EDUCATION MANAGEMENT AND MANAGEMENT SCIENCE

International Research Association of Information and Computer Science

IRAICS Proceedings Series

ISSN: 2334-0495

VOLUME 7

2014 INTERNATIONAL CONFERENCE ON EDUCATION MANAGEMENT AND MANAGEMENT SCIENCE (ICEMMS 2014), 7–8 AUGUST, 2014, TIANJIN, CHINA

Education Management and Management Science

Editor

Dawei Zheng
International Research Association of Information and Computer Science, China

CRC Press
Taylor & Francis Group
Boca Raton   London   New York   Leiden

CRC Press is an imprint of the
Taylor & Francis Group, an **informa** business

A BALKEMA BOOK

CRC Press/Balkema is an imprint of the Taylor & Francis Group, an informa business

© 2015 Taylor & Francis Group, London, UK

Typeset by V Publishing Solutions Pvt Ltd., Chennai, India
Printed and bound in Great Britain by Antony Rowe (A CPI-group Company), Chippenham, Wiltshire

Published by: CRC Press/Balkema
         P.O. Box 11320, 2301 EH Leiden, The Netherlands
         e-mail: Pub.NL@taylorandfrancis.com
         www.crcpress.com – www.taylorandfrancis.com

ISBN: 978-1-138-02663-6 (Hbk)
ISBN: 978-1-315-75214-3 (eBook PDF)

Table of contents

Education Management and Management Science – Zheng (Ed.)
© 2015 Taylor & Francis Group, London, ISBN 978-1-138-02663-6

Preface

We cordially invited you to attend the conference in Tianjin, China held August 7–8, 2014. The main objective of the conference is to provide a platform for researchers, engineers and academics as well as industry professionals from all over the world to present their research results and development activities in Education Management and Management Science. This conference provides opportunities for delegates to exchange new ideas and experiences face to face, to establish business or research relations and to find global partners for future collaboration.

The conference received over 400 submissions which were all reviewed by at least two reviewers. As a result of our highly selective review process 158 papers have been retained for inclusion in the proceedings, less than 40% of the submitted papers. The program of the conference consisted of invited sessions and technical workshops and discussions covering a wide range of topics. This rich program provided all attendees with the opportunity to meet and interact with one another. We hope your experience is a fruitful and long-lasting one. With your support and participation, the conference will continue its success for a long time.

The conference is supported by many universities and research institutes. Many professors play an important role in the successful holding of the conference, so we would like to take this opportunity to express our sincere gratitude and highest respects to them. They have worked very hard in reviewing papers and making valuable suggestions for the authors to improve their work. We also would like to express our gratitude to the external reviewers, for providing extra help in the review process, and to the authors for contributing their research results to the conference. Special thanks go to our publisher. At the same time, we also express our sincere thanks for the understanding and support of every author. Owing to time constraints, imperfection is inevitable, and any constructive criticism is welcome.

We hope you had a technically rewarding experience, and used this occasion to meet old friends and make many new ones.

We wished all attendees an enjoyable scientific gathering in Tianjin, China. We look forward to seeing all of you at next year's conference.

The Conference Organizing Committee
Tianjin, China

Education Management and Management Science – Zheng (Ed.)
© *2015 Taylor & Francis Group, London, ISBN 978-1-138-02663-6*

Organizing committee

GENERAL CHAIR

Prof. E. Ariwa, *London Metropolitan University, UK*

PUBLICATION CHAIR

Dr. R. Zheng, *International Research Association of Information and Computer Science, China*

TECHNICAL COMMITTEE

Dr. D. Pathak, *University of Pardubice, Czech Republic*
Prof. H. Ganjidoust, *Tarbiat Modarres University, Iran*
Dr. M. Dehghani, *Institute for Environmental Research, Iran*
Dr. A. Sharma, *Indian Council of Agricultural Research, India*
Prof. M.Yazdani-Asrami, *Babol University of Technology, Iran*
G. Mu, *Xi'an Jiaotong University, China*
Z. Jiang, *Biological, Chemical & Environmental Science Research Association, China*
G. Zhou, *Biological, Chemical & Environmental Science Research Association, China*
L. Jiang, *Science Technology Press, Hong Kong*
M. Xu, *Electronics and Information Engineering Research Institute, USA*

Education Management and Management Science – Zheng (Ed.)
© 2015 Taylor & Francis Group, London, ISBN 978-1-138-02663-6

The psychological research on the value of music education

Hongju Li
School of Arts and Communication, Beijing Normal University, Beijing, China

Xiaowei Zhu
Department of Education Science, Yangtze University, Jingzhou, Hubei, China

Xiaoqi Ding
Deshengmen Children's Palace, Xicheng District, Beijing, China

ABSTRACT: Music training is associated positively with performance on cognitive tests such as IQ test. However, the causal relationship between music education and nonmusical abilities is still controversial. In future, we should focus on the mediating variables and moderating variables, which may affect the relationship between music education and nonmusical abilities.

Keywords: music education; Mozart effect; mediating and moderating variables

1 INTRODUCTION

In 1993, Rauscher and his colleagues proposed the concept of Mozart effect on Science, the international core journals. Researchers first asked 36 university students heard Mozart's K448 double piano sonata and then let them complete the Stanford-Binet Intelligence Scale. They found that student's spatial cognitive ability increased significantly in 10 min after the listening. Therefore, Rauscher believed that Mozart's music may improve spatial abilities [1]. They seem to discover the visible and potential value of music education. However, Mozart effect has always been criticized.

2 THE RESEARCH OF MOZART EFFECT

Rauscher and her colleagues didn't mention some details in their experimental study. For instance, the limited 36 participants are divided into three conditions (experimental conditions: listen to Mozart's music, compare condition 1: listen to relaxing music, compare condition 2: no music) without being randomly ordered. Moreover, the repeatedly measure may lead to exercise effect or fatigue effect, both of them might influence participants' performance.

2 years later, Rauscher's research team modified their experimental method. In order to decrease the effect of exercise effect and other random error caused by measure frequency and period, they made a 5-day longitudinal study. In day 1, they tested the base line of student's spatial cognitive ability; In day 2, student completed spatial tasks under three conditions (condition 1: listen to Mozart's music, condition 2: listen to Philip glass's simple music, condition 3: no music); the following 3 days' experiment is similar with the former 2 days: only changing condition 2's music to story and dance music. The results showed no significant difference between the conditions. The reason is that with exercise effect, students' spatial task performance was becoming stable after first 2 days [2]. Nantais and Schellenberg [3] repeated the Mozart effect experiment and made a small adjustment: let some of the participants listen to Schubert's music. They found that participants in Schubert's music condition perform better than the compare condition, similar with participants in Mozart's music condition. That illustrated that other music can also have Mozart effect.

After Rauscher first put forward Mozart Effect, articles in this topic are continuing published each year. I used "Mozart Effect" as key word when searching the academic articles from 1993 to 2013 in database and found that every year there are a number of articles being published, which indicates that researchers do have continuous interests in this effect. The searching results also show us a research peak in 1999 and then researchers' passions began to slow down, as we can see about 10 new articles published a year in Figure 1. To sum up, there are nearly hundreds of research papers about Mozart Effect in the last 20 years.

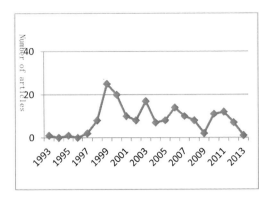

Figure 1. The figure illustrates the number of articles published in each year in the last 20 years.

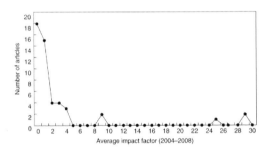

Figure 2. The figure illustrates the number of articles published as a function of average impact factor from 2004–2008.

The total numbers of published articles can only indicates the increasing concern on Mozart Effect. The Journal Impact Factor (IF) is a recognized objective indicators that can give us more information about the actual scientific value and meaning in the study area. As shown in Figure 2, Schulenburg recorded the average IF of the published articles between 2004 and 2008. From the picture, we can see that there are only a few articles' IF is higher than 1, which shows a big difference from the article's number we've seen in Figure 1.

3 THE RELATIONSHIP BETWEEN MUSIC TRAINING AND COGNITIVE ABILITIES

In 2007, Abbott pointed out in his article published in Science that many investigation and quasi-experiment studies found that music training has slightly but statistically significant and enduring effect on IQ [4]. Music education in individual's early development can have a widespread effect for a long time (Spinach et al, 2010)[5]. There are 9 articles in journal Music Perception in 2011, and

most of them concluded that there is a correlation between music education and nonmusical abilities.

Many researches confirm the correlation and even cause-and-effect relationship between music education and nonmusical abilities. Whereas other studies haven't found such relationship, even if there is a correlation between them, it is unknown which is the cause or the effect. Take the research of Ellen Winner (2000) as an example, who is a zero project expert of Harvard University. Ellen investigated the American students' SAT score and their art education in 4 years from 1988 to 1998, and use effect size to stand for the correlation intensity. The result showed that students who received art education presented increasing tendency in their SAT score, just as shown in Figure 3.

This indicates that accompany with the trend of the admission for the best applications, increasing competition make student tend to take one or more art course. However, it's far-fetched to explain that music education and nonmusical abilities are cause and effect. What Ellen concluded in the end is very different from the expectations of the public: art education is not the necessary condition to improve student's comprehensive learning ability and performance [6].

Even some researchers pointed out why there's a correlation between music training and cognitive function. It is because those who has high cognitive ability are more likely to choose music lesson and at the same time complete various tests perfectly, with no business of so-called Mozart effect [7]. Others believe that music training is related to specific cognitive ability.

Organization for Economic Cooperation (OECD) discussed art education's influences on

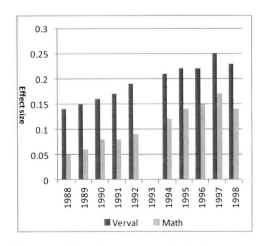

Figure 3. Effect sizes showing relationship between four year arts study in high school and scores. Effect sizes grow steadily larger from 1988–1998.

science, mathematics and other subject in 2011. On the conference, renowned scholars E G Schulenburg, E Winner has different opinion on whether there is relationship between music education and nonmusical abilities.

4 THE RELATION OF MUSIC EDUCATION AND EMOTION

Music can arouse some kind of mood. Studies have also found that positive emotions can promote the development of cognitive function. While negative emotional states such as anxiety or burnout will affect the cognitive function. Therefore, the researchers made an assumption that music training may not directly leads to the improvement of cognitive function, but participant's emotion states and awaken state which influenced by music may improve the cognitive function [8].

From this point, the researchers began to improve experimental research paradigm of the Mozart effect and launched a series of experiments. The results show that when using Schubert music as a control group, the participants showed Schubert effect; and when using funny Stephen King novel as a control group, their score was also no different with the Mozart music group in task space tests. During the interview after the experiment, researchers found that participant's number in the experimental group and the control group who like Mozart's music and who like stories are the same. Interestingly, Mozart effect is almost shown on subjects who mostly like Mozart's music, and subjects who like Stephen King novel is more likely to have Stephen King effect [3]. Although researchers didn't directly measure the emotion and aroused state in the experiment, but the above result indicates that music training is likely to influence cognitive function through emotion and aroused state.

Thompson [9] (2001) went to improve the method of experiment. They choose to directly measure the emotion and the aroused state. They utilized fast rhythm and major melodies to evoke happy emotion, slow rhythm and minor melodies to evoke sadness, respectively choosing Mozart D Major and Albinoni's Adagio in G minor (Albinoni Adagio). They found out the fast-paced major melody group got significantly higher spatial ability test scores than slow rhythm tunes melody group. On the basis of the experimental paradigm, researchers made some further adjustment; they use the same tracks with different rhythm and the mode to replace the original Adagio in G minor. That's because the songs are usually used in the funeral and systematic error is likely caused by the bereavement memory. Consistent with prior studies, fast rhythm major melody group had a higher score than slow rhythm minor melody group. These series of research results seem to point to the same conclusion: emotion and aroused state mediate the relation between music training and cognitive function.

However, emotion does not influence all the cognitive function. Studies have found that, when the researchers changed the test content (replace space ability with cognitive processing speed or working memory capacity), they did not find significant differences between groups. This shows that some cognitive function is not affected by emotion and aroused state. Of course, this conclusion is put forward under the assumption that emotion and aroused state really plays a role of mediation between music training and cognitive abilities.

In terms of other social or emotional function, when researchers investigated music training's effect on children's self-esteem and self-concept, they found no significant correlation between these variables [10]. In addition, some researchers studied the relationship between art education and children's social skills. They selected 6-year-old preschool age children who had just accept one year of art training as music training group and drama training. Also there is a control group. The results showed that only drama training group had a higher social skills improvement. Then in a broader similar research also found that there is no relation between music training and social skills [11].

Researchers believe that Mozart effect essentially is "stay sober" medicine effect. Music leads to the aroused state of individual, which promotes the intelligence test task performance (Schellenberg & Weiss, 2013) [12]. As a matter of fact, music arousal theory justifies the Mozart effect from the perspective of psychological mechanism. The effective of music education depends on the music stimulus or music education's impact on individual inner emotion. Music stimulates the mediation variables, emotion to affect intelligence, academic achievement, etc. Although this effect's lasting time is short, just about 20 min [4], but repeatedly exploration to music stimulation in the process of personal growth may gradually influence the cerebral cortex and cognitive ability, and finally short-term effect will produce long-term impact.

5 DISCUSSION

There is no agreement about the relationship between music education and nonmusical ability. Crucial additional variables may exist. Some scholars discussed the mediation function of cognitive executive function, but the result was not significant. Chabris (1999) believed that the existence of Mozart effect may be influenced by the participant's interest of music material [12]. In terms of the reality of Chinese art education, music learning interest may be an important moderating variable. This did not been concerned in the previous studies, and need to be explored by empirical research.

A lot of research literature has shown that the emotion arousal in music education plays a mediating role between and cognitive ability. More research needs to be done to test whether music education has the same psychological mechanism in China.

In the future, we should focus on the mediating variables and moderating variables, which may affect the relationship between music education and nonmusical abilities. With the aid of brain science research techniques, we may find the mechanism of music education and nonmusical abilities.

6 FUND PROGRAM

Supported by the fundamental research funds for the central universities.

REFERENCES

[1] Rauscher, F.H (1993). Music and Spatial Task Performance. Nature, 365, 611.

[2] Rauscher, F.H (1995). Listening to Mozart enhances spatial temporal reasoning towards a neurophysiological basis. Neuroscience letters, 185, 44–47.

[3] Nantais, K.M and Schellenberg, E.G (1999). The Mozart effect: An artifact of preference. Psychological Science, 10, 370–37.

[4] Abbott, A (2007). Mozart doesn't make you clever [OL]. Nature, http://www.nature.com/news/2007/070409/full/news070409-13.html.

[5] Spinath, B, Freudenthaler, H.H. & Neubauer, A.C (2010). Domain-specific school achievement in boys and girls as predicted by intelligence. Personality and Motivation, 18, 232–235.

[6] Winner, E (2000). The relationship between arts and academic achievement: No evidence (yet) for a causal relationship. Los Angeles, California, 17–31.

[7] Schellenberg, E.G (2011). Examining the association between music lessons and intelligence. British Journal of Psychology, 102(3), 283–302.

[8] Jones, M.H, and Estell, D.B (2007). Exploring the Mozart effect among high school students. Psychology of Aesthetics, Creativity, and the Arts, 1, 219–24.

[9] Thompson, W.F, Schellenberg, E.G., & Husain, G (2001). Arousal, mood, and the Mozart effect. Psychological Science, 12, 248–251.

[10] Portowitz, A, Lichtenstein, O, Egorova, L, & Brand, E (2009). Underlying mechanisms linking music education and cognitive modifiability. Research Studies in Music Education, 31,107–128.

[11] Schellenberg, E.G (2006). Long-term positive associations between music lessons and IQ. Journal of Educational Psychology, 98, 457–468.

[12] Chabris, C.F., Prelude or requiem for the Mozart Effect? Nature, 1999(8).

Education Management and Management Science – Zheng (Ed.)
© 2015 Taylor & Francis Group, London, ISBN 978-1-138-02663-6

Strategic positioning and thought of the rise of coastal trough region in new age—a case study on the coastal regions in Yancheng of Jiangsu

Yanmei Dong
School of Economics and Management, Nanjing University of Science and Technology, Nanjing, Jiangsu, China
Yancheng Teachers University College of Business, Yancheng, Jiangsu, China

Yingming Zhu
School of Economics and Management, Nanjing University of Science and Technology, Nanjing, Jiangsu, China

ABSTRACT: Multi-view strategic location analysis is carried out for Yancheng based on the studies on the opportunities and challenges of the rise of the coastal regions: seen from the global perspective, Yancheng is a costal competitive unit in the ocean strategy, the division unit in the secondary regional division of labor in 'growth triangle' of Northeast Asia, as well as the outward-looking economy port that is the closest to Japan and South Korea. Seen from the Chinese perspective, it locates in the deep-water port in the developing South Shipping Center and the eastern coastal characteristic economical plate. Seen from the perspective of Yangtze River Delta, it is now stepping in the diffusing effect of Yangtze River Delta, as well as the associated industries of Yangtze River Delta. Seen from the perspective of Jiangsu Province, Yancheng is a significant part of the central city and coastal hinge port group connecting towns. Here, Yancheng is defined as a potential new maritime portal in Huaijiang Region, as well as the capital of sustainable coastal ecological industry, which is now in the 'fourth growth pole' of the provincial development of Jiangsu. Besides, corresponding development thoughts are proposed.

Keywords: coastal trough; coastal area of Yancheng; rise

1   INTRODUCTION

Ever since the formal effectiveness of *United Nations Convention on the Law of the Sea* in 1994, the main maritime powers in the world, such as America, Japan, England, South Korea, Australia, etc. started to take the development and utilization of marine resources, as well as the development of marine business as a significant part of the national strategy, and made corresponding strategies and policies for marine economy development [1]. Ever since the 21st century, Chinese government also started a new round of strategic deployment for coastal development. From 2008 to 2009, the State Council released the development planning for some coastal areas in Guangxi, Fujian, Jiangsu, and Liaoning. Consequently, the new round of deployment made the coastal economy layout much more complete, and the development pattern, namely 'three large and four small regions' along the eastern coast, was gradually formed. The adjustment and implementation of national strategy brings unprecedented opportunities and challenges for the coastal development in China. However, Yancheng, which has the longest coastline in Jiangsu, is a depression in the coastal development. In 2012, Yancheng port has a cargo handling capacity of 31,186,000 tons, with a year-on-year growth of 48.4%, which only took up 7.6% of the cargo handling capacity in coastal ports in Jiangsu Province. Therefore, the coastal development in Yancheng has already become the bottleneck of coastal development in China, and the great-leap-forward development from 'beach Yancheng' to 'maritime Yancheng' has already become an inevitable choice for the rise of Yancheng.

2   THEORETICAL STUDY ON THE RISE OF COASTAL

2.1   *Late-developing advantage*

In 1952, Alexander Gerschenkron created the theory of 'late-developing advantage' in the article *Economic Backwardness in Historical Perspective*. It is a special advantage caused by the status of the late-rising country, and such advantage cannot be owned by advanced countries, or created by the late-rising countries with efforts. Instead, it co-exists with the relatively economic backwardness completely [2]. Afterwards, Abramovitz [3] proposed the 'pursuant hypothesis', while Brezis, Krugman [4] proposed the 'leap-flogging' model, which all pointed that the later-rising countries have technical last-developing advantages.

2.2 *Point-axis development theory*

'Point-axis' theory [5] was firstly proposed by Lu Dadao in 1984. Based on the central-place theory of Christaller, this theory mainly studies the strategic model of regional development from the perspective of spatial organization form, believing that at the beginning of the regional development, it shall concentrate limited strength to focus on the cultivation and element agglomeration for growth pole with favorable basic conditions, and then it shall spreads along the axis to drive the regional growth.

3 OPPORTUNITIES AND CHALLENGES CONFRONTED BY THE RISE OF COASTAL THROUGH

3.1 *Unique comprehensive regional advantage*

Located in the middle of eastern coastal regions in China, Yancheng has the Yellow Sea in the east, Huai'an and Yangzhou in its west, Nantong and Taizhou in its south, as well as Lianyungang in the north across the Guanhe. Seen from its economic location, Yancheng is seated in the intersection of China's eastern coastal economic belt and East Longhai economic belt, with the Yangtze River Delta economic area in its south, and it is the first stop of the development of Yangtze River Delta towards the north gradient. But for quite a long time, the natural moat of Yangtze River Delta restricts its connection to Shanghai, integration into the economic strength of Yangtze River Delta, as well as its economic connection to developed regions including Shanghai, South Jiangsu and Zhejiang, etc. Over the past years, Yancheng focuses on building the expressway network, such as its connection to the north and the south (Jingjiang-Yancheng, Yancheng-Nantong, Lianyungang-Yancheng expressway), to the east and west (Huai'an-Yancheng), etc. It breaks the natural moat of Yangtze River Delta by setting Sutong Bridge and Chongqi River-crossing Pathway, and it also opens an international route from Yancheng to Seoul, ocean trunk line, near-sea shipping line, as well as several domestic branches, which thoroughly changes the inferiority of Yancheng in economic position, and lays a solid foundation for Yancheng to accept the radiation of Shanghai and to integrate into the economic circle of 'Yangtze River Delta'.

3.2 *Gathering of various resources and huge development potential*

Yancheng is in possession of abundant characteristic resources, such as the shorelines, ports, beaches, agricultural products, oil, etc. The coastline is about 582 km, accounting for 60% of the entire province, as one of the prefecture-level cities, which has the longest coastline throughout the country. 12 natural ports are distributed on the coastline, which are all close to the deep-water channels, and connected to the open seas. The cultivable land per capita is 0.2 hectare, which is 1.5 times of the average level in Yangtze River Delta, and at present 21% of the land has not been exploited, and nearly 1/3 is beach. Furthermore, it grows at a speed of 1330 hectares/year. At present, when the land still restricts the regional development, these resources will provide powerful support for the coastal development of Yancheng. With abundant marine resources, it is a significant shellfish producing area and the largest sea eel catching and export base. The proved natural gas reserve is 80 billion cubic meters, and it is estimated that the total reserve is about 200 billion cubic meters, which is the largest onshore oil and gas field in eastern coastal region, with a broad exploration and development prospect.

3.3 *Weak radiating capacity of central cities and insufficient driving power*

Compared to the surrounding cities, the economic aggregate is relatively small in Yancheng, and it has a small radiating capacity to its surrounding regions. In 2011, the economic aggregate of Yancheng was not inferior when compared to that of Huai'an, Lianyungang, Suqian, Yangzhou and Taizhou. It was higher than the rest cities, except Xuzhou and Nantong (as shown in Fig. 1), while its urban GDP was only 38.2, 44.4, 47.5 and 72.1% of Xuzhou, Yangzhou, Nantong and Huai'an, respectively. Although the GDP of Taizhou was lower than that of Yancheng, its GDP per capita was about 1.4 times of that in Yanchen (as shown in Table1).

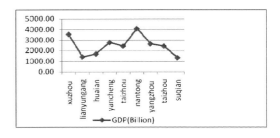

Figure 1. GDP of Yancheng and its surrounding cities.

Table 1. Major economic indicators of the downtown in Yancheng and its surrounding cities in 2011.

Indicator	Gross regional production (one hundred million)	GDP per capita (Yuan)	Total retail sales of consumer goods (one hundred million)
Yancheng	756.45	46903	297.03
Xuzhou	1981.47	63658	721.09
Huai'an	1048.48	39670	346.66
Lianyungang	507.23	48051	201.74
Suqian	446.148	30550	130.53
Yangzhou	1702.87	70688	563.21
Nantong	1594.08	69305.00	559.59
Taizhou	573.69	65089	206.97

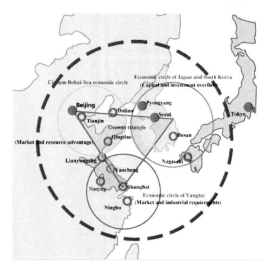

Figure 2. Secondary regional division of work system in 'growth triangular' in Northeast Asia.

4 MULTI-VIEW STRATEGIC LOCATION ANALYSIS

4.1 Global perspective

4.1.1 Coastal competitive unit in maritime strategy

In the new ocean age, the one who obtains sea supremacy will be able to integrate into the global division of work system. According to the researches, from 2000 to 2020, the world manufacturing industry base will be transferred to the eastern coastal region in China. After 2020, it may be further transferred to the Southwest Asian region along the coast of the Indian Ocean with India as the leading position. As a space unit in the eastern coast of China with abundant land and labor, Yancheng has a superior geographical location, and it is qualified for participating in the flexible competitive condition of global market system. It shall seize the opportunities, promote the ocean development strategy comprehensively, and seek for new development during the transfer of global manufacturing industry and the coastal regional competition process in the Western Pacific.

4.1.2 Secondary regional division of labor in the 'growth triangle' in Northeast Asia

During the course of history of Asian-Pacific economic development, the 'growth triangle' mode of Southeast Asia turns to be the successful example of transactional regional development. Singapore, Malaysia and Indonesia once formed the 'iron triangle' in Southeast Asia trough the functional complementation and regional division of labor. It has been reflected by facts that there is also 'growth triangle' in the Northeast Asian region. The technology and capital overflow of the economic circle of Japan and South Korea, the market and industrial foundation of Yangtze River Delta economic circle, as well as the resource and market potential of circum-Bohai-Sea economic circle form the regional supplementary relation. As a coastal zone in the north of Yangtze River Delta economic circle, Yancheng is equipped with the port shoreline, ecological beaches, land and labor, etc. As long as it captures the regional location, and integrate with it voluntarily, it will be able to integrate into the secondary regional division of labor system in the 'growth triangle' of Northeast Asia (as shown in Fig. 2).

4.1.3 The closest outward-looking economy port to Japan and Korea

In the 'growth triangle' of Northeast Asia, when compared to similar coastal ports that are newly developed, Yancheng port group with first-class port Dafeng port as the center has already become the international shipping seaport that is closest to Nagasaki of Japan (430 sea mile) and Pusan of South Korea (420 sea mile), which will provide extremely favorable portal condition for the coastal development of Yancheng.

4.2 Perspective of China

4.2.1 Deep-water port group in the north of the developing south shipping center

According to the five great port group planning approved by the State Council, Shanghai port group is the south shipping center, while Jiangsu and Zhejiang coastal port group are oriented as the two regional hubs. Compared to other regions in China, the capacity factor of ports in Yangtze

7

River Dealt Port is relative high. Ever since the 1990s, China's port handling capacity increased by 10% every year, and till 1999, the problem of insufficient handling capacity occurred, and in 2005, the duty factor of ports throughout the country was as high as 1.3, and the gap was about 500 million tons, and the Yangtze River Delta ports operated in excessive load. In 2010, the capacity factor was above 100%, and some ports displayed saturated and even excessive load operation [6]. It is estimated by experts that the transfer of world manufacturing industry and foreign trade will drive the constant increase of port handling capacity. Even if the current port handling capacity is doubled, the shipping capacity of Shanghai Shipping Center is still in shortage. In the north port group, Yancheng took up the 56% of the coastline of Jiangsu Province, and it also has the deep-water port resources, such as Dafeng port, Binhai port, etc. which will be favorable for cultivating the regional hub port in the north deep-water port group, so as to adapt to the increasing port handling capacity.

4.2.2 *Eastern coastal characteristic economic plate*

Yancheng is a characteristic brand innovation place: its self-developed brand 'Yanwu' was quite popular in China before after 1990. 'China's leather shoe' Senda was the first brand that was exported to foreign countries; 'Zonda' brand ranked the 98th among the 500 most valuable brands in 2005; 'Qianlima' Kia and 'Jiangdong' brand were extremely popular in the economical car and diesel engine industry respectively. Yancheng is the capital of eastern wet land, with the largest coastal beach and wetland resources in the western Pacific Ocean, as well as the ecological habitat of the rare animals like elk and red-crowned crane which are unique to China. Renowned as 'capital of eastern wetland', it has unique ecological resources. Yancheng is also a traditional agricultural salt base: traditional rice, cotton, silkworm cultivation base, the seal sault manufacturing base; it has advantages in resources and labor cost in related industries, such as deep processing of agricultural products, textile, silk weaving and salt chemical engineering, etc.

4.3 *Perspective of Yangtze River Delta*

4.3.1 *Stepping into the spreading effect region of Yangtze River Delta*

In the Yangtze River Delta economic circle, Shanghai, Southern Jiangsu and Northern Zhejiang form the 'core' of Yangtze River Delta region, while Yancheng has always been in the 'edge' of the Yangtze River Delta Region (as shown in Fig. 3).

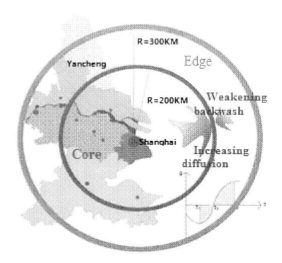

Figure 3.　Central-edge relation of Yangtze River Delta.

With the perfection of regional facilities and acceleration of regional collaboration, Yancheng gradually get closer to the core of Yangtze River Delta, and the 'spreading effect' is getting stronger than the 'echo effect'. It enters the T2 inflection point in the growth pole model, and the industry, capital and technology diffusion of the core area will inject powerful vitality for the coastal development of Yancheng.

4.3.2 *Associated industries of the core region in Yangtze River Delta*

Besides accepting the industrial expansion, Yancheng can also match the industrial function with core region of Yangtze River Delta. Firstly, it shall match with the touring base. The urbanization level of Yangtze River Delta is close to 70%, and GDP per capita exceeds 5000 USD. Shanghai metropolitan area, Suzhou, Wuxi and Changzhou metropolitan area, and Nanjing metropolitan area have dense population and convenient transportation. With the growth in the living standard, urban population has a huge demand on the ecology, popularization of science and leisure tourism. Yancheng has unique ecological wetland and idyllic scenery resource, and it can become the ecological touring base for Shanghai and southern Suzhou. Secondly, it shall match with the urban agriculture. Its cultural connections and cooperation foundation with Shanghai and southern Suzhou make it possible for the agricultural foundation construction, and it is quite potential in the construction of agricultural foundation for Shanghai metropolitan area, southern Suzhou and local central cities.

4.4 Perspective of Jiangsu

4.4.1 Convergence axis central city connecting cities and towns

In the regional space development strategy of Jiangsu, Yancheng locates in the middle of the convergence axis connecting towns, and it is one of the central cities cultivated by the development axis. With the rapid urbanization process in central and northern Suzhou, the urbanization level of Yancheng will accelerate rapidly, which will provide powerful support for the coastal development.

4.4.2 Significant part of coastal hub port group

Jiangsu's coastal ports consist of Nantongyang Port, Lianyungang Port and Yancheng port group, in which Yancheng port group has the largest number and best deep-water condition. Through comprehensive evaluation, Binhai Port and Dafeng Port have the most obvious advantages, in which Binhai port is set on deep-water, and 15 meters of isobath is about 3.4 kilometers away from the shore. It can hold 100000 to 200000 tons of quay berths, with the unique port construction condition in eastern China. Dageng port has 10 meters of isobaths, which can hold 50000 to 100000 tons. When considering the coordinated development of city-industry-port, Dafeng and Binhai port is oriented as the strategic core of coastal development in Yancheng, and it shall be cultivated and constructed with emphasis.

5 STRATEGIC POSITION AND DEVELOPMENT THOUGHT ANALYSIS

5.1 Potential 'new maritime portal' in Huaijiang region

As one of the most significant region in the rise of central China, Huaijiang is the potential growth zone in the edge of 'T-shape' development in China. For a long time, it lacks real marine outfall, and at present, load running of the lower reaches of Yangtze River, channel threshold and relatively high transportation cost, as well as the land bridge transport mode promotes Huaijiang to seek for appropriate export-oriented port. The creation of new sea portal in the downstream of Huaihe River shall be the choice for the globalization and marketization of Huaijiang region. Yancheng deep-water port locates in the downstream of Huaihe River, with the natural condition for constructing the ocean trade port. It will make Huaijiang region to be the first portal towards the world by the interactive development of port group and urban group. Recently, the regional hub port group with the first-class port Yancheng Port as the center in northern

Jiangsu is mainly developed comprehensively, so as to construct the portal city undertaking the combined transport of the river and the sea. In the future, it will construct an international deep-water port (more than 150000 tons) with ocean function, and form the super portal in Huai River basin.

5.2 Sustainable developing costal ecological industry capital

Located at the economic trough in eastern coastal region, Yancheng is connected to the Yangtze River Delta, and involved in the division of labor in 'growth triangle in Northeast Asia', which may make it convenient for accepting the world and regional industrial gradient transfer. With ample labor, land and coastline resources, it is confronted by the policy opportunities such as the rising of northern Suzhou, coastal development, etc. Through appropriate development and exploitation of the coastal deep-water port coastline and the largest ecological wetland on the western Pacific Ocean, it will give full play to its late-developing advantages, cultivate and form coastal ecological industrial city with the gathering of information logistics, developed harbor economy, wetland landscape and distinct ecological features. Recently, it mainly forms the coastal ecological industrial capital with intensive human power and technology, constant economic growth, and distinct ecological and cultural feature through the recycling of resources, saving of energy, clean production and construction of green industry.

5.3 Appropriate 'fourth growth pole' in the development of Jiangsu Province

At present, Jiangsu has already formed three metropolitan region growth poles, namely Xuzhou, Nanjing and Suzhou, Wuxi and Changzhou metropolitan region. Is there the fourth growth pole? Throughout the northern part of the Yangtze River, Lianyungang starts from the land bridge economic belt, Nantong has become the secondary center of Shanghai, Yangzhou has already integrated into the Nanjing metropolitan region, Taizhou is included into the riverside plate, Suqian is little but it is close to a corner, and Yancheng connects the middle and northern Suzhou, as the center of coastal economic belt and urban axis. Huai'an is the geographic center of northern Suzhou, with broad transportation junction. Therefore, Yancheng and Huai'an can complement each other and form the fourth growth pole. Yancheng city can construct large-scale Yanhuai metropolitan region with Yancheng as the portal city and Huai'an as the interior city with the support from deep-water port and coastal industry, so as to undertake the 'gradient transfer' of the industry, and drive the economic development in the middle and northern Suzhou. Recently, it mainly relies on Yancheng city, and regional 'double-core' model is formed in Dafeng port, while in the future, it will form complementary 'large-triangle' spatial structure, and construct Yanhuai metropolitan region by relying on 'Yancheng metropolitan region, Huai'an metropolitan region-Binhai metropolitan region and Fafeng Port city' as well as the secondary urban group and port group, thus to become the provincial fourth growth pole with comprehensive emissive power of social economy.

REFERENCES

[1] Zhou Dajun, Cui Wanglai, On Public Marine Policies. Beijing: China Ocean Press, 2009.
[2] Alexander Gerchenkron, Economic Backwardness in Historical Respective, Cambridge, Mass: Harvard University Press, 1962.
[3] M. Abramjoritz, Thinkong about Growth, Cambridge University Press, 1989.
[4] Grossman, G.M. and E. Helpman: Quality Ladders and Product Cycle, Quarterly Journal of Ecomomics, 1991, 106, pp 501–526.
[5] Lu Dadao, Regional Development and its Space Structure. Beijing: Science Press, 1995, 157–164.
[6] Research Report of Changjiang Securities. http://money.163.com/11/0225/07/6TNLLQB300253EOS.html.

FDI, Total Factor Productivity and income gap—the evidence from Jiangsu Province

Y.Y. Zhang, L.J. Meng & Xin Zhang
School of Economics and Management, Nanjing University of Science and Technology, Nanjing, Jiangsu, China

ABSTRACT: Based on the estimates of agricultural Total Factor Productivity (TFP) Jiangsu Province, we use panel data 2002–2011, 13 cities in Jiangsu province and collect the FDI and agricultural TFP into the same framework in order to analyze the impact of FDI and TFP on income gap between urban and rural areas. Research shows that in the sample period, FDI narrows the income gap between urban and rural areas in Jiangsu Province; while the agricultural TFP can narrow the income gap between urban and rural areas on the whole, but it is influenced by the level of regional economic development and urbanization of these two factors. In addition, the study also found that due to non-agricultural development result in a lag of rural labor force, making urbanization expanded the income gap between urban and rural areas; Trade dependence has increased the income gap between urban and rural areas in Jiangsu Province.

Keywords: FDI; TFP; income gap

1 INTRODUCTION

Income distribution problem is an enduring economic research topics, the income gap between urban and rural residents in China has the greatest impact. Research shows that there are many factors that can influence the income gap between urban and rural areas. In recent years, China's international economic cooperation is increasing and foreign direct investment scale expands constantly. The regional distribution, industry distribution and the various advantages for resources of FDI may affect the employment and income level. So FDI is becoming one of the hot spots in studying the influence on income gap. Pandej Chintrakarn etc (2012) using the panel cointegration analyze the relationship between FDI and income cap. They concluded that FDI had narrowed the income gap in the United States. Liu Yulin etc (2013) found that the relationship between FDI and income gap between urban and rural areas in China is inverted U-shaped. Most provinces in eastern and central had entered the right part of the parabola, while most western region is in the left half. Christian Lessmann (2013) used the panel data of 55 countries and they concluded that FDI widened the low and middle income countries, but not to narrow the income gap in high-income countries. As a coastal open city, Jiangsu has a superior location advantage, investment environment and huge market potential; it

has become an important FDI investment area. Statistics show that in addition to the 1980–1984 and 1993–1984, the income gap between urban and rural areas of Jiangsu province is expanding constantly. Whether the introduction of FDI has accelerated the expansion of urban-rural income gap is worth our study. In addition, the existing research considered the agricultural TFP impact on income gap, so we collect the FDI and agricultural TFP into the same framework. Under the condition of open environment, we research the impact FDI and agricultural TFP on income gap in Jiangsu Province; this has a certain significance for the Yangtze River Delta and other economic areas.

The article is organized as follows: the first part is an introduction. The second part analyzes the mechanism of FDI influence on income gap. The third part is the empirical part of this article. The last is the conclusion of this article.

2 THE MECHANISM OF FDI INFLUENCE ON INCOME GAP

FDI has become the main mode of foreign capital inflows, but it does not directly impact on the income gap in China, it through employment, wage level, industrial structure, foreign trade and TFP growth effect the income gap indirectly. First, FDI affects income gap through employment and wages.

FDI choosing different types of labor factor has different impact on the marginal benefit; Secondly, FDI can promote the optimization of the industrial structure, making the different income in different industry. Thirdly, FDI affects income gap by foreign trade. Total imports and exports of foreign-funded enterprises accounted for the proportion of total imports and exports of Jiangsu Province about 75% in 2011, it has become the important part of Jiangsu international trade. Fourthly, FDI affects income gap by TFP. Technology diffusion of FDI inflows will promote technological progress and productivity of the host country, and this will improve the region's income. But the effect is related to the degree of spillover effects.

3 THE EMPIRICAL ANALYSIS

3.1 The model

We use Malmquist productivity index method based on DEA to measure agricultural TFP, and then we establish econometric model analysis the impact of FDI and agricultural TFP on the income gap. In this paper, we use DEA constant returns to scale, input-oriented Malmquist index to measure Jiangsu Province agricultural TFP. In the process of measuring TFP, every county in Jiangsu province as a DMU.

From time s to time t, the technology T^s as reference, Malmquist index is defined as:

$$M_o^s(y_i^s, x_i^s, y_i^t, x_i^t) = \frac{D_i^s(y_i^t, x_i^t)}{D_i^s(y_i^s, x_i^s)} \tag{1}$$

Similarly, from time s to time t, the technology T^t as reference, Malmquist index is defined as:

$$M_o^t(y_i^s, x_i^s, y_i^t, x_i^t) = \frac{D_i^t(y_i^t, x_i^t)}{D_i^t(y_i^s, x_i^s)} \tag{2}$$

Among them: x_i^s, x_i^t and y_i^s, y_i^t respectively in case of DMU_i at t and s period input vector and output vector. $D_i^s(y_i^s, x_i^s)$ and $D_i^s(y_i^t, x_i^t)$ are respectively s and t period of distance functions, the technology T^s as reference. $D_i^t(y_i^s, x_i^s)$ and $D_i^t(y_i^t, x_i^t)$ are respectively s and t period of distance functions, the technology T^t as reference.

Using the Fisher ideal index of construction method Caves used the geometric mean of the two types as the Malmquist productivity index of productivity change from time s to time t. When it is greater than 1, the TFP grow.

$$M_o(y_i^s, x_i^s, y_i^t, x_i^t) = \sqrt{\frac{D_i^t(y_i^t, x_i^t)}{D_i^s(y_i^s, x_i^s)} \times \frac{D_i^s(y_i^t, x_i^t)}{D_i^t(y_i^s, x_i^s)}} \tag{3}$$

On the basis of calculation of Jiangsu agricultural TFP, we set up the following panel data model:

$$
\begin{aligned}
Gap_{i,t} = \alpha_i &+ \alpha_1 FDI_{i,t} + \alpha_2 TFP_{i,t} \\
&+ \alpha_3 Development_{i,t} + \alpha_4 Urban_{i,t} \\
&+ \alpha_5 Rur_{i,t} + \alpha_6 Open_{i,t} + \varepsilon_{i,t}
\end{aligned} \tag{4}
$$

The subscript I ($i = 1, 2, ..., 1$) is the different cities in Jiangsu. The subscript t ($t = 2002, 2003, ..., 2011$) is the different years. In the model, we control factors, such as: economic development level, urbanization level. Next, the variables and data processing will be specifically described.

3.2 Variable selection and description

All data are from Jiangsu statistical yearbook 2002–2011 (Table 1).

Table 1. Variable selection and description.

The variable name	Index selection
GAP	The wage income of urban residents/ Net income of rural residents
FDI	The actual FDI in Jiangsu cities
TFP	Malmquist TFP
Development	Per capita GDP in Jiangsu cities
Urban	Non-agricultural GDP/Total GDP
Rur	Employment number of industry/ Total employment
Open	Total trade/GDP

Table 2. The average agricultural TFP index of 13 cities in Jiangsu province from 2002 to 2011.

Area	effch	techch	pech	sech	tfp
Nanjing	0.963	1.194	0.973	0.990	1.150
Wuxi	0.982	1.199	1.000	0.982	1.178
Xuzhou	0.930	1.198	0.943	0.986	1.113
Changzhou	0.967	1.195	0.994	0.973	1.156
Suzhou	1.000	1.222	1.000	1.000	1.222
Nantong	0.951	1.179	1.000	0.951	1.121
Lianyungang	0.933	1.198	0.940	0.992	1.117
Huai-an	0.943	1.196	0.944	0.999	1.128
Yancheng	0.942	1.196	1.000	0.942	1.127
Yangzhou	0.931	1.200	0.948	0.982	1.117
Zhenjiang	0.955	1.196	0.997	0.958	1.143
Taizhou	0.951	1.186	0.949	1.001	1.128
Suqian	0.939	1.198	0.944	0.995	1.125
Average	0.953	1.197	0.971	0.981	1.140

Table 3. Panel data model regression results.

	M1	M2	M3	M4	M5
FDI	−0.0315**	−0.2639**	−0.2616**	−0.3021**	−0.2675**
	(−2.2318)	(−2.1765)	(−2.1359)	(−2.5431)	(−2.2768)
TFP	−0.1274***	−0.1406***	−0.1263	−0.0014	−0.0204
	(−2.7999)	(−3.1382)	(−1.5471)	(−0.2158)	(−0.2258)
Open		0.0049**	0.0049**	0.0047**	0.0045**
		(2.2518)	(2.2542)	(2.2203)	(2.1806)
Development			−0.0098	−0.0114**	−0.0083
			(−0.2137)	(−2.0143)	(−1.4378)
Urban				0.0278***	0.0488***
				(3.0917)	(3.7372)
Rur					1.9869**
					(0.0324)

Note: ***, **, *, respectively in 1%, 5%, 10% significant level.

3.3 Estimation results

This paper chose 13 cities in Jiangsu province as an independent production Decision Making Unit (DMU), in time to choose the 10 years from 2002 to 2011 to evaluate agricultural input and output values. When calculating the TFP, we refer to existing literature. We choose total number of employees in agriculture, forestry, animal husbandry and fishery, crop acreage, agricultural machinery power, agricultural chemical fertilizer and effective irrigation area as input variables and the first industry output value as output variable and deflate the value of variables. We use DEAP 2.1 measure Malmquist index of TFP, the results are shown in Table 2.

From the Table 2, we can get that the average agricultural TFP index of 13 cities in Jiangsu province from 2002 to 2011 is 1.14, among them, advances in technology (techch) is 1.197, technical efficiency (effch) is 0.953. The growth of the agricultural total factor derived from technological progress. On the space, the five fastest growth of agricultural TFP are Suzhou, Wuxi, Changzhou, Nanjing, Zhenjiang, all are located in south Jiangsu.

Based on the TFP, and we use Eviwes 6.0, the variable intercept, according to Hausman test to determine fixed effects or random effects. Model regression results are shown in Table 3.

We placed the control variable to the model one by one, we get 5 final models. When only contains FDI and TFP two variables, the coefficient of level significantly negative in 5% and 1% respectively (M1, M2); When join the Open, Development, Urban and Rur, coefficient of FDI is still significantly negative (M3—M5), so we get the conclusion, FDI narrow the income gap. The reasons are maybe the rural can gain more than the city from FDI inflows and FDI can promote agricultural industrialization. About the TFP, when join the Open, TFP is significantly negative, when add Development, Urban and Rur, coefficient of TFP by significantly is less significant, but still negative (M3—M5), The agricultural TFP is not the main factor influencing the income gap. What's more, according to the results of the model, trade dependence increased the income gap. This may be associated with the trade structure. Jiangsu mainly exports the raw product such as food and mineral raw materials. When the deterioration of trade terms occurs, there will be a lot of people out of work, who are mostly from rural areas. At last, Non-agricultural GDP and the number of primary industry employment are higher, the greater the income gap.

4 CONCLUSION

Based on the estimates of agricultural Total Factor Productivity (TFP) Jiangsu Province, we use panel data 2002–2011, 13 cities in Jiangsu province and collect the FDI and agricultural TFP into the same framework in order to analyze the impact of FDI and TFP on income gap between urban and rural areas. Research shows that in the sample period, FDI narrows the income gap between urban and rural areas in Jiangsu Province; while the agricultural TFP can narrow the income gap between urban and rural areas on the whole, but it is influenced by the level of regional economic development and urbanization of these two factors. In addition, the study also found that due to non-agricultural development result in a lag of rural labor force, making urbanization expanded the income gap between urban and rural areas; Trade dependence has increased

13

the income gap between urban and rural areas in Jiangsu Province.

REFERENCES

[1] Y.L. Liu. J. Li. "TFP, FDI and income gap between urban and rural," Economic Fabric. 2013(3).

[2] L. Christian "Foreign direct investment and regional inequality: A panel data analysis," China Economic Review, 2013, 24.

[3] H. Dierk and P. Nunnenkamp. "Inward and outward FDI and income inequality: evidence from Europe," Rev World Econ, 2013, 149.

[4] K. Taylor and N. Driffield. "Wage inequality and the role of multinationals: evidence from UK panel data," Labour Economics, 2005, 12.

[5] K. Sylwester. "Foreign Direct Investment, Growth and Income Inequality in Less Developed Countries," International Review of Applied Economics, 2005, 3.

[6] P. Chintrakarn and D. Herzer. "FDI and income inquality: evidence from a panel of US states," Economic Inquiry, 2012, 3.

Education Management and Management Science – Zheng (Ed.)
© 2015 Taylor & Francis Group, London, ISBN 978-1-138-02663-6

To strengthen the construction of college finance informationization, and raise the level of financial management in colleges and universities

Liu Liu

Guangxi University of Science and Technology, Liuzhou, Guangxi, China

ABSTRACT: Along with the development of higher education, the scale of higher education has been expanding fast, while profound changes have taken place both in the environment and the way of the financial work in colleges and universities. However, the financial work is much complex and meticulous, how should we do it well, which puts forward higher challenge, and highlights the importance of financial information. Based on the particularity of environment in colleges and universities, this paper analyzes the problems existing in the construction of financial informatization, and puts forward the necessity and urgency of the construction of financial informatization.

Keywords: college; finance; informationization

1   INTRODUCTION

As the scale of higher education expands, both the spending and the relevant funds rise as well, while personnel of the finance department personnel may not increase accordingly, which to some certain extent increases workload of the personnel. The trivial, complex and meticulous work may increase the psychological pressure of financial personnel, who feel their work is endless. Therefore, establishing financial informatization is necessary and urgent.

The construction of financial informatization in colleges and universities is an urgent and complex job itself, as the financial work becomes more complex, the request of the informatization is higher and higher, so it is necessary to form a unified, intelligence, integrated system to improve the quality and efficiency of university finance, to improve the quality of the overall management level and promote the development of teaching and scientific research of colleges and universities.

2   THE NECESSITY AND URGENCY OF THE FINANCE INFORMATIONIZATION CONSTRUCTION OF COLLEGES AND UNIVERSITIES

2.1   *Improving the quality and efficiency*

With the expansion of university scale, the computerization of the accounting becomes a quick trend inevitably, so we should only make full use of the information technology, which helps to expand the financial management work further. And it is beneficial to the internal control management, to improve the analysis ability and the financial analysis of colleges and universities, which also can improve the level of financial management, finish the huge financial data processing and analysis problems, improve the work efficiency.

2.2   *Help to information resources sharing*

On the basis of the accounting information system, with the modern Internet technology, integrating information resources in various aspects, processing and reprocessing the financial date deeply, discovering related information are helpful to the decision makers to make right decisions. On the other hand, the financial information should be shared between the various related units, such as tuition, teaching material and so on; which required to be connected with other departments so as to realize information resources sharing to increase timeliness and accuracy of the financial management information. So it cannot exist as a single isolated system, such as the financial management system, must be in combination with the school educational management information system, asset information, teacher information, student information to ensure providing a unified statistics. But in actual work, the phenomenon of information island is very serious, internal department fragmented, which is using the different application software, leading to the basic information can not be integrated finally, even a simple statistic result is not unified, differ in thousands ways.

2.3 *Help to the management decisions*

To make the right management decisions, we need to gather all kind of the financial information for comprehensive research and analysis, caring about the asset liquidity, making full use of the use of information we gathered to do quantitative assessment, so as to achieve reasonable allocation and usage of the money and understand the cash flow situation to prevent financial crisis to provide adequate basis for management decisions in the end.

2.4 *Help to the opening of the information*

By providing personalized information query, for example, the teachers and students authorized can query their own information, such as scientific research funds, salaries, tuition and so on. Setting up different personalized query can avoid the query flow of the students and teachers to finance department, and to some certain extent which can reduce workload of the financial personnel, and improve the work efficiency. So it is necessary to promote the opening of the financial information.

3 THROUGHOUT THE CONSTRUCTION OF COLLEGE FINANCE INFORMATIONIZATION, THE COMMON PROBLEMS EXISTING ARE AS FOLLOWS:

3.1 *The input of the information technology infrastructure is not enough*

The development of college financial information needs hardware like computers, servers, network, software such as financial software, which are basic conditions. But the input of the information technology infrastructure is not enough in our country, which prevents the development of the information construction seriously, and thus, affecting the service level and quality, making the operation mechanism chaotic, etc.

3.2 *The collection of college financial information is inaccurate*

The collection of college financial information involves different departments, so when it comes to the manpower, materials or other resources, the related departments often lack of corporations and communication, and often can not afford their responsibility in time. While there is no a formal way or standard for the gathering of financial information, which affects the efficiency of the accounting information collection, increases the information acquisition costs. So the financial system cannot exist as a single isolated system, which must contact with the students' information, asset information, educational information, to ensure that the statistics are accurate. But in practice, it is hard to make unified arrangement and planning, information isolated island phenomenon is common and serious. Since the financial date needs to be confirmed many times, this increases the financial personnel's workload and spend a lot time, and causes that the financial information could not be updated timely, some old information date even stays for many years.

3.3 *The team of financial informatization construction is slow*

The construction of financial management information needs a high efficient team, including leadership, the finance department staff and so on. But the personnel quality of university financial department is different in many ways, some cannot operate the computers well, some do not familiar with the accounting job. And due to the recruitment threshold of the colleges and low salaries, it is hard to recruit the high quality personnel to the financial department. The talent is a key factor in the construction of financial information in colleges and universities. The lack of financial talents in China is mainly manifested in two aspects: on the one hand, the leader cannot meet the requirements of the financial information management (such as strong coordination ability, innovation ability and insight). The other bottleneck is: it is hard to the staff being good at both accounting and computer tools in a long time.

4 THE SUGGESTIONS OF STRENGTHENING THE CONSTRUCTION OF FINANCIAL INFORMATIZATION

The financial information in colleges mainly uses computer technology to optimize the relevant financial data integration to make the financial management work becoming more convenient, quick, even without the limit of time and place, which not only can ensure the information date timeliness but also accuracy. Therefore, we should discover and solve the problems existing in the college finance informationization timely, and develop a solution strategy, to strengthen the financial management, promote the development of colleges and universities, which is the inevitable trend of era development, and is the necessary measure for colleges and universities to strengthen the financial management in the modern times.

4.1 Enhancing the leadership, increasing the cohesion

Only with the powerful management decisions of the leader, the department could operate smoothly, so does the construction of financial informatiztion, which needs the recognition and support from the decision-makers (such as the headmaster), if not, there would be not enough money to sustain the project.

4.2 Cultivating inter-disciplinary talents of high quality

To coordinate with the construction of college financial management informationization, a high quality management team also should be emphasized to settle the problem that the staff who do not familiar with accounting and computer tools. But how? since the team includes the department leader and personnel, one side, both of them should strengthen the learning of accounting and computer skills, another side, we should recruit some high quality staff who qualified in accounting and computers.

4.3 Strengthening the communication between departments

The financial informatization of colleges does not just involve the finance department only, which also involves the departments such as teaching, students, and technology etc. so, the accurate date should be unified in a formal way under the communication and cooperation between different departments, which is helpful to the information sharing and the construction process of informatization.

4.4 Establishing a high level system of financial information

To develop a suitable financial information system, the colleges and universities should combine with the actual situation of the school fully. At the same time, we should establish the consciousness of safe guard and strengthen the implementation of security measures, to construct a safe and efficient financial information network platform. And in daily work, learning the newest concept of management, improving the information system consistently, finding new problems and settling them, thus the information system can show its advantage and play the important role in the development of the high education.

REFERENCES

[1] Sophisticated management based financial information construction—Introduction to the use of the UF NC financial system, Finance and Accounting for international commerce, 2012(5), Qin Yuan Lu.

[2] Thinking of building financial decision support system in colleges, Dong Chen, Commercial Accounting, 2011(7).

[3] Reflections on college financial information construction. XiaoLi Huang, Entrepreneur World [J], 2011(7).

[4] Analysis of the plight and strategies to the university financial informatization construction. ZhiYu Huang, Jing Ye, Popular science & Technology, 2012(8).

[5] Analysis of the university financial informatization construction. HanRong Zhan, RongFang Chen, Communication of Finance and Accounting, 2012(4).

[6] Thinking about the financial informatization construction of the group company. Xin Meng, Finance & Accounting for communication, 2012(5).

[7] Thinking of the finance informationization construction of the western colleges. Yong Liang, Studies of Finance and Accounting in Education, 2012(2).

Education Management and Management Science – Zheng (Ed.)
© 2015 Taylor & Francis Group, London, ISBN 978-1-138-02663-6

Analysis on the asymmetric pricing behaviors of large retailers based on two-sided markets

Y. Han
School of International Economics and Business, Nanjing University of Finance and Economics, Nanjing, China

T. Han
School of Social Sciences, University of Southampton, Southampton, UK

ABSTRACT: As a new development in industrial economics, the theory of two-sided markets has been applied to study the behaviors of some industrial organizations with platform nature. Being a kind of two-sided markets, the behaviors of large retailers have some characteristics of platforms include asymmetric pricing as suppliers are charged slotting allowance and consumers are free except the price of commodities. This behavior can make the cross-group network externalities maximization, and bring the interests growth of suppliers and consumers when the retailers to achieve maximum benefits. In this paper, the asymmetric pricing of large retailers is described in order to explain the rationality of large retailers' pricing behavior, which does not belong to unfair competition and monopoly, so it is no need to be regulated by government.

Keywords: two-sided markets; platform firm; large retailer; asymmetric pricing; slotting allowance

1 INTRODUCTION

The theory of two-sided markets has been widely discussed by the academia and been a heated topic of the economics research after 21st century. Since this theory provides a new perspective to study industrial organizations and behaviors and explains some economic phenomena better than existing theories, it is widely applied to industrial organization problems of some special industries such as trade agencies and media in traditional industries and software, e-business platforms and credit card system in emerging industries (Chen & Xu, 2007). The research of two-sided markets has changed many viewpoints of classical theories. It will result in profound influences in both competition strategies of firms and governmental public policies.

Retailing is an industry with a very long history. During early times, the main function of retailing was subordinate, which enhanced sales for suppliers. Nevertheless, accompanied by the development of technology and economy, retailing has become a dominant industry in manufacturing sectors. This change inverted the relationship between retailers and suppliers as well. Taking advantage of their market power, some large retailers control the whole industry chain and charge suppliers slotting allowance for profit maximization. These behaviors of

retailers have resulted in controversies. Some scholars discuss that behaviors of retailers are abuse of market power and unfair competition that should be regulated by the government.

However, retailers are not common firms according to the theory of two-sided markets. As platforms, retailers have necessity and rationality to set prices and competition strategies under the rules of two-sided markets. Some domestic scholars have acquired contributing explorations in this respect. e.g., Shi Qi and Yue Zhonggang (2008) analyzed the market characteristics and policy implications of large retailers. Yue Zhonggang and Zhao Bo (2008) analyzed the problems of slotting allowance with the theory of two-sided markets. Shi Qi and Kong Qunxi (2009) gave opinions of access price, channel competitions and regulations of policies in payment with respect to retailers based on two-sided markets. Qu Chuang, Yang Chao and Zang Xuheng (2009) researched the competition strategies of large retailers in two-sided markets. Furthermore, Qu Chuang and Zang Xuheng (2010) discussed the strategies of slotting allowance for retailers by relating suppliers' scales to production diversities. As a new instrument and method to analyze market behaviors and performance of retailers, the theory of two-sided markets enriches and develops the theory of circulation economics. At the same time, it also provides a more scientific guidance to the formulation of policies for governments and adjustments of strategies for enterprises.

2 PROPERTIES OF PLATFORM AND ASYMMETRIC PRICING STRUCTURES OF LARGE RETAILERS

Retailers are intermediate traders who sell commodities to final consumers directly. Essentially, the fundamental function of retailers is the exchange of mediums that correlates suppliers with consumers through their business activities. Therefore, retailers have some basic features of platform firms. Because of these features, certain categories of retailers are treated as a kind of trade agency and one of the typical forms of two-sided markets including media, software platform and credit card system in many documents. Rochet and Tirole (2003) did a representative research. Similar to software, media and payment system, they attributed shopping malls to a kind of two-sided market as well, discussing that consumers and firms play the roles of correlated two sides. This view was accepted by many scholars and adopted in many sequent documents, especially in domestic documents. For example, Liu Qi and Li Mingzhi (2008) divided two-sided markets into four categories i.e. trade agencies, media, payment instruments, software platforms and attributed shopping malls to a kind of trade agency. Minli Huang (2007) paralleled shopping malls with e-business platforms, letting agencies, auction companies, search engines, securities exchanges etc. and attributed them to "market initialling" two-sided markets.

The properties of platform firms decide that the behaviors of large retailers must have features of two-sided markets. In other words, large retailers could set their own operating strategies by taking advantage of the interaction among two-sided users for the purpose of larger benefit. Similar to other platform firms, large retailers employ asymmetric price structure and pricing strategies as well. The users of large retailers can be divided into two parts. The first part is suppliers who can acquire services of distribution channels from retailers. The second part is consumers who can acquire diverse choices and services of commodities from retailers. Users of both parts have cross-group network externalities. The benefit that suppliers obtain from retailing platform depends on the scale of consumers, and the benefit that consumers obtain from retailers depends on the quantity of commodities and services offered by suppliers to such platform. Therefore, it becomes possible that the pricing mechanism of retailers is asymmetric. Generally, as the trade platform of two-sided

markets, only suppliers are charged slotting allowance by large retailers and consumers are free of charge on the other hand, i.e. they simply need to pay for the commodity prices.

This pricing behavior of retailers is decided by the quantity of users and the demand elasticity of suppliers and consumers. First of all, compared with the large amount of consumers, there are much fewer suppliers in the market with respect to the scale of users. According to Metcalfe's Law, the network externality is closely related to the scale of users, which shows an exponential growth with the increase in the scale of users. Hence, retailers prefer to choose the group of consumer (with more users) as the object of low-price strategy in order to ensure and enlarge the scale of this group such that the group of supplier (with fewer users) could be attracted effectively. This consequence would be much better than the low-price strategy based on suppliers. Secondly, there are significant differences between suppliers and consumers in terms of the demand elasticity. Because of lower switching costs, the demand elasticity of suppliers is lower than that of consumers. Thus, consumers have larger diversities of choices between different platforms. In two-sided markets, platform firms usually charge groups with low demand elasticity low prices and charge groups with high demand elasticity high prices (Ji & Guan, 2006). Accordingly, retailers charge suppliers high prices and charge consumers low prices in retailing markets for the sake of larger scale of consumers and ample network externalities that can create sufficient attraction to suppliers to make them be willing to pay higher costs for the participation in the platform, forming and maximizing the cross-group network externality between consumers and suppliers.

3 THE MECHANISM AND PERFORMANCE OF LARGE RETAILERS ASYMMETRIC PRICING FORMATION

We can use the instrument of welfare economics to derive the asymmetric pricing of retailers. Assume there is a large retailer with two-sided users of suppliers and consumers. As shown in Figure 1, the demand curves of suppliers and consumers are denoted by D_1 and D_2 respectively, where the slope of D_1 is less than which of D_2. This implies that the demand elasticity of suppliers is less than which of consumers. Setting P_0 as the initial price, the quantities of suppliers are the same (for the consideration of consistency, we can take different units, such as suppliers for 1, consumers for 1000), which are both Q_0 at this price level. In this case, the profit, which the retailer obtains from suppliers

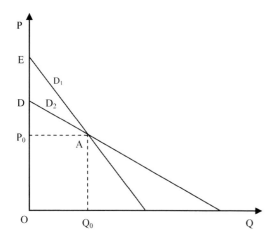

Figure 1. Same price of suppliers and consumers.

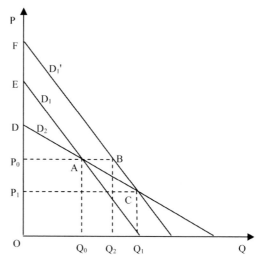

Figure 2. Constant price of suppliers and decreased price of consumers.

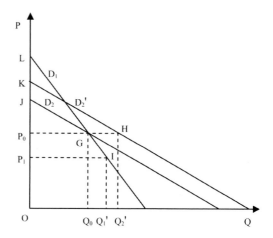

Figure 3. Constant price of consumers and decreased price of suppliers.

and consumers, are both the size of rectangular P_0 AQ_0O. Namely, the total profit is the value of size $P_0 AQ_0O$ multiplied by 2.

Firstly, let us analyze the strategy of high-price for suppliers and the low-price for consumers. The retailer keeps the price of suppliers constant at P_0 and diminishes the price of consumers from P_0 to P_1. Thereafter, the quantity of consumers will increase from Q_0 to Q_1. Meanwhile, due to the increased externality of suppliers caused by the increased quantity of consumers, suppliers will have a higher level of demand at the same price level. Thus, the demand curve will shift right from D_1 to D_1'. Correspondingly, the profit of the retailer changes to the value of size P_0BQ_2O + size P_1CQ_1O. Apparently, size P_0BQ_2O + size $P_1CQ_1O > 2 \times$ size $P_0 AQ_0O$. This formula illustrates the better situation under asymmetric prices now.

It should be noted that not only do the asymmetric prices bring about an increase in the profit of retailers, but they also contribute to a higher profit level of suppliers and consumers. As shown in Figure 2, the supplier surplus increases from size EAP_0 to size FBP_0 and the consumer surplus increases from size DAP_0 to size DCP_1, which means an increase in the total surplus i.e. the total welfare of the society.

Secondly, let us analyze the strategy of high-price for consumers and low-price for suppliers as a contrast. As shown in Figure 3, the retailer keeps the price of consumers constant at P_0 and diminishes the price of suppliers from P_0 to P_1. As a result of lower demand elasticity of suppliers, only lower increase in quantity could be achieved to Q_1' ($< Q_1$ in Fig. 1) with the same decrease in price. Meanwhile, due to the lower increase in the quantity of suppliers, the demand curve of consumers shifts only from D_2 to D_2' with a lower extent to D_1'.

Correspondingly, the profit of the retailer is the value of size $P_0HQ_2'O$ + size $P_1IQ_1'O$. Compared with the situation in Figure 1, the profit of the retailer indeed increases but is obviously less than the profit under the strategy of high-price for suppliers and low-price for consumers, i.e., $2 \times$ size $P_0 AQ_0O <$ size $P_0HQ_2'O$ + size $P_1IQ_1'O <$ P_0BQ_2O + size P_1CQ_1O. Moreover, compared with the situation in Figure 3, size $LIP1 <$ size $FBP0$ in terms of supplier surplus and size $KHP_0 <$ size DCP_1 in terms of consumer surplus. This demonstrates a worse situation for retailers, suppliers and consumers under the strategy of high-price for consumers and low-price for suppliers in comparison with the strategy of high-price for suppliers and low-price for consumers.

4 RECOGNITIONS OF RETAILERS' BUYER POWER AND REGULATION

Buyer power of retailers has been a focus of the research of the retailing industry. According to the theory of industrial organization, the definition of buyer power is "the ability to set and maintain the price lower than the competitive price". Further, the definition of buyer power of retailers given by Wang Zaiping (2007) is the credible threat that retailers can propose to suppliers. In other words, if suppliers decline the severe conditions offered by retailers, retailers could take action to suppliers, making the loss of suppliers much greater than that of retailers themselves. This definition actually implies an unfair circumstance to suppliers. Based on this recognition, some scholars hold a negative attitude to buyer power, arguing that the competition must be distorted under the existence of buyer power. Provided that the distortion happens, incumbent retailers must invade the benefits of suppliers and even consumers by using the privilege of growing market power. Therefore, buyer power is an object that should be controlled by the competition law.

However, the judgment of whether an enterprise commits monopoly or unfair competition requires to look at not only the behaviors, but also the reasons and results. The behavior should be intervened or regulated by the government only if the firm raises the price to acquire super-profit by taking advantage of some non-market factors (e.g. natural factors and human factors) because such behavior leads to a loss in the welfare of other communities and violates the principle of fair competition. Obviously, the behavior of asymmetric pricing and charging suppliers slotting allowance is not applicable to be intervened in this case.

Firstly, in terms of the reasons, retailers charge suppliers slotting allowance because of the group-cross network externality of two-sided markets. Retailers gather a large number of consumers by offering platforms and services, increasing the utility of suppliers. As compensation, it is reasonable for retailers to charge slotting allowance. This does not violate the market rule of fair trade but reflects the fair principle of the market. Secondly, in terms of the results, suppliers gain more profit from the platform system of retailers due to cross-group network externality in spite of the increased cost caused by slotting allowance. The welfare of suppliers will not decrease but even increase as a result of the slotting allowance provided that the

extra profit is greater than the increased cost. In this case, the behavior of retailers neither derives from non-market factors nor influences the profit of suppliers. On the other hand, it increases the profit of suppliers with the increasing welfare of retailers and consumers. Therefore, this behavior should not be judged as monopoly or illicit competition.

As a matter of fact, retailers cannot raise the price of suppliers without limitation based on the existence of buyer power. Being different to normal externality, the cross-group network externality in the two-sided markets is internalized. The internalization could be interpreted that the function of this kind of externality through the platform could be reflected by the market price. Moreover, each community could be constrained by the market when they are trading the result of externalities. In order words, the formation of prices in two-sided markets is still decided by the market mechanism. Behaviors of retailers cannot escape the constraints of the market. In this case, it is impossible for retailers to raise the slotting allowance higher than the extra profit of suppliers brought by the platform because the profit of retailers will deteriorate if retailers really perform this behavior.

Therefore, provided that the profit of suppliers is ensured, retailers should control the slotting allowance under the acceptable level of suppliers in order to achieve profit maximization. The government has no reason to intervene in behaviors of retailers due to their exposure in the market limitations and the effective function of the market mechanism. The market regulation will be damaged if the government reluctantly forces certain interventions in the slotting allowance charged by retailers. This situation could be considered as a "regulation failure" caused by "excess regulation" (Shi & Kong, 2009). Certainly, we cannot exclude the occurrence of monopoly and illicit competition of retailers caused by various non-market factors. Undoubtedly, in order to get rid of the adverse influences on the social welfare, the monopoly and illicit competition of retailers caused by factors beyond the two-sided market should be regulated according to certain laws and regulations. Nonetheless, this topic is beside the point because it does not belong to the market category.

5 CONCLUSION

This article discussed the price behavior of large retailers under the theory of two-sided market and analyzed the formation and the mechanism of asymmetric prices between suppliers and consumers. In addition, the article did a performance analyze and interpreted the market performance of the pricing behavior from retailers. Based on the research above, we can conclude that:

Firstly, large retailers are trade platforms with the characteristics of two-sided markets. The behaviors of large retailers are different from those of normal firms. Along with the expansion of scale and the changing role in the industry chain, large retailers are not mere trade mediums between production and consumption. They have a more important function to connect suppliers to consumers through platform services and generate cross-group network externalities. As platform firms, retailers have to decide their business strategies according to the features and rules of two-sided markets for the sake of sufficient reflection and maximization of the cross-group network externality between suppliers and consumers and therefore attain their profit maximization.

Secondly, large retailers have both necessity and reasonability to take asymmetric pricing strategies. In accordance with the characteristics of two-sided markets, platform firms should charge the users with larger network externality lower prices and charge the users with smaller network externality higher prices to compensate the loss caused by low prices and enlarge the market scale. Applying this strategy, such firms can ensure the formation and the maximization of the network externality. Specifically, in the two-sided users of large retailers, the scale of consumers with network externalities decides the effectiveness of the whole system. Hence, it is necessary for retailers to charge users with lower network externality certain fees.

Thirdly, the behavior of large retailers to charge slotting allowance is not a kind of injury to the interest of suppliers. On the contrary, it contributes to larger profit of suppliers. The slotting allowance paid by suppliers ensures that a large retailer supplier can provide more cheap goods and better service to consumers, which can attract and gather more customers and increase customer number, bringing more benefits not only for retailers, but also for suppliers. Visibly, slotting allowance is a necessary condition for suppliers to obtain interests through the trading platform and is compensated by cross-group network externalities. Therefore, slotting allowance does not harm suppliers but enhances the their welfare with the welfare enhancement of consumers and retailers.

Lastly, the buyer power of retailers is different from general monopoly power and should not be covered in government regulations because retailers are platforms formed in the two-sided market. In general, behaviors of firms are no longer affected by the market mechanism as a result of the formation of monopoly power, leading to decline in market efficiency and loss in social welfare. However, unlike the buyer power from retailers to suppliers, charge level is still regulated by market mechanism, which shall not be higher than the benefit of suppliers generated from network externalities in the platform. Since the market mechanism is still valid in the two-side markets, in general, the behaviors do not need to be intervened. The government should not mechanically apply regulation principles of one-side markets to large retailers. Instead, it should adjust and improve the existing laws and regulations according to the characteristics and requirements of two-side markets, providing an appropriate institutional and legal environment for platform enterprises including large retailers and promoting healthy development among them.

ACKNOWLEDGMENT

This research was financially supported by A Project Funded by the Priority Academic Program Development of Jiangsu Higher Education Institutions (PAPD) and A Project Funded by Jiangsu Modern Service Institute (PMS).

REFERENCES

[1] Chen, H. & Xu, L. 2007. Two-side: new perspective of enterprise competition environment, *Shanghai People's Publishing House.*
[2] Shi, Q. & Yue, Z. 2007. The two-side markets characteristics and policy implication of large retailers, *Finance & Trade Economics,* 29(2): 105–111.
[3] Yue, Z. & Zhao, B. 2008. Research on Slotting Allowances Based on Two-sided Markets Theory, *Journal of Shanxi Finance and Economics University,* 30(7): 12–16.
[4] Shi, Q. & Kong, Q. 2009. Access Pricing Channel Competition and Regulation Failure, *Economic Research Journal,* 44(9): 116–127.
[5] Qu, C. Yang, C. & Zang X. 2009. The competition strategy of large retailer in two-side markets, *China Industrial Economics,* 14(7): 46–75.
[6] Qu, C. & Zang, X. 2010. Supplier Size, Product differentiation and access pricing strategy, *Research on Financial and Economic Issues,* 31(12): 36–39.
[7] Rochet, J.C. & Tirole, J. 2003. Platform competition in two-sided markets. *Journal of the European Economics Association,* 1(4): 990–1209.
[8] Liu, Q. & Li, M. 2008. Theory of two-sided markets and platforms: a Survey, *Economic Problems,* 29(7): 17–20.
[9] Huang, M. 2007. Two-side markets and evolution of market conformation, *Journal of Capital University of Economics and Business,* 8(3): 43–49.
[10] Ji, H. & Guan, X. 2006. Research of two-sided markets and pricing strategy, *Foreign Economics & Management,* 28(3): 15–23.
[11] Wang, Z. 2007. Buyer power: welfare analysis and its public policy, *Journal of Shanghai University of Finance and Economics,* 8(4): 57–61.

Education Management and Management Science – Zheng (Ed.)
© 2015 Taylor & Francis Group, London, ISBN 978-1-138-02663-6

Construction and practice of multi-dimensional financial aid system oriented to post-1990s generation in financial difficulties—a case study of College of Hydrology and Water Resources, Hohai University

Yang Lu & Feifei Ji
College of Hydrology and Water Resources, Hohai University, Nanjing, Jiangsu, China

ABSTRACT: Post-1990s generation has increasingly become the major part of college students. Those in financial difficulty are characterized by the conflict between self-esteem and inferiority complex, sense of independence and dependence behavior, and the coexistence of learning and psychological problems. The current funding system has many shortcomings, which has been unable to meet the demand of growth and development of students in the new era. College of Hydrology and Water Resources of Hohai University has improved the funding evaluation system and expanded the funding support system oriented to post-1990s students facing financial difficulties, which strengthens the educational role of funding and builds a multi-dimensional funding system covering the whole process, across the board, and taking every aspect into consideration to provide a solid guarantee for the growth and development of students.

Keywords: post-1990s students; multi-dimensional funding system; construction and practice

1 INTRODUCTION

With the development of higher education in China, an increasing number of students have the opportunity to receive higher education, with the number of students in financial difficulties rising annually. Relying on the funding system oriented to students in financial difficulties and ensuring its smooth operation are the important guarantee for sticking to the implementation of the strategy of "rejuvenating the country through science and education" and "strengthening the nation through human resource development" as well as the important content of the practice of "scientific development" in education work 1. Currently, the university students are mainly composed of post-1990s generation. Their unique personalities have brought new challenges to the funding work in a university.

According to the characteristics of post-1990s students facing financial difficulties and with the goal of assistance and caring, providing educational service, and self-improvement and development, College of Hydrology and Water Resources, of Hohai University have combined the funding with the shaping of character, physical and mental development, capacity building, cultivation of and ideals and beliefs to explore the construction of a multi-dimensional funding system, which has achieved good results.

2 CHARACTERISTICS OF POST-1990S STUDENTS IN FINANCIAL DIFFICULTIES

2.1 *Great conflict between self-esteem and inferiority complex*

Many students come from economically disadvantaged families with low-income and huge burdens in remote areas. Due to differences in economic background and lifestyle, poor students often feel psychologically huge contrast, makes them unable to form an objective evaluation of the self, thus leads to strong inferiority. Meanwhile, despite financial difficulties, a lot of post-1990s students with financial difficulties, as the only child of their families, are spoiled by parents, with high self-esteem, and unwilling to fall behind others. Therefore, because of the conflict between the self-esteem and inferiority complex and afraid of losing face, they tend to give up the grant applications, and they even wear designer clothes so as to hide their inferiority.

2.2 *Serious contradiction between the sense of independence and dependence*

With the growing tide of reform and opening up, post-1990s generation accepts new things quickly. They are opinionate, independent and have strong sense of self, especially for students with financial difficulties, who have great desire to get rid of the plight short of money to pursue an independent life. However, due to lack of independence and social experience, their economic resources are still dependent on their parents. Some students, after receiving funding parties, are easy to get affected by passive waiting mood, without high self-motivation to overcome difficulties and demonstrating severe contradiction between dependence on others and independent consciousness.

2.3 The coexistence of learning and psychological problems

Post-1990s students from economically disadvantaged families are often sensitive and fragile, which will lead to inferiority and the lack of communication with the surrounding and parents who are busy working, thus developing into a mental state of loneliness. Virtual world often become the main channel to vent their demands. Some students with weak self-control become addicted to online games, resulting in poor academic performance. The lack of interpersonal communication, long-term self-repression and inner conflict and poor performance in seeking support can easily lead to psychological pressure and even distorted mental state. Thus, the economic problem is just a striking superficial issue of them, which is often accompanied by other learning and psychological ones.

3 DEFICIENCIES OF THE CURRENT FUNDING SYSTEM

3.1 Imperfect evaluation mechanism of the funding system

In the current funding system, the primary procedures to identify students with financial difficulties and provide subsidies for them are shown as follows: students submit applications and the panel of the class carries out democratic appraisal and then pass the applications to the college for final check. This process focuses on the identification and assessment of eligibility of students with financial difficulties, but lacks the corresponding performance evaluation mechanism. This kind of imperfect mechanism may lead a lot of students without financial difficulties to counterfeit materials for money, thus wasting the limited funding resources; on the other hand, some students squander the fund rather than apply it to study due to the lack of necessary binding, which is contrary to the original intention of funding.

3.2 Imperfect educational function of the funding system

In most colleges and universities, the current funding system emphasizes the financial aid, but lacks humanistic care. The mere financial aid is far from the requirement of promoting the growth and development of the post-1990s generation, improving their psychological quality and building interpersonal relationships. Let alone the cultivation of the integrity, social responsibility and dedication.

3.3 Lack of the incentives in the funding system

Currently, there has been a government–led and university-implemented funding system supplemented by the social support, thus various kinds of scholarships have been established, including national scholarships, national motivational scholarships, state grants, etc. The "free loans, grant-in-aid, tuition deduction" are the primary means, which may put students in a position of dependency and breed negative thoughts, thus lacking motivation. In addition, the free assistance is carried out based on the financial situation of students without considering their overall performance, so it fails to supervise and encourage students to study hard.

4 CONSTRUCTION AND PRACTICE OF THE MULTI-DIMENSIONAL FUNDING SYSTEM

The college of Hydrology and Water Resources, Hohai University has established a comprehensive funding system for post-1990s students with financial difficulties. This student-centered system focuses on the humanistic care, aiming to cultivate them into talents. The college is trying to improve and strengthen the role of education, thus building a multi-dimensional funding system to provide a solid guarantee for students.

The system focuses on three key points, the integrity, whole-staff-participation and comprehensiveness. To achieve the integrity, the funding activity stresses the process "Propaganda—Assessment—Publicity—Assessment—Updating" replacing the previous process with only two links, advocacy and assessment, which has strengthened the tracking and assessment, and improved the existing funding appraisal system; To achieve the whole-staff-participation, the social fund and part-time jobs, etc. are used to provide material support. In addition, teachers, parents, students and other forces are combined to form a linkage of "School—Family—Classroom—Community", which is conducive to encourage them to help each other and give them moral support; to achieve the

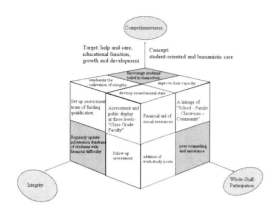

Figure 1. The diagram of the multi-dimensional funding system of the College of Hydrology and Water Resources, Hohai University.

comprehensiveness, combine the financial aid with students' character shaping, physical and mental development, capacity building to facilitate their growth and success.

4.1 Achieve an integrated evaluation system of funding

To effectively implement various policies about funding, the college has set up a special team with the deputy secretary as the leader, including instructors, class teachers, student representatives. This team is to be in charge of the application for scholarship, including the "award, loan and subsidies". This team should interpret the related documents and policies on a regular basis to ensure the standardization of assessment work.

Before the assessment, inform students the related polities via various channels, such as bulletin boards, class meeting, QQ group and so on, so that each student will have a comprehensive and objective understanding of the funding assessment system. In the assessment process, set up three teams constituted by members from three groups "Faculties—Grade—Class" and adhere to the "open, fair and impartial" principles. First the assessment team constituted by students to make a preliminary assessment for these applicants and then submit a recommended list; the second team composed of the instructors and student representatives is in charge of auditing this recommended list and then submits it to the faculty-level assessment team; the third team constituted by the college leaders, Deans, teacher representatives has the final determination for the recommended list. This kind of multi-dimensional assessment could ensure the scientification of assessment work.

In addition, to strengthen the tracking management, the college has established a electronic database for all the students with financial difficulties and updated it regularly every semester, in order to grasp the timely information; for students having been funded, their self-representation, communication with other students, counselors and parents could be used to conduct a comprehensive assessment of their state of life. Disqualify these dishonest students and guide them to use the fund in a rational way, thus achieving the integrity of the assessment.

4.2 Expand the funding support system to achieve the whole-staff-participation

The so-called support system means an individual could obtain material and spiritual help and support from others in his own social network. The funding support system includes not only the leaders, counselors, class instructors, classmates and friends, parents, etc., but also some social individuals and various organizations; both objective material support and subjective emotional support.

In addition to the existing subsides and scholarships, the college is actively expanding alumni resources and seeking help from social enterprises. Nowadays, the "1982 Alumni Scholarship" and "Jinshuiyanyu Scholarship" have been set up to provide financial aid for students with financial difficulties. In addition, various work-study jobs will be provided for them, such as the office assistants, laboratory assistants, assistant administrators, by which students could improve their abilities and make some money. Counselors and tutors communicate with students and their parents regularly to form a linkage mechanism of "School—Classroom—Community—Family"; in addition, a counseling team constituted by students could help them in terms of learning, living, etc. The diversified multi-channel support system could make use of the network to achieve the whole-staff-participation, thus providing material and emotional aid.

4.3 Play the function of education and promote the comprehensiveness

The funding system could promote the ideological and moral cultivation of students. In the document *On Further Strengthening and Improving the Ideological and Political Education of College Students* issued by the State Council, the ideological and political education should not only educate and guide students but also care about and help them. This means the ideological and political education should be used to solve the practical problems, thus helping these students with financial difficulties. In addition to the funding assessment and financing services, the college should emphasize the moral development, humane care and set some role models to develop good qualities like integrity, enhance their comprehensive ability and the spirit of self-strengthening.

4.3.1 *Take the high moral values establishment and people cultivation as the fundamental task to strengthen the moral integrity*

Honesty is the traditional virtue of the Chinese nation, so we should carry out people-centered and moral education-based education. Cultivating the integrity of the students with financial difficulties is conducive to promote the development of social credit system. Therefore, the college has implemented various activities, such as the lectures about integrity and legal knowledge, essay competition with the topic of integrity & gratitude expression, aiming to build flourishing campus climate. Focus on freshmen and seniors and help them firmly establish the consciousness of integrity by the class meetings, viewing videos, etc.; for the seniors, ensure the repayment of the loan.

4.3.2 *Take the humane care as the basic to promote mental health*

Students with economic difficulties are usually self-esteem, sensitive, and have poorly communicate with others, which causes to become unsociable and eccentric. Therefore, the timely psychological counseling is important so that it could reduce the psychological burden and shape healthy psychological psychology. The college adheres to the education philosophy of "the highest good is like water", which means to cultivate the dedication, cohesive force, tenacity and inclusiveness of students, especially those with financial difficulties. By some outdoor activities, psychological drama, microblogging communication, students would learn to be devoted to their study, cooperate with each other, and be strong and inclusive.

4.3.3 *Take the growth and development of students as the core and improve their ability*

As the old Chinese saying goes teaching one to fish is better than giving him fish, it is necessary to explore the potential of students when giving them financial assistance, for it could improve their comprehensive abilities. The college always adheres on emphasizing the growth and success of students and encourages them to participate in various activities such as recreational activities, social practice, aiming to develop their interpersonal skills and enhance their self-confidence as well as their innovative awareness, etc. In addition, simulated training of job market, intercommunion meetings are all carried out to improve their innovation ability, practical ability, employment ability, etc. thus achieving self-improvement.

4.3.4 *Set a good model to encourage them to strive constantly*

Low self-esteem always troubles those students with economic difficulties. To eliminate their inferiority and encourage them to strive constantly, it is necessary to help them to develop self-reliance and self-confidence. The college of hydrology and water resources has set up some scholarship to elect some good models who will encourage students to strive for the further, overcome the difficulties and never give up.

5 PRACTICAL RESULTS

There are 400 undergraduates with financial difficulties each year in College of Hydrology and Water Resources, accounting for about 1/3 of the total number. According to the statistics, 85.3% of households are located in rural areas, with a large number of family members and low-income parents. 75.9% of poverty-stricken students come from single-parent families or those with severely sick parents. 46.5% of them rank below the average in the same grade, and 15.6 percent have difficulty in learning (GPA is 3.0 or less and failing to pass CET 4). 65.4% rarely or never participate in any extracurricular activities, and 22.8% are in interpersonal difficulties.

According to the characteristics of post-1990s students and coupled with the actual situation, the faculty has established a multi-dimensional funding system and implemented it. Based on the original national scholarships and those at school levels,

Indicators	2011	2012	2013
The Number of Work-study Positions in Our College	15	18	24
Financial Assistance Coverage (%)	75.5%	80.3%	86.9%
Average Grade Point of Students with Financial Difficulties	3.32	3.46	3.59
Financially-challenging Students with Learning Difficulties	72	64	47
Financially-challenging Students with Psychological Problems (people)	4	3	0
Financially-challenging Students Obtaining Awards (at the school-level or above)	112	134	165

Figure 2. Achievements of the multi-dimensional funding system.

28

we have actively expanded resources of alumni and corporates, starting "82 Alumni Scholarship" since 2006 and the "Jinshui Yanyu Scholarship" since 2012, which are oriented to students with financial difficulties and currently have a total payment of more than hundred thousand Yuan. In addition, new work-study programs have been added each year to strengthen the financial aid, making the coverage of financial aid increase from 75.5% in 2011 to 86.9% in 2013.

Along with the financial aid, the faculty, relying on guidance activities of "seniors' class", strengthens the assistance to students with learning difficulties in the form of carrying out one-on-one mutual help and counseling services on the preparation of exams to enhance students' self-confidence and courage. The accumulated number of aided students in the near three years is more than 700, with the number of students with learning difficulties declining yearly.

In addition, the college also conducts comprehensive guide for students with financial difficulties, aiming to improve their mental health. Setting a good model could inspire them to strive constantly. For example, Min, a male student in our college, was from Shiyan of Hubei Province, and his parents are all famers, and his mother had a serious disease. The low-income, coupled with the medical expenses, makes the family bear huge economic burden. Therefore, this boy is self-abased and lacks communication with his classmates and friends. He is unwilling to participate in the group activities and not very active in study. Someday, he bursts potassium disease and a coma developed; he even wanted to drop out of school, as he believes that he cannot afford the high medical expenses. The college leaders timely communicated with his parents and paid all the medical expenses to the first time the full advance medical expenses, to put away his worries; his classmates and friends took care of him in the hospital and helped him ease the pressure. Then after the recovery, he was given the allowance and hired as the assistant of the college office, in charge of the letter distribution, for this could encourage him to communicate with others. In addition, he joined the summer internship in Qingdao, which was conducive to improve his professional skills and self-confidence. By this kind of multi-dimensional funding assistance, he has gone through the tracking evaluation and was awarded the scholarship. Besides, he has signed an employment agreement with the Water Resources Commission of Yangtze River.

In the nearly three years, due to this funding system, more than 400 students with financial difficulties in our college have obtained national scholarship, Xu Chilun scholarship and First Prize of National Mathematical Contest, the Fourth Place of the Third National Mountaineering Race, the Third Place of the Sixth National Energy Saving Contest. 23 students have been awarded the title of Provincial Excellent Youth League, Outstanding Party Member, Top Ten Outstanding Students, etc.

ACKNOWLEDGMENT

This paper is the research result of jiangsu province's project "Research and Practice on Integrated Revolution of Recruiting, Training and Employment of Professional Degree Postgraduate Students". The provincial project number is JGZZ13-020.

REFERENCES

[1] Xiao Liang. On the management of the funding for college students from poor families. Technological innovation Herald. 2012 (56).
[2] Wang Zhaohui study. Research on the mental funding of college students from poor families. Silicon Valley. 2009 (22).
[3] Zhao Anyong. Combination of funding of college students with financial difficulties and education. Higher Agricultural Education. 2012 (8).
[4] Shen Qianqian, Liu Diaozhao. Research on the Establishment of the Funding System of Post-1990 College Students with Financial Difficulties. Modern Property • Modern Economic. 2012 (11).
[5] Yang Xiaolie, Duan Tianqing. Make Every Effort to Build a Funding system and Promote Financial Assistance Work of Students—Financial Assistance Work of Gasu Polytechnic College of Animal Husbandry & Engineering. Teachers Forum. 2013 (2).

Explore new accounting standards and the quality of accounting information

Xi Chen

QiLu University of Technology, Jinan City, Shandong Province, China

ABSTRACT: February 15, 2006, the Ministry of Finance officially released the 39 accounting standards, including a basic standard and 38 specific accounting standards. This marks the country's socialist market economy, and in full coordination with the International Financial Reporting Standards, covering all types of enterprises of various economic activities, independent implementation of the new accounting standards system is basically set up. The impact of new accounting standards on the quality of accounting information systems, including both positive and negative aspects, but overall will help improve the quality of accounting information.

Keywords: new accounting standards; quality of accounting information; investigate

1 INTRODUCTION

February 15, 2006, the Ministry of Finance issued new accounting standards and regulations within the scope of practice of listed companies, to encourage other companies to perform since March 2007 l January 1, 2007. Enterprise Accounting Standards issued by the Accounting strengthened to provide useful information to investors and the public new ideas, such as the introduction of the "Replacement cost", "Net realizable value", "Present value" and "Fair value" and other more significant impact on the relevance of accounting information generated measurement attribute.

Quality of corporate financial accounting information is provided in the report of the basic requirements of the quality of accounting information, the financial report is to provide investors and other users of accounting information useful for decision-making should have the basic characteristics of the provisions of the basic criteria, which include reliability, relevance, understandability, comparability, substance over form, materiality, prudence and timeliness.

Practice, only with the characteristics of accounting information quality of accounting information is valuable. The developed system is based on the accounting standards to improve the quality of accounting information as a starting point. The new accounting standards issued by the accounting information to improve what extent have the success criteria related to the development of the system, the problem is also of great concern to the academic community.

2 THE IMPACT OF THE NEW ACCOUNTING STANDARDS ON THE QUALITY OF ACCOUNTING INFORMATION SYSTEM

The impact of new accounting standards on the quality of accounting information generated from the key quality characteristics of accounting information is the reliability and relevance of earnings manipulation and prevention of these three aspects were discussed.

2.1 Reliability

Reliability refers to accounting information disclosed in the company must be objectively true and fair view of the financial position, operating results and cash flows, as much as possible to reduce translation costs "asymmetric information" generated, increase the value of information. Both the "concept of fiduciary duty" or "decision-usefulness theory" emphasized the reliability of accounting information quality standards. Compared with the old system of accounting standards, more emphasis on the reliability of the new accounting standards of accounting information. Main features.

2.1.1 *"Basic criteria" aspect*
"Accounting Standards for Enterprises–Basic Standard" plays a role in the guidelines to govern the system, the criteria and the criteria, it is clear that the most important quality characteristics of accounting information is reliable, we should take transactions or events actually occurred as

a basis for accounting recognition, measurement and reporting accurately reflects the recognition and measurement requirements of the accounting elements and other relevant information, to ensure true and reliable accounting information, content and complete.

2.1.2 *"Specific criteria" aspect*

The new accounting standards for financial instruments, investment property, enterprises under common control, aspects of debt restructuring and non-monetary transactions are carefully using a fair value. Some people worry that the fair value will be abused and will reduce the reliability of accounting information. Actually it will not, compared with the scope of the international accounting standards in the fair value of the applicable accounting standards of Chinese enterprises in determining the fair value of the range of applications, a fuller consideration of the situation of our country, made prudent improvements. Use of fair value must meet certain conditions in Article 43 of the basic criteria clearly pointed out the use of replacement cost, net realizable value, present value, measured at fair value, should ensure that the identified accounting elements able to obtain and measured reliably. For specific guidelines on, for the use of fair value measurement, there are clearly defined limits. As in non-monetary assets exchange for use of the new standards of fair value in accordance with the provisions of the two prerequisites for non-monetary assets exchange process, namely whether the exchange has commercial substance, the existence of related party transactions between the parties. Again, the new standards require subsequent measurement of investment property, the cost model should be, if there is conclusive evidence that its fair value can be reliably achieved, can also be measured using the fair value model. These preconditions will effectively restrict the fair value measurement of earnings manipulation behavior. Thus, the application of fair value is strictly limited conditions, not allowed to be abused, so long as the strict implementation of the guidelines in accordance with the fair value will really be fair. In addition, the new accounting standards from the provision for impairment of assets, improving to determine the scope of capitalization and asset inventory valuation methods are reflected in the quality of accounting information reliability characteristics.

2.2 *Relevance*

New standards in its "Enterprise Accounting Standards—Basic Standard" under Article 4, the goal of financial accounting reports accounting information relating to the financial and accounting reports provide users with the enterprise's financial position, results of operations and cash flows, reflecting the corporate management fulfillment of fiduciary responsibilities, financial accounting reports to help users make economic decisions. It highlights the usefulness of the report of the decision, highlighting the relevance of accounting information, accounting information required to be provided with financial and accounting reports related to the user's economic decision-making needs. Embodied in the following aspects.

2.2.1 *The introduction of a variety of measurement basis of fair value, etc.*

The new accounting standards in the accounting measurement attributes include historical cost, replacement cost, net realizable value, present value and fair value. Diversification measurement attributes can effectively enhance the relevance of accounting information, to provide more information to aid their decision-making for investors, creditors and other stakeholders.

2.2.2 *To establish new standards in the core of the apparent balance*

The apparent balance refers to the accounting standard setters in the development of accounting standards governing certain types of transactions or events, you should first define and standardize the measurement of assets or liabilities generated by such transactions. Then, according to revenue recognized changes in assets and liabilities defined. Obviously, the new accounting standards system than before, more emphasis and focus on accurate measurement of assets and liabilities. For example, the new accounting standards for intangible research and development project expenditures handling requirements: Expenditure internal research and development projects, should be distinguished research expenditures and development expenditures were handled. The cost of capital and corporate in-house research stage investment, because it can not bring economic benefits, research costs are expensed handle future for the enterprise. When entering the development phase, with the transformation of research results, in fact, have a specific benefit from the development phase of the object (i.e. a particular commodity) as a carrier, enterprise research and development activities related to, conduct a commercial activity on its nature and basis relevant data can predict or measure the inflow of future economic benefits flowing. Therefore, the cost incurred in the development stage companies, in line with the conditions of capitalization should be capitalized. This changed our practice in the past too much emphasis on GAAP income statement, a reasonable proportion of income, profits first, in the measurement and amortization expense in the income statement of priorities, strict compliance with the principle of matching a certain extent conducive to business management authorities to develop long-term research and development programs, but also conducive to external users of accounting information for the future development of enterprises to make operations more objective predictions.

2.3 The new accounting standards for the control of profit manipulation

First, the new accounting standards system more compressed accounting estimates and accounting policy choice of projects, defining the spatial extent of corporate profits regulate, regulate and control the business of profit manipulation, cosmetic results of operations, improve the quality of accounting information. For example, inventory valuation issued, canceled "LIFO", all using the "FIFO method." The change in guidelines will make the enterprise no longer be able to change the method of inventory valuation adjustment of profits as a means to promote business inventories true reflection of actual historical cost and reduce the factors of manipulation.

Secondly, the new standards increase the disclosure of related party transactions. Corporate profits tend to use to manipulate the associated lack of fair trading, profits transferred from one party to another, so as to achieve the purpose of manipulating profits. According to the principle of substance over form, the new criteria for the definition of a related party made a clear expansion, including a series of three categories of enterprises with, joint control and significant influence. And whether or not the occurrence of related party transactions. Related parties with controlling companies should be disclosed in the notes in the report of the relationship between parent and subsidiary companies, is a multi-layered investment control, the association transaction should be disclosed to the bottom level enterprises. In addition, related parties transaction, cancel the amount or proportion of disclosure choices, seeking companies must disclose the transaction amount, a significant proportion of the transaction shall also disclose the transaction amount and type of transactions accounted for the total transaction amount; outstanding items required to disclose detailed information and amount; stressed only in providing sufficient evidence, companies can disclose related party transactions using the same terms and fair trade. Visible, the new standards increase the scope and content of disclosures related party transactions, will ever be more general and specific requirements of clarification, enhance the transparency of related party transactions.

Finally, the new accounting standards promulgated, will effectively curb the business combination under common control behavior of individual companies and the merger date fair value of the abuse and then adjusting the profit and owners' equity. Compared with the old merger guidelines, new standards for accounting for business combinations under common control do a new requirement that the assets and liabilities acquired in a business combination, it shall be measured at the carrying value of the merged party on the merger date, to abandon the use of fair value. This is mainly for the current merger of companies under common control most of the corporate merger, consolidation is a confirmation of the price of the two sides formally recognized at fair value, but in fact is not true "fair value."

3 CONCLUSION

By the previous discussion, the new accounting standards system in terms of reliability and relevance of accounting information have made a significant contribution to earnings manipulation and traditional means of curbing play a good role, which is bound to generate accounting information of enterprises positive impact. We can say that the implementation of new accounting standards will help improve the quality of accounting information, enhance financial transparency and provide more reliable and relevant accounting information for the financial report users.

REFERENCES

[1] Ministry of Finance: Corporate Accounting Standards (2006), Beijing: Economic Science Press, 2006 (2).
[2] Li Jidong Ma Shengxiang: The impact of new accounting standards for listed companies to manipulate profits analysis, Accounting, 2006 (7).
[3] Liu Yanguo moon: Analysis of the new accounting standards, accounting Anhui Water Resources, 2007(3).

Education Management and Management Science – Zheng (Ed.)
© *2015 Taylor & Francis Group, London, ISBN 978-1-138-02663-6*

Exploration of cultivating research students in excellent course construction

Jing Wang, Xiaoliang Chu & Ying Li

School of Information Science and Engineering, Ocean University of China, Qingdao, China

ABSTRACT: In excellent course construction, it is more important to inaugurate a way to cultivate research students through the teaching, in addition to course content, teaching methods and tools as well as website construction. The teaching object is the students whose major are the Optical Information Science and Technology. In order to cultivate students of practical ability, innovative and research ability, we develop a scientific and effective way combined with the major own characteristics and school conditions, by making the scientific research penetrate into teaching content of specific courses with the scientific Optics and Optoelectronics Laboratory conditions, and focusing on cultivating students' research ability and innovative consciousness in teaching. This teaching way has achieved very good results in the teaching practice with a certain promotion value.

Keywords: excellent courses; research student; research ability

1 INTRODUCTION

In the construction of excellent course, the mode of research and teaching mutual promotion is implemented through three aspects, i.e. research and professional teaching and that the graduation thesis, scientific research and teaching of the interest groups, professional course teaching and practical experiments. And the problem how to effectively mobilize the initiative and participation of students is paid attention to solve during implementation of the three aspects. From 2009 to 2011, many measures were taken such as research laboratory opened for students, the laboratory as the research base of the thesis and research projects as the support of the thesis. The implement of the measures has achieved good results.

2 COURSE PAPER AND CULTIVATION OF STUDENTS' BASIC SCIENTIFIC RESEARCH QUALITY

In the teaching of optical fiber communications, the teacher guides the students to independently raise questions, study and solve problems by creatively applying knowledge and synthetically using research methods, according to the curriculum content and knowledge accumulation of students. And the teaching form is taken for studying, creating and cultivating. The purpose of research teaching also includes other contents, such as encouraging students to master the research methods, cultivating the research spirit of students, inspiring the students with research enthusiasm. And the students will be trained to become a certain research ability person through the activity of research teaching. In other words, the purpose of research teaching makes the student not only master the basic knowledge and basic skills, but also learn how to apply the knowledge and theory studied from the course to solve the problems faced. So when the students face problems in the field, they know how to find literature, how to carry out theoretical analysis, and how to try experimental methods to solve the problems. The research teaching need teacher from concerning about students' mastery of subject knowledge, ability of imitating and reproducing the book knowledge to paying attention to the cultivation of the students' ability including a large amount of information collection, analysis, judgment, reflection and application.

The research teaching process is shown as Figure 1.

In accordance with this concept, we implemented the research teaching methods with the "research topic" as the main line, from 2009 to 2011. The topic of research paper can be self-made and also be determined under the guidance of the teacher. In the three school years, there are total of 109 completed course work including 103 replied. This allowed students to experience various steps of research work. We carried out in the student survey about the research teaching methods.

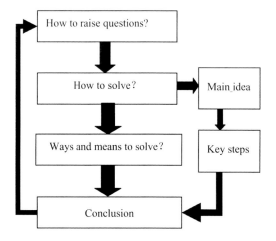

Figure 1. The process research teaching.

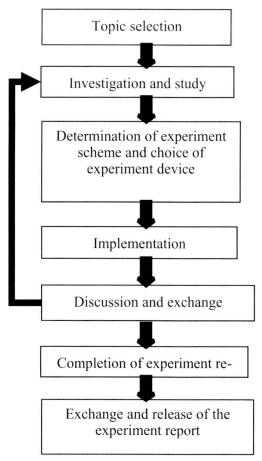

Figure 2. Methods of the self-design experiment.

The result showed that 96% students approved to carry out practical activities with research papers in the teaching, 92% students believed that "research learning" was large harvest through the whole term. At the same time, students gave many good suggestions for carrying out the research teaching methods.

3 TEACHING OF RESEARCH EXPERIMENT AND CULTIVATION OF STUDENTS' INNOVATIVE ABILITY

We implement the research experiment by two steps. One is to inspire students to design the experiment using existing laboratory experiment with addition few fittings; the other is to apply the software Photonic Transmission Design Suite (PTDS) to finish the various transmission experiment of optical fiber communications systems.

The methods used in the design experiment are shown as Figure 2. From Figure 2, we can see that the methods carried out need four steps. First, according to the experimental content and the request, the students need to consult literature, draws up the experimental scheme independently and choose the experiment instrument and effective method for determining data. Then, they can carry out the experiment and get the experiment result. Thirdly, the results were analyzed and experiment paper was written. At last, the results were exchanged and published through the reply and the discussion.

Based on developing application of the software PTDS, we has achieved a typical optical fiber communication system modeling and a variety design and transmission of optical fiber communication system simulating. We weigh the transmission performance of various communication systems through the indices such as bit error rate, eye diagram. The software is fully in English, however, students can design and complete experiment after they study the software. The experiments are explained and discussed in the class. And this make the student not only improve English level, but also summarize basic knowledge which they have learned in each chapter with better understanding. The teaching methods are popular with students and have good effect.

In short, in the research experimental teaching, the instructor's basic task includes indicating the experiment direction, examining experimental schemes of the students, organizing the students to explain and discuss the experimental process and helping the students to improve their experimental schemes. During the experimental process, the teacher pays attention to observe operation of the students and solve students' problems.

Furthermore, the teacher also needs to organize students to analyze and discuss the experiment results and the written papers, and evaluate the experimental results obtained. At the same time, the teacher finds the shortcomings of the results and gives the suggestions of improvement and help students to improve their ability of scientific experiment.

4 CULTIVATION OF STUDENTS' RESEARCH ABILITY COMBINED WITH THE UNDERGRADUATE THESIS

In the 7th semester, the students who are interested in this course begin to develop some preliminary work for the thesis with the guidance of teacher, such as consulting literature, exploring the theory, familiar with algorithms and programming. This lays a good foundation for the dissertation carried out in the 8th semester.

As second class with laboratory, the teacher focuses on training the students who are interested in the course and lets them participate in the teacher's scientific work. So the students can master the basic method of scientific research, carry out the research work and write papers which contact with the dissertation. All of this establishes a good foundation for a high level of thesis. Since 2001, there are 31 undergraduate theses about optical fiber communication and 8 science paper most of which are published in EI indexed journals.

All of undergraduate theses were related to the research projects, and so the research level of undergraduate theses was greatly improved. Moreover, the equipments of research laboratory were utilized more sufficiently. At the same time, the progress of scientific research was promoted by thesis published. Undoubtedly, the students are the biggest beneficiaries, because their research capacity has been greatly improved in further studies and they are well received by employer.

ACKNOWLEDGMENTS

We would like to thank financial supports from the Excellent Course of Shandong Province and Ocean University of China.

REFERENCES

[1] Rong Yuan, 2010, *Optical Fiber Communications (3rd edition)*, Beijing, Electronics Industry Press.
[2] Yan-Biao Liao, 2003, *Fiber Optics*, Tsinghua University Press.
[3] Gerd Keiser (writer), Yuquan Li (translator), 2004, *Optical fiber communications*, Electronic Industry Press.
[4] Qian Zhang, Chunyan Li, Zhi-Hong Fu, Dongping Xiao, 2010, National excellent Course construction of "Theory of Circuit", *Journal of electrical & electronic engineering education, 32 (1):26–28.*
[5] Yi Wang, Yan Liu, 2008, Realization of the excellent courses Construction, *China Metallurgical Education, 4:15–16.*
[6] Puyun Gao, 2010, Consideration on the excellent courses construction of Theoretical Mechanics, *Journal of Higher Education Research, 33 (3):64–65.*
[7] Shanli Duan, Zhenjuan Dong, 2005, Excellent Course should reflect a focus on "living" character—on Rethinking excellent Course Construction, *Chinese University Teaching 8:17–18.*

Education Management and Management Science – Zheng (Ed.)
© *2015 Taylor & Francis Group, London, ISBN 978-1-138-02663-6*

The effect and existing problems of nursing skills training in specialized secondary schools of Hebei province

Xin-zhong Gu
School of Medical Education, Hebei University, Baoding, Hebei Province, China

Fu-qing Yang
Department of College English Teaching and Research, Hebei University, Baoding, Hebei Province, China

ABSTRACT: Based on the review of 24 years of nursing skills training in specialized secondary schools in Hebei province, this article has made an analysis of the impact of skill training to the students and the development of nursing education, aiming to explore the existing problems in the training work. Meantime, the article has suggested some measures to the problems, as reference for the peers and management departments to improve the nursing education.

Keywords: nursing education; skills training; training objective; school construction

1   BACKGROUND AND PRESENT SITUATION OF THE TRAINING IN SPECIALIZED SECONDARY SCHOOL

Nursing skills training is a professional skill training, which requires students to be successful, skillful, and standard to operate the basic nursing skills according to clinical program, so that they can enter the internship and employment successfully. Nursing skills training in specialized secondary school of Hebei province (hereinafter referred to as the training) was carried out many years ago in the domestic. In 1990, based on the National Textbook "Basic Science of Nursing", "Medical College Nursing Skills Operation Standard and Scoring Method of Hebei Province" was compiled, and in the later application successively in 1991, 1993, 1998, 2001, 2003, 2005, 2006, 2007, 2012, and 2013, it has experience ten times of revision and perfection. In 2004, the province organized experts to design and make the "basic nursing skills manipulation"—a teaching CD which matches the standard and scoring method, Guaranteeing the training work to go smoothly (Health and Family Planning Commission of Hebei Province, 2013).

For more than 20 years China has experienced the period of building a well-off society, the state has paid much efforts to develop the nursing education vigorously. The Ministry of Health continuously release the "Chinese nursing career development plan" during the 11th and the 12th Five Year Plan, in 2008 The nurse regulations was promulgated and came into implementation. In China, nursing industry has been attached unprecedented importance, which greatly motivates and inspires the development of nursing education. During the 1900's, there are more than one hundred secondary health schools that enroll students in Hebei province, while only twenty more survived till this year (2014). The rise and fall of the schools is affected by the national policy and the management, in addition, the quality of education is undoubtedly the most basic condition for their survival.

Operating nursing skills is one of the core content of nursing education, the training of nursing skills has great effect on the development of the nursing education in the specialized secondary schools. Studying the successful experience of the training is bound to improve the effect of nursing education in Hebei province and even in the nation wide.

2   EFFECTS OF TRAINING

At the first stage of nursing skills training (years of 1990 to1994), some schools were actively involved into the training project. The graduates from these schools had done very good job on the internship posts. They were recognized not only by the training teachers and their fellow students but also by the hospitals and society. In the internship hospitals, they earned good reputation for their fulfillment and competence. More and more students who had undergone the training of nursing skills successfully get employed and get identification professionally. Therefore, the trained students benefited a lot from the training by successful employment, their parents get the follow-up benefit because the parents would never worry about the children's employment. On the other hand, the schools who were engaged in the training project also get a lot of benefits because they received good fame and got social identification at the same time. The achievements of the earlier stage of training attracted many more other schools to engage into the training project. Till 1996, the training work was in full swing in secondary vocational schools in Hebei Province.

2.1 Effects on the students

Nursing skills training being the core, the students—the training object, are the ones that can mostly reflect the effect and achievements of the training. Taking nursing theory as the premise, the students' basic skills as the core, this training project has also reached the students' psychological quality, communication ability, humanities and other aspects, hence constituting a comprehensive training. To a certain extent, the purpose of nursing skills training is to create a qualified nursing staff, so, this project has great and far-reaching effects on both the students' professional thoughts and skills. Those who can smoothly get through the training and assessment can easily get access to their internship and become qualified, hence successfully formally employed and get professional growth.

2.2 Effects on the school construction

The effects and achievements of the nursing skills training, ranging from the construction of laboratory to the strengthening of teaching staff, have generally seized the attention of the leaders of secondary professional schools. In the 1990's, secondary specialized schools were generally weak in nursing laboratory, even weaker than the basic subject labs such as biology lab and chemistry lab. With the extension of nursing skills training, on one hand, in order to lay a material foundation for the training project, each school has poured large amount of capital into the development of nursing education and nursing skills training to equip some new type of training rooms which are similar as the clinical nursing room. On the other hand, much attention has been paid to the professional development of the teaching staff. Because teachers play an essential and leading role in the training task, the growth of the students can not go without highly qualified teaching staff. Since the beginning of the training project, all of the schools attach great importance to the introduction of excellent expert teachers and the training of their internal teachers. From the government level, the Health Department and Education Department of Hebei province also organize teachers' skill training and communication, enforcing to strengthen the professional quality of secondary nursing school teachers.

2.3 Effects on the secondary nursing education

Through the strengthening of secondary nursing professional skill training, the school nursing teaching staff has been strengthened, and the laboratory equipments have been improved and bettered, which in turn have greatly improved the education quality, therefore, both the quantity and the quality of students have been greatly improved obviously. According to the statistics, the graduates from various secondary nursing schools changed from 734 in 1991, 5141 in 2001, and 15540 in 2011. During the 20 years of development of medical education, graduates from the major of nursing have increased by more than 20 times. No graduate in 1991 could find a job in Beijing or Tianjin cities, or went abroad to find a job, while 453 graduates went to Beijing and Tianjin, the number to the two large cities increased to 2381 in the year of 2011 (Health Department of Hebei Province,1991).

3 EXISTENT PROBLEMS AND COUNTER-MEASURES

3.1 The objective restriction of the number of trainees

The purpose of nursing skills training is to bring up a batch of qualified graduates, so all the graduates are required to engage in the training. When the schools began to expand enrollment, more teaching staff were required to meet the needs. However, the number of teachers and improvement of training equipment lag behind the expansion of enrollment, around 2000 the contradiction between the requirement of more teachers as well as better equipments and the increasing needs began to be highlighted. Education, as one of the social public welfare resources, is conditioned by its own law of development, which is difficult to meet the challenge from the marketization of education. Considering this situation, even though all the schools took every means to improve the school operation, adopted various kinds of measures and tried their uttermost efforts to carry out training programs, they still could not satisfy the objective needs of the students training, causing the quality of training to decline.

To solve this problem, which is also the biggest challenge ever since the beginning of the training program for more than 20 years, a top-down strategy is the best way out, that is, the government is expected to take macroeconomic regulation to control the enrollment according to the macro planning of education, preventing the public welfare of education from becoming commercialized in enrollment.

3.2 Contradiction between training content and clinical practice

The final aim of nursing skills training is for the students to do well in the clinical application, but the actual application in clinical medicine is changing rapidly with many new things happening daily, which evidently results in the contradiction between students training contents and practical application.

The solution to this problem lies in nothing but improving and perfecting the training contents. Meantime, the teaching staff should take the initiative in the annual training. Only when the teachers have grasped the essence and the new standards of the courses, can they transmit them to the students. Ever since the "Medical college nursing skills operation standard and scoring method of Hebei Province" was compiled and published, it has been undergoing the dynamical modification for 10 times, aiming to realize the maximum closeness to the clinical utility. In addition to the content and technical improvements, the revised standard and scoring method has also, from the point of view of clinical practice, given weight to the humanistic care in nursing work requirements (Zeng-xue Zhang, 2011).

3.3 Contradiction between aim and assessment

Though having gone 24 years, the training program having been adopted into the teaching schedules by almost all the schools, the prospect of carrying out the program is not optimistic yet. It is well known that the purpose of training is to make and cultivate many batches of qualified nursing talents, but the assessment of the responsible departments does not somewhat go with the purpose of the training. Generally, it is the Department of Health and Department of Education who join each other to carry out the assessment. Usually, they insist on the principle of "full participation and random sampling" to assess the effect of nursing skills training. While, affected by some factors, some new problems arise in recent years of evaluation and assessment. One is that the results of the assessment are publicized to the society. The publication of the results is a two-edge sword. On one hand, it enables the schools to value even highly the training program, while on the other hand, some schools may take unfair means of competition in order to pursue the excellent evaluation results unilaterally without considering the other aspects, which goes against the final aim of the training program. Another problem happens after the year of 2008, when the assessment was canceled for some reason. Since then, some schools were not very active to take part in the training and assessment program, more and more schools following the suit. Still another problem is that some schools put less and less stress on the training and assessment organized on the provincial level, they turn to the pursuit of the nationally organized "Olympic" type of excellent students training and competition. The purpose is so clear and obvious, because the workload is greatly reduced, the personnel and the financial investment is also accordingly reduced. No negative effect can produce even if the result is not good; while if proud result comes it can be vigorously preached.

REFERENCES

[1] Health and Family Planning Commission of Hebei Province (ed.) 2013. *Medical college nursing skills operation standard and scoring method of Hebei Province*: 41. Baoding, Hebei University Press.
[2] Health Department of Hebei Province (ed.) 1991, 2001, 2011. *The medical education statistics of Hebei Province* (internal reference).
[3] Zeng-xue Zhang. 2011. Evaluation of clinical medical professional skills training in higher medical colleges and Universities. In *Medical Research and Education*. vol. 28: 83–92. Baoding, Hebei University Press.

Research on the relationship between fashion culture and ideological and political education for the youth in China

Lei Xu & Ming Jiang
Shanghai Second Polytechnic University, Shanghai, China

ABSTRACT: As the culture extraction with features of the times, fashion culture is characterized with innovativeness, entertainment and consumption, etc. With the development of economy and global spreading of information, fashion culture has almost penetrated into every corner of our social life and created enormous influence over ideology of the youth. Being the teaching staffs in China, we should understand and accept the existence of fashion culture in an open-minded way, have a serious attitude towards its influence over the physical and mental development of the youth, investigate the effect of fashion culture on the growth of the youth in a new perspective and explore new measures for ideological and political education for the youth in the new era.

Keywords: fashion culture; the youth; ideological and political education

1   INTRODUCTION

Fashion is the popular dressing styles or behaviors within a certain period. As a kind of subculture phenomenon, fashion culture displays the most welcomed life style, value orientation, aesthetic taste and requirement for recreation and entertainment socially within a certain period.

With the development of economy and global spreading of information, the influence of fashion culture has almost penetrated into every corner of our society in China, surely including both the daily and school life and the ideology of the youth.

The youth is energetically curious about the world and accepts the fangled quickly, ideologically at the forming stage of philosophy, worldviews and values. Therefore, they are easily influenced by the trend of fashion, taking the pursuit of fashion as the important presentation of their prime of life and personalities. Fashion culture changes the attitudes toward life and values of the youth gradually and has created new opportunities and challenges for the ideological and political education for the youth at the same time.

To help the youth treat fashion culture with a correct and healthy psychology, it's important for the teaching staff to understand and accept the existence of fashion culture in an open-minded perspective, comprehend the features of fashion culture in detail and profoundly analyze the influence and impact experienced by the youth from fashion culture. Besides, the ideological and political education for the youth can't be comprehensively implemented unless the society assumes the educational responsibilities it's supposed to and families perform their educational obligations accordingly.

2   FASHION CULTURE

As a kind of subculture phenomenon, fashion culture is the cultural extraction of time, characters and displays the most welcomed life style, value orientation, aesthetic taste and requirement for recreation and entertainment socially within a certain period.

2.1   Dominant group

The dominant group of fashion culture is supposed to be comprised of the young and middle-aged people with independent financial capacity. However, the youth little by little has become the majority of fashion pursuers as they are accessing the latest information and technology with the popularization of information technology. They reach the goal of expressing individuality and the world in their heart through knowing, recognizing and accepting fashion. It's the result of conscious choice of the youth as well as the actual reflection of social life on the group of people.

The period in universities is a special stage for Chinese youth. They are filled with curiosity, feel pleasant to accept the new fangled, have a strong desire of self-presentation under the condition of living together with many others and look forward to joining the society and getting recognized by the society as soon as possible. Their attention can be easily attracted by the fashion trend and they have sufficient time and energy to spend focusing on, trying and chasing it as well, even getting addicted to have people's eyes on them by doing this. During such a process, the fashion culture has changed the attitudes toward life and values of those students at universities, bringing about new opportunities and challenges for the ideological and political education on campus.

2.2 *Characters*

The fashion culture is significantly characterized with innovativeness, entertainment and consumption, etc. by the times it belongs to and able to have people sensible desires fulfilled and bring them with sensory pleasure while it's not exactly a kind of delicate culture.

Continuous innovation is the essence of culture fashion. One of the main characters of fashion is fast and sustainable change. Only by changing all the time can fashion keep being focused on and imitated by people and then obtain an eternal activeness.

Entertainment is the main function of fashion culture. People will acquire pleasure and satisfaction mentally by experiencing the sense of superiority of the "minority" through the unique fangled when chasing the fashion, finally realizing self-entertainment both physically and mentally.

The carrier of fashion culture is certainly the commodity and the consumables produced in a great number, while fashion always displays itself in the form of some consuming trends, which is closely related to the consumer culture. Therefore, the fashion culture also has a magnificent character of consumption.

3 INFLUENCE ON THE YOUTH

Despite the enormous influence of the fashion culture, there are benefits as well as bad effects at the same time. As the teaching staff, it's unreasonable to simply ignore, reject or abandon it. Instead, attention should be paid to how it can influence the multi-dimension of the youth.

3.1

The fashion culture will facilitate the innovative thought development of the youth while it also may mislead their value judgment.

The fashion culture always comes up in the front of the youth with a completely new appearance and presents itself as unique by consciously opposing, rejecting and even mocking the tradition. Every change of fashion means an innovation, which is not only a breakthrough of behavior, but also an extension of life style, to some extent changing the attitude toward life and values of people. That the youth accept the fashion culture actually means that they accept the new "concept" and experience the new "trend". When after the fashion culture, the innovation idea of the youth has been inspired unknowingly and they gradually become the tester, acceptor and accelerator of fashion.

However, the standard of fashion isn't truth, goodness and beauty. To be extraordinary from tradition, fashion may get the strangest, useless and even the most ridiculous elements "dressed up" as the most fashionable. Even if it's the most mendacious, ugly and vulgar, it can also be called fashion when it's new and unique. Such a character, as mentioned above, would influence the value judgment of the youth and even sometimes make them confused about the standard of what is true, good and beautiful, weakening their concept of behavior and ethics norms.

3.2

The fashion culture plays a positive role of relieving the youth's study stress, but it may mislead their consumption concept.

Education for exams, ranking in class and expectations from parents have put too much pressure on the shoulders of the youth at such an early age. In the meanwhile, the entertainment of the fashion culture will arouse an emotional resonance of people by sensible figures and real-time pleasure, providing pleasant, exciting and satisfactory mental experience. Regarding the youth, it obviously becomes the most possible way of getting the study stress relieved.

However, the fashion culture is consumption-oriented. Under the circumstance of consumerism in the dominant place, the pursuit of fashion will end up as consumption of fashion commodity and even consumption of the symbolic meaning of commodity. While enjoying fashion, the youth may get into a bad habit of luxury consumption, competing consumption, conspicuous consumption or impulsive consumption, and thus form unhealthy consumption concept.

3.3

The fashion culture will facilitate the individuality presentation of the youth, but their personality concept may be misguided.

Fashion is an effective way of expression for the youth expecting to present their individualities. The fashion culture prepares a carrier for the expressive individualities of the youth and a way of releasing their rebellious psychology through various symbols, popular languages, behavior styles and dressing fashion, etc.

At the same time, the fashion culture may also weaken many spiritual conceptions including moral reason, aesthetic judgment, profound ideality and human solicitude, etc. having the youth gradually enhance and enlarge their own fashion desire unconsciously and unknowingly get indulged in hedonism and inclined to the halfhearted attitude of life, finally resulting in lack of ambition, giving up pursuit of sublime spirits and forming of ideal personalities, and weakening the desire of full development as human beings and ultimate solicitude for human.

Therefore, the fashion culture will also play a negative role in ideal personality development of the youth.

4 EDUCATIONAL MEASURES

To help the youth treat fashion culture with a correct and healthy psychology, it's important for the teaching staff to understand and accept the existence of fashion culture in an open-minded perspective, comprehend the features of fashion culture in detail and profoundly analyze the influence and impact experienced by the youth from fashion culture. Fully utilizing the social and family resources will realize the teaching staff's effect on education for the youth and comprehensively implement the ideological and political education for the youth.

4.1 Campus culture

What's the most important to ideological and political education for the youth is that utilize the effect of universities to introduce correct values. The school is always dominant in education. Teaching in class, campus culture and student activities, etc. are all "main fields" of the ideological and political education for the youth.

First of all, the teaching staff must persist in correct public opinion direction, show them what is correct philosophy and values when in class and help them master the method to dialectically treat the fashion culture, get rid of the habit of blindly pursuing the fashion culture and its negative influence. At the same time, through giving the reviews of the news, the teaching staff might show the youth the right way to see the social phenomena.

Second, take advantage of school culture and push forward the righteousness on campus and cultivate humanistic spirits through campus broadcasting and propaganda, etc. to help the youth gain knowledge, widen their minds and promote

their aesthetics through education. The teaching staff might use teaching on class and guiding in extracurricular activities to carry out aesthetic education on the youth, leading them to find the meanings and interests of classic and elegant culture and improve the ability of choosing culture willingly.

Thirdly, guidance shall be given to the club events organized by the youth to help them grow into good habits of self management, self education and self development and get them exercised in an atmosphere of diligence and positiveness so as to establish a world view and values that are modernization-, world- and future-oriented. Put ideological and political education into club activities on campus to realize educating through entertainment.

Last but not least, open psychological counseling hotline and online forums on fashion culture might improve the psychological health level of College students. The youth is vulnerable to psychological problems for they're experiencing changes in society, family and self. And over-pursuit of fashion is also easy to cause psychological obstacles to some extent and personality distortion. Psychological counseling hotline and online forums help the youth express their feelings, build mutual trust, go out of psychological perplexity, eliminate inferiority, lay down their burden, build confidence, and develop good attitude and personality.

4.2 Social atmosphere

The teaching staff must focus the social environment where the youth live and help create a healthy atmosphere of law, morality and public opinion for better growth of the youth.

Selected reading materials after class shall be suggested for them to help them extend their minds; answers shall be given to the existing social problems for the youth to help cultivate a critical spirit; a model figure with sense of the times shall be established for them to help them form an ideal personality.

Besides, new media of network, e.g. micro-blog (Chinese twitter), have great social influence now. It is popular among the youth, because their originality interactions are in accordance with the youth's high qualities, activeness and participation. The new social network atmosphere online makes new and higher demands for political ideology education. So educating staff should fully realize the social and cultural influence of new media and take advantage of them to teach political ideology online.

Furthermore, the patriotic education effect of moral education bases and social practice bases for the youth must be highly thought of at the same time. Therefore, more social resources shall be utilized to explore diversified education methods of ideology and politics.

4.3 *Family environment*

The teaching staff should help to build up a new family atmosphere and facilitate positive examples of the parents, because the family plays an important role to ideological and political education for the youth.

For the teaching staff, the effect of family atmosphere on education for the youth shouldn't be neglected. Instead, assistance shall be provided to parents in building up a healthy family atmosphere for better growth of the youth. The teaching staff shall communicate with parents as much as possible to understand the family background of the youth and their behaviors at home in order to know their key mental pressure and emotional issues, trying the best to solve problems at the early stage.

In the meantime, help parents to set a good example for the youth, correct what is wrong in their home education, assist them in choosing the best methods of education to reach goal and finally guide the youth to establish health philosophy and values.

5 CONCLUSIONS

Nowadays, the ideological development status of the youth is receiving great attention from the society day by day. It's quite urgent to find out how the education shall deal with the challenge brought up by the fashion culture to strengthen and improve the effective ways of morality education.

The ideological and political education for the youth should also be reviewed from multiple perspectives under the background of new era. An effective mode of ideological and political education for the new era shall be constructed in order to take positive measures to accelerate healthy growth of the youth.

ACKNOWLEDGMENT

This research was financially supported by Shanghai Second Polytechnic University. (XXKPY1316)

REFERENCES

[1] Barnes, R. and Eicher, J.B. (ed.) (1992). *Dress and Gender: Making and Meaning.* Oxford: Berg.
[2] Braham, P. (1997). 'Fashion: unpacking a cultural production', in P. du Gay (ed.), *Production of Culture/Cultures of Production.* London: Sage.
[3] Craik, J. (1993). *The Face of Fashion.* London: Routledge.
[4] Finkelstein, J. (1991). *The Fashion Self.* Cambridge: Polity Press.
[5] Flugel, J.C. (1930) *The Psychology of Clothes.* London: Hogarth Press.
[6] Hebdige, D. (1979). *Subculture: The Meaning of Style.* London: Methuen.
[7] Hu, Y. (2001). The Study of New Methods of Ideology and Polities Work in the Network Era. *Journal of Tsinghua University (Philosophy and Social Science),* 16:1, 16–21.
[8] Jiang, J. (2002). Internet and Ideology and Politics Education in Universities. *Journal of Higher Education,* 23:1, 83–85.
[9] Liu, M. (2000). Modern Mode of Ideology and Politics Education: A Network Construction. *Journal of Henan Normal University (Philosophy and Social Sciences Edition),* 2, 103–106.
[10] Steele, V. (1985). *Paris Fashion: A Cultural History.* Oxford: Oxford University Press.
[11] Wilson, E. (1985). *Adorned in Dreams: Fashion and Modernity.* London: Virago.
[12] Wu, W. (2005). Fashion Culture and Ideological and Political Education in University. *China Youth Study,* 9, 66–68.
[13] Zhang, Y. (2005). On Characters of Ideological and Political Education. *Thoughts, Concept and Education,* 2, 4–10.

Education Management and Management Science – Zheng (Ed.)
© 2015 Taylor & Francis Group, London, ISBN 978-1-138-02663-6

Research on occupational burnout of female employees in Chinese beauty industry

Lei Xu

Shanghai Second Polytechnic University, Shanghai, China

ABSTRACT: Occupational burnout is a state of emotional, mental, and physical exhaustion caused by excessive and prolonged stress within workplace, which is typically and particularly found within the human service professions. As a service industry, the Chinese beauty industry is growing rapidly, and more than 70% of employees are women. Compared with other employees, female employees in beauty industry suffer more pressure from the public and families, including the conflicting roles between work and home, which causes their occupational burnout and the frequent personnel change in the beauty industry. Improving management to intervene and/or prevent burnout would enhance female employee's work enthusiasm, which plays a key role in stabilizing human resource of beauty enterprises and industry in China.

Keywords: occupational burnout; female employees; Chinese beauty industry; improved management

1 INTRODUCTION

Occupational burnout or job burnout is characterized by exhaustion, lack of enthusiasm and motivation, feeling 'drained' (Potter 2005), and also may have the dimension of frustration and/or negative emotions and cynical behavior, and as a result reduced professional efficacy within the workplace (Maslach 1996).

To be defined more accurately, Burnout is a state of emotional, mental, and physical exhaustion caused by excessive and prolonged stress, thus the emotional exhaustion refers to the depletion or draining of emotional resources, from which refers to the indifference or distant attitude of work, and reduced professional efficacy refers to the lack of satisfaction with past/present expectations (Dierendonck 2005).

In recent decades, with the rapid development of economy, people are paying more attention to the pursuit of beauty resulting in the corresponding evolution of beauty industry in China. The latest statistics displayed that employees in Chinese beauty industry, including cosmetics, hairdressing, manicure, slimming etc., were more than 1.2 million, which made beauty industry one of the largest service industries. 76.4% of them were women, with an average age of 25.5.

The results of a survey "Employees' psychological health in 2012" on Chinese human resource website showed that 27.54% of the employees had a certain degree of psychological problems respectively, and up to 15% suffered from occupational burnout. Much more than others, 32% of female employees in beauty industry admitted their occupational burnout. Their dissatisfaction with jobs and frequent job-hopping had affected this industry's growth in China.

Therefore, the management should be improved to intervene and/or prevent the occupational burnout to enhance female employee's work enthusiasm, which will stabilize human resource of beauty enterprises and industry, and promote the development of Chinese beauty industry.

2 METHODOLOGY

In the research, I have interviewed 31 female employees from beauty industry aged between 20 and 45 in 17 cities of China on their own feelings about occupation and pressure. A questionnaire was designed to get the information about their personal details, education background, job situation, causes of pressure, ways to ease pressure, and attitude towards occupation. Based on the analysis of these fieldworks, several phenomena were discovered.

3 PHENOMENA

First of all, beauty firms suffer from notoriously busy revolving doors. Although they were still working in the beauty industry, almost everyone whom I interviewed, once at least, had the idea to switch to another job or leave this industry. 12 women, 40% of them, had the experience of switching from one company to another, while 3 had done it more than twice.

Second, occupational burnout is typically and particularly found within the human service professions. Among the 31 female employees I interviewed, 14 women served in beauty shops or salons and 8 were saleswomen. Half of them thought that their jobs were very boring and their values were not realized. This proportion was much higher than that of those who managed in beauty enterprises or manufactured in cosmetics factories.

Third, people who are most vulnerable to occupational burnout are those who are strongly motivated, dedicated, and involved in the work. As work for them is the source of importance in which they derive meaning in life, it is significant that they find meaning by achieving their goals and expectations. A phenomenon has been discovered from the interviews that the female employees who have got Bachelor degrees in spare time but not yet at management positions showed the most anger towards their employers and the inclination of quitting the job.

Last but not least, usually occupational burnout is associated with increased work experience, increased workload, but female employees' burnout increases with confliction between work and family, rather than work experience. In my interviews, the burnout of these women reached the peak when they were 25 to 30 years old, which were about 5 to 10 years after they began to work, and the first 3 years after they had babies. After that special time, their burnout was obviously eased with different levels.

4 INFLUENCE

Occupational burnout is a type of stress condition and as such results in concentration problems or decreased problem solving abilities. Ignored or unaddressed burnout can have significant consequences, including: excessive stress, fatigue, insomnia, depression, anxiety, negative spillover into personal relationships or home life, vulnerability to illnesses, etc.

Occupational burnout also affects social relationships and attitudes making interactions at home and at work difficult either because of the social withdrawal of the burned-out person or of making him more prone to conflict (Potter 2005). Withdrawing is a type of defense mechanism but in fact this has a negative effect because of the importance of social interactions for one's well being.

Burnout problems may lead to general health problems because of the stress becoming chronic, symptoms like headaches, colds, insomnia may appear together with overall tiredness, although burnout itself is not an ailment and is not recognized as a neurosis (Potter 2005).

Furthermore, as far as the enterprises and industry are concerned, occupational burnout of employees, resulting in absence from work, lack of creativity, reduced work enthusiasms, efficiency and quality, will give negative impact on management, the stability of staff, and their growth.

5 CAUSES

Generally speaking, occupational burnout is caused by many factors: lack of recognition, inadequate pay, conflicting roles, demanding clients, etc., while female employees in Chinese beauty industry were standing more pressure than others. These pressures can be summarized as follows:

5.1 Popular prejudices

In China, the social status of employees in beauty industry were considered as lower than that of other people of different occupations, because they did the blue-collar jobs or service. There was an old saying in China that "a good scholar can become an official". The official was the best occupation in the Chinese traditional concept. As a result, people weren't proud of laborers and those who were in service positions. They even had prejudices against them sometimes.

Moreover, in the long-term work, adhering to the industry principle "customers are God", employees in beauty industry were prone to have inferiority complex. Their suppression of own emotions and personality might cause the indifference and numbness.

At the same time, some of the negative news, such as problems of raw material, plastic accidents, high profits, etc. aggravated the public bias against beauty industry and the physical and mental pressure of the employees.

5.2 Dysfunctional workplace dynamics

In the workplace, inequality is the most important reason for female employees losing their motivation to work. Although "equal pay for equal work" is a basic principle in enterprises, in practice, paying the same time and efforts, female employees didn't receive the same treatment as their male colleagues receive.

Another reason why burnout is so prevalent in female employees in beauty industry is due in part to the high stress environment, emotional involvedness, and outcomes that might be independent of the effort exerted by the working individual. The complex interpersonal relationship could also lead to occupational burnout of employees.

Compared to employees in other industries, beauty industry employees face more complex interpersonal relationship. They had to face the relationship with bosses, superiors and colleagues, and at the same time, the guests, clients, and whom they served.

Feeling isolated, working with an office bully, undermined by colleagues or micromanaged by the boss might bring more stressed. Lack of trust from bosses, cooperation from colleagues, and envy between employees could let the female employees lose the sense of security and organization, then involved in burnout.

5.3 Work-family imbalance

What made female employees most depressed in occupations is that they could not keep their work and family in balance. Almost all of my interviewees felt that jobs took them so much time and energy that they couldn't spend adequate time with their families and friends.

Because of the specialty of beauty industry in China, employees had the different work schedule from others. Considering their customers need their service in spare time more than work time, beauty industry employees had to work in holidays or weekends, and some of they worked from 9 am to 10 pm every 4 days of a week. These made them lose the chances of family gathering in weekends or holidays and communication with their families.

Besides, Chinese hold the thoughts that women should pay more attention to family, take care of families, and educate children. Otherwise, women were unfortunate if they did great job but had unhappy families.

Facing the conflict between work and families, pressure form jobs, families and traditional opinions brought female employees guilt and frustrations, making them blame work, complaint jobs, and then have psychological burnout.

5.4 Mismatch in values

When employees' values differ from the way their employer does business or handles grievances, the mismatch may eventually take a toll. The desire to achieve high goals and expectations may collide with physical, emotional and mental exhaustion resulting from an inability to achieve them, which can lead to a type of burnout that may involve even a reflection on the failure to find meaning and growth in life (Dierendonck 2005).

Compared with male employees, opportunities of promotion for female employees were less in beauty enterprises and industry in China. If there was a vacancy of senior manager, male employees were more possibly promoted than females, although female staffs were competent enough with better work experience. The factors, such as age, gender and plan to give birth, set so many obstacles to women's careers, and caused the promotions of women were far more difficult than those of men, especially the promotions for senior managers.

Unfairness and unrealized personal values could produce emotional exhaustion and alienation, which might affect female employees' enthusiasm for work, satisfaction for jobs and commitment for organizations, and easily lead them to job burnout.

6 PREVENTION AND INTERVENTION

For the purpose of preventing occupational burnout, various stress management interventions have been shown to help improve employee health and wellbeing in the workplace and lower stress levels. Training employees in ways to manage stress in the workplace have also proven effective in prevention of burnout.

One study suggested that social-cognitive processes such as commitment to work, self-efficacy, learned resourcefulness and hope might insulate individuals from experiencing occupational burnout (Elliott 1996). Increased job control is another intervention shown to help counteract exhaustion and cynicism in the workplace (Hatinen 2007).

Besides the psychological adjusting by employees themselves, improving management to intervene and/or prevent burnout would enhance female employee's work enthusiasm, which plays the key role in stabilizing human resource of beauty enterprises and industry in China. These ways can be summarized as follows:

6.1 Adjusting values

Active measures should be taken to guide female employees' values to their jobs, which might help them to understand the importance of their work. Efficient management could help them establish self-esteem, confidence and sense of honor of their jobs, and overcome the negative emotions.

Female employees should realize that the meanings of work for women are same to men. Everyone have the rights to work, be promoted or success in workplaces. The traditional views of service industry and beauty industry in China are gradually changing. Work might be the best way to realize values of them. Success doesn't only mean standing in management position, but also mean continuously learning from work, constantly improving oneself, and creating more values in workplaces.

6.2 Humanizing management

Beauty enterprises might satisfy their female employees with better working environment, children care centers, and flexible holidays plan.

It is proved that a good working environment improves satisfaction of female employees, because women pay more attention to around than men. Decorated offices, reduced noise, cleansed air and increased fitness equipment give women chances to exercise, relax, rest and relieve fatigue, thereby reducing the incidence of burnout.

Many female employees have met the situation that they have to take care of their children while working. A care center for children could help them very much. And the flexible holidays plan might help them get more chances to take part in family gathering party, to communicate with families, and to keep up with their children's growth.

6.3 Providing fairness

Providing female employees reasonable incomes and promotion opportunities are the best way for beauty enterprises to show their fairness, and to intervene occupational burnout of female employees.

Provided "equal pay for equal work", removed their career obstacles, offered skills train, shared the latest information, built reasonable promotion system could help female employees with their professional development, and meet their needs of achievement. And the establishment of effective communication systems, smooth communication channels, communication between employees and employers without delay ensure that information is shared timely and the staffs got assistance on time.

With these measures, fair working environments might be established gradually, which could enhance female employees' respect, trust and support to the enterprises.

6.4 Offering help

To the female employees who are concerned about occupational burnout already, employers and managers might take direct actions to help them.

Psychological counselors could be hired in firms to give employees regularly psychological counseling and suggestion. In every possible way, help should be offered to employees to solve psychological problems, release stress, and relieve tension. Paid attention to their mental health, found the psychological problems as early as possible, intervened and treated burnout on time are the efficient measures to intervene the occupational burnout of female employees.

If possible, enterprises might offer an Employee Assistance Program (EAP), which would prevent that job burnout would undermine employees' health.

7 CONCLUSIONS

Occupation burnout is a complex psychological problem, caused by various reasons. Burnout experts believe that in order to reduce occupational burnout, a strategy of combining both organizational and individual level activities may be the most beneficial approach to reduce the main symptoms. Improving upon job-person fit by focusing attention on the relationship between the person and the job situation appears to be a promising way to deal with burnout (Maslach 2001).

Compared with other employees, female employees in Chinese beauty industry suffered more pressure from the public and families, which caused their occupational burnout and the personal flow frequently in beauty industry. Improved management, humanized regulations, provided fairness in work, and offered direct help or intervention would alleviate their burnout, ensure their physical and mental health, and ensure the healthy development of the industry.

ACKNOWLEDGMENT

This research was financially supported by Shanghai Second Polytechnic University. (XXKPY1316)

REFERENCES

[1] Dierendonck, D., Garssen, B., & Visser, A. (2005). Burnout Prevention Through Personal Growth. *International Journal of Stress Management*, 12(1), 62–77.

[2] Elliott, T., Shewchuk, R., Hagglund, K., Rybarczyk, B., & Harkins, S. (1996). Occupational burnout, tolerance for stress, and coping among nurses in rehabilitation units. *Rehabilitation Psychology*, 41(4), 267–284.

[3] Hatinen, M., Kinnunen, U., Pekkonen, M., & Kalimo, R. (2007). Comparing two burnout interventions: Perceived job control mediates decreases in burnout. *International Journal of Stress Management*. 14(3), 227–248.

[4] Maslach, C., Jackson, S., & Leiter, M. (1996). *Maslach Burnout Inventory Manual* (3rd ed.). Palo Alto, CA: Consulting Psychologist Press.

[5] Maslach, C., Schaufeli, W.B., & Leiter, M. (2001). Job burnout. *Annual Review of Psychology*, 52, 397–422.

[6] Potter, B. (2005). *Overcoming Job Burnout: How to Renew Enthusiasm for Work*, Ronin Publishing.

Education Management and Management Science – Zheng (Ed.)
© 2015 Taylor & Francis Group, London, ISBN 978-1-138-02663-6

Public watercolor art curriculum teaching method in science and engineering universities

Jing Xu
Shanghai University of Electric Power, Shanghai, China

ABSTRACT: The relationship between watercolor and public artistic education and the features, teaching methods of public watercolor curriculum are discussed. Innovative concepts are raised to public watercolor teaching modes by exploring the effect of emotion, innovation thoughts and moral education that watercolor art will take on science and engineering university students, and according to the characteristics of science and engineering university students.

Keywords: science and engineering universities; public artistic education; watercolor teaching method

1 INTRODUCTION

For the past few years, with the deepening of the education system reform and changing of the education concept in colleges and universities, the focus on the humanistic quality education is on increasing trend day by day; public artistic education is facing an unprecedented new opportunity. In the year 2001, Ministry of Education promulgated "National School Artistic Education Development Plan (2001–2010)". In the plan, it is proposed that "Schools of all types and at all levels shall deeply strengthen art curriculum teaching, provide enough time guarantee for art curriculum under the guidance of the state regulations and requests, carry out rich and colorful cultural and artistic activities through art courses, to develop and improve students' aesthetic ability and artistic appreciation the same time when they learning basic art knowledge and skills, promote students to develop in a comprehensive and harmonious way." In year 2006, it is also proposed in "Public Art Curriculum Guidelines for National Common Colleges and Universities", that "Public art are elective courses set to cultivate high quality talent people for the need of socialist modernization drive, art courses are in an irreplaceable role in improving the aesthetic quality, cultivate the innovative spirit and practice ability, and shape a healthy personality". In January 2014, Ministry of Education issued "Opinions on Promoting the Development of Artistic Education", it pointed out clearly that "Artistic education has unique and important effect on talent cultivation. Art education will cultivate students' ability to feel, to express, to appreciate and to create the beauty, guide students to establish a correct concept on

beauty-appreciation, cultivate noble moral sentiment, raise profound national emotions, stimulate the imagination and innovation consciousness, and promote all-round development and healthy growth for students". This further deepened the function and value of artistic education. As a main channel for promoting students' humanities art appreciation, public artistic education in colleges and universities is increasingly playing an important role.

It is decided by the function of public artistic education that set up watercolor courses in science and engineering universities should get their focuses and directions. The curriculum and arrangement should be designed to meet with the characteristics of science and engineering students, course structures and teaching contents should be well arranged to focus the cultivation of emotion, interest and enthusiasm, to promote a pleasant and healthy development in both physical and mental for the students.

2 CHARACTERISTICS OF STUDENTS IN SCIENCE AND ENGINEERING UNIVERSITIES AND THE MEANING OF SETTING UP PUBLIC ARTISTIC EDUCATION IN SCIENCE AND ENGINEERING UNIVERSITIES

To students, university stage is a very important period for self-consciousness and personality building, public artistic education during universities aims to improve the nature beauty, social beauty and artistic beauty, and to promote health development of humanities accomplishment and moral sentiment of students. Integrate science education, humanities education and art education, to build the future vision and personality of college students.

In science and engineering universities, students paid more attention to study and train professional skills, while relatively neglect the knowledge gaining in humanities and social sciences. At the point of the innovative talents cultivation, carrying out art and humanity education in science and engineering universities, merge art education and science education together, emphasize on the promotion of artistic accomplishments during vivid art activities while mastering professional skills and knowledge, is conducive for students to carry out work exploratory, initiative, cooperative and creative. In the process of teaching, through art style creation, aesthetic edification and imagery thinking training, can effectively help them expand their ideas, breakthrough mindsets caused by technical rationality and logic thinking inertia, training their divergent thinking ability, improve their imagination and creativity. "By conducting art education, can set up a bridge between strict logic and vivid image, inspire the students' ability of imagination, insight, and make thinking more agile and flexible".[1] From the humanity appreciation to the mental development, from the way of thinking to the professional innovation, the significance of artistic education for the personality building and all-round development to science and engineering students is clear and obvious.

3 WATERCOLOR COURSE AND PUBLIC ARTISTIC EDUCATION

Public artistic education use art as a media for emotional education, it is in the general education field, through different forms of activities to stimulate the enthusiasm and creativity to beauty of students. First of all, public artistic education focus on the cultivation of aesthetic ability and emotional education, guide students to establish correct aesthetic concepts in the teaching activities, and enrich aesthetic experience. "The ultimate purpose of aesthetic education is to analyze and evaluate art works and art phenomenon based on the regularity of art, to get inspired and understand the profound thoughts within, to get edified during the progress of art appreciation, and achieve 'edutainment'".[2] Second, the creation of artistic image is also a way to explore, to feel, to observe and to create beauty. Michael J. Parsons and H. Gene Blocker once pointed out in their writings: "Art can not only express feelings, make the person's creative impulse to the biggest display, and can improve the students' insight, understanding, expression and communication skills and ability to solve practical problems." To purify our hearts in inspiration, to enhance the ability of understanding and insight in observation, to train the ability of imagination and creativity in creation, to express emotion and relax body and mind in beauty expression, and meanwhile promote the comprehensive accomplishment in the artistic edification. Finally, to construct arts and humanities quality education that adapted to the times and professional characteristics of contemporary college students are also a key point to reform the current status of education in our country. Public art education in science and engineering universities should be in accordance to the students' professional orientation to explore a set of teaching modes that fit for the features of students. Teaching methods should fully respect special rules of public artistic education, and focus on the common features of course teaching, its target positioning students' professional quality, innovative thinking, and coordinated ability, etc.

As a main form of public artistic education—watercolor has a positive impact on student's image thinking ability, artistic understanding ability, imagination and expression. First, the lyricism of watercolor art gives students a unique aesthetic pleasure, the poetic language features exquisite pure and fresh, lively aesthetic artistic conception and interest. In the process of appreciate, learn, analyze and compare to watercolor arts, the artistic appeal will act on students' thoughts and feelings, to let them feel the beauty of watercolor arts through contents and forms. Second, watercolor art both have the spirit of traditional Chinese painting and modelling method of western painting. During the appreciation of art masterpieces, we can feel extensive and profound artistic conception of traditional Chinese painting, and the beauty in form and color of western painting, and will serve as a modest spur to induce the eastern and western art concepts and contextual framing. Third, the process performance of watercolor is a way to coordinate with eyes, thoughts and hands, the author should taking everything into consideration to deal with all unpredictable situation when water and color mixed together, during painting, recreation should be taken according to the randomly generated water and color form, and the author need to continuously breaking original thought and pattern, to develop the brain potential and exercises visual thoughts. Thus, the sensitivity, reaction degree, imagination and creativity of thoughts can be exercised during continuous thinking and inspiring. Albert Einstein said, "Imagination is more important than knowledge, because knowledge is limited, but imagination summarize everything in the world, driving the progress, and is the source of knowledge evolution."[4] This kind of innovative consciousness, creativity and imagination is just in all the necessary premise of creative work. Fourth, tools for watercolor painting are easy to carry, suitable for a fast sketch and improvisational performance, it can effectively help students to understand natural color from the view of artistic, to integrate natural color and form rule, so as to improve the observation ability, aesthetic ability, modeling ability and artistic performance ability of students.

4 TEACHING METHODS OF PUBLIC WATERCOLOR ART COURSE

Different from direct talking teaching methods, in public artistic course teaching, consideration should be given to the capacity and the actual situation of different students. On the content setting, should connect aesthetic appreciation education and practice of experiential education, combine universal teaching and advanced teaching together, lay emphasis on achievement exhibition and extension and complement of classroom teaching.

4.1 Integrating aesthetic education and experience education

At present, aesthetic education is mainly run as elective humanities art courses, common teaching methods focus more on beauty understanding, to cultivate students' ability to feel and appreciate the beauty, exert a subtle influence of the visual aesthetic functions of art itself to students. Practical teaching focuses on experience, which need to fully mobilize students' life experience and cultural appreciation ability, through imagination and legend actively participate into the creation and practice of art works, to experience the joy and hardship. Considerable initiative and openness are included in this kind of experience, so it is easy to drive students' potential of creativity and imagination.

Curriculum of public watercolor aesthetic should be set from multi-angle and multi-level in order to guide students to understand the width and depth of beauty. We can choose hotspot of art which students interested in as a breakthrough point, to discuss the aesthetic concept of contemporary art with them, and can also choose several classical masterpieces for comparison, to analyze from the style, art form, expression features, so as to broaden students' field of view and aesthetic thought, make them have the ability to appreciate art works from different angles. During the progress of appreciation, viewers can enjoy pleasant of beauty in spirit from a watercolor artwork with both nice modality skills and ideology contents. Creativity in concept can inspire human

thoughts, this will be great helpful in the cultivation and increase of students' aesthetic quality.

The practice process of public watercolor teaching focuses on user experience, make it possible for students to use paint brush to feel the attraction of art performance. Teacher should guide students to select topics they interested in to exercise, to fully understand the behavior of watercolor from color mixing and paint, to spur students' ability to paint. Only by doing it themselves can students have a vivid impression on the visual attraction of art language.

Also there are several advantages that the teaching method of integrating aesthetic and experience have, such as: 1. Link theory with practice. Take advantages of art practice through different course contents to improve students' art expression and aesthetic ability. 2. Get high participatory. The progress of art education is open, joyful and relaxed, aiming at gain benefit of the art and comprehension while practice and thinking.

4.2 Integrating universal teaching and advanced teaching

The teaching method of watercolor art should not only take into consideration the universal teaching content especially for students majoring in science and engineering, but allow for those students who have basic painting skill and keen interest in painting. The arrangement of aesthetic education course should be rational designed, and the teaching method should diversified and flexible enough, professional theory of art shouldn't be too complex, an easy and relax teaching method can be taken, and aesthetic appreciation can be taught during the interactive with students. "Aesthetic does not need a defined result, you just need to continuously gain novelty and pleasure from the awareness of the world." [5] Practical course focuses on the experiment of skills and technology, to taste watercolor art from paint, color and forming. Start from simple doodles, to cultivate judgment and cognitive ability, paint coordination from color, shape and space perspective. The purpose is to let students with no basic painting skills to start painting quickly and easily, prevent them from losing confidence for a try. Teaching forms like review and response can be taken to let students know the merit and demerit of their artwork clearly.

Meanwhile, we should also pay attention on talented participator, to form a specialist art group from students, to create a platform for those who already have basic painting skills and keen interest in art. Forms like art creating studio can be taken in daily creation, communication and studying, link theoretical teaching with practical education, to fully express their imagination and innovation ability through watercolor. With targeted counseling, further broaden and improve their professional skills on art, promote their create enthusiasm and initiative.

The advantages to treat universal teaching and advanced teaching differently are the following: 1. Get high pertinence. Integrate different teaching mode and advantages, consider students' actual situation at all level, grasp their keen interests, and teach them targeted. 2. Get strong flexibility. Students can freely select course content according to their personal situation, create art works in a flexible and relax environment, to further show their potential on art creation.

4.3 Integrating art activities and achievement exhibition

List art activities on the usual education schedule, emphasis on cultivate the ability to explore, create and appreciate beauty through vivid and interesting art activities. For instance, activities such as visiting museums, art galleries can be taken regularly to broaden students' field of vision, and widen their innovation thoughts. Held watercolor art exhibitions between teachers and students, collect outstanding students' artworks to album, arouse the enthusiasm of students learning, and let them learn with a sense of achievement.

There are advantages of lay emphasis on the extension and complement of classroom teaching, and integrating art activities and achievement exhibition, on one hand, this can record the whole progress of students learning; on the other hand, work exhibition will give sense of identity to students, and promote their enthusiasm and confidence in art creating.

5 CONCLUSION

Under the background of great development in culture, art education in colleges and universities is bound to usher in new opportunities and challenges, how to grasp the opportunity, pioneering and innovative, needs the joint efforts and struggle of every art education workers. Building reasonable education method is the key to develop an optional mechanism for art, and set up teaching contents that are fit to further promote students' humanity and artistic accomplishment, to make a new attempt and exploration in the cultivation of all-round developed talented people, and establish an effective reference for science and engineering universities to set up humanity and art education.

REFERENCES

[1] Zhang Changbo. Implementation measures and way of art education in Science and Engineering Universities. Journal of Changzhou Institute of Technology (Social Science Edition). Aug. 2011.
[2] Luo Jian. Colleges and universities to carry out aesthetic education value exploration. Majestic and grand.
[3] Mlchael J. Parsons, H. Gene Blocker. Aesthetics and Education. Chengdu: Sichuan people's Publishing House. 1998.
[4] Yang Enhuan. Aesthetic Education. shenyang: Liaoning University publishing house: 1987.
[5] Xiang Huailin. Brightening and Understanding: Pointing to the Truth of Artistic Aesthetics. Journal of ChongQing University (Social Science Edition) Vol. 18 No. 4 2012.

Education Management and Management Science – Zheng (Ed.)
© *2015 Taylor & Francis Group, London, ISBN 978-1-138-02663-6*

The use of derivatives, corporate risk management and firm financial performance: Evidence from non-financial listed companies in Zhejiang province, China

Q. Feng

Business School, Beifang University of Nationalities, Yinchuan, Ningxia, China

ABSTRACT: Since 1970s, financial derivatives have been utilized increasingly for enterprise risk management. Debates on whether derivatives optimally hedge risks and improve firms' financial performance then arose and still remain controversial. To examine this, an empirical study was conducted by analyzing Zhejiang non-financial listed companies in 2012. The test shows that there is positive association between the usage of financial derivatives and to a firm's value and performance, which indicates that using derivatives to hedge risks could improve a firm's financial performance.

Keywords: derivatives; risk management; financial performance; non-financial listed companies

1 INTRODUCTION

Derivatives, as financial instruments for corporate risk management, have been increasingly used over past decades to hedge a firm's business risks worldwide. In recent years, companies in Zhejiang province in China, as national pioneers, have been greatly involved in capital market and faced dramatic changes in operating environment. For non-financial companies, derivatives, therefore, are utilized to optimally hedge risks and improve firms' financial value and performance through hedging instead of speculating. This paper examines the impact of using derivatives by Zhejiang's non-financial companies on the scope of corporate risk management, and firms' performance. Being one of the most dynamic economic regions in China, Zhejiang has an active capital market where enterprises actively involve to hedge business risks by derivatives. Therefore, a study on Zhejiang experience can provide typical evidence for the whole from the perspective of the usage of derivatives, companies' risk management, and firm's value.

2 LITERATURE REVIEW

According to risk management theories, firms optimally hedge to reduce the cost of financial distress and volatility. Leland (1998) shows that through hedging, firms can reduce the probability of financial distress and increase their debt capacity. From the perspective of corporate risk management, however, it has generated intense debate over whether financial derivatives could lower business risk and enhance firm's earnings. Proponents who maintain that the use of derivatives could reduce market friction cost and affect the level of firm's cash flow and lower earnings volatility (Geczy et al, 1997). These hedging activities could encourage prudent risk management and thus increase earnings. This view is supported by the empirical findings of Allayannis & Weston (2001). They verified that foreign currency hedging increased company financial value by 4.7%. On the other hand, opponents argue that the excessive pursuit for utility maximization by management themselves could lead to new agency problems and thus cause immeasurable harm to business operations.

Recent empirical evidence (Liu & Ye, 2008) of American insurance industry suggests that the utilization of derivatives has no clear effect on firm's financial performance. They indicate that due to the outcomes probably, insurance companies always use derivatives for speculation purposes instead of hedging business risks. However, unlike financial companies, financial derivatives have been mainly used to hedge business risks in non-financial companies. Jia & Chen (2009) empirically examined all non-financial companies listed on Shanghai and Shenzhen Stock Exchanges in 2007. They find that derivatives for risk management tool to hedge risk do have, to some extent, positive effect on improving firm's financial performance.

3 HYPOTHESES

According to recent risk management findings, firms which hedge their business risks could reduce the cost of financial distress and increase debt capacity (Allayannis & Weston, 2001). Through hedging, volatility of cash flow is smoothened and hence the ability of debt financing increases. When external financing is more costly, hedging can also ensure that the firm has enough cash flow to finance attractive investment internally (Froot et al, 1993). Otherwise, firms may abandon investment projects because of high cash flow volatility. Under such controversial scenario, the use of derivatives will ease such contradiction and encourage firms to invest more for future expansion and get improved performance.

Based on the hedging benefits, I hypothesize:

H: the use of financial derivatives for risk management improves financial performance of non-financial listed companies in Zhejiang province.

4 RESEARCH APPROACH

4.1 Samples selection and data source

Most non-financial firms use derivatives for hedging risk instead of speculation. This paper chose all non-financial companies listed on Shanghai and Shenzhen Stock Exchanges in Zhejiang province in 2012 as samples since the capital market in Zhejiang is one of the most dynamic in China, where enterprises have been increasingly hedging business risks by derivatives. Therefore, study of Zhejiang experience can provide typical evidence for the whole. I further classify those companies into two categories: companies using derivatives and those who do not, with 38 and 77, respectively. The utilization ratio is 33.04%. Data was collected from annual financial reports including statement footnotes in 2012. Quantitative analysis was conducted using SPSS 17.0.

4.2 Variables

4.2.1 Dependant variables

Reviewing literatures about firm's financial performance valuation, the ratios of ROA and ROE, which implies companies' abilities of earning, debt paying and future sustainable development are two primary indicators. No single indicator would strongly explain the performance (Pan & Cheng, 2000). The higher ROE, the better income a company can get. And the higher ROA, the more efficient assets employing, thus can increase business profit and reduce the cost of capital. I use the ratio of net income to weighted average stockholder's equity for ROE and the ratio of net income to annual average total assets for ROA.

4.2.2 Independent variables

In practice, non-financial companies use derivatives for their risk management. To test the hypothesis,

I use one dummy variable indicating the application of derivative. Hence, RMdummy takes the value 1 if the firm reports managing business risks using derivatives and discloses the fair value or nominal value of the specific product and 0 otherwise.

4.2.3 Control variables

Several control variables are used for the multivariate analysis, based on the extensive literature on financial performance determinants.

1. Firm size: Empirical studies show that firm size is negatively correlated to firm financial performance (Leland, 1998; Pan & Cheng, 2000). To control for firm size, I use the annual average total assets of the firm.
2. Leverage: Proper capital structure can reduce cost of financing and increase return on equity (Leland, 1998). As a proxy, I use the ratio of the book value of total debt to total assets.
3. Growth: The ability of company's growth has positive effect on firm's current value and future development. Research findings show that financial forecast based on historical data of revenue will give investors a big picture for firm market valuation (Pan & Cheng, 2000; Defond & Hung, 2003). Thus, I use the ratio of operating revenue growth rate to represent the whole.
4. Composition of assets: Different firm asset compositions will lead to various types and levels of business risks, finally affecting operational performance. Big proportion of fixed assets induces high depreciation cost, low asset turnover; then, low profit. Moreover, the accounting of goodwill in intangible assets also could relate to earnings volatility in the long run. Thus, I use net book value of fixed and intangible assets to total assets as representative control variables.

4.3 Models

To test the hypothesis, I estimate the models:

$$\text{VALUE} = \beta_0 + \beta_1 * \text{RM} + \beta_2 * \text{Size} + \beta_3 * \text{Growth} + \beta_4 * \text{Fixed} + \beta_5 * \text{intangible} + \upsilon \quad (1)$$

$$\text{ROA} = \beta_0 + \beta_1 * \text{RM} + \beta_2 * \text{Size} + \beta_3 * \text{Growth} + \beta_4 * \text{Fixed} + \beta_5 * \text{intangible} + \upsilon \quad (2)$$

$$\text{ROE} = \beta_0 + \beta_1 * \text{RM} + \beta_2 * \text{Size} + \beta_3 * \text{Growth} + \beta_4 * \text{Fixed} + \beta_5 * \text{intangible} + \upsilon \quad (3)$$

where, VALUE is a set of variables including ROA and ROE; the dependent variable is ROA in the second model and ROE in the third model. RM, dummy variable, takes the value 1 if the firm reports managing business risks using derivatives and 0 otherwise. Hence, the coefficient β_1 captures the general impact of using derivatives for risk management. SIZE is the annual average total assets of the firm; and the ratio of operating revenue growth rate stands for Growth. Fixed and Intangible represent the proportions of fixed assets or intangible assets to total, respectively.

5 EMPIRICAL RESULTS

5.1 Descriptive statistics

Descriptive statistics provided in Table 1 indicate that the maximum and the minimum of ROA and ROE. With values of 52.80, 0.20, 36.50 and 0.20%, the results imply the earnings ability varies dramatically among sample companies, facilitating to examine the impact of using derivatives on firm's valuation. The descriptive statistics on control variables are in comprehensible patterns.

Table 2 shows the performance of firms using derivatives is better than those with no derivatives.

5.2 Correlation analysis

To test whether there is correlation between independent variables and the dependent variable, I conduct Pearson correlation analysis. Table 3 shows there is certain relation between some variables, i.e. the variable of capital structure is significantly positive related to the variable of intangible assets proportion at 5% level. Existing research studies find that multicollinearity would not have great influence on regression results and not reduce the predictive power or reliability of the model as a whole as long as the correlation coefficient is less than 8%. The values of correlation coefficient in this test are acceptable.

5.3 Multiple regression analysis

Regression analysis involves identifying the relationship between a dependent variable and one or more independent variables.

Regression test of model 1:

Based on results in Tables 4, 5, the regression results of model 1 show that:

1. The F-statistic is the test statistic for the analysis of variance (ANOVA) approach to test the significance of the model or the components in

Table 1. Descriptive statistics.

	Min.	Max.	Mean	Std deviat.
ROA	0.002	0. 528	0. 105	0. 087
ROE	0.002	0.365	0.082	0.062
Intangible	0.145	0.969	0.564	0.180
Fixed	0.059	2.30E0	0.281	0.375
Size	1.18E8	6.46E9	8.91E8	1.12E9
Leverage	0.156	1.05E2	1.39E1	2.26E1

Table 2. ROA & ROE descriptive statistics.

	N	Min.	Max.	Mean	Std deviat.
ROA1	38	0.012	0.528	0. 135	0. 102
ROA0	67	0.002	0.365	0.088	0.071
ROE1	38	0.002	0.365	0.089	0.071
ROE0	67	0.008	0.188	0.070	0.042

Table 3. Correlation analysis.

	N = 115	ROA	Intan.	Fixed	Size	Lev.	ROE
ROA	Pearson	1	0.20	0.14	0.51**	0.17	0.72**
	Sig.		0.23	0.40	0.00	0.31	0.00
Intan.	Pearson	0.20	1	−0.54**	−0.19	−0.36*	−0.22
	Sig.	0.23		0.00	0.26	0.03	0.19
Fixed	Pearson	0.14	−0.54**	1	0.34*	0.23	0.55**
	Sig.	0.40	0.00		0.04	0.17	0.00
Size	Pearson	0.51**	−0.19	0.34*	1	0.52**	0.47**
	Sig.	0.00	0.26	0.04		0.00	0.00
Lev.	Pearson	0.17	−0.36*	0.23	0.52**	1	0.26
	Sig.	0.31	0.03	0.17	0.00		0.19
ROE	Pearson	0.72**	−0.22	0.55**	0.47**	0.22	1
	Sig.	0.00	0.18	0.00	0.00	0.19	

*.Correlation is significant at the 0.05 level (2-tailed).
**.Correlation is significant at the 0.01 level (2-tailed).

Table 4.   Regression test results of model 1.

Coefficient[a]

Model 1	Estimate	Std. Err.	Std. Esti	t-tatistic	Sig.
Model	−0.005	0.027		−0.176	0.861
Intan.	0.092	0.040	0.395	2.289	0.029
Fixed	0.020	0.019	0.179	1.041	0.035
Size	1.96E-11	0.000	0.525	3.113	0.004
Lev.	8.230E-7	0.000	0.000	0.003	0.098

[a]Dependent variable: ROA.

Table 5.   Regression test results of model 1.

Anova[b]

Model 1	Sums q	DF	Means q	F	Sig.
Regression	0.024	4	0.006	4.904	0.003[a]
Residual	0.040	33	0.001		
Sum	0.064	37			

R	R^2	Adjusted R^2	Std. Estimated Err.
0.611[a]	0.373	0.297	0.0350

[a]Predictor variables: leverage, fixed, size, intangible.
[b]Dependent variable: ROA.

Table 6.   Regression test results of model 2.

Coefficient[a]

Model 2	Estimate	Std. Err.	Std. Esti.	t-statistic	Sig.
Model	0.044	0.065		0.678	0.502
Intan.	0.051	0.096	0.089	0.525	0.063
Fixed	0.133	0.046	0.489	2.907	0.006
Size	3.10E-11	0.000	0.339	2.049	0.048
Lev.	0.000	0.001	−0.039	−0.231	0.089

[a]Dependent variable: ROE

Table 7.   Regression test results of model 2.

Anova[b]

Model 2	Sums q	DF	Means q	F	Sig.
Regression	0.153	4	0.038	5.423	0.002[a]
Residual	0.233	33	0.007		
Sum	0.387	37			

R	R^2	Adjusted R^2	Std. Estimated Err.
0.630[a]	0.397	0.323	0.0841

[a]Predictor variables: leverage, fixed, size, intangible.
[b]Dependent variable: ROE.

the model. The closer to 0 of the value of sig., the fitter the model to data set. In this test, the F-statistic of the linear fit versus the constant model is 4.904, with a sig. value of 0.3%. The model 1 is statistically significant.

2. The value of R and R^2 are, respectively, 0.611 and 0.373 in this test showing good fitness explanation of the model to the dataset. And there is little possible multicollinearity among the predictor variables. Each statistic tests for the significance of each term given other terms in the model. Moreover, in terms of residual examination, the value of all residuals is low, indicating each residual is independent from others and each independent variable is not self-correlated to each other.

3. From tables above, what we found is that the control variables are not significantly related to explanatory variable. Then, I can conclude: the usage of derivatives could help improve firms' financial performance.

Regression test of model 2:
Results of model 2 in Tables 6, 7 show that:

1. The values of F-statistic of model 2 are significant at the 1% significance level, with a sig. value of 0.2%. The model 2 is statistically significant.
2. The values of R and R^2 are, respectively, 0.630 and 0.397 in this test. The model 2 can well explain the dataset. And there is little possible multicollinearity among the predictor variables.
3. According to regression results in model 2 above, like model 1 analysis, the conclusion is same: the usage of derivatives could improve firms' financial performance.

6 CONCLUSIONS

Derivatives, financial instruments increasingly being used for risk management by non-financial firms, have been utilized to optimally hedge risks and improve firms' financial value and performance. To examine the relationship among derivatives usage, firm's risk management and their financial performance, all Zhejiang's non-financial companies listed on Shanghai and Shenzhen Stock Exchanges in 2012 are selected as samples since the capital market in Zhejiang Province is one of the most active one in China and companies there have stronger demand for risk management. Given the empirical study in this paper, a regression model is set to test the hypothesis which is that using derivatives to hedge risk could improve firm's financial performance. Empirical study verified the hypothesis and there is positive association between the usage of financial derivatives and to a firm's value. Further research will be conducted on the scope of total non-financial companies in China.

ACKNOWLEDGMENTS

The paper is supported by Chinese National Social Science Foundation, *Study on IT and internal control information disclosure of listed companies in China*, 11CGL020; and Key Laboratory Program of State Ethnic Affairs Commission of China, *Research on implications on derivatives regulation and cooperate risk management in China*.

REFERENCES

[1] Allayannis, G. & Weston, J. 2001. The use of foreign currency derivatives and firm market value. *Review of Financial Studies* 14 (1): 243–276.
[2] DeFond, M. & Hung M. 2003. An empirical analysis of analysts cash flow forecasts. *Journal of Accounting and Economics* 35: 73–100.
[3] Froot. K.A. & Scharfstein, D.S. & Stein, J.C. 1993. Risk management: Coordinating corporate investment and financing policies. *Journal of Finance* 48: 1629–1658.
[4] Geczy, C.C. & Minton, B.A. & Schrand, C.M. 1997. Why firm use currency derivatives. *Journal of Finance* 52: 1323–1354.
[5] Jia, W.Y. & Chen, B.F. 2009. Empirical research on the influence of risk management on the listed companies value and performance of our country. *Communication of Finance and Accounting* 17: 77–80.
[6] Leland, H.E. 1998. Agency cost, risk management and capital structure. *Journal of Finance* 53: 1213–1243.
[7] Liu, Y. & Ye, D.L. 2008. The value and performance effects of derivatives application in American insurance companies. *Securities market Herald* 3: 9–14.
[8] Pan, Y. & Cheng, X.K. 2000. Principal component analysis on performance evaluation in publicly listed companies. *Accounting Research* 1: 31–35.

Education Management and Management Science – Zheng (Ed.)
© *2015 Taylor & Francis Group, London, ISBN 978-1-138-02663-6*

Evaluation of Chinese cattle and sheep industry's subsidies policy

Xiao Han, Fang Liu, Jie Yu & Zhongwei He
Economic and Management School, BUA, Beijing, China

ABSTRACT: In this paper, Inner Mongolia cattle and sheep farmers are surveyed for the effect of cattle and sheep subsidy policy using DID model analyses. As a result, income subsidies for farmers have significant role in promoting the participation of farmers' per capita income and the contribution rate reached 71%. Under the influence of natural and market risks, the influences are continuity, but to continue the popularity of the subsidy policy, it needs to improve.

Keywords: cattle and sheep industry; DID model; subsidy policy

1 INTRODUCTION

At present, our country's cattle and sheep industry subsidy policy are formulated by two aspects: one is a yellow box policy, mainly represented by subsidies for improved varieties policy. The other is green box policies, such as support funds for cattle and sheep industry standardized scale farming policy, major animal epidemic compulsory immunization subsidy policy, the grassland ecological protection subsidies reward mechanism, etc. In this paper, the subsidies involved are all yellow box policy, namely, subsidies for improved varieties policy. According to new development plan, China is expected to issue 1.7 billion Yuan to support the development of cattle and sheep industry. With the implementation of the subsidy policy, cattle and sheep supply ability and the overall income of farmers produced certain positive effect. Along with the increasing production cost and gradually reducing the breeding female benefit, breeding stock of cattle and sheep and its production growth is slowing, the gap between demand and supply increases, and the demand in some regions is unsustainable. The price of cattle and sheep is rising; the phenomenon will not change fundamentally in the short term.

Qualitative analysis methods are mostly used at present stage for the effect of subsidies for cattle and sheep evaluation research, and rarely involve econometric analysis method. But on the evaluation of the effect of pigs, cows, and other fields of subsidy policy, the use of measurement method research has become quite mature. From the point of research methods, domestic policy effect evaluation measurement method is more abundant. Such as (Wang Jiao Xiao Haifeng, 2006), using the econometric model, food subsidies have little effect on grain yield, but significant effects on

farmers' income. Above studies show that cattle and sheep subsidy policy in China have made great achievements, but policy objective remains to be determined and perfected. Through search of the references, it was found that there is no literature evaluating special system for cattle and sheep policy effect, even slightly involving multi-purpose qualitative analysis. Most scholars research on cattle and sheep policy with no support of certain macro data and survey data. In this paper, DID model from other areas of innovation is introduced into the cattle and sheep policy effect evaluation.

2 DATA SOURCES

2.1 *The questionnaire design*

This research is based on questionnaire survey in pastoral areas of Inner Mongolia-farmers' basic situation, including family information, operation and management status, cost benefit. Through the above information, cattle and sheep subsidies cognitive situation and farmers' satisfaction of subsidies were studied.

2.2 *The questionnaire distributed region selection and processing*

Cities and counties were randomly selected for questionnaire survey according to the main cattle and sheep breeding area of Inner Mongolia. The selection principle is that the region belongs to key area of cattle and sheep production; secondly, considering the distance, choosing by distance such as near and far away. Finally, to focus on screening, some who received subsidies and others did not receive subsidies to farmers, according to the proportion of 1:1. During the summer vacation in

2013, research group of Inner Mongolia region, sent/received questionnaire count is about 500/487, with questionnaire effective rate being 97.4%.

3 EFFECTS OF THE CATTLE AND SHEEP SUBSIDY POLICY IN INNER MONGOLIA

3.1 *Cattle and sheep market volume significantly increased*

In 2010, there was 2.387 million tons of meat production in Inner Mongolia, including cattle and sheep production, ranking fourth and first in

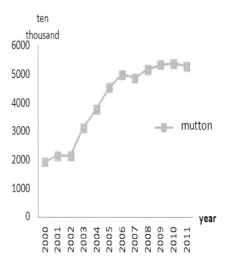

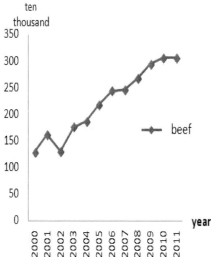

Figure 1. Cattle and sheep market volume variation in inner mongolia from 2000 to 2011.

the country, respectively, 7.6% and 22.4% of the total output in turn. Inner Mongolia mutton production is the second in China which is 1.9 times Xinjiang's mutton production. Through the analysis of 10 years variation trend of the Inner Mongolia cattle and sheep production, there are three points which are listed as follows: firstly, cattle and sheep breeding stock's proportion of the country raised year by year in Inner Mongolia. After that period, it fell in 2007 and was gradually restored; Secondly, the proportion of Inner Mongolia cattle and sheep production increased year by year in China, and fell in 2007 after that, reached a new peak in 2012; Thirdly, Inner Mongolia sheep and cattle's breeding stock accounted for the proportion of the national growth rate slower than the growth of sheep and cattle production accounts for the proportion of Inner Mongolia, starting in 2004, the proportion of beef and mutton production is greater than the proportion of breeding stock, the difference between them are more and more significant (Fig. 1).

4 SUBSIDY POLICY IMPLEMENTATION EFFECT RESEARCH BASED ON DID MODEL

4.1 *Build and variable selection of the model*

This paper evaluates the existing cattle and sheep subsidy policy impact on farmers' income based on DID model, accurate assessment of China's cattle and sheep subsidy policy implementation effect, guide the decision-making behavior of farmers. DID model by large sample can be divided into two kinds: reference sample named M, and experimental sample named N.

A virtual variable P, when the P = 1 for farmers to sample N, P = 0 said farmers M belongs to the reference sample. Virtual variable time, expressed in T here, when T = 1, said cattle and sheep benefits after implementation, when T = 0, is not to subsidies. "e" as random perturbation terms the disturbance. In order to analyze the effect of the subsidies, the simplest equation is:

$$Y = \beta_0 + \alpha_0 T + \beta_1 P + \alpha_1 TP + e$$

That is to say, when subsidies before, during and after the implementation of income Y can be expressed as:

$$Y =$$

The reference sample M before and after the policy subsidies is $\Delta E(Y)1$ can be expressed as follows:

$$\Delta E(Y)1 = diff1 = (\beta_0 + \alpha_0) - (\beta_0) = \alpha_0$$

$$\Delta E(Y)2 = \text{diff}\,2 = (\beta_0 + \alpha_0 + \beta_1 + \alpha_1) - (\beta_0 + \beta_1)$$
$$= \alpha_0 + \alpha_1$$

The government subsidies to farmer's income promotion's net effect are expressed as:

$$\Delta E(Y)2 = \Delta E(Y)2 - \Delta E(Y)1 = (\alpha_1 + \alpha_0) - (\alpha_1) = \alpha_0$$

DID model basic formula TP coefficient alpha 1 is DID's estimate or double difference estimates. Alpha 1 is government subsidies utility value; it is the core of model DID in the process of the application data. The model and demanding coefficient value can accurately measure the government subsidies net effect. Subsidy effect analysis in Inner Mongolia area, the multiple control variables is introduced to ensure the DID optimal fitting model and reality. The introduction of control variables DID model can be represented as:

$$Y_{ab} = \beta_0 + \alpha_0 T_b + \beta_1 P_a + \alpha_1 T_b P_a + \sigma X_{ab} + e$$

"a" is on behalf of farmers, b is on behalf of the period. Y_{ab} is "a" in "b" period of farming income; X_{ab} series of control variables such as age (age), level of education (this indicator used by the education time goes by to represent; edu), farmer number (people) and breeding farm size (scale), etc.

Since 2006, Inner Mongolia based on pilot by dairy subsidies for growing superior seed varieties extend to cattle and sheep subsidies for growing superior seed varieties. And un-subsidized farmers were selected as the control group from subsidies households in 2009. Research information are selected in 2008 year (year) without subsidies, choose 2008 and 2012, the two time points, can be more precise analysis net utility of cattle and sheep subsidy policy.

4.2 Analysis of estimate value

The results show that before the subsidy policy is implemented, the experiment sample farmers earned relatively 1689 Yuan less than reference sample farmers farming income. Subsidies after implementation, reference farmers are 16464 Yuan higher than the experimental sample. The net utility of subsidies before and after the implementation is 18153 Yuan, the difference-in-difference estimates value is 18153. Although before and after subsidies two samples' farming income increased, but the growth rate is far different. The experimental sample farmers income growth was significantly higher than the reference sample farmers (Table 1).

Table 2 shows that level of education is positively correlated to the farming income, and it is very significant. Family population and wages for

Table 1. Farming income before and after the subsidy policy.

Farming income	Reference sample	Experimental sample	Diff
2008	21481	19792	−1689
2012	31754	48218	16464
Diff	10273	28426	18153

Table 2. The results of control variable model.

Variable	Coefficient
α_1: DID estimated value	18153.47***
β_0: affected reference sample farmers income change	21481.47*
α_0: subsidies farmers income changes	10273.12*
β_1: two samples income changes before and after the subsidies	−1689.14**
X_1: legal representatives age	0.54*
X_2: education year	345.12***
X_3: number of family	10.14

farm income is, respectively, positive and negative effects, but not significant, the statistical results show that household population and wages for farm income has little to do. And the fit of the model as a whole is 0.8124. It explains model results are perfect.

It is concluded through DID the general expression of the model:

$$Y = 21481 + 10273.12\,T - 1689.14\,P + 18153.47\,TP$$

Subsidies after the experimental sample farmers compared with the reference sample farmers, farming income increased significantly. The results show that the experimental sample farmer's 71% of growth rate in income should be promoted by the government.

5 CONCLUSIONS

The above conclusion for cattle and sheep subsidy policy enlightenment is as follows: firstly, the subsidies information should ensure universal access to farmers, and adopt multi-channel propaganda, train farmers to apply for subsidies and concern consciousness, avoid the information subsidies internalization. Secondly, government should provide the technical training for farmers, improve quality, enhance the ability to encourage both farming experience, and also have a high degree of farmers into farming, synchronously improve breeding scale and technology. Third, continue to increase subsidies, improve the social security for farmers, and eliminate worries.

This paper's corresponding authors are He Zhongwei and Liu Fang.

ACKNOWLEDGEMENT

Thanks to the support of this thesis Fund: 2013 Department of Agriculture soft science topics (201309); 2013 National Natural Science Foundation of China (71373025); Colleges and Universities in Beijing Talent introduction and Training projects (CIT & TCD20140314); Beijing Agricultural College Advantaged Technology team (Beijing agricultural industry security theory and policy research innovation team) projects.

REFERENCES

[1] Wang Jiao, Xiao Haifeng. China's grain direct subsidy policy effect evaluation. China's rural economy, 2006 (2): 12–15.

[2] He Zhongwei. Price conduction mechanism between China's pig industry production and marketing research, the empirical analysis based on VAR model. Agricultural technology economy, 2012 (8): 38–45.

[3] Fan Gongxia. Inner Mongolia mutton sheep industry competitiveness research. Inner Mongolia University of agricultural, 2012.

Education Management and Management Science – Zheng (Ed.)
© 2015 Taylor & Francis Group, London, ISBN 978-1-138-02663-6

The relationship between Corporate Social Performance and Corporate Financial Performance: A literature review of twenty years (I)

Gang Fu
College of Economics and Management, Sichuan Agricultural University, Chendu City, Sichuan Province, China

Ping Zeng
Accounting Department, Sichuan Business Vocational College, Chendu City, Sichuan Province, China

ABSTRACT: To explore the real relationship between corporate social performance and financial performance, this research takes 63 studies from 1990s as samples to analyze synthetically the conception of CSP, basic theories, methodologies, industries and control variables, stakeholder groups, the measures of CSP and CFP and the correlation between CSP and CFP. We find that (1) dominant researchers assert the positive or neutral relationship of CSP and CFP, (2) basic theories are vague and unclear, (3) the measures of CSP and CFP is a "complex" phenomenon, (4) control variables play an important role in those studies, and (5) many researchers perform the across-industry studies, and view the industry as a key control variable.

Keywords: Corporate Social Performance; Corporate Financial Performance; control variable; stakeholder, research methodology

1   INTRODUCTION

Exploring the relationship between Corporate Social Performance (CSP) and Corporate Financial Performance (CFP) has been a lively debate since Milton Friedman's (1962, 1970) challenge that "a corporation's social responsibility is to make a profit."

The research on the link between CFP and CSP has traditionally involved two different empirical issues: First, the direction (if exists) of the causality; and second, the sign of those identified causal relationship (Shane, Spicer, 1983; Strachan, Smith, Beedles, 1983; Vance, 1975; Wier, 1983; Makni, Francoeur, Bellavance, 2009; Giannarakis, Theotokas, 2011); some researchers have found contradictory results, that is, the positive and inconclusive relationship (Anderson, Frankle, 1980; Freedman, Jaggi, 1982; Fray, Hock, 1976), or the positive and negative relationship (Chen, Metcalf, 1980; Cochran, Wood, 1984; Coffey, Fryxell, 1991). The majority of researchers have found the positive relationship (Belkaoui, 1976; Bowman, 1978; Hart, Ahuja, 1994; Orlitzky, Schmidt, Rynes, 2003; Recchetti, 2007; Mahoney, LaGore, Scazzero, 2008; Mcguire, Sundgren, Schneeweis, 1988; Simpson, Kohers, 2002; Stanwik, Sarah, 1998, el.). Therefore, the impression that 'in the aggregate, results are inconclusive' regarding any theoretical conclusions about the relationship between CSP and CFP has persisted until today (Orlitzky, Schmidt, Rynes, 2003).

This paper specifically focusses on the previous studies. The remainder of the paper is organized as follows. In the next section, we introduce how to get the literature, and the number and characteristics of the literature. The third section reviews the conception of CSP. The fourth section is about basic theory on the research of CSP and CFP. The fifth section reviews the methodology, referred indusries and control variables of twenty years. The sixth section is about stakeholders, the measures of CSP and CFP. The following section discusses the relationship between CSP and CFP. The last section includes conclusions and enlightenment.

2   THE LITERATURE OF TWENTY YEARS

The relationship between CSP and CFP has been a topic of interest and controversy for more than half a century, and serious empirical research on the association between financial and social performance indicators has been going on for several decades (Peston & O'Bannon, 1997). Griffin & Mahon (1997) analyzed 51studies of twenty five years and found 16 studies in 1970s, 27 studies in 1980s, and 8 studies in 1990s. In the analysis,

Table 1. The literature since 1990s.

Writers	Journal or Meeting	Year
From 1990 to 1999 (8)		
Moses L. Pava, Joshua Krausz	Journal of Business Ethics	1996
Jennifer J. Griffin, John F. Mahon	Business & Society	1997
Lee E. Preston, Douglas P. O'Bannon	Business & Society	1997
Ronald M. Roman, Sefa Hayibor, Bradley R. Agle	Business & Society	1999
Janet F. Phillips	Journal of Health Care Finance	1999
Richard A. Johnson, Danieal W. Greening	Academy of Management	1994
Peter A. Stanwick, Sarah D. Stanwick	Journal of Business Ethics	1998
Sandra A. Waddock, Samuel B. Graves	Strategic Management Journal	1997
From 2000 to 2009 (39)		
Marc Orlitzky, Frank L. Schmidt, Sara L. Rynes	Organization Studies	2003
Abagail McWilliams, Donald Siegel	Strategic Management Journal	2000
Homer H. Johnson	Business Horizons	2003
Lois Mahoney, Robin W. Roberts	Accounting Forum	2007
Marc Orlitzky	Journal of Business Ethics	2001
Geoff Moore and Andy Robson	Business Ethics: A European Review	2002
Michael L. Barnett and Robert M. Salomon	Academy of Management	2002
Benjamin A. Neville, Simon J. Bell, Bulent Menguc	European Journal of Marketing	2005
W. Gary Simpson, Theodor Kohers	Journal of Business Ethics	2002
Geoff Moore	Journal of Business Ethics	2001
Bernadette M. Ruf, Krishnamurty Muralidhar, Robert M. Brown, Jay J. Janney, Karen Paul	Journal of Business Ethics	2001
Marc Orlitzky	University Auckland Business Review	2005
Jordi Surroca, Josep A. Tribo	Academy of Management	2005
Meng-Ling Wu	The Journal of American Academy of Business	2006
John Peloza	California Management Review	2006
Douglas A. Schuler, Margaret Cording	Academy of Management Review	2006
Michael L. Barnett and Robert M. Salomon	Strategic Management Journal	2006
Leonardo Becchetti	Finance & The Common Good/Bien Commun	2007
Jeffrey S. Harrison, Joseph E. Coombs	Academy of Management Best Conference Paper	2006
Gary Woller	ESR Review	2007
Gerwin Van der Laan, Haans Van Ees, Arjen Van Witteloostuijn	Journal of Business Ethics	2008
John Peloza and Lisa Papania	Corporate Reputation Review	2008
Jean-Pascal Gond, Guido Palazzo	Academy of Management Annual Meeting	2008
Pieter van Beurden, Tobias Gossling	Journal of Business Ethics	2008
Jonas Nilsson	Journal of Business Ethics	2008
Stephen Brammer and Andrew millington	Strategic Management Journal	2008
Hasan Fauzi, Lois S. Mahoney, Azhar Abdul Rahman	Issues in Social and Environmental Accounting	2007
Peter May, Anshuman Khare	Journal of Environmental Assessment Policy and Management	2008
Richard Peters, Michael R. Mullen	The Journal of Global Business Issues	2009
Edward Nelling, Elizabeth Webb	Review of Quantitative Finance & Accounting	2009
Scott J. Callan, Janet M. Thomas	Corporate social responsibility and environmental management	2009
Lois Mahoney, William LaGore, Joseph A. Scazzero	Issues in Social and Environmental Accounting	2008
Darren D. Lee, Robert W. Faff, Kim Langfield-Smith	Australian Journal of Management	2009
Rim Makni, Claude Francoeur, Francois Bellavance	Journal of Business Ethics	2009
Hasan Fauzi	Issues in Social and Environmental Accounting	2009
Leonard A. Jackson, H.G. Parsa	International Journal of Business Insight & Transformation	2009
Hasan Fauzi	Globsyn Management Journal	2009
Robert Neal, Philip L. Cochran	Business Horizons	2008
Clyde, Eirikur Hull and Sandra Rothenberg	Strategic Management Journal	2008

(Continued)

Table 1. (*Continued*).

Writers	Journal or Meeting	Year
From 2010 to 2011 (16)		
Anis Ben Brik, Belaid Rettab, Kamel Mellahi	Journal of Business Ethics	2011
Margaret L. Andersen, John S. Dejoy	Business and Society Review	2011
Roberto Garcia-Gastro, Miguel A. Arino, Miguel A. Canela	Journal of Business Ethics	2010
Jordi Surroca, Josep A. Tribo, Sandra Waddock	Strategic Management Journal	2010
Jong-Seo Choi, Young-Min Kwak, Chongwoo Choe	Australian Journal of Management	2010
David J. Flanagan, K.C. O'Shaughnessy, Timothy B. Palmer	Corporate Reputation Review	2011
Margaret L. Andersen	The Academy of Accounting and Financial Studies	2010
William S. Chang	International Journal of Business and Management	2010
Pushpika Vishwanathan	Academy of Management Annual Meeting	2010
Kranti Dugar, Brian T. Engelland, Robert S. Moore	Society for Marketing Advances	2010
Johan Gromark, Frans Melin	Journal of Brand Management	2011
Marc Orlitzky	Business Ethics Quarterly	2011
Margaret L. Anderson, Lori Olsen	Academy of Accounting and Financial Studies Journal	2011
Zhi Tang, Clyde Eirikur Hull, Sandra Rothenberg	Academy of Management Annual Meeting	2011
David P. Baron, Maretno Agus Harjoto, and Hoje Jo	Business and Politics	2011
Supriti Mishra, Damodar Suar	Journal of Business Ethics	2010

Ronald, Sefa and Bradley (1999) revised the data from Griffin & Mahon (1997) and found 12 studies in 1970s, 28 studies in 1980s, and 10 studies in 1990s, which research the relationship between CSP and CFP.

The literature comes from the EBSCO host database. We search the articles based on the following set: the first combination is 'social performance' and 'financial performance' as 'title'; the second combination is 'social' and 'financial performance' as 'title'. We have 63 studies on the correlation of CSP and CFP after we kick 'book review' and another 5 articles. Of 63 studies, there are 8 studies in the period of 1990 to 1999, 39 studies in the period of 2000 to 2009, and 16 studies in the period of 2010 to 2011 (see Table 1). When we read these articles carefully, we review the writers, journal or meeting, Year and Issue, methodology, industry, stakeholders, control variables, measures of CSP and CFP, the sign and conclusions of each study.

3 WHAT IS CORPORATE SOCIAL PERFORMANCE?

Griffin (2000) indicated that one of the most puzzling inconsistencies in their article is the use of the word CSP. Much of the debate on corporate social performance is of a normative nature, building upon the idea that moral principles should or should not guide corporate decision making (Gerwin, Hans,

Arjen, 2008). Corporate social performance is one of three constructs to refer to business involvement in social issues, which are the outcomes of socially responsive behavior. There are different definitions of corporate social performance. After reviewing the literature, we find three category definitions, which are *comprehensive view*, *activity view* and *outcome view*. Many researchers favor the **comprehensive perspective** given by Wood (1991). He defines CSP as 'a business organization's configurations of principles of social responsibility, processes of social responsiveness, and policies, programs, and observable outcomes as they relate to the firm's societal relationships' (Wood, 1991a: 693). Wood describes CSP as comprising three major components—the first component is the level of corporate social responsibility which is based on legitimacy with society, public responsibility within the organization, and managerial discretion by each individual within the organization. The second component is the processes of corporate social responsiveness which includes environmental assessment, stakeholder management and issues management. The third component refers to the outcomes of corporate behavior and includes social impacts, social programs, and social policies (Peter, Sarah, 1998). CSP is a critical factor to consider for all organizations since CSP components such as "social issues, environmental pressures, stakeholders concerns are sure to affect corporate decision making and behavior far into the future" (Wood, 1991b). Surroca, Tribo and Sandra

(2010) view Corporate Responsibility Performance (CRP) as CSP, and indicate "CRP is conceptualized as the broad array of strategies and operating practices that a company develops in its efforts to deal with and create relationships with its numerous stakeholders and the natural environment" (Waddock, 2004). It reflects the idea that responsibilities are integral to corporate actions, decisions, behaviors, and impacts, whereas the concept of corporate social responsibility connotes the discretionary responsibilities of business (Surroca, Tribo and Sandra, 2010).

The second perspective is the *outcome view.* Douglas & Margaret (2006) define CSP as a voluntary business action that produces social (third-party) effect. Scott (1992) defines CSP as "social outcomes of firm behaviors" with little elaboration. Griffin (2000) also approves this view and has suggested "outcomes represent the joint product of organizational performance and environments response", and "outcomes ascertain satisfaction or effectiveness" (Lenz, 1981).

The third perspective is the *activity view*. Some researchers mix Corporate Social Responsibility (CSR) and corporate social performance and they always view CSR as CSP. For example, McWilliams and Siegel (2001) define CSR as actions that appear to further some social good, beyond the interests of the firm and that which is required by law. The key in this definition is that CSR activities are on a voluntary basis, going beyond the firm's legal and contractual obligations. It involves a wide range of activities such as being employee-friendly, environment-friendly, mindful of ethics, respectful of communities where the firms' plants are located, and even investor-friendly (Benabou & Tirole, 2010).

After seeing the connotation of CSP, we might find that the firm may require some sacrifice in the short-term profits with the process of performing or bearing social activities. Moreover, more people are concerned as to why firms would perform social responsibility, and if the sacrifice of short-term profits would be compensated by the improvement in firm's long-term financial performance. Or, are they purely feel-good activities initiated by corporate insiders? (Choi, Kwak, Choe, 2010)).

4 SLACK RESOURCE THEORY, GOOD MANAGEMENT THEORY AND STAKEHOLDER THEORY

4.1 *Slack resource theory*

Slack resource theorists argue that better financial performance potentially results in the availability of slack (financial and other) resources that provide the opportunity for companies to invest in social performance domains, such as community relations, employee relations, or environment. If slack resource is available, then better social performance would result from the allocation of these resources into the social domains, and thus better financial performance would be a predictor of better CSP (Waddock, Graves, 1997).

4.2 *Good management theory*

Good management theorists argue, alternatively, that there is a high correlation between good management practice, and CSP, simply because attention to CSP domains improves relationships with key stakeholder groups, resulting in better overall performance. For example, good employee (including women and minorities) relations might be expected to enhance morale, productivity, and satisfaction. Excellent community relations might provide incentives for local government to provide competition-enhancing tax breaks, improved schools (and a better workforce over the long term), or reduced regulation, thereby reducing costs to the firm and improving the bottom line (Waddock, Graves, 1997). Further, positive customer perceptions about the quality and nature of a company's products, its environmental awareness, and its government and community relations, are increasingly becoming bases of competition (Cf. Prahalad & Hamel, 1994), blurring the lines between good management practice and 'social' performance. Such positive perceptions of the firm by outside stakeholders may lead to increased sales or reduced stakeholder management costs. The work of McGuire (1988, 1990) also supports the good management theory in that it provides empirical support for financial performance as the dependent variable.

4.3 *Stakeholder theory*

Stakeholder theory posits that firms possess both explicit and implicit contracts with various constituents, and are responsible for honoring all contracts (Bernadette etc., 2001). As a result of honoring contracts, a company develops a reputation that helps determine the terms of trade it can negotiate with various stakeholders (Bull, 1987; Cornell & Shapiro, 1987; Jones, 1995). While explicit contracts legally define the relationship between a firm and its stakeholders, implicit contracts have no legal standing and are referred to in the economic literature as self-enforcing relational contracts (Bernadette, etc., 2001). Since implicit contracts can be breached at any time, implicit contracts become self-enforcing when the present value of a firm's gains from maintaining its reputation is greater than the loss if the firm reneges on its implied contracts (Telser, 1980).

Moreover, stakeholder theorists view shareholders as one of the multiple stakeholder groups managers must consider in their decision-making process (Donaldson & Preston, 1995; Wood & Jones, 1995, etc.). These stakeholder groups include internal, external, and environmental constituents. Like shareholders, the other stakeholders may place demands upon the firm, bestowing societal legitimacy. Firms must address these demands or else face negative confrontations from non-shareholder value, through boycotts, lawsuits, protest (Bernadette, etc., 2001). From a stakeholder theory perspective, corporate social performance is assessed in terms of a company meeting the demands of multiple stakeholders. Firms must at some level, satisfy stakeholder demands as an unavoidable cost of doing business. So, firms should adopt different approaches to satisfying stakeholder demands, ranging from cost minimizing to societal maximizing (Freeman, 1984).

5 METHODOLOGY, INDUSTRY AND CONTROL VARIABLES

5.1 Methodology

The research conclusions on the relationship between CSP and CFP are inconsistent or contradictory, which has been recognized by most researchers in academy. Some researchers (Cochran and Wood, 1984; Ullmann, 1985; Woddock and Mahon, 1991) argue that methodological differences are one reason of inconclusive relationship between CSP and CFP. "The methodological limitations of prior studies obscure the true association between CSP and CFP thereby resulting in inconsistent results" (Darren, Robert, Kim, 2009). In the analysis, we divide the methodology into three types: Empirical Analysis (EA), Normative Analysis (NA) and Meta-Analysis (MA). Of 63 studies (see Table 2), 41 studies use empirical analysis method, such as Anderson & Olsen (2011), Gromark & Melin (2011), Chang (2010), Waddock & Graves (1997), Griffin & Mahon (1997), Phillips (1999), McWILLIAMS & SIEGEL (2000) etc., 16 studies adopt normative analysis method, such as Pava & Krausz (1996), Johnson (2003), William (2010), Orlitzky (2002). And there are 6 meta-analysis articles, such as Orlitzky, Schmidt & Rynes (2007), Orlitzky (2001, 2011), Moore &Robson (2002), Wu (2006), Sefa & Bradleyr (1999), Vishwanathan (2010).

Table 2. Methodology of the relationship between CSP and CFP.

Method	EA	NA	MA	Total
No.	41	16	6	63

5.2 Industry

The industry's level of differentiation may also affect firm's performance, as competition in a highly-differentiated industry is unlikely to be price-based and, thus, is likely to be profitable for all concerned (Porter, 1980, 1986). However, some industries lend themselves to higher levels of differentiation than others, and there is evidence that industry-level factors, such as overall levels of differentiation, impact performance (McGahan and Porter, 1997). Given the evidence that suggests that firms do better in industries in which companies allocate more resource to differentiation activities, it seems reasonable to expect that industry differentiation will impact firm performance (Clyde & Sandra, 2008).

Therefore, numerous researchers take industry factor into account in process of investigating the relationship of CSP-CFP, and a few researchers especially have researched a certain industry. For example, Moore & Robson (2002), Moore (2001) investigate supermarket industry in the U.K., Barnett & Salomon (2002, 2006) focus on socially responsible investment funds, Simpson & Kohers (2002), Woller (2007) research banking industry, and Jackson & Parsa (2009) perform a typology for service industries. Most researchers have performed the cross-industry research, in which they take the industry factor as a key control variable. Of 41 empirical analyses, 22 studies apply industry as control variable.

5.3 Control variable

Many researchers have identified a number of variables believed to impact how a firm's social performance relates to its financial performance (Andersen, Dejoy, 2011). Callan and Thomas (2009) find that control variables must be properly specfied to avoid bias and that some of these measures are quadratically related to CFP. Using appropriate control variables is critical to obtaining reliable results, as pointed out repeatedly by previous research (Callan, Thomas, 2009), so many researchers apply different control variables to investigate the relationship between CSP and CFP. Of 41 empirical studies, control variables include firm size, risk, industry R&D expenses, advertising expenses, average age of corporate asset and others (see Table 3). In control variable categories, size, risk and industry are used much frequently, respectively, being 73%, 56% and 54% of total studies Different researchers use different indicators for same control variable catalogue. For example, the measures of firm size include total sales, total assets, the number of employees, the log of assets and average turnover over the period (McWilliams, Siegel, 2000; Mahoney, Roberts,

Table 3.   Control variable categories.

Name	Size	Risk	Industry	R&D	Advertising	Age	Others
No.	30	23	22	10	6	3	9

2007; Johnson, Greening, 1994; Peter, Sarah, 1998 etc); the measures of risk include debt/total asset, financial leverage, beta, long-term debt to total assets, average gearing over the period, an interest coverage ratio, quick ratio, cash ratio, P/E and MV/BV (Gerwin, Haans, Arjen, 2008; Jordi, Josep, 2005; McWilliams, Siegel, 2000; Waddock, Graves, 1997 etc); the measures of industry refer to Standard Industry Classification Code, dummy variable (1 for manufacturing, 0 for non-manufacturing) (McWilliams, Siegel, 2000; Bernadette etc., 2001; Anis etc., 2011 etc); the measures of R&D expenses refer to R&D expenditures to sales, amount of R&D expenses (McWilliams, Siegel, 2000; Margaret, John, 2011 etc); the measures of advertising expenses include the amount of advertising expenses and the ratio of advertising expense to sales (McWilliams, Siegel, 2000; Stephen, Andrew, 2008 etc). And other measures are complex, such as type of ownership, country, year, capital expenditure, the level of pollution emissions, dividend ratio (Supriti, Damodar, 2010; Mahoney, Roberts, 2007 etc).

ACKNOWLEDGMENT

This research was financially supported by the program (13XJC630005) from Ministry of Education of the People's Republic of China.

REFERENCES

[1] Moses L. Pava, Joshua Krausz, 1996, The Association Between Corporate Corporate-Responsibility and Financial Performance: The Paradox of Social Cost, Journal of Business Eethics, 15: 321–357.

[2] Marc Orlitzky, Frank L. Schmidt, Sara L. Rynes, 2003, Corporate Social and Financail Performance: A meta-analysis, Organization Studies, 24(3): 403–441.

[3] Jennifer J. Griffin, John F. Mahon, 1997, The Corporate Social Performance and Corporate Financial Performance Debate: Twenty-Five Years of Imcomparable Research, Business & Society, 36(1): 5–31.

[4] Lee E. Preston, Douglas P. O'bannon, 1997, The Corporate Social-Financial Performance Relationship-A Typology and Analysis, Business & Society, 36(4): 419–429.

[5] Ronald M. Roman, Sefa Hayibor, Bradley R. Agle, 1999, The relationship Between Social and Financial Performance-Repainting a Portrait, Business & Society, 38(1): 109–125.

[6] Janet F. Phillips, 1999, Do Managerial Efficiency and Social Responsibility Drive Long-term Financial Performance of Not-for-Profit Hospitals before Acquisition?, Journal of Health Care Finance, 25(4): 67–76.

[7] Abagail Mcwilliams and Donald Siegel, 2000, Corporate Social Responsibility and Financial Performance: correlation or Misspecification?, Strategic Management Journal, 21: 603–609.

[8] Homer H. Johnson, 2003, does it pay to be good? Social responsibility and financial performance, Business Horizons, 11: 34–40.

[9] Lois Mahoney, Robin W. Roberts, 2007, corporate social performance, financial performance and institutional ownership in Canadian firms, Accounting Forum, 31: 233–253.

[10] Marc Orlitzky, 2001, Does Firm Size Confound the Relationship Between Corporate Social Performance and Firm Financial Performance, Journal of Business Ethics, 33: 167–180.

[11] Geoff Moore and Andy Robson, 2002, The UK supermarket industry: an analysis of corporate social and financial performance, Business Ethics: A European Review, 11(1): 25–39.

[12] Michael L. Barnett and Robert M. Salomon, 2002, Unpacking Social Responsibility: The Curvilinear Relationship Between Social and Financial Performance, Academy of Management, B1–B6.

[13] Benjamin A. Neville, Simon J. Bell, Bulent Menguc, 2005, corporate reputation, stakeholddrs and the social performance-financial performance relationship, European Journal of Marketing, 39(9): 1148–1198.

[14] Richard A. Johnson, Danieal W. Greening, 1994, Relationships Between Corporate Social Performance, Financial Performance, and Firm Governance, Academy of Management, 314–318.

[15] W. Gary Simpson, Theodor Kohers, 2002, The link between corporate social and financial performance: evidence from the banking industry, Journal of Business Ethics, 35: 97–109.

[16] Geoff Moore, 2001, Corporate Social and Financail Performance: An investigation in the U.K. Supermarket Industry, Journal of Business Ethics, 34: 299–315.

[17] Peter A. Stanwick, Sarah D. Stanwick, 1998, The Relationship Between Corporate Socaial Performance, and Organizational Size, Financial Performance, and Environmental Performance: An Empirical Examination, Journal of Business Ethics, 17: 195–204.

[18] Bernadette M. Ruf, Krishnamurty Muralidhar, Robert M. Brown, Jay J. Janney, Karen Paul, 2001, An Empirical Investigation of the Relationship Between Change in Corporate Social Performance and Financial Performace: A Stakeholder Theory Perspective, Journal of Business Ethics, 32: 143–156.

[19] Sandra A. Waddock and Samuel B. Graves, 1997, The Corporate Social Performance-Financial Performance Link, Strategic Management Journal, 18(4): 303–319.

[20] Marc Orlitzky, 2005, Social responsibility and financial performance: Trade-off or virtuous circle?, University Auckland Business Review, 38–44.

[21] Anis Ben Brik, Belaid Rettab, Kamel Mellahi, 2011, market orientation, corporate social responsibility, and business performance, Journal of Business Ethics, 99: 307–324.

Education Management and Management Science – Zheng (Ed.)
© 2015 Taylor & Francis Group, London, ISBN 978-1-138-02663-6

Improvement of teaching method for petroleum industry introduction to improve the learning interest of non-professional student

Jian Yan
Petroleum Engineering College, Xi'an Shiyou University, Xi'an, China

ABSTRACT: The petroleum industry introduction is a specialized multi-subject course, for petroleum engineering and non-engineering students. The objective is to broaden the students' professional knowledge, cultivate practical ability, and to cultivate talents for the oil industry. Because this course has strong basis, wide range of knowledge, non-petroleum engineering students lack interest in learning. Also, because teaching method is sometimes undeserved, students can only master little about what have been taught. In this paper, combining with the teaching practice, several teaching methods for training and improving students' learning interest of this course are put forward, such as the harmonious relationship between teachers and students, alleviate the pressure of study; combined with the actual needs, cultivate interest in learning; flexible teaching methods, improve the learning interest, which will arouse students' enthusiasm, initiative and creativity fundamentally. The study results lay a good foundation in improving the teaching effect.

Keywords: petroleum industry; non professional; teaching method; improvement

1 INTRODUCTION

Petroleum industry introduction is currently one of the general education courses for the petroleum university. It is a professional multi-subject course which introduces the main field of petroleum industry, that is, some basic concept, theory and process of oil exploration, development, storage and transportation, oil refining and petrochemical industry. At the same time, it briefly introduces the status and role of petroleum industry in the development of the national economy, the oil industry and the development trend in the future. It is a stepping-stone to learn petroleum engineering education theory for petroleum engineering students, and it is a new measure to broaden the knowledge, cultivating ability for non-petroleum engineering students.

During the teaching process in recent years, teachers reflect generally that oil engineering students have strong learning interest, good command, but the non-professional students have poor grasp of basic concept and theory. The reason may be: first, students think that this course knowledge has nothing to do with his future work, so there is no need to learn. Second, many liberal arts and science class students think learning of engineering knowledge is difficult. In addition, it is also related to our teaching experiences, teachers do not grasp the teaching method of specialty courses for non-professional students. Therefore, this text combines the rich teaching experiences of several teachers and my own experience in the process of teaching, put forward in step by step methods to inspire and cultivate the students' interest.

2 IMPROVING METHODS

2.1 *Harmonious relationship between teachers and students, and alleviate the pressure of learning the course*

Harmonious relationship between teachers and students can make students accept the knowledge in the pleasant atmosphere, thinking, and grasp the method. It is the precondition to cultivate interest in learning. Non harmonious relationship between teachers and students causes the student to resist the teacher, and to the studying of the course. Of course, the efficiency of learning is very poor. The relationship between teachers and students can be improved through the following ways: improving education, moral relations and relations in heart. Education relationship is that there are different status and role between teachers and students. We need to change the state that teachers dominate everything and ignore the students' subject position. Moral relationship refers to the fact that the teachers and students should do in accordance with the requirements of ethics. Change the situation of mutual distrust into mutual respect. Heart relationship is that whether the mind identity and compatibility between teachers and students achieve harmony. We should change the phenomenon of emotional confrontation and mutual misunderstanding. Now most of the teachers in teaching petroleum industry introduction are young, mostly, straight out of college. As such, the age gap with students is small. First, the inner condition of the communication between teachers and students; second, teachers should care about students, understand students, have consideration for students everywhere, and become a real friend of students. Only in this way can students become masters of learning.

2.2 Combining with practice, cultivate learning interest

After giving students a preliminary image that it is easy to be close, the need is to begin cultivating students' interest in learning courses. This step is critical, but it is also hard to take forward. Because professional students want to study professional knowledge relative to their major, who can lay a good theoretical basis for the future work. They lack the learning interest of professional knowledge, and even have a contradiction mood. Suggest: Don't jump into the course. Firstly, tell the students the need to learn this course: the first is the need of the international political environment. Because oil has special strategic value, it causes the competition of global oil supply resource; also, it is an important foundation of the world politics and foreign policy. In the rapid development of world economy, the demand for oil also will increase, but oil reserves are drying up. The country who masters the oil will have advantageous position in the international political arena; the scarcity of world's oil resources can lead to the severity of the scramble for oil resources. We can take the three world's oil crisis as an example, Explain the importance of controlling oil resources. The second is the need of China's national conditions. Now the contradiction between the demand and output of crude oil in China has become more and more prominent— external dependency degree reaches more than 50%, energy security is becoming more and more low. And even more alarming is that most oilfields on domestic land have entered high water cut development period, difficulties of development increases, and potential resources reduce, new resource discovery can only depend on the development of overseas and on sea, oil fields abroad are tied down by the country and the local government, while for offshore oil field, it is known that the bottom of the sea often contain rich mining resources, including oil resources, the South China Sea, recognized as one of the most promising deepwater area containing oil and gas resources in the world, but there still exists certain disputes with the neighboring countries on boundary. In addition, our country still lacks technology for developing the deepwater oil resources. We also need to put in more efforts in the future. So caring about the development of petroleum industry can enhance students' consciousness of safeguarding China's territorial sovereignty. The third is the need of the employment environment. The rapid development of social and economic energy development put forward higher demand for oil, which needs more high-quality talent into the oil industry, in addition to the professional graduates, more and more non professional (including machinery, materials, computer, communications, electronics, instruments, economy, management, foreign language, etc.) graduates enter the petroleum enterprises and institutions, and as the non-professional graduates in petroleum colleges and universities, whether in the recruitment or in the future work, because they contact prior to the oil industry and have more obvious competitive advantage than those from other colleges and universities. Through the necessity, we gradually let the students begin to like and to know more about our petroleum industry spontaneously.

2.3 Flexible teaching methods, improve the interest in learning the course

After inspiring the desire of learning, if we still adopt the conventional teaching ways, the desire of students could be nipped in the bud. So, we must adopt flexible teaching methods to start with the fire of desire.

2.3.1 Rich multimedia, reduce visual fatigue
Multimedia teaching has become a very good teaching tool and method in current scenario, but the dull words and the single color only increase the students' visual fatigue, and another bad habit—lethargy. So, in order to reduce the visual fatigue of students, in the process of multimedia lessons, some images, animation, or video can be interspersed, and text must be refined, with different colors to highlight the key points that are important for emphasis and to grasp. But avoid too much color, which will cause another kind of visual fatigue—have dim eyesight.

2.3.2 Introducing the life common sense, make the knowledge understanding easier
For non-professional students, especially for the liberal arts and science classes, some technique and theory are relatively difficult to understand. Because they do not need to grasp the professional

knowledge, understanding is ok, and some life common senses might as well be able to be introduced into the classroom, to reduce the difficulty of learning. For example, when introducing the petroleum exploration means, you can ask the students how to distinguish whether the watermelon is ripe in our daily life. There are usually two steps, the first step is to use the finger tapping the surface, if the sound is ringing, it is ripe, if the sound is dull, it is unripe. This method has certain risk. The second step is to open a small window with a sharp knife, directly take out of the pulp, see its color, taste its flavor, and then determine whether it is really ripe. This approach is very straightforward, almost no risk. This is very similar to the process of oil exploration, the exploration has indirect and direct methods. Indirect method is doing some exploration work on the surface of earth, including geological, geochemical and geophysical exploration method, to determine whether there is oil underground, with certain speculation; the direct method is to drill, direct contact with the formation, including drilling, test, perforation and test to determine whether there is oil, and evaluate properties of formation rock and fluid and production capacity, etc.. This process is a forecast to the verification process. Through such simple life common sense, we can make it easier for students to understand exploration methods and means.

2.3.3 *Combination of theoretical study, physical matter and scene, increasing the perceptual knowledge*

Through the interpretation of the theory and image display in the classroom, students may have a preliminary understanding of teaching things, but due to a lack of physical display and touch, often it is easy to forget them. So, in the course of lectures one can try to show some physical objects, such as oil reservoir. If you don't know the oil industry before, one may think that the crude oil stored in the ground like a pool. But the fact is that oil is stored in microscopic pore in underground rock, and the flow capacity is very small, we call it seepage. At this time, we can show them some cores or core slices used in our scientific works. Let the student personally touch and let them really understand the oil storage place. In addition, our school have internship facility. We can take students to

visit it and assist in teaching. When the perceptual understanding deepens, we believe that they can hardly forget the knowledge that they see.

2.3.4 *Increase learning activities to improve students' innovative consciousness*

In addition to the interpretation in the classroom, we can also add some knowledge competition, arouse more interest of learning in the competition, at the same time, in order to obtain the final pride, students can pay more attention to learning the course. Also, if some students feel more interest, the form of a "small project" can be used to enhance the innovation consciousness of students. These small projects can come from the teacher's scientific research project. It may also be the projects that they feel interest in, with the guidance and help of teacher, or do independent design and complete the project. Some outstanding works can be properly rewarded.

3 CONCLUSION

In a word, the teaching of petroleum industry introduction for the non-professional students has certain difficulties. We should avoid rushing, guide students step by step, stimulate desire of the students to study the course after eliminating the conflicting emotions of students, improve the learning interest and innovation consciousness through the improvement of teaching methods and means. We believe that most of the non-professional students will like the course gradually after using the above step by step methods.

REFERENCES

[1] Bo, X, 2012. Setup and practice of General Curriculum in Petroleum University. *Education Forum*, (2): 134–135.
[2] Xiaojuan, R. 2007. Petroleum industry introduction. *Sinopec press.*
[3] Lexiang, Y. 2000. Further research on teaching to stimulate students' interest in learning. *Petroleum Education*, (9): 27–28.
[4] Lexiang, Y. 2000. On Cultivation of students' creative thinking ability. *Petroleum Education*, (7): 33–34.

Education Management and Management Science – Zheng (Ed.)
© 2015 Taylor & Francis Group, London, ISBN 978-1-138-02663-6

Review of Counterproductive Work Behavior of the new generation employees

Y.P. Sun
Business School, Hohai University, China

ABSTRACT: This paper focuses on two concepts: the new generation employees, and the counter-productive behavior. It summarizes the research process of counterproductive behavior, research content, domestic research progress, and also describes the division of research and the definition of the new generation of employees at home and abroad. It finally draws a conclusion in counterproductive behaviors division, dimension, affecting factors of the new generation of employees.

Keywords: new generation; counter-productive behavior; review

1 INTRODUCTION

Counterproductive Work Behavior (CWB) is prevalent in all types of businesses and other organizations, most of which conduct dual characteristics as covert and destructive. Not only current theory research focuses on this issue, the business community has also set out to learn from the practice of anti-productive behavior. The highest attention on counter-productive behavior research in the West appeared after 2000. However, in China, although some scholars have begun to study this issue, the study is still in the stage of the introduction of Western theories and localization of Chinese context. Most existing researches focus on reviews of Western theories, wishing to apply them in Chinese practice. Meanwhile, with the new generation of employees gradually stepping into the workplace, their work behavior, especially counter-productive work behavior, has attracted great attention. In view of this, this paper starts with the definition of the new generation and the research of employee's counter-productive behavior, reviewing the concept, structure dimensions, affecting factors and research progress in this area, aiming to draw attention on the research of CWB characteristics of the new generation of employees.

2 NEW GENERATION EMPLOYEES

The so-called generation, refers to people born in the same era, with special social and cultural background and historical events of this era that they share closely and stable attitudes and values of life (Kupperschmidt, 1998; Smola and Sutton, 2002).

The study of intergenerational divide mainly started in the West—they summarize the 20th century and the population in order to summarize the differences between people of different ages. Although there are differences amongst the different scholars in divided era, but the Western sociologists believe that each generation probably lasts about 20 years (Schaeffer, 2000), and every age has a specific social context.

Currently, the Western mainstream intergenerational theory has mainly divided people into: Baby Boomers, X Generation, and Y era (also known as the younger generation). This classification is mainly due to the recognition of a particular social context and historical events, the World War II, economic recession, high-tech features as a landmark for generations divided crowd. This standard of divided intergenerational studies in other countries has also become intergenerational issues in accordance with Japanese scholars. Miyuki Takase referenced to this classification standards and the specific circumstances of the country, forming a nursery block generation (1946–1959), the new generations of mankind (1960–1974 years), the nursery block generation (1975–present) division.

From life cycle perspective, each generation has a certain age range. According to the average child-bearing age of Chinese women, each generation should be about 24 years. This standard, as the traditional "generation" division provides a theoretical basis. As early as 1999, the research group of intergenerational relations with China on the basis of this generation division, three generations of Chinese values, expectations and mutual understanding of the empirical study, reflects the relationship between the longitudinal intergenerational.

This generational divide is not common in the country. In order to make intergenerational divide of more practical significance, Taiwan scholars in addition to referring to the West after World War II intergenerational classification, also introduce the Japanese name "new breed" by comparing the net and gross agricultural production of Taiwan's industrial production. To be born in the postwar world has significant changes in the industrial structure of the population to become "new man". After that, every 10 years forms a generation, thus forming a unique intergenerational division standard. Intergenerational divide the continent, but also follow some basic criteria that was divided according to certain social background. Sun, JM (2010), summarized the study of social change in Mainland iconic events, including the Great Leap Forward, the agricultural revolution, reform and opening up, social transformation, etc. In these social background, each generation formed their unique personality traits and values, and accordingly formed a division of intergenerational. This classification of intergenerational continuity to the overall population of samples to the social background of criteria for the classification, the differences can reflect different times and basic characteristics of the population.

In addition, the concept of intergenerational division as the 70s, 80s, 90s etc. is also popular domestically, and has gradually become an issue of academic study. Although this limits to ten years for the intergenerational divide, is not very scientific, but it contains a profound social and cultural background. The study of this generational divide, at first, focused on a representative on behalf of the population, such as the study after 80s, initially concentrated in the research on the literary youth of 80s, the expansion of 80s sports stars, as well as final thoughts on the formation of the overall population of 80s. At present, this research has evolved from a point to the surface, 80s, 90s personality traits, values. The scholars conducted a survey of statistics, so that the characteristics of these intergenerational groups become increasingly clear.

3 COUNTERPRODUCTIVE WORK BEHAVIOR

In the 1980s, more and more psychologists and researchers were concerned about organizational citizenship behavior, and they tried to find relation between citizenship behavior and their performance. And in this in-depth course of the study, scholars have gradually found that some negative behavior also affects organizational performance. Indeed, the foreign counter-productive behavior of research

began in the 20th century and initially interpreted as tissue negative behavior. Scholars try to use Aggression, Deviance, Delinquency, Retaliation to summarize the various disruptive behavior generated by the organization. With constant attention to this area, scholars have found that original formulation of these terms are still not clear enough, or even cross, thus resulting questionnaire have duplicate entries. In the end, counter-productive behavior came to summarize the term for similar behavior.

Counter-productive behavior of the present study focused on the following aspects: concept and structure; antecedents; intervention mechanism. The concept and structure of counter-productive behavior constituting the basis for the field of research, including the production of anti-behavioral characteristics are summarized and their classification; study of antecedent variables (factors) mainly through the study of the impact of anti—relationships between factors of production behavior and behavior among its mechanism of action better description of the process and the behavior of each factor. For example: Marcus & Schuler summarize the classification framework of CWB antecedent variables, previous studies can take advantage of this framework. This is a two-dimensional frame structure in which one dimension is the individual factors—situational factors, including the former personality, self-control variables such as individual differences of employees, which includes job satisfaction, institutional norms or situational variables situational perception variables; another dimension is induced factors—factors that inhibit the induction factor refers to the pressure, sense of fairness, such as counter-productive behavior induced variable, which is corrected straight, tissue culture suppress these behavior variables. The two dimensions constitute the analytical framework of the four quadrants. In addition to the research framework antecedents, many scholars have studied different variables relations with CWB, and the mechanism of action has been studied. Professor Greenberg's study found that reducing wages will lead to increased employee theft rate, but if the issue managers fully communicate with employees, it will significantly reduce the theft rate. Sense of injustice will result in retaliation in employees' intent, and the intent will result in retaliatory counter-productive behavior. In addition, the environment, sense of responsibility, emotional stability, personal moral standards, self-monitoring, self-superiority have been shown to occur with the CWB. They play an important role in predicting the occurrence of CWB.

However, the study of counter-productive behavior, in the end aims to improve organizational performance. Conducting prevention and intervention when counter-productive behavior is taking place or foreseeable is the ultimate goal of the study. Current research for prevention and intervention of counter-productive behavior is still relatively small. From the research point of view, such studies will be the basis of the full study on the antecedent variables. Certain achievements in research-based antecedents will help discuss how to prevent and intervene. According to the results already obtained from the selection, training, performance management, job design, one can develop and create a corporate culture and other aspects of the institutional environment, and explore the prevention and intervention means of counter-productive behavior. For example: To investigate the academia is how to get more qualified staff in the selection of personnel from the source to avoid counter-productive behavior; while strengthening management in the organization, to create a good organizational climate, to prevent these acts from both institutional and cultural generation. Ultimately, CWB future research will focus on the basis of previous studies, the exploration and excavation of a counter-productive behavior of early warning and intervention mechanism, providing solutions for a variety of businesses and social organizations.

4 RESEARCH STATUS IN CHINA

Domestic research of CWB started quite late. From around 2010, there have been scholars noticing this trend, which was mainly through literature review. It will sort out the progress of CWB study abroad and try to summarize significance for the local business or organization. Sun J.M. & Xun, W. have also worked on related foreign theories commentary, summed up the foreign study of five kinds of theoretical models, that frustration—the attack on the self-control theory, causal push theory of planned behavior theory and stressors-emotional theory. Finally, he noted that the concept and structure of future counter-productive behavior should enhance scientific research methods, in close connection with Chinese cultural background and organizational contexts and times. Peng domestically studied more representative and continuity, both from the field research, research paradigm, counter-productive behavior of the formation process of the four aspects of the CWB and intervention mechanisms conducted theoretical overview, conducted research on counterproductive behavior, obtained 66 domestic CWB, were divided from its extent and degree of damage is less than

the basic sort of counter-productive behavior of the overall hierarchy.

Thus, forming the three characteristics of domestic research: 1. In theory, the basis of tradition and continuity of speech abroad, and localization improvements; 2. counter-productive behavior of understanding, comprehension and China CWB different scenarios under different cultures, produced without the same CWB, but the effectiveness of these reliable CWB also need a lot of research to confirm. 3. Most of the studies still use foreign sophisticated measurement scale, measured in different populations and in different industries, thus getting to know the situation of domestic CWB.

5 CONCLUSION

This paper focuses on the literature review of the concepts of new generation of employees, and counter-productive behavior. It summarized the historical progress and the paradigm of counter-productive behavior research, together with its domestic research progress. Meanwhile, it also describes the division of Intergenerational new generation of employees abroad and domestic, and on this basis defines the new generation of employees, which has settled a good foundation for exploring of the occurrence, influencing factors and management of counter-productive behavior.

REFERENCES

[1] Kupperschmidt B.R. Understanding generation X employees. JONA. 1998, 28 (12), PP: 36–43.
[2] Schaeffer J., Kemper Reports (Winter-Spring). Kemper Distributors, Inc., Chicago, IL. 2000.
[3] Miyuki T., Keiko O., Noriko Y. (2007) Generational differences in factors influencing job turnover among Japanese nurses: An exploratory comparative design. International Journal of Nursing Studies. 2009, 46, PP: 957–967.
[4] Sun J.M. & Xun W. Value difference between generations in china: a study in China. Journal of Youth Study. 2010, 13(1), pp: 65–81.
[5] Spector P.E., Fox S. An emotion—centered model of voluntary work behavior: some parallels between Counterproductive Work Behavior (CWB) and Organizational Citizenship Behavior (OCB). Human Resources Management Review, 2002, 12: 269–292.
[6] Marcus B., & Schuler H. Antecedents of Counterproductive Behavior at Work: A General Perspective. Journal of Applied Psychology. 2004, 4, 647–660.
[7] Greenberg. Employee theft as a reaction to underpayment inequity: the hidden costs of Pay cuts. Journal of applied Psychology, 1990 (12): 21–36.

Adhere to the scientific development and construct technician cradle—the connotation construction and practical exploration of the adult vocation skill training center of Shanghai Second

Shujuan Lu & Ming Jiang

Shanghai Second Polytechnic University, Shanghai, China

ABSTRACT: Higher vocational colleges should focus on cultivating high skill talents according to social needs, implement vocational skill training, and being of benefit to the regional economic and social development. After analyzing the characteristics of higher vocational colleges serving society and implementing social vocation skill training, we summarize the advantages of vocational skill training in higher vocational colleges. At the same time, in order to further enhance the connotation and the training level and ability of the adult vocation skill training center of Shanghai Second Polytechnic University, we explore and sum up its development train of thoughts.

Keywords: higher vocational colleges; vocational skill; development train of thoughts; connotation construction; practice

1 THE DEVELOPMENT OF SKILLS TRAINING IN HIGHER VOCATIONAL COLLEGES

Social service function is one of the three big functions in higher vocational colleges, and it is reflected by adapting to the demand of the social and economic development and the demand of industrial structure adjustment and the improvement of the laborer's quality, and carrying out vocational skills training according to the need of enterprise and labor market.

In 2010 the "State Council on Strengthening Occupation training to promote employment opinions" proposed "encouraging institutions of higher learning to vigorously carry out occupation and employability skills training." Higher vocational colleges should carry out occupation skill training focusing on social needs of highly skilled personnel training, and favor the regional social development, and economic development. If higher vocational colleges can solve the Employment Anxiety for urban and rural workers and enterprises, they will get recognition and support from the society and enterprises and will naturally become the main force of regional economic development, will win the praise of the community, and will also highlight school-running characteristics. The vocational education objective proposed by educator Mr. Huang Yanpei, is preparing for the living and social services of individuals and for the productivity improvement of the country and world.

2 THE CHARACTERISTICS OF THE SKILLS TRAINING IN VOCATIONAL COLLEGES

2.1 *The pertinence of the training objectives*

According to the new challenges and the need of the market, the vocational quality should be improved constantly. So according to the requirements of the job and trainees, the vocational colleges carry on the targeted training.

2.2 *The practicability and synchronicity of the training content*

The contents of the skill training are practical. In order to satisfy the specific needs of enterprises, according to the vocational position classification knowledge and skill requirements, the vocational colleges reasonably adjust the proportion between the theoretical knowledge and practical skills; The skill training content is synchronous with vocational demands and vocational colleges take into account the ability of the trainees and consider the demand of the market to keep the training contents synchronized with the job development.

2.3 *The training process combined with vocational practice*

The essential character of vocational skill training is job-oriented, and the vocational training contents in vocational colleges are always in accordance with the actual needs of the employment direction, closely integrated with the occupation field practice. Vocational quality cannot be obtained only through the teaching of theory and must be gained through direct, realistic, perceptual practical activities, which determines that we should adhere to the principle of "Combination of production and academy", take the "ability based" training mode, combine the teaching process with the vocational sites.

3 COMPARED WITH THE COOPERATIVELY-RUN SCHOOLS AND INDUSTRY TRAINING ORGANISATIONS, VOCATIONAL COLLEGES HAS THREE ADVANTAGES IN VOCATIONAL SKILLS TRAINING

3.1 *The advantage in teachers and teaching*

Higher vocational college have theory teacher, operation training teacher and double-position teachers. Colleges and universities attach great importance to teaching and research, conducting teaching competition, complete strict evaluation system, use advanced modern teaching means, own self-designed teaching materials. These have great advantages in teaching.

3.2 *The advantages in facilities and investment*

Higher vocational institutes have established a series of laboratories, training room, the equipment form a complete set and can be used efficiently through coordination. Each school has special funds to purchase equipment, update outdated equipment, and some equipment can be transferred for training. In 2008, our university invested 800000 to strengthen our skills training facilities. It also has obvious advantages compared to private and industry training institutions.

3.3 *The advantage in keeping up with the pace of industry update*

The global financial crisis will provide the opportunities to buy advanced foreign enterprises and technology. It also makes the enterprise to update the technology that is close to the world technology standard. On the premise that the government pays more and more attention to the occupation skill training, higher vocational colleges will find these changes quickly and let them penetrate into the teaching. The occupation skill training in higher vocational colleges is not attached to a particular industry, and just follows the need of industrial adjustment and development.

4 INSIST ON SCIENTIFIC DEVELOPMENT, EXPLORE THE PRACTICE OF SHANGHAI SECOND POLYTECHNIC UNIVERSITY(SSPU) ADULT VOCATIONAL SKILLS TRAINING CENTER CONNOTATION CONSTRUCTION

SSPU adult vocational skills training center (hereafter referred to as the "center"), founded in 1997, has experienced two stages of development. From 1997 to 2002, "the center" ran in colleges and universities system, because of the combination of schools. In 2002, "the center" gradually changed. Especially in the independent market, how to survive? How to develop? Through baptism we gradually realize that we should analyze scientifically the present situation to determine the development train of thought, focus on the quality and regulations to win the market, put service to heart and be people-oriented. All this is the core of development.

4.1 *Analyze scientifically the present situation to determine the development train of thought*

Firstly, Service-oriented. According to the characteristics of Baoshan District's economic and industrial structure, "the center" have put Baoshan district as the main area for training and service. In order to develop the vast hinterland of the future, "the center" will use the strategy of dislocation competition according to the characteristic and advantage of its training program.

Secondly, Development strategy. Through study and research, we realize that vocational skill training is valued and funded by the government and is needed by the enterprise; training needs to link theory with practice and have definite standards and regulations. Vocational skills training should be undertaken by the training agency of the university with engineering background. Therefore, our development strategy is carrying out vocational skill training mainly, extensively contacting and undertaking the new social projects, and taking the initiative to strengthen cooperation with the enterprise.

Thirdly, the main project. According to the idea of dislocation and sustainable development, we determined to put maintenance electrician training as the main project which will enable students to have a sustainable development.

4.2 Focus on the quality and regulations to win the market

4.2.1 Training scheme, teaching staff, teaching environment, teaching method, training conditions, tracking service and school-enterprise cooperation are the rigid construction of the training regulations and quality

1. Training scheme. A complete scheme must have clear training objective, market demand, job opportunities, work task, personnel specification, curriculum, teaching process, training places, joint enterprise and teachers. There are two points for planning training scheme. Training plan has two points, one is how to make the training course connected with the job requirement. Two is to guarantee the appropriate proportion between the theoretical courses and practical training courses.
2. Teachers. To own excellent teachers is important for the quality and regulations of training. "The center" should make full use of the social resources. Through the use and evaluation of teachers, "the center" strengthens the teachers' service and contact, and gradually builds a more stable team of teachers.
3. The teaching environment and training conditions are the indispensable hardware construction for the training regulations and quality. "The center" owns its classrooms and training facilities.
4. The application of project teaching method to improve the teaching effect. "The center" claims that students should learn and research independently under the guidance of teachers to gain learning experience and understanding in many aspects. The project teaching method creates a wide stage for the development of students' subjective initiative, which can help them in applying the skills that they have owned, expressing their thoughts creatively and solving problems. The project teaching method cultivate the students' consciousness of mutual cooperation and research, learning to communicate, learning to help each other, learning to share, and promote the students' team cooperation spirit, exploration and innovation ability.
5. Training technicians through industry-academy cooperation. We should actively promote the contact with the enterprise and industry, and arouse the enthusiasm of enterprises and industries to participate in vocational skills training.
6. Pay attention to feedback and follow-up service. "The center" manages and supervises the whole process of teaching, and to achieve the optimization of training effect constantly adjusts the training content, method and means.
7. The combination of curricula education and non-curricula education. "The center" focusses on training the students' vocational skills and satisfy the students' desire to improve the curricula education level.

4.2.2 Managing all aspects of the skills training normatively is the soft construction of the training regulations and quality

Improving attendance is one of the most serious problems for the training organizations. The students who have worked accounted for about 70%, and most of them have families of their own. Therefore "the center" uses some methods to increase the attendance, such as psychological counseling.

The standardized management of teaching process and the comprehensive monitoring of teaching quality are important for the training organization, but are easily ignored. Affected by the formal teaching traditions, most teachers in "the center" abide by the training rules and attach great importance to the teaching quality. The administrators of "the center" hold regular seminars and listen to the opinions of the students, and at the same time give feedback to the teacher to improve the problems existing in the teaching process.

4.3 The core of the development is serving to heart and people-oriented

4.3.1 Serving to heart

Some students who attend the junior and intermediate skills training are unemployed and migrant workers. They want to gain knowledge and skills, get certificates, get a relatively better and stable work, support the family. Some students who attend the senior skills training have worked, they want to gain the advanced certificate, learn more skills and get higher salary. The administrators work enthusiastically and guide the students to have the right direction and the correct way of learning. The administrators of "the center" can cooperate with each other in all directions, help to solve the problems that the teachers and students have come up against, help the needy students to pay the tuition by installment, care about the students who have not passed the exam, remind the students who have passed the junior and intermediate skills training to attend the higher training, appeal the students to help the one who encounter special difficulties. Through the efforts of everyone, now all the training classes have gradually formed the correct teaching style and study style and have owned the harmonious teacher-student relationship, and "the center" is becoming more and more prosperous.

4.3.2 *Psychological counseling*

For the changes in the students and teachers' psychology, thought and self-consciousness, the administrators make efforts to analyze and persuade. There are few students who expect that they can get the certificate without labor; there also few teachers who just care about training wage. We want to use the relevant theories of management psychology to explain, analyse, regulate and control, to make everyone handle the relationship appropriately.

4.3.3 *Pay attention to learning continuously*

To serve efficiently, the staff of "the center" must improve their own quality continuously. Only through continuous learning they can understand, grasp, be familiar with relevant laws and regulations, such as method and mechanism for government purchasing of training, the scope, methods and procedures of occupational skill testing authority. Only when the staff and teachers of "the center" master the relevant policies and knowledge, they can serve and improve the training quality well.

ACKNOWLEDGMENT

Source of the project: Shanghai education scientific research project: B12061.

Shanghai Second Polytechnic University construction of key disciplines (foster), subject name: vocational and technical education, subject number: XXKPY1316.

REFERENCES

[1] Zhenhua Li. Explore the virtual vocational skill training model in Higher Vocational Colleges, Chinese Vocational and Technical Education, 2013.04.
[2] Ningling Miao. Present situation and thinking of the social service of Higher Vocational Colleges in China, Journal of Harbin Vocational & Technical College, 2014.01.
[3] Fang Han, Xinyan Wang. Study on higher vocational colleges integrating the local economic to construct the multi-level vocational skill training system, Manager' Journal, 2014.02.

Education Management and Management Science – Zheng (Ed.)
© 2015 Taylor & Francis Group, London, ISBN 978-1-138-02663-6

Current situation, problems and solutions of highly skilled personnel training mode through school-enterprise cooperation

Ming Jiang

Shanghai Second Polytechnic University, Shanghai, China

ABSTRACT: Highly skilled talents are needed urgently by the current information society. School-enterprise cooperation is an effective model of training highly skilled talents. This paper analyzes the profound meaning and status of highly skilled personnel training mode through school-enterprise cooperation and finds out its main problems. On this basis, it puts forward methods and strategies of perfecting highly skilled training personnel mode through school-enterprise cooperation, which has reference value for the reform and development of skilled personnel training mode in Chinese vocational colleges.

Keywords: high-skilled personnel; training mode; status; questions; countermeasures

1 INTRODUCTION

Since the 21st century, along with the rapid increase of Chinese industrialization level and overall level of information, the social demand for numbers and quality of highly skilled personnel has increased significantly. To meet the development, vocational colleges as important bases for training highly skilled personnel and enterprises as the main stage to display talents of highly skilled personnel in the future should adapt to the times, strengthen all-round cooperation in the process of training highly skilled personnel, explore needed training model adaptable to social development, and provide strong support for training highly skilled personnel in China.

2 MEANING AND STATUS OF HIGHLY SKILLED PERSONNEL TRAINING MODE THROUGH SCHOOL-ENTERPRISE COOPERATION

School-enterprise cooperation is an emerging high-skilled training model. It refers to universities and companies establishing simulation training center consistent with professional skills training and training objectives, simulate the modern enterprise's management philosophy, culture, production environment and production process in the teaching process to make students participate in the practical enterprise production and management process directly, and build a number of enterprises-related skills training inside and outside universities as training bases to combine school teaching with enterprise management. It can realize seamless docking highly skilled training mode between students' capacity and enterprises' demand. This model enables students to feel the real business environment, production processes and management model in school and create good conditions for transforming knowledge and skills into professional skills and vocational ability after graduation.

School-enterprise cooperation is an important measure for vocational colleges to pursue their own development, implement with the market, improve the quality of education greatly and educate practical and technical personnel for enterprise. It combines students' knowledge with practice. Through this way, it can solve both the problem of insufficient practice space in enterprises and the problem of lack of training equipment in college, make colleges and enterprise realize the importance of complementing each other's advantages, resource sharing, improve the quality of training skilled personnel, and reach the combined goal of production, study and research. This learning mode includes both students' obtaining basic knowledge, vocational knowledge, technical skills and human qualities in indirect environment and student's training of practice, technology and professional quality. It is a new vocational education model for personnel training in China.

In recent years, highly skilled personnel training mode through school-enterprise cooperation has been widely used in China. Some vocational colleges have attempted it boldly and achieved good results. For example, in 2011, the leading provider of household goods Jiashangcheng company set up a school-enterprise cooperation practice base in Technician College of Guangdong in order to provide students with business simulation training platform, and won the agreement of school leaders, teachers, the majority of students and parents. It not only offered students more opportunities to practice in famous enterprises and enhanced their ability but also enabled enterprises to recruit personnel from colleges quickly. It can be seen as a successful example in school-enterprise cooperation. But in general, this mode is still at an early stage of exploration with narrow coverage and without an effective mode of training. It needs to be improved through reform.

3 THE MAIN PROBLEM OF HIGHLY SKILLED-PERSONNEL TRAINING MODEL THROUGH SCHOOL-ENTERPRISE COOPERATION

At present, although some vocational colleges have made some progress in training highly skilled personnel through school-enterprise cooperation, this mode is still far from perfect from the perspective of the country. There are many problems.

1. The school-enterprise cooperation mode has no sound and comprehensive legal protection system yet. At present, although some vocational colleges have attempted it boldly, it is still in a nascent state. Despite the fact that the nation has formulated relevant policies and measures to support and encourage close co-operation between schools and enterprises, real legal safeguards have not still been generated along with serious shortage of external driving force. In addition, the internal dynamic is unbalanced. Many places do not allow actual implementation. This training mode only shows spontaneous state without wide promotion because of the absence of comprehensive legal system.

2. The school-enterprise cooperation mode lacks strong support from operational mechanism. From the scope of the world, they all get the government's active support and participation and form a scientific and standardized mechanism. For example, in German vocational education model of dual system, schools and enterprises are all principals of education with different division. Enterprises are responsible for training and schools are responsible for cooperation and service. Both sides are under regulation of the federal laws and local government. Schools train students according to contracts of vocational training. Teaching Factory in Singapore is also under government intervention. It introduces advanced teaching facilities and real business environment into the school and combines with teaching tightly while achieving the integration of schools, training centers and enterprises.

Although the cooperation training model in some Chinese vocational colleges is admitted by the national education sector, there is no specific national supportive mechanism and successful experience. Therefore, the result is unsatisfactory. There is difficulty in overall progress.

3. The enterprises' enthusiasm in participating in school-enterprise cooperation is not high enough. Currently, one of the outstanding problems is vocational colleges are usually active, whereas related enterprises often passive with lack of inherent power of cooperation. Therefore, the effect can be imagined. The reason is that the benefit is not apparent for enterprises. In fact, the current school-enterprise cooperation in our country is in primary stage, and has shallow cooperation of sending talents to enterprises from school. The establishment and maintenance of cooperation relies on a network of major leadings. Although this kind of cooperation mechanism based on personal feelings is simple, flexible and easy to be operated, it is lack of long-term stability and not conducive to mobilize the enthusiasm of enterprises. Moreover, both sides have not been constrained by relevant laws in terms of capital investment, distribution of benefits, information exchange and management system etc. Enterprises feel more risk and less profitable, so there is lack of enthusiasm in participating in cooperation with schools.

4. A serious shortage of double teachers in school-enterprise cooperation. In this mode, the theory and practice teaching are equally important and should be coordinated closely, integrated deeply and promoted mutually. It requires higher quality of teachers. They can not only teach theory in class easily but also be proficient in practice as an engineer. In fact, in recent years, due to the vigorous development of vocational education, many teachers have no practical experience in enterprises. Therefore, their teaching contents are out of touch with the actual situation. Even those old teachers with some practical experience in the past have degraded seriously in professional competence for lack of practice in enterprises again due to the heavy work load. Therefore, the problem of lack of double teachers is serious now.

4 COUNTERMEASURES OF IMPROVING AND PERFECTING THE HIGHLY SKILLED PERSONNEL TRAINING MODE THROUGH SCHOOL-ENTERPRISE COOPERATION

1. Accelerating the process of legislation and perfecting related laws and regulations. Laws

and regulations are fundamental base and the effective way of perfecting school-enterprise cooperation training mode. In order to expand the mode successfully across the country, the nation should formulate "Promotion Law of training highly skilled personnel through school-enterprise cooperation", make principles in terms of mechanism setting, the division of power, training objective, information exchange, management mechanisms, funding, income distribution, labor aspects of access and safeguard measures etc., make legal definition of social status and welfare of skilled personnel, reduce the risk of cooperation effectively, improve positivity of both sides, and promote the healthy development of the cooperation. At the operational level, multi-sectors should formulate "measures of school-enterprise cooperation contract management" and define implementation of the management contract explicitly. The contact should include vocational training, cooperation mode, training objective, training content, training time, training career, acceptance criteria, standards of treatment as well as rights and obligations of each partner. So both sides have regulations to abide by.

2. The government should participate in it positively to provide powerful driving force for promotion of school-enterprise cooperation mode. From perspective of the world, the governments' vigorous promotion is extremely important. Especially in China, all levels of governments' active participation are bound to receive huge boost. Therefore, the Ministry of Education should develop relevant policies for specifying projects of training, practical training course content, time and method of different professions clearly. That will ensure that school-enterprise cooperation have rules to follow. Governments at all levels should take effective measures to guide enterprises play an important role in the establishment of market demand, talent specifications, knowledge and skills structure, curriculum, teaching content and assessments. Besides, they should coordinate vocational colleges to adapt changes of enterprises' needs, adjust professional direction, determine the scale of training reasonably, amend personnel training programs timely, and improve the adapting ability of the graduates effectively. By the government's positive guidance and regulation, the school-enterprise cooperation can be win-win for both.

3. Profound integration of the school and enterprise can mobilize the enthusiasm of enterprises participating in school-enterprise cooperation mode. In order to reverse the dilemma of negative participation of enterprises, we must strengthen reform and innovation of cooperation model, promote profound integration of the school and enterprise, help both sides' interests be tied closely, and urge enterprises to change their role from spectator to participant. School-enterprise cooperation can bring the following benefits to an enterprise: the selection of qualified technicians from graduates, retraining older workers in colleges, transition to a learning-oriented enterprise, enhancement of core competitiveness, development of new products with talent and equipment of colleges, upgrading old equipment, the introduction of new technologies, establishment of good image, expanding the positive publicity, allowing students to understand their products. And entity enterprises can bring direct economic benefits. Order training mode means that both sides sign the employment order and personnel training agreement so that the enterprise gets involved in development of training program and management of training personnel procedure, provides equipment and practical place, students' tuition, grants and loans, and attracts graduates to work in the enterprise directly. The order training mode forms a legal commissioned relationship between schools and enterprises. It can not only enhance the pertinence of training, but also stimulate the enthusiasm of students.

4. Multi-channel training teachers can provide qualified teachers for school-enterprise cooperation. It is not easy to solve serious shortage of "double" teachers. We must make concerted efforts. First, we recruit teachers with high educational degree, strong theoretical level and work experience in related enterprise, and avoid recruitment of graduates directly. Second, we employ engineers with rich theoretical knowledge and practical ability to teach in the college. Third, the college should arrange teachers' tasks reasonably, and encourage teachers to go to the enterprise to expand the depth and breadth of skills and knowledge. Fourth, personnel exchange should be increased. The college arranges a certain number of teachers to enterprises for visit or practice each year. The enterprise sends a number of engineers to colleges for theoretical study or exchange of experiences. In the process, both can achieve mutual promotion and common enhancement. Fifth, vocational colleges should create good environment for teachers. For example, the generous wages and benefits, good career prospects and so on, so that the existing "double" teachers can focus on their work. It is an effective way of preventing the loss of talents.

ACKNOWLEDGMENTS

Source of the project: Shanghai education scientific research project: B12061.

Shanghai Second Polytechnic University construction of key disciplines (foster), subject name: vocational and technical education, subject number: XXKPY1316.

REFERENCES

[1] Jianxun You, Liubin Chen. How to enhance highly skilled personnel training mechanism: from "cooperation" to "integration". The Education and occupation, 2010 (33).

[2] Haijun Cui, Ronghong Liu. Analysis of innovative training mode of higher vocational colleges: "school-enterprise cooperation". Journal of Qingdao Vocational and Technical College, 2011 (6).

[3] Airu Jin. Study of school-enterprise cooperation model in vocational colleges. North China Electric Power University (Baoding) Master Thesis, 2009.

[4] Junhui Sun. Improvement and explore on vocational school-enterprise cooperation innovative training model. Shandong University master's degree thesis, 2012.

[5] Junsheng Ning, Xixi Wang. Reflections on the professional colleges cooperative innovation mode. Education and occupation, 2011 (29).

[6] Yibing Huang. Innovation of higher vocational education school-enterprise cooperation model. Anhui Business College of Vocational Technology, 2012 (4).

[7] Jinsong Yang. Exploration and practice of school-enterprise cooperation training mode. Education and occupation, 2010 (22).

Education Management and Management Science – Zheng (Ed.)
© 2015 Taylor & Francis Group, London, ISBN 978-1-138-02663-6

Adult students' self-regulated learning ability evaluation research

Bo Yang & Wenyan Song
Chongqing Radio and TV University, Chongqing, China

ABSTRACT: It is particularly important to scientifically evaluate students' self-regulated learning ability, because the level of students' self-regulated learning ability directly affects individual autonomy learning effect and network education quality during online education. According to the adult students' self-regulated learning ability, we establish tertiary index system which uses the fuzzy comprehensive evaluation method to establish fuzzy set and fuzzy membership degree matrix for comprehensive evaluation. And it also illustrates the specific application of the fuzzy comprehensive evaluation by examples.

Keywords: adult students; self-regulated learning; self-regulated learning ability; fuzzy comprehensive evaluation

1 INTRODUCTION

Self-regulated learning in the scientific research field is not a fashionable concept [1], but is a natural need which people rely on for their own learning [2], and are a self-motivation means and a way of growing up for learners to decide what and how to learn [3]. Currently, for students' autonomous learning and autonomous learning ability research, domestic scholars often focus on common formation mechanism, influencing factors, strategies for upgrading, [4] [5] [6] but the research on how to make effective evaluation of the students self-regulated learning ability is relatively rare. This paper attempts to establish a quantitative evaluation model approach. It can effectively quantify, evaluate the adult students self-regulated learning ability to make them have an objective understanding about their self-learning ability, which lays a good foundation to specifically enhance the self-regulated learning ability and self-regulated learning effect.

2 CONNOTATION AND INDEX SYSTEM OF SELF-REGULATED LEARNING ABILITY

Self-regulated learning ability in the network environment is that according to learners' intentions, objectives, plans, under the guidance of teachers, relying on their own ability and adopting various learning media, the learners take the initiative to acquire knowledge and skills. It is a comprehensive, integrated learning ability, including the ability to self-planning, self-control, self-evaluation and collaborative learning. According to the connotation of self-regulated learning ability and constitution, it establishes three evaluation index systems, as shown in Figure 1. The first level is self-regulated learning ability (U); the second level of self-regulated learning ability (U) includes five sub-capacity U_i (which $i = 1, 2, 3, 4, 5$), they are separated network technology capabilities (U_1), self-planning capacity (U_2), self-regulatory capacity (U_3), self-evaluation capabilities (U_4), team collaboration (U_5); the third level

First level	Second level	Third level
Self-regulated learning ability U	Network technology capabilities U_1	Web browsing and querying skills U_{11}, Interaction and communication skills U_{12}
	Self-planning capacity U_2	Making a reasonable plan of study U_{21}, Arranging appropriate learning content U_{22}
	Self-regulatory capacity U_3	Self-monitoring U_{31}, Self-regulation U_{32}, Self-control U_{33}
	Self-evaluation capabilities U_4	Self-awareness U_{41}, Self-examination U_{42}, Self-test U_{43}
	Team collaboration U_5	Actively participate in group learning activities U_{51}, Mutual communicate, help and learn U_{52}

Figure 1. Three evaluation index system of self-regulated learning ability.

is further broken down to the ability of the second stage (U_{ij}), it is divided into 12 objects, respectively, Web browsing and querying skills (U_{11}), interaction and communication skills (U_{12}); making a reasonable plan of study (U_{21}), arranging appropriate learning content (U_{22}); self-monitoring (U_{31}), self-regulation (U_{32}), self-control (U_{33}); self-awareness (U_{41}), self-examination (U_{42}), self-test (U_{43}); actively participate in group learning activities (U_{51}), mutual communication, help and learning (U_{52}).

3 THE CONSTRUCTION OF SELF-REGULATED LEARNING ABILITY EVALUATION MODEL

Because the factors that affect students' self-learning ability are uncertain or difficult to quantify to make students' self-regulated learning ability with a certain ambiguity, when in the evaluation, fuzzy comprehensive evaluation method based on the principles of fuzzy variables and the maximum membership degree principle can carry on overall evaluation.

Fuzzy comprehensive evaluation has the following key steps:

3.1 Establishment of fuzzy sets

① Establish factors layer factor set $U = \{U_1, U_2, U_3, U_4, U_5\}$, the corresponding set of weights $A = (a_1, a_2, ..., a_x)$. $a_k (k = 1, 2, ..., x)$ Which represents the proportion of factor U_k in U, and $\sum_{k=1}^{x} a_k = 1$.

② Define index set of index layer as $U_k = (U_{k1}, U_{k2}, ... U_{km})$, the corresponding set of weights $A_k = (a_{k1}, a_{k2}, ... a_{km})$. $a_{ki} (i = 1, 2, ..., m)$ Which represents the proportion of U_{ki} in U_k, and $\sum_{i=1}^{m} a_{ki} = 1$.

③ Establish evaluation set $W = \{W_1, W_2, ... W_n\}$, $W_j (j = 1, 2, ... n)$ which indicates all different comments from high to low levels. This article takes n = 5, "W_1", "W_2", "W_3", "W_4", "W_5" separated represent "best" "better", good", "poor", "poorer".

3.2 Establish a fuzzy membership matrix

$$R = \begin{bmatrix} r_{11} & r_{12} & \cdots & r_{1m} \\ r_{21} & r_{22} & \cdots & r_{2m} \\ \cdots & \cdots & \cdots & \cdots \\ r_{p1} & r_{p2} & \cdots & r_{pm} \end{bmatrix}$$

Specific calculation process is to take "m" classes as samples to assess their self-learning ability. At first, collecting self-learning ability evaluation index data $r_{ij} (i = 1, 2, ... 5; \ j = 1, 2, 3)$ of the "m" classes as the element and making comprehensive evaluation separately to each the third level index set by $B_i = A_i R_i (i = 1, 2, ..., 5)$, draw first level judge vector B_i; the first level judge vector constitute the secondary index evaluation matrix for the second index, then get the second judges vectors $B = AR$. The first level index is the overall evaluation of self-learning ability; the second level index is network technology capabilities, self-planning capabilities, self-control capabilities, self-evaluation capacity, and team collaboration capabilities.

4 MODEL APPLICATION EXAMPLES

4.1 Assessment process

One class applies WebQuest test teaching model to enhance the self-regulated learning ability of

Second index	Third index	At the beginning					In the end				
		best	better	good	poor	poorer	Best	better	good	poor	poorer
network technical capacity U_1(0.28)	web browser inquiry skills U_{11}(0.55)	6 (0.15)	8 (0.2)	16 (0.4)	8 (0.2)	2 (0.05)	8 (0.2)	12 (0.3)	18 (0.45)	2 (0.05)	0 (0)
	interaction and communication skills U_{12}(0.45)	5 (0.125)	12 (0.3)	15 (0.375)	4 (0.1)	4 (0.1)	7 (0.175)	14 (0.35)	11 (0.275)	5 (0.125)	3 (0.075)
Self-planning capacity U_2(0.18)	Making a reasonable plan of study U_{21}(0.46)	0 (0)	7 (0.175)	12 (0.3)	15 (0.375)	6 (0.15)	6 (0.15)	13 (0.325)	13 (0.325)	5 (0.125)	3 (0.075)
	Arranging appropriate learning content U_{22}(0.54)	0 (0)	6 (0.15)	14 (0.35)	14 (0.35)	6 (0.15)	5 (0.125)	15 (0.375)	14 (0.35)	4 (0.1)	2 (0.05)
Self-regulatory capacity U_3(0.17)	Self-monitoring U_{31}(0.40)	0 (0)	4 (0.1)	12 (0.3)	12 (0.3)	12 (0.3)	5 (0.125)	8 (0.2)	15 (0.375)	6 (0.15)	6 (0.15)
	Self-regulation U_{32}(0.25)	0 (0)	3 (0.075)	15 (0.375)	10 (0.25)	12 (0.3)	6 (0.15)	9 (0.225)	14 (0.35)	7 (0.175)	4 (0.1)
	Self-control U_{33}(0.35)	0 (0)	0 (0)	24 (0.5)	15 (0.25)	11 (0.25)	3 (0.075)	12 (0.3)	15 (0.375)	5 (0.125)	5 (0.125)
Self-evaluation capabilities U_4(0.15)	Self-awareness U41(0.35)	0 (0)	3 (0.075)	20 (0.5)	12 (0.3)	5 (0.125)	5 (0.125)	10 (0.25)	12 (0.3)	10 (0.25)	3 (0.075)
	Self-examination U_{42}(0.35)	0 (0)	2 (0.05)	25 (0.625)	7 (0.175)	6 (0.15)	6 (0.15)	9 (0.225)	10 (0.25)	10 (0.25)	5 (0.125)
	Self-test U_{43}(0.3)	0 (0)	2 (0.05)	25 (0.625)	6 (0.15)	7 (0.175)	6 (0.15)	9 (0.225)	10 (0.25)	11 (0.225)	4 (0.1)
Team collaboration U_5(0.22)	Actively participate in group learning activities U_{51}(0.55)	6 (0.15)	10 (0.25)	12 (0.3)	7 (0.175)	5 (0.125)	10 (0.25)	15 (0.375)	6 (0.15)	5 (0.125)	4 (0.1)
	Mutual communicate, help and learn U_{52}(0.45)	7 (0.175)	12 (0.3)	13 (0.325)	5 (0.125)	3 (0.075)	11 (0.25)	15 (0.375)	7 (0.175)	4 (0.1)	4 (0.1)

Figure 2. Self-learning ability of some adult class surveying summary table at the beginning and the end.

adult students, and 40 students self-learning ability are surveyed at the beginning and the end of test. Each indicator survey in evaluation model obtains the data as Figure 2. Experts give each index weights in column brackets of the second index and the third index. For example, self-learning ability in the weights of the network technical capacity is 0.28; Web browser queries skills U_{11} in the weights of network technology capabilities is 0.55, interaction and communication skills U_{12} is 0.45. The columns of the beginning and the end are the evaluation of the third level index and the weights of evaluation value are in the brackets. Such as at the beginning of the test, the 40 students Web browser inquiry skills U_{11}, there are 6 students evaluated as "best", weights 0.15; there are eight students evaluated as "better", weights 0.2; there are 16 students evaluated as "good", weights 0.4; there are 8 students evaluated as "poor", weights 0.2; there are 2 students evaluated as "poorer", weights 0.05.

According to data in the table, get the beginning of the Evaluation vector of network technology capabilities, self-planning capabilities, self-control capabilities, self-evaluation capacity, and team collaboration capabilities as follows:

$$B_1 = A_1 \circ R_1$$

$$= (0.55, 0.45)\begin{pmatrix} 0.15 & 0.2 & 0.4 & 0.2 & 0.05 \\ 0.125 & 0.3 & 0.375 & 0.1 & 0.1 \end{pmatrix}$$

$$= (0.139, 0.245, 0.389, 0.155, 0.072)$$

$$B_3 = A_3 \circ R_3$$

$$= (0.4, 0.25, 0.35)\begin{pmatrix} 0 & 0.1 & 0.3 & 0.3 & 0.3 \\ 0 & 0.075 & 0.375 & 0.25 & 0.3 \\ 0 & 0 & 0.5 & 0.25 & 0.25 \end{pmatrix}$$

$$= (0, 0.059, 0.389, 0.27, 0.281)$$

$$B_4 = A_4 \circ R_4$$

$$= (0.35, 0.35, 0.3)\begin{pmatrix} 0 & 0.075 & 0.5 & 0.3 & 0.125 \\ 0 & 0.05 & 0.625 & 0.175 & 0.15 \\ 0 & 0.05 & 0.625 & 0.15 & 0.175 \end{pmatrix}$$

$$= (0, 0.059, 0.581, 0.211, 0.149)$$

$$B_5 = A_5 \circ R_5$$

$$= (0.55, 0.45)\begin{pmatrix} 0.15 & 0.25 & 0.3 & 0.175 & 0.125 \\ 0.175 & 0.3 & 0.325 & 0.125 & 0.075 \end{pmatrix}$$

$$= (0.161, 0.273, 0.311, 0.153, 0.102)$$

Evaluation vector for self-regulated learning ability at the beginning:

$$B = (0.28, 0.18, 0.17, 0.15, 0.22)$$

$$\times \begin{pmatrix} 0.139 & 0.245 & 0.389 & 0.155 & 0.072 \\ 0 & 0.162 & 0.327 & 0.361 & 0.15 \\ 0 & 0.059 & 0.389 & 0.27 & 0.281 \\ 0 & 0.059 & 0.581 & 0.211 & 0.149 \\ 0.161 & 0.273 & 0.311 & 0.153 & 0.102 \end{pmatrix}$$

$$= (0.074, 0.177, 0.39, 0.22, 0.139)$$

Equally, get the end the Evaluation vector of network technology capabilities, self-planning capabilities, self-control capabilities, self-evaluation capacity, and team collaboration capabilities as follows:

$$B_1' = A_1 \circ R_1' = (0.189, 0.323, 0.371, 0.084, 0.033)$$

$$B_2' = A_2 \circ R_2' = (0.137, 0.352, 0.339, 0.111, 0.061)$$

$$B_4' = A_4 \circ R_4' = (0.141, 0.249, 0.268, 0.242, 0.1)$$

$$B_5' = A_5 \circ R_5' = (0.25, 0.375, 0.161, 0.114, 0.1)$$

Evaluation vector for self-regulated learning ability in the end:

$$B' = (0.28, 0.18, 0.17, 0.15, 0.22)$$

$$\times \begin{pmatrix} 0.189 & 0.323 & 0.371 & 0.084 & 0.033 \\ 0.137 & 0.352 & 0.339 & 0.111 & 0.061 \\ 0.114 & 0.241 & 0.369 & 0.148 & 0.128 \\ 0.141 & 0.249 & 0.268 & 0.242 & 0.1 \\ 0.25 & 0.375 & 0.161 & 0.114 & 0.1 \end{pmatrix}$$

$$= (0.173, 0.315, 0.303, 0.13, 0.079)$$

4.2 Result analysis

According to the principle of maximum membership, on the whole, by the application of WebQuest teaching model to promote students' autonomous learning ability has played a more significant effect, the overall self-regulated learning ability of this class has improved from "good" to "better". Specifically, the students' ability of the network technology and the team cooperation ability have a modest improvement, and self planning ability, self-control ability, and self-evaluation ability have also improved a lot. Therefore, the use of fuzzy comprehensive evaluation method to evaluate the students' autonomous learning ability can help us know clearly about the change situations of students' self-regulated learning ability, and make targeted diagnosis of teaching activities.

ABOUT THE AUTHOR

Yang Bo (1974–), male, Han, Economics and Business Administration College of Chongqing University, PhD; Finance and Economy Department of Chongqing Radio and Television University, vice professor, vice president; Main research directions: modern distance education, innovation and entrepreneurship education.

Song Wenyan (1984–) female, Han, Finance and Economy Department of Chongqing Radio and Television University, Main research directions: modern distance education, innovation and entrepreneurship education.

Contact: Tel: 15086821550,

Email: yangbo74@163.com

Address: Finance and Economics Management Department of Chongqing Technology and Business Vocational Institute, Siyuan Road 15, Hechuan District, Chongqing.

Code: 401520.

REFERENCES

[1] Knowledg, M. Self-directed learning: A guide for learners and teachers. New York Association press, 1975.
[2] Tough, A. The adult's learning projects. Toronto: Ontario Institute for Studies in Education, 1971.
[3] Garrison, D.R. Self-directed learning: Toward a comprehensive model. Adult Education Quarterly, 1997, 48(1):1–18.
[4] Lihong Zhou, Xiao meng Wu, Yan Yin, Research Network learner autonomy status—Learner of Peking University School of Online Education as Case China Educational Technology, 2010 (6):46–54.
[5] Yaohui Hong, Jiangbo Xia. Open education students autonomous learning ability generate resistance and crack strategies explore. Modern distance education, 2007 (3):9–11.
[6] Junke Zhuang, Baoxun He, Network autonomous learning behavior system-level framework and autonomous learning behavior Tower. Chinese Educational Technology, 2009 (3):41–45.

Education Management and Management Science – Zheng (Ed.)
© *2015 Taylor & Francis Group, London, ISBN 978-1-138-02663-6*

Characteristics of general education in Taiwan's universities and the inspirations to the universities in the Mainland

Shuqin Ren
Shandong Polytechnic, Jinan, Shandong, China

ABSTRACT: General education is actively carried out in colleges and universities in Taiwan in order to meet the needs of personnel training. Since the new century, with the general education evaluation activities carried out in an orderly manner, the curriculum system of the general education, with clear concepts, rich contents and perfect support, is becoming more perfect and has been inspiration to various colleges and universities in the Mainland.

Keywords: Taiwan; general education; inspirations

1 INTRODUCTION

General education in Taiwan's colleges and universities develops more maturely compared with the culture and quality education or the general education of the colleges and universities in the Mainland. Promoted in 1984, absorbing the advanced educational experience of the USA, Japan and other developed countries, general education has become an important work of the current educational reform in Taiwan's colleges and universities.

The general education was advocated in the early 1980s in Taiwan's universities. Education should be a 'whole person' education, and the 'whole person' equals 'profession' stamens and 'general education' petals. The 'whole person' education is the final goal of the general education of Taiwan. The 'whole person' education argues that the fundamental purpose of education is to cultivate the whole person, which reflect not only in the development of the individual's knowledge and skills, but also the experiences and promoting of the moral and spiritual states; not only pay attention to the full development of the human beings, but also emphasize the coordination and interaction between people and society, people and nature [1].

2 CHARACTERISTIC OF TAIWAN'S GENERAL EDUCATION

More than ten years ago, Shen Junshan, the former president of National Tsing Hua University, once said, in Taiwan, the practice of general education is more difficult than the argument of it. The difficulties of practice included that no one wanted to

manage, no professionals wanted to teach and no students wanted to spend their minds to listening. During that time, in Taiwan, this was a common phenomenon in the implementation of the general education, which caused that many students took the general courses as 'nutrition' courses. At present, this phenomenon has ceased to exist. Taiwan's general education has embodied certain characteristics.

2.1 *Clear concepts of courses*

Special attention has been paid to the development of the general education in Taiwan's colleges and universities. There are specialized management institutions, such as general education centers and so on, which are responsible for the planning of the general education courses. Since the 21st century, with the constant improvement of the general educational evaluation mechanism, the 'whole person' educational concept has reflected gradually in the distinctive general education in the colleges and universities, and it pays attention to develop holistic personality in its curriculum [2]. As mentioned above, the biggest problem in the courses offered is the problem of 'fracture', including 'the fracture between tradition and modern' and 'the fracture between departments or fields'. Clear courses concepts come from the common view of the general educational core spirits. The core spirit of the general education in Taiwan's colleges and universities is to cultivate students' proper cultural qualities, life wisdoms, analysis abilities, communication skills and lifelong-learning-and-growth motivations. A student with the ideal general educational personality will not only have the basic knowledge of the humanity, social and natural science. More important, he or she can think critically, understand the meaning of self-being, respect the values of different lives and civilizations, be curious about the university, and know how to explore it.

2.2 *Rich contents of courses*

As we mentioned earlier, there are various general educational courses in Taiwan's colleges and universities, and students can take suitable elective classes from different areas, which increases the elasticity of the students' elective. For example, in the eight fields of National Taiwan University, there are 462 courses of general education and students take 18 credits from 5 to 6 fields. The courses emphasize the fundamentality of the contents, that the contents must contain the basic elements of the human civilization; emphasize the subjectivity of the students, that is, to exert the principal role of the students by discussion, intellectual enquiries, comparison and critical learning; emphasize the conformability and fusing characteristic, that is, to reconcile knowledge of different fields and realize the core value of general education across different subjects [3].

2.3 *Perfect supports of courses*

Implementation of general education is the process of taking the curriculum plan into practice and is an important condition to truly improve the quality of general education. Generally speaking, the main factors influencing the curriculum implementation are the characteristics of the curriculum plans, the teachers, the schools and the external environment of the schools. We can say that all the factors are good for the effective implementation of the general education in Taiwan. For one thing, the general educational curriculum plans under the guidance of the 'whole person' education are all designed relatively perfect, despite the differences among schools; for another, since the 1990s, colleges and universities in Taiwan all set up general educational centers as the specialized agencies, which plan and evaluate the general education as a whole, provide students diversified cognitive views and provide communication and dialogues between humanities and science and technology, etc. Some centers even publish journals of general education, to promote the scientific research with teaching and to drive the teaching with scientific research. To improve the quality of general education constantly, the Ministry of Education in Taiwan specially established the general education committee and general education evaluation institution, responsible for coordinating and planning to deal with the related matters. It also set up an improving

program office of general education, responsible for the coordination of the improvement work. Two big educational evaluations promote the constant perfect of the curriculum system of general education in Taiwan. In addition, 'award for outstanding teachers of general education' was set up in the year 2007, in order to encourage excellent teachers devoted to general education. In recent years, the ministry of education in Taiwan subsidized the colleges and universities to improve their abilities of teaching and researching, as well as improve the general education, in the name of 'plan to promote the fundamental education', 'plan to promote the competitiveness of the universities' and so on. Also, seminars for general education and teachers are often held to provide the training opportunities for general education teachers. All these measures promote strongly the standardization and institutionalization of general education, and guarantee the effective implementation of the courses.

3 GETTING INSPIRATIONS OF TAIWAN'S GENERAL EDUCATION TO THE UNIVERSITIES IN THE MAINLAND

3.1 *To grasp accurately the essence of vocational education and to make the goal of general education of higher vocational colleges clear*

In recent years, the higher vocational education developed rapidly in the Mainland. The Ministry of Education put forward clearly the employment-oriented guiding principle for the higher vocational colleges, in order to solve the problem of employment that the graduates are facing. However, many colleges lack the dialectic comprehension of the employment-oriented essence. In the process of personal training, they put undue emphasis on the professional skills, ignoring the cultivation of the students' comprehensive qualities. The basic attribute of education is to realize people's overall development. Vocational education should also undertake the mission of promoting students' overall development as well as cultivating the quality of professional skills. The experiences that higher vocational colleges in Taiwan promote the general education show us a very commendable enlightenment: vocational education should not only pay attention to the value of skills and tools, but also should pay attention to the cultivation of students' comprehensive qualities, pay attention to the unity of the functions of professional education and human's all-round development. General education centers of Kunshan University and other colleges and universities did some research among the graduates. The feedback showed that receiving more general education or not in the universities is different for the self-development of the graduates.

Maybe at the beginning of the employment, general education is not important. But when students get to work, general education will play an important part in students' potential development, promotion and so on. Some graduates feel that professional education is of great help to the work that requires independent character. But as the nature of jobs changes, or one changes from a professional job to a managerial job, the role of professional knowledge and technology becomes less important; the sight, mind, sense of responsibility, communication skills, cooperation skills, expression skills and judgment that general education cultivates become more and more important. Due to the personal feelings and experiences of graduates, the understanding of the students at school changes a lot about the general education.

Therefore, higher vocational colleges should vigorously carry out and promote general education, strengthen the construction of campus culture and promote the fusing among humanities, science and technology, professional and general knowledge. Higher vocational colleges should make efforts to train senior technical talents who have both outstanding professional skills and qualities and knowledge of humanities, who have both noble professional ethics and strong sense of innovation, and who have the courage to bear the social responsibility.

3.2 To carry out the concepts of general education and to reconstruct the public and fundamental courses system in higher vocational colleges

Vocational and technical colleges in Taiwan use general courses as the main channel to implement general education. The curriculum is carefully designed and systematically planned, emphasizing on the extensive, classic and multivariate sources, and widely covers the main fields of human culture and knowledge, realizing the integration among the categories of humanities, social and natural knowledge. Nowadays, in the higher vocational colleges in Chinese mainland, the general courses contain two parts: public compulsory courses and public elective courses. Public compulsory courses include college Chinese, English, political theory, ideological and moral lessons, sports, computer, etc., while public elective courses are set independently by colleges without systematic planning and further argument [4]. These general courses are gradually formed according to the requirement of development of social politics, economy, culture and science and technology under certain historical condition. They lack clear concept, unified goal and overall planning. The courses mostly adopt national textbooks, have unified outline and requirement, and do not have enough of

pertinence, flexibility and humanism. At the same time, the structure of the courses and the proportion are not reasonable: the courses of basic skills political introduction are more, while the courses of historical culture, ethics, social analyses, art and science are less; informative courses are more while methods and thinking courses are less; compulsory courses are more while elective courses are less; one-subject-based courses are more while comprehensive courses are less. Judging from the concepts and goals of the general education, public courses in the higher vocational colleges in the Mainland are difficult to be called general education in the true sense and they cannot have the function of general education. Therefore, based on the idea and pattern of general education, the higher vocational colleges should fuse between Chinese and western, ancient and modern, science and humanities, have an overall plan for the general educational system, and reconstitute the system of public basic course. The higher vocational colleges should have integration, selection and promotion to the contents of general education, make their effort to show the scriptures and essence of human culture to the students, establish a complete general education platform, build the professional education based on the wide and extensive general education, and realize the fusion of professional education and general education.

3.3 To establish effective implementation mechanism of general education and to promote general education comprehensively

From Taiwan's experiences we can see that, to develop general education one must establish and improve the implementation and safeguarding mechanism, have full guarantee of organization, human, material and financial resources and other resources, fully implement and advance from the aspects of courses structure, environment influences, life experiences, social practice and so on, and form the whole educational atmosphere and effects. First, it is necessary to strengthen the publicity of general education, create actively the atmosphere propitious to the implementation of general education, and make all the teachers and students form a profound knowledge of general education. Thus, they will highly approve and actively support the implementation of general education [5]. Second, it is necessary to establish powerful protection and enforcement mechanism, including establishing the system of decision-making, consulting and implementing, strengthening the construction of teaching staff and so on, and set up school consultative committees and general education centers, responsible for the deliberation, promoting, scrutiny and the specific organization and implementation of the general education work. Third, besides the formal courses, we should also make plans of the campus landscape by painstakingly constructing positive atmosphere in colleges. We should carry out colorful cultural activities in the campuses like social practice, seminars, culture and art festival, science-and-technology awareness month, reading and discussion and so on. We should create excellent school spirit, learning spirit and teaching spirit and promote general education through the influence of the environment.

REFERENCES

[1] Jerry G. Gaff. New Life for the College Curriculum: Assessing Achievements and Furthering Progress in the Reform of General Education. San Francisco: Jossey-Bass. 1991. 12.

[2] Palomba, C.A., & Banta, T.W. (1999). Assessment essentials: Planning, implementing, and improving assessment in higher education. San Francisco: Jossey-Bass.

[3] United States Department of Labor. The Secretary's Commission on Achieving Necessary Skills. (1991). What work requires of schools: A SCANS report for America 2000. Washington, D.C.: The Department.

[4] Leskes, A., & Wright, B.D. (2005). The art & science of assessing general education outcomes: A practical guide. Washington, D.C.: Association of American Colleges and Universities.

[5] McCue, Michelle (2012-07-02). "Ja-neane Garofalo And Larry Miller Star In Comedy general education; In Theaters & VOD August 24". wearemoviegeeks.com. Retrieved 2012-08-27.

Education Management and Management Science – Zheng (Ed.)
© 2015 Taylor & Francis Group, London, ISBN 978-1-138-02663-6

A new model of forecast enrollment using fuzzy time series

Hongxu Wang & Jianchun Guo
College of Tourism Management, Qiongzhou University, Sanya, China

Hao Feng & Fujin Zhang
College of Science and Engineering, Qiongzhou University, Sanya, China

ABSTRACT: In this paper, a concept of forecasting enrollments based on fuzzy time series is led by Song and Chisson who put forward the first model of forecasting enrollments based on fuzzy time series. Since twenty years, many scholars have proposed series' method of forecasting enrollments based on fuzzy time series, and the forecasting accuracy rates of historical data have been improved. We apply the historical data percentage of year to year changes as the universe of discourse to change the construction of inverse fuzzy number, and propose a model of forecasting enrollments based on fuzzy time series. We used 22 years of freshmen's enrollments data of the University of Alabama to illustrate the forecasting process. The result shows that proposed method can get higher forecasting accuracy rates for forecasting enrollments than the existing method.

Keywords: year to year changes; percentage; inverse fuzzy number; fuzzy time series; forecasting model

1 INTRODUCTION

Song and Chissom combined fuzzy forecasting technology with time series, and proposed the first forecasting model of fuzzy time series [1–2]. Since twenty years, many scholars have proposed many different models and the range of study is extensive. Such as forecasting enrollments with fuzzy time series [1–10], stock index forecasting [12], forecasting stock markets [13], exchange rates forecasting [14], data mining [15–16], temperature prediction using fuzzy time series [17–18], and forecasting for car road accidents [19] and so on. This paper applies the knowledge discovery, using the historical data of year to year percentage changes, the inverse fuzzy number of Jilani and Ardil [6], and the forecasting model technology of Saxena, Sharma & Easo [4]. We change the construction of inverse fuzzy number and propose a new forecasting model of fuzzy time series.

We apply the case of historical freshmen's enrollments data of University of Alabama (1971–1992) to illustrate the forecasting process with the new model of fuzzy time series.

2 PREPARE KNOWLEDGE

This paper applies the concerned concept in literature [4] and [1, 2, 3, 5–11], and here we omit it.

3 THE NEW METHOD OF FORECASTING

We apply the forecasting questions of historical freshmen's enrollments data of University of Alabama (1971–1992) as the case to introduce the applications' step of the new method, and compare with the existing method. The specific application step can be drawn as follows.

Step 1. Listed the table of historical data. The freshmen's enrollments data of University of Alabama (1971–1992) shown in Table 1 [1].

Step 2. Integrate historical data. The year to year percentage changes are defined as $c_i = [(e_i - e_{i-1})/e_{i-1}] \times 100\%$, where e_i and e_{i-1} is the i year' historical data and $i - 1$ year's historical. Thus, we can get the universe of discourse and fill in Table 1.

Step 3. Establish forecasting formula and forecasting

The formula of Inverse fuzzy number:

$$u_i = \frac{0.004 + 1}{\dfrac{0.004}{c_{i-1}} + \dfrac{1}{c_i}}, \ 1973 \le i \le 1992 \quad (1)$$

Forecasting formula:

$$f_i = e_{i-1} \times (1 + u_i\%), 1973 \le i \le 1992. \quad (2)$$

The simulation forecasting of freshmen's enrollments data of the University of Alabama (1971~1992), the results shown in Table 2.

Table 1. The historical enrollments of the University of Alabama.

Year	Enrollments e_i	Year to year	Percentage c_i	Year	Enrollments e_i	Year to year	Percentage c_i
1971	13055			1982	15433	1981–82	−5.83%
1972	13563	1971–72	3.89%	1983	15497	1982–83	0.41%
1973	13867	1972–73	2.24%	1984	15145	1983–84	−2.27%
1974	14696	1973–74	5.98%	1985	15163	1984–85	0.12%
1975	15460	1974–75	5.20%	1986	15984	1985–86	5.41%
1976	15311	1975–76	−0.96%	1987	16859	1986–87	5.47%
1977	15603	1976–77	1.91%	1988	18150	1987–88	7.66%
1978	15861	1977–78	1.65%	1989	18970	1988–89	4.52%
1979	16807	1978–79	5.96%	1990	19328	1989–90	1.89%
1980	16919	1979–80	0.67%	1991	19337	1990–91	0.05%
1981	16388	1980–81	−3.14%	1992	18876	1991–92	−2.38%

Table 2. Actual enrollments and forecasting enrollments of the University of Alabama.

Years	Enrollments e_i	φ_i (%)	f_i	$e_i - f_i$	$(e_i - f_i)^2$	$|e_i - f_i|/e_i$
1971	13055					
1972	13563					
1973	13867	2.243791786	13867	0	0	0.000000
1974	14696	5.940484135	14691	5	25	0.000340
1975	15460	5.202703656	15461	−1	1	0.000065
1976	15311	−0.964552284	15311	0	0	0.000000
1977	15603	1.933023649	15607	−4	16	0.000256
1978	15861	1.650895338	15861	0	0	0.000000
1979	16807	5.898614011	16797	10	100	0.000595
1980	16919	0.669668738	16920	−1	1	0.000059
1981	16388	−3.212787786	16375	13	169	0.000793
1982	15433	−5.810169348	15436	−3	9	0.000194
1983	15497	0.411175583	15497	0	0	0.000000
1984	15145	−2.330696403	15136	9	81	0.000594
1985	15163	0.120505481	15163	0	0	0.000000
1986	15984	4.601784825	15861	123	15129	0.007695
1987	16859	5.469758316	16858	1	1	0.000059
1988	18150	7.67801204	181548	2	4	0.000110
1989	18970	4.552739394	18976	−6	36	0.000316
1990	19328	1.894391504	19329	−1	1	0.000052
1991	19337	0.050194688	19338	−1	1	0.000052
1992	18876	−2.951482215	18766	110	12100	0.005828
AFER						0.08508%
MSE					1384	

4 A COMPARISON OF EXISTING METHOD

Compare this paper's method with the method proposed in existing papers in literature [1], [3], [4], [5], [8], [10], [11], the data taken from Table 6 in literature [4], shown as Table 3. In Table 3, the formula of AFER and MSE is drawn as follows [4]:

$$AFER = \frac{|e_i - f_i|}{n} \times 100\%,$$

$$MSE = \frac{1}{n} \sum_{i=1}^{n} (e_i - f_i)^2$$

where, E_i, F_i are the enrolments of i and the forecasting data of i.

From Table 3, we can see, the forecasting results of the model proposed in this paper can get the smallest AFER and MSE than other models of forecasting fuzzy time series.

Table 3.   A comparison of the forecasting results of different forecasting models.

Year	Enrollments	Saxena, Sharma, Easo [4]	Stevenson, Porter [5]	Song, Chissom [1]	Song, Chissom [3]	Hwang, Chen, Lee [10]	Chen [8]	Chen [11]	Proposed model
1971	13055	–	–	–	–	–	–	–	
1972	13563	13486	13410	14000	–	–	–	14000	
1973	13867	13896	13932	14000	–	–	–	14000	13867
1974	14696	14698	14664	14000	–	–	14500	14000	14691
1975	15460	15454	15423	15500	14700	–	15500	15500	15461
1976	15311	15595	15847	16000	14800	16260	15500	16000	15311
1977	15603	15600	15580	16000	15400	15511	15500	16000	15607
1978	15861	15844	15877	16000	15500	16003	15500	16000	15861
1979	16807	16811	16773	16000	15500	16261	16500	16000	16797
1980	16919	16916	16897	16813	16800	17407	16500	16833	16920
1981	16388	16425	16341	16813	16200	17119	16500	16833	16375
1982	15433	15657	15671	16789	16400	16188	15500	16833	15436
1983	15497	15480	15507	16000	16800	14833	15500	16000	15497
1984	15145	15214	15200	16000	16400	15497	15500	16000	15136
1985	15163	15184	15218	16000	15500	14745	15500	16000	15163
1986	15984	15995	16035	16000	15500	15163	15500	16000	15861
1987	16859	16861	16903	16813	15500	16384	16500	16000	16858
1988	18150	17965	17953	19000	16800	17659	18500	16833	181548
1989	18970	18964	18879	19000	19300	19150	18500	19000	18976
1990	19328	19329	19303	19000	17800	19770	19500	19000	19329
1991	19337	19378	19432	19000	19300	19928	19500	19000	19338
1992	18876	18984	18966	–	19600	15837	18500	19000	18766
AFER		0.34%	0.57%	4.38%	3.11%	2.44%	1.52%	3.11%	0.085%
MSE		9169	21575	775687	407507	226611	86696	32148	1384

5   CONCLUSION

In this paper, we combined the existing knowledge, especially modified the forecasting method of Saxena, Sharma and Easo [4], and apply the existing knowledge to establish the new method. This new method proved the smallest *AFER* and *MSE*.

We will look forward to developing the fuzzy time series forecasting method which not only have higher accuracy rate simulation forecasting but also higher forecasting accuracy rate for the unknown years.

REFERENCES

[1] Q. Song, B.S. Chissom. Forecasting enrollments with fuzzy time series—part 1. Fuzzy Set and Systems, Vol. 54, pp: 1–9, 1993.

[2] Q. Song, B.S. Chissom. Fuzzy time series and its models. Fuzzy Set and Systems, 1993, Vol. 54, pp: 269–277, 1993.

[3] Q. Song, B.S. Chissom. Forecasting enrollments with fuzzy time series—part 2. Fuzzy Set and Systems, Vol. 62, pp: 1–8, 1994.

[4] Preetika Saxena, Kalyani Sharma, Santhosh Easo. Forecasting enrollments based on fuzzy time series with higher forecast accuracy rate. Int. J. Computer Technology & Applications, Vol. 3, No. 3, pp: 957–961, 2012.

[5] Meredith Stevenson and John Porter. Fuzzy time series forecasting using percentage change as the universe of discourse. Proceedings of World Academy of Science, Engineering and Technology, Vol. 55, pp: 154–157, 2009.

[6] T.A. Jilani, S.M.A. Burney, C. Ardil. Fuzzy metric approach for fuzzy time series forecasting based on frequency density based partitioning. Proceedings of World Academy of Science, Engineering and Technology, Vol. 34, pp: 1–6, 2007.

[7] T.A. Jilani, S.M.A. Burney. M-factor high order fuzzy time series forecasting for road accident data. IEEE-IFSA 2007, World Congress, Cancun, Mexico, June 18–21, Forthcoming in Book series Advances in Soft Computing, Springer-Verlag, 2007.

[8] S.M. Chen. Forecasting enrollments based on high-order fuzzy time series. Cybernetics and Systems: An International Journal, Vol. 33, pp: 1–16, 2002.

[9] K. Huang. Heuristic models of time series for forecasting. Fuzzy Sets and Systems, Vol. 123, pp: 369–386, 2001.

[10] Jeng-Ren Hwang, Shyi-Ming Chen, Chia-Hoang Lee. Handling forecasting problems using fuzzy time series. Fuzzy Sets and Systems, Vol. 100, pp: 217–228, 1998.

[11] S.M. Chen. Forecasting enrollments based on fuzzy time series. Fuzzy Sets and Systems, Vol. 81, pp: 311–319, 1996.

[12] Huarng K.H., Yu H.K. A type 2 fuzzy time series model for stock index forecasting. Physical A: Statistical Mechanics and Applications, Vol. 353, No.1–4, pp: 445–462, 2005.

[13] Chen T-L, Cheng C-H, Teoh H-J, High-order fuzzy time-series based on multi-period adaptation model for forecasting stock markets. Physical A, Vol. 387, pp: 876–888.

[14] Leu Y.H. Lee C.P. Jou Y.Z. A distance-based fuzzy time series model for exchange rates forecasting. Expert Systems with Applications, Vol. 36, No. 4, pp: 8107–8114, 2009.

[15] Fu T-C. A review on time series data mining. Engineering Applications of Artificial Intelligence, Vol. 24, pp: 164–181, 2011.

[16] Hu X, Xu P, Wu S.Z., Asgari S, Bergsneider M. A data mining framework for time series estimation. Journal of Biomedical Informatics, Vol. 43, pp: 190–199, 2010.

[17] Chen S.M. Hwang J.R. Temperature prediction using fuzzy time series. IEEE Transaction on Sys-tems, Man, and Cybernetics-Part B: Cybernetics Vol. 30, pp: 263–275, 2000.

[18] Lee L.W. Wang L, Chen S.M. Handing forecast-ing problems based on two-factors high-order time series. IEEE Transactions on Fuzzy Systems, Vol. 14, No. 3, pp: 468–477, 2006.

[19] Jilani T.A. Burney S.M.A., Ardil C. Multivariate high order fuzzy time series forecasting for car road accidents, International Journal of Computational Intelligence, Vol. 4, No. 1 pp: 15–20, 2007.

Education Management and Management Science – Zheng (Ed.)
© 2015 Taylor & Francis Group, London, ISBN 978-1-138-02663-6

The application of 3D technology in vocational education

Fawei He
Chongqing Water Resources and Electric Engineering College, Yongchuan, Chongqing, P.R. China

ABSTRACT: 3D technology as a new technology in modern design is featured in its intuitive, understandable and realistic image results. In vocational technical education, 3D technology can be exploited to develop ideal physical model on computer, establishing a virtual and open platform for practice, creating relatively realistic 3D images, and creating realistic virtual physical environment, so as to help students study, and to make breakthroughs in difficulties in technical education; for example, to understand the initial structure and working method of closed machines like pumps and valves. In this way, the effect of teaching can be improved.

Keywords: vocational education; 3D technology; mechanical principles; simulation; visualization

1 INTRODUCTION

3D technology is derived from The United States in the 90s of the previous century. To start with, it was used in Ames Lab of NASA (National Aeronautics and Space Administration). Later on, similar researches were carried out in the UK, Germany and some other European countries, and resulted in great achievement. In Asia, 3D technology in Japan developed rapidly. Most of these researches are used in aviation trainings, satellite maintenance, space station trainings, molecule modeling and surgical operations. The application of 3D technology relies on professional talents, and must be through professional means such as programming. In early ages, it was rather difficult to use this technology in teaching.

In the recent years, with the development of computer technology, its abilities of calculating, graphing and data management becomes increasingly powerful. Accordingly, a large scale of 3D software came into being, such as UG, ProE and SolidWorks, as well as CAXA Solid Design which is developed by China on its own. Such software are easy to learn and put into use. With computer, it is now possible to create a realistic, three-dimensional, perceptive environment. A virtual and open practice platform can be established with 3D technology, and this platform can be used to carry out some human activities. In vocational education, 3D modeling software is also used to model machine parts, so that their structure and working theories can be observed and analyzed. Present practical experiences of human activities on the virtual platform are helpful to make learners of different level study.

2 TRADITIONAL TEACHING METHODS AND THEIR LIMITATION

In this article, traditional teaching methods refers to those such as writing on the blackboard, using multimedia, going out of school for practice, etc. These methods are necessary and natural to some degree, but all have some limitation.

2.1 Class teaching

Result of this method depends highly on the teacher and the content of class. The advantage of class teaching consists in its flexibility, and it is very suitable for classes requiring high logic such as higher mathematics and engineering mechanics. However, it is unlikely to be very proper for engineering classes requiring students to understand the inner structure of machines. So as to understand the structure of machines, at least assembly drawings are needed, and it is quite impossible to draw in class. Even with prepared pictures which are two-dimensional graphs, it is difficult for students to understand due to their limited graph reading abilities, let alone following the teacher to understand their working principles and outcome in different circumstances, as well as their positions and use for the entire machine. Hence, the existing problem is obvious, and it is for this reason that such teaching method is now gradually being replaced by advanced ones.

2.2 Multimedia teaching

Multimedia teaching is a teaching method popularized in the recent decade. It combines text, pictures, animation as well as video in teaching, and makes classes lively, which makes it a satisfying new teaching method used by most teachers. To some degree, multimedia solves some difficulties in teaching, while in case the teacher uses it improperly and relies too much on it, the interaction between teacher and students disappears, which makes it hard for the teacher to play his leading role and tough for the students to positively receive knowledge, and creates embarrassing situations as transforming class into slideshow play. Besides, only with multimedia, it is hardly possible to make students clearly understand the structure and working principles of machine parts such as overflow valve, whose shell is close and inner structure is quite complicated.

2.3 Practice teaching

Practice-teaching integration is now advocated, as to let students "study in practice, and practice in study". Practice teaching model which is centered on project teaching and guided by practice is applied. And in practice, it emphasizes on professional abilities and a practical application, according to the requirement of working positions. There's no doubt that it is highly positive, and useful for raising students' practical skills. However, this teaching method is not perfect. Many difficulties exist in practice, and due to the fact that students may scatter in multiple factories whose conditions and facilities may vary on a large scale, so that their practice could also be quite different. It is unlikely that the teacher designs different practice guidance materials for each one of scores of students. And still the students may still find it hard to understand those tiny and complex inner structures of the machine parts, as well as their working principles. In addition, teachers in factories are commonly equipped with plentiful experience but weak theoretical knowledge as well as problematic teaching techniques. Besides, what the students see in factories are merely what exist in these certain factories, which may not be representative, and can only provide limited help for students' study.

Another thing to point out is that, in practice, some parts such as valve and bearing could be rather tiny, their inner structure are complicated, and sometimes they even have close shells. It is not quite possible to take apart real machine parts for learners to observe their inner structures, which when unseen the learners may find it tough to understand their working principles. Even if some parts could be opened, their small scale can still make it hard for learners to see their inner structures. And when there are many students, those who stand behind can hardly see the structures shown. On the other hand, big parts could be too heavy to move, so that the teaching is quite limited.

3 EXAMPLES OF THE APPLICATION OF 3D TECHNOLOGY IN VOCATIONAL EDUCATION

3D technology here doesn't refer to the advanced programming technology that NASA adopted in virtual aviation or surgical operations, which is hard for normal people to master. Instead, it means using computer 3D modeling software to generate relatively realistic 3D pictures on screen, and to create virtual physical environments that are realistic to a certain extent, so as to help users experience geometrical, technological and physical properties of the machinery, which then can be used in teaching. In this way, students are put in a realistic vocational environment. As a supplement to class teaching, this method can lead to great results. The following are examples.

As is shown in Figure 1, the author used the 3D model of a gear pump to demonstrate its working principles in teaching. For the convenience of teaching and to make the image clear, the pump and some parts are not drawn according to the real structure. The pump is shown as semi-translucent, so that the fuel entry hole and exit hole can be seen clearly. Besides, non-critical parts such as transmission shaft and screw holes are not shown. It can be seen from the picture that when the driving gear (upper toothed gear in Fig. 1) rotates counter-clockwise, the driven gear (lower toothed

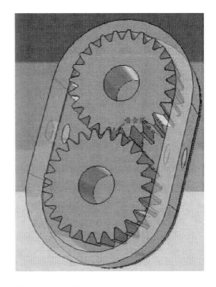

Figure 1. The working principle of gear pump.

gear in Fig. 1) would rotate clockwise. Because the space between the right of the gears and the pump becomes larger, negative pressure was created, so that the fuel flows in from the right. As the pump rotates, the fuel is taken to the left side through the space between gears and pump and pressed out, and this is the working principle of a gear pump. In teaching practice, 3D software can also be used to generate animation of the pump to show its working principle. With animation, its working condition can be understood more easily.

When the gear pump operates, another problem that must be solved is the entrap phenomenon. Entrap phenomenon occurs when the two gears rotate. The volume between the mesh of the two gears (close area indicated in Fig. 1) changes over rotation, while fuel is considered incompressible in normal temperature, and this leads to unstable operation, and creates shaking and noise. In order to solve this problem, the usual way is to make a groove on the shell of the pump to hold the excrescent fuel. Without 3D modeling, and considering students' limited recognition abilities of 2D graphs, this simple question could be hard to understand. Figure 2 shows the 3D model of the pump without the lower gear, and the dark red area is the groove on the shell of the pump. This groove can hold superfluous fuel brought about due to the volume change when the gears operate, so that the fuel pressure becomes stable, and entrap phenomenon is avoided. Without 3D modeling, students can find it difficult to understand this principle, while with 3D model, the teaching becomes easy.

Meanwhile, 3D modeling can also be applied in systems such as convex wheel and four-rod mechanism, to understand their operating status. It is even more useful in hydraumatic. For example, create a hydraumatic according to the hydromatic working principle, and it is easy to show how the fuel flows and the outcome of it when the valve plug of the hydraulic valve changes its position. Figure 3 is a hydraulic loop graph, and its working principle is as follows: with the reversing valve, the fuel pumped out by the pump makes the hydraulic cylinder move forward, backward and stop. How does the reversing valve work? The students may find it hard to understand only by reading Figure 3, and for those whose fundamental knowledge is poorer, it is even harder. Figures 4–6 show the 3D model of the reversing valve according to Figure 3, in which P refers to fuel inlet connected with hydraulic pump, and A and B are fuel outlet connected with hydraulic cylinder. T refers to returning opening connected with fuel tank.

It can be seen from Figure 3 that when the valve plug of the three-position four-way reversing rotary valve is in the middle, fuel inlet P, outlet A and B, and returning opening T are not connected. In this condition, there is no fuel coming in or out, and the valve is not working. Why is the returning valve obstructed? The real condition can be seen from Figure 4. In Figure 3, when the returning valve is positioned on the left, and P is connected with A, while B is connected with T; fuel comes in from the left of the cylinder and return on the right, and the

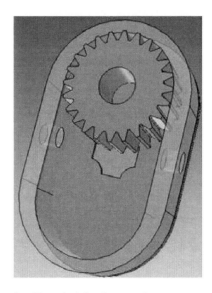

Figure 2.   The principle of entrap phenomenon.

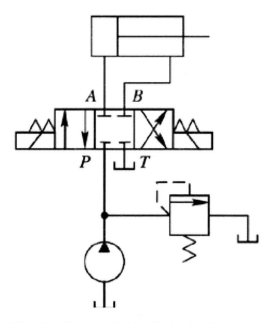

Figure 3.   Working principle of hydraulic valve.

101

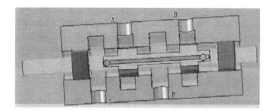

Figure 4. When the returning valve is positioned in the middle, four of the holes are all disconnected, and the cylinder stops working.

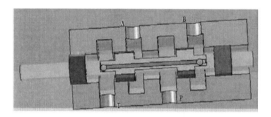

Figure 5. When the returning valve is positioned on the left, and P is connected with A, while B is connected with T, the cylinder moves forward.

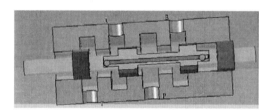

Figure 6. When the returning valve is positioned on the right, and P is connected with B, while A is connected with T, the cylinder moves backward.

piston comes out. Figure 5 shows this condition. Still back to Figure 3, when the returning valve is positioned on the right, and P is connected with B, while A is connected with T; the fuel comes into the cylinder from the right side and the piston goes back. Figure 6 shows this condition.

Without 3D modeling, and with only the machinery graph of the three-position four-way reversing rotary valve, it is quite difficult to clearly see the complex structure because it is complex and small, and there are too many lines in the graph, and let alone understanding its working principles. In teaching, the students find it hard to understand the working condition of the valve even in such simple circumstances, and when the structures are more complex, the students can only give up.

4 SUMMARY

If 3D design is combined with teaching resources, and common parts, mechanism and structures can be made into 3D graphs, the inner structure of machines can be shown vividly. They can even be rotated for all-round observation. In this way, the problem of hard observation due to close shell, too small scale or too large to move can be avoided. With the animation simulation functions of 3D software, animation can be made and played on many media players. Combining modern internet, 3D models can also be shared online. Hence, 3D technology is very useful in vocational education.

REFERENCES

[1] Huang Hongsheng: On the Value of Virtual Technology. Technology Progress and Plans, Vol. 25 (2008).
[2] Du Yigang, Tian Hongqiang, Zhang Zhiwen. Design and Simulation based on Pro/E. Industry and Automation, Vol. 7 (2013).
[3] Zhang Jing, Huo Hong, Chen Yanhua, *CAXA Design* (Beijing Press Industry Press, 2012).
[4] Chen Yimu, Yuan Hong, Hou Dezheng: Renovation in Graphing and Measurement Teaching Assembly Technology, Vol. 12 (2012).
[5] Gao Qian, The Application of Solidworks in Virtual Product Design. CAD/CAM and Industry Informatization. Vol. 4 (2009).

Education Management and Management Science – Zheng (Ed.)
© *2015 Taylor & Francis Group, London, ISBN 978-1-138-02663-6*

Reform practice of cooking in higher education—take Sichuan Tourism University as an example

Xiang Li
Sichuan Tourism University, Sichuan, China

ABSTRACT: This article starts from the higher education of cooking, and takes Sichuan Tourism University as an example. This paper introduces the development history of China's higher education of cooking and the current situation of the cooking major in Sichuan Tourism University. It analyzes Sichuan Tourism University's plan of applying the cooking and nutrition education major in the future by SWOT. Then, it puts forward the measures that should be taken in future development of the project and plan of cooking major.

Keywords: cooking; higher education; reform practice

1 INTRODUCTION

On March 18, 2013, vice-minister of the Ministry of Education, Lu Xin delivered an important speech, Speed up the Development of Modern Vocational Education to Provide Technical Support for Building China into a Moderately Prosperous Society in all Respects, at a work conference on vocational education and adult education. The minister stressed that we should construct well the modern vocational education system, and make positive contribution to build China into a moderately prosperous society in all respects.

2 DEVELOPMENT HISTORY OF CHINA'S COOKING IN HIGHER EDUCATION

Our cooking education started in the 1980s. In 1983, Jiangsu Business College (now Tourism Culinary Institute of Yangzhou University) established the cooking department, and set a cooking major; in 1985, the Ministry of Education approved the establishment of Sichuan Higher Institute of Cuisine by the report of Commerce Department, which became a unique school in China western which trains senior cooking personnel. Guangdong University of Business Studies, Heilongjiang University of Commerce, Jilin commercial College and other colleges began to set cooking major one after another later, and our cooking education entered its initial stage. At the end of the last century and the beginning of this century, as the national economy is developing rapidly, people's standard of living gradually transit from the subsistence to

being well-off, and the cooking education has begun to enter a rapid development stage. Cooking major is also becoming a popular program, and drawing more and more people's attention and favor. Especially Chengdu, gourmet capital, became a place where many cooking experts and food culture experts exchange experience, perform and transmit scriptures and deliver treasure. And a large number of cooking masters and experts skilled in cooking gathered there to study the development of cooking in higher education. Chengdu also attracts a large number of students who come here to study hard, and decide to take cooking as their career. On April 18, 2013, by the approval of Ministry of Education, former Sichuan Higher Institute of Cuisine and former Sichuan Agricultural Management Institute combined and established the Sichuan Tourism University which is a unique university for whole-day tourism education.

3 CONSTRUCTION SITUATION OF COOKING MAJOR IN SICHUAN TOURISM UNIVERSITY

Early in 1985, Sichuan Higher Institute of Cuisine (predecessor of Sichuan Tourism University) set the special major of cooking process, and added the cooking and nutrition education major in 2004. In 2002, Sichuan Tourism University was integrated with Sichuan Normal University and set tourism management major (direction of cooking and nutrition education) among undergraduate majors, cooking technology, culinary nutrition hygiene, cooking chemistry, nutritional meal distribution and production, food microbiology, food sensory evaluation, pedagogy, educational psychology, education, educational skills and other relevant professional courses, so it accumulated a lot of experience in undergraduate education. Besides, from 1999 to 2006, the college has been integrated with Sichuan University of Science & Engineering, Chengdu University of TCM, Chengdu University and other colleges in order to set up the food science and engineering major, pharmacy major, bio-engineering major and other undergraduate programs.

4 DEVELOPMENT PLAN OF COOKING MAJOR

4.1 Continue to apply for the undergraduate major of cooking and nutrition education

In order to consolidate advantages and characteristics of cooking major of Sichuan Tourism University in the country, we should try to fight for the support from Education Department of Sichuan Province and Ministry of Education, and try to get the approval of the enrollment of this major to finally build a complete education system which includes cooking majors of secondary vocational education, vocational education and undergraduate education.

4.2 Define the educational position of cooking major

4.2.1 Undergraduate education
Cultivate applied talents with high technology. For one thing, the talents should have comprehensive professional ability and comprehensive quality in the corresponding field, and their skills should be applied and compounded at the aspect of professional theoretical knowledge and practical technical skills to adapt to the technology improvement of the industry (profession) or the technical positions and the need of diversification of the ability structure of knowledge; Secondly, they should have very strong practical ability and practical skills, which is a very important ability characteristics to distinguish applied talents from other talents; Thirdly, they should have innovation consciousness and innovation ability, which are the common requirements for the talents of different levels as the development of economic and progress of technology improve the technical content of work. So, the cultivation of applied talents must be based on ability, and this kind of ability is not only a professional ability but also a comprehensive ability; not only a reproducible skill but a creative skill; and not only an employment ability, but also a certain entrepreneurial ability.

4.2.2 Vocational education
Cultivate high technology talents who are engaged in specific production work at the enterprise line. Strengthen technical level of the college students, and emphasize the practical ability and operational skills.

4.3 Development goal

Undergraduate education: Get the evaluation of qualified vocational education and awarding degree within five years, strive for awarding master of cooking and nutrition education major within ten years, and build the major into a domestic excellent major.

Vocational education: Strengthen connotation construction on the basis of stable scale, develop various types of high-quality class, and take international professional education as characteristics to improve teaching quality.

5 MEASURES

5.1 Design good curriculum around the professional orientation

Before assessment of faculty's right to grant undergraduate college degree, undergraduates should highlight features under the premise of specification, further strengthen the features after degree-granting is achieved, intensify undergraduate quality construction and accelerate professional development. Specialties should intensify teaching reform, strengthen the connotation construction, and improve the level of personnel training.

5.2 Solve the key problem of personnel training-teachers

First, to solve the vacancy of engineering basic course in cooking and nutrition education undergraduate, we can adopt ways of introducing from elsewhere and additional training. On the other hand, we can increase efforts to train young teachers, regularly send theory teachers to practice bases to take some project tasks by testing exercise, learn practical skills and organizational management, carry out the latest scientific research and research projects for enterprises to solve problems, create real economic value by improving scientific research, in order to promote teaching and improvement of the quality of teachers.

5.3 Strengthen course reform and course construction

Deconstruct and reconstruct course content, improve teaching materials, reform teaching methods and learning styles, standardize assessment methods, and improve the quality of teaching courses.

Cooking is a major with theory and practice, a major with the blending of tradition and modern, a promising major that needs experience. To make cooking as a career, you need to study hard day after day; year after year. Numerous myths of wealth have been created in food and beverage industry, and entering the culinary industry will lay a foundation for your success tomorrow. Cooking is closely related to us, small to "mundane" and large to "governing a large country is like cooking a small fish". Today, cooking shows the Chinese people's wisdom in a unique perspective; in the future, cooking will show the great tomorrow of our motherland with a new look. 2004(1):20–22.

REFERENCES

[1] Sun Yaojun, et al. Current Situation and Thinking of Culinary Professional Higher Education Henan Vocational and Technical Teachers College (Vocational Education Edition), 2004(1): 20–22.

[2] Zheng Xiufang, et al. Construction of new office colleges applied talents training mode, Science Technology and Industry, 2011(4): 101–103.

[3] Fund: This article is an initial result of 2011 Sichuan Province "Higher Education Quality Project" construction project, "Comprehensive Reform of Cooking Technology and Nutrition Major".

Education Management and Management Science – Zheng (Ed.)
© 2015 Taylor & Francis Group, London, ISBN 978-1-138-02663-6

Sports value outlook and physical training behavior of middle school students in China's western ethnic minority regions

Hu Yang
Sports Institute, Sichuan Agricultural University, Yaan, China

ABSTRACT: *Objective*: To effectively carry through China's "Sunshine Sports Project", guide middle school students in China's western ethnic minority regions through active engagement in exercise, and comprehensively improve their physical and psychological health.

Methods: Literature review, questionnaire survey, interview and consultation and comparative study are employed to research the sports value outlook as well as the status quo and characteristics of 2000 randomly selected middle school students in China's western ethnic minority regions middle school.

Results: The majority of middle school students in China's western ethnic minority regions have good sports value outlook and their fitness value orientation ranks top in China. There exists a certain positive correlation between their recognition of sports value outlook and their status of exercise engagement and there are differences in the characteristics of physical training behavior between boys and girls. Major factors affecting the active engagement in physical training for middle school students in China's western ethnic minority regions are their interest in and awareness of physical training.

Suggestions: A soft environment favorable for middle school students to form a right sports value outlook should be nurtured; physical education teachers should adopt different approaches and measures to cultivate students' awareness of physical training; physical education classes should be well conducted and extracurricular sports activities should be actively organized and developed; physical training forms and contents fit for students of different grades and sexes should be created in view of students' different characteristics.

Keywords: Sunshine sports; middle school students; physical training; value orientation; behavioral characteristics, China's western ethnic minority regions

1 INTRODUCTION

China's ethnic minority regions middle school sports development has been an important issue in the development of physical education. It is the theme of our thinking on how to develop the middle school sports, promote the middle school students to take part in physical exercise. Physical training has such functions as healing people's physical and psychological wounds, improving their health and relieving their psychological pressure. It is of great practical significance to research the sports value outlook and physical training behavior of middle school students from China's western ethnic minority regions, for it can provide important references for the further implementation and development of China's "Sunshine Sports Project".

2 ANALYSIS OF THE SPORTS VALUE OUTLOOK FOR MIDDLE SCHOOL STUDENTS IN CHINA'S WESTERN ETHNIC MINORITY REGIONS

2.1 *Good sports value outlook and top fitness value orientation for middle school students in China's western ethnic minority regions*

The sports value outlook refers to people's understanding of the value of sports to individuals or groups and their according designation of value orientation of sports behavior [1]. People depend on it for their views on sports behavior, or individual evaluation of the virtues and vices, the right and wrong as well as the importance of sports behaviors. It is also people's psychological bases for sports participation. A survey of the sports value outlook of middle school students in China's western ethnic minority regions reveals that 86.49% of the surveyed fit completely, comparatively or basically into entries of the sports value outlook for middle school students. The overall average value is 3.822, generally fit into the "comparatively" type of the contents of the sports value outlook. It is shown that middle school students in China's western ethnic minority regions have good sports value outlook. In the ranking of the average value for sports value outlook contents, the list from the highest to the lowest includes sports value outlook on fitness, entertainment, spiritual adjustment, interpersonal relationship and education. Fitness ranks top in the five aspects of sports value outlook for middle school students, with its average value being 3.98. Among the sixteen surveyed entries, the frequency ratio for the students to choose the "completely" type of fitness entries is 39.9%, 38.6% and 35.5%, respectively, ranking first, second and fourth in the sixteen entries, respectively. It is reflected that middle school students have focused their sports value outlook on fitness. Middle school students are still in puberty. Boy students hope to become stronger and taller, while girl students pay more attention to their physical appearance and size as well as their body weight and fat. Physical training is undoubtedly a means actively adopted by both boy and girl students to maintain charm.

2.2 Physical training-taking students' better understanding of sports values than that of nonphysical training-taking ones

It is found that physical training-taking students constitute 84.1% of the total number of students surveyed, while non-physical training-taking students constitute 15.9%. The frequency for physical training-taking students to choose the "fit" type of sports value outlook entries is higher than that of non-physical training-taking students, with the former group of students constituting 88.21% of the total number of students surveyed, and the latter group of students constituting 77.45%. The average value of sports value outlook for physical training-taking students is also higher than that of non-physical training-taking students, with the former being 3.91 and the latter being 3.34. The average value of physical training-taking students for such five aspects of the sports value outlook as fitness, entertainment, interpersonal relationship, spiritual adjustment and education is higher than that of non-physical training-taking students. It is revealed that understanding of sports values for physical training-taking students is better than that of non-physical training-taking students. Certain links exist between students' exercising behavior and their personal understanding of sports values and some experts believe there exists a certain

positive correlation between sports value outlook and sports behavior.

2.3 Role of the sports value outlook on entertainment in stimulating students' active engagement in physical training

It is found that entertainment, fitness, interpersonal relationship, spiritual adjustment and education value are listed from the highest to the lowest in terms of contents of the sports value outlook for physical training-taking students, while fitness, entertainment, spiritual adjustment, interpersonal relationship, and education value are listed from the highest to the lowest in terms of contents of the sports value outlook for non-physical training-taking students. It can be seen that entertainment plays an important role in students' active engagement in physical training. Although middle school students have recognized the fitness function of sports, it is hard for them to view the sports value outlook on fitness as the major incentive to taking physical training, for most of them are in good health and in an important learning stage where they are given many assignments but little spare time. The unsound psychology of middle school students who are experiencing rapid physical and psychological development makes it hard for them to rationally regulate their behavior. Entertainment thus becomes the major incentive for middle school students to engage in physical training.

3 ANALYSIS OF THE PHYSICAL TRAINING BEHAVIORAL CHARACTERISTICS FOR MIDDLE SCHOOL STUDENTS IN CHINA'S WESTERN ETHNIC MINORITY REGIONS

3.1 Characteristics of the sports pursued by physical training-taking students

The majority of middle school students in China's ethnic minority regions are engaged in such sports as basketball, table tennis, football and badminton. Few students are engaged in such sports as aerobics and Wushu, for these sports are not well developed in these regions and raise a comparatively higher demand for students' physical qualities, making it hard for students to engage in these sports. On the other hand, factors like the condition and location of the sports facilities affect students' choice of sports they want to pursue. It is found out that school sport facilities in China's western ethnic minority regions are dominated by basketball, football and volleyball courts, athletic fields, table tennis platforms and badminton courts, while few schools boast aerobic centers and Wushu courts. This is why some sports are chosen by a large number of students while other sports are chosen by only a small number of students. Boy students constitute 60.4% of the total number of 1200 students surveyed, while girl students constitute 39.6%, which indicates that girls are not that actively engaged in physical training as boys are. Factors accounting for this phenomenon include that girls are more shy than boys and more afraid of dirty and difficult things and they are constitutionally conditioned differently than the boys. There exists a marked difference between the sports boys and girls choose to pursue. Boys usually choose such sports with many physical contacts and strong exercising intensity as basketball and football, while girls tend to choose such sports with few physical contacts but laden with entertainment and amusement and helpful in maintaining individual image as table tennis and badminton.

3.2 Characteristics of the physical training frequency and duration

The weekly physical training frequency and duration of middle school students, to some extent, reflect the implementation status of the Sunshine Sports in a given region. It is found that the majority of middle school students are engaged in physical training three or four times each week, constituting 49.8% of the total number of students surveyed. And they spend 31 to 60 minutes on each physical training, constituting 43.9% of the total number of students surveyed. Few students engaged in physical training more than seven times each week and spend more than 91 minutes on each physical training, constituting 3.5% and 8.6% in the total number of students surveyed, respectively. The physical training intensity of boys is slightly higher than that of girls. In the first place, when it comes to the physical training frequency, the ratio of boys taking physical training three or more than three times each week is larger than that of girls, with the ratio of boys being 78.4% while that of girls being 63.1%. In the second place, when it comes to the physical training duration, the ratio of boys training each time for more than 31 minutes is higher than that of boys, with the ratio of boys being 83.5% while that of girls being 69.5%. Constitutional differences between boys and girls can partially account for the fact that the physical training intensity of boys is higher than that of girls. For one thing, compared with boys, girls are more likely to feel tired after physical training and thus they tend to lower the physical training frequency and duration. For another, compared with boys, puberty girls are more conscious of their images and tend to be afraid of the physical training-induced sweat and dirtiness, which is also one of the major factors

affecting the physical training frequency and duration of girls.

3.3 Characteristics of the physical training sites

The physical training sites can also reflect the physical training behavioral characteristics of students. School playgrounds are the major source of physical training sites for middle school students. It is found that the ratio of students choosing school playgrounds or stadiums as their physical training sites is 69.7%, ranking first among the total number of students surveyed. While the ratio of students choosing school dorms or corridors as their physical training sites is 13.9%, ranking second among the total number of students surveyed. School sports apparatuses and facilities are comparatively comprehensive and concentrative and students spend comparatively longer time in school than at home. So a large number of students choose to engage in physical training in schools. Next to schools, nearby sports venues or students' own homes have become important physical training sites of third and fourth importance, respectively, with the ratio of the former higher than that of the latter. For one thing, the limited family space and simplistic sports apparatuses greatly affect middle school students' decision to engage in physical training at home. For another, those middle school students who choose to engage in physical training in nearby places are usually accompanied by their classmates or friends, which, to some extent, adds to the number of students choosing to engage in physical training in nearby places. Certain differences exist between the choice of physical training sites for boys and girls, which has something to do with the different sports boys and girls choose to engage. Besides, the different psychological traits of boys and girls also account for the difference in the choice of physical training sites.

4 FACTORS AFFECTING THE ENGAGEMENT IN PHYSICAL TRAINING OF MIDDLE SCHOOL STUDENTS IN CHINA'S WESTERN ETHNIC MINORITY REGIONS

4.1 Students' extracurricular physical training time usurped by the heavy task of learning

Both teachers and parents consider learning as the major task for middle school students. As long as there is nothing wrong with students' bodies, few parents will volunteer to involve their children in extracurricular physical training. Be it on weekdays or weekends, teachers and parents cram students with assignments, leaving students little spare time. So, lack of physical training time is the major factor affecting students' participation in physical training [2].

4.2 Lack of the link of health education and lack of the cultivation of students' fitness awareness

A survey of the teaching status of middle school physical education teachers reveals that 84.1% of the teachers surveyed arrange no teaching contents related to the cultivation of students awareness of and interest in life-long sports. Classes of middle school physical education teachers are usually conducted in a very casual way. Even if some teachers are serious and responsible in teaching classes, most of the teaching contents are related to competitive sports. In the broad context of the Sunshine Sports development, schools in China's ethnic minority regions seldom issue relevant policies and adopt effective measures to guarantee the Sunshine Sports development, lacking right guidance on cultivating students' awareness of and interest in physical training. It is found that only 20% of the schools surveyed have taken relevant measures to guarantee the implementation and development of Sunshine Sports in their schools. To make matters worse, most of the measures become empty talks. Students are confronted with the tedious and heavy learning tasks as well as the temptations of such social vices as online games. Society, schools, families, teachers and parents fail to offer students health education and guidance, causing students to show no interest in and develop biased understanding of physical training and causing schools to fail in nurturing a favorable physical training atmosphere. Lack of interest in physical training, unfavorable physical training atmosphere and the conception that physical training is dirty and difficult are factors restricting students' participation in physical training.

4.3 Impediment posed by lack or shortage of sports facilities to students' enthusiasm in physical training

Participation subjects, objects and media are necessary conditions for physical training. As students are the participation subjects, they must have certain sports abilities to make it possible for them to engage in physical training; otherwise, physical training will cease to exist. Time and space are the participation objects. Spare time is needed for students to engage in physical training. Space here refers to sports venues and apparatuses. Most schools in China's ethnic minority regions are newly-built or newly-renovated, with each school equipped with some sports venues and apparatuses. It is found out that 72.7% of the teachers surveyed

believe that sports venues and apparatuses in their schools can basically satisfy students' demands for physical training; 20.5% of the teachers surveyed maintain that sports venues and apparatuses in their schools basically or completely cannot satisfy students' demands for physical training and only 6.8% of the teachers surveyed think sports venues and apparatuses in their schools can completely satisfy students' demands for physical training. Sports venues and apparatuses in the schools surveyed are simplistic and unified, generally consisting of one athletic field, one football field and several basketball courts. For many sports, there are no necessary facilities and apparatuses, which greatly hinder students' choice of their interested sports and further narrows their range of choices, decreasing their enthusiasm in engaging in physical training. Physical training media refer to the sports students choose when engaging in physical training. Students are required to have grasped certain sports skills before they are allowed to engage in physical training, which also decreases their enthusiasm in engaging in physical training.

5 CONCLUSIONS AND SUGGESTIONS

5.1 Conclusions

Physical training-taking students have a better understanding of sports values than those non-physical training-taking students, and both groups of students are different in their selection sequence for contents of sports value outlook. But middle school students generally have good sports value outlook and the fitness orientation is their first choice.

Sports chosen by middle school students to engage in physical training are dominated by basketball, table tennis, badminton, running and rope-skipping. Physical training is mainly undertaken with the companionship of classmates or friends in physical training sites outdoors with a frequency of every three to four times per week and every 30 to 60 minutes per physical training.

Due to the difference in students' physiological and psychological traits as well as the effect of external conditions and environment, certain differences can be found in the behavioral characteristics of physical training for boy and girl students.

At present, sports facilities of schools in China's western ethnic minority regions can basically satisfy students' demand for physical training, but the soft sports cultural environments in these schools are poor, with little guidance offered by teachers to students on establishing the right sports value outlook.

Factors affecting students' engagement in physical training are listed in terms of their weight from the highest to the lowest as physical training time, interests and hobbies, companionship, venues and facilities, students' level of sports skills, students' conception of sports, school physical training environment, weather and students' physical conditions as well as others.

5.2 *Suggestions*

Students' living environments should be purified and the sports cultural construction should be enhanced. The living environments for middle school students include the school environment, the family environment, and the community environment. Governments of all levels, schools and families should adopt various policies and measures to purify students' living environments, making middle school students keep off the unhealthy influence of online games and violence and creating a healthy, civil and harmonious living environment. Meanwhile, radios, media, newspaper reading sections, banners and sports competitions should be employed to enhance the sports cultural construction to guide students through establishing the right sports value outlook.

The long-term mechanism for the "Sunshine Sports" should be established. Under the right guidance of the responsible departments, schools should establish, according to their practical situations, the organization system to carry through the Sunshine Sports and cultivate students' awareness in life-long physical training, which includes rules and regulations, the guarantee mechanism, the supervision mechanism, and the evaluation mechanism [3].

Sports functions and regional characteristics should be combined to guide students' active engagement in physical training which can not only promote the growth of middle school students but also enhance their abilities to respond to stimuli and adapt to environment. Schools, teachers and families should take into consideration the China's ethnic minority regions to educate and guide students to stimulate their enthusiasm in engaging in physical training [4].

Guidance on students' extracurricular physical training should be enhanced, for extracurricular physical training is a major form and means for students to engage in sports. Schools and teachers should adopt various means and measures to guide students' extracurricular physical training, helping students' find a right way to scientific fitness improving their interest in and passion of active engagement in physical training [5].

REFERENCES

[1] C.M. Huang, H.Z. Geng. 2009. Research into the Sports Value Outlook of Middle School Students in Urumqi. *Journal of Jilin Sport University* 2(4): 154–155.

[2] Y.F. Wang, J.Q. Zhu. 2005. Summary of the Issues on College Students' Sports Value Outlook and Sports Behavior. *Zhejiang Sports Science* 27(6): 81–84.

[3] C.L. Xu. 2007. Merge and Conflict between Eastern and Western Sports Cultural Value Outlooks. *Introduction to the Sports Culture* (10): 52–54.

[4] Y.M. Liu, ... et al. 2002. On the Multi-Dimensional Characteristics of Sports Behavior. *Journal of Shandong Sport University* (4): 6–9.

[5] Z. Li. 2012. Sports Value Outlook and Physical Training Behavior of Middle School Students in Disaster-Afflicted Regions—Citing as an Example Sichuan Regions Suffering from Frequent Natural Disasters. *Master* (10): 245–246.

Education Management and Management Science – Zheng (Ed.)
© *2015 Taylor & Francis Group, London, ISBN 978-1-138-02663-6*

The support environment mechanism of clusters ecosystem

Xi Zhan Yu & Chuan Bo Zhang
Shangdong University of Science and Technology, Qingdao, Shandong, China

ABSTRACT: It is a necessary foundation for cluster ecosystem to exist and evolve whether industry cluster ecosystem could structure external support environment mechanism. The software and hardware environment are formed by the factors such as the policy and system, social culture, infrastructure and innovation resources. They match each other to make an impact on the innovation activities, so as to determine and restrict the nature and evolution direction of the system. Due to commonness and individuality in the process of cities developing, environment support system of the industrial cluster ecosystem also contains two systems in general and special. To improve the effectiveness of the system environment, some lessons could be drawn from the method to construct the intelligent city.

Keywords: industry cluster; ecosystem; innovation environment; support mechanism

1 INTRODUCTION

There is a close connection among the cluster ecosystem and the complex external environment. The better external environment is more important for the ecosystem to raise the systematic innovative ability and efficiency. Only in a suitable environment, could the ecosystem be formed by the innovative elements and innovative resource, such as companies, universities, research institutions and agencies. Under the trend of economic globalization, regional competition has focused on the performance of competitive development environment. As a result, the external environment has already become the key constraint for economic development.

2 THE CONNOTATION OF THE REGIONAL INNOVATION ENVIRONMENT

The concept of regional innovation environment is comprehensive. In the study of the region with high rate of growth and development ability, such as the industry cluster of small and medium-sized enterprises in Italy, France and other countries, the scholars observed the occurrence condition of regional innovation and then gave a definition of innovation environment. After carrying out research on some regions in Europe and USA's Silicon Valley, GREMI took the lead in putting forward the innovative environment concept: that is, the total constitution of the innovative subjects (innovation agencies or organizations), non-subject factor (innovative material conditions) and system and policies to coordinate various elements relationship.

Today, the international competition ultimately comes down to the innovative abilities. If only the innovation environment existed, could the innovation and diffusion of knowledge be carried out. Therefore, it has become one of the most important factors affecting and restricting the economic development whether the innovative environment could be formed.

3 SUPPORT STRUCTURE OF INDUSTRIAL CLUSTERS ECOSYSTEM

3.1 *Infrastructure environment*

It mainly includes natural resources, ecological environment, infrastructure, information infrastructure, technology-based information, etc. Natural resources and ecological environment are composed of the better natural conditions, a good ecology, a suitable living environment, etc., and production infrastructure including transportation, communications, water supply, power supply, gas supply and other infrastructure, life infrastructural facilities including infrastructural facilities such as dwelling house, heat addition, knowledge information infrastructural facilities including library, data base, science and technology intelligence retrieval system etc.

3.2 *Innovative resource environment*

It mainly includes some factors such as talented person, education and research institutes, etc. A measure to the innovativeness of regional talent market environment mainly relies on the indicator system regional talent supply and mobility of talented personnel. As the important part of regional innovation environment, the education training institutions include basic education, occupation education, training of intermediary agencies and others. As the driving force of regional innovation, the scientific environment includes enterprises, research institutes, universities, etc.

3.3 Policy and institutional environment

It mainly includes four key elements such as policy law and regulation, management system, marketplace and finance service. Technological innovation is a complicated social and economic process, involving many policy, legal and management issues. A good financing environment and a complete regulated intermediary market system also are important factors to boost the regional innovative activities.

3.4 Social and cultural environment

It mainly includes the regional customs, labor culture level and psychological quality, value idea, social atmosphere and social relationship networks etc. It could not make a direct impact on whether actors generating enthusiasm for the pursuit of innovation to form the spirit of innovation, also become the important factors to establish cooperation relationship and to build the open and liberal atmosphere to exchange ideas in people.

The innovation resources and infrastructure factors commonly constitute the regional hardware innovation environment, and the social and cultural environment and the policy system of environment constitute the software. Numerous studies show that there are four types of regional innovation environment by the matching relation between soft and hard innovative environments (as Fig. 1). The different influences would be made on the regions by these innovative environments.

1. Synergy. The software and hardware innovative environments are all in the high level states.

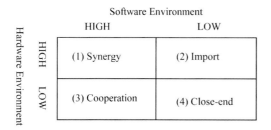

Figure 1. The matching figure of environment factors.

These include the more infrastructure facilities, perfect system, loose and open society interpersonal network, etc. The characteristics of this environment are with stronger innovation power and ability, more obvious innovative synergies, stronger opening and better communication and cooperation mechanism. In this environment, the innovative main bodies are able to get sufficient incentives and innovation can be fully implemented. This is the most ideal system environment for innovation activities.

2. Import. In the system, the hardware environment is at high level and the software is at low. Its main characteristics are with more infrastructure facilities and innovation resources, but poorer policy and institutional environment, more deficient innovation policy, not form a loose innovative culture atmosphere and the lack of harmonious social relationships. As a result, it is difficult to take the lead in achieving innovation for the less enthusiasm for cooperative innovation among main bodies and more incomprehensive incentives. But these are more infrastructure facilities, the area is able to attract innovation resources and introduce the innovations outside the system, so as to implement innovative development.

3. Cooperation. In the system, the hardware environment is at low level and the software is high. The infrastructure is not perfect and the necessary facilities to support innovation activities are deficient, but these are many better innovative policies and more harmonious and opening innovation atmosphere. In this environment, due to the deficient infrastructures, it is difficult to introduce the innovation resources outside the area and the exchange and cooperation could not be extensively formed in the innovation activities. Therefore, the innovation activities could only be carried out among the system main bodies but in a wide range of fields. As closely related to the traditional culture, the innovative activities have the obvious region characteristics.

4. Close-end. The software and hardware innovative environments are all at the low level. The system is in the state of few natural resources, poorer ecological environment and backward transport infrastructure. The innovation elements are scarce and scattered, and the harmonious social network relationship is able to be formed for frequent friction in various mechanisms and the bodies' relationships. In the close-end environment, there are hardly any communication and exchange, thus it is in a closed state. System innovation is not only difficult to get the effective support of relevant subject inside, also hard to attract innovation

resources and introduce innovations from the outside. Therefore, innovation activities are difficult to be carried out and the innovative effect is very poor.

4 INNOVATIVE ENVIRONMENTAL SUPPORT SYSTEM OF CLUSTERS ECOSYSTEM

4.1 General support system

In the clusters ecosystem, the universality of innovative support environment lies in that its constituent elements and supporting system are same as others. What constitute the general support environment of the industrial includes several parts as the following.

4.1.1 Innovative support system

In the industrial clusters ecosystem, the innovative cooperation mechanisms can be improved by way of building the regional innovation system. In the world, most famous science and technology parks are geographically close to universities and research institutes, and have been developed by them. Therefore, in the process of industry cluster ecosystem development, great attention must be paid to the introduction of intellectual resources, such as university and research institutions.

4.1.2 Industry development system

It is the main objective and the primary task for cluster ecosystem to optimize regional industrial structure by way of industrial innovation, so as to promote urban transformation and enhance the region competitiveness. So the industries of comparative advantage should be chosen and the regional characteristic pillar industries should be cultivated and developed according to local real conditions. On the basis of the pillar industries, great attention should be paid to cultivating industry chains and value chains. Then these chains are united to gradually form the industry network. As a result, a more comprehensive industry environment is being developed to give investors a much better condition.

4.1.3 Policy support system

As it is difficult to achieve the optimal level of social demand through innovation activities only by the market mechanism, the government must formulate and implement some related policies and regulations to build innovation environment, which can guarantee the effective innovation vitality. Currently, these are several main policies and regulations that can be adopted to support innovation, such as the industrial policies to promote developing high-tech industry and upgrading industrial technology, relevant regulations on protecting intellectual property rights, fiscal and tax policies on encouraging technological innovation, the policies on supporting the development of small and medium-sized enterprises, supportive policies on encouraging to establish research and development institutions, etc.

4.1.4 Science and technology service system

Its main task is to establish and improve the service system of production-study-research cooperative innovation, of which the main bodies are the science and technology intermediary agencies. In order to solve the difficult problems such as technical resources sharing and information asymmetry among enterprises, universities and research institutes, various types of technology intermediaries should be actively guided and supported to develop.

4.1.5 Financial service system

Its main faction is to gradually establish and improve the multi-channel and diversified investment mechanisms of innovation, which is guided by government, given priority to enterprises and supplemented with social capital. To achieve joint-investment and interest-sharing, the innovation investment mechanism should be actively put in practice, which takes government finance, banking institutions and enterprises as the main investment bodies.

4.1.6 Human resources system

In general, the regional competitive advantages and innovative potential could be obtained to enhance the region development by way of assembling the skilled workers and the highly qualified scientists and engineers. Whether high-quality labor forces can have unrestricted flow is also one of the important indexes to measure qualities of regional innovation environment. Therefore, a sound talent introduction and incentives should be established to optimize the innovative development environment.

4.1.7 Infrastructure environmental system

To build the developed and perfect infrastructures is one of the most important factors in the process of constructing and planning clusters ecosystem. It means that the one-stop service should be formed on the basis of adequate supply of hydropower, road unobstructed, convenient communication, municipal facilities, entertainment and leisure facilities, etc. Accordingly, the infrastructure environmental system will evolve to digital, networking and humanization, finally to achieve self-reliant operation. Thus, culture and humanistic environment should be created to foster and encourage innovation. These are the equal, free and relaxed working condition and open information communication environments conducive to the spread of new ideas and technologies in the region.

4.2 Special support system

In industrial cluster ecosystem, special support system is decided by the particularity of its environment. As the cluster ecosystem has been developing on the basis of local urban innovation, the special support ecosystem environment is accordingly made up by the city innovation system. Analysis indicates that the city already contains the industry cluster. As the central city is the regional center of politics, economy, culture and education, it has some obvious agglomeration advantages over infrastructure, industrial reserves, resources supply, market information and other aspects. Learning from the economic developing practice of developed countries and the eastern coast of China's regional, the central city can produce the economic and technological radiation effects, so as to greatly promote the regional economy growing by leaps and bounds.

5 THE WAY OF IMPLEMENTATION

To different cities, the resources, such as talents, capital, technology and information, are relatively scarce so that the city is very difficult to have strong comprehensive strength and play the leading role of the central city. Therefore, the central city should be constructed as the breakthrough of developing industrial clusters, and great attention must be paid to the radiation and leading role of central city in surrounding areas. However, the central city must meet some standards, such as stronger money supply and economic strength, better quality of urban environment, rich intellectual resources and talent pool, good public service system, advanced urban infrastructure, well-developed public education system, etc. Central city is essentially not only a science and technology centre but also a regional innovation system that is made up of the innovation activity subjects and objects. Among them, the principal part includes local governments, education and research agencies, financial institutions and intermediaries, etc. and the objects consist of infrastructure, urban environment, science and technology service system, etc. To make central city into regional industry important base of innovation and accumulation, can be achieved through the construction of innovation-oriented city.

The practice and experience of constructing innovative city at home and abroad can draw some enlightenment and reference to the central city.

1. With their own advantages, actively create special innovative features. For some cities with significant innovation advantage, such as New York and London, they all focus on developing the knowledge innovation as the endless motivating force to industry innovation. On the contrary, the poor innovation resources cities, represented by Shenzhen, choose their special way to innovation from their own actual situation. Then, they have produced stronger creative abilities in some key fields.
2. Emphasis on the cooperative innovation. The interaction and cooperation among government, enterprises, universities and research institutes in some ways run through developing the innovation-oriented city. Therefore, to build up a good partnership among these innovation subjects, has become one of the requirements for carrying out innovation activities. As a result, the open innovation system could be formed for knowledge production and application.
3. Emphasis on the non-technical innovation resources. At the same time as the advanced resources have been made into the advantage of wealth, the mechanism innovation and system innovation should be carried out in time to form sustainable economic advantage.
4. Emphasis on the roles of government guidance and support for innovation. The cities at home and abroad have enhanced the role of government in promoting scientific and technological innovation. The government builds the external environment to encourage innovation by implementing various policies and measures and give support to the enterprises under the market rules by government procurement.

6 CONCLUSION

In the industry cluster ecosystem, the support environment includes software and hardware. According to their matching relations, there are four types of innovation environment which could have different influence on innovation. Similarities and different characteristics in the process of city developing, give rise to nature of industrial cluster ecosystem which is not only general, also its particularity. In order to improve the effectiveness of the system environment, lessons can be drawn from the intelligent city construction.

REFERENCES

[1] A.E. Douglas. Symbiotic Interactions. Oxford: Oxford University Press, 1994: 3.

[2] Yu Xizhan, Sui Yinghui. The Industrial Clusters Ecosystem Based on City Innovation. Science & Technology Progress and Policy, 2010 (21): 56–60.

[3] Li Jianjun. Silicon Valley Model and Its University—industry Innovation System. Renmin University of China Doctoral Dissertation, 2000 (6): 50–51.

[4] Si Shangqi, Cao Zhenquan. Research on the Cooperation Mechanism of Institutions and Enterprises—A Symbiotic Theory and Analytical Framework. Science of Science and Management of S & T, 2006 (06): 15–19.

[5] Yu Xizhan. Industrial Clusters Ecosystem and its Applied Research in Resource City Innovation. Shandong University of Science and Technology Doctoral Dissertation, 2011 (12): 45–56.

[6] Persaud A. Enhancing synergistic innovative capability inmultinational corporations: An empirical investigation. Product Innovation Management, 2005, 22(5): 412–429.

Education Management and Management Science – Zheng (Ed.)
© 2015 Taylor & Francis Group, London, ISBN 978-1-138-02663-6

Exploration of the path construction of brand of building equipment group in Guangxi province

Chunyi Duan, Yun Zheng & Jian Gong
Guangxi Polytechnic of Construction, Nanning, China

ABSTRACT: This paper analyzed the current situation and problems of construction equipment in Guangxi professional training, raised the value of innovation and brand building, positioned professional groups from the demonstration framework, focused on professional, typical foster public platform construction, specialty construction and other aspects of economic development planning between the professional group for construction equipment brand building, construction and implementation. Finally, approached a professional group brand, and got some success in the study of brand building.

Keywords: vocational; construction equipment professional group; brand building

1 INTRODUCTION

With the 2008 Guangxi Beibu Gulf Economic Zone Development Plan which figured in the national development strategy, Guangxi accelerated the pace of urbanization, equipment and technology in modern urban construction, such as water, electricity, air conditioning, fire protection systems became more complex. The proportion of the construction equipment investment in the total investment was growing. Building equipment installation supported the development of construction industry, employing 11% of the region's construction industry practitioners. The next few years, construction equipment and installation industry professional technical and management personnel would need to add 10 million people; 3.5 million people in training senior workers, 10,000 technicians and senior technicians were trained [1].

Therefore, we adhered to the service for the purpose of employment-oriented courses in vocational colleges, took the principle of combining production path of development, on the basis of comprehensive understanding of construction equipment and specialty construction socio-economic background, professional groups and professional characteristics and formation mechanism were of great significance to build a professional group of construction equipment brand building strategy design and professional group of construction path. In recent years, Guangxi Polytechnic of Construction had done a lot of useful exploration in school-enterprise cooperation to jointly promote the construction equipment specialty. There were a number of well-established patterns. To focus on building water supply and drainage

engineering technology which could drive professional group, we strengthened the connotation of higher vocational colleges. We achieved the central financial support for higher vocational schools and enhanced the professional capacity of the service industry development projects' planned operational requirements [2].

This paper took the practice of Construction Equipment Group specialty construction in Guangxi Polytechnic of Construction for example, with the support of central finance for the project as the basis for construction equipment brand building vocational colleges' base paths, which were studied, and promote specialty group practice of brand building.

2 INNOVATIVE VALUE AND POSITIONING OF THE PROFESSIONAL GROUP BRAND BUILDING

Construction equipment, construction of specialty group, is actually a direct business-oriented post group, social and economic state of the industry chain. Its orientation was different from the professional disciplines in colleges in the traditional sense of the academic and vocational education with separate fundamentals. It was a breakthrough and innovative academic education. The professional group theory meaning was the concept of vocational education innovation.

First of all, professional brand building has important meanings. The higher occupation education had entered from extensive development scale expansion to rely on enhancing the quality and developed stage characteristics connotation.

College construction equipment professional must base on the post and integrate with the enterprise industry closely and combine with the development of regional economy and society tightly. Starting from the construction of specialty group formed a major cluster of its advantages and their own characteristics and brand. High vocational education should pay attention to theoretical knowledge. Students should take the applicability of the principle. Basic course teaching should meet the occupation development and professional knowledge needed for the standard, for the purpose of applying sufficient measure to strengthen the application of professional course teaching as the focus. We must highlight the work of pertinence and practicability. Combined with the construction of professional equipment needs or related courses, rationally integrate the professional settings according to the course of the occupation and professional which were focused on higher vocational education. This would make more prominent the technical ability training and exercise of occupation post [3].

Secondly, as a new type of education in higher vocational education, we must have a new mode of education. From the teaching point of view, integration of learning and work study combination was the route that we must take; on the school enterprise cooperation in the aspect of school running, the combination of production was the only way. So that the social demand for talents of occupation were to give a more profound connotation for the professional group of brand building. Professional group brand construction was as a new platform for the development of higher vocational colleges, constituted a new form of teaching organization and the new training model which was the innovation practice mode of higher vocational education.

3 THE PLANNING OF CONSTRUCTION EQUIPMENT PROFESSIONAL GROUP BRAND BUILDING

The formation of professional building equipment group was mainly for the internal push derived from the drainage of engineering technology and the external pull based on community's urgent need. Construction equipment professional groups included water supply and drainage, fire protection, construction equipment, intelligent buildings, building electrical engineering.

3.1 *The framework of professional group*

Based on the positioning of the professional group of brand building, Guangxi Polytechnic of Construction had carefully combed, aggregated and reorganized the existing professionals. It eliminated outdated professionals, developed new professionals, identified association and integrated to form a professional group of construction equipment. Meanwhile, we should develop the core professional according to the local economic structure, industrial structure and regional pillar industries and the direction, to form a core professional construction equipment focused professional group [4].

3.2 *The demonstration of the key professional*

The service function of single professional was small, so that only the formation of a professional group could generate economies of scale, and reflect the local economic development truly. From the microscopic point of view, only to build key professional could play a key professional radiation function. Built relevant professional groups, which focused on building professional, were to fill the need to form a professional group. From a macro point of view, the overall effect was a professional group that focused on building and contributing to improving the professional level of the building. There was a mutual restraint and mutually reinforcing relationship between key professional and relevant professional groups. Guangxi Polytechnic of Construction built comprehensive considerations and institute's industry background, location advantages, professional school conditions, teacher, the professional relationship between radiation functions and professional development and the local economy, etc., to determine construction equipment professional group that focused on brand building professional on water drainage engineering technology as the core. Surrounded by a professional with a focus on building characteristics and advantages of forming a professional brand, professional training model building process was of the formation of the core curriculum system construction, the paradigm of teaching organizational model, which along with the related specialty construction, so that the rules of the construction of chapter-based professional group could be found.

The goals of focusing on key professional were: (A) Developing and introducing a number of teaching leaders as backbone with solid theoretical basis, highlighting the ability of the professional teaching practice; (B) Build a number of training base which combined financial education, training, vocational skills identification and technology development functions; (C) Develop a number of distinctive professional combined features reflected in engineering curriculum, the formation of advanced educational philosophy and production closely and high professional employment rate. On the basis of the professional brand building, forming a key construction professionals as a leader, as the support of relevant building professional construction equipment group, constructed jointly by local school and focused on building a professional basis related to the intrinsic link closely. Resources could be shared by professional groups and the overall advantages of professional groups. So, focus on building a professional was the most important thing of professional group construction and local industry and regional economic development services.

3.3 Typical cultivation of construction equipment professional group

On the basis of actual situation of vocational colleges that with capital to establish a limited ability to raise relatively, small professional characteristics and advantages, and generally number was a lot in professional total. The number of professional expertise to invest in construction and professional group within the group should adhere to the "spirit" principle, not too much, 3 to 5 was appropriate. This was conducive to the rapid formation characteristics and advantages of the professional group, the typical professional to play an exemplary role in the radiation group. Therefore, the professional group of professional was set for construction equipment drainage, fire protection, construction equipment, intelligent building, and electrical and other construction engineering technology [5].

4 THE IMPLEMENTATION OF BRAND BUILDING PATHWAYS PROFESSIONAL GROUP

4.1 The main professional group of brand building

First, the professional brand building included teachers, teaching building facilities, curriculum and textbook construction, practical teaching system construction in four areas.

4.2 Professional group brand building security system

Strengthen the professional development from a holistic approach and seek to optimize comprehensive benefits, and therefore, in addition to the main aspects of construction, which should meet the requirements, should also establish a professional group of brand building security system. Seek professional group development steadily. Professional group brand building security system construction including evaluation feedback system, corporate advisory committee and professional industry, graduates, professional assessment, organizational management, vocational ability evaluation system, professional development funds, relied on industry enterprises and other aspects [6].

4.3 Strengthen the anchor construction industry

Professional vocational colleges' construction equipment group was close to industries and enterprises. In brand building needs of regional economic development and market demand, there was a clear interaction between the anchor building vocational colleges, vocational education and social sector [7]. Structure changed in the industry jobs and improved the quality of workers and technical requirements, market and internationalization had a direct impact on school-oriented vocational and training objectives and mode. Adjusted the curriculum, teaching methods and means were achieved. Creating a technological advantage, providing technical support would also enhance the industry level, promote industrial upgrading and enhance competitiveness of the industry. Its a resource extension institution, base construction, employment and sustainable development talent to win vitality and vigor.

ACKNOWLEDGMENT

Project source: This paper is one of the results of Guangxi higher education teaching reform projects in 2013. Project name: Exploration and practice of developing construction equipment professional group brand construction under the background of higher vocational education. Project number: 2013JGB354. Approval number: GUI teach [2013] 28, higher education.

REFERENCES

[1] Wuzhong, effects and counter measures of urbanization construction in Guangxi College Graduates Employment. Guangxi University, 012: 6–54.
[2] Zhuang zhong Xia, Some Thoughts on teaching professional project construction equipment. Guang dong Technical College of Water Resources, 2011, 9 (4):49–52.
[3] Shi Wei. Path construction equipment to build a professional group study. Professional time and space, 2010, 6 (3):103–105.
[4] Liurui Jun, Qu Fang, Zhao Dan. Study and practice of professional groups to explore the construction of vocational college's connotation. Liaoning Higher Vocational Technical Journal, 2012, 4 (1):1–3.
[5] Wang Qinghua. Construct mechanisms of doctrine and specialty construction. Shijia zhuang Vocational and Technical College, 2012, (24):1–25.
[6] Wu Yunxiang. Vocational training system construction equipment construction professionals. Chinese construction education, 2012, 1 (1):32–35.
[7] Zhuangzhong Xia, Qiuhan Qi, Yin six apartments, vocational construction equipment building program combines professional engineering Study. Guang dong Technical College of Water Resources, 2008, 12 (4) 6–9.

Group AHP and evidential reasoning methods in the evaluation of green building

Xiao Yu He & Lu Ping Yang
School of Civil and Environmental Engineering, Anhui Xinhua University, Anhui, China

ABSTRACT: In this paper, we research the problem of group AHP and evidential reasoning methods used in residential construction assessment of green building. First, we have elaborated the analysis of group AHP to determine the index weight of evidential reasoning algorithm steps and then based on the "Green Building Evaluation Standard" establish the green building index system applied in the principle of selecting indicators index system which constituted 4 level indicators and 20 secondary indicators. Finally, we use the assessment system to a new residential building in Hefei live green building assessment.

Keywords: green building evaluation; group AHP; evidential reasoning

1 INTRODUCTION

The research of green-related buildings emerged in the past 10 years with the proposed green, ecology, environmental philosophy. Accordingly, the construction sector in ecological building, green building also began to sprout, and green building evaluation also emerged. In 2004, the Central Economic Work Conference, first proposed to vigorously develop energy-saving residences. The Eleventh Five-Year National Economic and Social Development Program released clearly states that the task of building the field is to develop land-efficient buildings (Wolff, S.V & Zhang, Q.F 2007 China building energy manual). How to select a comprehensive and reasonable green building evaluation system from China's "green building evaluation standards" in order to establish a comprehensive and rational and easy to operate evaluation system is worthy of thoughtful questions. There are many uncertain factors in evaluation system for green buildings. The group decision-making methods are a very good method through mutual sharing of information and interaction, and ultimately determine the collective action plan to solve the problem. Practice has proved that the Evidence Reasoning (ER) can more effectively solve complex scientific assessment system. The method gets good application in the fields of quality assessment, decision-making and other projects. This paper explores the AHP and ER methods used in green building evaluation.

2 THE GROUP APH

The group decision is an effect of group reflecting the group decision-making in which certain groups, members according to some negotiation rules, through mutual sharing of information and interaction, ultimately determine the collective action plan (Chen, G.J & Chen, Z.J 1993 Comprehensive evaluation of the three gorges project on the ecological and environmental impact).

2.1 Basic concept

The group AHP is a common method of group decision. The core problem is how to consider expert authority with its foundation still being the traditional AHP. The basic idea of group AHP is to let element be hierarchical and make decisions according to the decomposition, comparative judgment, and integrated way of thinking (Liu, X.B. 2009. Decision Analysis and Decision Support Systems).

2.2 Application of evidential reasoning steps

The steps of using AHP are as follows:

1. Establish a hierarchical model
   Dividing the factors associated with the problem into several layers, each factor also affected the upper and lower layers.
2. Make up judgment matrix
   From the layer 2 in the model, compare matrix with the pairwise comparison method according to the 1–9 scale comparison for subordinate (or impact) of each factor on the floor with a layer of various factors.
3. Calculate the weight vector and do the consistency test
   For the structure of the judgment matrix, find the eigenvectors corresponding to the maximum eigenvalue. Normalize them and get the

weight vector. In this paper, use the summation methods to calculate the weight vector approximately. The formula is as follows:

$$w_i = \frac{1}{n} \sum_{j=1}^{n} \frac{\alpha_{ij}}{\sum_{k=1}^{n} \alpha_{kj}}, i = 1, 2, \ldots, n \qquad (1)$$

w_i—weight vector; α_{ij}—judgment matrix elements; n—number of elements.

In this paper, it makes up judgment matrix according to the complete consistency which can save workload and avoid consistency problems.

4. Calculate the total weight of each index

Calculation of the total weight of each index is also called the total level sorting, which is complex from the bottom to up layer by layer. Calculate the total weight of the index in each layer relative to the top layer, by the use of matrix table and weight summation method.

3 D-S EVIDENCE REASONING

Evidence reasoning theory created by Dempster and Shafer has been widely used in data integration and expert systems (Beynon, M. & Curry, B. 2000. The Dempster-Shafer theory of evidence). Because evidence reasoning provides a natural and powerful method to express and synthesize the uncertain information, it is developed into a capable decision method for uncertain, incomplete, and unclear data. The method is ER (Yang, J.B. & Singh, M.G. 1994. An evidential reasoning approach for multiple attribute decision making with uncertainty).

3.1 Basic concept

The basic concept of theory of the evidence include identification frame, reliability structure and so on. Identification frame (represented by a collection of H) includes all estimated possible outcomes of one un-composed basic index; identification frame elements are mutually exclusive. Reliability structure means using the probabilistic to express evaluator preferences. It will give out the vague comment for judgment expression of the probability distribution according to each element in identification frame. In order to make the belief functions play the role to associate to each index using rules of evidence synthesis to synthesize belief functions.

3.2 Application of evidential reasoning steps

Set up the sub-index of E is $e_i (i = 1, 2, \ldots L)$, then index E can be expressed as $E = (e_1, e_2, \ldots, e_L)$, the weight is $w = (w_1, w_2, \ldots w_L)$. The formula is

$$0 \leq w_i \leq 1 \text{ and } \sum_{i=1}^{L} w_i = 1 \qquad (2)$$

Get the synthetic E belief functions through basic belief functions; the steps are as follows:

1. Use vector $H = (H_1, H_2, \ldots, H_N)$ present comment level of every index, define the corresponding utility value presented by $u(H_N)$.
2. $\beta_{n,i}(B_l)$, is presented in the index e_i, the level of plan B_l comments is the belief assignment of H_n, For the evaluation object $B_l = (I = 1, 2, \ldots, M)$, the basic belief functions to make sure its index $e_i = (i = 1, 2, \ldots, L)$ is as follows.

$$\beta_{n,i}(B_l) \geq 0 \text{ and } \sum_{n=1}^{N} \beta_{n,i}(B_l) \leq 1 \qquad (3)$$

$$S(e_i(B_l)) = \left\{ (H_n, \beta_{n,i}(B_l)), n = 1, 2, \ldots, N \right\},$$
$$i = 1, 2, \ldots, L, \ I = 1, 2, \ldots, M \qquad (4)$$

Matrix $D_g = (S(e_i(B_l)))_{L \times M}$ can present the basic belief assignment of all the basic indexes.

3. Set up $m_{H,i}$ present one remaining probability after estimating e_i. Set up $m_{n,i}$, present the index E probability distribution. $m_{n,i}$ and $m_{H,i}$ show as following:

$$m_{n,i} = w_i \beta_{n,i}(B_1) \; n = 1, 2, \ldots, N \qquad (5)$$

$$m_{H,i} = 1 - \sum_{n=1}^{N} m_{n,i} = 1 - \sum_{n=1}^{N} \beta_{n,i}(B_1)$$
$$i = 1, 2, \ldots, L \qquad (6)$$

$m_{H,i}$ is the sum of $\bar{m}_{H,i}$ and $\tilde{m}_{H,i}$ which means $m_{H,i} = \bar{m}_{H,i} + \tilde{m}_{H,i}$, during the formula $\bar{m}_{H,i} = 1 - w_i, \tilde{m}_{H,i} = w_i \left(1 - \sum_{n=1}^{N} \beta_{n,i}(B_1) \right)$ probability distributions $m_{H,l(L)}, m_{n,l(L)}, \bar{m}_{H,l(L)}$ and $\tilde{m}_{H,l(L)}$ are synthesized by recursive evidential reasoning iterative method. Get all the basic probability assignment and assemble them finally.

4. Synthesis probability distribution of index E of a_t $m_{n,l(1)} = m_{n,1}(n = 1, 2, \ldots, N), \bar{m}_{H,l(1)} = \bar{m}_{H,1}, \tilde{m}_{H,l(1)} = \tilde{m}_{H,1}, m_{H,l(1)} = m_{H,1}$, use recursive ER iterative method to synthesize probability distribution $m_{H,l(L)}, m_{n,l(L)}, \bar{m}_{H,l(L)}$ and $\tilde{m}_{H,l(L)}$, assemble all basic probability assignment as follows:

$$\{H_n\} : m_{n,l(i+1)} = K_{l(i+1)}[m_{n,l(i)}m_{n,i+1} \\ + m_{H,l(i)}m_{n,i+1} + m_{n,l,(i)}m_{H,i+1}], \\ n = 1, 2, \ldots, N$$

$$\{H\} : m_{H,l(i)} = \tilde{m}_{H,l(i)} + \bar{m}_{H,l(i)}$$

$$\tilde{m}_{H,l(i+1)} = K_{l(i+1)}[\tilde{m}_{H,l(i)}\tilde{m}_{H,i+1} + \bar{m}_{H,l(i)}\tilde{m}_{H,i+1} \\ + \tilde{m}_{H,l(i)}\bar{m}_{H,i+1}]$$

$$\bar{m}_{H,l(i+1)} = K_{l(i+1)}[\bar{m}_{H,l(i)}\bar{m}_{H,i+1}]$$

$$K_{l(i+1)} = \left[1 - \sum_{t=1}^{N} \sum_{\substack{j=1 \\ j \neq t}}^{N} m_{t,l(i)}m_{j,i+1} \right]^{-1},$$
$$i = 1, 2, \ldots, L-1 \qquad (7)$$

5. Set up β_n as the H_n reliability of index E of B_l, calculate the synthetic reliability of index E of B_l. It can be got by synthesizing all the index e_i which are all the sub-index related to index E:

$$\{H\}: \beta_H = \frac{\tilde{m}H,I(L)}{1-\tilde{m}H,I(L)} \tag{8}$$

$$\{H_n\}: \beta_H = \frac{m_{n,I(L)}}{1-\overline{m}_{H,I(L)}} \tag{9}$$

Synthetic reliability:

$$S(y(B_l)) = \left\{(H_n,\beta_n(B_l)), n=1,2,\ldots,N\right\}$$

6. Calculate the expectation of B_l and all the distribution of estimated reliability: set up $u(H_N)$ as the effectiveness of H_n, get the expectation of $S(y(B_l))$ as:

$$u(s(y(B_l))) = \sum_{n=1}^{N} \beta_n(B_l)u(H_n) \tag{10}$$

7. Calculate the effectiveness range of B_l: When you meet uncertain or unknown factor which will cause evaluation is incomplete, the $\beta_n(B_l)$ will present the boundary value of the minimum possibility of B_l which is rated as H_n, at the same time the boundary value of the maximum possibility is $\beta_n(B_l) + \beta_H(B_l)$. This leads to the generation of effectiveness range. The defined evaluation level shows the minimum evaluation level is H_1, the maximum evaluation level is H_n. And the max and min average expected utility of B_l can be got by the following formula:

$$u_{\max}(B_l) = \beta_1(B_l) + \beta_H(B_l)u(H_1) + \sum_{n=2}^{N} \beta_n(B_l)u(H_n) \tag{11}$$

$$u_{\min}(B_l) = \beta_N(B_l) + \beta_H(B_l)u(H_N) + \sum_{n=1}^{N-1} \beta_n(B_l)u(H_n) \tag{12}$$

$$u_{avg}(B_l) = \frac{u_{\max}(B_l) + u_{\min}(B_l)}{2} \tag{13}$$

From the above formula: if $u(H_1) = 0$, then $u(S(y(B_l))) = u_{\min}(H_1)$.

If all the original evaluation $S(e_i(B_l))$ of the belief matrix is complete, then $\beta_H(B_l) = 0$ and $u(S(y(B_l))) = u_{\min}(B_l) = u_{\max}(B_l) = u_{avg}(B_l)$.

4 EVALUATION OF GREEN BUILDING BASED ON THE GROUP AHP AND EVIDENCE REASONING METHODS

4.1 Green building assessment index system

The green building assessment is reflected on the whole life cycle of the building such as initial construction, the construction period and the applicable period. This paper takes energy-saving and resource utilization B_1, land saving and outdoor environment, B_2 materials saving and material resource utilization B_3 and Operations Management B_4 four level indicators and their under-20 secondary indicators: Total energy of architectural design C_1 Comprehensive utilization of energy C_2, Renewable heating and the total generation resources C_3, Lighting power density C_4, Non-conventional water resources utilization C_5, Total construction graphic design C_6, Exterior windows and building wall can be opened area C_7, Thermal storage treatment C_8, Storm water treatment and utilization programs C_9, Water safety C_{10}, Waste land use C_{11}, Outdoor permeable floor area ratio C_{12}, Site environmental noise C_{13}, Building wind speed around the pedestrian area C_{14}, The utilization of reusable building materials C_{15}, The utilization of reusable building materials C_{16}, Building structures system C_{17}, The weight of the construction site construction materials C_1, Resource management incentives C_{18}, Intelligent building systems and information network system C_{19}, Automatic monitoring system technology C_{20} green building evaluation index system constituted according to the rules in the standard "Green Building Evaluation Standard" (2006.Green building evaluation standard). For detailed information, see Table 2.

4.2 Case study

Use the selected green building evaluation system indexes to evaluate one new residential building in Hefei. In this paper, 8 experts were invited who are engaged in structural engineering and mechanical engineering, building environment and energy saving, construction quality, building foundation, concrete and pre-stressed concrete technology, heating ventilation and air-conditioning engineering, power Engineering and Engineering Thermo physics, new technology and testing of concrete structures. Get the weight of indexes in every level by AHP, improved expert assessment indexes show different preferences phenomenon. Set up five levels for comments as worst, poor, average, good and best, assigned the value as (0, 0.25, 0.5, 0.75, 1).

The experts evaluate the new residential building according to the 1–9 scale comparison, get the basic probability assignment and weight of every indexes; the result is shown in Table 1.

Integrate evidence of sub-level two in each level one index by the formula (5)–(7), get the basic probability assignment of the 4 level one indexes, then integrate evidence of level one, get the whole residential basic belief assignment. Get all the estimation belief assignment by the formula (8) and (9). In this paper, all the belief vectors of new

Table 1. Index weights and basic probability assignment.

Index and weight		New residential building				
Level one index ω_i	Level two index ω_{ih^i}	Worst	Poor	Average	Good	Best
B1 (0.36)	C1 (0.12)	0	0.25	0.5	0.25	0
	C2 (0.12)	0.7	0.3	0	0	0
	C3 (0.11)	0.875	0.125	0	0	0
	C4 (0.11)	0	0	0	0.5	0.5
	C5 (0.11)	0.375	0.375	0.25	0	0
	C6 (0.09)	0	0	0	0	1
	C7 (0.09)	0	0	0	0.375	0.625
	C8 (0.09)	0.125	0.625	0.125	0	0
	C9 (0.08)	0	0	0	0.625	0.375
	C10 (0.08)	0	0	0.125	0.375	0.5
B2 (0.24)	C11 (0.24)	0.625	0.125	0	0	0
	C12 (0.31)	0	0.25	0.5	0.25	0
	C13 (0.22)	0	0	0	0	1
	C14 (0.22)	0	0	0	0	1
B3 (0.25)	C15 (0.37)	0.75	0.25	0	0	0
	C16 (0.35)	0.125	0.125	0.625	0.125	0
	C17 (0.28)	0	0	0	0.125	0.725
B4 (0.15)	C18 (0.38)	0	0	0	0.75	0.125
	C19 (0.31)	0	0	0	0.25	0.625
	C20 (0.31)	0	0	0.25	0.5	0.25

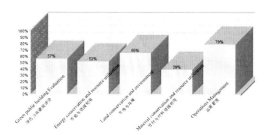

Figure 1. Utility value of 4 level one indexes.

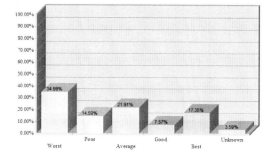

Figure 2. The belief assignment of Materials saving and Material resource utilization.

residential are got by the evaluation software Intelligent Decision System (IDS):

$$S(y(\alpha)) = \{(H_1, 0.2083), (H_2, 0.1266), (H_3, 0.1367), (H_4, 0.1861), (H_5, 0.3121), (H, 0.0303)\}$$

Because incomplete evaluations existed, the max utility value is 0.5819, the min utility value is 0.5516 and the average utility value is 0.5667. The average utility value shows the evaluation is good. Use IDS software. Get the 4 level one utility value as shown in Figure 1.

In the 4 level, the Operations Management is the best, the Materials saving and Material resource utilization is the worst. For further study of experts evaluation for the Materials saving and Material resource utilization, use IDS. Get the

belief assignment of Materials saving and Material resource utilization as shown in Figure 2.

From Figure 2, the belief values of worst is 34.99%, the values of good and best are 7.55% and 17.35%, which means there are some problems existing in the respect of Materials saving and Material resource utilization.

5 CONCLUSIONS

This paper studies the application of group AHP and evidence reasoning method for green building evaluation. By the results obtained, one can indicate that the AHP and evidence reasoning method can make a reasonable evaluation for residential building.

REFERENCES

[1] Beynon M, Curry B, Morgan P. 2000. The Dempster-Shafer theory of evidence: an alternative approach to multicriteria decision modelling. Omega (The International Journal of Management Science), 28: 37–50.

[2] Chen G.J. & Chen Z.J. 1993. Comprehensive Evaluation of the Three Gorges Project on the ecological and environmental impact. Beijing: Science Press, 98.

[3] Green building evaluation standard GB/T 50378-2006. Ministry of Construction PRC, 2006.

[4] Yang, J.B. & Singh, M.G.1994. An evidential reasoning approach formultiple attribute decision making with uncertainty. IEEE Trans. Syst., Man, Cybern, 24(1): 1–18.

[5] Liu, X.B. 2009. Decision Analysis and Decision Support Systems. Beijing: Tsinghua University press, 6.

[6] Sebastian, V.W. & Zhang, Q.F. 2007. China Building Energy Manual. Ministry of Construction PRC.

Education Management and Management Science – Zheng (Ed.)
© 2015 Taylor & Francis Group, London, ISBN 978-1-138-02663-6

A Multi-Transactions-Oriented Interface Management Model

X. Tu
School of Economics, Wuhan University of Technology, Wuhan, Hubei, China
College of Information Engineering, Hubei University for Nationalities, Hubei, China

Q.F. Yang & P. Song
School of Economics, Wuhan University of Technology, Wuhan, Hubei, China

X.Z. Yang
School of Economics, Wuhan University of Technology, Wuhan, Hubei, China
College of Information Engineering, Hubei University for Nationalities, Hubei, China

ABSTRACT: Interface management is an effective method to enhance interface efficiency among integration units. In order to improve the interface management efficiency, this study proposes a multi-transactions interface concepts, uses of formal method for the management process modeling of multi-transactions interface, establishes a Multi-Transactions-Oriented Interface Management Model ($MTOIMM$) for a complex system, includes five aspects: Subjects Model (SM), Transactions Model (TM), Issues Library Model (ILM), Solutions Library Model (SLM) and Multi-Transactions Interface Model ($MTIM$). Finally, verifies the effectiveness of the $MTOIMM$ used in the interface management.

Keywords: interface management; multi-transactions interface; formal method

1 INTRODUCTION

In practice, failure to manage interfaces may result in additional work or in low project performance. In order to reach a common goal and mutual cooperation between project participants, one needs to effectively manage interface (Guo 2009). Interface is described as the role and the contact between project participants. Interface issues must exist in the interface management., Interface Issues are described as the state of substance, energy and information exchange process, and Interface Issues are the sum of all the negative effects (Liu & Shen 2012). Guo (1999) summarized the resulting factors of Interface Issues including sticky information, target differences, cultural differences and background differences, and pointed out the important role of effective interface management for business innovation and market performance. Vandevelde & Vandierdonck (2003) considered that different personalities, cultural differences, differences of organizational model, location factors and language problems led to the Interface Issues.

Interface Issues severely affected and restricted the development of enterprises., The survey found that when there are serious management problems in the R & D marketing interface, 68% of the R & D project would be a complete failure in the commercial, 21% of the project would be partial failures (Souder & Chakrabarti 1978). Chen et al. (2009) regarded that Interface management failure may lead to low productivity, low product quality, and increased production costs. In order to effectively manage the Interface Issues, Guan et al. (1999) explained the relationship between the interface management and innovative performance. Liu et al. (2007) constructed the measure model based on the harmonious matrix in order to effectively measure Interfacial Issues. From the relationship between the starting node enterprises, the interface of the supply chain network organization is divided into market-based interface, network-based interface and enterprise-type interface (Du & Yang 2008). Ren (2010) researched enterprise cooperation and innovation from horizontal and vertical interface. Barczak (1995), Wheelwright (1992), & Moenaert (1995) found that in order to complete the requirements of the interface, high harmonious team is more effective than the low harmonious team.

When there is effective interface management application and implementation, effective project management is evident (Siao & Lin 2012). The key of interface management is to accurately describe interface management process and interface management issues, and thus effectively control the interface issues in the interface management process. Therefore, we propose the concept of multi-transaction interface, and try to apply formal method to describe the general interface management process based multiple transactions interface, provide the methodology for accurately positioning and solving interface issues. The remainder of this paper is organized as follows. Section 2, proposes Multi-Transactions Interface concept, uses of formal method for the management process modeling of multi-transactions interface, establishes a Multi-Transactions-Oriented Interface Management Model (*MTOIMM*). It verifies the effectiveness of the model in Section 3. Section 4 offers some concluding remarks.

2 MULTI-TRANSACTIONS-ORIENTED INTERFACE MANAGEMENT MODEL

Multi-Transactions-Oriented Interface Management Model (*MTOIMM*), includes Transactions Model (*TM*), Subjects Model (*SM*), Issues Library Model (*ILM*), Solutions Library Model (*SLM*) and Multi-transactions Interfaces Model (*MTIM*) etc., and is formally described as *MTOIMM = (TM, SM, ILM, MTIM)*.

2.1 *Related concepts*

Definition 1: One transaction is the operation process which happened in the experience in order to complete a common task by the $n(n > 1)$ subjects.
Definition 2: Multi-Transactions Interface is the state of interaction of materials, energies and information among many subjects in order to complete several relevant transactions.

2.2 *Transactions model*

TM is the core of *MTOIMM*. It describes all the transactions in a complex system. There are Meta Transaction and Long Transaction in the complex system.

2.2.1 *Meta transaction*
Definition 3: *Meta Transaction (MT)* is the simple transaction which involves only 2 subjects in the complex system.

MT = (ID, Name, <Sender, Receiver>, Description, <Material | Energy | Information>, Precursor, Successor, StartTime, EndTime, TimeConsuming, {Feature}).

Some explanations about the *MT*:

1. MT is the minimum component of TM.
2. In terms of the number of subjects constituting transaction, MT is the simplest type of transaction.

3. The Sender and Receiver belong to SM.

MT is shown in Figure 1.

2.2.2 *Long transaction*
Definition 4: *Long Transaction (LT)* includes *m* *MT* and *n* *LT*, $m + n > 1$.

LT = (ID, Participants, Method, Amount, Subtransaction, Description, {feature})

Participants are a subjects set in the LM, that is, $\{SubID_1, ..., SubID_n\}$. *Method* is the Participators actions. *Amount* is the number of the subjects in the LT. *Subtransaction* =$\{MT\} \cup \{LT\}$.
Introductions to *LT*:

1. The characteristics of *LT* are equal to the summation of the characteristics of all *MTs* and long transactions included in this long transaction plus the transaction characteristics owned by this long transaction itself.
2. Each LT include n other LTs, and one LT could be included in *n* other LTs. It can be described in a tree-structure as follows: each LT can have *n* direct precursor LTs and *n* direct successor LTs.
3. The inclusion relations do not exist among several LTs or between LTs and MTs included in one LT.
4. Method in LT must be the motions of all participants involved in one transaction.
5. One LT could not include only one MT and another LT, $m+n > 1$.

2.2.3 *Transactions model description*
The transaction model describes the set of all transactions in a complex system. The logical relations among all transactions of *TM* can be described as special tree structure, as is shown in Figure 2.

2.3 *Subjects model*

SM describes the universality and individual features of all transaction participant subjects involved in a complex system. It is a carrier of the

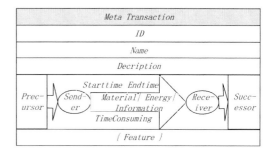

Figure 1. Meta transaction.

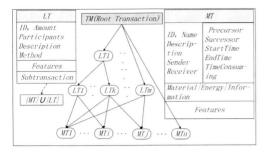

Figure 2.   Transactions model.

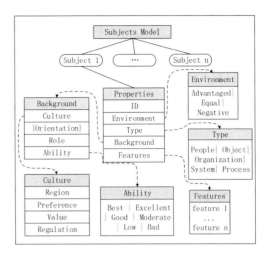

Figure 3.   Subject model.

Interface Management. The *Subject* described as *Subject = (ID, Environment, Type, Background, Features)*.

2.3.1   Subject environment

The *Environment* describes the relation state between subject and its external environment. *Environment = <Advantaged | Equal | Negative>*, which, respectively, indicates advantaged, equal and negative effect the external environment has on subjects.

2.3.2   Subject type

Type describes the universal characteristics between the subjects, which correspond to the different interface features. The subject type is classified as people, object, organization, system, process or processes, etc., described as *Type = < People | Object | Organization | System | Process>*.

2.3.3   Subject background

Background is an important factor that has effect on interface state. It describes the cultural features, role features, ability features and orientation features, *Background = <Culture, Orientation, Role, Ability>. Culture = <Region, Preference, Value Regulation>, Orientations = {Orientation}, Ability = < Best | Excellent | Good | Moderate | Low | Bad >.*

2.3.4   Subject feature

Features describes the individual features set of subjects, *Feature={feature}*. Except the universal feature of type and the invisible feature of background, each subject possesses its individual feature in specific environment, and those individual features often have certain invisible interface issue.

2.3.5   Subjects model description

SM describes the set of all subjects in a complex system which possesses individual features, that is, *SM = ({Subject})*, as is shown in Figure 3.

2.4   Issues library model

ILM describes the set of interface issues in a complex system; it is a strong support for effective interface management. *ILM* is structured by basic interface Issues. *ILM* is used to make new interface issues into library dynamically and dynamically upgrade new interface management plan in order to structure perfect Issues Library Model. Hence, *ILM* is an expertise database with large amount of knowledge and experience in interface management field. It can solve various kinds of Interface Issues of inner interface management field in complex systems. *ILM* possesses the following characteristics.

1. Independent. *ILM* is independent without belonging to any complex systems.
2. Sharing. *ILM* is opened to public and can be shared by different complex systems.
3. Universal. *ILM* is proposed for interface management issues of universal management interface.
4. Instructive. *ILM* possesses great instructive significance in solving interface management issues.
5. Quasi-perfectness. The issues library in *ILM* is not perfect, but it can improve itself.

2.4.1   Interface issues model

A single issue is described as *Issue = (ID, Incentive, Description, Background)*. The incentive of issues is the core of Interface Issues Model (IIM). We adopted Analytic Hierarchy Process (AHP) to make an analysis on the incentives of interface issues and got several incentive factors included: induced by subject's features, by requirement of transactions and by features of transmitted information, energy and material. That is to say, the incentive of Interface Issues can be classified into subject incentive, transaction incentive and information incentive, formally expressed by *Incentive* = < *SI, TI, II* >, as is shown in Figure 4.

1. *SI* refers to the Interface Issues induced by attribute features of participant subjects in interface, which included the difference of goal, type, morphology, role, preference, culture, background and ability etc. among subjects, formally expressed by *SI* = <*GD, TD, MD, RD, PD, CD, BD, AD*>.
2. Transaction incentive refers to the Interface Issues induced by the features of transactions in interface, which includes timeliness and volume rate etc., formally expressed by *TI* = <*TL, VR*>.
3. Information incentive refers to the Interface Issues induced by material, energy or information transmitted by transactions in interface, which includes standard difference and attribute difference of transmitters, formally expressed by *II* = <*SD, AD*>.

2.4.2 *Issues library model description*
ILM is described as: *ILM* = *(Issues, addIssue( ), updateIssue( )), Issues* = { *Issue*}, that is to say, *ILM* is the set of interface issues and includes "addIssue" method can input the new issue into library; "updateIssue" method can upgrade old issues automatically, as is shown in Figure 5.

2.5 *Solutions library model*
SLM is a solutions set of the interface issues in the issues library, described as *SLM*=({*Solution*}).

I	II	III
Incentive	SI	GD TD MD RD PD CD BD AD
	TI	TL VR
	II	SD AD

Figure 4. Incentive of interface issues.

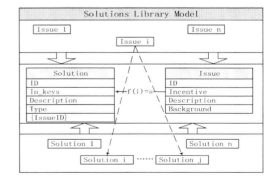

Figure 5. Issues library model.

Solutions library is quasi-perfect as issues library. In project management practice, *SLM* dynamically inputs the new solutions into library, in order to build the perfect library. *Solution* = *(ID, InKeys, Description, Type,{IssueID})*.

A mapping relationship between the issues library and the solutions library, described as *IS* = *(I, S, f)*. *I* = {*Issue*}; *S* = {*Solution*}; $f = (I \rightarrow S)$. If $i \in I$ and $\{s_1, ..., s_n\}, \in S$ then $f(i) = \{s_1, ..., s_n\}$, as is shown in Figure 6.

2.6 *Multi-transactions interfaces model*
MTIM is structured on the basis of the above 4 models.

2.6.1 *Multi-transactions management interface*
Definition 5: Multi-Transaction Interface is:

1. *T* is a finite set of all transactions in a complex system,
$T = \{t_1, ..., t_n | \nexists t_j \in t_i. \text{substransaction}\}$;
2. *S* is the set of all participant subjects in *T*, $S = \{S_1, ..., S_n | n > 0\}$;
3. If $t_i \in T$ & $t_j \in T$ and $S_i = \{t_i, Participants\}$ & $S_j = \{t_j, Participants\}$, then $S_i \subseteq S$ & $S_j \subseteq S$. If $S_i = S_j$, then t_i and t_j belong to the same interface;
4. $t_i \in \{t_1, ..., t_m\}$, $S_i \in \{S_1, ..., S_m\}$ $S_i = \{t_i, Participants\}$. If $S_1 = \cdots = S_m$, then the *m* transactions belong to the same Long Transaction (*LT*), which is called Multi-Transactions Interface.

Structure Multi-Transactions Interface as follows:

Step 1: Make $T = T_1 \cup ... T_i \cup T_n$, the T_i is the transactions set which have the same participants.
Step 2: $T_i = \{T_{i1}, ... T_{ij} ..., T_{in}\}$. T_{ij} is a *LT*, analyze the interface issues of T_{ij} according to SM and *ILM*, and get TI_i, which is one transaction interface.
Step 3: Analyze each transaction in T_i through step 2, and get MTI_i, which is the Multi-Transactions Interfaces Model.

Figure 6. Solutions library model.

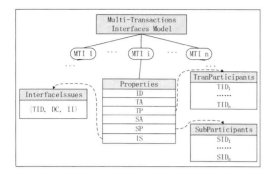

Figure 7. Multi-transactions interfaces model.

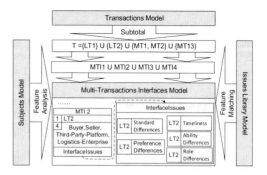

Figure 8. *MTOIMM.*

Step 4: Analyze each transactions set in T through step 3, and get $MTI = \{MTI_1, \ldots, MTI_n\}$.

Transaction Interface (TI) can be described as $Ti = (ID, \{Participant\}, \{Issue\})$. Multi-Transactions Interface can be described as $MTI = <ID, TA, TP, SP, SA, IS>$. TA is the amount of long transactions, TP is the transaction set $\{TID_1, \ldots, TID_n\}$, SP is the subjects set $\{SID_1, \ldots, SID_n\}$, SA is the number of participants, IS is the interface issues set $\{IS_1, \ldots, IS_n\}$, and $IS = <TID, DC, II>$. TID is id of the transaction caused interface issue, DC is the issues description, II is an interface issue, and $II \in ILM$.

2.6.2 *Multi-transactions interfaces model description*

$MTIM$ describes all multiple transactions interfaces in a complex system, $MTIM=(\{ MTI \})$, as is shown in Figure 7.

3 MODEL CHECKING

The structure process of $MTOIMM$ is the structure process of SM, TM, ILM and $MTIM$. We define an $MTOIMM$ of B2B online payment system to test the validity of the model, as is shown in Figure 8.

In Figure 8, $MTI2$ can be described as: there is one interface participant transaction $LT2$ and four interface participant subjects, namely, Buyer, Seller, Third-Party-Platform and Logistics Enterprise; the probable interface issues in this MTI are:

1. *Buyer* prefers *Platform 1* while *Seller* prefers *Platform 2*, so the probable interface issue is *Preferences Differences (PD)*;
2. What is transmitted among subjects is information, so the probable interface issue is *Standard Differences (SD)*;

3. MT9 is timely, so the probable interface issue is *Timeliness (TL)*;
4. The ability eigenvalue of logistics-enterprise is moderate, so the probable interface issue is *Ability Differences (AD)*;
5. Role characters are different among subjects, so the probable interface barrier is *Role Differences (RD)*.

4 CONCLUSION

This paper proposed a Multi-Transactions-Oriented Interface Management Model ($MTOIMM$) to complex systems, abstracted the universal interface management problem in complex systems and described interface management process by formal method. The paper put forward 5 layer models of $MTOIMM$: Subject Model, Transactions Model, Issues Library Model, Solutions Library Model and Multi-Transactions Interface Model, described interface management model of complex systems by structuring those 5 models, which provided methodology for the analysis and control of interface issues for interface management in complex systems. Finally, the paper verified the effectiveness of $MTOIMM$ through the modeling of B2B online payment system. According to the instance analysis, it is definite that $MTOIMM$ plays an effective role in analyzing and controlling interface management problems in a complex system.

ACKNOWLEDGMENT

This research was supported by: (1) National Natural Science Foundation of China, under Grant 71233006; (2) National Natural Science Foundation of China, under Grant 71073122; (3) the Fundamental Research Funds for the Central Universities.

REFERENCES

[1] Barczak, G. 1995. New product strategy, structure, process, and performance in the telecommunications industry. *Journal of Product Innovation Management*, 12(3), 224–234.

[2] Chen Q., Reichard G., Beliveau Y. 2009. Object model framework for interface modeling and IT-oriented interface management. *Journal of Construction Engineering and Management*, 136(2): 187–198.

[3] Du, Y. & Yang, J.J. 2008. Study on Interface Management of Supply Chain Network Organizations. *Soft Science*, 06:63–67.

[4] Guan, J.C. & Jin, P.A. 1995. Empirical Study on Interface Management of Firm. *Economic Theory and Business Management*, (6): 67–69.

[5] Guan, J.C., Zhang, H.S., Gao, B.Y. 1997. An Empirical Study of R & D/Marketing Interface Management. *Chinese Journal of Management Science*, 7(2): 8–16.

[6] Guo, B., Chen, J., Xu, Q.R. 1998. Interface Management: a New Thend in the Management of Enterprise Innovation. *Studies in Science of Science*. (1): 60–68.

[7] Guo, B., Chen, J., Xu, Q.R. 1997. Interface Management in the Process of Enterprise Innovation. *Quantitative & Technica Economics*, (7): 38–41.

[8] Guo, B. 1999. Empirical Research of Enterprise Interface Management. *Science Research Management*, 20(5): 73–79.

[9] Guo, M.L. 2009. Study on the Financing Problem of the Middle-and-small-sized Enterprise in Perspective of Interface Management. *Guide to Business*, (10): 71–72.

[10] Liu, B. & Shen, J.Q. 2012. Definition of the Concept of Interface and Interface Management. *East China Economic Management*, 26 (9): 109–111.

[11] Liu, X.M., Xu, F.W., Zhang, Y.S. 2007. Evaluation Method Research of Effective State of Enterprise Innovation Interface, *Science Research Management*, (5): 31–35.

[12] Liu, X.M. & Xue, F.W. 2005. Validity Study Based on the Harmonious Interface. *Technology Innovation and Management*, (3): 28–31.

[13] Moenaert, R.K., De Meyer, A., Souder, W.E., & Deschoolmeester, D. 1995. R & D/marketing communication during the fuzzy front-end. Engineering Management, *IEEE Transactions on*, 42(3): 243–258.

[14] Ren, R. 2010. The Integration of Cooperative Innovation and Organizational Levels Based on Interface Management Theory. *Economic Management Journal*, 10: 180–186.

[15] Siao, F.C., & Lin, Y.C. 2012. Enhancing construction interface management using multilevel interface matrix approach. *Journal of Civil Engineering and Management*, 18(1), 133–144.

[16] Souder, W.E., & Chakrabarti, A.K. 1978. The R&D/marketing interface: results from an empirical study of innovation projects. Engineering Management, *IEEE Transactions on*, (4), 88–93.

[17] Vandevelde, A., & Van Dierdonck, R. 2003. Managing the design-manufacturing interface. *International Journal of Operations & Production Management*, 23(11), 1326–1348.

[18] Wheelwright, S.C., & Clark, K.B. 1992. Organizing and leading "heavyweight" development teams. *California management review*, 34(3), 9–28.

Education Management and Management Science – Zheng (Ed.)
© 2015 Taylor & Francis Group, London, ISBN 978-1-138-02663-6

The curriculum design of EM in the major of ME of the USA and its enlightenments

L. Gong, Q. Xue & Z.J. Zhang
School of Mechanical Engineering, Beijing Institute of Technology, Beijing, China

ABSTRACT: It is an important objective to cultivate Mechanical Engineering (ME) students with high and comprehensive quality in the major design of mechanical engineering in English teaching. The Engineering Management (EM) curriculum in the universities of the USA with mechanical engineering ranked top 10 are collected by analyzing the training objective for mechanical engineering students. Moreover, by summarizing the characteristics of curriculum design of these colleges, the authors propose some suggestions for the design of relevant courses in the colleges of China.

Keywords: mechanical engineering; English teaching; engineering management; curriculum design

1 INTRODUCTION

At present, numerous colleges in China have set up English teaching majors and courses to improve students' communication ability and professional competence in English and lay the foundation for more international communication, academic visiting and advanced study, etc. As early as in 2001, the Ministry of Education in China pointed out that conditions should be created for teaching common required and professional courses using English or other foreign languages in undergraduate education, according to the requirement that education should be oriented toward modernization, facing the world and the future.

Mechanical Engineering (ME), as one of the important engineering disciplines, has been taught in many colleges using English or by setting pilot of English teaching in China. Based on the requirements of USA Accreditation Board for Engineering and Technology (ABET) for engineering students, the research analyzes the design and content of Engineering Management (EM) courses in the colleges of the USA with mechanical engineering ranked top 10. Moreover, suggestions for the content and design of relevant courses in colleges of China are proposed to provide basis for cultivating mechanical engineering students with high comprehensive quality.

2 OBJECTIVE FOR ME TEACHING AND DEMANDS FOR EM COURSES

2.1 *Objective for ME teaching in the USA*

USA ABET is one of the originators of Washington Accord and its engineering accreditation is recognized worldwide. ABET formulates engineering accreditations for different disciplines and levels annually to put forward the requirements for fostering students clearly. It presented explicit criteria for master level programs and teachers of mechanical engineering in the Criteria for Accrediting Engineering Programs, 2014–2015; for the abilities that students should present at their graduation of baccalaureate education, ABET formulated the general criteria for engineering disciplines. The details are as follows (ABET, 2014):

1. an ability to apply knowledge of mathematics, science, and engineering
2. an ability to design and conduct experiments, as well as to analyze and interpret data
3. an ability to design a system, component, or process to meet desired needs within realistic constraints such as economic, environmental, social, political, ethical, health and safety, manufacturability, and sustainability
4. an ability to function on multidisciplinary teams
5. an ability to identify, formulate, and solve engineering problems
6. an understanding of professional and ethical responsibility
7. an ability to communicate effectively
8. the broad education necessary to understand the impact of engineering solutions in a global, economic, environmental, and societal context
9. a recognition of the need for, and an ability to engage in life-long learning
10. a knowledge of contemporary issues
11. an ability to use the techniques, skills, and modern engineering tools necessary for engineering practice.

2.2 Objective for mechanical engineering education at baccalaureate level in China

According to the Specifications for Mechanical Discipline Teaching at Baccalaureate Level in Higher Education Institutions presented by the national mechanical engineering teaching advisory committee under the Ministry of Education in China, the curriculum of mechanical engineering students should meet the following criteria (Liu Z.Q. et al. 2008):

1. basic knowledge of humanities and social sciences, economic management, and natural science, particularly a good humanistic quality
2. systematic knowledge of required technological theories in the discipline
3. familiar with at least one or two orientations or professional knowledge in the discipline as well as their frontier and developing trend
4. a basic skill of drawing, calculating, testing, investigating, literature reviewing, and basic process operating, etc, and a good ability in applying computer
5. grasping a foreign language, be able to read professional books using the language, and the ability of listening and speaking
6. self-study, expression, and problem analysis and solving, and a high consciousness of innovation
7. good teamwork spirit and cooperative ability.

2.3 Demands for engineering management courses

The above-mentioned criterion show that the teaching discipline at baccalaureate level both in China or the USA require students to have broad basic knowledge, and abilities in economic management.

However, considering the small proportion and credit of engineering management courses, it has been a concern on how to design the relevant courses and their content in the cultivation of mechanical engineering students, especially English teaching mechanical subjects.

3 CURRICULUM DESIGN RELATING TO MECHANICAL ENGINEERING SUBJECT IN THE COLLEGES OF THE USA

Aimed at the core issue in this research, the authors studied the colleges with mechanical engineering subject ranked top 10 (owing to the fact that three colleges are tied for the 10th place, there are 12 colleges investigated), according to the ranking presented in US News 2014. Additionally, information including the curriculum design, curriculum type, credit, and content, etc, was analyzed. The analysis in detail is as follows.

3.1 Curriculum design

The engineering management courses in the major of mechanical engineering that rank top 10 in the colleges of the USA are listed in Table 1.

3.2 Content of the courses

In order to better understand the objective of opening the courses, the authors analyze the content of each course in detail. Considering some courses are traditional economic and engineering management courses, such as principles of macroeconomics, principles of microeconomics, and engineering economics, etc. Because some courses

Table 1. Engineering management courses in the major of mechanical engineering in the top ten ranked colleges in the USA.

Rank	University	Courses	Type	Credits
1	Massachusetts institute of technology	Management of engineering	Elective	3
1	Stanford University	Technology entrepreneurship	Elective	4
		Technology policy		5
3	California Institute of Technology	Management of technology	Elective	9
3	University of California—Berkeley	Introduction to product development	Elective	3
5	Georgia Institute of Technology	Engineering economy	Elective	1
5	University of Illinois—Urbana-Champaign	Microeconomic principles	Elective	3
		Macroeconomic principles		3
5	University of Michigan—Ann Arbor	Global manufacturing	Elective	3
8	Cornell University	Entrepreneurship for engineers	Elective	3
8	Princeton University	Global technology	Elective	3
10	Carnegie Mellon University	Business communications	Elective	9
10	Purdue University—West Lafayette	Technology and values	Elective	3
10	University of Texas—Austin (Cockrell)	Engineering communications	Elective	3
		Engineering finance		

are similar in content, the authors introduced the concrete teaching contents of some courses.

1. Engineering management
   It mainly contains brief introduction of engineering management, standards for financial management innovation, technical strategy, and optimal management practice. Class discussion is to learn the analysis method by cases studies. This course pays attention to the development of personal skills and management tools.
2. Technology management
   This course concentrates on the improvement of students' interest for applying current technologies in the production of effective products. Students are expected to solve cases in Harvard Business School (HBS) and illustrate the core issues. The cases are provided by teachers. The course improves students' learning and working abilities in enterprises and business school by carrying out team projects. The content of team projects includes technology development, financial foundation, process of business integration, product development and investment management, learning curve, risk assessment, technological trend and methodology, and motivation and reward.
3. Global manufacturing
   Global manufacturing includes examples of global manufacturing, integration of production, process, and business, product innovation strategy, customization, individualization, reconfigurable product, batch and lean production, mathematical analysis for mass customization, traditional manufacturing system, reconfigurable manufacturing system, reconfigurable machine, analysis of system configuration, business response model, global strategy of enterprise, and globalization enterprise.
4. Technology and value
   The course is mainly to teach as follows: the influences of science and technology on individual and social value system; engineer's responsibility; implementation method for guiding the development of future science and technology by applying human value; reflection on social issues such as war, energy, overpopulation, resource consumption, and environmental degradation; and multidisciplinary collaboration.
5. Engineer's entrepreneurship
   Engineer's entrepreneurship is designed to improve students' ability in identifying, evaluating, and opening new enterprises. Its knowledge includes intellectual property, competition, business plan, technological forecasting, finance and accountant, and capital resource. The course requires students to formulate financial documents and plans, and analyze human resource models using strictly quantitative methods.

Complexity analysis method, ownership structure, and law and business document have to be used in the formulation. The course is performed by means of discussion, homework, and the composition and report of complete business plan.
6. Engineering communication
   For the students majored in engineering, the course improves effective writing and communication ability in the following aspects: the collection, organization, and evaluation of data; the drafting, composition, and revision of documents; effective oral report; the understanding, analysis and organization of effective arguments; the development of critical thinking ability; the understanding of social and global influences of engineering communication, research, and practice.

3.3 *Analysis of curriculum design*

The design of engineering management courses in mechanical engineering subject in the 12 colleges in the USA exhibits the following characteristics:

1. Broad and comprehensive knowledge.
   Except several colleges, such as California Institute of Technology (it has three semesters annually and the calculation mode for credit is different from others), most courses in other colleges have 3 credits and the courses cover multi-field knowledge, including economy, management, and finance, etc. The courses are different in content.
2. Project—or case-based teaching
   Most courses apply the project—or case-based teaching method, which enables students to master essential knowledge systematically.
3. Mastery of advanced technologies
   Great importance is attached to the introduction and analysis of advanced technologies. It makes engineering students grasp advanced technologies as well as learn the professional knowledge, particularly those relating to information technology.
4. Operational capacity
   Most courses pay attention to students' practical ability. This is consistent with the education objective formulated by ABET. To improve students' cooperative ability, team work with grouped students is carried out.

4 ENLIGHTENMENTS FOR CURRICULUM DESIGN RELATING TO MECHANICAL ENGINEERING IN THE COLLEGES OF CHINA

Currently, regarding mechanical engineering in the colleges of China, the courses relating to engineering management have been set as well. However, the teaching is generally basic and simple in content. Considering English teaching courses, they have to be more international in particular. Aimed at this problem, the authors present the following suggestions:

1. Setting courses with varied content
   The curriculum has to be designed with wide content. The knowledge including management basis, project operation, financial management, engineering economy, and technology development, etc. can be integrated in 1–2 courses. Considering the objective for baccalaureate level programs, the depth of each aspect can be adjusted. But the courses should cover a wide range of knowledge.
2. Practical teaching
   The curriculum stresses on the case—or project—based teaching and divides students into groups to train their practical ability and teamwork spirit in teaching.
3. Driven by advanced technologies
   Advanced technologies, particularly those relating to information technology which have been applied or integrated in mechanical engineering should be included in the teaching, to make students comprehend the application range of different technologies.
4. Aim for grasping and applying knowledge
   The teaching is to achieve the goal that students are able to master and apply the learned knowledge. Meanwhile, based on the professional courses, students have to comprehend the influences of engineering technology on economy and society and obtain the ability of applying all kinds of technologies, skills, and modern engineering tools in practice.

At present, an engineering management course is open for mechanical engineering subject taught by English in Beijing Institute of Technology. The course has 2.5 credits; covers fields consist of fundamentals of management, project management, technology management, and engineering economics, etc. This course applies the case-based teaching method.

It enlarges students' knowledge horizon and improves the operation ability.

5 CONCLUSION

The design of engineering management courses in the major of mechanical engineering in the colleges which rank top 10 in same field in the USA are collected and studied by analyzing the cultivating objective for students in this field. The characteristics of the curriculum design are summarized. The research provides the suggestions for designing courses relating to mechanical engineering, especially English teaching ones, for students at baccalaureate level in the colleges of China. It lays a foundation for the training of mechanical engineering students with high and comprehensive quality.

ACKNOWLEDGMENTS

This work was supported by Beijing Higher Education Young Elite Teacher Project.

REFERENCES

[1] ABET Engineering Accreditation Commission. 2013. *Criteria for accrediting engineering programs.* Baltimore: ABET.
[2] Liu Z.Q. & Wu Y.H. & Han G.C. 2008. On Higher Education of Mechanical Engineering in China Based on ABET Accreditation in America. *2008 International Conference on Science, Technology and Education Policy.* 192–195.
[3] http://grad-schools.usnews.rankingsandreviews.com/best-graduate-schools/top-engineering-schools/mechanical-engineering-rankings. 2014.
[4] http://student.mit.edu/catalog/search.cgi?search=2.96 &style=verbatim. 2014.
[5] http://www.mce.caltech.edu/academics/ugrad. 2014.
[6] http://www.engin.umich.edu/college/academics/bulletin/courses/mecheng. 2014.
[7] https://engineering.purdue.edu/ME/Academics/Undergraduate/index.html. 2014.
[8] http://courses.cornell.edu/content.php?catoid=14&navoid=3134. 2014.

Education Management and Management Science – Zheng (Ed.)
© 2015 Taylor & Francis Group, London, ISBN 978-1-138-02663-6

The great significance and implement methods on university students' sports spirit cultivation

Z.H. Tang & C.L. Deng
Department of Physical Education, UESTC, Chengdu, Sichuan, China

ABSTRACT: The methods including documents, investigation, experimental observation, action research, and summarization are adopted in the paper, in order to research significance and implement methods on university students' sports spirit cultivation. The great significance of cultivating university students' sports spirit aims to carry forward national spirit, strengthen personality education, deepen quality-oriented education, create special campus culture, and summarize practice problems of cultivating university students' sports spirit from both university sports teaching and university sports students' associations. The purpose is creating special campus sports culture, shape university students' personality education, deepen connotation of quality-oriented education, and further improve university sports developmental level.

Keywords: university students; sports spirit; cultivation; significant; methods; China

1 INTRODUCTION

Sport is an inevitable outcome of human society development, and it is developing with human society. It also has great influence on human society development and course of history, which promote human society's continuous improvement. During sports development, a value which makes great contribution to human society development is gradually formed. And this value is treated as an excellent nourishment for the mind—sports spirit. It is an integral part of the spirit of the Chinese nation culture connotation, and its perception, integration and extraction throughout the whole human society development process. Therefore, it is very important to research the role of physical education spirit on the human society development. The research use sports spirit as center, student associations as carrier, cultivate university students' sports spirit, provide exemplary, instructional, and informative suggestions in researching how to produce special sports culture, making university sports culture construction in a sustained, rapid and sound manner. And more importantly, there is a great significance in shaping university students' personality education, deepen quality-oriented education and make students understand when they are in the competitive social environment, sports spirit will play an irreplaceable role.

2 RESEARCH OBJECT AND METHODS

2.1 Research object

Use methods and path to cultivating UESTC students' sports spirit as research object.

2.2 Research methods

Documents: Research new educational theory and development trend in teaching reform at home and abroad, especially relational theory of quality-oriented, recreational education and cultivating sports spirit; learn from existing results to structure this issue's theoretical framework and methodology, transform education opinion, and adjust teaching thought.

Investigation: Investigate teachers' teaching situation and former students' living condition; investigate, summarize and extend the new thought, methods and experience during teachers cultivation of students' sport spirit; investigate students' associations development situation, provide living examples for effective organizational activities, feedback and adjust activities strategy.

Experimental observation: Through operating variable, observe variation, measure front and back contrast, analyze causal relationship, make cognition from sensibility to rationality, and then guide our work.

Action research: Use the latest teaching theories to solve latest problems. Practice and summarize research results, timely feedback and amend action plans. The method that research in action and then action in research is always the main research strategy. Because the research of teaching efficiency is affected by all kinds of facts and aspects, so we have to adjust our strategy based on the latest problems in order to improve research timeliness.

Summary: This issue requires every teacher to exchange and analyze opinions and experience of cultivating university students' sports teaching, write down personal experience and opinion, put forward more scientific and reasonable activity plans based on summarizing and analyzing teachers, teaching and research group and other aspects.

3 RESEARCH RESULTS AND ANALYZE

3.1 *Related research status at home and abroad*

1. The research of sports spirit explanation. From 1980s to 1990s, Chinese government was making four modernization construction. Material civilization was strengthened and spiritual civilization was emphasized at the same time, physical education spirit was treated as important content at that moment. Some scholars even put forward that sports spirit should be popularized in the whole society in order to promote social progress. After that sports spirit was generally used and rapidly became an academic hot point. Large numbers of scholars gave all kinds of explanation of sports spirit. Representative examples include Chinese Sports Spirit is the Spiritual Wealth of the Whole Nation (2000) by Xi Qionghuan, Sports Spirit and the Reform and Opening-up Policy (2002) by Hu Xiaoming, Cultural Connotation and Value Construction of Sports Spirit (2007) by Huang Li etc. The scholars' research results provide an academic foundation of sports spirit's connotation, characteristic and value, provide beneficial reference to solve problems which recent sports spirit lack, but there is no relevant research about university sports spirit which is very important in today's society.
2. The research of cultivating university sports spirit. A good university pays more attention to create student's personality; besides those two factors, a famous university also pays more attention to cultivate sports spirit. The thematic study about sports spirit began to be active until 2003. Representative examples include The First Exploration of Humanistic Significance in University Campus Sports Culture (2003) by Yuan Minglian, Campus Sports Culture and University

Spirit (2006) by Chen Hong, University Sports Teaching and Students' National Spirit Cultivating (2006) by Sun Bing, University Time Spirit and University Sports Cultivation (2008) by Xue Feng, Sports Spirit Need to be Cultivated in University (2011) by Yang Yan, Practical Significance and Methods of Cultivating University Sports Spirit (2012) by Huang Xiao Bo etc. This research provides beneficial references and valuable opinions on researching university sports spirit's concept, characteristic, function and constructional path.
3. Relevant research evaluation and analysis. All of above research achievements about university sports spirit remain at explaining sport spirit and its importance but never grasp its law and essence. That is to say, university sports spirit can't be treated as key content of university development, and some knowledge of sociology. There is no academic and path analysis when researching university sports spirit. In recent years, different universities have different theories on school management, value pursuit, objective of personnel cultivation and attention to sports. So university sports spirit is different in different universities. All in all, there is few achievements aiming at university sports spirit. So this issue uses sports spirit cultivation in UESTC as study object, systematically study university sports spirit, research successful cases about cultivating sports spirit on students' associations, explore strategy of promoting and cultivating sports spirit.

3.2 *The important significance on cultivating university students' sports spirit*

1. Necessity to promote national spirit. Sports spirit is an important content and concrete representation of national spirit. Promoting sports spirit is exactly promoting national spirit [1]. Sports spirit was treated as valuable nourishment for the mind, and sports spirit not only contains lofty beauty of personality and profound intellectual beauty, but also deposits an eternal soul which symbolizes a great nationality. It is core content of national spirit. Society needs innovation talents, and sports spirit is the source power in innovation development.
2. Necessity to strengthen personality education. Personality education is a cultivation process which harmoniously develops human being's ability of cognition, emotion, volition, action, making one wholesome with thorough psychological quality. State physical Culture and Sports Commission has generalized Chinese sports spirit after years of working: win honor for the country, selfless contribution, scientific realistic,

observe law and discipline, solidarity and friendship and work tenaciously [2]. Through personality education, university students will have integrated development in school sports activities.

3. Necessity to deepen quality-oriented education. Sports spirit is gems of wisdom which belongs to countless talents; it contains national spirit, aesthetic taste, life philosophy, standard of behavior, values and ideology etc. [3] And it has become a catalyst for one's growth of talent. Sports spirit helps students feel enigma of life, provides insight and then change one's destiny. It also helps students light creative fire, enlighten wisdom, strengthen original creativity.

4. Necessity to create special campus culture. UESTC is a famous key university in China and it belongs to "211 project" and "985 project". UESTC covers all electronics disciplines, of which electronic information science and technology is core, engineering course is main aspect of penetration of technology; science, engineering, management and arts harmoniously develop in UESTC. For a long time, UESTC paid high attention to sports. Through hard work, UESTC has own qualification to recruit high-level athletes in basketball, track and field, swimming, football and tennis. The "spirit" and "task" was initiated by sports spirit. It is our task to use sports as special culture, cultivating students' sports spirit. Through years of practices experience, different kinds of students' associations were formed, an effective activity program of sports teaching and sports association are found, the special sports activities which suit university students are created, and finally university development has its own sports characteristic, and this characteristic is actually a concrete expression of sports spirit [4].

3.3 Implement approach to educate university students sports spirit

1. Sports teaching approach. Explore effective way to cultivate university students sports spirit during sports teaching. The specific paths are: 1. use class teaching as main route, activity courses, after-class training, contests as basic approach; 2. use sports history and Olympic knowledge as theory approach, use sport-specific learning and competitive sports games as practical approach; 3. during sports teaching, use sports spirit events as teaching content, adopt all kinds of teaching organization pattern; 4. during sports teaching, construct good sports culture atmosphere, build long-term mechanism of sports spirit cultivation.

2. Sports association approach. There are two aspects of cultivating sports spirit through university students' sports association: 1. Build special competitive sports associations, use colorful associations activities to cultivate students' sports knowledge, technology and technical ability, cultivate students' comprehensive abilities of organizing, coordinating, communicating and leading, cultivate students' sports spirit, make students form lifelong exercise habit. 2. Build national features sports associations, combine with sports curriculum, introduce national features sports into class, create national features sports associations including figure TiJian, rope skipping, Tai Chi, Baduanjin, five-animal boxing, cultivate students' national sports spirit [5].

3.4 Actual result of implementing those two approaches to cultivate students' sports spirit

1. Students initially learn sports spirit knowledge through research and practices, students' sports behavior, self-confidence, sense of pride cohesion are strengthened, students' sports spirit of patriotic dedication, solidarity and cooperation, fair competition, striving happiness and health are cultivated, the sports spirit of UESTC is cultivated.

2. According to actual situation of UESTC, different departments and different subjects have sports culture association activities, explore forms of sports association activities, edit text books of sports association activities, explore positive effects and effective building of which student's sports associations affect students' physical fitness, mentality, thought, ability, intellectual education, moral education. Integral elevation of teachers' quality, optimize teachers' contribution.

3. The whole school's general special sports activities are made through sports teaching and colorful sports association activities. There are technical sports activities which suit the class, association and personal specialty, from part to entirety, popularization to improvement, improvement to characteristic, and finally form sports program including basketball, five-a-side football, swimming, track and field, table tennis, badminton which have reached the top-level in all of universities of Sichuan province.

4 CONCLUSION

In order to change university students' current situation and create university special culture, some methods have been found to cultivate university students sports spirit, explore the implementation scheme to create special campus culture, including effective filter of university special sports culture, successful cases of implementing, effective actions during implementation, distillation process of sports brand culture. From practice aspect, through exploring and practice, use UESTC as sports teaching center, students' sports association activities as carrier, cultivate university students' sports spirit. From theory aspect, reveal how to build campus sports culture under new education situation, create theoretical foundation of special campus culture, provide blueprint including practice to theory for other universities. I wish relevant scholars can pay some attention to this issue, research and practice together, better combine practice and theory, further improve overall development of school sports education.

ACKNOWLEDGMENTS

This paper is high education personnel training quality and teaching reform project in Sichuan province (2013–2016), education teaching reform research project in UESTC (2013XJYYL002): one of phased objective on "use sports teaching as center, students' associations as carrier, cultivate university students' sports spirit in researching and practicing".

REFERENCES

[1] F.Q. Meng. 2004. Sports Spirit and University Students' Personality Shaping. *Jilin Sport University Journal* (3):30–46.
[2] W.M. Zheng. 2009. Chinese Sports Spirit and University Students' Quality Education of the Time. *Shandong Youth Management Cadre Institute Journal* (4):58–59.
[3] J. Li. 2012. The Sports Humanities Education on College Students' Spiritual Construction. *Wuhan Sport University Journal* (10):93–96.
[4] X. Qiu. 2007. *The Research of Campus Sports Culture Effect on University Students' Humanistic Spirit.* Guangxi Normal University.
[5] H. Li. 2009. On the Integration and Development of National Traditional Sports and College Student National Spirit Education. *Chengdu Sport University Journal* (7):92–94.

Education Management and Management Science – Zheng (Ed.)
© 2015 Taylor & Francis Group, London, ISBN 978-1-138-02663-6

On curriculum development of vocational college teacher training

Zhenwu Zhou
Shanghai Second Polytechnic University, Shanghai, P.R. China

ABSTRACT: Qualified teachers are the prerequisite of developing feature vocational education, which requires that we intensify the training of teachers' professional techniques and build up a teaching rank with high qualities. The paper shows the basic process of curriculum development in vocational college teacher training.

Keywords: curriculum development; vocational college; teacher training

1 INTRODUCTION

The basic objective of teacher training is to update teachers' professional knowledge, change their teaching attitude, improve their teaching skills and optimize their professional conduct so that they can better adapt themselves to the requirements of teaching. Developing a training curriculum is such a basic skill that a trainer should possess and such a task that he must accomplish that we should not only increase our awareness of the importance of curriculum development, but earnestly and seriously study the general rules and strategies of vocational college teacher training. Only by doing so can we adapt ourselves to the development of continuing education in vocational schools.

2 THE NATURE OF DEVELOPING TRAINING CURRICULUM FOR VOCATIONAL COLLEGE TEACHERS

The nature of developing training curriculum for vocational college teachers is just a process, in which the trainer collects and works over teachers' needs and many other factors regarding curriculum resources, and then, integrates them together according to some specific purposes and requirements.

The training need of vocational college teachers is to shorten the gap between what they are required and what they have acquired. The current training needs of vocational college teachers can be examined from the following aspects:

First, it is the social development that requires vocational college teachers to be trained. These factors, such as scientific and technological progress, the increasing high-tech contents during production, the newly rising professional technical

posts and the growing contents of intelligence, require technical personnel at new posts, which have also put forward more demands for vocational education. As a matter of fact, most of our teachers who have graduated from general institutions of higher learning are not completely familiar with vocational education and vocational teaching, so it is imperative to intensify pre-and-post professional training for teachers.

Second, it is teachers' own professional development that is required. As a professional, the developmental connotation of a teacher is multifaceted and covers many fields. Unlike the one-off shaping of industrial products, teacher training is a long-term learning process because the society keeps developing and new technologies and new modes of training emerge constantly, all of which requires that the vocational college teachers improve their professional competence.

Third, it is teachers' own individuality development that is required. Curriculum development of vocational college teacher training must make a minute analysis of teachers' individual demand structure and satisfy their future individuality development. First, teachers' individual demands change with some factors. A teacher, undertaking a teaching task, is assumed to be academically qualified. Hence, he will always set a higher qualification as his pursuing goal and once this goal has been achieved another one will be established. During the process of performing vocational instruction, teachers should not only impart students the theoretical knowledge in one field, but also strengthen their actual practicing abilities. What's more, teachers are assumed to study and master new knowledge and techniques on their subjects timely, to comprehend vocational college students' psychological characteristics, to improve their teaching methods and to change their roles constantly. Therefore, the constantly changing requirements, which are teachers' inner impelling forces of continuing education, require that the curriculum development set its program, orientation and contents according to these objective conditions. Second, teachers' need of individuality development is stratified. Because of the different structures of knowledge, specialty and professional titles, as well as the different social experiences and working environment, teachers' needs always demonstrate different features. As a result, curriculum development is assumed to cater to different levels and diversities of teachers' individuality development so that it will keep its distinct characteristics. Furthermore, teachers' demand for individuality development is advanced step by step. The makings of a fine teacher is fostered and accumulated step by step, some of which have already been possessed during the study in universities, others developed gradually during the process of teaching practice. Thereby, in this era of life-long education, curriculum development should be carried out step by step, moved on from the shallower to the deeper, pushed forward at a gradual pace and advanced in succession. Emphasis of curriculum development should be laid upon how to satisfy teachers' individuality advancement from the contents, accesses and approaches and how to achieve the goal of diversity.

Training resources refer to the summation of relevant factors supporting and helpful to teaching and learning activities, which include these factors such as manpower, financial and material resources, time, space, information and so on. Among all these training resources we lay special emphasis on trainers' curriculum development and exploitation, because most of the trainers are well-experienced in actual practice.

3 THE FUNDAMENTAL PRINCIPLES INVOLVING IN CURRICULUM DEVELOPMENT OF VOCATIONAL COLLEGE TEACHER TRAINING

3.1 *Displaying the concept of the current vocational education reform*

Vocational education has undergone great changes along with the rapid development of society and economy, which include the following three aspects. For one, the vocational education has transformed itself from knowledge-based to skill-based. In this age when science and technology have developed by leaps and bounds, the knowledge advantages of a knowledge-based teacher compose no longer the whole qualities of a fine teacher in educational front because the multiplying new concepts and thoughts have made the old ones out-of-date, and the real intention of specialty teaching, on the other hand, is not to impart mechanically but to apply knowledge and skills comprehensively. Second, vocational education has shifted its function from a substitute to an instructor. Quite different from the traditional classroom teaching that is based on teachers' textbook explaining and students' individual understanding, vocational education targets at students' active participation and independent practice, which requires that the teachers not only teach but, more importantly, instruct as well. Third, vocational education has changed from examination-oriented to ability-oriented. In accordance with the demands of reform, vocational education should be employment-oriented and give top priority to students' ability development and make full use of their potentials. Students can experience the essence of knowledge while working in business, and practice their knowledge acquired from textbooks in an actual job.

3.2 *Promoting teachers' professional development and forming a reasonable knowledge structure*

It is impossible for teachers in vocational colleges to effectively deal with all the problems arising from their teaching by virtue of their own experiences and abilities. They should employ some unique professional skills, that is, the professional education knowledge and skills, to support their teaching, which is also the symbol of teachers' professional specialization. While a vocational college teacher who has met the standards of professionalism should possess some general pedagogic knowledge (such as the general rules and strategies of classroom management and knowledge learning), some knowledge of subjects (including the information of evolution in subjects, teaching materials and concepts and the developmental trend of the subjects), relevant knowledge of teaching methods, (referring to the teaching approaches and tactics of understanding and transforming relevant subject knowledge), the knowledge of educational context (that is, teachers should have the ability to control the influences from students' family, the college and the society upon their teaching conduct), he must have a solid pedagogic foundation and the abilities to perform scientific research, develop curriculum, supervise classroom teaching, evaluate the quality of teaching, self-question and the abilities to advance continuously.

3.3 *Intensifying practice ability teaching and paying attention to the introduction of new technologies and outcomes*

The target of vocational education is to cultivate laborers on production front to be technically professional and creative and enthusiastic in work.

So, qualified teachers are the prerequisite of developing feature vocational education, which requires that we intensify the training of teachers' professional techniques and build a teaching rank with high qualities.

3.4 *Adapting to the local developmental level of vocational education curriculum reform*

The main function of vocational education is to serve local economy, which means that vocational education should be closely bound with local economic development and provide service to local economy. Accordingly, the curriculum development of vocational college teacher training must adapt to the local developmental level of vocational education curriculum reform.

4 THE BASIC PROCESS OF CURRICULUM DEVELOPMENT IN VOCATIONAL COLLEGE TEACHER TRAINING

4.1 *Determining the need of training*

The first step of curriculum development is to determine the need of training, namely, the necessity of training. Before determining the need of training, the following factors are mainly taken into considerations: training objectives, cost, value, materials, the managers and the trainees.

Analyzing the need of training is the first step of determining the need of training. Lots of approaches can be employed during this process. Hierarchical approach, for example, can analyze the different levels among some set range; hierarchy of post can study certain different grades and posts; individual approach makes analysis of individual performance and assessment feedback; hierarchy of time period can analyze both current and future developmental trend. To some specific level, other approaches can be introduced, such as face-to-face talking, questionnaire, observation, and visiting. Generally, these approaches are intersected with one another, without an absolute demarcation line, depending on the concrete situations.

Taking the training of curriculum criterion as an example, analyzing the need of training can be done by the following steps:

Visit the relevant leaders, teachers and researchers to learn their needs and opinions;
Go to the teaching front to have on-the-spot observation and make investigation and research;
Ask those teachers performing actual teaching self-assessment and put forward their requirements;
Study the written materials such as syllabus, minutes, exercises and teaching materials.

Based on these, then, specially invite the trainees to answer the following questions:

Is it necessary to have training?
What are the main problems?
What are the best training approaches?
What kind of materials and information should be collected?

After this, the curriculum developers must conduct a further careful, thorough and painstaking analysis before determining the options, the topic and the approaches.

4.2 *Setting the objective of curriculum*

The objective of curriculum is the result that must be obtained after training, which serves not only as the guidelines directing the learners and the goals for them to achieve during the process of learning, but also as the target set by the developers embodied by these terms such as: learn, master, improve and reach. Setting objective is subject to the needs of both the organization and the individual learner, and on the other hand, it changes constantly because of the restrictions of training resources. As a result, the curriculum developers cannot just lock their objectives, but take all factors into account and make necessary revision anytime. They must deliberate on the following issues:

What are the pros and cons concerning the acceptability of curriculum objective?
How to provide maximum satisfaction to learners' needs during the process of development?
In what form and by what ways can the various training resources be developed and exploited to the fullest extent?
How to determine the developing procedures? What should be given top priority?
Can the goal be reached?

4.3 *Working out the training program and content*

To work out training program, the training purpose should be clarified at the first place, followed by the training contents, sequences, and types of the teachers, the training targets, forms and training schedule. Finally, it is the turn of carrying out these measures. While working out the training program, importance should be laid on curriculum contents. The developers can make full use of the library, the reading-room, the network, multi-media devices and some other audio-visual apparatus to collect from all sides the information relevant to the curriculum contents and seek advice from training specialists and scholars and in particular, to be adept in finding, seizing, collecting and sorting the trainees' own experiences, all of which may be shifted into developing outcomes. Apart from that, the developers can also borrow experiences from similar developed courses. As for the displaying form of curriculum contents, it can be a text, a video or a picture. No matter in what form, it should be brief, vivid, active and practical.

4.4 *Selecting training strategies*

During the process of developing curriculum for training vocational college teachers, another important link that must be taken into account is what tactics need to be used to ensure the realization of training goal. Attention should be paid to the following aspects when selecting training strategies:

Make the trainees well prepared for being learning agents. Everything should center on the trainees for the purpose of maximizing their initiative and change the situation that the lecturer keeps his non-stop teaching till the end to achieve the goal of two-way interaction in classroom teaching and apply those effective methods such as discussion, talk and deliberation to the training of vocational college teachers.

Conform with the training contents. The method of teaching or discussion, for example, can be adopted in pure knowledge imparting; demonstration, displaying and performance are better in skill training; and situation simulating is more suitable for emotional attitudes.

Conduct within the permitted objective conditions. These objective factors such as time, venue, the number of the trainees and teaching devices sometimes limit the training. The way of observation and discussion, for instance, will not be suitable for overall-staff-training, for it covers all staff and sometimes is urgent due to the factors of short period, large contents and relatively large number of trainees engaged in the implementation of curriculum.

During the whole process of actualizing the teacher training, the developers must pay great attention to and be engaged in it, collecting the feedback, assessing the qualities timely, amending the existed deviation and predicting the possible problems. Among which, focus should be laid on these questions, such as:

Has the anticipated goal been achieved?
Have the contents been finished?
Have the relevant measures been carried out?

By doing this, the developers can be more successful in completing the set developing program and providing some valuable experiences for a new round of curriculum development.

ACKNOWLEDGMENT

Project source: Key subject construction (cultivate) of SSPU.
Subject name: Vocational and technical education. NO: XXKPY1316.

REFERENCES

[1] Pan Haiyan: A Necessary Handbook for Professional Trainers of Teachers' Continuing Education. Huazhong University of Science and Technology (HUST) Press, Wuhan, 2006.
[2] Ministry of Education, P.R.C. Suggestions on the Employment-oriented Principle and Deepening the Higher College Vocational Education Reform. 2004.
[3] To Cho-Yee. The Scientific Merit of the Social Sciences: Implication for Reseach and Application. England: Trentham Books Limited, 2011.

Education Management and Management Science – Zheng (Ed.)
© 2015 Taylor & Francis Group, London, ISBN 978-1-138-02663-6

Research on integral social planning for age accommodated community

Xiaoyun Li & Rong Yi
Jiangxi Normal University, Nanchang, Jiangxi Province, China

ABSTRACT: Against the increasing population of elders and the skyrocketing need of nursing for the aged, this paper, based on the dynamic necessities of the aged, initiates a system for planning of elder-accommodated integral community, featuring factors of housing, healthcare, security, culture and services specialized for the aged. It exploits as well the functions of social engagement and mutual assistance to promote the integration of different ages and to further advance the creation of age-friendly community.

Keywords: age-friendly community; integral social planning; age fusion

1 INTRODUCTION

The definition and contents of community planning haven't been unanimously agreed upon among academics because of various perspectives and methodologies applied when assessing the problem. Domestic scholars Wang Ying and Yang Guiqing defined it as a social entity centering on the inhabitation of human being within erected borders (Wang Ying et al. 2009). In recent years, as city planning transforms gradually from physical function oriented to public policy oriented, community planning, which enriches its contents, has gained an increasing role in the development of cities (Peel et al. 2007; Sinclair, 2008; Lawson et al. 2010; Zhang Chun et al. 2009; Yuan Ye 2013; Liu Yanli et al. 2014). However, the current theories and practices of city planning of our country inevitably result in community planning's leaning to physical and technical achievements, which apparently leads to a top-down model short of a bottom-up mechanism characterized by social engagement and democratic decision-making. Moreover, problems of wealth inequality and social differentiation appear impending given the macro background of social transformation. Consequently, a reassessment of demands by individuals and their diversifications and a call to construct a system of integral social planning covering housing, healthcare, education, culture, etc., accordingly are on the doorstep.

Individuals differ in age, profession, education, well-being, etc., from which are derived their personalized demands. Therefore, the aged undoubtedly possess both the commonalities of general public and their featured personalities. Growing old, the aged will experience a progressing degeneration of health, sensory abilities, sport skills, etc., eliciting changes in physiology, psychology and social economy, etc. The changes encourage dynamic and diversified social needs from the aged and only when those needs are satisfied can they enjoy a qualified living. From above, we can conclude that age-accommodated integral community planning should draw on the physical, psychological and social changes of the aged and planning strategy needs to be drafted along with their individualized demands on housing, healthcare, security, education, entertainment, workout, etc. Only when those tasks are fulfilled can we create a cozy and familiar living environment for the aged.

2 STRATEGY OF INTEGRAL SOCIAL PLANNING FOR AGE ACCOMMODATED COMMUNITY

2.1 *Providing housing of diversity and multi-level*

2.1.1 *Multi-level proportioning of housing*

Being the most paramount spatial attachment to the aged, housing naturally becomes their primary needs. For the moment, many cities have proposed pension targets of "9073" or "9064", which means 97% or 96% of the aged are going to be sheltered under common roof. On the contrary, the demands for age-appropriate housing, though small at the moment, are rising on a yearly basis. Proportioning of multi-level housing shall strike forward for the goal of fulfilling both the aged group who live independently and those who need extra care, healthcare in particular.

Divergences in well-being, income and education of the aged will without doubt lead to needs for multi-level housing. So the proportioning of different types of housing has to be built on the basis of analyzing hidden influencing elements like the acreage, floor and layout of houses, public service infrastructure, transportation, etc. Through the control of land use, spatial arrangement, construction of public service infrastructure, together with a flexible incentive mechanism to encourage the participation by organizations, private house owner, real estate developer, we can provide the aged with multiple choices for housing.

2.1.2 *Housing security for the aged with low and medium income*

Not only providing affordable housing for the group with low and medium income but also making specialized housing security policy for the aged shall be included in the housing security system. To meet the housing needs of the aged, Hong Kong Housing Authority launched "Public Housing Rental", according to which single elderly persons are eligible to "Single Elderly Persons Priority Scheme", two elderly persons or more who undertake to live together upon flat allocation are subject to "Elderly Persons Priority Scheme" and elderly living with families may choose the "Harmonious Families Priority Scheme". This offers a sound example for the policy-making and supervision of affordable housing in mainland. Priority for the aged, as is stated in the Beijing City Affordable Housing Management Regulation (Trail) that families with members over 60 plus (60 included) eligible for priority when applying, can be traced from eligibility criteria for applying affordable housing in some cities. But such kind of priority stops at the point without detailing available house type.

Overall, we have to draw on the experience of developed nations and districts and put together the government, market and the aged during affordable housing policy making. On one hand, investment in affordable housing needs to add up; on the other hand, policies need to be further detailed and individualized for demands of single elderly, elderly couples and those living with families. And at last, affordable housing should be adapted to the aged with inconvenience. The repairing and adaptation expenses of old houses for the aged with low and medium income should be shared among the government, market and the aged.

2.2 *Maintaining healthy and secure inhabitation environment*

2.2.1 *A healthy and secure land use policy*

Community should put in place a compact and mix land use policy, combining organically the housing, green area and public service infrastructure, to shorten the distance and reduce reliability on public transportation when going out, which encourages walking and bicycle use and then improves elderly well-being. And a mix land use policy will

offset the demands for cars, meaning less carbon emission and cleaner and safer environment for elderly. What's more, it enhances elderly physical and psychological healthiness by providing them with diverse space for communication.

2.2.2 *Healthy and secure housing environment*

Housing environment affects men's health and safety in a stark direct way. A filthy housing environment may be the cause of preventable diseases such as respiratory, neurological and cardiovascular disease and cancer, or may be the safety hazards for physical injury and fire fighting. Therefore, a healthy and secure housing environment requires a healthy composition of building density, FAR, greening rate, layout design, construction materials, etc. However, embroiled in problems of high build density, inadequate sunshine, insufficient ventilation, under-developed affiliate infrastructure and condensed public space, etc., the aged living currently in the old districts of cities are risking their health and safety. Taking those into account, we are obliged to facilitate the elderly adaptable transformation with current housing environment to eradicate health and safety hazards for the aged by erecting handrails in corridors and lavatory, installing emergency call set, putting in place fire extinguisher devices, laying out anti-slippery tiles, improving lighting capacity, etc.

2.2.3 *Healthy and secure transportation*

Healthy and secure transportation environment firstly has to rely on a highly efficient and convenient public transportation system, in which bus stations stand within walking range of the aged. Such a system complies with elderly expectations for convenience and shields them from carbon emission of cars. Secondly, a parallel flexible, accessible and secure pedestrian surrounding is indispensible for healthy and secure transportation environment. This kind of pedestrian surrounding withholds functions of protecting elderly from car pollutions and activating their social interactions.

2.3 *Culture and education for the aged*

2.3.1 *Communities respecting the aged*

A pleasant life for elderly is guaranteed by communities bearing esteem for the aged, which as well bond groups of different ages. In order to cultivate such communities, we have to employ multiple measures. At first, we shall make respecting elderly a merit widely aspired through festivals themed in culture of respecting elderly and by setting examples. Moreover, respect by the younger needs to be evoked by urging them to voluntarily assist the aged, thus breeding mutual respect among generations. Lastly, it's equally essential for elderly self-esteem and improvement. The aged are obliged to establish an image of self-esteem and exuberance compatible with the current society through self-education and social engagement.

2.3.2 *Education and entertainment for the aged*

Enclosed in a scientific curriculum arrangement, which makes a role of importance for elderly education and entertainment, are three essential functions. Firstly it meets with elderly demands for health, leisure, quality of life. For this part, health, calligraphy, dance and gardening courses are recommended. It also includes courses of computer, law, finance, English, etc., to fulfill elderly desires for professional skills and higher education. Thirdly, it consists of a mechanism for elderly engagement both in class and in the arrangement itself. What also grab attention apart from curriculum arrangement are community policies accelerating the build-up of non-profit and commercial entertainment programs to absorb variable requirements from elderly of unpredictable backgrounds.

2.3.3 *Facilities of elderly education and entertainment*

Existing communities are currently tackling problems of inadequate resources for elderly education and entertainment, and poor-conditioned facilities. To effectively alter this situation, two tactics need to be put in place. One is to maximize the utilization of nearby public services of schools, libraries, museums, squares, youth centers. The other is to upgrade the facilities, comprising renovations of education and activity centers and introductions of digital devices and modern technologies, within communities.

2.4 *Flexible community service system for the aged*

Families, communities and professional organizations are equal providers of community services for the aged. Differences of elders' physical and psychological state, income and social status lead to changed expectations for community services: Expending numbers of the aged raises demands; rising income results in the improvement of quality; and accelerating aging adds incentive to business of housekeeping and medical care. Dynamic as elderly demands are, community services should adapt accordingly. With regard to self-reliant elders, who fare well in terms of health, facilities for entertainment, education and sports, which help reignite their passion, are welcome; and to those partly relying on external care, communities should provide them with basic entertainment and medical care, and create an elder care center where they can stay when their families are at work. Concerning the group without ability of caring for oneself, intensive medical care and psychological intervention are expected from communities to ensure their immediate access to professional medical center when necessary.

3 INTEGRATED AGE-FRIENDLY COMMUNITY

Current researches on community planning mostly devote to specific age group rather than that of age integration. To break the status quo of fragmentation by age, together with the goal of realizing integrated age-friendly community, the planning is obliged to figure out the common grounds among different age groups.

As well, community planning advocates cooperation of different age groups. Studies show that the majority of the aged preferred a community with young group instead of one set apart from it. All evidences urge community planners to thoroughly analyze the interrelations and shared demands among kaleidoscopic age groups, creating an atmosphere in which the necessity and importance of an integrated community are presented to all. Meanwhile, filial morality ought to be widely advertised in order to raise awareness for respecting elders and inspire interactions between the aged group and the others.

4 CONCLUSION

Community planning in an aging society has raised our thoughts on allocation of spatial resources among age groups. Benefits have to be proportionately applied to all groups. Planners are tasked to make age integration criteria of community planning and construction, to coordinate contradicting interests and to achieve sustainable development of communities.

ACKNOWLEDGMENTS

Sponsored by Jiangxi Normal University Doctor Fund "Research on Jiangxi urban elderly friendly community comprehensive social planning research, No. 5619)"; Humanities and Social Science Project of Jiangxi Colleges and Universities "Research on Aging-friendly Community Planning and Construction of Small Town in Jiangxi".

ABOUT THE AUTHOR

Address: College of City Construction, Jiangxi Normal University, NO. 99 of ZiYang Road, Nanchang. Jiangxi Province. (330022)
Contacts: Li Xiaoyun
Tel: 18679655198, 0791-88120430
Email: eric812@126.com.

REFERENCES

[1] Lawson, L., A. Kearns. Community engagement in regeneration: are we getting the point?. Journal of Housing and the Built Environment, 2010 25(1): 19–36.
[2] Liu Yanli, Zhang Jinquan, Zhang Meiliang. Community Planning Compilation And Implementation In China. Planners, 2014 30(1): 88–93.
[3] Peel, D., G. Lloyd. Community Planning and Land Use Planning in Scotland. Public Policy and Administration, 2007 22(3): 353–366.
[4] Sinclair, S. Dilemmas of Community Planning. Public Policy and Administration, 2008 23(4): 373–390.
[5] Wang Ying, Yang Guiqing. The Construction of City Community in the Period of Social Transition. Beijing: China Architecture & Building Press. 2009.
[6] Yuan Ye. Evaluation of Community Plan Implementation: An Empirical Study of Public Space in Caoyang Estate in Shanghai. Urban Planning Forum, 2013 207(2): 87–94.
[7] Zhang Chun, Lv Bing. Collaborative Approach to Harmonious Community Planning: Case Study of Jiaodaokou. Urban Studies, 2009 16(4): 101–105.

On the aid of sentence modification to enhance non-English majors' grammatical proficiency in writing

Junqiang Zhao & Shunliang Shi
School of Foreign Languages, Lanzhou University of Technology, Lanzhou, Gansu, P.R. China

ABSTRACT: In the process of college English teaching, it is thought-provoking for the content of grammar to naturally permeate the skill training of reading, listening, speaking and writing, for the boring grammar teaching changed to lively and vivid grammar application to ensure the accuracy of language. This paper aims to explore how to utilize this syntactic modification practice as the specific language output to make the learners realize and understand the more difficult linguistic form, to improve grammar proficiency in writing and pragmatic competence of non-English majors.

Keywords: sentence modification; English writing; grammar proficiency

1 LITERATURE REVIEW

In addition to rich content and tight organization, diction command and sentence structure also play a decisive role in a good piece of article. The same event and the same point of view can be expressed by different sentence patterns. A good piece of article is clear and consistent in argument, integral in content, natural in cohesion, accurate and vivid in language. Even if an article with the simple sentence being excessive, sentence pattern being monotonous, less syntax errors and clearer plots will also be boring to some extent. If it is full of long, complex sentences, it will be very difficult to understand. Based on simple sentences, with the appropriate, diversified compound sentences and complex sentences, the flexible and varied sentence patterns can improve the writing proficiency. If the simple sentences, complex sentences are used alternately to often transform sentence patterns with the proper use of transitional words, the writing will be more coherent.

A common problem that Chinese EFL learners often have in English writing is the monotonous, boring sentence structure, and what the majority of students are lacking in writing is not the systematic writing theory and method, but the most basic sentence writing ability and the exact understanding of vocabulary usage. Therefore, how to practice writing sentences in English is a problem worthy of study. Some scholars emphasized the sentence changes to improve English writing ability from the perspective of English grammar, specifically grammar teaching idea. Chen Dongchun [1] laying the emphasis on English grammar, discussed

the grammar translation teaching approach to detect the master grammar knowledge by writing and text translation. This paper showed that the students' listening and speaking ability, as well as reading ability have improved, while the knowledge of grammar and writing proficiency did not improve, even declined. Ren Lihua [2] talked about the importance and methods of university English grammar teaching, the students often not recognizing the close relation between the mastery of grammar and reading and writing ability. Gao Yuan [3] expressed that without the knowledge of grammar, the meaningful communication is not possible. More knowledge of grammar is needed to cultivate the ability of high-level language comprehension and expression. Li Yuxian [4] believed that there was a more or less positive correlation between different grammar teaching modes and English achievements. Fan Jiaolan [5] asserted that grammar teaching could accelerate the mastery of language structure, with grammar teaching being the most effective method to improve the accurate use of language. From the literature review of the previous research, different scholars have explored the ways and strategies of university English grammar learning from different dimensions, but rarely there is study on how to improve the command of grammar and writing proficiency by rewriting and modifying the sentences.

The diversified and progressive practice mode to rewrite sentences allows learners to get rid of the traditional monotonous, boring grammar exercises. The practice of rewriting sentences is the systematic practice of thinking process. If properly designed, the flexibility in the use of language, the free expression of ideas can be achieved; if designed in haste, the rewriting exercise can become mechanical repetition, or mere the game of words, failing to fulfill the objective of inspiring innovative thinking and training language skills.

2 THEORETICAL BASIS

If the acquisition is a process of information processing, the first condition is the attention of target (i.e., language form). Schmidt [6] puts forward the noticing hypothesis, believing the subconscious learning is not enough to explain the second language acquisition; with attention being the necessary and sufficient condition for intake; without attention, there will be no learning. Semantic understanding does not necessarily lead to acquisition. Students must "discuss the significance of interaction" [7], so that they would pay attention to their lack of linguistic knowledge to attend to language form to enhance the acquisition effect through the sentence reconstruction operation.

Gass [8] proposed a second language acquisition model, explaining how language information is perceived, understood, taken in, and then integrated into the grammar system. There are many factors which influence the acquisition, including morpheme and syntax (morphosyntactic) analysis quality. From her point of view, the higher quality of this analysis, the fuller understanding of language, the more likelihood of the language form being ingested, with the language understanding being a necessary step for acquisition. Gass & Torres [9] believed that, in the input and interaction, for students to gain an understanding of the language, the higher the complexity of the structure, the more difficult it is to understand. In this case, external assistance or external intervention will help them improve acquisition efficiency.

Cai Yun [10] believed that the teaching practice should adhere to the meaning exchange, notice the language form and design, the corresponding input and output operation at the same time, by the explicit analysis of sentence structure to improve the students' understanding of the syntactic structure function and changing rules. The output is to be encouraged to let students repeatedly construct, inspect and correct these structures in order to improve the learning efficiency.

Obviously, in the adult second language acquisition, relying on the semantic understanding is inadequate or inefficient. For the high level of complexity in language form, attention and understanding are very important. According to this understanding, the active intervention should be given to the teaching of writing process, with the students taking initiative to understand the knowledge of grammar and sentence structure in the writing process, enjoying the wonderful transformation, designing appropriate input and output operations to improve writing acquisition efficiency, instead of negatively waiting for natural acquisition to occur.

3 DEMAND ANALYSIS

The training of English writing ability is an effective way to develop students' thinking ability and expressing ability. It is also one standard to measure the effect of English teaching. According to our investigation, most students lack interest in writing, unable to plan their writing, with the outline not listed in the process of writing, without any changes in the composition, the first draft usually being the final draft. The teacher assigns a proposition for students, and the traditional writing mode allows students to learn writing in an isolated environment, the writing content and writing process being often ignored. The teacher is usually the first to analyze the proposition, and then lets the students imitate the model to write an essay. This kind of teacher-centered teaching method makes the students' composition structure and content similar, without more thoughts.

Three types of the traditional writing teaching model each have its own merits, with each other complemented. There are 312 students in the two grades surveyed on the basis of students' expectations for the writing teaching. It can be seen from the statistical results, the moderate students with stable psychological challenges are less sensitive to the university English writing teaching, the data of general satisfaction and satisfaction are nearly flat; while students with better foundation are more clear in attitude, the data for being not satisfied and dissatisfied up to 11%, there being a 13 percentage gap between the general satisfaction and satisfaction, which is the indication for careful thinking of these students. In a word, current situation of writing teaching is not optimistic.

In addition, what were the usual difficulties in writing was investigated. It was found that the most difficult is grammar, vocabulary and sentence structure following closely behind. So many difficulties students encounter in the process of writing can be generally described as confused thinking, lack of vocabulary, poor command of grammar, barren sentences etc. In view of this, it is necessary to study syntactic strategies for diversity, to rewrite the sentences as the starting point, to invigorate the vocabulary and grammar usage, to dredge the idea of writing, to let students experience the writing as a kind of creation, as a kind of art.

4 SYNTACTIC CHANGE STRATEGIES

Syntactic change is closely related to English writing quality. The two change strategies are to be introduced in this section, which aims to explore how to improve the students' ability to effectively deal with the sentences.

4.1 Synonymous sentence transformation

Synonymous sentence transformation practice is to process, transform, change the original sentence, and thereby create a new sentence. This method can cultivate divergent thinking in the course in the flexible use of grammatical knowledge, making writing just like a fish in water, doing a job with the skill and an easy feeling. Synonymous sentence transformation is the best way to practice language use, also for the smooth approach for text comprehension.

For example, the sentence "Tom is a curious boy; he is not only interested in whats but also in whys" can be expressed differently. Some versions are listed below:

1. Tom curious about whats and whys is a boy.
2. There is a boy called Tom being curious who is interested not only in whats but also in whys.
3. The reason why Tom is not only interested in whats but also in whys is that he is a curious boy.
4. Tom is a boy with curiosity, not only interested in whats, but also in whys.
5. Tom is a boy with curiosity, being not only interested in whats, but also in whys.
6. Being a curious boy, Tom is interested not only in whats but also in whys.
7. Tom having interests not only in whats but also in whys is a curious boy.
8. Tom being interested not only in whats but also in whys is a curious boy.
9. Tom who is interested not only in whats but also in whys is a curious boy.
10. As a curious boy, Tom is interested not only in whats but also in whys.
11. Tom is a curious boy, interesting himself not only in whats but also in whys.
12. Not only whats but also whys Tom is interested in, he is a curious boy.
13. Whats and whys are interesting to the curious boy called Tom.
14. To know whats and whys, Tom is interested in everything.
15. To know whats and whys of everything makes Tom a curious boy.
16. It is Tom who is curious about whats and whys.
17. Curiously, Tom is interested in both whats and whys.
18. It is the fact that Tom is not only interested in whats but also in whys that makes Tom a curious boy.

In these sentences, different patterns express the same meaning. Synonymous sentence pattern transformation can truly practice grammar knowledge flexibly. In these sentences, clauses, emphatic patterns, non predicate verbs, preposition phrases, noun clauses are involved. For example, the eleventh sentence, *interesting* is now used as the present participle and the generally it is used as an adjective. This sentence lets a person find something fresh and new. The seventeenth sentence is concise, comprehensive and imposing, with the adverb *curiously* placed at the beginning of the sentence.

4.2 Sentence combination

When writing English sentences in Chinese thinking, sentences written must be monotonous, with the lack of coherence. A series of simple sentences or complex sentences are not idiomatically acceptable. For example, the meaning "this story was very funny, Bill kept laughing while reading" can be generally expressed in English "the story was very funny and Bill kept laughing while reading it" with the two simple sentences connected, but the simple and seemingly fused sentences are in fact loose and lifeless in structure, with traces of Chinese thinking. If the original simple sentences are combined in English thinking like "The story very funny, Bill kept laughing while reading it." or "The story being very funny, Bill kept laughing while reading it." Or "The story was very funny, Bill keeping laughing while reading it." These two sentences being strong, smooth and natural are combined and compact.

For another example, in the sentence "the driver is responsible for the accident. His car knocked down a tree and a man on his bike." The two clauses, in fact, can express the cause and effect logic. If it is written in a way as "The driver is responsible for the accident. His car knocked down a tree and a man on his bike." The basic meaning of the Chinese sentence is passed on, but from the writing point of view, there is still much room for the improvement. If it is written as "The driver being responsible for this accident, his car knocked down a tree and a man on his bike." The combination of the two clauses, the present participle in front of a comma indicates the result. If it is written as "The driver is responsible for this accident, his car having knocked down a tree and a man on his bike." With two clauses combined, the present participle after a comma indicates the reason that he is responsible for the accident. These sentences are better than the use of adverbial clauses expressing the causes.

In the sentence with the meaning "I have a good friend. He regularly sent money to parents, he thought it was the right thing to do", If the Chinese expression meaning to write the sentence "I have an English friend." "He sends money to his parents on a regular basis." "He considers it as a right thing." All these are the basic literal expressions without syntax errors, but it appears malnourished. When the sentences are combined, the writing proficiency can be improved. It can be rewritten as "I have a friend who considers sending money to his parents on a regular basis as a right thing." Or "I have a friend considering sending money to his parents on a regular basis as a right thing." One sentence with the present participle and gerund successively appearing is constructed based on the original three sentences, the ups and downs in the sentence making the reading smooth.

5 CONCLUSION

In the university English teaching, it is a question worth considering that in what way, how much the grammar knowledge will be skillfully processed in language acquisition. Learners with certain knowledge of grammar can practice this specific language output, paying conscious attention to the understanding of more complex language forms, enjoying the pleasure of language learning in the creation and process of writing, enhancing the initiative and enthusiasm in writing. This sentence modification practice can improve the writing proficiency of non-English Majors' mastery of grammar, and realize truly the creative study and application.

AUTHOR INTRODUCTION

Zhao Junqiang (1980.04-), male, born in Tongwei Gansu, lecturer, mainly interested in functional linguistics. Shi Shunliang (1962.11-), male, born in Lanzhou Gansu, professor, mainly engaged in functional linguistics, discourse analysis.

REFERENCES

[1] Chen Dongchun. *On the Status of College English Grammar*. Journal of Tianjin Foreign Studies University, 2004, (5): 73–76.
[2] Ren Lihua. *On the Importance and the Teaching Methods of College English Grammar*. Journal of Changchun University of Science and Technology (Social Science Edition), 2005, (1): 107–108.
[3] Gao Yuan. *Raising Grammar consciousness, Strengthening Grammar Teaching—On the Reform of College English Teaching*. Foreign Language and Literature Studies, 2006, (1): 34–48.
[4] Li Yuxian. *Teaching Model of College English grammar*. Journal of Southwest Agricultural University (Social Science Edition), 2010, (3): 139–141.
[5] Fan Jiaolan. *On Raising Grammar consciousness and Strengthening Grammar Teaching*. Journal of Changchun University of Science and Technology (Social Science Edition), 2011, (4): 173–174.
[6] Schmidt, R. *Consciousness and foreign language learning: A tutorial on the role of attention and awareness in learning*. In R. Schmidt (ed.). Attention and Awareness in Foreign Language Learning. Honolulu: University of Hawaii Press, 1995.
[7] Long, M. *The role of the linguistic environment in second language acquisition*. In W.C. Ritchie & T.K. Bhatia (Ed.). Handbook of Second Language Acquisition. San Diego, CA: Academic Press, 1996.
[8] Gass, S. Integrating research areas: A framework for second language studies. Applied Linguistics, 1988, (9): 198–217.
[9] Gass, S. & M. Torres. Attention when? An investigation of the ordering effect of input and interaction. SSLA, 2005, (27): 1–31.
[10] Cai Yun. *The Influence of Input and Output on Second Language Acquisition*. Modern Foreign Languages, 2009, (1): 76–84.

Education Management and Management Science – Zheng (Ed.)
© 2015 Taylor & Francis Group, London, ISBN 978-1-138-02663-6

Construction of aesthetic education objective in physical education

Chuan Zhou
Department of Physical Education, UESTC, Chengdu, Sichuan, China

ABSTRACT: As a part of education, aesthetic education should be implemented in all stages of education. In physical education, aesthetic education also has its own explicit objective, which is like other kinds of education. Based on documentary method, this paper elaborates the significance, function, basis, frame, and utilization of aesthetic education in physical education. The research result of this paper is as follows: the significance and function of aesthetic education in physical education is attaching the students with the beauty in bodily form, spirit, action, and language; the construction of aesthetic education in physical education is based on history, reality, and theory. In the end, this paper proposes detailed assumption on the construction of aesthetic education in physical education, with stress on the importance and application value of aesthetic education in physical education.

Keywords: physical education; aesthetic education objective; construction

1 INTRODUCTION

The tenet of school physical education is to cultivate talents, aiming to develop the sport consciousness and ability through comprehensive implementation of national educational policy. In this way, students can obtain integrated development in both mind and body and, by means of the combination with moral education, intellectual education, and aesthetic education, students will become the constructor and successor of socialist cause. Physical education permeates in moral education, intellectual education, aesthetic education, and labor education. It is an important platform for aesthetic education. Many components of aesthetic education are contained in physical activities, including the beauty of behavior, language, bodily form, action, etc. Sport is an effective way to promote normal development of human beings, develop correct posture, train coordinated actions, and improve physical fitness. Many sports activities, including catch-up, free-standing exercise, aerobics, rhythmic gymnastics, sports dancing, synchronized swimming, and figure skating are performed to rhythmic music. Some activities, such as track and field, gymnastics, martial art, swimming, and ball games, have no clear rhythm, but strong rhythm sensation. These most beautiful actions are completed despite high speed and difficulty. When Liu Xiang won the applause of millions of people in the 110-meter hurdling, he showed his beauty of bodily form, which is engraved in our impression. The beauty of bodily form always makes so much influence. How do students develop aesthetic sentiment, aesthetic consciousness, and aesthetic method, improve the ability of feeling, appreciating, and expressing beauty, and dwell on healthy body with sound mind? How to achieve educational quality that affects whole physical lessons and the objective of education?

2 SIGNIFICANCE AND FUNCTION OF AESTHETIC EDUCATION IN PHYSICAL EDUCATION

2.1 *Beauty of bodily form*

Beauty of bodily form is presentation of the natural beauty of sports. The growth and development of human body is based on the rule of beauty. Modeling beauty of body, line, and posture, body beauty of muscles, bones, and skin color and vitality beauty of life are manifested in human body. Students of PE major should be armed with shapely bodily form, strong bones, good muscles and healthy skin color. Sitting upright, standing straight, and walking naturally, they should enhance the exercises to make their body healthier and more beautiful and fully energetic, and to make all functions of their body develop coordinately in an all-around way [1]. Mayakovski, a poet from Soviet Union, once complimented copper-colored skin and massy muscle with enthusiastic lines.

2.2 *Beauty of spirit*

Vladimir Lenin said, "A sound mind is in a sound body." Mao Zedong said, "Body is the vehicle of knowledge and the home of morality." These words of wisdom show that the beauty of spirit and the beauty of body are interwoven. The beauty of spirit makes the sport activities brighter and hotter. Sport activities feature progress, competition, confrontation, undertaking, overcoming difficulties and enduring defeat, which are good for developing the ideal beauty of pursuing winning and national honor, the will beauty of courage, persistence, decision, perseverance and modesty, and the emotional beauty of patriotism and internationalism [2].

2.3 Beauty of action

Exquisite skills of sports and beauty of body and spirit enhance each other's beauty, creating the perfect image of athletes. The beauty of skills consists of coordinated actions, good rhythm and agile reaction. Through skills and strategies, these exquisite sports are painted to be appealing pictures, and written about by wonderful poets. For example: ball games show the rapid reaction ability; gymnastics represents coordination and accuracy, and beautiful gesture which is free, graceful, strong, and decisive; body building reflects strong muscle and power; aerobics requires strong and cheerful rhythm; and track and field sports are the basis of various sports, which affect in many ways.

2.4 Beauty of language

Language is the tool for communication. It is the external reflection of spiritual wisdom, cultural quality, morality, and aesthetic taste, as well as the symbol of social civilization. Under no circumstances should we speak rudely or abusively. We should make use of our advantages in the beauty of bodily form and action and professional language, and evoke beautiful feeling by beautiful languages which features simplicity, orderliness, and vividness. In this way, beauty effect is created and language becomes the means of beauty transmission.

3 BASIS OF THE CONSTRUCTION OF AESTHETIC EDUCATION OBJECTIVE IN PHYSICAL EDUCATION

3.1 History basis

In the specialized education organization which appeared in slave society, physical education and aesthetic education are both important contents of school education. In the "six arts" (including rites, music, archery, driving a chariot, learning and mathematics) of the Zhou Dynasty, archery and driving a chariot are of physical

education elements. The second art, music, is a kind of aesthetic education. Since ancient times, whether gifted scholars, elegant young ladies, politicians, militarists, or scientists, they all deem "four arts" important, i.e. lyre-playing, chess, calligraphy and painting their required qualities [3]. The "four arts" are the historical crystal of the Chinese nation's spiritual civilization. During the long history of Chinese, the evolution process of aesthetic education is lengthy. As a young discipline, aesthetics has a history of more than 200 years. However, with the development of science and technology, together with natural science and social science, aesthetics is involved in daily life, application of science, art and sports, creating some new subdisciplines and reflecting characteristics of the time and the nation. Based on the traditional culture of the nation, Confucian school's "ten unkindness against great deeds", Taoist school's "extraordinary refined", and Buddhist "bright and clean" all oversee the life in a very high level, only for clean disposition, free will beyond life. These are traditional aesthetics ideology of the Chinese nation.

Along the river of history, the concept of beauty is expanding and the carrier of beauty is spreading. Today, from the aspect of the form of beauty or the relationship between the subjective and objective of aesthetics, aesthetic education has crossed natural beauty, social beauty, art beauty and science beauty, which is beyond the "four arts". However, no matter how the carrier of beauty expands, people's mind and body are inseparable. Therefore, in physical education, through the special physical method, sports improve people's body in action and posture, and make it strong and vigorous, which creates beauty of strength. With the continuous development of material and cultural level, high level of civilized life is required. In order to meet the aesthetic demand of physical sport, which is to have people enjoy sports and matches like entertainment programs, many sports with strong aesthetic features are created. The construction of aesthetic education objective should spread to the relationships between natural world and scientific world, social beauty and mental beauty, art beauty and methodology. Aesthetic education should be applied to physical education, and vice versa. Aesthetic education should be spread with physical education as carrier.

3.2 Reality basis

In modern society, it is no longer danger that threatens and corrodes the beauty, but "oppression" of material and spiritual civilization. Spiritual civilization level manifests in the personality level. These "oppressions", including variety of mental illness, increasing number of mental patients, psychological imbalance of terrorists, and political situation transformation, result in our soliciting, conquering, and even damaging attitude to the nature; lead to continuously bad relationship among people; and bring about conflict between human and self (internal human and external human), which leads to increasing number of schizophrenia patients. However, the existing art education emphasizes skills rather than arts, which makes it difficult to develop creative intelligence, or to seize the spirit, form, and significance of the connotation to appreciate art on the same wavelength.

3.3 *Theory basis*

In general, aesthetic education is all education that is implemented based on the rules of beauty, rather than the education about certain art skill. The primary task of aesthetic education is to give educators the ability to appreciate beauty. Educators should be equipped with some basic physical education knowledge, sport experiences, life experiences, and cultural accomplishment, and improve their aesthetic accomplishment in practice. They should be able to apply the beauty they feel and extract from the textbook to the education, teach through lively activities, and give education before students notice. It is clear that the origin of aesthetic knowledge in physical education lies in maintaining the rules of beauty in the internal and external unification [4]. Only in this way, emotion and theory, experience and judgment, generality and individuality, aesthetics and health can be unified in the aesthetic activities. Thus, the special function of aesthetic education in physical education is brought to play. In fact, the ultimate objective of aesthetic education is to develop people with perfect personality, namely, the beauty of personality. The beauty of personality is based on the beauty of mind and spirit, and manifests in the beauty in bodily form. The objective of physical education is to cultivate qualified talents with physical and mental health.

4 THOUGHT AND FRAME OF THE CONSTRUCTION OF AESTHETIC EDUCATION IN PHYSICAL EDUCATION

In the aesthetic education, appearance of teachers is the first element that shows beauty to students. Only by enhancing the sense of mission about aesthetic education, as well as becoming the model of beauty, can physical education teachers give full play to their leading role. Therefore, physical education teachers should possess comprehensive cultural literacy and artistic accomplishment no

matter in morality, knowledge, physique, skill, dedication, or temperament. In this way, they could become scholarly mentor and beneficial friends to their students and lead their students to the beauty of idea, action, sentiment, posture, and movement. The second element is the explanation and demonstration in the class. Clear, brisk, and accurate explanation with coordinated actions of teachers will create the feeling of beauty for the students. On the contrary, halfhearted explanation, inappropriate terms, unrefined term explanation, and casual demonstration will hardly bring about the sense of beauty for the students. As a result, for teachers who lack cultural literacy and artistic accomplishment, the beauty of physical education is hard to manifest, let alone the construction of aesthetic education objective. To manifest the beauty of physical education, related aesthetic education objective must be established to accomplish the contents of courses. Therefore, teachers should develop their own good personality and overcome uncivilized action and style.

As an integral part of school education, aesthetic education is interconnected and mutually promoted with moral, intelligence, and physical education [5]. The function of aesthetic education is to develop the ability of feeling and appreciating beauty through aesthetic activities. It is important to apply beauty to physical education. In other words, we should focus on students' training of physical exercises and basic skills, as well as develop moral quality and aesthetic abilities [6]. From philosophy perspective, physical education and aesthetic education objective form a unified whole. Physical education contains aesthetic education objective, while aesthetic education objective is a form of physical education. As Karl Marx says, "all matters are unified, interdependent, and compatible." Aesthetic education objective is an integral part of the teaching objectives of physical education. The beauty of bodily form, language, and spirit can be manifested in physical education. In physical education classes, teachers utilize various teaching methods and styles to show the beauty of spirit and cultivate their students' will and morality. Based on the new curriculum standard, the beauty of spirit is a new style of modern education. Traditional education focuses on the grades, despite the students' will. On the contrary, new curriculum standard emphasizes on spirit accomplishment rather than grades. What we should do is improving grades through enjoyment in the spirit. This is the development tendency of modern education.

Therefore, during the expansion and extension of aesthetic education, the construction of aesthetic education objective aims to follow the idea that "applying beauty to physical sport, laying

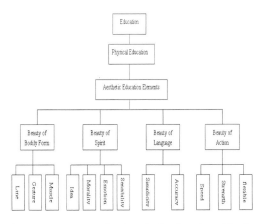

Picture 1. Framework of aesthetic education objective in physical education.

physical education in beauty, sublimating physical education in beauty, and letting beauty blossom in sports". Applying beauty to physical sport means applying theoretical and practical education of aesthetic education in physical education and dissolving aesthetic education in physical education. Laying physical education in beauty means revealing aesthetic phenomena and regulations in physical education, studying on how to raise physical education activities to the aesthetic level and achieve the best effect, and dissolving aesthetic education in physical education. Sublimating physical education in beauty and letting beauty blossom in sports means taking students from physiology level, utility level, morality level, to aesthetic level. Students would forget the self, coordinate mind and body, enrich intelligence, and release potential abilities to achieve beautified personality. The frame of aesthetic education objective in physical education is shown in Figure 1.

5 CONCLUSION

In the construction of aesthetic education objective in physical education, the objective and task of aesthetic education must be defined in the first place. The ultimate objective of aesthetic education is to develop the beauty of personality, i.e. person with perfect personality. The main tasks of aesthetic education are to ensure that the students manage the regulations of beauty, and develop the ability to feel, appreciate, and create beauty. Second, basic knowledge of aesthetics should be managed and correct aesthetic idea and method should be established. For students, 1–3 class hours of aesthetics basic knowledge should be defined in theoretical physical education. For teachers, basic aesthetic knowledge should be included in the training activities as updated information of knowledge structure. Third, take part in practical activities. Aesthetic education targets should be established and various training activities implemented. In this case, sport dancing is one of the best activities that combine physical education with music (aesthetic education). Last but not least, study and develop the beauty in physical education. Special research group should be organized to study the development of aesthetic education in physical education. Some chapters on aesthetic knowledge should be added in the theory lessons to gradually develop disciplinary system of "physical education aesthetics".

REFERENCES

[1] J.H. Tian. 1995. Aesthetic Qualities Required for Physical Education Teachers. *Journal of Yanbei Teachers College* (3):55–56.
[2] D.B. Zhang. 1995. On Physical Aesthetics. *Natural Science Journal of Harbin Normal University* (1):102–104.
[3] X.D. Yang. 2002. Music, Archery, and Driving a Chariot in "Six Arts" V.S. Physical Education and Aesthetic Education. *Nankai Journal* (6):98–102.
[4] Q.H. Wu. 2004. Aesthetic Education Function of Physical Education. *Journal of Wuhan Institute of Physical Education* (1):22–23.
[5] P. Guo. 2002. Physical Education V.S. Moral, Intelligence, and Aesthetic Education. *Physical Education Culture Guide* (1):51–52.
[6] X. Cheng. 2001. Approaches of Aesthetic Education in Physical Education. *Journal of Nanjing Sport Institute* (1):39–40.

Education Management and Management Science – Zheng (Ed.)
© 2015 Taylor & Francis Group, London, ISBN 978-1-138-02663-6

Preparation of lesson plans in higher education

Xiaotao Guan

School of Civil Engineer and Architecture, East China Jiaotong University, Nanchang, Jiangxi, China

ABSTRACT: In the colleges and universities of our country, the role of teaching guidance of lesson plans in the colleges and universities is often neglected; the important link of preparation of lesson plans is ignored. The meaning and importance of lesson plans are introduced in this paper, and it is pointed out that in the environment of new curriculum reform, in order to stimulate the teachers' enthusiasm to teach and the students' initiative to learn, it is important to prepare good lesson plans, play their due role in regular teaching and enhance the teaching efficiency and quality.

Keywords: higher education; lesson plans; preparation; teaching efficiency and teaching quality

1 INTRODUCTION

Lesson plans are usually called teaching designs, and they are practical teaching instruments for teachers to carry out teaching activities smoothly and efficiently, conduct specific design and arrangement concerning teaching content, teaching procedure, and teaching methods by taking class hour or subject as a unit and according to the actual situation of teaching syllabus and textbooks and students. Currently, people's understanding of lesson plans for colleges and universities is mistaken. Lesson plans are considered the patents of secondary schools; it is okay for university teachers as long as they have lecture notes. For some it is even enough to have electronic notes such as "PPT". Many of our university teachers only have lecture notes without lesson plans, and some teachers confuse between lesson plans and lecture notes. And some new teachers simply do not know what lesson plans mean, or the meaning and effect of lesson plans.

2 CONNOTATION OF LESSON PLANS

2.1 Connotation of lesson plans

Lesson plans are the teaching framework carefully designed to achieve the teaching syllabus. They are meant to achieve a certain stage of the expected course objectives for the teacher of different levels, and the students of different majors, during which, the systematic point views and methods are used; the basic rules of teaching process are followed to systematically plan and arrange the teaching activities. That is, the teachers formulate teaching plans of the implementation of teaching for each knowledge point (Group) according to the teaching hour. According to characteristics of physical and mental development of students, curriculum knowledge system, different education teaching requirements, the preparation of lesson plans is also different. Preparation of lesson plans in the schools of primary and secondary education follows the pattern of "small steps, high time-frequency", while

Table 1. Differences between lesson plans and lecture notes.

No.	Lesson plans	Lecture notes
1	Ideas and plans of teachers' teaching process.	Teachers 'lecture notes.
2	Teaching organization and management information is carried.	Knowledge is carried.
3	Formation of thinking is subject to the management logic of teaching process.	Formation of thinking is subject to the knowledge logic in teaching process.
4	Systematical project content is concerned.	Intellectual project content is concerned.
5	They concern how the teacher teaches.	They concern what the teacher teaches.
6	Short length.	Long length.

the preparation of lesson plans in universities or professional schools follows "big strides, low time-frequency" mode. In other words, the lesson plans in universities are decided by the subject, after determining the subject capacity, the teaching design is implemented by teaching students in accordance of their aptitude, flexibly allocate the class hours and promoting it by skipping.

2.2 *Relationship of lesson plans and lecture notes*

Lesson plans are not equivalent to the lecture notes. Both are interrelated but different from each other. Lesson plans and lecture notes are intrinsically linked: the lecture notes are the specific requirements to enrich and refine the lesson plans, and they specify the lesson plans according to the teaching content; the lesson plans and lecture notes are the relationship of deciding and being decided. Differences between lesson plans and lecture notes are shown in Table 1.

3 ROLES OF LESSON PLANS IN UNIVERSITIES

There are four main roles of lesson plans in the teaching process: the first one is the basic program of every teaching to clarify the teaching goal and the use plan of education resources; the second one is the basis of teaching activities, teaching activities must be carried out orderly and effectively according to teaching preparation; the third one is the results of teaching research, lesson plans are the research results of the combination of textbooks, students, teaching methods; the fourth one is the tool to implement the teaching activities, lesson plans in the teaching process is the frame of reference, which can prompt teaching content, focus, difficulties, goals, ideas, help teachers effectively complete each teaching activity.

Lesson plans are the visual results of teaching design, are an important basis for classroom teaching, the guarantee of effective classroom, and the basis for reflection after class. They play a vital role in teaching and management in universities. Efficient and practical preparation of lesson plans can guide teachers to teach, design and normalize the teaching behaviors, improve teachers' professional quality, improve teaching efficiency, and ensure the effect of classroom teaching so as to ensure the realization of the goal of professional talent cultivation.

Optimization of teaching process plays the most direct role in improving the teaching efficiency, and teaching design plays a decisive role in teaching process. Therefore, only optimizing the teaching design, improving teaching methods

can we improve teaching efficiency, the teaching targeted at quality education can give full play to the teachers' role of teaching and guiding, in-depth study of the students' cognitive process, seek the best point of teaching and learning, improve the teaching quality comprehensively.

4 PREPARATION OF LESSON PLANS IN UNIVERSITIES

Lesson plans are teachers' teaching design and ideas. Experienced teachers can use their own lesson plans, but the lesson plan must be constantly enriched, and prove to be feasible through their own use. For some experienced teachers, if the teaching purpose and difficult points of the content taught have not changed, they can write the lesson plan in a simple way. For new teachers, especially the content which has not been taught, they should write the lesson plans in detail. In general, the preparation of lesson plans have the following basic elements: the course name, course type, teaching time, teaching objectives, teaching requirements, teaching focus and difficulty, teaching content and process design, teaching strategies, assignment, teaching evaluation and so on.

In addition, there are some special elements for some special courses such as art classes, physical education, etc. They are teaching grouping, security matters, teaching conditions, etc., and sometimes, the lesson plans are to increase some references or other elements according to the needs, which are regulated as required by each department, the teaching and research section according to the teaching needs of different professionals and different courses.

4.1 *Design of lesson plans*

Design of lesson plans should solve three problems: the first question is what to teach, which is to solve the problem of teaching objectives; the second question is how to teach, which is to solve the problem of teaching strategies in teaching process; the third question how well the teaching is, which is to solve the problem of teaching evaluation. Therefore, the basic content of the design of lesson plans should include the design of teaching objectives, the design of teaching strategies and the design of teaching evaluation. The flow chart is shown in Chart 1.

4.2 *Design of teaching objectives*

Design of teaching objectives refers to the "overall course objectives" to be reflected in the teaching and the objectives in "chapter, section or practice teaching unit, and the expected effect". In the design

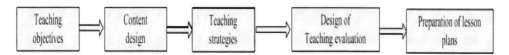

Chart 1. Flow chart of design of lesson plans.

of teaching objectives attention must be paid to three issues. First, in order to avoid the abstraction of the teaching objectives, the course objectives cannot be regarded as teaching objectives. Second, it is needed to have the objectives of "process and method, emotional attitude and values", and the objectives of "knowledge and skills" are not enough, so as to avoid the single function of classroom teaching. Third, the description of teaching objectives should be scientific.

4.3 Teaching content and process design

Teaching content and process design are the core and main parts of preparation of lesson plans. They are to formulate the process of teaching and learning, which includes the lead-in of new lesson, teaching content, teaching process design and so on. They are the sum of knowledge and information of determining the course teaching of the unit by analyzing teaching syllabus, teaching materials and reference data. In the teaching process design, it is needed to show full respect for the creativity of teachers' teaching and learning autonomy of students. It does not have to be standardized.

4.4 Teaching strategies

Teaching strategies refer to the teaching methods (lectures, demonstrations, experiments, implementation, discussions, case analysis, simulation or real live implementation guidance, etc.), teaching assisting means (teaching aids, models, charts, real objects, modern teaching facilities and equipment, and special teaching environment, etc.), teacher-student interaction, blackboard design in accordance with the teaching objectives. It mainly focuses on effectively arousing the enthusiasm of students, promoting positive thinking of students, stimulating students' potential to design. Higher education focuses on learning methods, situation creation, issue guidance, media usage, feedback control and other teaching strategies, and special attention should be paid to the dynamic generation and real-time flexibility of teaching.

4.5 Design of teaching evaluation

Teaching evaluation includes formative and summative evaluation. The classroom teaching design

mainly uses the former, which is a process evaluation, aiming at timely diagnosing and adjusting the "teaching" and "learning" between teacher and students so as to achieve teaching objectives; the latter is mainly suitable when teaching activities are done in a teaching period (such as a large teaching unit, one semester or one academic year), which is suitable for macro teaching design. Summative evaluation is the reflection on the teaching design of all aspects after the teaching. It is to summarize the success and failure of the teaching design and implementation, and the teachers analyze and summarize the teaching design, teaching focus and difficulties, the application of teaching methods, the teaching effect and other classroom teaching processes after finishing the teaching in the unit, which provides experience and teaching materials for the future.

4.6 Key points of preparation of excellent lesson plans

Lesson plans are brief and orderly records for the lesson content, and they are the samples of teachers to teach in class. In simple terms, lesson plans are the teachers' memorandums. A good lesson plan should have the following characteristics: scientificity, innovation, artistry, and operability. How to prepare good lesson plans? There are two main points to prepare good lesson plans: first, arouse the enthusiasm of teachers to write lesson plans, improve teaching efficiency and teaching plan should be different depending on person, course, teaching content. It is needed to break the fixed, rigid model of traditional lesson plans, to write personalized, innovative lesson plans. Second, the writing of lesson plans should reflect abstraction of the content, fuzziness of the form, and uncertainty of the structure. It is needed to remember that the content should not be in greater detail, the form should not be too trivial, the structure should not be too closed and stylized, so that the lesson plans can adapt to the new situation, accommodate the new content, establish the new strategy, leave some leeway for the interactions and resonance between teachers and students, the generation of new knowledge, the establishment of new situation. Simply speaking, lesson plans should set aside some white space. In short, lesson plans should have openness and flexibility so that they form a special kind of "tension" between the lesson planning and classroom teaching, which is conducive to keeping a broader way of thinking and open-up concept for teachers in teaching, more easily to be integrated into the new content, adapt to the new situation, change the original design, and realize the ecological classroom teaching.

Writing of lesson plans is a basic skill for teachers, and is the basis for the teachers to teach. Every teacher should pay attention to the lesson plans so as to prepare excellent lesson plans, improve teaching ability, teaching efficiency and teaching effect.

REFERENCES

[1] Cao Haibin, Shao Jianxin, Hou Juan. Brief Discussion on How to Write Lesson plans in Colleges. "Science and Technology Information", 2009, (33).
[2] Wang Hongmei. The Lesson Plans are Necessary Prerequisites for Teaching. "Academy Education", 2013, (18).
[3] Zhang Yexu, Fu Guiyan. How to Write Lesson Plans. "Jilin Pictorial" (Education of Hundred Schools B), 2014, (1).
[4] Huang Shaoyuan. Education Guide and Teaching Plans—On Lesson Plans in Universities. Journal of east China Jiaotong University, 2006, 23(z1).

Education Management and Management Science – Zheng (Ed.)
© 2015 Taylor & Francis Group, London, ISBN 978-1-138-02663-6

The analysis of external factors about sports marginalization of migrant workers in China

Luojing Zhu
College of Physical Education, Central South University of Forestry and Technology, Changsha, China

ABSTRACT: The present paper analyzes external factors about sports marginalization of migrant workers. The four dimensions as follows: the enterprise, sports administration, social forces and relative laws and regulations. The paper shows that there are four factors: 1) The municipal government can't afford to provide adequate public sports facilities for migrant workers. 2) The enterprises of China have no intention of provide sports facilities for migrant workers. 3) The relevant laws don't effectively ensure sports right of migrant workers. 4) Moral, cultural and the comparatively weak non-governmental organization constrain sports right protection about migrant workers.

Keywords: migrant workers; marginalization; external environment factors

1 INTRODUCTION

After every Spring Festival, hundreds of millions of farmers enter the cities in China. These farmers are called migrant workers. According to data of National Bureau of Statistics of China migrant workers amounted to 262.61 million people in 2012 [1]. They hope to find job in the cities because urban incomes are higher than rural incomes. Objectively speaking, migrant workers have made an important contribution towards China's urban development. Migrant workers, however, are faced with embarrassing situations. On the one hand, government highly appreciates migrant worker's important contribution for urban development; on the other hand, most of migrant workers are still at the bottom. Most of them have no sense of belonging in the city because of household registration system and dual social security system caused by the separation structure between urban and rural. In the social realm including the world of sports, migrant workers face painful experiences of rejection and marginalization. Sports marginalization of the migrant workers is a hot topic in the present the sociology of sports in China. This paper attempts to analyze the external factors of sports marginalization of migrant workers. There are four external factors as follows: the enterprise, sports administration, social forces and relative laws and regulations.

2 EXTERNAL ENVIRONMENT FACTORS ABOUT SPORTS MARGINALIZATION OF MIGRANT WORKERS IN CHINA

2.1 *The employers have no intent to provide sports welfare for the migrant workers*

The survey by author shows that migrant workers having no consciousness of sports right is the first constraint (29.2%). This reflects the lack of consciousness of rights and interests of migrant workers. There are two constraints being tied for second. One of the two is that the company system doesn't involve sports right of workers. This reflects the union being disadvantaged in the company in China. Another constraint is that the superior leaders don't require lower organizations to provide sports service. The third constraint (16.7%) is that the Labor law and the Trade Union law of PRC do not have relevant clause about the protection of sports rights. Meanwhile, the Sports law has no direct legal binding effect to enterprises (see Table 1).

The author, however, found that there are other constraints. For instance, many migrant workers don't join the union. In addition, there are multi-level contracting and illegal sub-contracting existing in the construction industry. There aren't direct labor relations between migrant workers and enterprises because of multi-level contracting.

Table 1. The survey of constraints of enterprises providing migrant workers with sports service coming form union officials. Data based on the survey of 8 enterprises in China.

Constraints of enterprises providing migrant workers with sports service	Percentage
Migrant workers have no consciousness of sports right	29.2%
The company system don't involve sports right of workers	25.0%
The superior leaders don't require lower organizations provide sports service	25.0%
The Labor law and the Trade Union law of PRC have not relevant clause	16.7%
Enterprise cost	4.2%

2.2 The municipal government can't afford to provide adequate public sports facilities for migrant workers

1. The municipal government can't afford to provide adequate public sports facilities because of the limit of city's public sports resources.

The urban public service resources can't meet population growth. In fact, the growth rate of urban population is often faster than the growth rate of public sports resources. On the one hand, the proportion of urbanization increased from 17.4% to 53.7% between 1978 and 2013, according to China statistics yearbook [2]. By 2020, the proportion of urbanization will be 60% at such a high pace. At the same time, the city population increased by 300 million people. Among them, the migrant workers comprise about 2/3 of the increasing population [3]. On the other hand, data from the statistical statement of Fifth National Sports Venues Investigation show that per capita sports venues areas are 1.03 m². Since 1995, growth has averaged 5.92% a year. And 91.82% of sports venues are distributed in the city [4]. The data shows that per capita sports venues area has a certain increase. However, we can't ignore the fact that sports venues of National Education System take 65.6% of all of our country's sports venues. Moreover, sports venues of National Education System are closed to the public. In other words, per capita actual area of sports venues is far less than the statistical statements.

As the rate of population increase is much faster than growth rate of public sports resources, the public sports resources is more and more serious as more and more migrant workers are entering the city. With the shortage of the public sports resources, how do the municipal government meet both the demand of public sports resources of native citizens and migrant workers? That is one of the objective factors of restraining migrant workers from sharing public sports resources.

2. The funds investment proportion of the executive branch of sports is not reasonable.

The executive branch of sports attaches great importance to athletics sports, and neglects sports for all in China. The public welfare funds of sports lottery is one of important sources of public sports funds. But, the allocation of public welfare funds of sports lottery is not reasonable. For example, Article 19 according to the Interim Measures on public welfare funds of sports lottery: General administration of sport of China is in charge of the allocation of the public welfare funds of sports lottery. 60% of total funds are to be spent into sports for all. 40% of total funds are to be spent into "the program of striving for Olympic glory" [5].

I always question the allocation principle. The nature of sports lottery is public welfare. The public welfare funds of sports lottery comes from hundreds of millions of lottery players. As a usual rule, the public welfare funds should come from people and be used for people. It's not unfair that general administration of sport of China allocate 40% of total funds for tens of thousands of national team. After all, the national team obtains large amounts of funds through media sponsorship, corporate sponsorship and government appropriations, etc.

2.3 The relevant laws don't effectively ensure sports right of migrant workers

1. The Labor law and the Trade Union law of PRC don't effectively ensure basic right of migrant workers From the point of view of Maslow's hierarchy of needs theory [6], it will constrict the higher level of requirement—sports requirement of the migrant workers because the Labor law and the Trade Union law of PRC can't effectively protect their low level of requirement about physiological needs and security requirements. For example, having enough leisure time is not a sufficiency but an essential condition of sports participation. Unluckily, migrant workers undergo work long hours with high labor intensity everyday. Article 36 according to the labor law of PRC: The state shall implement a working hour system under which laborers shall work for no more than eight hours a day and no more than 44 hours a week on the average [7]. Nonetheless, data show that 84.4% of migrant workers work for more than 44 hours per week [8].

Besides, according to Article 16 of the labor law of PRC [9]: A labor contract is the agreement reached between a laborer and an employing unit for the establishment of the labor relationship and the definition of the rights and obligations of each party. A labor contract shall be concluded where a labor relationship is to be established. In reality, only 43.9% of the migrant workers signed labor contracts in a recent official survey [10]. It means that more than half of the workers couldn't obtain protection from laws.

2. PRC's sports law lack of operability in the field of protection sports rights of the migrant workers.

The concrete performances are as follows: To begin with, the sports law of PRC was drawn up as basic laws of sports in 1995. Great changes have taken place in the field of social sports since 1995 in China. The well-worn laws can't adapt to a new situation. Secondly, the clauses of sports law are too general. There are many fuzzy words in the sports law such as enhance, encourage and increase, so on. It looked more like guideline document. Finally there is no specific law enforcement to supervise implementation of sports right. For example, the survey shows that sports grounds area about 245 million square meter are encroached upon [11]. However, we don't see the sports administration doing anything.

2.4 Analyzing of social factors about influencing sports rights of migrant workers

1. Social exclusion of urban resident to migrant workers still exists.

According to a blue paper released by the Chinese academy of social sciences in 2012 [12]: on the one hand, migrant workers solved the problem of labor shortage in the city. On the other hand, large number of migrant workers entering into the cities led to scarcity of public resources in urban areas. Meanwhile, it led to discontent of some urban residents. The survey shows that the attitude of the urban residents and rural residents has certain disagreements in many aspects.

2. The migrant workers hardly find recognition of cultural identity because of conflict between the modern sports of city and the traditional sports of country.

Nowadays, modern sports are very popular activities in the city. And modern sports contain some modern elements. In contrast to modern sports, the migrant workers from rural areas tend to like folk-custom sports. The folk-custom sports are created by the farmers. It is both physical culture and life culture. This kind of folk-custom sports is rooted in the daily behavior and the faith of rural people. The collision of cultures will occur when the migrant workers enter city after the Spring Festival. Therefore, the migrant workers hardly find recognition of their cultural identity.

3. The labor union organizations and NGOs fail to effectively protect the sports rights and interests of migrant workers because of their status as appendage in China.

Theoretically, the labor union organizations and NGOs could fill the gaps when the relevant laws go unenforced and government's functions are often absent. In the real world, however, there are many factors restricting the supervising capability. On the one hand, the labor union organizations and NGOs can't care about sports right protection because they are busy with other rights protection, such as back pay, work conditions, living conditions and migrant worker's children education. On the other hand, the labor union organizations and NGOs fail to effectively protect the sports rights and interests of migrant workers because of their status as appendage in China.

3 CONCLUDING REMARKS

The sports right protection of the migrant workers relies on not only improving the external environment but also improving their own quality, enhancing rights protection consciousness and increasing the degree of migrant workers' organization in China. The sports right protection of migrant workers can't be guaranteed unless the above conditions are met. Perceived at a profound level, the urban-rural dual structure is the institutional cause of the problem.

ACKNOWLEDGMENTS

1. The research is supported by Social Science Fund of Hunan Province. (Grant No. 13YBB228).
2. The paper is supported by Research Fund of Central South University of Forestry and Technology. (Grant No. 2013YB07).

REFERENCES

[1] National Bureau of Statistic of the People's Republic of China. Monitoring report of migrant workers 2012. http://www.gov.cn/gzdt/2013–05/27/content_2411923.htm, 2013-5-27.
[2] The head of national bureau of statistics answer to reporters' request about the situation of national economy operation in 2013. http://www.stats.gov.cn/tjgz/tjdt/201401/t20140120_502414.html, 2014-01-20.

[3] Heli. McKinsey & Company forecast that migrant workers could make up half the population of city. http://www.canadaren.Com, 2008–03–28.

[4] General Administration of Sport of China. The Statistical Bulletin of The Fifth Sports Facilities Survey. http://www.sport.gov.cn/n16/n1167/n2768/n32454/134749.html, 2005–2–8.

[5] General Administration of Sport of China, Ministry of Finance, People's Bank of China. Interim Measures of Sports Lottery Welfare Fund Management, 1998–9–1.

[6] Maslow's Need Hierarchy Theory. http://baike.baidu.com/view/690053.htm, Baidu Encyclopedia.

[7] The President of the People's Republic of China No. 28. Labor law of the People's Republic of China. 1995–1–1.

[8] National Bureau of Statistic of the People's Republic of China. Monitoring report of migrant workers 2012. http://www.gov.cn/gzdt/2013–05/27/content_2411923.htm, 2013–5–27.

[9] The President of the People's Republic of China No. 28. Labor law of the People's Republic of China. 1995–1–1.

[10] National Bureau of Statistic of the People's Republic of China. Monitoring report of migrant workers 2012. http://www.gov.cn/gzdt/2013–05/27/content_2411923.htm, 2013–5–27.

[11] General Administration of Sport of China. The Statistical Bulletin of The Fifth Sports Facilities Survey. http://www.sport.gov.cn/n16/n1167/n2768/n32454/134749.html, 2005–2–8.

[12] Ruxin, Xueyi Lu, Peilin Li. Blue Book of China's Society. Social Sciences Academic Press. 2011–12–19.

APPENDIX

Table A. The list of the chairman of the labor union of respondents.

The enterprise surveyed	Position of respondents	Types of enterprises
Hunan province food limited company	The Chairman of the Labor Union	Food
The fifteenth bureau of China Railway Construction Corporation	The Vice-chairman of the Labor Union	Construction
The fifteenth bureau of China Railway Construction Corporation	The Chairman of the Labor union	Construction
Branches of SANY	The Chairman of the Labor Union	Manufacture
The third branches of CCCC second harbour engineering company LTD	The Chairman of the Labor Union	Construction
Changsha BAHUAN Food Company	Manager, The Chairman of the Labor Union, Concurrently.	Food
Guangxi CHENGYUANFU Building Materials Limited Company	Manager, The Chairman of the Labor Union, Concurrently.	Building Materials Industry
Changsha Xincheng Composite Material Limited Company	Manager, The Chairman of the Labor Union, Concurrently.	Manufacture

Table B. The list of the chairman of the labor union of respondents.

The enterprise surveyed	Position of respondents	Types of enterprises
The second Branches of China Railway Tunnel Co., Ltd	Project manager, Secretary, concurrently.	Construction
Hunan Province XIANGPAI Food Limited company	Manager, The Chairman of the Labor union, concurrently.	Manufacture
The third Branches of the Fifteenth Bureau of China Railway Construction Corporation	Project Manager	Construction
Changsha MINZHI Food Company	Manager	Manufacture
Shanghai Tunnel Engineering Co., Ltd	Project Manager	Construction
The first Branches of the nineteenth Bureau of China Railway Construction Corporation	Project Manager	Construction
The second Branches of the eleventh Bureau of China Railway Construction Corporation	Chief of Electromechanical	Construction
Guangxi CHENGYUANFU Building Materials Limited Company	Manager, The Chairman of the Labor union, concurrently.	Building materials industry
Changsha Xincheng Composite Material Limited Company	Manager, The chairman of the labor union, concurrently.	Manufacture
Changsha Bahuan Food Company	Manager, The chairman of the labor union, concurrently.	Food

Education Management and Management Science – Zheng (Ed.)
© 2015 Taylor & Francis Group, London, ISBN 978-1-138-02663-6

Networked movie courseware theoretical exploration and development model research

Xiaoping Li, Yongliang Hu, Qiong Xu, Zhenghong Wang & Lin Zhang
Beijing Institute of Technology, Beijing, China

Xiaojun Wang
Beijing University of Posts and Telecommunications, Beijing, China

ABSTRACT: The traditional multimedia courseware are often limited to the courseware formats, they always lack some flexibility. Meanwhile, teachers have high dependence on the traditional multimedia courseware, the use of which led to less effective benign interaction with students. Thus, we explore the proposed networked movie courseware theory, which is an organic product of educational technology, interactive television and ninth art and other related theories. This theory can effectively solve the interaction problems between teachers and students as well as greatly improve the students' enthusiasm. Then, we explored to propose the networked movie courseware development architecture, while proposing the specific developing process. At last, we described its application prospects in detail. This theory can greatly integrate existing online instructional resources and make effective secondary design of it, so that it ensures the effective reuse of resources and improve the utilization as well.

Keywords: online education; movie courseware; instructional design

1 INTRODUCTION

In the rapid development of educational information technology and new media, the new instructional ideology has been gradually formed and promoted the birth of new problems and challenges in resource constructions and online education fields. In the traditional classroom, most instructional methods and achievements can be reused in the next classroom teaching and the teacher always stood in the leading position. But if we transplant the traditional classroom teaching mode to online education, it may not be able to get a better teaching effect. The reason lies in the online education students are much more complex, it's difficult to grasp their learning psychology as well as effectively control each student's knowledge understanding degree, teaching methods selected by teachers became much more subjective and unilateral.

Meanwhile, the traditional multimedia courseware lacks effective interaction and feedback mechanism with the students. Although the text, movie and other transmission ways have extended and expanded the propagation mode of teaching information, but the traditional multimedia courseware instructions have not changed its passive position during the students' learning process.

From the teacher's perspective, multimedia courseware usually acts as teaching tools. On the one hand, they can help to improve teaching effect; on the other hand, the teacher's instructional contents are always constrained by courseware's content and format, which lacks flexibility and impromptu artistic creativity. From the students' point of view, they just regard the multimedia courseware as a teaching tool. They can only passively receive information while can't effectively participate in the instructional media's formation and revision process. So, it's hard to form effective feedback between them.

Meanwhile, the traditional instruction methods lack the equal relationship between teachers and students. It emphasizes too much on the teachers' subjective position, ignoring the students' enthusiasm of constructing knowledge. Although the traditional teaching way transfers an amount of information, having a strong controllability, but the teacher-centered learning process have paid more attention to students' knowledge and skills, ignoring their interests, abilities and life experience. The one-way information disseminations have caused the students to think less, failing to teach them in accordance with their aptitudes. As American educational psychologist Lindgren said, "Education, as the other social process, its effectiveness depends on communication." Constructivism learning theory[2] also thinks that knowledge is useful only after learners' accomplishment of construction. Therefore, to improve the teaching effect needs us to fully mobilize the enthusiasm, initiative and creativity of learning as well as unify teachers' leading role and students' subjective role.

In view of the above problems, we explored proposing networked movie courseware theory[6]. The theory is formed under the framework of educational technology, using movie technology, game skills and other technologies comprehensively to realize the optimization of teaching resource construction. The theory is intended to maximize the online teaching effect with smaller cost. Through the use of the learners' own experience and snooping psychology, we take the method of classification design for different learning groups, and analyze the learning groups as well as research deeply on instructional environment design, combining the education psychology of cognitive learning theory and constructivism learning theory. We explored the innovative ideas of inter-disciplinary and cross-domain, while taking the cognitive views in the film theory[8], conflict theory as well as suspense design into instructional design, promoting the integration of innovation. By means of new media instructional ways, we can complete the transposition between online education and traditional classroom under the new instructional ideology. Thus, we converted the target design into film expectation, taking empathy conversion into the teachers' image design as well as adding the game theory into the physical teaching. Finally, we can make full use of the students' curiosity to fulfill knowledge exploration.

Through the application of the visual-centered cultural atmosphere, we can change the learners' sense and experience. The learning process can be completed in a slightly delightful atmosphere by the use of the beautiful scenes, stunning visual effects and ingenious plots. This method can establish the new observation and excitement points which are different from that of the traditional instructional resources. The immersive learning experience can be provided by introducing the ninth art, high-tech means, 3D virtual environment and 3D characteristic models. It needs us to further discuss the confluent innovation of the teaching practice and the ninth art, meanwhile, emphasizing on the learning method of simulating real world experience in the learning process. We need to let the students participate in the instructional design process actively, while formulating a multi-level, multi-angle, multi-role, multi-function optional situational learning model with the help of interactive storyline, role and task switching and other interactive forms.

2 NETWORKED MOVIE COURSEWARE SYSTEMATIC ARCHITECTURE

2.1 Theoretical basis

Just as science and technology are the necessary conditions for the birth of movie art, both movie art and educational technologies are the important components of movie courseware, while the former is expressiveness, the latter is the means and carrier. The renewal of movie education ideas will follow the times and the realization will change comprehensively from ways to the connotations. The following theories displayed in Figure 1 played an important role in forming the networked movie courseware theory.

The concept of movie courseware is formulated on the basis of movie art media theory, educational technology theory[3] and other related theories[1,4,5,7]. It connects the essence of movie art, while expressing intuitive visual modeling, rich sound and picture languages, figuratively content displaying in education forms. Meanwhile, it extends the manifestation of movie art to different disciplines and expertise areas. For the differences between different disciplines, it proposed the design philosophy of "liberal of engineering, dramatic of liberal arts."

2.2 Application platform construction

The application platform of networked movie courseware is intended to complete the second design on the basis of traditional curriculum design. It is the technical basis of implementing networked movie courseware teaching, while acting as a teaching system containing environment

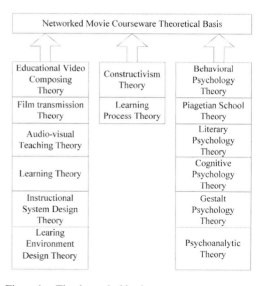

Figure 1. The theoretical basis.

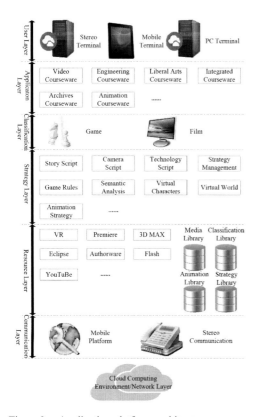

Figure 2.　Application platform architecture.

navigation means, you can get the virtual rewards together with formative assessments so that to arouse the students' interests.

Through the integration of new teaching techniques, we need to deeply explore the law of instructional resources construction. With the newest high-tech as the guideline, we should enhance the overall design of the networked instructions under the engineering background, starting the overall planning of networked boutique resource constructions in an international perspective, formulating a multi-angle, multi-role, multi-function optional scene-based learning model, showing constant, practical, dramatic, mobile, movie interactive features, constructing an opening instructional environment for the purpose of culturing students' knowledge, skills, intelligence, emotions comprehensively. For the curriculum composition, there is no need to stick to one pattern, which does not say what kind of plate and features it must have, you can independently grasp the courseware structure according to the characteristics of each course under the basic instructional resource principles. At the same time as achieving the instructional targets, we should have a certain ability to interact with students so as to complete the formative assessments. Its general development framework is shown in Figure 3.

3.2　Development process research

The basic development process of networked movie courseware is shown in Figure 4, which contains

design, process design and technical design. The application platform utilized layered architecture. From bottom to up are the network layers, communication layer, resource layer, strategy layer, classification layer, application layer and user layer, achieving a logical division and physical implementation stratification as shown in Figure 2.

3　NETWORKED MOVIE COURSEWARE DEVELOPMENT MODEL

3.1　Design path research

In the networked movie courseware theory, we proposed the 3s of design philosophy: *surprise*, *suspend* and *satisfaction*, completing "no conflicts thus no drama, no psychological expectations thus can't complete the instructional objectives, no suspense thus can't reach network study effect" classic design idea. Through the integration of the ninth art theory, you can use the motivational strategies in the electronic game theory to arouse students' learning interests, while developing networked virtual simulation experiments to improve their practice experience. With the help of appropriate

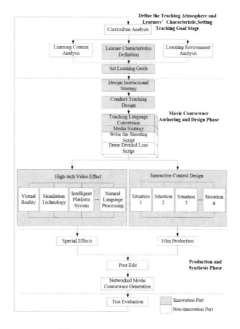

Figure 3.　13磅 development framework.

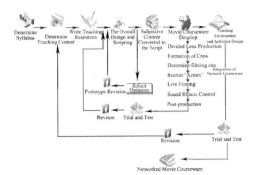

Figure 4. The basic developing process.

determined curricula, instructional contents and other aspects like the traditional courseware development, also including the unique part of storyboard production, recruitment of "actors" and so on. Here, we assume that teachers have in-depth understanding of the students' learning characteristics, and the course has formed mature syllabus and knowledge architecture.

4 NETWORKED MOVIE COURSEWARE APPLICATION VALUE

First, we carried out a bold interdisciplinary integrated innovation with the help of movie art producing technology. We solved the connection between the new media and the traditional networked instructional design through the integration of the learners' personal experience, taking the movie empathy thoughts to complete curriculum secondary design, which makes the ordinary course classic and artistic. Use the movie's tension and impact force to adjust teaching rhythm, while trying to achieve the effect of an excitement each two minutes as well as a turning point each fifteen minutes, then we could tightly and loosely complete the course. With the use of movie pragmatic approach and artistic means, you can successfully complete the instructional goals. The course can be an artwork by knowledge points' construction and sublimation. The learning process has transferred into the process of enjoying new media art, which completed the transformation of the online education concept.

Second, we made a bold attempt on the students' self-experiences and spy psychology, completing building a unified model for different levels, occupations, fields learners by using classified design method for different people, which is based on the education cognitive rules and knowledge acceptance rules in the network. Facing the complex anonymous audience, just dividing the learners by the education level in the instructional design would lead to the courseware's single function, so

that it's hard to reach the goal of lifelong and comprehensive learning. Then, we need to quantify and refine the traditional teaching, fully using students' self-experience and spy psychology, taking classification method, defining the teaching atmosphere and learner's characteristics and instruction targets as well as effective learning path for different people. Integrating interactive scene instructional design and self-experiences, formulating diverse learning strategies, completing instructional down-to-earth design, completing the character empathy, so as to expand featured instruction service.

5 CONCLUSIONS

We proposed the networked movie courseware and the design principle of "liberal of engineering, dramatic of liberal arts" have been intended to take full advantage of a variety of high-tech means to complete the whole human knowledge repetition, completing human knowledge reproduction on the internet. It needs us to make full use of the internet advantage to realize knowledge reconstruction. This will promote the emergence of new discipline classification in media instructional design, which will greatly promote the development of education. A new theory and model from a proposal to maturation is not an easy thing, it must go through countless practice and theoretical improvement. In the future, we will further improve and refine the networked movie courseware theory, further design more distinctive online movie courses, further study the networked movie courseware platform development. At the same time, let this new theory be more widely applied to various fields of social education and have a profound impact on the development of education technology.

REFERENCES

[1] Butzin, S.M. 2004. Project CHILD: A proven model for the integration of computer and curriculum in the elementary classroom. *Asia—Pacific Cyber education Journal* I: 1.30.
[2] Fosnot, C.T. 2013. *Constructivism: Theory, Perspectives, and Practice.* Teachers College Press.
[3] He, K.K. & Li, W.G. 2009. *Education Technology.* Beijing: Beijing normal university publishing group.
[4] Huang, H.L. 2007. *Movie Audience Theory.* Beijing: Beijing normal university publishing group.
[5] Gagne, R.M., Wager, W.W., Golas, K.C., Keller, J.M. & Russell, J.D. 2004 (5th ed.). *Principles of Instructional Design.*
[6] Li, X.P. 2013. *A Textbook on Networked Movie Courseware.* Beijing: Beijing institute of technology press.
[7] Prensky, M. 2001. *Digital Game-based Learning.* New York: McGraw-Hill.
[8] W. Schurian. 1998. *Movie Psychology.* Chengdu: Sichuan people's publishing house.

Education Management and Management Science – Zheng (Ed.)
© *2015 Taylor & Francis Group, London, ISBN 978-1-138-02663-6*

The influence of emotional intelligent and adversity quotient on job motivation: An evidence case from insurance company

Mian Lin
Asia University, Wufeng, Taichung City, Taiwan, P.R. China

Xing Li
Longdong University, Xifeng, Qingyang City, Gansu, China

ABSTRACT: Insurance companies relate to individual's life in society. In order to promote job motivation, insurance company employees should keep adversity quotient and emotional intelligence at a high level. Theory suggests that individuals who are both higher in emotional intelligence and adversity quotient are likely to exhibit higher performance. This paper studies the influence of emotional intelligence and adversity quotient on job motivation. It utilized the questionnaire, the research results show that relationships among adversity quotient, emotional intelligence, and job motivation were statistically significantly correlated. Insurance companies should strengthen adversity quotient and emotional intelligence, which will change job motivation for better.

Keywords: emotional intelligent; adversity quotient; job motivation; insurance company

1 INTRODUCTION

Under fierce competition, insurance companies are facing many problems; the most important one is the high employee turnover rate. In the worst cases, employers attrition rates approaches 40 per cent (Huang, 2010), the largest proportion of employee turnover rate is the position of insurance salesman. Sales staff are not only working longer hours than others, but also working under more pressure. The finding shows the insurance salesman always has high rate of employee turnover. Thus, the insurance salesman should have a higher degree of adversity quotient and emotional intelligence. These two factors should be the necessary personal traits.

A decade ago, Paul Stoltz (1997) introduced a new yet interesting & intriguing concept-Adversity Quotient (AQ), which tells how well one withstands adversity and his ability to triumph over it. In fact, more researches recently have shown that measurement of AQ is a better index in achieving success than IQ, education or even social skills. Therefore, the people with high AQ will have a stronger ability to change the outcome when they are facing difficult environment. Results pointed out that the higher adversity quotient value has the stronger resistance to pressure at work. To measure AQ, Stoltz developed an assessment instrument called

Adversity Response Profile (ARP). It was divided into control, ownership, reach, and endurance. So, the study is based on the four dimensions to investigating AQ.

Emotional intelligence (Coleman, 2008) can be defined as the ability to monitor one's own and other people's emotions, to discriminate between different emotions and label them appropriately, and to use emotional information to guide thinking and behavior. Daniel Goleman (1998)introduced a model, focuses on EI as a wide array of competencies and skills that drive leadership performance. Goleman's model outlines five main EI constructs: Self-awareness, Self-regulation, Social skill, Empathy, Motivation. Thus, this research depends on analysis of those five dimensions.

The Work Preference Inventory (Amabile, 1994) is designed to assess individual differences in intrinsic and extrinsic motivational orientations, the WPI has meaningful factor structures, adequate internal consistency, good short-term test-retest reliability, and good longer term stability. Herzberg, Mausner and Snyderman (1959), analyzed the foundations of job motivation based on motivation and hygiene factors. In conclusion, job motivation mainly covers the intrinsic and extrinsic two aspects. This study discusses both dimensions.

Based on the previous literature review and background, this study focuses on the insurance company salesman, according to the empirical analysis to explore the relationship among adversity quotient, emotional intelligence and job motivation, Implications for managerial practices and future research are discussed.

2 THEORY BACKGROUND AND HYPOTHESIS

2.1 *Adversity quotient*

Dr. Paul Stoltz (1997) defines Adversity Quotient as "the capacity of the person to deal with the adversities of his life. As such, it is the science of human resilience." He developed an assessment instrument Adversity Response Profile. Control dimension means that those with higher adversity quotient have significantly more control and influence in adverse situations than do those with lower adversity quotient. Even in the situations that appear overwhelming or out of their hands, those with higher adversity quotient find some facet of the situation they can influence. Those with lower adversity quotient respond as if they have little or no control and often give up. Ownership dimension means higher adversity quotient hold themselves accountable for dealing with situations regardless of their cause. Those with lower adversity quotient deflect accountability and most often feel victimized and helpless. Reach dimension means higher adversity quotient keep setbacks and challenges in their place, not letting them infest the healthy areas of their work and lives. Those with lower adversity quotient tend to catastrophize, allowing a setback in one area to bleed into other, unrelated areas and become destructive. Endurance dimension means higher adversity quotient have the uncanny ability to see past the most interminable difficulties and maintain hope and optimism. Those with lower adversity quotient see adversity as dragging on indefinitely, if not permanently.

2.2 *Emotional intelligent*

Goleman's (1998) model outlines five main EI constructs: Self-awareness means the ability to know one's emotions, strengths, weaknesses, drives, values and goals and recognize their impact on others while using gut feelings to guide decisions. Self-regulation involves controlling or redirecting one's disruptive emotions and impulses and adapting to changing circumstances. Social skill means managing relationships to move people in the desired direction. Empathy means considering other people's feelings especially when making decisions.

Motivation means being driven to achieve for the sake of achievement.

2.3 *Job motivation*

Deci (1985) found that giving people unexpected positive feedback on a task increases people's intrinsic motivation to do it, meaning that this was because the positive feedback was fulfilling people's need for competence. Extrinsic motivation refers to motivation that comes from outside an individual. The motivating factors are external, or outside, rewards such as money or grades. These rewards provide satisfaction and pleasure that the task itself may not provide. The Work Preference Inventory (Amabile, 1994) is designed to assess individual differences in intrinsic and extrinsic motivational orientations, aim to capture the major elements of intrinsic motivation (self-determination, competence, task involvement, curiosity, enjoyment, and interest) and extrinsic motivation (concerns with competition, evaluation, recognition, money or other tangible incentives, and constraint by others). The instrument is scored on two primary scales, each subdivided into two secondary scales. WPI scores are related in meaningful ways to other questionnaire and behavioral measures of motivation, as well as personality characteristics, attitudes, and behaviors.

From the arguments mentioned above, we specify the following hypotheses to focus our empirical investigation.

Hypothesis 1: Adversity quotient is positively related to job motivation.

Hypothesis 2: Emotional intelligence is positively related to job motivation.

Hypothesis 3: Adversity quotient is positively related to emotional intelligence.

Hypothesis 4: Adversity quotient and emotional intelligence had significant effect on job motivation.

3 METHOD

3.1 *Sample and data collecting*

This research adopts the investigation method of the questionnaire. Insurance salesmen from Zhe Jiang province in Mainland China were investigated on principle of convenience sampling to understand how adversity quotient and emotional intelligence influenced the job motivation in insurance industry. A total of 210 questionnaires are used in this study, recycling of 160, recovery 76.19%, excluding 47 invalid questionnaires, 113 valid questionnaires, with the available rate of 70.63%.

3.2 Measurement

Adversity quotient: In our investigation, we applied Short Form Adversity Response Profile (ARP) Questionnaire developed by Stoltz. This measurement, consisting 40 items, was a self-report scale. All items ranged from 1 (strongly disagree) to 5 (strongly agree). The Cronbach's alpha for this scale was 0.87.

Emotional intelligence: The twenty-five items measuring emotional intelligence as developed by Goleman (1995), also combine the model from Salovey and Mayer (1990), and the response format was a 5-point scale, ranging from 1 (strongly disagree) to 5 (strongly agree). The Cronbach's alphas of emotional intelligence was 0.89.

Job motivation scale: The scale system according to the Herzberg, Mausner, & Snyderman (1959) two factors theory and reference the Work Extrinsic and Intrinsic Motivation Scale (WEIMS) is an 18-item measure of work motivation theoretically grounded in self-determination theory (Deci & Ryan, 2000), and the response format

was a 5-point scale, ranging from 1 (strongly disagree) to 5 (strongly agree). The Cronbach's alphas was 0.85.

4 RESULTS

Through the statistical analysis results, as Table 1 below shows, was significantly related to adversity quotient, emotional intelligence and job motivation. In other words, if insurance salesman's adversity quotient level is higher, the emotional intelligence performance is better, and helps to boost insurance sales work motivation.

To further explore the prediction of job motivation variables, use multiple regression analysis method for insurance salesman. Tables 2–4 show that, on the whole, the adversity quotient and emotional intelligence are the prediction of job motivation index. Individually speaking, emotional intelligence "social skill" and "self-awareness" is to predict the intrinsic job motivation variables; secondly, adversity quotient "social skill" and

Table 1. Descriptive statistics and product-moment correlation among variables.

Dimensions	1	2	3	4	5	6	7	8	9	10	11	12	13	14
Control														
Ownership	0.43**													
Reach	0.35**	0.38**												
Endurance	0.41**	0.25*	0.53**											
AQ	0.73**	0.63**	0.73**	0.75**										
Self-awareness	0.21	0.23*	0.22*	0.22*	0.26**									
Self-regulation	0.15	−0.03	0.04	0.21*	0.13	0.15								
Motivation	0.16	0.16	0.19*	0.23**	0.24**	0.32**	0.34**							
Empathy	0.23**	0.21**	0.13	0.23**	0.28**	0.63**	0.28**	0.53**						
Social skill	0.24**	0.16	0.16	0.33**	0.33**	0.63**	0.64**	0.61**	0.73**					
EI	0.25**	0.18	0.18	0.36**	0.33**	0.63**	0.61**	0.73**	0.83**	0.86**				
IM	0.21*	0.01	0.13	0.15	0.19*	0.41**	0.23**	0.28**	0.33**	0.50**	0.46**			
EM	0.22*	0.11	0.13	0.30**	0.29**	0.21**	0.03	0.24*	0.23**	0.30**	0.26**	0.56**		
JM	0.23*	0.07	0.19*	0.30**	0.30*	0.31**	0.13	0.24*	0.23*	0.50**	0.46**	0.84**	0.86**	

Notes: AQ = adversity quotient, EI = emotional intelligence, IM = intrinsic motivation, EM = extrinsic motivation, JM = job motivation $p < 0.05$* $p < 0.01$**.

Table 2. The results of multiple stepwise regression analysis: adversity quotient and emotional intelligent predicting intrinsic job motivation.

Predictive variables	R^2	ΔR^2	β	F	t
Social skill	0.23	0.23	0.47	33.71**	6.11**
Self-awareness	0.15	0.35	0.34	18.76**	5.61**

Notes: $p < 0.05$* $p < 0.01$**.

Table 3. The results of multiple stepwise regression analysis: adversity quotient and emotional intelligent predicting extrinsic job motivation.

Variables	R^2	ΔR^2	β	F	t
Social skill	0.08	0.08	0.33	13.21**	3.31**
Endurance	0.08	0.15	0.32	10.56**	3.61**

Notes: $p < 0.05$* $p < 0.01$**.

Table 4. The results of multiple stepwise regression analysis: adversity quotient and emotional intelligent predicting job motivation.

Variables	R^2	ΔR^2	β	F	t
Emotional intelligent	0.18	0.18	0.43	23.26**	4.83**
Adversity quotient	0.06	0.25	0.22	5.16**	2.71**

Notes: $p < 0.05$* $p < 0.01$**.

"endurance" are to predict the extrinsic job motivation index.

The results showed that adversity quotient positively predicted job motivation significantly; emotional intelligent positively predicted job motivation significantly; adversity quotient is positively related to emotional intelligent; adversity quotient and emotional intelligent had significant effect on job motivation. We can see that hypothesis of H1, H2, H3 and H4 are supported by empirical data.

5 CONCLUSION

The purpose of this study is to explore the relationship among adversity quotient, emotional intelligence and job motivation of Chinese insurance salesman. Through questionnaire survey and data statistics result, an important conclusion was summarized by the study. There is a significant positive correlation of adversity quotient, emotional intelligence and job motivation. In addition, the emotional intelligence of "social skill" and "self-awareness" support have significant effect on intrinsic job motivation; adversity quotient "social skill" and "endurance" have significant effect on

extrinsic job motivation. Overall, adversity quotient and emotional intelligence have significant predicting effect on job motivation. Accordingly, insurance salesman should enhance the level of adversity quotient and emotional intelligence, to strengthen job motivation. This study suggests that to enhance insurance salesman adversity quotient, promote emotional intelligence level, to improve work motivation and promote the insurance business performance.

REFERENCES

[1] Amabile, T.M., Hill, K.G., Hennessey, B.A., & Tighe, E.M. (1994). The Work Preference Inventory: Assessing Intrinsic and Extrinsic Motivational Orientations.
[2] Coleman, Andrew (2008). *A Dictionary of Psychology* (3 ed.). Oxford University Press.
[3] Deci, E.L., & Ryan, R.M. (2000). The "what" and "why" of goal pursuits: Human needs and the self-determination of behavior. *Psychological Inquiry*, 11, 227–268.
[4] Goleman, D. (1998). *Working with emotional intelligence*. New York: Bantam Books.
[5] Herzberg, F., Mausner, B., & Snyderman, B.B. (1959). *The Motivation to Work*. New York: John Wiley & Sons.
[6] Huang, LL. (2010). *A Narrative Analysis of the Insurance Broker*.
[7] J.D. Mayer, P. Salovey. What is emotional intelligence P. Salovey, D.J. Sluyter, Eds. *Emotional Development and Emotional Intelligence*. New York, NY: Basic Books, 1997:3–31.
[8] Salovey, P., & Mayer, J.D. (1990). Emotional intelligence. *Imagination: Cognition and Personality*, 9(3), 185–211.
[9] Stoltz, P.G. (1997). *Adversity Quotient: Turning Obstacles into Opportunities*, Wiley.

Education Management and Management Science – Zheng (Ed.)
© 2015 Taylor & Francis Group, London, ISBN 978-1-138-02663-6

Break through the traditional model and create a new way of teaching

G.Z. Hou
Army Aviation Institution of PLA, Beijing, China

X.Z. Wang & J. Wu
North China Institute of Science and Technology, Sanhe, Hebei, China

ABSTRACT: Aiming at the existing problems in computer basic education, some reform measures are put forward from teaching preparation and teaching implementation based on the deep research on the reform of computer basic teaching. Thus we will break through the traditional model and create a new way of teaching.

Keywords: computer basic education; teaching reform; new way of teaching

1 INTRODUCTION

With the high-tech industry booming in today's world, the network process advances rapidly. From a social perspective, the process of social informatization accelerates and the computer ability requirement of the employee is generally improved. From the education situation, computer education has been set up in primary and secondary schools. It is urgent to integrate the professional course teaching in the higher education and computer technology. And the network teaching environment has been popularized. With the penetration of information technology into the social life, computer literacy has become one of the important signs to measure students' work ability and an important part of employment competitiveness.

In this case, computer basic teaching is facing new opportunities and new challenges. The society puts forward new requirements for computer basic education in university. The teaching content and teaching means of the original computer basic course in university have been unable to meet the demands of the new epoch development, which perform mainly in the following aspects.

1. The computer basic education has the characteristics of being relatively stable, which makes computer basic teaching lag behind the development of computer technology.
2. The students' computer level is uneven. How to solve the needs of students at different levels and how to select and organize teaching contents are the problems to be solved in the present computer basic teaching.
3. The existing assessment methods cannot fully mobilize the enthusiasm of the students, don't

make full use of the network resources and multimedia technology. The assessment of students is limited to the assessment of "master". And it is unable to effectively check "application".
4. The teaching process makes the teacher as the center. Students are busy memorizing the knowledge. And it is difficult to form the systematic knowledge structure, which hinders the students' innovation ability training.

Obviously, the existing teaching methods cannot meet the requirements for students in modern social development. The reform of computer basic teaching [1–3] is imperative. Thus, we will break through the traditional model and create a new way of teaching. The reform of computer basic teaching can be conducted from the following two aspects.

2 TEACHING PREPARATION

In the teaching content selection, the computer basic teaching needs to orient to application and pay attention to being practical. The content of computer basic teaching cannot copy the teaching content of computer professional, and is not a simple cut, which makes the teaching materials of computer basic teaching become the simplified edition of computer professional materials. We should give full consideration to the characteristics of teaching object and potential demand for occupation. And we select teaching content conceptually and expansively. For example, for accounting major students, teachers can expand the conventional teaching contents of Excel in preparing the teaching content in order to enable students to adapt to the use of Excel in financial data processing in a short time. For students of graphic design professional, teachers should focus on the graphics processing software in the preparation of teaching content. Teachers should not only explain the using method of the software representative, but also explain the relevant content in processing advanced graphics. In a word, in order to prepare suitable teaching contents, break through the limitations of syllabus firstly. Don't let the syllabus become a new block for students to acquire new knowledge and technology. Secondly, break through the limitations of textbooks. Textbook is the basic tool for teaching. A textbook can only focus on one direction and one field. The teaching process is a complex process. As the promoter of this process, a teacher cannot be bound by a textbook. Although it cannot be widely quoted, teachers should know the sequence of events and have one's words at hand. Teachers should be able to select the teaching contents for specific audiences from so many sources. Thirdly, break though the limitation of thinking. The current trend of computer teaching is the training of the students' application ability. Therefore, the teaching content is determined by the objective of training the practical ability of students in the teaching preparation. But how to carry out this process is not set. The teacher should benefit by mutual discussion. The collective preparation for lessons should be conducted for each knowledge point and each chapter. And the optimal teaching, contents are discussed so as to avoid a blind situation. The above description is only for teaching preparation. In the real teaching implementation, we should also follow the following aspects.

3 TEACHING IMPLEMENTATION

A teacher is taken as the center in the implementation of traditional teaching process. The thinking of students can only follow the teacher passively. In the new era, we should change the teaching mode. Take the students as the center, take the teachers as the guidance, and pay attention to skills training. Specifically, follow the below mentioned points.

3.1 Arouse the interest of students and guide the exploration learning

American cognitive psychologist Bruner pointed out that the best stimulation to study is the interest of learning materials. The best way to make the students interested in a subject is to make the students feel it is worth learning. There is a close relationship between student interest in a subject and his achievement in this subject. The interest is the internal driving force for study. Therefore, teachers

should fully mobilize the learning enthusiasm of students, improve student interest in learning, encourage students to think actively and ask questions, guide the students to find a solution to the problem, make the students as the subject of study to construct their own knowledge structure in the process of teaching. In this process, the teacher should guide the students to explore the basic principle in the practicing process of solving the problem. The teacher plays a great role of a guide in this process.

3.2 Develop practice method and increase practice strength

The teaching of computer basic course is to cultivate computer application ability of students. This ability is not listening, not looking at, but practicing. A variety of computer extracurricular activities, such as computer lectures, knowledge contests, program design contests etc., should be carried out actively. The open teaching mode of second classroom is constructed. These activities are not limited by the class, which is in favor of the cultivation of students' personalities and abilities. By making the students participate actively, they will have a sense of accomplishment, be interested in the course, think and apply the knowledge initiatively, so as to achieve the purpose of strengthening the memory.

3.3 Select teaching content and optimize curriculum structure

At present, the college students are more or less having some computer application foundation. In this context, if we continue to teach computer introduction, Windows operating system, Office operation, Internet basic knowledge and use, it will reduce the personnel training specification, waste valuable learning time of students, depress learning enthusiasm of students. Therefore, teachers must consider many factors comprehensively and choose proper teaching content. In order to make students receive a better understanding of the teaching content, teachers can test the computer skills of students at the beginning of the class. According to the test results, students will be divided into different learning groups. Then, the teacher proposes different learning requirements according to different groups during the class.

3.4 Integrate of the school resources and create perfect practice environment

These reform measures can open a new window for the computer basic teaching at present. But in order to make the related reform measures achieve good results, the colleges and universities need to provide a good hardware support for the teaching professional, so as to meet the needs of computer courses for different levels and different professionals and take this as the foundation and promote the network examination system vigorously. The system includes student examination system, teacher evaluation system and database management system. This system takes a lot of question bank, examination papers and user information as data records to store. It's easy for teachers to produce a paper from the database randomly. The teachers can conduct online examinations at any place and time. And the scores, statistics, analysis and evaluation can be got immediately. On the other hand, for the students who need comprehensive practice, the school should make use of not only internal resources but also external resources and provide the opportunities of the curriculum design and graduation practice for students. Let the students contact the real work environment and further improve the students' practical ability.

3.5 Introduce certification system of computer industry

The development trend of the present computer education is combining occupation training with daily teaching. From this point of view, computer rank examinations related to national or computer ability tests set by other well-known enterprises are introduced into computer teaching in universities [4–6]. They are taken as evaluation mode of related curriculums, which can make the students learn with clearer goals, so as to improve the students' internal driving force of autonomous learning. On the other hand, this method of evaluation can also save time and economic cost consumed by occupation ability training on students.

4 CONCLUSION

As a place where we can understand the unknown world and explore the objective truth, the university undertakes the task of cultivating and training of creative talents. Therefore, computer knowledge and application ability will be taken as an important part in the personnel training plan of university. In order to fulfill this task, the reform of computer basic education is in there and has one's finger on the trigger. In this paper, we propose some ideas related to computer reform. We will put these ideas into practical teaching process. The practice is to verify the effect of these reform measures, so as to break through the traditional model and create a new way of teaching.

ACKNOWLEDGMENTS

This work was financially supported by the Fundamental Research Funds for the Central Universities (3142014087, 3142013098, 3142014125, 3142014096, 3142014085), Natural Science Foundation of Qinghai Province of China (2012-Z-935Q, No.2012-N-525), Key Prevention and Control Technology of Major Safety Accidents in Production (2013).

REFERENCES

[1] Fan, M.Z. 2010. Practice and thinking on the reform of basic computer teaching. *Computer Education* (1): 90–92.
[2] Huang, X. 2011. Discussion on the teaching reform of computer base in university. *Journal of Hunan University of Science and Engineering* 32(4): 59–61.
[3] Han, G.H. 2014. Research on the reform of basic computer teaching. *Software Guide* 13(2): 162–164.
[4] Hu, P., Fu, C.J. & Li, M. 2011. Thinking on the computer grade examination and computer basic teaching reform. *Chinese out of School Education*: 167–168.
[5] Ling, Y. 2010. NCRE and computer foundation teaching reform of higher education. *Computer Education* (21): 138–141.
[6] Huang, Z.G. 2013. Exploration on the teaching reform of computer base based on the status of computer grade examination. *Science and Technology Information* (23): 54–82.

Education Management and Management Science – Zheng (Ed.)
© 2015 Taylor & Francis Group, London, ISBN 978-1-138-02663-6

Teaching experience on the stability design of portal frame

Zaihua Zhang
College of Civil Engineering, Hunan City University, Hunan, China

ABSTRACT: The thesis has introduced the teaching practice and experience of steel frame stability in light portal frame design course teaching process in detail, which tells the concept of effective cross section from the view of lightweight structure design and analyzes the stabilization of steel frame components from the view of the structure overall stability. It has also introduced some background knowledge of related formula, which can make the students more quickly master the basic concept of portal frame stability and clarify the idea of studying the steel frame stability design.

Keywords: light-weight portal frame; stability analysis; teaching experience

1   INTRODUCTION

After the design principle of steel structure course completion, civil engineering specialty in our university sets the curriculum of steel structure design for students majoring in direction of housing construction engineering as a required professional course, 2 credits with 32 class hours. The curriculum focusses on the design of steel roof truss, the light portal frame structure, multi high-rise steel frame structure and space grid structure. Among them, the design of light portal frame structure is generally taught eight class hours, including three major parts of the structure type and layout, steel frame design and envelope structure design. Of these, the related concepts and calculation of portal frame beam-column stability design is an important and difficult content in the course's teaching, where the calculation formula is complex and the corresponding concept and calculation are various. So, how to teach well is a very critical factor for the student to master the portal frame structure design.

Combined with the author's teaching practice of the steel structure design, this thesis introduces and summarizes some teaching experience of the steel frames stability design problems during portal frame structure design teaching.

2   INTRODUCE THE CONCEPT OF EFFECTIVE CROSS SECTION FROM THE PERSPECTIVE OF STRUCTURE LIGHTWEIGHT DESIGN

Portal frame is a kind of light steel structure and there are two approaches to realize lightweight design. One is to reduce the weight of the secondary components; the other is to make full use of the bearing capacity of the main structure. For portal steel structure, H-type section is the common section form of main structure beam-column. For the flexural member of H-type section, the bending moment generated by load mainly is borne by the flange and the web mainly bears the shear. Increasing the height of the web can significantly improve the flexural capacity of flange, so it is easy to get the best economic effect to adopt the bigger proportion web of height (width) and thickness. But it may also make height (width) and thickness proportion of web exceed the limit required by the local stability defined by the small deflection theory, which causes the local buckling of web. The flange can form reliable constraints to web as long as flange plate doesn't appear buckling at this time, which makes the web plate of convex surface produce film tension effect (as shown in Fig. 1a). At this time, the section whole is not a failure. Meanwhile, it also has the ability to continue to carry, and the section borne load is the effective section. Adopting effective section to design fully utilizes the bearing capacity of materials and well realizes the lightweight design.

At present, the method of direct strength about component buckling analysis hasn't spread. The effective section analysis method is the main analysis method of design, so it is a basis of portal steel structure design to establish a clear concept of effective section and correctly calculate the effective section and geometric parameters of section. The following points of effective section can be easily obtained from the introduction of above concept:

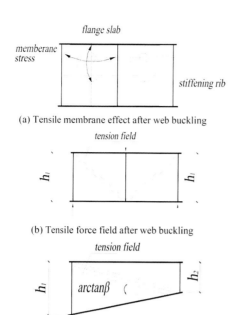

(a) Tensile membrane effect after web buckling

tension field

(b) Tensile force field after web buckling

tension field

$arctan\beta$

a

(c) Tensile force field after non-equal altitudes web buckling equal altitudes web buckling

Figure 1. The tensile force field's formation after web buckling.

2.1 The premise of usage of effective section

Effective section is formed based on the concept of tension field after web buckling[1]. It refers to after buckling of web plate for solid web steel girder, the tension direction of web forms into tension field, which anchors in the upper and lower flange and both sides of web stiffeners, and makes the girder into truss form and continues to bear the load. In order to form tension field, the flange plate should be able to provide very good constraints on web, so it is not allowed buckling. That is to say that the free outer up width-to-thickness ratio of flange plate should meet $b'/t_f \leq 15\sqrt{235/f_y}$. Figures 1b and 1c show the tension field distribution of web with equal and unequal altitude, respectively. We can find that, when the web's altitude is unequal, the dipangle of tension field will lower and the height changes greater the dip angle smaller, that the ability of resistance to shearing force of web is lower. So we suggest that only when $\beta \leq 0.06$ can the web with unequal altitude use the intensity after shear buckling. That is to say the variation of web height changes along the axis direction cannot exceed 60 mm per meter. Meanwhile, if the component bears alternating load, the convex direction of web may change alternately, which is not conducive to form stable tension field. So the Standard (CECS102-2002)[2] provides that only components under static load are considered

to use the effective section design. In addition, from the perspective of construction, if the height of some web is too high, it is hard to make the web flat, which will bring the increase of web deflection and is a disadvantage of the stability of web. Thus, the web's largest ratio of height and thickness is stipulated not exceeding $250\sqrt{235/f_y}$.

2.2 Objects of effective section usage and confirmation of effective section

From Figure 1a we can see that the effective section is formed after external convex curve produced by compressive stress of web beyond a certain value. That is to say that only the web of compression zone has the concept of effective section; if the web is tensioned, the area is certainly effective, so the concept of $h_e = \rho h_c$ is obtained. h_e is the effective height of web at press section, h_c is the height of web of compressive zone, ρ is the effective height coefficient of web at press zone. So we can establish distribution of effective height of web as shown in Figure 2.

For the figure, $\sigma_2 = \beta \sigma_1$, the web buckling properties can be concluded from the web's stress distribution character. If the compression buckling stability coefficient k_σ and height thickness ratio parameter of web λ_p is introduced, the effective height coefficient ρ of web can be defined:

$$k_\sigma = \frac{16}{\sqrt{(1+\beta)^2 + 0.112(1-\beta)^2} + 1 + \beta} \quad (1)$$

$$\lambda_p = \frac{h_w/t_w}{28.1\sqrt{k_\sigma}\sqrt{235/f_y}} \quad (2)$$

$$\rho = \begin{cases} 1 & \lambda_w \leq 0.8 \\ 1 - 1.9(\lambda_p - 0.8) & 0.8 < \lambda_w \leq 1.2 \\ 0.64 - 0.24(\lambda_p - 1.2) & \lambda_w \geq 1.2 \end{cases} \quad (3)$$

If the effective section height h_e of web is concluded, the distribution of effective section can be confirmed from the following formula:

$$\begin{cases} h_{e1} = 2h_e/(5-\beta); & h_{e2} = h_e - h_{e1} & \beta \geq 0 \\ h_{e1} = 0.4h_e & h_{e2} = h_e - h_{e1} & \beta < 0 \end{cases} \quad (4)$$

(a) compression on the total cross-section
$\beta > 0$

(b) compression on the portion cross-section
$\beta < 0$

Figure 2. Web effective height distribution.

3 ANALYZING THE STABILITY DESIGN OF COMPONENT FROM THE ANGLE OF STRUCTURAL OVERALL STABILITY TO DEEPEN THE UNDERSTANDING OF THE OVERALL STABILITY ANALYSIS METHOD

Portal frame structure is a whole composed of all kinds of components, whose stability analysis is to determine the buckling critical load of the whole structure. The current design adopts the method of approximation, which transforms the stability of the steel frame into the columns, But the overall consciousness is necessary in the process of component stability analysis.

When we talk about the stability of the components for the design theory, components are isolated uniform section bars. The factors which have effect on overall stability of them mainly include the load and section type, the initial defects, calculation length, of which calculation length is mainly confirmed by the constraint condition of rod end, the arrangement of bar support. For the beam-column stability analysis of portal steel structure, the bars analysis should be considered on the whole system of steel bar.

For the stability analysis of portal frame column, the calculation length analysis in plane generally has look-up table method, the first and second order analysis method; the look-up table method usually focuses on in the class. Look-up table method is applicable to the single span of hinge of the column bottom or multi-span portal frame with swing column, in this case, the swing column bearing axial force is designed to uniform section column. Meanwhile, the bending bar bearing bending moments applicable of change situation of bending moment is designed to wedge section. For the calculated length coefficient of wedge section column bending moments effecting in-plane, the constraint conditions of inclined beam of column top to column should be analyzed, namely the bending stiffness of beam-column and the relative ratio relation should be clear. In general, the oblique beam is designed to wedge section according to enveloping graph shape of bending moment. At this time, the flexural rigidity specification of cant beam is considered according to equivalent section beam (shown in Fig. 3)[3]. The height of equivalent beam adopts the small section height of oblique beam, calculation length of beam has become ψS, ψ is the length calculation factor of oblique beam ($\psi \leq 1.0$), corresponding oblique beam line stiffness is represented as $K_2 = I_{c0}/2\psi S$. For the steel frame column, the line stiffness is determined by the big head section of column in the standard form ($K_1 = I_{c1}/h$), the plane calculation length of steel wedge column is looked up in

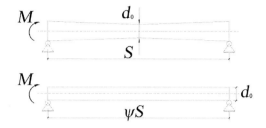

Figure 3. In-plane calculation length of oblique beam.

the table based on the relation of the line stiffness ratio of beam-column and section stiffness ratio of big and little head of column.

$$\frac{N_0}{\varphi_{xr}A_{e0}} + \frac{\beta_{mx}M_1}{\left[1 - \varphi_{xr}\left(N_0/N'_{Ex0}\right)\right]W_{e1}} \leq f \qquad (5)$$

$$\frac{N_0}{\varphi_y A_{e0}} + \frac{\beta_t M_1}{\varphi_{by}W_{e1}} \leq f \qquad (6)$$

After having confirmed the calculation length of wedge section column, CECS102-2002 provides that the stability calculation in-plane of column is made according to formula (5). The most characteristic of calculation expression is that the bending moment and axial force, respectively, adopt the internal force and sectional characteristic of big and little head rather than the internal force and sectional characteristic of same section. From the analytical expressions, the section characteristics corresponding to axial force are confirmed based on small head section, and section features corresponding to bending moment are confirmed based on big head. It is nondescript seemingly that the stability analysis method in the same formula, the axial force items are subject to little heads while the bending moment items are subject to big heads. But if you understand these from the concept of whole stable, you will find it is completely necessary that both big and little heads are adopted when the wedge components are calculated: the stability calculation of variable cross-section column is the overall stability calculation of components rather than the section calculation, if it is the completely elastic ideal straight-bar, φA is same for big and little head. But because the steel is elastic-plastic body, considering the component with residual stress and geometrical defects, and the second order effect, the smaller slenderness the bearing capacity reduces more calculated according to big head. That is to say the little head $(\varphi A)_0$ is less than the big head $(\varphi A)_1$, so it is safer that the first item (the axial force) is calculated according to small head relative to big head. Certainly, the second item's

181

(bending moment) M_1 and W_1, respectively, adopt the maximum bending moment and the effective section modulus of the plane of bending moment.

Stability analysis of bending moments out-of-plane of wedge section steel column adopts the analysis expression (6). In the expression, N_0 is the axial pressure design value of the small head, M_1 is bending moment design value of big head, β_1 is the equivalent bending moment coefficient. For out-of-plane stability analysis according to expression (6), the calculation length and the equivalent bending moment coefficient are the two key parameters. The analysis expression has two stability factors φ_y and φ_{by}, the coefficient of calculation length is grasped from the corresponding buckling mode: φ_y is the stability coefficient of axial compression members bending moments out-of-plane. For the little head, the corresponding calculation length l_y is the space between column Out-of-plane supports, φ_{by} is the bend torsion overall stability factor. The corresponding calculation length should, respectively, consider the effect of free torsion and constraint torsion of wedge section, so μ_s and μ_w, the two calculation length coefficient are introduced to consider separately:

$$\mu_s = 1 + 0.023\gamma\sqrt{lh_0/A_f} \tag{7a}$$

$$\mu_w = 1 + 0.00385\gamma\sqrt{l/i_{yo}} \tag{7b}$$

Considering the effect of free torsion and constraint torsion of wedge section, bending and twisting overall stability factor of components are determined according to the following formula:

$$\phi_{by} = \frac{4320}{\lambda_{y0}^2} \cdot \frac{A_0 h_0}{W_{x0}} \sqrt{\left(\frac{\mu_s}{\mu_w}\right)^4 + \left(\frac{\lambda_{y0} t_0}{4.4 h_0}\right)^2} \left(\frac{235}{f_y}\right) \tag{8}$$

For this formula, special attention should be paid to λ_{yo}, it is the slender proportion confirmed by the winding y-axis turning radius of the section composed by compression flange and compressive web 1/3 height.

4 CONCLUSION

Light type portal steel frame is one of the most widely applied steel structure form at present, whose stable design involves a lot of concepts and calculation formulas, which is a big difficulty in design teaching of steel structure. Having some background knowledge of related formula can help students to understand the nature of problems so as to better grasp the application of formulas. Based on this, the author puts forward some experience according to recent years' teaching practice hoping to offer some help to the course teaching of portal frame structure design.

REFERENCES

[1] Youquan Chen, Chaowen Wei. Design and interpretation for construction problems of portal frame light building steel structure. China building industry press, 2009, 08.
[2] Technical specification for steel structure of lightweight buildings with gabled frames, Beijing, The Standardization Institute of Chinese Construction, 2003.
[3] Shaofan Chen, Stability calculation of light steel structure variable cross-section portal steel frame. Building structure, 1998, Vol 8th pp: 10–14.

Education Management and Management Science – Zheng (Ed.)
© *2015 Taylor & Francis Group, London, ISBN 978-1-138-02663-6*

The market concentration research of jujube processing industry in Hebei province

Wei Chen & Liang-liang Wang
School of Economics and Management, Hebei University of Science and Technology, Shijiazhuang, China

ABSTRACT: To evaluate the market concentration and enhance economies of scale in jujube processing industry, according to current development situation of jujube processing industry in Hebei province, adopting a combination of qualitative and quantitative methods, selecting data that are sales of the top 50 processing enterprises, applying CR4 and HHI Index model to analyze jujube processing industry concentration. The conclusion is the low market concentration, the more quantity of processing enterprises and single product, the low scale of the enterprise market share, no significant economies of scale. The main reason is that small and scattered processing enterprises, no obvious economies of scale; low degree of organization; lack of core competitiveness. Finally, putting forward the countermeasures: mining resource endowments, expanding comparative cost advantage, improving the degree of organization and core competencies.

Keywords: industry concentration; jujube processing industry; CR4 index; HHI index

1 INTRODUCTION

Hebei province is the focus region of jujube production in the country, as well as the major cultivation areas of jujube. Jujube industry is one of the important industries of rural economic development in Hebei province. There are 29 million hectares under jujube plantation in Hebei. Annual output is 1.26 million tons, ranking first in the country. However, with the upgrading and adjustment of agricultural structure, jujube production growth compared to 2003 decreased by 15.81% in 2013; jujube processing rate has been about 23%, which makes the competitiveness and profitability of jujube processing industry very low. According to the current situation and problems in jujube industry in Hebei, Peng Jian-ying [1] and Ge Wen-guang [2] analyzed the strengths and weakness of jujube industry. Finally, putting forward the constructive suggestion, Jia Wei-guo, Chen Huan-di [3] and Wang Chang-wei, Wu Zhi-hua [4] analyzed the market structure of Subei poplar processing industry and the market performance of China's rice processing industry, respectively, by SCP analysis method. Zhang Yu-kun, Pang Qing [5] analyzed market concentration of the tea by CR4 index and HHI index. SCP analysis method is an effective method of analysis of market structure. Currently, no one studies jujube industry with SCP method. Therefore, CR4 and the HHI index in SCP analysis methods are adopted to analyze the

market concentration of jujube industry in Hebei province, which is helpful to improve the jujube industry market structure. Improving jujube industry market performance has important practical significance.

2 THE ANALYSIS OF CURRENT SITUATION AND CONCENTRATION

2.1 The current situation of jujube processing industry

According to China Agricultural Statistics Yearbook, China's jujube output has reached more than 540 tons in 30 years from 1979 to 2012. Hebei, Shandong, Shanxi, Xinjiang, Shanxi, Henan are the main producing areas, contributing rate more than 90%, which is about 24% in Hebei, about 23% in Shandong, about 20% in Shanxi. The current domestic large jujube research institutions and production enterprises are more than 40, which is located in Beijing (4), Hebei (8), Shanxi (6), Shandong (8), Shanxi (1), Xinjiang (6).

Hebei province has established seven major jujube processing markets. Processing enterprises are mostly distributed in the processing markets around. There are varieties of jujube juice, jujube wine, jujube tea, of which Allied, "Qian Tong" jujube wine, Huanghua big jujube possess high technological and added value, and is an influential brand. However, it also highlights the characteristics of the processing industry: The high output of annual high, the low processing ability; the more processing enterprises, but few leading companies; rich product varieties, and the less high-tech and value-additions of the product, the smaller brand influence. These features result in diseconomies of scale, low market concentration in jujube processing industry in Hebei.

2.2 The analysis of jujube processing industry concentration

Market concentration is an important indicator of the market structure concentration, and is an important factor to influence market behavior and performance. The main index Measured market concentration are CRn index and HHI index.

2.2.1 CR4 index analysis

Jujube processing enterprises in Hebei is so widely dispersed and small in scale that collecting data becomes very difficult. Given the availability and accuracy of data, the paper selected data that operating income from the top 50 companies with annual sales of 50 million or more for three years 2011–2013, with the sum of previous 50 sales instead of the total sales. Specific data are shown in Table 1.

Concentration ratio calculation model: $CR_n = \left(\sum_{i=1}^{n} X_i\right)\Big/X$, $\sum_{i=1}^{n} X_i$ represents the sum of the top n output, X represents the total output. CR4 (%) is 35.14 in 2011, 34.37 in 2012, 31.34 in 2013, respectively (Table 1). According to Bain types of industrial monopoly and competition, CR4 (%) is the low market concentration from 30–35. This can explain why the concentration of the jujube industry in Hebei province is low.

2.2.2 HHI index analysis

HHI index is the total number and size distribution based on the manufacturer's industry, HHI index calculation model: $HHI = \sum_{i=1}^{n}\left(X_i/X\right)^2 = \sum_{i=1}^{n} S_i^2$, x_i is the market share of i vendor; x is the total market share. The advantage of the index is an absolute and relative method of mathematical method, which explains that the influence of market concentration caused the distribution of firm size. Selecting data of value in the top 20 processing enterprise, calculated HHI = 0.0672 in 2011 by Excel, in order to

Table 1. The concentration of the jujube industry—CR4 index.

	2011	2012	2013
The sum of top 4 (million)	272	344	403
Totle (million)	747	1001	1286
CR4 (%)	35.14	34.37	31.34

facilitate the observation, it is multiplied by 10000 amplification, namely, 672. Similarly, calculated HHI = 634 in 2012, HHI = 607 in 2013. According to Market Structure Classification based on HHI index, jujube processing industry is in competitive type I, which means more vendors and low market share of the top 20.

2.3 The changes analysis of the market concentration

The main factors influencing the industry market concentration is economies of scale. Through the last three years' CR4 index and HHI Index, market concentration coefficients and manufacturers scale are reducing, namely, the economies of scale is also smaller. Although the absolute value of the former four enterprise sales increases, but the growth rate has lagged behind the growth rate of the industry, which is specifically reflected in the 2011 is 35.14% down to 31.34% in 2013. HHI Index is constantly decreasing, indicating that the top 4 market share of the total market share is reducing, narrowing the gap in the scale of industry vendors, reducing the dominance of manufacturers. Namely, the development of the leading enterprise is instability, and even the number of the leading enterprises is shrinking.

3 CONCLUSION AND ANALYSIS

3.1 Conclusion

Overall market concentration is low and gradually decreasing. From the CR4 index, market concentration is gradually reducing, and the overall concentration is low, economies of scale are not significant.

The more quantity of enterprises and single product. There are more than 2000 jujube processing enterprises in Hebei province, of which the rough-oriented enterprises accounted for more than 80% of the total processing companies. The processing varieties are mainly candied fruit, dried fruit, etc.; the convergence of processed products caused the low price competition, low profits.

The lower share of leading processing enterprises. Based on the HHI index, the market share of the leading processors is decreasing, and the dominant market power is also reducing, leading to the slow development of the processing industry.

3.2 Analysis

3.2.1 Small-scale and scattered, low economies of scale

There are more than 2000 large and small plants in the main growing areas in Hebei, most processing enterprises produce rough-oriented jujubes.

The leading enterprise accounted for about 10%. Deep processing plants are very few. The family-run factory is the main form of rough guide, processing single products, rough. Although leading enterprises have higher production equipment and advanced management methods, which can meet the basic needs of the market, it is difficult to meet the need of a higher level, give full play of medicinal and nutritional value of the dates, and product differentiation is low, competition intense, economies of scale insignificant.

3.2.2 *The low degree of organization*

According to the survey, only individual regions unify management, which is limited to pests and diseases. 85% of planters don't join farmer cooperatives, and jujube production is still mainly a decentralized management of a household. Therefore, they cannot match with the big market, with weak ability to obtain information, poor bargaining power, low ability to resist risks. In addition, the competitive relationship formed between the large, medium and small enterprises are very complex in jujube processing industry. A large number of small and medium-sized enterprises failed to form a business relationship with the specialization and cooperation, leading to large, medium and small enterprises in the product structure is very similar to form a "large and complete whole, small and complete whole" competitive landscape. A large number of small-scale processing enterprises compete with the few medium-sized enterprises in the two markets of raw materials and products, resulting in excessive competition situation, reducing the market performance of the entire industry.

3.2.3 *Leading enterprises lack the core competitiveness*

First, observation from the well-known trademarks in China (Table 2), Hebei Province although has a famous Golden Silk jujube and Zanhuang big jujube, leading enterprises still do not register well-known trademarks, and other provinces have already surpassed Hebei in the brand, which have affected the competitiveness of processed products in reputation and brand.

Secondly, the products and technology innovation of processing enterprises in Hebei are inadequate. As can be seen from Table 3, the number of patent applications per year of Hebei province has been far below the entire country in recent years. The invention patents of jujube processing enterprises have been growing slowly, and the growth rate is far behind the national average, which causes the slow improvement of production technology, little change of differences in product quality and species, declining in the competitiveness of enterprises.

Table 2. Jujube products—China's well-known trademarks.

Brand	Registrant	Location
Haoxiangni	Haoxiangni jujube Industry Co., Ltd.	Henan
Hetian jujube	Hetian jujube Industry Association	Xinjiang
Tian Yuan	Shanxi tianyuan Jujube Industry Co., Ltd.	Shanxi
Han Po	Shan Xi Hanpo Food Co., Ltd.	Shanxi

Source: State Administration for Industry and Commerce.

Table 3. The number of patent applications in jujube processing enterprises.

	2009	2010	2011	2012	2013
Hebei province	8	15	23	20	31
The whole nation	33	69	112	125	152
Percent	24.2%	21.7%	20.5%	16%	20.4%

Source: China Patent Information Center.

Finally, the jujube processing enterprises lack high-end talent. According to incomplete statistics, jujube processing enterprises have only about 10% of the total number of employees with specialist qualifications; the total number of undergraduate degree less than 5% in Hebei province. At present, only Hebei Agricultural University, Hebei Normal University of Science and Technology has fruit trees professional, which cultivate very few trained people each year. The talent they need has failed to satisfy the business development and widen sales channels.

4 THE MEASURES TO IMPROVE THE INDUSTRY CONCENTRATION

4.1 *Mining resources endowment, expanding the comparative cost advantage*

Hebei is located in the eastern part province in North China, belongs to the temperate continental monsoon climate, precipitation relatively concentrated, distinct seasons, light enough in natural conditions. Hebei province has about 20 cities and counties in large-scale planting jujube, annual output more than 10,000 tons. Jujube processing enterprises should make full use of land resources to expand the scale of planting and processing. From a technical point of view, Hebei Province has Hebei Agricultural University, Chinese jujube Research Center and other institutions of higher learning and scientific research units, processing enterprises lack of innovation capacity should cooperate with these research units to optimize processes, and develop niche products.

Jujube processing enterprises in more favorable areas should also improve the processing technology, introduce advanced equipment, reduce costs, obtain price advantage. In turn, they make full use of the profits to invest up mining resource endowments, achieving a virtuous circle.

4.2 *Improving the degree of organization*

Improving the degree of organization's goal is to promote effective competition in industry, enhance the competitiveness of industry to form moderately concentrating large, medium and small enterprises symbiotic market structure. First, exerting the ability regulation of jujube corporate trade associations. Jujube processing enterprises should actively join in the professional committees, gifting a stronger ability to regulate the association, so that the professional committees play a positive role in the organization of jujube processing enterprises processing. Then, establishing a group of industry consolidation ability of large enterprises by the role of the market mechanism, appropriately raising market concentration, so as to improve the level of the economies of scale. Finally, the government should actively support the development of small and medium-sized enterprises, guide them toward professional development. The small enterprises become "small but exquisite" and form a rational division of labor, mutual cooperation and complement of competition with big business.

4.3 *The implementation of brand strategy, enhancing core competitiveness*

Processing enterprises should further strengthen the brand consciousness, focusing on jujube products brand registration and protecting intellectual property. Large-scale leading enterprises should actively participate in the jujube brand assessment activities, enhancing corporate and brand visibility.

Enhancing innovation and marketing capabilities. First, we must improve the technological innovation capability of enterprises.

Jujube processing enterprises should cooperate with Hebei Agricultural University, Chinese Jujube Research Center, and other institutions of higher learning and scientific research units in various ways. Setting up own scientific research and technology development organizations, improving the value-added and technological content of products. Second, jujube processing enterprises should enhance marketing capabilities. Jujube processing enterprises should establish modern marketing concept, learning and using of modern marketing skills, strengthening the construction of network marketing. First, enterprises should make an investigation to determine market demand of jujube products and then make product positioning and market positioning based on market demand. Finally, plan development of relevant marketing planning of the product in detail. With the help of the mass media, improve the awareness of products and the influence of services.

ACKNOWLEDGMENT

Forestry Soft Science Fund Project "The Construction Research of Modern Jujube Industrial System in Hebei Province".

REFERENCES

[1] Peng Jian-ying. The industrial structure and layout of jujube in Hebei Province. Practical Forestry technology. 2012 (11): 57–59.

[2] Ge Wen-guang, Ma Li-ran, Wang Jie. Present situation, advantages and development counter mersure of jujube industry in Hebei province. Research of Agricultural Modernization. 2011, 32 (6): 713–716.

[3] Jia Wei-guo, Chen Huan-di. Analysis on the market structure of poplar processing industry of northern Jiangsu province. Problems of Forestry Economics. 2007, 27 (4): 371–375.

[4] Wang Chang-wei, Wu Zhi-huan. The analysis of rice processing performance in China based on the theory of SCP. Cereals and oil processing. 2009 (8): 25–28.

[5] Zhang Yu-kun, Pang Qing, Liu Qiong. Analysis industry concentration from the development of our country tea industry. Chinese Economic and trade Herald. 2009 (13): 56–57.

Education Management and Management Science – Zheng (Ed.)
© 2015 Taylor & Francis Group, London, ISBN 978-1-138-02663-6

The governance of higher education in the era of big data

Q. Meng
School of Education, Bohai University, Jinzhou, Liaoning Province, P.R. China

X.D. Xu
School of Education, Huazhong University of Science and Technology, Wuhan, Hubei Province, P.R. China

ABSTRACT: The development of big data has brought challenges and opportunities to higher education. Data, as a strategic source of organization, can provide support for organizational decision-making. Data governance can improve the quality of higher education, make university decision-making process more transparent and improve the efficiency of university decision-making. Data governance can be divided into several stages: data access, data extraction, integration and analysis, explanation and prediction. Although data governance, as a part of higher education governance, is still in the embryonic stage, supporting the decision-making is the driving force of data governance.

Keywords: big data; higher education; data governance

1 INTRODUCTION

In 2012, the term "big data" began to enter the public view. New York Times column says: "The era of big data has come, in the commercial, economic and other fields, the decision will increasingly be made based on data and analysis, rather than based on experience and intuition." (Lohr, 2012) Data will become one of the world's most powerful things in the twenty-first century. "It's a revolution," says Gary King, director of Harvard's Institute for Quantitative Social Science. "The march of quantification, made possible by enormous new sources of data, will sweep through academia, business, and government. There is no area that is going to be untouched." Higher education is facing the opportunities and challenges in this data revolution.

2 THE AGE OF BIG DATA

Big data is anther hot term after clouding computing and Internet of Things (IOT). The effect of big data is emerging in various fields. The development of big data makes the collection, storage, analysis and application of data easier and more efficient and convenient, has exerted great influence on national, government and universities decisions, and may greatly improve the management efficiency and public service capacity. On March 29, 2012, American government put forward 'Big Data Research and Development Initiative' to improve the ability to extract knowledge and insights from large and complex collections of digital data, to accelerate the pace of discovery in science and engineering, to strength the national security and transform teaching and learning. In July 2012, the United Nations issued a white paper 'Big Data for Development: Challenges & Opportunity'. All these indicate that data has become a kind of strategic resource.

Data governance is not only the key to organizational development, but also a new perspective of organizational research. Government agencies and companies have been aware of the positive role of data in improving management and the internal and external relations. According to Info Dev's report, data governance can meet the following requirements: (1) providing a way for public to access government information, (2) through encouraging the interaction between citizens and officials to improve the democratic participations, (3) providing opportunities for the development of organization, especially the immature organization.

Before big data, government could force people to provide information, so they can get large amount of data. In the age of big data, people ask government to release data. Big data doesn't directly give us the answer to the question, only provides us clues to the answer. With the development of big data, information is no longer occupied by powerful institutions, and data sharing is possible. The development of big data can help universities to explore and use the underutilized dark data (Laney, 2012). Just like other organizations, universities produce large amount of data, the age of big data for higher education has come. Data governance can benefit university development.

3 DATA GOVERNANCE FOR HIGHER EDUCATION

Decisions based on the data significantly improve organization productivity. The development of higher education and technology progress are inseparable. Plans such as MOOCs have changed the traditional way of teaching and learning. In each evolution, data which we usually ignore has revolutionary influence on higher education. Today, we only use data for ensuring students can choose the right course and teacher, rather than using students and course information to plan university strategic development.

3.1 Data governance is conductive to improve the quality of university decision

Chinese higher education has changed from elite education to mass education, from scale-oriented development to quality-oriented development. Society and government pay much attention on the curriculum quality, teaching quality, research equipment quality, the number of students, university ranking and so on. The accountability of higher education requires universities to focus on their performance. Meanwhile, severe international higher education competition and China's goal to world-class university have prompted us to improve quality in the first place. Big data is an effective way to improve the quality of education.

American universities have applied Learning Analytics technology to analyze students' learning behavior. Teachers can monitor students' learning and point out their mistakes; thus, teachers can develop highly personalized learning program for every student to increase the graduation rate. Rio Salado College developed 'Progress and Course Engagement System' to automatically track students learning status. System analyzes three data: the number of student Logon, the number of student reading materials and course score. The report has three warning levels: green, yellow and red. The prediction accuracy is about 70%. There are other similar systems, such as Northern Arizona University 'Grade Performance System', Purdue University 'Course Signals System', Ball State University 'visualizing collaborative knowledge work'.

3.2 Data governance makes university decisions more transparent

Data governance uses Information Communication Technology (ICT) to achieve information sharing and exchange. Data governance can be understood as a transparent process using technology to transfer information to public effectively and quickly. This kind of information sharing can provide decision-making opportunities for bottom or vulnerable group. The quality and potency of information decides the extent of stakeholders participating in decision-making. For bottom-to-up university, data governance can ensure everyone participates in decision-making and improve the quality of administrative decision-making.

Data governance conforms to the characteristics of decentralization and deliberative democracy, provides a way to improve the transparency of university governance. Data technology not only brings the renewal of teaching and learning method, but also changes management method.

The University of Texas System Productivity Dashboard is the pioneer for the application of big data. This system is open for public. This platform provides materials for public to evaluate UT fairly, for example, student graduation rate, tuition and affordability ratio compared with other universities, teacher to student ratio, passing ratio for mechanical, law, pharmacy and other professional certification, the scores of entrance examination for SAT, ACT, GRE, GMAT, and time for completing degree. These data help students, parents and other stakeholders to evaluate UT independently. Universities leaders are considering establishing college or school level information systems which are also completely open to public.

3.3 Data governance is necessary to improve the efficiency of university decision

As Fry et al. (2001) advised, if university wants to win in the market of higher education, they must take technology as a strategic tool for planning. Big data pays attention to customers' experience and needs, so it is considered a powerful tool for analysis and improving organizational efficiency. Big data helps university see clearly what advantages are and what disadvantages are, is more conducive to university decision-making.

Data governance makes all levels of university to commit to the goal of university through greater participation, online discussion and learning method reform. Data governance which is different from traditional governance lies in its attention to each person's demand, to reduce human intervention as far as possible, to establish fair, transparent and rapid response system.

Data governance has become a research topic of higher education for its high efficiency, accountability and transparency. It is consistent with the characteristics of good governance: transparency, responsibility, responsiveness and effectiveness.

Data governance can make universities more intelligent and smarter. Some scholars have predicted that big data governance will bring revolutionary change in higher education (Manyika, 2011).

4 ANALYSIS FRAMEWORK OF HIGHER EDUCATION DATA GOVERNANCE

Data governance has three stages: access and extraction, integration and analysis, interpretation and prediction.

4.1 Access and extraction

It is a basic work to extract useful data from a large number of metadata. Data is from different producers, including university-level data, school-level data, program data, curriculum data, student data and learning management system data or course management system data. Then, we can discard unwanted data according to a standard and retain the useful or relevant data.

4.2 Integration and analysis

Although data is messy and dynamic, we can still obtain useful data. In view of data being huge and its heterogeneity, it is the key step to integrate and analyze data. How to structure the different data and code them in computer language is crucial for data processing. We may find the hidden information among data when we check relevant data, data redundancy and data loss. The lack of cooperation and non-structural database (non-SQL) are the problems big data should face.

4.3 Explanation and prediction

If data cannot be effectively interpreted, there is no value for decision makers. In this stage, we should test all hypotheses and review analysis process, use data mining technology and description or inference statistic method such as correlation, regression, market analysis and chart analysis to explain data. Data interpretation can help decision makers understand customers' needs more clearly, understand the key indicators of influencing university development. Based on the report, leaders can formulate the effective strategies and policies by using decision trees and strategy maps.

5 CONCLUSION

At present, university data governance technology is not perfect. There are many technological problems to be solved, for example, data volume is too large, data has high homogeneity and data updates too quickly. There are also some running problems, such as there are not enough teachers or administrators qualified for dealing with big data, students' personal privacy and security issues when we integrate students' data into other data systems. But its importance for university development is self-evident, just like Sloan Digital Sky Survey (SDSS) for human development.

For public, accessing data is more important than getting results. They prefer to draw results from data themselves, rather than being given results directly. The application of big data is still in the embryonic stage, the combination of big data and higher education need leaders' recognition and support for decision based on data, also depends on the future development of big data. Supporting decision-making system is the dynamic of big data development.

REFERENCES

[1] Centre for Democracy & Technology. November 2002. *The E-government Handbook for developing countries: A Project of InfoDev and The Centre for Democracy & Technology*. Washington, DC: World Bank.

[2] Douglas L. October 2012. *Big Data Strategy Components: Business Essential*. Stanford: Gartner Report.

[3] Heather F, Steve K. and Stephaine M. 2001. *A Handbook for Teaching and Learning in Higher Education*. Kentucky: Routledge.

[4] James M. 2011. *Big Data: The Next Frontier for Innovation, Competition and Productivity.* http://www.mckinsey.com/insights/business_technology/big_data_the_next_frontier_for_innovation. Retrieved May 3, 2014.

[5] Rajeev S. 2014. *E-Governance Center of Excellence: The Future of Efficient, Effective, and Transparent Governance.* http://egov-coe.ncc.gov.ph/index.php?option=com_content&task=view&id=55&Itemid=1. Retrieved May 3, 2014.

[6] Steve L. February 11, 2012. The Age of Big Data. *New York Times*.

[7] Tulasi B. 2013. Significance of Big Data and Analytics in Higher Education. *International Journal of Computer Applications*, 68(14): 21–23.

Education Management and Management Science – Zheng (Ed.)
© 2015 Taylor & Francis Group, London, ISBN 978-1-138-02663-6

Study on the application of smartphone technology in college education

L. Chen & C.H. Deng
School of Electrical Engineering, Wuhan University, Wuhan, P.R. China

Y. Yuan
Pan Asia Laser Art Company of Hubei Daily, Wuhan, P.R. China

ABSTRACT: Along with the rapid development of information and communication, the application of smartphone has attracted more attention, and in a sense, smartphone technology may bring lots of changes. In this paper, regarding the influence of applying smartphone technology to college education, some studies are performed. Firstly, the background and current development of smartphone is presented, and then its potential effects for teachers and students are analyzed. Furthermore, a few helpful mobile applications which teachers and students can use inside and outside of the classroom are summarized. At the end of this article, the prospect forecast of smartphone technology for improving college education is suggested.

Keywords: smartphone technology; college education; mobile application software

1 INTRODUCTION

Smartphones emerged around 2000, and sales have consistently increased with each succeeding year. Growth in demand for advanced mobile devices boasting powerful processors, abundant memory, larger screens and open operating systems has outpaced the rest of the mobile phone market for several years [1]. At presently, smartphones have become very popular, and in a sense, the smartphone technology may bring lots of changes to our lives from different kinds of aspects. In consideration of that college education is critical to the success of our workers and our economy, the study of the relationship between smartphone technology and college education is meaningful.

In this paper, regarding the influence of applying smartphone technology to college education, some studies are performed. The article is organized in the following manner. Section II presents the background and current development of smartphone technology. In Section III, the smartphone's potential effects on teachers and students are analyzed, and a few mobile applications which may be suitable for them are summarized. Finally, the prospect forecast of smartphone for improving college education is suggested.

2 BACKGROUND AND CURRENT SITUATION OF SMARTPHONE

Herein smartphone features are firstly explained in brief. A smartphone is smart because it has an operating system which manages the phone's hardware and software. It is no different from the operating system of a desktop computer. Another characteristic of a smartphone is the presence of at least one home screen, and it is the main menu display that shows the apps and widgets. In addition, various types of novel apps, they are also one of the reasons why a smartphone is smart and different.

In a certain sense, Apple Company redefined smartphone. In 2007, Apple Inc. introduced the iPhone, one of the first mobile phones to use a multi-touch interface. The iPhone was notable for its use of a large touchscreen for direct finger input as its main means of interaction, instead of a stylus, keyboard, or keypad typical for smartphones at the time [2]. With regarding to another mainstream smartphone operating system, Android is an open-source platform founded by Andy Rubin and backed by Google [3]. Although Android's adoption was relevantly slow at first, it started to gain widespread popularity in 2010. Under this platform, Samsung Company plays an important contribution and has a large market share.

Table 1. Historical sales of smartphones (in millions of units).

Years	Android	Ios	Windows phone
2012-Q1	81.07	33.12	2.71
2012-Q2	104.8	26.0	5.4
2012-Q3	122.5	23.6	4.1
2012-Q4	144.7	43.5	6.2
2013-Q1	162.1	37.4	7.0
2013-Q2	177.9	31.9	7.4

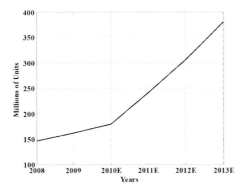

Figure 1. Worldwide smartphone shipments from 2008 to 2013.

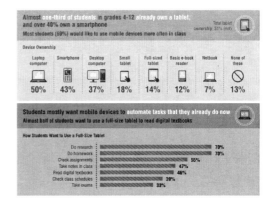

Figure 2. Mobile device service condition examination for school students.

Meantime, there are some alternative smart operating systems, such as BlackBerry, Windows phone, etc. As shown in Table 1, it indicates the historical sales in recent years (2012–2013).

Figure 1 shows the recent smartphone market growth from 2008 through 2013 [4]. It can be observed that, smartphones will be undoubtedly a huge market for consumer electronics.

3 APPLICATION OF SMARTPHONE TECHNOLOGY TO COLLEGE EDUCATION

3.1 Impacts of smartphone technology to college teachers and students

For the effects of introducing a smartphone on college education, mobile learning is a key. Students can use smartphones to connect to the digital world more conveniently, either through Wi-Fi or 3G/4G. Due to the use of mobile learning, the following advantages can be achieved, such as: 1) Students may learn in a way they are comfortable. 2) Students can get answers quickly. 3) Audio and video can bring learning to life. 4) Smartphones allow for social learning.

As shown in Figure 2, it indicates a mobile device service condition examination for school students [5]. As a matter of fact, along with the size of smartphone screen becomes larger, taking Samsung Galaxy Note for an example, Samsung Galaxy Note III smartphone with 5.70-inch 1080 × 1920 display powered by 1.9 GHz processor alongside 3 GB RAM and 13-megapixel rear camera. This hardware configuration may contribute to have a good experience of operating a tablet, and using this smartphone may be more beneficial to strengthening mobile learning.

For teachers, using smartphones can enhance the contact between course content and student learning. Teachers should really reconsider allowing smartphones in class. Although smartphones might look like toys at first, they can be very efficient, especially now that many app developers are creating educational apps. On the basis of smartphones, difficult problems existed in the learning process can be feed backed timely through mobile-network information communication. Even more, a specific mobile application software can be designed pointedly. Regarding this specific APP's required functions, lecture notes, assignments, and other supporting materials might be included. Furthermore, directing at those difficult knowledge points and problems, detailed solutions could be presented.

3.2 Example analysis of a specific smartphone app for college course education

On the whole, a philosophical concept of information exchange should be realized in the smartphone APP. Herein taking one of my teaching courses for example, some concrete explanations are conducted. Automatic control theory is one of the major courses for electrical engineering and automation. Through learning automatic control theory, students should not only grasp the basic operation principle of automatic control system and relevant analytical methods, but also learn how to carry out the mathematical modeling, performance estimation and controller design for process of production. The course mainly covers classical and modern control theories. For classical modern theory, it is included that introducing the mathematical model of control system, explaining the time-domain/frequency-domain analysis and root locus method suitable for continuous linear system, and describing the design of PID controller as well as the system correction method. Regarding modern control theory, the course content including state-space equation, Lyapunov stability method, system controllability/observability, state feedback and pole configuration will be systematacially taught.

Aiming at the aforementioned knowledge nodes, to improving students' learning efficiencies, the designed APP should configure the following functions (or technical contents): 1) summarizing each chapter's points of knowledge, and explicating relevant basic concepts, 2) contrastively analyzing time-domain/frequency-domain performance indexes, classical/modern control method, as well as state-space/transfer-function expression, 3) including different kinds of test questions, which are consisting of table completions, true or false items, and calculations, 4) having a statistical function. It can refine the features from the question of a student, and then return the best answer of the question from the knowledge base with relational algorithms. 5) Creating a concise and user-friendly interface. As indicated in Figure 3, it shows a possible user interface for the smartphone APP called as Automatic Control Theory Supported Learning.

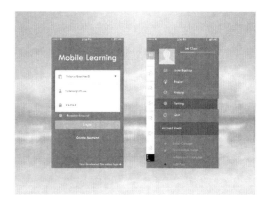

Figure 3. A user interface for the smartphone APP called as Automatic Control Theory Supported Learning.

3.3 Collection of a few helpful smartphone apps for college teachers and students

In this section, a collection of smartphone apps that teachers and students can flexibly use inside and outside of the classroom are presented, their functions are also introduced in brief.

In respect to students, the following four APPs may guarantee to help them manage their learning-time and improve their learning-qualities more responsibly [6–7]. 1) Recordroid-Students who are tired of transcribing their teacher's lessons can use Recordroid to create audio recording notes. Files created will be sent to their desktop computer via e-mail, so they can review what the teacher said at home. 2) iHomework-This app can make organization a cinch. Use it to keep track of assignments, readings, courses, teachers, and schedules. Even better: students can set alarms for their homework and be notified of upcoming deadlines. 3) Youdao Dictionary-Youdao dictionary mobile version is suitable for various platforms. It has typical acoustic translation, smart screen word identification, real time synchronization with network up-to-date vocabulary, and other powerful features. Youdao dictionary could even translate the text in the picture using OCR technology. By using this app, the distance between the questions and answers (aiming at language study) can be shortened for students. 4) iTunes U—the App is a great resources for anyone, whether the university supports it or not. If it does, there's a good chance for students who can find lectures and class material available for download straight to their iPhones or iPads.

For teachers, smartphones can be powerful teaching tools [8]. According to the educational apps, we can now find reference books, flashcards, calculators, and much more on a smartphone. 1) Margins-This app can help teachers prepare for class faster and easier. It may let teachers take notes about what they read. This is very helpful when relating news and media to the class. 2) Evernote—This app is hands down one of the best note taking apps available, with support for tags, separate notebooks, iCloud sync, and more. Teachers can even take photos of a blackboard or syllabus and tag it accordingly for specific courses. 3) Grade Rubric—This app can help teachers who use a grading rubric for assignments. It is actually an option to auto-generate an email with detailed grade report for students. 4) Quick Graph+. This app is a lot more convenient and makes for one less thing to carry around. With the ability to plot graphs and so much more, it's a good option for those taking math courses. Quick Graph+ supports cartesian, polar, cylindrical and spherical coordinate systems as well as hyperbolic and inverse functions. 5) Google Search-This app provides way more options than the web-based engine, and using it can seek out different types of research papers and relevant materials more fully for teachers.

4 EPILOGUE

According to the International Data Corporation (IDC) Worldwide Quarterly Mobile Phone Tracker, the smartphone market passed an important milestone in 2013 when worldwide shipments surpassed the 1 billion mark for the first time, driven by continued momentum from Android and iOS. Currently, the Apple App Store is still the biggest app store showing more than 750,000 apps, and almost 11% of them are educational apps. It may be believed that, along with the further progress of information and communication, and the social requirements for educational apps have been developed by the related education enthusiasm, the educational apps' quantities and qualities can be enhanced to a certain extent. It is hoped that our world may become better owing to the popularization of smartphone technology.

REFERENCES

[1] Dong-Hee Shin, Youn-Joo Shin, Hyunseung Choo, Khisu Beom. 2010. Smartphones as smart pedagogical tools: Implications for smartphones as u-learning devices. *Computers in Human Behavior* 27: 2207–2214.

[2] Ryan Block. 2007. The iPhone is not a smartphone. Engadget.com.

[3] Butler, M. 2011. Android: Changing the Mobile Landscape. *IEEE Pervasive Computing* 10(1): 4–7.

[4] Xun Li; Ortiz, P.J.; Browne, J.; Franklin, D.; Oliver, J.Y.; Geyer, R.; Yuanyuan Zhou; Chong, F.T. 2010. A Case for Smartphone Reuse to Augment Elementary School Education. *International Green Computing Conference*: 459–466.

[5] David Nagel. 2013. Students Use Smart Phones and Tablets for School. http://thejournal.com/articles/2013/05/08/report-students-use-smartphones-and-tablets-for-school-want-more.aspx#6uOYLPWdhIwyiZUq.99.

[6] The Green Electronics Trade. 2011. Smartphones as Tools for Education: Getting Smart with Smartphones. http://www.ecyclebest.com/smartphone/articles/smartphones-as-tools-for-education.

[7] Allyson Kazmucha. 2013. Best iPhone and iPad apps for college students: Evernote, Notability, iTunes U, and more. http://www.imore.com/best-iphone-and-ipad-apps-college-students-evernote-notability-itunes-u-and-more.

[8] Young, Jeffrey R. 2011. Top Smartphone Apps to Improve Teaching, Research, and Your Life. *Essential Readings Condensed for Quick Review* 76(9): 12–15.

The influence of meta-cognitive strategies on English acquisition from private universities in China

J.H. Wang & X.D. Xu
School of Public Administration, Huazhong University of Science and Technology, Wuhan, China

X.L. Zhao
School of Foreign Language Studies, Southwest Forestry University, Kunming, China

ABSTRACT: Metacognitive strategies are taken as the most advanced and core factor of human's mental power, which is the key to exploring one's intelligence, potentiality and language proficiency. The thesis is devoted to examining students' use of metacognitive strategies. The findings show metacognitive strategies have positive influence on English acquisition.

Keywords: metacognitive; english acquisition; private university

1 INTRODUCTION

Since the late 1970s, classroom instruction has shifted its focus from teacher-centered to student-centered approach based on the ground that if learners do not intend to learn, it does not make any difference no matter how well the teachers teach. Consequently, learners had become the dominant figures in the language classrooms where learning tasks had been conceptualized and approached from the learners' own point of view (Rubin, 1987). Thus effective strategies students adopt are conducive to improving language development and fostering their learning ability.

2 LITERATURE REVIEW

Researchers have realized that apart from these strategies, there exist Some high-leveled skills which make learners more adept at and successful in the language development. This refers to metacognitive strategies. However, metacognitive strategies are taken as the most advanced and core factor of human's mental power, which is the key to exploring one's intelligence, potentiality and one's language proficiency. For the time being, though there are a plethora of research studies relating to language learning strategies from home to abroad, such as those from Rubin (1975), Naiman, Frohlich and Todesco (1975) and those from Wen Qiufang (1993) and WangLifei (2003). Among all the researches on metacognitive strategies carried out by experts and professors, Oxford (1990) stressed

that "metacognitive strategies are indispensable to successful language learning." In China, Wen Qiufang (1995) stated that the monitoring role of metacognitive strategies is quite effective, which also means that the importance of metacognitive strategies can never be over emphasized. From then on, researchers have been passionate about their studies on metacognitive strategies.

3 PRIVATE UNIVERSITY STUDENT LEARNING SITUATION

The 1990s witnessed a dramatic social and economical transformation and jump of China, in which the reform of educational system is quite eye-catching. It has exerted a far-reaching impact on China's higher education ever since. A number of private universities sprang up and jumped into people's focus at that time. (Private university, also named as non-government run university or non-governmental run university, are not operated by governments though many receive public subsidies, especially in the form of tax breaks and public student loans and grants. Depending on their location, private universities may or may not be subject to government regulation.) After some years' development, now private universities have become quite influential not only because of their large scale, but also because of the high quality of some of them. It goes without saying that the emergence of private universities plays a positive role in extending the higher education and improving the capacity of state-owned universities to keep a satisfactory performance under the circumstances of fast development. However, considering from another angle, the extension of higher education has leveled down the standards of universities, and it naturally leads to the difficulties in English instructions, because nearly most private university students are low-achievers of schools and they are more dependent on parents and teachers. Meanwhile, the students in private universities are from different provinces and the middle-school education levels in different provinces are varying, so accordingly, the students' English levels are also varying when they are enrolled. In addition, their English levels are influenced by other factors, such as family background, personal experiences, etc. All these factors mentioned above have hindered their English development considerably. Besides, most private universities do not have their own teachers. Soon after classes are over, teachers usually leave schools. So the students very often have little time to stay with their teachers and few opportunities to communicate with them. Consequently, these students, more often than not, have to do self studies. Therefore learning strategies and learning autonomy are important to them.

4 METHODOLOGY

In this section, the author employed a statistical analysis to investigate whether there were differences of metacognitive strategies use between these two groups. The groups were classified by the standard of TEM-4 scores. 67 students on the top 30% were regarded as "high", while 68 in the bottom 30% were taken as "low". This classification grounded on the fact that TEM-4 scores of the participants are lower compared with the students from government—run universities. So it was maneuverable to adopt this division scale.

From Table 1, we can see: (1) The mean value of high achievers is higher than that of low achievers

Table 1. The difference between high and low achiever.

Elements	Level	Number	Mean	SD
Planning	High	67	3.332	0.639
	Low	68	2.795	0.585
Monitoring	H	67	3.334	0.617
	L	68	2.789	0.686
Evaluation	H	67	3.215	0.588
	L	68	2.817	0.677
Metacognitive strategies	H	67	3.265	0.491
	L	68	2.791	0.565

(Difference is significant at the level of 1%).

in planning, monitoring as well as evaluation. So it goes without saying that high achievers do much better than low achievers in metacognitive strategy use and its three subcategories; (2) the standard deviation of high achievers is bigger than that of low achievers in planning, whereas the standard deviation of high achievers is smaller than that of low achievers in monitoring and evaluation as well as the whole metacognitive strategies.

5 DISCUSSION ON OVERALL SITUATION OF METACOGNITIVE STRATEGY USE

The statistical analysis manifests that the use of metacognitive strategies and its subcategories is in its middle and even slightly lower level among private university students. Because we can easily observe that the statistics of each subcategory of metacognitive strategy is in a medium level, which indicates that most students, generally speaking, are not quite skillful at utilizing metacognitive strategies. Judging from the educational background of the students who have enrolled in private universities, compared with the students in government-run universities, there appeared the following phenomena: (1) when students are enrolled to universities, they have not laid a solid foundation of English. Accordingly, they can not meet the requirements of the schedule; (2) the English teaching methods may not be very systematic, for a large number of private universities lack their own teachers and mainly depend on part-time teachers; (3) most students are forced to enter universities by their parents, so their English learning motivation is comparatively low, and thus they always tend to depend on class hours to learn English and are less active to make plans and to communicate with others in English; (4) most students fail to form a good habit of language learning, which may be another devastating cause for their English deficiency and poor academic performance. In order to confirm the students' learning situation, the author also had an interview with the participants. During the interview, the author found that although it seemed that some students had their own plans to follow, they only finished the tasks assigned by their teachers. They could just make some simple plans without ruminating about some difficult ones. All in all, the overall situation of metacognitive strategy use is not satisfactory and needs to be improved. About the importance of metacognitive strategy use in English learning, a group of Chinese scholars have put it into practice, and they all get satisfying findings. For example, Ji Kangli (2002) and Zhang Yanjun (2004) respectively have conducted the general metacognitive strategy training. Ji's findings demonstrate that after training all the trainees would have a more frequent use in all the categories of metacognitive strategies, and that they would succeed in developing a correct sense of learning, independent thinking and autonomous leaning ability. Ji also suggests that it is feasible to integrate metacognitive strategy training program into the English teaching syllabus. Zhang reports that after training both the strategy use frequency and performance are improved and that training can heighten the autonomous learning ability.

6 PEDAGOGICAL IMPLICATION

Probably any researcher or educator who takes up the teaching profession as a career may possibly know a famous statement about the significance of good learning strategies: "Give a man a fish and he eats for a day; teach him how to fish, and he may eat for a life time." This statement adequately uncovers the importance of learning strategies. The same meaning is echoed in Wenden and Rubin (1987)'s statement "One of the leading educational aims of the research on learner strategies is to train a person to become an autonomous learner, who possesses the appropriate skills and strategies to learn a language in a self-directed way." So if a learner is provided with the answer directly, he may not bother to think further. Nevertheless, if they are required to search for the answers by themselves, they are forced to conduct their own thinking and learning. From this, we can observe that learning strategies weigh heavily in the process of learning a foreign language, for they are, more often than not, taken as tools for active involvement, which give rise to language competence. Appropriate learning strategies result in improved proficiency and greater confidence. Thus, learning strategies definitely have a deep impact on students' English improvement. In this paper, the author is inclined to focus his attention on the importance of metacognitive strategies. Based on the research and discussions above, some invaluable implications are given: First, based on the findings of the present research, for teachers or educators, they should help students form a good habit of self-study, which might begin with instructing them aims or learning objectives. If students do not have aims, they may lose directions and even do not know why and how to study. So it is quite necessary for teachers to help students set learning goals and guide them to make plans according to individual's different learning conditions. Second, teachers or educators should also instruct students

self-monitoring strategies. Teachers should always remind the students of whether they have mastered the knowledge or not, require them to focus their attention on the learning methods and the features of their own learning style, guide them to note down the difficulties in learning and supervise them to eradicate these difficulties so that they are able to do self-regulation and self-monitoring quite well in the process of learning English.

7 LIMITATIONS AND SUGGESTIONS

In spite of careful preparing and thoughtful thinking, the current research is far from perfect because of some limitations. Thus, some limitations for the present research are presented; meanwhile, some suggestions for further research are tentatively offered as well.

First of all, the study is carried out in a relatively ideal environment, since it is specifically designed for this paper. So there is a long way to go to make up this phenomenon. Second, every researcher knows that learning process involves many other factors in the individuality of the participants, such as motivation, social environment, learning style, personality, anxiety, family background, intelligence quotient and so on. However, in this research, these variables are not taken into consideration due to many reasons, such as the research level of the author, the limitation of time as well as the lack of research funds. Third, researches need to be conducted with larger scale considering the further accuracy of research studies.

REFERENCES

[1] Anderson, J. R. (1983). *The Architecture of Cognition.* Cambridge, Mass: Harvard University Press.

[2] Andrew, D. Cohen. (2000). *Strategies in Learning and Using a Second Language.* Beijing: Foreign Language Teaching and Research Press.

[3] Rubin, J. (1975). What "the good language learner" Can Teach Us. *TESOL Quarterly, 9,* 41–51.

[4] Stern, H. (1975) What Can We Learn from the Good Language Learner. *Canadian Modern Language Review,* 3, 304–318.

[5] Wenden, A. & Rubin, J. (1987). *Learner Strategies in Language Learning.* Englewood Cliffs, NJ: Prentice Hall.

[6] Wen Qiufang & Johnson R K. (1997). Second Language Learner Variables and English Achievement: A Study of Tertiary Level English Majors in China. *Applied Linguistics,* 1, 27–4.

Education Management and Management Science – Zheng (Ed.)
© *2015 Taylor & Francis Group, London, ISBN 978-1-138-02663-6*

The reform of Mechanical Principle teaching for cultivating outstanding mechanical engineers

Zhanguo Zhang
College of Mechanical Engineering, Beihua University, Jilin City, China

ABSTRACT: Outstanding mechanical engineers are the applied talents working in frontline of production in the field of mechanical engineering, who are different from the research-type talents. This paper established the roles and teaching objectives of Mechanical Principle course in cultivating outstanding mechanical engineers according to their cultivating objectives and core abilities. Taking training students' engineering practice ability and innovation ability as main line, the paper discussed the key links of this course, including theory teaching, practice teaching and extracurricular science and technology activities. The article can provide some references for implementing the Program for Educating and Training Outstanding Engineers and deepening the teaching reform of Mechanical Principle course.

Keywords: outstanding mechanical engineers; Mechanical Principle course; teaching reform

1 INTRODUCTION

The Program for Educating and Training Outstanding Engineers is an important measure to cultivate and create a large number of engineering and technical personnel with strong innovation ability and to meet the needs of China's development of economic society. It is also the inevitable choice to strengthen Chinese enterprises' core competitiveness, to build an innovation-oriented country and to walk a new road of industrialization.

The Program for Educating and Training Outstanding Engineers focuses on training students' ability to solve the practical engineering problems by combining theoretical knowledge and practical engineering projects. It aims to incorporate the idea of industry-education-research cooperation into students' entire cultivating process of engineering education, and to make the cultivating process far closer to society, market and production by simultaneous activities including teaching, practice, research and application.

Mechanical Principle is an important technical basic course for cultivating outstanding mechanical engineers. According to the industry standards and professional standards for cultivating outstanding mechanical engineers set in the Program for Educating and Training Outstanding Engineers, Mechanical Principle course should orient towards engineering and give prominence to the ideas of deepening foundation, emphasizing practice and focusing on application. Its teaching content should be simplified, the class hours of its theory teaching should be decreased properly and its practice teaching should be strengthened. Therefore, how to improve students' initiative and autonomy of studying the course so as to make students not only master its main contents, but also experience a certain degree basic skill training and scientific research training in short time to train and improve their practice ability and innovation ability has become one of the most important topics of the course teaching reform.

2 THE CULTIVATING OBJECTIVES AND CORE ABILITIES OF OUTSTANDING MECHANICAL ENGINEERS

The cultivating objectives of outstanding mechanical engineers are established as following: Guarantee students to have the preliminary ability of applying their acquired knowledge and skills to solve the basic problems in the field of mechanical engineering such as design, manufacture, marketing and service of mechanical product and construction, operation and maintenance of engineering project. Also, they should have great development potential and reach the level of mechanical assistant engineer.

Cultivating outstanding mechanical engineers must take the social needs as guidance, actual projects as background, engineering practice and scientific research training as main line, and focus on training their learning ability, engineering practice ability, innovation ability, management ability, communication skills, social adaptability and so on. The core abilities in the above are engineering practice ability and innovation ability. The engineering practice ability represents a good ability to deal with onsite problems and a certain ability to design and develop new products. The innovation ability represents the technical innovation ability which is needed in the processes of technology integration, transplant, implementation and renovation.

3 THE ROLES AND TEACHING OBJECTIVES OF MECHANICAL PRINCIPLE COURSE

In the curriculum of cultivating outstanding mechanical engineers, Mechanical Principle course not only plays the connecting link role for students studying related technical basic courses and specialized courses, but also plays the role of enhancing the adaptability and developing innovation ability for students engaging in mechanical product design and development work in the future.

According to the cultivating objectives and core abilities of outstanding mechanical engineers defined in the previous section of this paper, the teaching objectives of Mechanical Principle course are established as following: make students master the motion analysis, dynamic analysis and basic design theory and methods of mechanical system; make them master the modern design methods such as computer-aided design, computer-aided analysis, optimization design, reliability design, reverse design and so on; make them have the basic ability of designing kinematic scheme of mechanical system; train their engineering thinking, innovation consciousness and innovation ability.

4 THE THEORY TEACHING REFORM OF MECHANICAL PRINCIPLE COURSE

4.1 The reform of teaching contents

The reform of theory teaching contents should focus on training students' basic skill, analysis and design ability of mechanisms, independent studying ability, innovation consciousness and innovation thinking. On the one hand, lecture on "traditional" basic concepts, basic theories and basic methods, meanwhile, one must combine concrete engineering objects to think and study on the topics of the course. On the other hand, for training students' analysis and design ability of mechanisms using the modern mechanical design methods and computer technology, the section of the application of virtual prototype technology in analysis and design

of mechanisms should be added to teaching contents of the course by striving to integrate closely the analysis and design methods with today's computer technology.

4.2 The reform of teaching methods

Mechanical Principle course has strong theoretical and practical meanings. In its teaching process, teachers should focus on both imparting knowledge and training ability. Teachers should avoid tedious theory deduction, but rather emphasize the practical application. At the same time, teachers not only should impart basic concepts, basic theories and basic methods of the course to students, but also should make students learn to contrast, think, summarize and discover new problems, so as to obtain new knowledge and improve their innovation ability. For this purpose, we should discard the spoon-feeding pedagogy, but adopt the heuristic and discussion-based pedagogy, the perception model pedagogy, the pedagogy of setting up doubt, the quasi-practice pedagogy and so on.

4.3 The reform of teaching means

Varied teaching means should be utilized synthetically; the traditional teaching model should cooperate with the multimedia teaching model. Specifically, teachers should carry out the classroom teaching by the way of combining blackboard writing and computer-aided instruction courseware, enlighten and guide students' thinking with physical product, model and animation, strengthen the classroom interaction to train and improve students' independent thinking ability and make students quickly establish the perceptual knowledge to the frequently-used mechanisms, links and kinematic pairs. For example 1, while lecturing on the common mechanisms, teachers should demonstrate their basic structure and movement with their models or animations when teachers are explaining. For example 2, students have much difficulty in understanding the evolution of eccentric wheel mechanism. If the evolving process is made into an animation to make students personally experience evolving process and operate the model of eccentric wheel mechanism, the classroom teaching of this section will become more intuitive, visual and vivid. For example 3, while lecturing on the graphical methods of mechanism's analysis or design, teachers should ensure that explaining and drawing are synchronized which can make students deepen the understanding of the methods.

Second, teachers should strive to combine the methods of mechanism's analysis or design and the computer technology. In Mechanical Principle course, many graphical methods have the characteristic that their concepts are clear, such as the graphical methods of mechanism's kinematic analysis, force analysis, design of cam contour curve. The analytical methods of these teaching contents are often considered to be difficult to provide vivid concepts. In teaching, with the drawing function and visualization technique of computer, many computing processes and results of the analytical methods and numerical methods can be visualized and show the clarity of the graphical methods in concept. Even some parameters can be changed so as to draw inferences from one or more cases. Doing so push the students' study and research on Mechanical Principle to a higher level.

5 THE PRACTICE TEACHING REFORM OF MECHANICAL PRINCIPLE COURSE

5.1 The reform of experiment teaching

On considering students' cognitive rule and innovation thinking process, the hierarchical, modular and serialized experiment teaching content system which is interrelated with theory teaching contents should be constructed relying on mechanical basis laboratories. The experiment items can be set to three levels. The first level is base level, including the cognitive experiment and confirmatory experiment which help students deepen their comprehension of the contents of classroom teaching and focus on training their observing and analyzing abilities. The second level is improving level, including the comprehensive experiment and self-designed experiment in which practice contents are mainly aimed at the key points and difficult points of the course, knowledge with good coherence or integrated application value, which focus

on training students' experiment operating ability, manipulative ability, innovation consciousness and innovation thinking. The third level is the innovative experiment which focuses on training students' engineering practice ability, innovation ability and primary scientific research ability. The experiment items of Mechanical Principle are shown in Table 1.

According to the construction concepts of software driving hardware and resources sharing, taking the cultivation of students' engineering practice ability and innovation ability as objective, taking the building of experiment teaching system and its connotation construction as key points, we should integrate and optimize the existing mechanical basic laboratories to achieve the construction objectives of menus-style experiment items, hierarchical experiment contents, personalized training to students and informative laboratory management. We should strengthen the construction of key laboratories and set up open-ended and personalized experiment teaching platform for cultivating students' engineering practice ability and innovation ability, so as to provide a wider stage for their innovation practice. For the excellent students having surplus acumen in learning, besides completing the compulsory experiment items prescribed in the syllabus, they can also choose some optional experiment items according to their interest. The experiment task of chosen experiment item can be assigned in class, its experiment scheme can be designed outside class and its experiment results should be assembled, debugged and accepted by teacher's guidance, evaluation and assessment in laboratory. In this elective way, the degree of freedom with which student selects experiments can be increased significantly; the effect of teaching students according to their aptitude must be achieved.

Table 1. The experiment items of Mechanical Principle.

Experiment items	Class hours	Experiment type
Mechanisms cognizing	2	Cognitive
Rotor balancing	2	Cognitive
Mechanism kinematic diagram drawing	2	Confirmatory
Enveloping principle and geometric parameters measuring of the involute gear	2	Confirmatory
Dynamic speed governing of the mechanical system	2	Comprehensive
Mechanisms assembled by the crank, slider, guide bar and cam	2	Comprehensive
Kinematics and dynamics simulations of mechanisms based on secondary development of the software ADAMS	Open	Comprehensive
Modular machine design and assembly	Open	Self-designed
Creative combination and test analysis of mechanisms	Open	Self-designed
Mechanism combination and innovative design	Open	Innovative
Creative combination of fischertechnik based on the secondary development	Open	Innovative

5.2 *The reform of course design*

As a practical training link for college students, course design is an important part which can make them deepen the understanding of theoretical knowledge and improve the ability to solve practical engineering problems by using learned knowledge synthetically. At present, both the course design of Mechanical Principle and the course design of Mechanical Design are carried out separately, they all are aimed at digesting respective course contents. The former mainly trains students' abilities to design kinematic scheme of mechanical system, to design and analyze mechanisms and design creatively. The latter mainly trains students' abilities to select material of machine part and design mechanical structure.

Based on the needs of cultivating outstanding mechanical engineers, we should integrate the two course designs. Taking off from the whole process of mechanical system design, the course design should blend the scheme design, kinematic and dynamic analysis and dimension synthesis of executive mechanism, the scheme design of transmission system, the structural design of key machine part and so on together, so as to ensure that the training contents are complete, systematic and integrated.

The course design topics can be from which given by teachers or set by students themselves. We should promote that the course design is combined with the practical application and development of society and technology by involving the practical engineering problems, research projects and subject competition projects with the contents of course design.

6 DEVELOPING EXTRACURRICULAR SCIENCE AND TECHNOLOGY ACTIVITIES EXTENSIVELY

In order to fully exert the educating function of second classroom and to guide and help students to perfect their knowledge structure, to train their innovation consciousness and practice abilities, to improve their comprehensive quality, for the excellent students having surplus acumen and interest in innovation practice, we should lead them to actively participate in extracurricular science and technology activities set up in the second classroom. These extracurricular science and technology activities mainly include scientific research training, National Undergraduate Innovative Test Program, State Undergraduate Innovative Training Program and subject competition and so on. The activities can well train students' innovation design abilities of the principle of function and kinematic scheme, train their design and analysis abilities of

mechanisms, and further develop their engineering practice ability, innovation ability and research ability. In turn, the representative results of scientific research can be integrated into case teaching, experiment teaching and course design, namely, can nurture and promote the teaching work of Mechanical Principle course as a high-quality education resource.

The scientific research training is an effective way to train students' abilities to do research independently, which can effectively help students have more industry background knowledge and technology frontier knowledge. Meanwhile, it can cultivate their innovation consciousness, train their innovation thinking and improve their innovation ability. In order to create the autonomous and research-based learning conditions for students, universities should make the most of laboratory, engineering training center, engineering research center, undergraduate scientific and technological innovation practice base, research platform and innovation platform to construct the multi-dimensional and open-ended scientific research training bases. At the same time, universities should establish the effective and positive development mechanisms that scientific research activities are brought into the scientific research training bases in accordance with the principles of leaving more room in time, having sufficient sites in space, sharing full degree of freedom in mechanism and providing adequate insurance in condition. Universities should also advocate the team-working learning model that incorporates teachers, senior students and junior students, and advocate the way that combines long-term culturing with short-term training to train students' innovation spirit and practice ability. The scientific research training projects should have engineering pertinence, diversity and hierarchy, that mainly include the teachers' scientific research projects, the laboratory equipment modification projects, the subject competition projects at all levels and the projects set by students themselves that interest them. The measures described above can provide them with ample opportunity to study independently, good conditions for their independent studying, abundant research resources and a broad range of opportunities for communication.

According to the guiding ideology of encouraging everyone involved, strengthening process training, promoting the intersection between different disciplines and balancing the construction of study atmosphere, universities should strive to construct the state-level, province-level and school-level implementing systems of the National Undergraduate Innovative Test Program and State Undergraduate Innovative Training Program, which can guide undergraduates early to involve in scientific research and innovation practice and can strengthen their innovation ability. During carrying out the two above programs, universities should strengthen the projects' process management, promote interdisciplinary communication and collaboration, implement multi-step reply and exchange mechanism, increase the rate of projects accomplished on schedule year by year, so as to realize the desired objectives of the programs.

The subject competition is an effective link which can make students show themselves and improve their abilities, is an important supplement to practice training in curriculum. Universities should actively create conditions to carry out school-level event of subject competition, to expand competition items, to design subject competition with wide coverage, to nurture brand competition and gradually to form state-level, province-level and school-level subject competition systems. In the meantime, universities should take the "Challenge Cup" National Undergraduate Extra-curricular Academic Science and Technology Works Competition, National Undergraduate Mechanical Creative Design Competition, China University Robot Contest and so on as carriers which can drive the development of innovation practice activities, and should make the subject competitions reach the level that everyone can be involved and competition events are organized every month. Only then can we improve effectively students' innovation spirit and practice ability.

REFERENCES

[1] Z.Z, Zhang. 2009. How to train students' practice and innovation abilities in Mechanical Principle teaching. *Career Horizon* (7):98–99.

[2] Z.Z, Zhang. 2007. Exploration and practice of Mechanical Principle practical teaching reform. *New Course* (11):40–42.

Education Management and Management Science – Zheng (Ed.)
© 2015 Taylor & Francis Group, London, ISBN 978-1-138-02663-6

Research on the development of Chinese higher education based on fractal theory

Xiao Wang & Xiaolan Zhu
School of Civil Engineering, Nanjing Tech University, Nanjing, Jiangsu, China

ABSTRACT: Based on fractal theory, the authors research the development and future trend of higher education from the perspective of time fractal, spatial fractal and social fractal. Service attributes of higher education are summarized and the life experience of higher education is pointed out, which provide the basis for analysis of other relevant phenomena and problems of higher education.

Keywords: fractal; higher education; life experience

1 INTRODUCTION

1.1 *The fractal theory*

Fractal theory is a new branch of modern mathematics, showing the change of nature's self similarity. Its essence is a new world outlook and methodology, which provides a new view of nature, time, space and movement. It advocates understanding infinite from the limited knowledge, seeing whole from its part and laying emphasis on process and transition state, and has useful method in the field of science and sociology.

1.2 *Research background*

With the rapid development of Internet technology, the era of knowledge economy is coming. And the target of deepening reform of the cultural system has been proposed in the eighteenth plenary. China's higher education has ushered the opportunity and challenge of reform. Researching on higher education and its characteristics of fractal can help to further understand the rules and trends of higher education, which is helpful to promote institutions of higher learning to find the rational orientation, reform the teaching content, teaching methods, teaching management, and cope with competitive crisis.

2 THE FRACTAL ELEMENT OF HIGHER EDUCATION

Fractal element, which is an invariant of constant iteration, is used to describe the characteristics of fractal. Along with the development of higher education, the fractal element is the morphology of teaching and learning.

The dominant structure of teaching relationship is made up of the morphology of teaching and learning, the educated, educators, teaching content and teaching methods. The hidden structure of teaching relationship is made up of the relationship about human, knowledge and culture. As a unity, the relationship between teaching and learning reflects the particularity and the universal nature of higher education. Different teaching relationship is exactly different social forms, can't be separated from the teaching and learning morphology. Educational activities change with the hierarchical or substantial changes of educators, the educated, teaching content and teaching means [1].

3 THE FRACTAL OF HIGHER EDUCATION

Time fractal can be used to understand the maturation and promotion of China's higher education. Spatial fractal can be used to know social development and the future trends of China's higher education. Social fractal can be acquired from the use and management of social resources.

3.1 *The time fractal of higher education*

The maturation and promotion of our country's higher education can be seen from the concept of higher education appearing, educational activities changing from weak to strong and improvement of relationship between teaching and learning. Modern education, which is closely linked to the urgent demand of survival, is a process of self education, self adjustment and adaptation repeated in practice. Higher education was firstly put forward by RenY in-schooling in 1902. In the time of social change, education is a tool of statecraft, which is particularly prominent. Higher education has gone through a tortuous development, led the ideological liberation and become value oriented.

In the promotion of the third industrial revolution, cloud technology, networking and big data make the information, data knowledge available everywhere. External environment led to the fact that various teaching forms have been produced, such as distance education, mobile learning and MOOCs. The teaching content is unprecedented and open, teaching means multi sensory and vivid, the space of teaching environment can be extended infinitely.

3.2 *The spatial fractal of higher education*

Communicational range and regional development of higher education has undergone major changes. Since the reform and opening up, China's political, economic system reform and the rapid development of economy, social information provide rapid development of higher education in a favorable social environment. But at the same time, geography, social politics, economy, culture, population, nationality and the education system itself have an impact on the spatial distribution and structure of higher education.

The era of knowledge economy, online education bring new educational experiences to society. The scope of educational communication is no longer restricted by time and space. Convenient medium because of Internet provide convenience for the exchange of information. Online education provides the possibility to achieve regional balance, universal education, and lifelong education.

3.3 *The social fractal of higher education*

The development of social politics, economy, culture makes the social masses take great part

in higher education. The utilization of social resources by relying on technical support becomes more important. Higher education administration and school management needs reform and teaching management idea based on knowledge management will meet disruptive change.

Colleges and universities in the era of knowledge economy have become an open platform. Profound changes have taken place in the educated, educators, teaching contents and teaching methods. The formation of educators has no fixed pattern. The education is no longer limited to the students in the school. The teaching content becomes more open and free with the development of Internet. Educators are free to choose the curriculum. Experiential and virtual learning constitute rich teaching methods.

It is not just the teaching-learning relationship, profound changes have taken place in teaching concept. Indoctrination in education traditionally is subverted. The human society will move from the material-resources type in the past to knowledge-resources type in the future. Knowledge of the possession and allocation and knowledge production, dissemination and consumption will become the basic activities of the new economic era.

The teacher is a lifeline of knowledge, a leader of knowledge of life, a producer of knowledge production, a partner of survival of knowledge [2]. Organizers of personnel training are learning partners with learners, or have relationships such as collaboration and service. For a school, the main competitive power lies in the mining and use of tacit knowledge. Along with the change of higher education industry, guiding the transformation from tacit knowledge to explicit knowledge occupies more prominent position. The main duty of teachers is to provide planned, purposeful, systematic education, and guide the students to make tacit knowledge explicit.

4 THE FRACTAL CHARACTERISTIC OF HIGH EDUCATION

Fractal dimension is a quantitative description of the fractal characteristic. The concept has a great influence on modern science and philosophy, which combines the nature, society and thought in a series of dialectical relationship at a higher level. In practical, there are a variety of definitions and corresponding calculation methods of fractal dimension, such as the Hausdorff-Besicovitch dimension, information dimension, correlation dimension and so on. Fractal development of higher education is a continuous nonlinear process; its fractal function is more complicated. The accurate description of higher education fractal characteristics contribute to grasping the law of

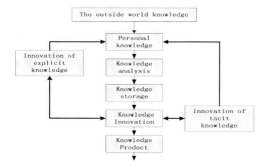

Figure 1.   The knowledge transformation model in colleges and universities.

education and teaching learning relationship. Using the mathematical method of quantitative calculation, combined with the limits of resources, to rationally determine the dimensions of various functions and accurate position. All the above have a significance to find a reasonable position in the education market.

5 SERVICE AND EXPERIENTIAL ATTRIBUTES OF HIGHER EDUCATION

5.1 Service attributes of higher education

In 2001, China joined the World Trade Organization. One of the trade agreements defines education as the service industry. Four service functions of higher education—personnel training, scientific research, social services and cultural heritage have changed in meaning. The significance and responsibility of serving the society have the dominant advantages, coupled with great vitality on the market in the education industry, led to the change in ideology and organization of Universities and colleges.

5.1.1 Social service

Higher education to provide social service is the inevitable outcome of market competition and market orientation is the inevitable result of the establishment of social service function of higher education [3]. Universities have their own legal status in comparison and competition with other institutions in society.

Education serving society reflects providing service to country, society and family service. In addition to cultivating all kinds of talents needed in the age of knowledge economy, universities also use their knowledge advantage to serve the majority of learners from society, which is the main content of social services provided by universities in the era of knowledge economy. Colleges and universities provide distance learning opportunities no matter whether the person has formal education or no degree education and provides learning opportunities in different time. In the third industrial revolution, more and more universities sign contracts with social organizations, feed knowledge base, science and technology enterprises jointly, including outputting knowledge-based products-books, audio, video, provide training, cooperative schools etc.

5.1.2 Personal service

To provide personal service and even tailored education service is the development trend of higher education in the future. Individuals through having the opportunity to get education, to learn the basic knowledge, then analyze, innovate and then acquire personal wealth and social status. Education can be seen as a way of labor reproduction, is not only a kind of conspicuous consumption, but also production and investment hidden with potentially enormous value.

Modern education is not only facing the dissemination of knowledge, but also even more is the needs of customers. Education can provide all aspects of education ordering service to customers. To individuals who accept the education services, the cost of education investment on the one hand, is paid for enjoying current education service, on the other hand, is an investment for future.

5.2 Learning services

Colleges and universities provide learning services to masses, which is the core service.

5.2.1 Establish platform

Establish a blended community communication platform, whose core is colleges and universities.

5.2.2 Services provided by colleges and universities

As the core of community platform for communication and cooperation, colleges and universities provide a variety of inquiry learning mode and experiential learning services for diversified education theme, to meet the personalized and life needs. The core competence of university embodies the ability to show personalization and integrate resources, is reflected through customers' feedback about satisfaction.

Figure 3 is the concrete teaching activities provided by colleges and universities based on the Internet. Feedback information is received in the process of education.

5.3 Experience of higher education

Modern education is not limited to a specific age, specific academic background, and is not limited to time and place. Because of the values of individual, influence of intelligent features, psychological expectation, people have rational and non-rational demand of individual demand for higher education.

Rational demand for the higher education is knowledge demand, economic demand, political demand and spiritual demand. Knowledge requirement means that the educated is willing to get more advanced knowledge and desires to receive higher education. Economic demand is the preparation for occupation, the educated accept the higher education, make themselves more competitive in the society, so as to obtain more work opportunities and have a good working environment, and enjoy more affluent material life. People's social status is determined by his social position (i.e. social

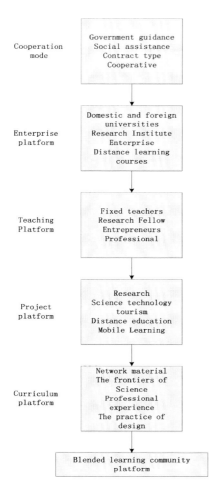

Figure 2. Blended learning community platform.

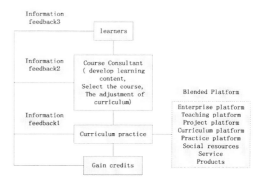

Figure 3. Experience of learning services.

stratification), and the role of education is to make people to flow in the upper social classes, which constitute the political demand. Higher education is an important way for people to continue to improve their quality of life and to realize own value, this is the spirit demand. Factors of non-rational demand come from others and traditional values. The educated can be free to choose the opportunity for higher education and training mode according to individual needs, access to credit and recognition.

The barrier of higher education has been broken; many of the participants to the university bring a lot of experience for the services provided by universities. To the educated, such services are opportunities to enjoy the life. The fact that higher education meet the needs of every individual, universities have the responsibility to improve social ideological style, improve the public intellectual accomplishment, pure national taste, provide true principles for popular enthusiasm, provide an assured target for popular aspiration, extend the ideological content and maintain the thought in waking state in the era, promote the use of political power. Spiritual lifestyle is to determine the level of an individual's quality of life.

University, as a noble spirit habitat, to meet the spiritual needs of the educated individuals, provides a unique, suitable environment, creates conditions for people to learn thinking, imagination, be aware of wisdom and beauty. This is a precious wealth that university still won the independent dignity and meaning of existence in the spring tide of market economy, is also the life force and the soul of the foundation of university.

6 CONCLUSION

This paper describes fractal characteristics of higher education from the view of time fractal, spatial fractal and social fractal, proposes the reform of knowledge-teaching relationship and educational management based on the conversion from tacit knowledge to explicit knowledge. At the same time, facts reveal the future trend of higher education—social services combined with personal services. In this paper, the specific form of services provided by colleges and universities is also described. Finishing school is a life experience full of spiritual wealth.

REFERENCES

[1] Fang Yaomei. The epistemological significance of fractal theory in Higher Education. Research on Higher Engineering Education, 1999, 4: 31–34.
[2] Fan Zeheng. The core activities and way of realization of knowledge management in Colleges and universities and. The modern education management, 2010, 1: 43–46.
[3] Yu Rong. Reflections on social service provided by universities in the era of knowledge economy. Journal of Chaohu College, 2009, 4: 108–113.

Education Management and Management Science – Zheng (Ed.)
© 2015 Taylor & Francis Group, London, ISBN 978-1-138-02663-6

Interpreting *Starbucks*' image-creating and its functions in Chinese markets: A semiotic perspective

Siyuan Huang

Faculty of Business Administration, University of Macau, Taipa, Macau

ABSTRACT: *Starbucks*' phenomenon has been the debatable topic in the academic field. Numerous scholars from marketing, psychology, anthropology, linguistics, etc. have touched on this issue from their own perspectives. Yet it is found that few research is conducted on how *Starbucks* creates its image and what ontological and social pragmatic functions *Starbucks* as a sign has from the semiotic perspective. The present study avails itself of semiotic theories, supplemented by marketing and discourse theories to inquire into this issue so as to expose how *Starbucks* as a sign emerges and functions ontologically and socio-pragmatically. This study reveals that *Starbuck*'s image creation is the process of signifying. When the status of *Starbucks* as a Sign is established, it has multiple functions and social pragmatic implications. What *Starbucks* sells is not just a substantive coffee, but goods-and-services, and above of all, the distribution of social goods.

Keywords: *Starbucks*; image-creating; ontological function; social pragmatic function; semiotic approach

1 INTRODUCTION

Starbucks, the largest coffeehouse in the world, has been universally acclaimed as a miracle in the commercial empire. Its operation philosophy and operating paradigm as well as the ways to get enormous profits have been a debate in the academic field. A large number of scholars in the domains of marketing (i.e. Thompson and Arsel 2006), psychology (i.e. Michelli 2006), anthropology (i.e. Gaudio 2003), linguistics (i.e. Yang 2012), etc. have done many in-depth analyses of this commercial wonder from their own perspectives. They mainly discussed such issues as how the brand creates its own unique image, how consumers converse with each other in *Starbucks*, how *Starbucks* gets tremendous profits by designing a set of special mugs and labeling shocking prices, and so on and so forth. As for how *Starbucks* obtains its huge profits in Chinese markets, a Chinese scholar Yang (2012) gave a detailed analysis, but his analysis is carried out from the pragmatic perspective, and he, based upon pragmatic theories in linguistics, just analyzed how *Starbucks* realizes its amazing economic interests by employing some clever or even cunning strategies. Actually, we hold that *Starbucks*' victory is not the matter of 'coffee selling', rather it is the victory of 'a cultural sign'. Then how does this 'cultural sign' settle itself in Chinese markets and become an idiosyncratic icon? The present study will attempt to use semiotic theories, supplemented

by marketing and discourse theories, to investigate into this issue so as to discover the process of *Starbucks*' image-creating, the ontological functions that *Starbucks* have and the social pragmatic functions that *Starbucks* is used to fulfill.

2 INTRODUCING RELATED THEORIES

2.1 *Situating signs and discussing Roland Barthes' semiotic model*

Semiotics, simply speaking, means the scientific study of signs. Signs are meaningful units, which are constituted, according to Saussure, by two components: signifier (a sound pattern) and signified (a concept) or according to Peirce, by three components: the represent amen (the form), an interpretant (the sense) and an object (the thing that the sign refers to). For Saussure, a sign, just like two sides of a coin, must have a signifier and a signified, and for Peirce, a sign, just like a triangle, must have the form, the sense and the object. The same signifier might stand for a different signified, hence a new sign. In the meanwhile many signifiers could stand for the same concept, so, many pairs of signifier and signified constitute different signs.

From the above discussion, we can arrive at such conclusions. First, with the change of either signifier or signified, the conventional sign will be changed into another one. Second, the creation of signs cannot be separated from the context.

Third, signs—creating is actually meaning-making. However, meaning here only refers to denotative meaning. Can signs have connotative meaning? How can signs create connotative meaning based upon the denotative meaning?

Roland Barthes (1915–1980), a leading French structuralist and Saussure's follower, used Hjemslev's theory of connotation and "defined a sign as a system consisting of E, an expression (or signifier), in relation (R) to C, a content (or signified): E R C. Such a primary sign system can become an element of a more comprehensive sign system. If the extension is one of content, the primary sign (E_1 R_1 C_1) becomes the expression of a secondary sign system: E2 (= E1 R1 C1) R2 C2. In this case, the primary sign is one of denotative while the secondary sign is one of connotative semiotics. (NÖth 1990:310–311) What Barthes implies is that a sign contains the denotative sign (i.e. primary sign) and the connotative sign (i.e. secondary sign). The former expresses denotative meaning and the latter connotative meaning. In Saussure's sense, Barthes' model can be restated as follows. First, the denotative sign consists of signifier 1 + signified 1, and the meaning that this primary sign communicates is conventional. Second, signifier 1 + signified 1 can serve as a new signifier, expressed as signifier 2, which has its new signified, expressed as signified 2; signifier 2 + signified 2 forms into a new sign. It is obvious that the meaning of this new sign is larger than the denotative sign. In Barthes' sense, this meaning is 'ideological', which, in our stance, equates to "political". What is the appropriate definition of 'ideological" or "political"? This will be further dealt with in the following.

2.2 Christie and Martin's multi-layed model of semiosis

When elaborating on the relationship among genre, register and language, Christie and Martin (1997:13), by using Hjilmslev's notions, set up a model in which such notions as connotative semiotic, denotative semiotic, stratified context plane, expression plane, stratified content form and expression form are in use. In this study, these notions and the model is borrowed and modified to be put into use.

2.3 The working definition of politics and function

In this study, Gee's (2000:2) definition about politics will be adopted. According to Gee, "ideological" or "political" is related with the way that "social goods" are distributed and "social goods" refer to power, status, worth, intelligence, etc. As for "function", we combine Peirce's notions "icon", "index" and "symbol" with "interaction" to indicate that

Starbucks as a sign has a number of social pragmatic functions.

2.4 Comment

The above theoretical introduction reveals that semiotic analysis, supplemented by discourse and marketing theories, helps us, first, to "become more aware of the mediating role of signs and of the roles played by ourselves and others in constructing social realities" (Chandler 2002:14); second, to realize that meaning is not just 'contained' in an object or a thing or an event; third, to uncover what is really hidden behind the sign and how a reality, for example, a commercial miracle such as Starbucks, is constructed and maintained. Thus, in the following, these theories will be employed to analyze Starbucks' image-creating, its ontological and social pragmatic functions in Chinese markets.

3 ANALYZING STARBUCKS' IMAGE CREATION, ONTOLOGICAL AND PRAGMATIC FUNCTIONS

In 1999, Starbucks started its business in China. Since then, it has been growing tremendously and successfully, making China become the "Second Domestic Market" of Starbucks. There is no doubt that Starbucks performs a miracle, not only establishing its outstanding position in the commercial field, but also embedding its coffee icon or its image in China. From the semiotic perspective, it can be said that Starbuck's image creation is the process of signifying.

3.1 Starbucks: from materialization to signifying

Starbucks originally is the name of a man who is a protagonist in an American novel. To make this signifier (sound image)—Starbucks signify another signified (concept), namely to have people psychologically associate the sound image Starbucks with a concept 'coffee brand', the founders and their working teams must have done a large number of things so as to speed up the process from materialization to signifying. They must have formulated their own special long-standing strategic goal, the core business philosophy and the business pattern. In addition, they must have opted for the shops, the coffee, the staffs, the language, etc. When their efforts pay off and when people recognize "Starbucks" as a coffee brand, a new sign Starbucks appears. In Barthes' sense, this is the primary sign, functioning as denotation. This primary sign Starbucks as a coffee brand then can be an expression, which has its own content. This expression and content forms the secondary sign, which implicates more beyond a coffee brand. Up to the present state, *Starbucks* has finished its dematerializing process and reached to the signifying stage. In the process of semiotization, what elements are involved in?

Based on Christie and Martin's multi-layed model of signs (1999:7), we propose that two planes and four categories are vital in the process. Two planes refer to denotative semiotic plane and connotative semiotic plane. Four categories are the form of expression, the form of content, the local context and the macro context. The relationship between denotative semiotic plane and connotative semiotic plane is a kind of realization, that is, the connotative is realized by the denotative. Connotative semiotic plane is a stratified context plane, and denotative semiotic plane is a stratified expression plane. The stratified context plane is composed of two categories: the macro context and local context. For *Starbucks,* the macro context includes "the long-standing strategic goal" and "the core business philosophy", and the local context "the business pattern". The stratified expression plane is made up of two parts: stratified content form and the expression form; the former contains two elements: one is the meaning of the sign, the other is the ways including the ostensive means such as the place, the outside and interior decoration of the shop; the breed, quality, processing and price of coffee, the quantity and quality of working staffs; the conversational pattern, etc. When the denotative semiotic entailing the connotative semiotic (i.e. *Starbucks*) is accepted by a (speech) community, this denotative semiotic is not just an empty form, rather it has turned into a form of life and a way of being. Also in the process of dematerialization of *Starbucks,* quite a number of factors play a part. Thus, it can be concluded that when the signifier *Starbucks* signifies a new concept—a coffee brand with its special and diversified features, they go through the process from materialization to signifying. This process makes this *Starbucks* as a coffee brand totally different from that *Starbucks* as the person's name. So it is evident that the emergence of this new sign creates a new image. When this new image attaching to this sign is accepted by a speech community, it will take roots and become routinized. To this stage, it has become a cultural sign. As a cultural sign, *Starbucks* performs multiple functions and can do many things for the society.

3.2 *Starbucks as a sign: ontological functions*

When *Starbucks* enters into the Chinese markets and is accepted by Chinese, it has changed into a sign. The formation of this sign, just as what has been discussed in the above, involves the participation of expression form and stratified content form on the denotative semiotic level, and most importantly, expression plane and stratified context plane on the connotative semiotic level. Additionally, humans' psychological association matters here, because the two planes and the four categories are just parts and parcels of a sign, which need humans' mental processing to assemble them together. Thus, it can be tentatively said that this sign is not a "pure sign", actually, while it fulfills the role of icon, it has the symbolic function, and also indexical function, in spite of the fact that in certain contexts, the profiling function may be iconic, or symbolic or indexical. As for icon, symbol and index, Peirce gave a detailed illustration (1985). Based upon Peirce's definitions, the three functions can be further explained as follows. First, it can be inferred that when it serves the function of an icon, the image of *Starbucks* possesses qualities which "resemble" those of this kind of coffee shop. On top of that, it caters for the social psychology of the present-day Chinese people. When we Chinese visualize the sign with our own eyes or in our minds' eyes, we will have an analogous structure and simultaneously associate it with a kind of life or habit or disposition or quality. So Starbucks functions as a symbol. That is to say, *Starbucks* is not just a product, it represents a kind of consuming or fashionable culture. What it sells is not just coffee, instead it sells goods—and—services of high quality. And the consumers go there not just only for drinking a cup of coffee, but for a special higher level of physical and mental embodiment, which contributes to the construction of their way of life, humans' interactional patterns, consuming views, etc. Apart from the symbolic function, *Starbucks* has the indexical function, that is, to indicate something which is conceptually closely connected. Since *Starbucks* is a popular brand coming from America, some of us tend to associate it with such notions as "fashionable", "modern", "idiosyncratic", "free", "taste", "elegant", "a place in which social identity and status can be manifested", etc. Then it can be assumed that *Starbucks* has such indexical meanings as "being fashionable, being modern, being idiosyncratic, being free, being taste, being elegant, a site in which our social identity and status can be implicitly manifested, etc." In this way, when we go into *Starbucks*, all of the elements involved project ourselves as "a socially-situated identity" (Gee 2000:12) and meanwhile project ourselves as engaged in "a socially-situated activity" (Gee 2000:12).

Therefore, it can be proposed that owing to the ontological multifunctionality of Starbucks as a sign, even though it is quite expensive, it still sells well in Chinese markets.

3.3 *Starbucks as a sign: social pragmatic functions*

When *Starbucks* has been established as a sign, it has a number of social pragmatic functions.

The first one is to communicate information. *Starbucks* as a sign can have life only through communication. In communication, the iconic and the symbolic functions of *Starbucks* can be produced and extended. In the meanwhile, through communication, the messages embedded into *Starbucks* are shared.

The second one is to help the growth of cognitive abilities. Humans live in the world of signs. Without signs, there will be no world. In order to understand the world, we have to rely upon signs. So signs are the media through which humans cognize the world. In other words, signs are the media through which our cognitive abilities are cultivated. And the growth of our cognitive abilities is based upon the concurrent semiotic actions. The emergence of *Starbucks* as a sign makes us realize that there is such a special kind of coffee empire existing in the world and the effect that it brings, which broadens our vision, increases our cognitive abilities and accumulates our knowledge so as to help us to have a better understanding of the world.

The third is "to scaffold the performance of social activities" (Gee 2000:1). *Starbucks* as a culture, a social group and an institution shapes social activities, including drinking coffee, chattering, having party, relaxing, interaction, doing business, etc. In the meantime, *Starbucks* as a culture, a social group and an institution "gets produced, reproduced, and transformed" (Gee 2000:1) through these various sorts of human activities. In the process, the participants' social goods as power, status, wealth, intelligence, etc. are well distributed.

The fourth is "to scaffold human affiliation within cultures and social groups and institutions" (Gee 2000:1). When we log into the authoritative website of *Starbucks*, the arrangement of the website leaves us a deep impression. There are five commercial items on the top: Coffee, Menu, Coffeehouse, Responsibility, Card, Shop, each of which has its subcategories. Take Coffee as an example. The label Coffee has four columns: Browse by Profile, Browse by Form, Espresso, Starbucks Reserve Coffee, each of which contains a long list of description. For instance, under the heading Espresso, the following types of drinks can be seen: Macchinato Beverages, Latte Beverages, Mocha Beverages, Cappuccino Beverages, Americano Beverages, Espresso Beverages. From the design of the website and the language used, we find at least two "who-doing-whats" voices (Gee 2000:24). One is the seller telling the consumer their goods- and -services. The other one speaks with the authoritative voice, which implies that they are in possession of more knowledge about coffee making, coffee breeds, etc. This reflects a kind of relationship between an expert and laymen. It is obvious that the design of the website and the language used in *Starbucks* can show the identity of the participants and the activity that the participants are doing.

4 CONCLUSION

Starbucks is not just a commercial phenomenon, but also a cultural phenomenon. In the above, we analyzed this phenomenon from the perspective of semiotics. The analysis reveals how *Starbucks* grows up from a person's name to an unknown coffee shop which in turn is transformed into a commercial wonder. Our view point is that *Starbucks*' success is the success of *Starbucks* as a Sign. In other words, without the formation of *Starbucks* as a Sign, there will be no success for *Starbucks*. When the status of *Starbucks* as a Sign is established, it has multiple ontological functions and several social pragmatic functions. *Starbucks* as a Sign creates not only a commercial empire but also a cultural model. What it sells is not just a substantive coffee, but goods-and-services, and above all, the distribution of social goods (i.e. power, status, wealth, etc.).

REFERENCES

[1] Chandler, D. 2002. *Semiotics: The Basics*. London: Routledge.
[2] Christie, F.& Martin, J.R. 1997. *Genre and Institutions: Social Processes in the Workplace and School*. London: Cassell.
[3] Gaudio, R.P. 2003. Coffeetalk: Starbucks™ and the commercialization of casual conversation. *Language in Society* 32(5): 659–691.
[4] Gee, J.P. 2000. *An Introduction to Discourse Analysis: Theory and Method*. Beijing: Foreign Language Teaching and Research Press.
[5] Michelli, J. 2006. *The Starbucks Experience: 5 Principles for Turning Ordinary Into Extraordinary*. New York: McGraw-Hill (1. edition).
[6] NÖth, W. 1990. *Of Semiotics*. Bloomington and Indianapolis: Indiana University Press.
[7] Peirce, C.S. 1985. Logic as semiotic: The theory of signs. In R.E. Innis (ed.), *Semiotics: An Introductory Anthology*. Bloomington: Indiana University Press.
[8] Thompson, C.J. and Arsel, Z. 2004. The Starbucks brandscape and consumers (anticorporate) experiences of glocalization. *Journal of Consumer Research*. 31(3): 631–642.
[9] Yang, L. 2012. Perlocutionary act—An analysis of the economic factors underlying the special linguistic phenomena of Starbucks. *Shandong Foreign Languages Teaching* 33(5): 31–35.

Education Management and Management Science – Zheng (Ed.)
© 2015 Taylor & Francis Group, London, ISBN 978-1-138-02663-6

Reform and practice in teaching data structures

Xin Liu
School of Information Engineering, Shandong Youth University of Political Science, Jinan, China
Key Laboratory of Information Security and Intelligent Control in Universities of Shandong,
Shandong Youth University of Political Science, Jinan, China

ABSTRACT: This paper analyzed the characteristics of data structure courses and many factors that influence teaching effectiveness. In order to overcome these deficiencies, a series of effective teaching reform measures were proposed, i.e., implementing teaching content convergence, adopting diversified teaching methods, performing hierarchical teaching, reforming experiments and evaluation methods, establishing the curriculum group and so on. It is pointed that these measures can effectively stimulate students' enthusiasm for learning and significantly improve their practical ability, thus achieving the teaching goals of improving students' analytical and programming skills.

Keywords: computer education; data structures; teaching reform; hierarchical teaching; case teaching

1 INTRODUCTION

Data structure is a subject designed to study the operation objects and the relationship between them, which emerge in the problem of non-numerical programming. Data structure lays the foundation for program design as well as realization of compilers, operating systems, database systems, and other applications [1]. Currently, the data structures course has been considered as one of the core curriculums of computer-related specialty. In addition, it has been set as an important professional elective of electronic information specialty in more and more colleges and universities.

However, the current data structures teaching has the following prevalent shortcomings: (1) bad connection with preceding courses, affecting the learning effect; (2) backward teaching methods and poor teaching effectiveness; (3) single teaching content, not taking into account the different learning needs of students at different levels; (4) excessive validation experiments, lacking of comprehensive and practical content; (5) unscientific assessment methods, failing to call up the learning enthusiasm of students. In order to overcome these problems, and to achieve the goal of "take the practical ability as the core, and focus on improving students' analytical and programming skills", we propose a serials of teaching reform measures, which are involved with teaching content, teaching methods, experimental content, evaluation methods. It is emphasized that when combining these measures with the construction of curriculum group, network teaching platform and teaching team, we can effectively mobilize the enthusiasm of students and improve the effectiveness of teaching.

2 THE IMPROVING WAYS OF TEACHING

2.1 Establishing good connection between the data structures course and underlying courses of programming languages

Data structures course is a computer-related professional basic course. Students generally feel this course is not easy to get started, and difficult to understand. In addition to strong and highly abstract theoretical characteristics of the course, for those students who have poor basis of high-level language, the algorithm descriptions of the course have become a huge obstacle to comprehension. Data structures textbooks usually choose C language as the description language for algorithms, and make high requirement for functions, pointers and the structures, which are also teaching difficult points in the course of "C Programming". Unfortunately, there is often poor curriculum connection between data structures and its prerequisites C programming, resulting increased students' learning difficulty in data structures. In this regard, our approach is to encourage regular exchanges of teachers who teach the data structures curriculum and those who teach the C programming curriculum, and to set parts which closely tie with data structures as the focus content of C programming. Meanwhile, in our computer professional training program, the C programming curriculum is scheduled in the first year and its teaching process is divided into two semesters. Thus, when we offer the data structures course in the third semester, students have a better grasp of the C language. In the other hand, after the introduction section, the teachers are suggested arrange 4–6 hours devoted to strengthen the mastery of the pointers and the structures.

In recent years, the current domestic mainstream data structures textbooks use simplified C-like language as a tool for describing algorithms. The benefit of this approach is that many syntax details of the underlying C language can be shielded, thus allowing students to focus on the understanding of the algorithm design ideas. However, students can not see the executable source program, and therefore they are often psychologically skeptical to the question that if the algorithms can be implemented properly. Moreover, most of textbooks do not show "how to translate the algorithm described by C-like language into executable source program", which to some extent, increases students' perplexities. In this regard, our experience is to carefully select comprehensive cases, and require students to use the C programming technique to implement specific problems in these cases. So, we not only can stimulate students' learning initiative and interest, but can exercise their programming ability.

2.2 Ensuring the teaching effect of the introductory chapter

Many teachers often explain the introductory chapter on the cheap. In fact, the importance of introduction is not only the concepts of algorithm, data structure and time complexity, but to stimulate students' interest in data structure by combining with application examples. At the same time, by providing specific examples to the students, the teachers can help them understand the relationship between the data structures course and the other courses in the system structure of computer science. Only by catching on the significance of data structures in practice, can students better mobilize their enthusiasm.

2.3 Improving the demo effects of complex algorithms by using multimedia

In the traditional "writing + PPT" teaching mode, students can seldom have a clear understanding on the obscure process executions of many complex algorithms, thus reducing their learning enthusiasm. This problem is caused by the deficiency of traditional teaching method in expressing. To improve this situation, teachers can find some high quality web courses (such as the national level excellent course websites) on which some high quality demo teaching software or Flash are provided freely, and take these modern multimedia means as a useful supplement to traditional teaching.

2.4 Implementing hierarchical teaching

The data structures course is a professional basic course of computer specialty, as well as an important elective course for electronic information specialty. Obviously, it is not appropriate to teach students who are of different specialties or of different levels with the same content and textbook, just for convenience. In fact, even for students of the same specialty, their learning motivation is not exactly the same. Specifically, the students who have poor professional foundation do not intend to engage in the program design work in the future, and their learning motivation is just get credit. Some students hope for direct employment after graduation, their demand is weakening theory and emphasizing applications. For other students, their goal is to become graduate students and continue their studies in the near future, so they wish teachers to give them in-depth explanation on design and analysis of algorithms. To address this problem, the best way is to carry out the mode of hierarchical teaching [2]. According to the above needs of students, teachers can divide their students into three levels: A, B, and C. (Level A means minimum requirements, and level C means the highest requirements), and select teaching material and content suitable for students of different levels, and formulate the corresponding teaching program for different levels.

2.5 Carefully selecting experimental content

Conventional experimental content focuses on validating a specific knowledge (such as single linked list, stack in order, etc.) in the textbook. However, its drawback is the lack of comprehensive experimental content, and it is not closely linked with actual applications. Although students can pass to finish the experiment content, but the harvest is nothing but the basic technique of converting pseudo code description of algorithms on the textbook to executable programs. In essence, there is often small difference between the pseudo code description of algorithms and their actual executable programs. So verification experiments cannot significantly improve students' ability of programming, and they are easy to lead to the confusion of whether the data structures course is useful in future's practical work. Therefore, teachers should carefully select each experimental content, and they are encouraged to select cases which are combined with practical application or subsequent courses and to adopt case driven teaching method [3]. Of course, it put forward higher requirements to teachers, i.e., their usage of data structures should reach higher level. The specific experiments are shown in Table 1–3. Specifically, the confirmatory experiments in Table 1 are provided for students in level A, and the expansion experiments in Table 2 are designed for students in level B, and the comprehensive experiments in Table 3 are prepared for students in level C.

2.6 Evaluating experimental results rationally

In past experimental classes, teachers are busy helping individual students to correct syntax/logical errors in their program and maintaining the classroom discipline, such as preventing students to use the Internet to watch video or chat. In the evaluation of experimental results, teachers overly pay attention to students' experimental results, such as whether they can complete the experimental task, whether the experimental reports are handed in on time, while ignoring to evaluate the experimental process. As a result, many students are not willing to carefully write and debug a program, but copying others' experimental results and experimental reports. At the end of an experiment course, only a small number of students get the harvest. Consequently, teachers should reform the traditional experimental evaluation mechanism and manage to improve students' passion for experiment course. We introduce the scientific evaluation mechanism suggested by [7], which can be seen as the combination of self-assessment, peer assessment and experimental reports. The self-assessment means that students at the end of each semester are required fill out a self-assessment questionnaire

(as shown in Table 4), teachers can use this form to investigate the needs of students and their points of interest. In the phase of the so-called peer assessment, teachers demand each group of students to record their experimental process into a video file by screen recording software, and then organize the students to watch and evaluate the experimental performance of other groups. In addition, the experimental report is a very important link. Teachers should enhance the writing requirements of experimental reports, promoting students to improve their ability of summary, analysis, and scientific writing, which will lay the solid foundation of the graduation design in the last semester.

2.7 Reforming the examination mode

In previous teaching, students' total marks are computed by way of "regular grade (20%) + final exam grade (80%)", and final exam grade accounts for a larger proportion of the total. However, the drawback is that students become more utilitarian. They usually do not seriously participate in teaching activities, but copy homework after class, and keen to cram for the final exam. To address this

Table 1. Experiments for students in level A.

No.	Title
1	Basic operations of single linked lists
2	Basic operations of stacks
3	Binary tree traversal (recursive algorithm)
4	Memory structure of graphs
5	Static search tables
6	Insertion sort

Table 2. Experiments for students in level B.

No.	Title
1	The Josephus problem [5]
2	Simulation of a queuing service desk [5]
3	Huffman encoder [4]
4	Graph traversal [5]
5	The design of hash table [5]
6	Bidirectional bubble sort [4]

Table 3. Experiments for students in level C.

No.	Title
1	Addition operation on polynomial of one indeterminate [4]
2	Rearrangement of train compartments [6]
3	Setting of signal amplifiers [6]
4	Preparation of teaching plan [4]
5	Counting the number of words by using binary tree [4]
6	Machine scheduling [6]
7	Management of Parking lot [4]

Table 4. Self evaluation questionnaire.

No.	Problem
1	Which attitude do you take of the expe-riments of data structures? ( ) A. like B. doesn't like C. does not matter
2	Which way do you use to complete the experiments? ( ) A. accomplish them independently B. accomplish them by cooperation C. always unable to accomplish them
3	Which way do you take to solve the technical problems in the experiment? ( ) A. think by myself B. discuss with others C. ask the teacher
4	Which type of experiments do you like most? ( ) A. replication experiment B. practical applications and case C. comprehensive experiment
5	Do you make preparations before each experiment? ( ) A. yes B. no C. occasionally
6	What if the class time is insufficient to complete your experiment content? ( ) A. use extra time to complete B. continue to do in the next experimental class C. give up
7	How do you evaluate your programming ability? ( ) A. good B. only middling C. poor
8	Which experiment is the most difficult?
9	Which experiment is in no need to open?
10	Which experiment is the most attractive?

issue, we suggest taking the way of "assignments (10%) + course report (10%) + open jobs (20%) + experiment (20%) + final exam (40%)". The purpose of course report is to leave the chapters suitable for self-study as assignments, and require students take their group as a unit to report in class, which helps to promote their active learning. The open jobs are designed to make up for lack of assignments and experiment teaching at ordinary times, students are given a more comprehensive problem and are required to complete it in their spare time, and to verify the designed algorithm by machine validation.

3 OTHER METHODS FOR IMPROVING THE TEACHING EFFECT

3.1 Improving the teaching effect by means of course group

According to the importance of data structures course and the relation between it and the subsequent courses, we form a course group, including C programming, data structures, object oriented programming and algorithm design and analysis. Whenever encountering overlapping of knowledge (such as selection sort, bubble sort, etc.) in courses of the group, teachers are encouraged to make an "enhanced" instruction, so as to promote students understand and grasp the important knowledge points in repeated strengthening processes. Also, the abstract datatype is an important concept in "data structures", but in view of the limitation of the underlying C language, it is difficult to use the structural body of C language to describe this complex concept. In the teaching of subsequent course of object oriented programming (such as Java), we recommend to redefine the basic abstract datatypes in data structures by using of the concept of "class", so as to improve students' understanding of this important concept.

3.2 Enriching the teaching means by using of network teaching platform

Experience shows that it is not only very hard for students to completely understand and grasp the teaching content through classroom, but very difficult to teach themselves after classes. After the deployment of network teaching platform in our school, we have proactively participated in the application of the "Network Course Project of Data Structures". Through the constant construction, our network course of data structures can provide teaching courseware, experimental guidance, job instruction, course forum, extracurricular extended reading, and innovative experiment project content for students, thereby making complement to classroom teaching.

3.3 Setting up course teaching group

Usually, only one or two teachers are responsible for the data structures course, while they are seldom willing to exchange their teaching experience and to conduct teaching research. With the expansion of our computer science and electronic information specialty enrollment, and the need to improve the teaching effect, we constitute the "Data Structures Teaching Team" in the core of doctors. At present, the team has been carrying out series of promotional activities to update teachers' knowledge, skills and attitudes, such as teaching observation among one another, hierarchical teaching, studying practical cases, coaching students to attend the national graduate school entrance exam, and participating in training courses by University Teachers Training Center Network (http://www.enetedu.com/) sponsored by the Minister of Education.

4 CONCLUSION

In this paper, we provide in-depth analysis of the existing problems in teaching of the data structures course, and put forward a series of measures to promote teaching reform. Through the positive research and practice, students' initiative study consciousness is fully aroused, and their programming ability is significantly improved. It should be noted that our reform measures also pose higher requirements for teachers' professional quality.

REFERENCES

[1] Yan, W. & Wu, W. 1997. Data structures (C language version), Beijing: Tsinghua University press.
[2] Wei, T. & Wang, L. 2013. Hierarchical teaching mode of data structure. *Journal of Changchun University of Science and Technology (social science edition)* 26(7): 203–205.
[3] Xu, X. 2010. Case teaching in data structure. *Computer Education* (24): 61–64.
[4] Xu, C. 2009. Data structure course of the case (C language version), Beijing: Peking University press.
[5] Wang, L. 2002. Experimental course for data structures (C language version), Chengdu: Sichuan University press.
[6] Sahni, S. 2004. Data structures, algorithms, and applications in C++, 2nd Edition. Summit: Silicon Press.
[7] Shen, C. 2009. Research and practice of the teaching of e-commerce security. *The Science Education Article Collects* (7): 26–27.

Education Management and Management Science – Zheng (Ed.)
© 2015 Taylor & Francis Group, London, ISBN 978-1-138-02663-6

An analysis on negative transfer of native language in second language acquisition

Dongmei Wang
Foreign Languages Department, Beijing Information Technology College, Beijing, China

ABSTRACT: Native language transfer is a common phenomenon in learning a second language. According to contrastive analysis, Native language transfer tends to produce two-side influences on second language acquisition, namely, positive transfer and negative transfer. This paper firstly dwells on the negative transfer of native Language in second language acquisition at different aspects, then puts forward the conclusions that native Language transfer could be regarded as learning strategy and communication strategy, finally gives some suggestions for second language learning and teaching.

Keywords: second language acquisition; native language; negative transfer

1 INTRODUCTION

As the second language/foreign language learners have developed a habit of native language behavior, so in the language learning process, the formation of new language habits is bound to be affected by the impact of language to the old habits. According to Lado's comparative analysis, it is assumed that when certain features of native language are similar or identical with target language, positive transfer is easy to be produced; while negative transfer is produced by the differences between native language and the target language, the greater the difference, and the greater the interference. Generally, positive transfer is helpful to foreign language learning, and negative transfer hinders the foreign language learning [1].

2 CLASSIFICATION AND PERFORMANCE OF NEGATIVE TRANSFER

Negative transfer involves many aspects, can generally be divided into two categories, namely, language negative transfer and pragmatic negative transfer [2]. The former is reflected in pronunciation, vocabulary, syntax, etc. and the latter can be divided into pragmatic language and social negative transfer migrate according to their pragmatic knowledge.

2.1 Language negative transfer

2.1.1 Pronunciation negative transfer
In the process of learning foreign language, no matter how Chinese people speak English fluently, they are more or less with a little Chinese accent, and impossible to speak like native speakers as pure authentic. As in pronunciation, it must be negatively affected by ways and rules of native language. Specific reasons are as follows:

Look from the pronunciation language system, English rely on the intonation to distinguish the meaning of the words, that is intonation language. Chinese word meaning is distinguished depending on the difference between the tones, that is tone language. These two languages vary considerably in the number and combinations of phonemes.

Firstly, the English word can not only be open-syllable words with vowel at the end, but also be closed syllable words ending in consonants. But Chinese words often end in vowels. Therefore, beginners often unconsciously add a vowel after the English consonants, so work is read into a worker, bet is read into better. Secondly, some English phonemes in Chinese do not exist, so thank, this, sing, and shy is often read as [s · nk] [dis] [sin] [sai]. Thirdly, in English, lost blasting, voiced consonants, linking, stress, etc., also tend to make Chinese student confused, and lead to the pronunciation negative transfer.

2.1.2 Vocabulary negative transfer
In second language acquisition, vocabulary negative transfer affects students' learning a foreign language, which is a big barrier for students. Vocabulary negative transfer is more complex, but overall, can be divided into two kinds. [3]

Firstly, the words are expressed in different ways. In general, every language has its unique set of words, ways and habits of expression. And for foreign language learners, especially for junior level, often translated the native language into the target language from the perspective of meaning. Examples of English words with native semantic are caused by equivalent translation, such as: raining chicken, heavily ill, some eat things, people mountain people sea, small school, and How's your body?

Secondly, the connotations of the word, Lenovo and emotion meaning are different. Vocabulary as the basic unit of communication contains cultural significance. [4] The same words in different cultures represent the different meaning and emotion. For example, "raining chicken" in Chinese expression can evoke the sense of mercy, but hasn't image Lenovo expression in English. Chicken's meaning in English is a coward, but in Chinese refers to women undertake the sex business.

Words' emotional meaning is also worth of attention. Describing student's achievement, I use "propaganda" to describe the word as I regard it has the same meaning. And later I realized that "propaganda" in English meaning political group's influence on public and exaggerated or false information, which is different from the "propaganda" in Chinese meaning. Its contextual meaning varies considerably.

2.1.3 *Syntax negative transfer*

Although the basic structure of sentences both in English and in Chinese are the subject (S) + Predicate (V) + object (O), but the internal structure of differences between the two forms of language often lead to the negative transfer.

Firstly, Chinese language is belong to no marking language, and no change in shape. English is much more complicated than Chinese in terms of relative changes in word order and form. [5] For English learner, they tend to make mistakes in change of number, tenses, personal pronouns, subject-verb agreement and other aspects. There are some examples to illustrate: You is my favorite teacher; You teach our how to do a human; She see me very happy.

Secondly, the characteristics of Chinese are the significant subject. The basic structure of the sentence reflected the relationship between the main themes and title, rather than the relationship between the subject and the predicate. So In the process of learning English, Chinglish sentence is often produced. For example: There are many tourists visit there. Here the students are affected by the Chinese, and regard "many tourists" as the theme, "visit there" as a state title. So there is an error in the syntactic.

Thirdly, in general, Chinese sentence focus on parataxis. As long as the meaning can be expressed clearly, whether the form is complete is not important, so subject, and related words are often omitted. You should research on the matter, in which subject "we or you" must be added, because English stress on SVO logic and grammar intact. But the subject can be omitted in Chinese expression. "If you don't go, I won't either" in which, "if" is omitted in Chinese expression. Chinese students are influenced by parataxis and list a sentence together simply.

2.2 *Pragmatic negative transfer*

With the second language acquisition and development of the study of pragmatics, it was discovered that the transfer does not just stop at the language level, and foreign language learners will be affected by mother tongue and native culture, and misuse mother tongue pragmatic rules when they use of the target language, which is called "Pragmatic Negative Transfer". Generally, compared with the pronunciation, grammar knowledge, communicative language is getting people's more attention. "Negative Pragmatic Transfer" can be divided into two categories:

2.2.1 *Pragmalinguistic transfer*

It means that when foreign language learners use the target language to communicate, they are influenced by the native expression. For example:
Manager: Thanks a lot. That is a great help.
Secretary: Never mind.
Manager: (Embarrassed)

When trade manager praised secretary had done the excellent work, secretary wanted to express "You're welcome", but she is influenced by Chinese expression and misuse of the English expression to responded resulting in each other's embarrassment.

2.2.2 *Socio -pragmatic transfer*

"socio-pragmatic transfer" means the learner communicate in the habit of native language for the lack of understanding of the target language culture resulting in a verbal behavior does not meet the target language and cultural habits.

When Chinese people meet each other and often ask in Chinese native language: have you eaten? and where to go? Greetings as these are not really the question. If you translate them into English equal "greeting" as "hello", this translation will make them feel surprised and would mistake you want invite them to dinner, or to interfere with their private affairs. As for the taboos, Chinese people often ask each other's age, occupation, income, marriage and family status. But if foreigners ask how old are you? or how much do you earn every month? The other will think it is a gross interference in their personal privacy, causing unhappiness. As for compliment, Westerners generally regard it as affirmation of their ability, so they tend to accept it, and generally say "thank you" while Chinese people are generally more subtle, modest, tend to self modest positive. For example, a teacher praises a Chinese student for her good spoken English, students answer "no. Oh, really? Thank you". This answer shows that the students in the foreign language learning process, they understand Western culture, it should be said that in this case, he should say "Thank you", but the Chinese people with shy personality make him began to deny this unconsciously. [6] Chinese and Western phenomenon should be said that the performance by the migration of Chinese culture. It shows that this phenomenon is influenced by the Chinese and Western culture by the Chinese culture negative transfer.

3 UNDERSTANDING OF NATIVE LANGUAGE NEGATIVE TRANSFER

From the above discussion, we can see that the native language's interference in second language acquisition involves many aspects, for this objective phenomenon, we have to face up to its existence, and recognize its impact on SLA. On the other hand, we should recognize that the error caused by the native language is inevitable, language learners should be freed from the psychological barrier of fear mistakes and this will help them to take the initiative in foreign language learning, to achieve the best learning a foreign language effect. Moreover, the error occurs is not necessarily a bad thing, during the process of experiencing negative transfer of the native language, learners are experiencing a continuous learning process in fact. We can also regard the negative transfer as learning strategies, and communication strategies to promote learners constantly internalized their own Knowledge structure, and gradually transit to the process of the target language.

Negative transfer can also be used as a communication strategy to help learners to overcome communication problems. Krashen thinks the native knowledge is paving the way for learners, when learners can't master knowledge of the target language completely, and can't use the target language knowledge language to communicate, native expression will be used to fill this gap. For example, the sentence in the above-mentioned "how about your body?". It can be understood as an authentic expression, and student's grasp of English is not enough, which is imprinted with the Chinese expression in communicative activities, but will convey meaning. When you want to offer help for the people who buy things and you will say" What do you want". This expression is affected by the Chinese. The exact expression should be "Can I help you" or "What can I do for?". But foreigners will still understand your kind and friendly, and communication can proceed smoothly.

4 ENLIGHTENMENT FOR FOREIGN LANGUAGE TEACHING

In foreign language teaching and learning process, the negative transfer phenomenon is an unavoidable problem. Ellis (1985:29) believes that the average percentage of error interference is 33%, a number of studies have shown that adult learners of English in China have reached 51%. Therefore, one of the tasks of foreign language teaching is to guide students to gradually learn to use the language of thinking, to reduce or eliminate the influence of the ultimate learning interference of the native language. So, the following suggestions are made:

Firstly, in foreign language teaching process, teachers can take a comparative analysis on the part of the error to predict and correct them. For big differences between Chinese and English, teachers can guide students to be more attention, and can increase the intensity of interpretation, to make students to practice more, so that students understand the meaning to master the correct usage.

Secondly, Krashen believes that the real solution depends on inputting a lot of comprehensible materials. Teachers should provide a lot of comprehensible material and create a good learning environment. Teachers make full use of class time to provide linguistic data. we must broaden the language and input methods and channels to encourage students to use a variety of after-school media, ways to learn English, so that students make a lot of practice, and experience the difference in English and Chinese so that students gradually formed the habit of thinking in English.

Thirdly, language is an exchange of ideas, and tool of information delivery, the main purpose is to use the foreign language to communicate, including two forms in verbal and written, therefore, teachers should pay attention to the language output. [7] Students in this process can be conscious of their language problems, and correct their improper language habits. Teachers can also get feedback to analyze the causes of their errors. In addition, teachers should encourage students to communicate with the foreign teacher, and organize English extracurricular activities to enhance students' English Pragmatic Competence in communicative process.

5 CONCLUSION

Chinese students' English learning process is based on the fact that English is as the target language, and is also a constantly adjustment, restructuring and construction process. In this process the impact of the native language is inevitable. If the teacher can combine the error analysis, language input, and output effectively; even compare cultural and language acquisition mechanism, the teacher can use the existing knowledge to help students to reduce the impact of the negative transfer and achieve the transition from Chinese to English in thinking.

REFERENCES

[1] Bhela, B. Native language interference in learning a second language: exploratory case studies of native language interference with target language usage. International Education Journal, 1999(1):1–16.

[2] Corder, S.P. Introducing Applied Linguistics. Harmondsworth: Penguin, 1973.

[3] Ellis, R. Understanding Second Language Acquisition. Oxford: Oxford University Press, 1985.

[4] Ellis, R. The Study of Second Language Acquisition. Oxford: Oxford University Press, 1994.

[5] Lado, R. Linguistics Across Cultures. Ann Arbor: University of Michigan Press, 1857.

[6] Odlin, T. Language Transfer. Cambridge: Cambridge University Press, 1989.

[7] Selinker, L. Interlanguage. International Review of Applied Linguistics, 1972(10).

Deepening the graded teaching reform in higher mathematics

R.L. Tian
Department of Mathematics and Physics, Shijiazhuang Tiedao University, Shijiazhuang, China

X.W. Yang
School of Traffic, Shijiazhuang Institute of Railway Technology, Shijiazhuang, China

ABSTRACT: Based on the graded teaching of higher mathematics and aiming at the characteristics of a comprehensive university, the dynamical graded teaching is carried out and the teaching idea which centers on students is established according to subjects. The graded teaching is taken mainly from teaching content, teaching method and examining mode and the effectiveness obtained from the reform of graded teaching is analyzed.

Keywords: higher mathematics; graded teaching; teaching reform; basic subject

1   INTRODUCTION

Higher education in China is already headed from elite education to mass education and the traditional teaching mode cannot meet the requirement of the tendency in education reform, especially in basic education—mathematics. Therefore, it is very important to study the new method of teaching mode for higher educational workers. Higher mathematics is the most important basic subject and as the development of higher education, the reform in teaching content and subject system of this subject should be carried out by higher educational workers. Many universities including Tsinghua University took some useful methods for cultivating students using mathematic subjects more effectively and finding ways to stimulate the enthusiasm of students to learn mathematic subject and improve the innovative abilities [1]. The graded teaching is one of the methods [2–6].

Shijiazhuang Tiedao University works hard to found the well-known engineering university with special characteristics. According to its aims of personal training and the ideas that the teaching of basic subjects can both meet the requirement of quality education and reflect the engineering characteristics of the university, the aims of teaching higher mathematics is defined as making students master basic idea, basic method, basic theory and knowledge of higher mathematics and finish mathematical computation with the help of mathematic software. The trained students can grasp the sound basis and use higher mathematics to solve engineering problems, which play an important role

in their future works. The whole idea of teaching reform of higher mathematics in Shijiahzuang Tiedao University is that the graded teaching is carried out according to professional development and basic differences of students.

The graded teaching reform was carried out comprehensively and purposefully by Shijiahzuang Tiedao University and the obvious effectiveness was obtained. Because of rapid development in recent years, Shijiahzuang Tiedao University has already grown into a comprehensive university in arts and science which mainly engages in engineering. In order to meet new demand, the graded teaching reform of higher mathematics must be deepened.

In order to stimulate teachers to reform in teaching content, teaching method and examination mode and education quality must be improved gradually, Shijiazhuang Tiedao University carried out the activity—Class Teaching Reform from 2012 to 2013. Through the study and practice on class graded teaching reform in higher mathematics for one year, there was great improvement in teaching quality and effectiveness.

2   THE NEW METHOD FOR GRADED TEACHING REFORM AND PRACTICE OF HIGHER MATHEMATICS

2.1   *Improve existing graded method and carry out dynamical graded teaching*

According to specialties of Shijiazhuang Tiedao University, engineering course was divided into three grades—A, B, C. Compared with the former graded arrangement, the graded teaching of arts was added. In every grade, some parallel classes were arranged to offer the same mathematics simultaneously. Every student can change their grades and classes for learning proper courses.

2.2 *Update teaching content*

The teaching material of higher mathematics was revised. The new teaching material can change the tendency of enhancing technique and neglecting idea, emphasize the basic idea of mathematics, strengthen the training of basic mathematical methods and mix the ideas and methods of modern mathematics into class teaching. The teaching contents come from practical problems and make students abstract mathematical concept, build mathemetical model and search mathematical methods to solve practical problems. The aim is to improve their interest in learning mathematics and make students grasp the mathematical methods to strengthen their application ability. The new teaching material has several obvious characteristics including,

a. The basic idea and method of differential and integral calculus are highlighted to make students grasp the inner relations between every part completely. For example, the differential calculus can be regarded as the study on local property of function and the integral calculus as whole property; Definite integral serves as the main part of function integral of one variable and indefinite integral is used as the auxiliary tool of definite integral, which can stand out in the relation between indefinite integral and definite one; The content of 3D vectors and special rectangular coordinate can be conceived as n-dimensional vector.
b. It can reflect the property of engineering mathematics and meet the real requirement of engineering students. The explanations of its content are from concrete to abstract and from special to common. The words are described exactly and in a thought-provoking manner.
c. It pays more attention to the train of mathematical application ability and plays down some computational skills. The idea of mathematical model is spread throughout. The foundation and application of mathematical concepts are emphasized. The application example exists in every chapter to explain the foundation and solution of mathematical model.
d. In order to meet the teaching requirement of different grades, every chapter has rich diversity of exercises. The basic exercises are prepared to grasp chapter contents and the comprehensive ones is used for students who have great

interests in mathematics or want to take part in the entrance exams for postgraduate schools.

2.3 *Advocate the elicitation and participatory teachings*

The art of teaching is to help students find new things. Therefore, according to different teaching progress and aiming at concrete conditions of students, teachers should use different methods. When teaching new knowledge, systemic teaching method should be used; when summing up chapter content, teachers should take the method of teaching techniques; when emphasizing key and difficult points, some visible and concrete example can be used to help students learn related knowledge; when training the thinking of students, teacher can use scenario methods including analogical scene, mode scene and visual scene and so on; when teaching exercises, the participatory method should be used to help students learn abstract knowledge directly and visually and improve class teaching effect.

2.4 *With students as the main subject, highlight personal development*

Aiming at the new teaching materials for art subjects, new teaching plans and syllabus were worked out with adding mathematical development history. The teaching aim is to enhance reasonable and logical thinking and improve mathematical quality of students. For test class, when arranging teaching content and building material system, mathematical model was used as main part and mathematical application was greatly provoked. During the course of study, design, practice and innovation, the students who have profound basis, good practical ability and innovation sense were trained specially.

3 THE EFFECTIVENESS OF DEEPENING THE REFORM IN GRADED TEACHING OF HIGHER MATHEMATICS

3.1 *The learning activity of students was improved*

In the former teaching class of higher mathematics, some students were tired of study and the main reason was lack of power and interest, which means that study is a kind of burden to them. Throughout the reform in class teaching, teachers and students found the perfect relationship which went into the class and made students feel like home. The strong atmosphere of learning can be created. Therefore, students can understand the aim of studying clearly and their activity and consciousness were improved.

3.2 *The learning grades of students in test class were greatly improved*

The effectiveness of graded teaching should be measured on two hands, the first is the improved magnitude of learning grade and the second is the speed of students' making progress. From the class reform in graded teaching for test class, the results of students in different grades were improved and the learning progress was obvious.

3.3 *The teaching quality was heightened in a large area*

After carrying out the reform in graded teaching, the teaching idea of teachers has changed thoroughly. In class, teachers not only impart knowledge, train skill, but also pay attention to dynamical learning course and method of students and enhance the two-way communication with students, which can strengthen the subjectivism of students, adjust the activity of learning each other between teachers and students. The teaching quality was heightened in a large area.

3.4 *The working idea and quality of teachers were improved*

Teachers comment on the teaching contents and methods of other teachers, which can help teachers update the teaching idea, improve the teaching quality and promote the increase of professional standard. The working quality of teachers improved greatly, which plays an important role in heightening educational level.

4 CONCLUSIONS

The building of graded teaching of higher mathematics must be based on continuous creativity, updating teaching content and using several teaching methods to encourage the interests of students in study. With students as the main subject, teachers should give them individualized instruction and help them develop personality, which can give full play to subjective initiative and creativity. Teachers should optimize mathematical courseware, build uniform standards and try their best to make abstract, static knowledge concrete, dynamical, which can improve the effect of teaching class. The examination mode is reformed properly to encourage students to cultivate the good habit of studying. The teaching quality control is enhanced and the new form of teaching activity is used to help teachers update teaching idea, improve teaching level and find the perfect relationship with students, which makes students enjoy their class and study.

REFERENCES

[1] Liu Juan, Ma baolin, Wu Liang, 2006, Clarifying some misconstruction of level-based teaching of "higher mathematics", Journal of Urumqi Adult Education Institute, 14(2): 02–0091–03.

[2] Zhao Jinyu, 2012, The present condition analysis and solving scheme based on teaching at different levels, Journal of Liaoning Higher vocational, 14(9): 39–40.

[3] Yang Jinyuan, Chen julong, Zhang Xiulan, Zhan Xueqiu, Lin Feng, 2012, Research and practice of classifying teaching of college mathematics in general colleges of engineering, Journal of Jilin Institute of Chemical Technology, 29(6): 40–43.

[4] Hu Guikai, Peng Ping, 2010, Study and practice on the hierarchical teaching of higher mathematics under teacher-selecting system, Journal of East China Institute of Technology, 29(3): 287–290.

[5] Wu Guogen, Chen Huodi, Xu Dinghuo, 2006, Carrying on the graded teaching improving the quality of teaching-the theory dan pratice of the graded teaching of the course advanced mathematics, College Mathematics, 22(4): 18–22.

[6] Yao Xiangfei, 2008, Probing on hierarchical teaching pattern of engineering higher mathematics, Higher Education Forum, 3: 85–87.

Education Management and Management Science – Zheng (Ed.)
© 2015 Taylor & Francis Group, London, ISBN 978-1-138-02663-6

Strategies of pragmatic distance embodied in non-English majors' English writings

Junqiang Zhao & Xiaoli Zhang
School of Foreign Languages, Lanzhou University of Technology, Lanzhou, Gansu, China

ABSTRACT: Being uneven and intractable, the situation of non-English Majors' English writing deserves our attention. It is thought-provoking to deal with strategies of pragmatic distance embodied in non-English majors' English writings. It is necessary to analyze how to deal with pragmatic distance strategies in non-English majors' English writing from the angle of relevance theory and to explore the relationship between strategies of pragmatic distance and genders of writers by the statistical analysis of vocabulary and syntax reflecting pragmatic distance strategies, thus this paper being the enlightenment for the further writing proficiency of non-English majors.

Keywords: non-English majors; English writing; relevance theory; pragmatic distance; gender difference

1 LITERATURE REVIEW

Directly restricting the language exchange, pragmatic distance can make people perceive a kind of expected language expression. If both sides of communication coincide with each other on the pragmatic distance, the language communication can be smooth without mental disorders. However, if there is the gap in the expectation, the exchange will suffer from psychological disorders [1]. What's more, the communication would be broken down. Actually, when talking, people can choose between pragmatic distance deployment, making the language express whatever appropriate both for their own pragmatic distance. Pragmatic distance would be an essential factor in communication as it models the relationship that exists between two speakers of a language [2]. Yule [3] pointed out that the speaker can choose direct speech, indirect speech or a mixed form in the narrative discourse to express the different pragmatic distances, so as to obtain some communicative effect.

In the field of pragmatics, usually politeness and pragmatic distance are closely related. Leech [4] defined the politeness principle mainly from the means of semantic expressions; Brown & Levinson [5] introduced the face theory to define politeness principle of languages from the perspective of some social factors. Previous research focusing on manners investigated the relationship between language politeness and social distance from the angle of static social distance, while the research for politeness from the dynamic pragmatic distance is scarce.

The domestic study on pragmatic distance is not so much, mostly some cited research done from the perspective of politeness principle. Wang Jianhua [6], He Jinyan [7] and Shi Hongmei [8], respectively, conducted research on the utterance politeness and pragmatic distance, the pragmatic distance principle and the social pragmatic distance, the utterance politeness and the politeness principle in translation. From the other angles, mainly Xu Xueping [9] discussed the adaptation theory and pragmatic distance; Liu Yang [10] made an analysis of the use of English and Chinese plural first person deixis and pragmatic distance; Qi Jiafu [11] studied the relationship between pronouns and pragmatic distance; Liang Xiaobing [12] analyzed the experience of regulating the pragmatic distance in library service circulation windows.

From the review of the previous research related papers, there is a rare attempt to study the pragmatic distance strategies in non-English major' English writings. More scarce is the contrastive analysis of the pragmatic distance strategies by two genders in English writings. In English writing, authors use different discourse types, various lexical and syntactic means, and the timely adjustment of pragmatic distance to communicate with readers. If the pragmatic distance is handled properly, the credibility and affinity of writings can be increased to ensure the effective communication between authors and readers.

2 THEORETICAL BASIS

As a new hot spot in pragmatic research at present, relevance theory has been applied in many fields by more and more scholars, while few researchers interpret pragmatic distance strategies used in English writings from the perspective of relevance theory proposed by D. Sperber & D. Wilson [13]. This paper analyzes how to achieve the optimal relevance in the use of pragmatic strategies in English writings, to find a theoretical basis for the use of pragmatic distance strategies, and to prove the pragmatic distance strategies can be used on the basis of relevance theory criterion. In short, the relevance theory can be applied to explain the use of pragmatic strategies.

3 STUDY PROCEDURE

3.1 Corpus collection

The writing topic of non-English majors is stipulated for an English composition in the given time, and then the articles, respectively, by 20 males and females are sorted out as the research corpus.

3.2 Study question

Gender differences in the use of pragmatic distance strategies are explored in the English writings by non-English majors, respectively, from the two specific aspects of lexical and syntactic categories with their statistical frequencies. With the help of concrete examples, the pragmatic distance strategies used in English writings are described and analyzed.

Relevance theory with its dynamic explicit and inferential communication mode of thinking can be used to explain the selection and treatment of pragmatic distance strategies embodied in English writings by non-English majors, so that the flexible and explanatory power of relevance theory can be confirmed.

3.3 Study methodology

Because of the complexity and diversity of language expressions restricting interpersonal relationship and reducing pragmatic distances in specific speeches and language strategies being complex, unsystematic, we also cannot quantify the pragmatic distance in the relatively same way as how we treat spatial distance. However, according to the intimacy and elasticity of interpersonal relationship, the pragmatic distance can be described. This research is partially a descriptive analysis of the pragmatic distance embodied in different examples in English writings; on the other hand, it is a statistical description of specific lexical, syntactic strategies of pragmatic distance.

The methods are of the theory description and example explanation, as well as the quantitative statistics and qualitative analysis. The reduction or expansion of pragmatic distance can be statistically analyzed in the choice of vocabulary and syntax.

3.4 Data analysis

A series of persuasive data can be used to describe the related focuses of attention. Table 1 is the overall description of English writings by the male and female students with the comparable statistics concerning text complexity, lexical density and reference density. It is shown that there are no significant differences between the genders, with the first and third person reference being used more often by the two genders, and overall the females using person reference slightly more than that of the males.

The detailed use of pragmatic distance strategies is obtained in Table 2. Out of the total 541, lexical strategies are much more than syntactic ones. In terms of lexical strategies, person deixis and hedges are used the most, time deixis the least; as for syntactic strategies, mainly elliptical sentences are used to regulate pragmatic distance.

Communication with written language is the process of reasoning by the author and the reader. Authors of two genders employ a variety of ways to express ideas, having differences in the use of pragmatic distance strategies. Tables 3 and 4 show detailed information of the differences in vocabulary and syntax, respectively. In writings, the cognitive level of the reader can be considered, and then different strategies of pragmatic distance can be used to convey ideas.

It is seen from Table 3 that the use of lexical pragmatic strategies of girls is more than that of boys, with both genders using person deixis the most, time deixis the least to regulate the pragmatic distance. It is worth mentioning that, in the use of modal auxiliaries, there was a significant difference between the two genders.

Table 1. Writings' information.

Items	Two genders	Males	Females
Text complexity			
Av. word length	4.53	4.57	4.49
Av. segment length	154.53	156.40	152.65
Lexical density			
Lexemes per segment	76.72	78.30	75.15
Lexemes % of text	49.65%	50.06%	49.23%
Reference density			
1p reference	2.378%	2.21%	2.55%
2p reference	0.227%	0.22%	0.23%
3p reference	3.511%	3.45%	3.57%

Table 2. Pragmatic distance strategies (541 in all).

Lexis	Items	Person deixis	Spatial deixis	Time deixis	Abbreviations	Modal auxiliaries	Hedges
92.2%	Percentage	31.7%	11.0%	4.6%	8.6%	19.6%	24.4%
499	Frequency	158	55	23	43	98	122

Syntax	Items	Elliptical sentences	Imperative sentences		Direct speech	Indirect speech	
7.8%	Percentage	92.9%	2.4%		0.0%	4.8%	
42	Frequency	39	1		0	2	

Table 3. Lexical distance strategies used by 2 genders.

Lexical items	Person deixis	Spatial deixis	Time deixis	Abbreviations	Modal auxiliaries	Hedges	Total
Males							
Percentage	28.5%	12.1%	3.8%	9.2%	25.1%	21.3%	100%
Frequency	68	29	9	22	60	51	239
Females							
Percentage	34.6%	10.0%	5.4%	8.1%	14.6%	27.3%	100%
Frequency	90	26	14	21	38	71	260

Table 4. Syntactic distance strategies used by 2 genders.

Syntactic items	Elliptical sentences	Imperative sentences	Direct speech	Indirect speech	Total
Males					
Percentage	88.5%	3.8%	0.0%	7.7%	100%
Frequency	23	1	0	2	26
Females					
Percentage	100.0%	0.0%	0.0%	0.0%	100%
Frequency	16	0	0	0	16

In syntax, as Table 4 shows, both male and female writers cannot regulate pragmatic distance with skill and ease to effectively express the thought. Direct speech, sonorous and forceful, is far beyond the reach of writers. The boys have relatively more divergent and flexible thinking in syntactic strategies, the girls only use elliptical sentences.

3.5 *Sentence analysis*

With the concrete examples in students' writing, the use and effect of pragmatic strategies can be analyzed. There are obvious grammatical mistakes in the following two sentences, but the basic meaning can be understood.

Example 1: "In my opinion, we should not on fare evasion, if you think the charge is not rational, you can response to higher level. So we should fare on what we get."

Example 2: "In my opinion all the people should combat unhealthy phenomenon. We should supervise each other and related branch should make some rules. Especially we need to pay more attention to ourselves' character building and set up correct values."

"In my opinion" defining the credibility of propositions is only to represent the authors' personal opinions, the fuzzy estimation avoiding misleading readers and shortening the pragmatic distance. The too-frequent use of modal auxiliary "should" indicates that the authors impose personal ideas on readers, thus the commanding voice only pulling far the pragmatic distance. In Example 2, "should" is replaced with "need to" in the end, a gentle tone for the easy acceptance and recognition of the readers. In addition, the multiple uses of "we" in two examples makes authors and readers friendly and close in terms of pragmatic distance.

Writing process is the thought interaction between the author and the reader, and is also a process of finding and determining a kind of relation, with the flexible adjustment of the pragmatic distance in the communication. Words and sentences are not the only elements in writing, with the related discourse understanding being focused on, and the more subtle pragmatic distance being understood on a macro level.

4 CONCLUSION

The relevance theory is enriched with this research analyzing gender differences of the pragmatic distance strategies in non-English majors' English writings. Readers can spare as little effort as possible to find the association between the writing discourse and the cognitive context, with reading acuity being improved. Being a unique perspective for a more comprehensive and accurate understanding of English writings, the study sphere of pragmatic distance is extended from a spoken dialogue to the written text.

AUTHOR INTRODUCTION

Zhao Junqiang (1980.04 -), male, born in Tongwei Gansu, lecturer, mainly interested in functional linguistics. Zhang Xiaoli (1980.10 -), female, born in Tongwei Gansu, lecturer, mainly engaged in functional linguistics, teaching methodology.

REFERENCES

[1] Wang Jianhua. Utterance Politeness and Pragmatic Distance. Journal of Foreign Languages, 2001, (5): 25–31.
[2] Kasper, G., Blum-Kulka, S., Interlanguage Pragmatics. Oxford University Press, Oxford, 1993.
[3] Yule, George. Pragmatics. Shanghai: Shanghai Foreign Language Education Press, 2000.
[4] Leech, G.N. Principles of Pragmatics. London and New York: Longman, 1983.
[5] Brown, P. & S. Levinson. Politeness: Some Universals in Language Usage. Cambridge: Cambridge University Press, 1987.
[6] Wang Jianhua. Pragmatic Distance Principles. Journal of Dong Hua University, 2002, (4): 31–32.
[7] He Jin Yan. On Discourse Politeness and Pragmatic Distance. Inner Mongolia University, 2012.
[8] Shi Hongmei. Social Pragmatic Distance, Utterance Politeness and Polite Principles in Translation. Academic Exploration, 2014, (2): 137–141.
[9] Xu Xueping. Adaptation Theory and Pragmatic Distance. Foreign Language and Literature Studies, 2005, (2): 91–95.
[10] Liu Yang Usage and pragmatic distance of the plural first person deixis in English and Chinese. Journal of Wuhan Institute of Technology, 2010, (8): 78–81.
[11] Qi JiaFu. Cross cultural address pronouns and pragmatic distance. Journal of PLA University of Foreign Languages, 2011, (6): 21–25.
[12] Liang Xiaobing. Pragmatic distance regulation in library window circulation. Success, 2013, (6): 178–180.
[13] D. Sperber, D. Wilson, Relevance. Communication and Cognition, Blackwell, Oxford, 2nd ed., 1995.

Education Management and Management Science – Zheng (Ed.)
© 2015 Taylor & Francis Group, London, ISBN 978-1-138-02663-6

Research and practice on Industry-University-Research collaborative training mode for innovative talents of forestry engineering specialties

Xia Zheng, Xingong Li & Yiqiang Wu
Central South University of Forestry and Technology, Changsha, China

ABSTRACT: Constructing Industry-University-Research collaborative training mode for innovative talents of Forestry engineering specialties is the key to solve the problems of lag in teaching idea on traditional Forestry engineering specialties, scarcity of Industry-University-Research collaborative platform, disjointed Industry-University-Research docking, and deficiency of students' creative consciousness and ability. Methods and measures of constructing Industry-University-Research collaborative training mode for innovative talents of Forestry engineering specialties were described in this paper from 5 aspects of Industry-University-Research specialty of innovative education teaching system and innovative teaching team construction system, the importance and necessity of Industry-University-Research collaborative training mode for innovative talents was analyzed as a reference for the construction of the universities innovative talents training mode in China.

Keywords: forestry engineering specialties; innovative talents; Industry-University-Research collaborative; training mode

1 INTRODUCTION

Forestry engineering is one of the engineering specialties that has a strong practicality; it requires its students have good practice skills and strong innovative ability. Traditional Forestry engineering specialties have the disadvantage of lagging behind in teaching idea, scarcity of Industry-University-Research collaborative platform, disjointed Industry-University-Research docking, and deficiency of student's creative consciousness and ability. Developing Industry-University-Research collaborative training mode for innovative talents of Forestry engineering specialties, constructing Industry-University-Research collaborative training mode for innovative talents that ensure equal attention was paid to knowledge and ability, theory and practice, improve the students' ability of innovation practice of the Forestry engineering, it is imperative to provide high-quality talents for the development of Chinese forestry engineering industry. In this paper, research and practice on Industry-University-Research collaborative training mode for innovative talents was carried out in an indepth way based on four majors of Wood science and project, Forest chemical industry, Forest engineering, and Furniture design and manufacturing, constructing Industry-University-Research collaborative innovative talents training mode that adapt to development of science and technology innovation

of Forestry engineering specialties in China, from the start of specialty construction, curriculum construction, theory teaching research, practice of teaching platform construction and teaching team construction, comprehensive study the training method of Forestry engineering students' innovative practice ability aims to cultivate talents with innovative practice ability who can make contributions to the development of forestry in China.

2 CARRY OUT THE SPECIALTY CONSTRUCTION, BUILD CONSTRUCTION SYSTEM OF INDUSTRY-UNIVERSITY-RESEARCH COLLABORATIVE

2.1 Constructing Industry-University-Research collaborative training mode

From the perspective of Industry-University-Research collaborative, Comprehensive reform and construction of Forestry engineering were developed overall, correctly handle relationship between specialty and theory, specialty and the production practice, specialty and innovation, to reposition for plan and target of talent cultivation, innovative personnel training mode that mainly engaged in "General quality training— Professional quality training—Create quality training" was constructed, personnel training and quality evaluation system which participated by schools, businesses and social was carried out. The system not only ensure the Forestry engineering professional basic course as it should be, and the necessary adjustments was done for the structure of professional basic course, make the course more close to produce results. Given the premise that the engineering professional basic course was ensured, and the necessary adjustments were done for the structure of professional basic course, persist in the curriculum construction combined with regional characteristics and national unity, stick to the knowledge cultivating ability and improve the quality of the comprehensive unification, enlarge the proportion of Industry-University-Research collaborative guided by the nation and industry major requirements, from the perspective of technology innovation and the economic benefit, business decisions and innovation process were showed in the process of cultivation to dig the education value of Industry-University-Research collaborative and improve the student's handing ability and research ability.

2.2 Implement the strategy of professional brand construction

Open your old file and the new file. Switch between these two with the Window menu. Select all text of the old file (excluding title, authors, affiliations and abstract) and paste onto bottom of new file, after having deleted the word INTRODUCTION (see also Section 2.5). Check the margin setting (Page Setup dialog box in File menu) and column settings (see Table 1 for correct settings). After this, copy the text which have to be placed in the frames (see Sections 2.3 and 2.4). In order to avoid disruption of the text and frames, copy these texts paragraph by paragraph without including the first word (which includes the old tag). It is best to first retype the first words manually and then to paste the correct text behind. When the new file contains all the text, the old tags in the text should be replaced by the new Balkema tags (see Section 3). Before doing this, apply automatic formatting (AutoFormat in Format menu).

3 BEGINNING COURSE CONSTRUCTION, BUILDING INDUSTRY-UNIVERSITY-RESEARCH COLLABORATIVE COURSE CONSTRUCTION SYSTEM

3.1 Implementing strategy of quality courses construction, promoting development through high quality

To implement strategy of quality courses construction, following rules of ensuring quality, emphasizing effect, guiding effect of demonstration and inducing effect of radiation, under course construction's thoughts of first-class teaching ideas, first-class teaching contents, first-class teaching platform, first-class teachers group, first-class textbooks, first-class teaching measures, first-class teaching methods, quality courses for forestry engineering will be elaborated, which will lead and stimulate whole improvement of specialized courses' construction level.

3.2 Building Industry-University-Research collaborative course construction system mode for cultivation of professional innovative talents

Course construction is specific implementation and embodiment of reforming talents training mode, teaching ideas and talents training's goal, and is core of educational reform, too. Based on undergraduate teaching contents as well as reform and practice of curriculum system for Forestry engineering orienting to the 21st century, course system will be overall optimized further, four curriculum mode of basic education module, professional theory module, practice teaching module and professional emphasis module, will be built, ratio of specialized courses of practical skills will be enlarged through which teaching centre of specialized courses transfer to enterprises research and development and production center; which give a chance for students to learn and acquire knowledge at first line of research and production. Meanwhile, emphasize innovative teaching content; increase proportion of innovative quality education and engineering skills education; update teaching measures and methods constantly; models will be built during productive process of design and manufacture, simulating whole productive process in parallel in real time to forecast product performance and product manufacturing technology, which give a chance for students to know and experience process of technology research and development and industrial production.

3.3 Creating collaborative teaching theory of perception + intervention + exploration, building Industry-University-Research collaborative theory teaching system

Create collaborative teaching theory of perception + intervention + exploration which has been systematically and deep researched and explored for more than 10 years. The collaborative teaching theory properly solved interrelationship between teach and study, which, with inspiring students' thinking at the core, fully mobilized students' learning initiative and motivation; Inspired students thinking independently; Developed students' ability of logical thinking and apperceive ability to knowledge; Emphasized heavy intervention and positive interaction between teachers and students during teaching procedures. Encourage teachers and students to intervene enterprise and first line of research and production; Guide students to follow 'two ins, two outs' teaching module which are walking into workshops and labs, walking out of classroom and library; Lead and guide students apply knowledge; Find, analyze and solve problems independently in research site and production place; Accumulate knowledge, cultivate ability and practice thought.

4 CONSTRUCT THE INNOVATION EDUCATION PRACTICAL TEACHING SYSTEM TEXT AND INDENTING

4.1 Constructing two systems

Constructing university practice teaching course system. Elaborate design practice teaching courses and the content that combined industry characteristic and front development of discipline. Establishing a multi-level and step by step practice teaching content system from simple to complex, incremental to the comprehensive according to the demand of the discipline development and applied talents training. Constructing the system of college students' innovation experiment program. Establish scientific research institutes, enterprise, college innovation experiment project on the basis of national, provincial and university innovation experimental project, set up a college students' innovation experiment program system.

4.2 Establishing four main platforms

Establish the practical education platform. The practical education platform is the key point for training applied innovative talents. And according to the principle of high starting and efficiency, as well as outstanding characteristic, the project group established the practice education platform with the features of open, sharing and layout-reasonableness by comprehensively constructing laboratory. It is the construction of the practice education platform that lay a solid foundation for practical education mode, applied innovative talents training, and the exploration of diversified internships for undergraduates.

Establish the competition platform for scientific and technological innovation. Not only in order to providing students a stage to show their personalities, a chance to cooperate and communicate, as well as developing abilities, but also for forming the consciousness of thinking and practicing, a competition platform for scientific and technological innovation combining college, university, province and even the country should be built. Moreover, this platform can furthest discover the potential of undergraduates.

Establish achievements transformation platform. Through this platform, we will encourage students to apply for patents, and appraise the achievements with development prospects. On the other hand, excellent competition entries will be recommended to participate in competitions of school, province and even the country. Furthermore, we can also pick and subsidize the innovative teams as well as programs with entrepreneurship values, with support of capital and entrepreneurship training.

Establish the international education platform. Establishing the international education platform aims to rise international talents. Thus, we have been practicing and exploring in varied areas to build such an education platform, including providing undergraduates chances to attain interns in foreign companies, and also post-graduates choices to do research in universities in America, France, or even Japan, and the opportunities to attend international meetings, as well as, inviting famous foreign professors to do give lectures and holding international meetings.

5 CONCLUSIONS

As the Industrial structure of Forestry has adjusted and upgraded, the positioning of the talent cultivation target of Forestry engineering professionals has presented new goals. The scientific-technical progress and the improvement of living standard not only promoted the adjustment and upgrading of the Industrial structure of Forestry, but also proposed a new requirement about appointment of qualified personnel. To foster the senior engineering and technical personnel who has a certain ecological consciousness, engineering, economic and management consciousness, engaged in the Forest Industry Engineering or related areas about the design of the production process, product development, technical services, research and management work is a top priority. We should take the road to create new teaching mode to train talented personnel in forest engineering major by constructing a complete system in terms of the building of professions, curriculum, educational science and teaching group as to solve the problems of backwardness in teaching concept of related forest engineering majors, lack of platforms in combination of production, teaching and research, the separation of production, teaching and research, and the shortage of creativity of the students.

ACKNOWLEDGMENTS

Thanks for the subsidy from the educational reform project in department of education of Hunan province: "Wood science and engineering characteristics of professional curriculum system and innovation personnel training mode" and "Forestry engineering specialty construction and the research and practice of innovative talents cultivation mode".

REFERENCES

[1] Cheng J.F. 2006 Theory of cooperative education and its practice in China—Research on education mode of combining study and work (Shanghai: Jiao Tong University Press).

[2] Pang M.Y. 2008 Several theoretical problems of cooperative education of Production, Teaching and Research (China University Teaching) pp 15–17.

[3] Niu D.H. 2008 The value orientation of the construction of modern University System named Ecology university (China Higher Education Research) PP 65–66.

[4] Gan J.Q. 2008 Review of the study on the operational mechanism of academic power in Chinese universities (Heilongjiang Researches on Higher Education) PP 55–58.

[5] Hao Y. 2004 The Obstacles and Approaches of Converting Scientific Researches into Positive Results at Universities (Tsinghua Journal of Education) PP 36–39.

[6] Zeng L.Q. 2005 Preliminary study on the operational mechanism of cooperative education (China Higher Education Research) PP 43–47.

[7] Wang D.G. 2004 Present situation and study on countermeasure of cooperative university education of Production, Teaching and Research (Journal of Higher Education Research) pp 12–15.

[8] Jiang J., et al. 2006 The Connotation and Characteristics of Age in Cooperative Education of Production, Teaching and Research (Research in Teaching) pp 23–26.

Education Management and Management Science – Zheng (Ed.)
© *2015 Taylor & Francis Group, London, ISBN 978-1-138-02663-6*

Analysis of FDI spatial aggregation trend in the western region

Jing Xu
School of Economics, Northwest University for Nationalities, Lanzhou, China

ABSTRACT: In this paper, the total annual FDI of 12 provinces and cities in the western region is studied. The spatial statistical analysis method is used to establish spatial autocorrelation model and trend surface model. According to analysis, it is concluded that FDI in the western region has spatial autocorrelation, most provinces and cities are in high-high aggregation or low-low aggregation areas and high FDI areas have limited radiation effects on surrounding areas; total FDI in the western region increases slowly and is mainly concentrated in urban centers under good development.

Keywords: FDI; the western region; spatial aggregation; spatial autocorrelation; trend surface

1 INTRODUCTION

Foreign Direct Investment (referred to as FDI) is a hot topic for domestic scholars. In terms of FDI-related research in the western region, there are three main ideas: First, it is the impact of FDI on economic development in the western region. Zhang Shenglin and Wu Haiying proposed that FDI in the western region affects economic growth mainly by short-term demand and long-term equilibrium effects;[1] Dou Dengquan proposed that FDI in western region has technology spillover effect on domestic enterprises but has some suppression function in local economic growth.[2] Second, it is the comparative study of different regions. Guo Zhiyi and Yang Xi proposed that FDI has significant technological spillover and capital crowding effect in the eastern region and crowding effect in the central region, without obvious capital crowding effect and technology spillover effect in the western region.[3] He Wenhua has compared spatial differences in the distribution of FDI in the eastern, central and western regions.[4] Third, it is the spatial effect of FDI distribution. He Xingqiang and Wang Lixia proposed that there is significant spatial effect among the FDI in the sample provinces and cities.[5] Tian Suhua and Yang Yechao analyzed the determinants affecting the change of location of FDI in China.[6] There is little research literature about the geographical effects of FDI in the western region. Therefore, spatial autocorrelation model and trend surface model is established in this paper to analyze the spatial aggregation trend of FDI in the western region, which is full of practical significance.

2 DATA COLLECTION AND RESEARCH METHODS

2.1 Data collection

In this paper, the research is focused on the total FDI statistics in 12 provinces and municipalities (Chongqing, Sichuan, Guizhou, Yunnan, Guangxi, Shaanxi, Gansu, Qinghai, Ningxia, Tibet, Xinjiang, Inner Mongolia) in the western region. According to the theory of China's economic development after reform and opening up, 1998, 2002, 2008 and 2012 are selected for the study.

2.2 Research method

2.2.1 Spatial autocorrelation

Spatial autocorrelation index can reflect the FDI spatial pattern and change features in regions through the aggregation or dispersion degree in space, wherein the global spatial autocorrelation index is used to determine whether the phenomenon or property values have aggregation characteristics in space. The common indexes representing the global spatial autocorrelation include global Moran's I, global Geary's C and global Getis-Ord G, all of which can measure the global spatial autocorrelation by comparing the degree of similarity of the observation values in adjacent positions. After comparison, it is more reliable to adopt the Moran's I index results to determine whether there is spatial aggregation in a certain region, especially when the estimated area is at the edge of the region. The property value can be used to directly determine whether it is high or low value aggregation. Local Moran's I is used to identify possible different spatial association modes in different locations, thereby observing local instability of the space and discovering spatial heterogeneity.[7–8]

Spatial autocorrelation index should be manifested through spatial weight matrix. According to the close relationship of the spatial position, three types of spatial coefficient matrix are built: based on neighboring relation, based on the distance relationship and selection of the nearest K points.[9]

The four quadrants of Moran scatter plot are corresponding to the four types of local spatial patterns of all provinces and cities and their neighboring provinces and cities. The first quadrant represents high-high correlation; the second quadrant indicates low-high correlation; the third quadrant represents low-low correlation; and the fourth quadrant indicates high-low correlation.[10] The LISA significant level is combined with Moran scatter plot to obtain "Moran significance level map" to visually display space hotspots.

2.2.2 Trend surface analysis

Trend surface analysis is made by using regression analysis principle and generalized least squares to fit a two-dimensional function to simulate geographic distribution of elements in space. Usually, the linear and quadratic trend surface fitting are adopted. If it cannot meet the requirements, cubic trend surface or quadratic trend surface and even higher trend surface will be used.[11]

3 ANALYSIS OF FDI SPATIAL AGGREGATION IN THE WESTERN REGION

FDI statistics data is collected to perform logarithmic transformation on the raw data in order to reduce the amplitude, and ArcGIS and Geoda software are adopted, respectively, for global and local Moran's I analysis. Different spatial weight matrix will affect the global Moran's I value, wherein the variation of Moran's I value obtained by the nearest point K, namely, the space-attribute value has a little impact on it. In this paper, the nearest point K is used to create space coefficient matrix.[12] The original null Hypothesis (H0) refers to that there is no spatial autocorrelation among FDI in 12 provinces and cities in the western region and the alternative Hypothesis (H1) refers to that there is spatial autocorrelation among FDI in 12 provinces and cities in the western region.

3.1 Overall spatial characteristics

According to Table 1, from 1998 to 2012, global Moran's I index of FDI in 12 provinces and cities in the western region are positive. Based on normal distribution assumption, the significance level

Table 1. Global Moran's I index of total FDI in the western region.

Year	Global Moran's I index	Z-score	The value of P
1998	0.29	2.10	0.036
1999	0.29	2.13	0.033
2000	0.24	1.78	0.075
2001	0.28	1.94	0.052
2002	0.37	2.54	0.011
2003	0.41	2.79	0.005
2004	0.36	2.58	0.010
2005	0.34	2.43	0.015
2006	0.27	2.08	0.037
2007	0.22	1.72	0.085
2008	0.17	1.46	0.143
2009	0.19	1.54	0.123
2010	0.24	1.90	0.058
2011	0.29	2.13	0.033
2012	0.28	2.03	0.042

is $\alpha = 0.1$. For Moran's I index test results, all have pass significance test except Z in 2008 and 2009, namely, there is significant positive correlation in terms of FDI in the areas of the western region, showing spatial aggregation of FDI in similar observation values. The provinces and cities with high FDI level are mostly adjacent to those with a high level of FDI, forming a high-high aggregation area; and the provinces and cities with low FDI level are mostly adjacent to those with a low level of FDI, forming a low-low aggregation area.

3.2 Local spatial characteristics

GeoDa software is applied to obtain Moran scatter plot of the FDI spatial distribution status and to further get Table 2.

According to Table 2, HH aggregation is mainly located in the southwest region, LL aggregation is mainly located in the northwest region, and most regions belong to HH or LL aggregation, presenting HH and LL polarization aggregation trend; the regions of the three types of aggregation are relatively stable, and the aggregation type of some individual provinces and cities change.

4 TREND SURFACE ANALYSIS

4.1 Establishment of model

4.1.1 Parameter estimation

The polynomial coefficients $\alpha_0, \alpha_1, ..., \alpha_p$ are determined based on actual observation value Z so as to minimize the residual sum of squares. ArcGIS software is applied to convert the panel

Table 2. Corresponding areas of FDI Moran scatter plot in the western region in main years.

Year	1998	2002	2008	2012
HH	Shaanxi, Chongqing, Sichuan, Yunnan, Guangxi, Guizhou	Shaanxi, Chongqing, Sichuan, Yunnan, Guangxi	Shaanxi, Chongqing, Sichuan, Yunnan, Guangxi, Inner Mongolia	Shaanxi, Chongqing, Sichuan, Yunnan, Guangxi, Inner Mongolia
LH	Ningxia, Inner Mongolia	Ningxia, Inner Mongolia, Guizhou, Gansu	Ningxia, Guizhou, Gansu	Ningxia, Guizhou
LL	Gansu, Qinghai, Xizang, Xinjiang	Qinghai, Xizang, Xinjiang	Qinghai, Xizang, Xinjiang	Gansu, Qinghai, Xizang, Xinjiang
HL				

elements into point elements and extract geographic plane coordinates (x,y) as the independent variables x and y of trend surface model, wherein the dependent variable z is the average amount of FDI in provinces and cities during 1998 and 2012 (Unit: USD 100 million). To prevent the value of (x,y) is too large so as to affect parameter estimation and model establishment, the unit of x and y is set to $10^4 m$. Eviews7 software is applied for linear, quadratic and cubic regression analysis to eliminate heteroscedasticity and get a polynomial equation as follows: Linear trend surface:

$$z_i = -1009.499 + 3.565 x_i + 3.187 y_i$$

$(R_1^2 = 0.6173,\ F_1 = 7.2584,\ P_1 = 0.0196)$

Quadratic trend surface:

$$z_i = 3880.897 - 23.4355 x_i - 10.23792 y_i \\ + 0.035 x_i^2 + 0.035 x_i y_i + 0.015 y_i^2$$

$(R_2^2 = 0.9474,\ F_2 = 21.6051,\ P_2 = 0.0009)$

Cubic trend surface:

$$z_i = -9457.390 + 86.438 x_i + 35.862 y_i - 0.253 x^2 \\ - 0.23 x_i y_i - 0.114 y_i^2 + 0.00024 x_i^3 \\ + 0.00036 x_i^2 y_i + 0.00038 x_i y_i^2 + 0.00018 y_i^3$$

$(R_3^2 = 0.9904,\ F_3 = 23.0430,\ P_3 = 0.0227)$

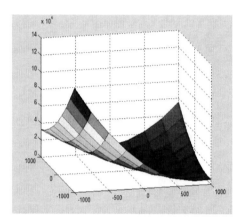

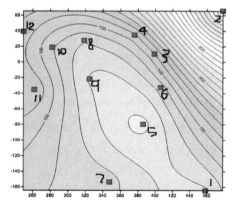

Figure 1. Trend surface and isoline.

4.1.2 Model inspection

4.1.2.1 Inspection of goodness of fit

Based on R^2 inspection method, linear, quadratic and cubic trend surface determination coefficients are obtained, namely, $R_1^2 = 0.6173$, $R_2^2 = 0.9474$ and $R_3^2 = 0.9904$. It can be seen that quadratic and cubic trend surface models are of high goodness of fit.

4.1.2.2 Significance F test

Based on F test method, the values F of linear, quadratic and cubic trend surfaces are obtained, namely, $F_1 = 7.2584$, $F_2 = 21.6051$ and $F_3 = 23.0430$. Under the confidence level of $\alpha = 0.05$, according to the F distribution table: $F_1 \rangle F_{1\alpha} = F(2,9) = 3.8625$, $F_2 \rangle F_{2\alpha} = F(5,6) = 4.3874$ and $F_3 \rangle F_{3\alpha} = F(9,2) = 19.3848$. It can be seen that quadratic and cubic trend surfaces are feasible. However, *P* value of quadratic trend surface model is much less than 0.05, with greater statistical significance. Therefore, the quadratic model can better reflect the trend surface of FDI in the western region.

4.2 *Trend surface and isoline*

Based on quadratic trend surface equation, Matlab software is used to draw quadratic trend surface Figure 1-a, surfer software is applied to draw isoline Figure 1-b, wherein: 1-Xinjiang; 2-Inner Mongolia; 3-Ningxia; 4-Shaanxi; 5-Qinghai; 6-Gansu; 7-Tibet; 8-Chongqing; 9-Sichuan; 10-Guizhou; 11-Yunnan; and 12-Guangxi.

It can be seen from Figure 1-a, the two ends of the curved surface are upward, indicating the presence of trend. In Figure 1-b, the axis *x* represents the geographic coordinate *x* and axis *y* represents geographic coordinate *y*, the isoline interval is 5 billion dollars, most of the areas are within the 150 isoline, namely, that these areas fail to absorb FDI value significantly, wherein Inner Mongolia, Ningxia and Yunnan witness great increase, Shaanxi and Guangxi witness slight rise, Qinghai, Sichuan and Chongqing decline, and other areas suffer little change.

5 MAIN CONCLUSION

First, high FDI areas in the western region have limited radiation effects on surrounding areas so as not to drive the growth of FDI in the surrounding low FDI regions. The aggregation types are relatively stable and only some individual provinces change; second, FDI investment in the western region witnesses some increase but it is mainly focused in urban centers under better development. Regional internal growth is very uneven and slow. Third, three is a positive radiation effect among the cities and provinces in HH regions and negative radiation effect among the cities and provinces in LL regions.

REFERENCES

[1] Zhang Shenglin, Wu Haiying; Empirical Analysis of the Impact of Foreign Direct Investment on the Economic Growth of the Western Regions; Social Sciences in Ningxia; 2005, (1).

[2] Dou Dengquan; Empirical Research of the Impact of Foreign Direct Investment on the Economic Growth of Western Regions; Academic Forum; 2011, (7).

[3] Guo Zhiyi and Yang Xi; Differences in Function Mechanism of Foreign Direct Investment on Economic Growth in Eastern, Central and Western Regions of China; Nankai Economic Studies (2008), (1).

[4] He Wenhua; Difference between FDI and Economic Growth in Regions: Study Based on Chinese Provincial Panel Data; Forward Position in Economics, 2009 (2) to (3).

[5] He Xingqiang, Wang Lixia; Study of Spatial Effect of FDI Location Distribution in China. Economic Research Journal; 2008, (11).

[6] Tian Suhua, Yang Yechao; Determinants of the Location Change of FDI in China—Empirical Research Based on D-G Model; The Journal of World Economy; 2012 (11).

[7] Ma Ronghua, Pu Yingxia, Ma Xiaodong; GIS Spatial Association Pattern Discovery; Beijing: Science Press; 2007:101–111.

[8] Wang Yuanfei, He Honglin; *Spatial Data Analysis Methods*; Beijing: Science Press; 2007:94–119.

[9] Zhang Xinfeng; Research of Spatial Autocorrelation Data Analysis Methods and Application; Lanzhou: Lanzhou University; 2009.

[10] Wei Hao; Spatial Effects of Foreign Direct Investment of 29 Provinces and Cities during 1985 and 2007; Journal of International Trade; 2009, (9).

[11] Zhao Guoqing, Luo Hongxiang; Spatial Aggregation of FDI in China and the Trend Surface; World Economy Study; 2012, (1).

[12] Xu Bin; Analog Analysis of the Impact of Spatial Weight Matrix on Moran's I Index; Nanjing: Nanjing Normal University; 2007:20–23.

Education Management and Management Science – Zheng (Ed.)
© 2015 Taylor & Francis Group, London, ISBN 978-1-138-02663-6

Teaching exploration and students' innovative ability training based on "Testing Technology" course

Y.B. Zhou, Z.R. Wang, H.D. Wang & Y.J. Du
School of Mechanical and Automotive Engineering, Hubei University of Arts and Science, Xiangyang, China

ABSTRACT: The "Testing Technology" course is an important professional foundation course of mechanical engineering. According to the characteristics of the mechanical undergraduate education, combining with teaching practice, from the teaching content, teaching methods, textbook and courseware construction, this paper analyzes the importance of the "Testing Technology" course, and introduces the methods of teaching innovation and students' innovative ability training. Recently, teaching practice shows that, the teaching exploration and reform in the testing technology have achieved obvious results, the improvement in students' innovative ability has got fully manifested in disciplines race, patent application, and paper publishing, etc.

Keywords: teaching reform; undergraduate education; testing technology; innovative ability

1 INTRODUCTION

Manufacturing scale and level are the main marks to measure a nation's comprehensive strength and modernization degree. To make our country shift from "big manufacturing country" to "power manufacturing country", we must implement the strategy of informatization driving industrialization. Without an advanced instrument and meter industry support, it can't complete this task. With the progress of manufacturing technology, the role and importance of testing technology have increased significantly in mechanical systems and manufacturing processes (Yin et al. 2010). Testing technology has become a necessary part that has penetrated into each link in the process of manufacturing, and has become an integral part of mechanical science and advanced manufacturing.

"Testing Technology" is an important professional foundation course of mechanical engineering; over the years, multiple majors of our school of mechanical and automotive engineering set it as an undergraduate specialized course. It is of great significance for building mechanical undergraduates solid professional theoretical basis, broadening the research idea, improving the undergraduate quality, and constructing a foundation for further study, research, and dealing with mechanical engineering technology problem.

As course content updated quickly, less class hours and other objective reasons, it must be constantly perform the teaching reform and innovation to make students to master professional theory, the real development forefront and necessary professional skills. As the chief course instructors, combining with the teaching practice of recent years, I will introduce my own teaching practice, respectively, from teaching content, teaching methods, teaching materials and coursework construction.

2 TEACHING CONTENT

Testing technology is one of the advanced sciences in developing and perfecting; therefore, the corresponding teaching content also has the expansibility and the uncertainty (Zhou et al. 2011). Teaching teachers must follow the discipline and academic development, accumulating a lot of literature, to guarantee the timely update of the teaching contents, construct a solid professional foundation for students, to promote the improvement of teaching quality.

At present, the domestic similar course teaching still face the following problems: (1) The knowledge structure appears fossilized, existing curriculum content is difficult to keep up with today's testing technology development level, unable to reflect the developing law of testing technology, and failure to grasp the common problems of modern testing technology. (2) Lots of time spent on the teaching of theories, lack of logical thinking, reduces the classroom learning enthusiasm, and students lose interest. (3) The teaching knowledge is scattered, lack of a backbone of all knowledge; students lack knowledge application wholeness, are difficult to form a knowledge system, lead to the understanding of the testing technology only remain on the perceptual. (4) Theory separates from practice, and ignores the comprehensive use of knowledge, neglects cultivating the students' ability of solving actual problem using knowledge.

To solve the above problems, we scientifically arrange and reasonably adjust the teaching content. First, modern testing technology development trends are introduced and the testing technology development present situation, characteristics and trends are discussed. Then, we introduce testing technology of common knowledge and the law which is the basis of the various sensors and effect, focus on the analysis of the basic effects of different sensor, and make the student have a new understanding of sensor. Then, we have carried on the comprehensive introduction of intelligent sensor, including its structure and system, functions and characteristics, software design and manufacture, etc. Finally, we introduce several new sensor technologies, appearing in recent years, such as MEMS sensors, biosensors, magnetic sensor, machine vision, nondestructive testing, multiple sensor fusion, etc.

The whole set of course system has grasped the key, difficult and new technology tendency, based on the basic concept and principle of testing technology, emphasizes on the sensing new principle, new technology, and new effects, combines sensing and information processing at the same time, makes the students have a better grasp and understanding of the abstract theoretical knowledge information processing.

Due to involvement of multiple professional students in class, we add related interdisciplinary parts to the course according to the background of students, such as car safety condition monitoring, information decoupling, and multiple information fusion, networked sensors, robot control, etc. In addition, the teaching link is also equipped with practical training, arranged with the content of self-study and hot research topic. By writing papers related to this course, train the student skills of consulting literature material and summary writing, and enable students to master the latest developments and the international front, cultivate the students' professional learning interest, practice ability and innovation ability, and construct the foundation for the thesis topic selection and research in the future.

3 TEACHING METHOD

The object students of testing technology course are mainly sophomore and junior students, which have a preliminary knowledge structure, and have a certain ability of self-study. During the class teaching, the teacher's role is not only delivering knowledge, but is also paying more attention to inspire students' thinking, and the teaching should focus on the scientific methodology, teaching methods, reasoning methods, the method of data collection, how to make a conclusion from the facts, and how to analyze the facts and comprehend facts.

Testing technology is a comprehensive course of science and technology, involving a large number of basic theoretical knowledge and the content of the sensing technology research progress. According to the characteristics of the undergraduates, we have carried on the reform on teaching methods, using a variety of teaching methods such as heuristic method, discussion method, and project-driven method, to stimulate students' interest, guide students thinking, and inspire their learning interest and research thought.

3.1 Heuristic method

The heuristic teaching method requires teachers to focus on students' thinking process, not the answer. First, to introduce students to a kind of research contents and progress of the research project or direction roughly, and put forward some inspirational question, according to the specific content, guide students to discuss, to think, let the student to ask questions, collect related data, investigate, design research plan (Yang 2013). The teachers guide students' research methods and finally solve the problem with the students themselves. The heuristic teaching method should not be looked as to ask questions in class simply and unilaterally, but let the students to practice more. In the process of teaching, the teachers should timely and reasonably enlighten and induce the students' learning activities, change the teachers thinking to students thinking, exercise the student's ability of understanding and judgment, fully mobilize students' learning initiative and enthusiasm.

Such as the introduction of a photoelectric absolute encoder, first to let the students understand that it is a kind of angle encoder, which can convert an angular displacement into a binary code, and it is convenient to connect with computers, capable of non-contact, small in size, high resolution, good reliability, and convenient use, etc. Then introduce, respectively, the absolute encoder binary code disc and cyclic code disc, inspire the student to discuss the following questions: Why design these two coding schemes? What are the advantages and disadvantages of each? What is the relation of these two schemes? How to design a new kind of coding scheme on the basis of existing code disc? What is the application of this coding scheme? What key problem can be solved? And so on.

3.2 *Discussion method*

This kind of teaching method first lets the students study a thematic content independently, present some research problems at the same time, let the student go deep to find the information, and try to find the methods or research ideas to solve the problems; then let the students give their views, and implement team or group discussions in view of these opinions, the teacher give a summary analysis finally. This is a kind of teaching method between teachers and students, students and students, mutual communicate and mutual evaluation with each other. The discussion can cultivate students' ability of thinking, language expression and sum up, guide the students to study and put forward theoretical perspectives freely, and change students from passive acceptance into active participation in the course teaching (Zhou 2013). Ask questions from the students, inspire thinking, change the main content to a particular study subject, and give students a certain time to think and discuss, in the process, it mobilizes the attention and interest of the students, greatly improves the teaching effect, lets the student obtain the experience of the research process, and exercise the students' scientific research ability.

Discussion teaching method, on the other hand, is a process of teachers and students as equals, and teaching benefits teachers as well as students. Allowing the students and teachers discuss and debate on the academic question also require the teachers to keep on learning and understanding of international academic frontiers, and be familiar with the common sense of this discipline. In the process of classroom teaching, through equal discussion and thinking with classmates, teachers' academic level will be further improved.

In the teaching of sensor characteristics, we arrange a discussion topic in sensor temperature compensation method research. In the process of discussion, the students, according to the characteristics of different sensors, talked about a variety of sensor temperature compensation methods. In the discussion, the students found that the twice interpolation method can calculate any temperature sensor characteristic curves; eliminate the influence of temperature factors on sensing properties. But the actual sensing properties are often linked to a variety of environmental factors (such as temperature, humidity, pressure, etc.). Therefore, we can guide the students to think further about how to compensate the sensor of multi-dimensional factors.

Adopting this teaching method, students in the process of learning can really change from "want we to learn" to "I want to learn". Upon completing the project, students find problems, timely consult teachers, through the communication with the teacher or discussions with the other students to solve problems, create a more harmonious teaching atmosphere. Timely summary after class, the teacher gives some questions or discussion topics, helps students to better grasp the relevant knowledge, and improves the learning initiative of students.

3.3 *Project-driven method*

Project-driven teaching method starts with cultivating the students' scientific research ability, and by combining practical projects to achieve the teaching goals. According to the teaching content, the teacher arranges some practical projects for students to complete, or lets the students to attend to their own research projects directly, creates the scientific research practice opportunities for students, and develops the students' scientific research emotions and improves the scientific research quality in the practical scientific research (Liu 2013). The teacher provides guidance and help to students when necessary, at the same time, introduces their own research results into teaching, and the teaching content can be updated timely.

During the teaching of the testing technology course, we pay more attention to combine the course content with the teachers' scientific research project. Also, some students are involved in the relevant research work of the project, in the process of completing the task, students take the initiative and take to creativity, make dull formula derivation and problem solving become an interesting test process, cultivate the research ability, improve the learning effect, and have played an important role in utilizing research projects to improve teaching quality.

Over the years of teaching practice, we have received obvious teaching effects from combining scientific research projects with the curriculum teaching process of testing technology. Some students combined with the course content, by participating in the teachers' scientific research project, have obtained certain research results and have published many academic papers (Rao 2012, Wang 2012). As an effective teaching method, project-driven teaching method has a good compensation effect to solve the disjointed problem of teaching theory and practice, can give full play to students' individual learning initiative, greatly improves the student's comprehensive quality, and effectively cultivates students' abilities of autonomous learning, analyzing problems, solving problems, and scientific research innovation.

4 TEXTBOOK AND COURSEWARE CONSTRUCTION

The textbook is an important part of undergraduate course teaching, the quality of the textbook directly affects the effectiveness of course teaching (Xu & Ma 2014). At present, many domestic textbooks of "testing technology" course are available, the course teaching is basically ordered in knowledge system. Firstly, introducing basic concept and basic features of the testing technology, and then introduce some kind of sensor according to different sensing principle and related applications, such as resistive sensor, piezoelectric sensor, capacitance sensor, electromagnetic sensor, photoelectric sensors, etc. Some general sensing principle also comes with a few simple demonstration and validation tests. This teaching method takes knowledge as the core of the course teaching, lack of innovation and design experimental projects in the true sense, too much emphasis on imparting theoretical knowledge, and neglects the cultivation of the ability to apply knowledge, thus brings about many negative impacts.

We carefully choose the good textbook for teaching, and students are required to refer to several relevant literatures. In addition, combined with years of experience in teaching and scientific research achievements, we also write a supplementary handout, and add the new developments of testing technology in recent years. We strive to enable students to systematically study this course, master the basic knowledge of testing technology, and understand the latest research developments. This handout is trial and continues to be amended, and in the future, we will strive to publish in the form of textbooks.

Due to the fact that time of classroom teaching is not much, in order to increase the course information, we have also designed and developed electronic lesson plans and network courseware for teaching utilizing the multimedia application software platforms such as Flash, PowerPoint. These make the teaching content richer and diverse, also transfer the traditional chalk board type teaching to the modern electronic teaching, and make the past abstract, difficult content, due to no physical objects (such as specific sensor), become concrete, vivid, and visual.

5 CONCLUSION

The "Testing Technology" course has an important status and role in mechanical or related professional curriculum system. The teaching practice of recent years shows that to teach this course well, teachers must timely update teaching content, adapt the teaching methods suitable for the teaching content, accelerate the construction of teaching material and courseware, make full use of various resources, perform corresponding course reform, which can make students in the process of learning to master basic theory, cultivate innovative ability, and construct a solid foundation for future professional work.

ACKNOWLEDGMENT

This work is supported by the Hubei Province Educational Science "Twelfth Five-year" Plan Project (2013B204), the Hubei Province High School Provincial Teaching Research Project (2011350), and the Hubei University of Arts and Science Teaching and Research Project (JY2013033).

REFERENCES

[1] Yin, G.F. et al. 2010. An overview of advanced measurement technology and its development trend for modern manufacturing. *China Measurement & Test* 36(1): 1–8.
[2] Zhou, D.Q. et al. 2011. Primary teaching reform on testing technology. *Journal of Jiangsu Teachers University of Technology* 17(4): 86–90.
[3] Yang, Y. 2013. Review of heuristic teaching. *Journal of Juamjusi Education Institute* (12): 290.
[4] Zhou, G.Z. 2013. The strategic research of new discussion-based teaching. *Mechanical Vocational Education* (12): 57–58.
[5] Liu, Y.X. 2013. Analysis of the project-driven teaching mode. *Journal of Yangtze University* (*Natural Science Edition*) 10(22): 155–156.
[6] Rao, J. et al. 2012. An on-line monitoring equipment of BUE based on machine vision. *Mechanical Management and Development* (6): 10–11.
[7] Wang, C. et al. 2012. Design of an on-machine vision measurement device for surface roughness of turning parts. Equipment Manufacturing Technology (7): 46–47.
[8] Xu, K.J. & Ma, X.S. 2014. Textbook construction for sensor and measurement technologies. *Journal of Electrical & Electronic Education* 36(1): 115–117.

Education Management and Management Science – Zheng (Ed.)
© 2015 Taylor & Francis Group, London, ISBN 978-1-138-02663-6

Research on construction of public relations laboratory

Hongyan Li
Shanghai Second Polytechnic University, Shanghai, China

ABSTRACT: Building public relations laboratories is related to quality of practice teaching directly. It is important to increase the intensity of the laboratory construction and improve practice ability of undergraduate students. In this article, some problems existing in public relations laboratory are analyzed according to the nature of public relations professional disciplines and companies' actual operation and needs. At last, a framework for construction of public relations laboratory and the corresponding construction strategy are put forward.

Keywords: public relations; construction of laboratory; practice; countermeasure

1 INTRODUCTION

Public relations laboratory is the base for teaching and research and an important indicator of level of public relations management. Its function is not only to meet the needs of experimental teaching, but also provide guarantee for development of scientific research and students. However, the public relations profession usually puts more focus on the theory study, and the traditional practice teaching use more forms of case teaching, so the public relations professional laboratory building lags behind obviously and the practice teaching is needed to be strengthened urgently.

2 THE GENERAL PRINCIPLE OF PUBLIC RELATIONS PROFESSIONAL LABORATORY BUILDING

2.1 Changing the concept of weight theory and light practice

Modern society requires the public relations talent to meet the principle of "thick foundation, strong ability, wide adaptation, emphasizing innovation". Experimental teaching is an important means to develop students' creativity, development capabilities, independent analysis capability and solving problem abilities. Laboratory building is a prerequisite to carry out experimental teaching, so it must be incorporated into the overall plan. Only in this way, the laboratory building can get real attention.

2.2 Constructing a reasonable practical teaching system

Public relations laboratory should overcome blindness and simplistic tendency of opening experiments, build a scientific practice teaching system based on modern information technology and computer network technology. This can ensure the effectiveness of laboratory building.

2.3 Tapping the existing potential of laboratories to create a common experimental platform

Public relations laboratory construction must be planned overall and distributed rationally to maximize the use of limited resources and develop existing potential of laboratory. The complementary functions of different experiments can be achieved by building a common platform.

3 IDEAS OF BUILDING PUBLIC RELATIONS LABORATORY

3.1 Increasing infrastructure construction

Infrastructure construction of public relations lab includes lab space, hardware and software etc.

1. The construction of a comprehensive open lab environment
   Building public relations laboratory needs to combine computer technologies, communication technologies, multimedia technologies and network technologies together to construct open and comprehensive experimental teaching environment focusing on experimental teaching and learning.
2. The development of public relations professional experimental software
   Public Relations professional software mainly refers to software corresponding to innovative laboratory, including database website of researching public relations market.

The software is consistent with real enterprises. It will stimulate students' interest in learning to participate in teaching practice activities.

3.2 *Strengthening the construction of experimental teaching system*

1. Changing teaching concepts
   Traditional ideas of main theoretical teaching and supplementary experimental teaching should be changed. Theory teaching and practical teaching should be seen as an organic whole. Theory course should focus on development of skills and change the misconception of high level of teaching theory and low level of experimental teaching. Staff with high experimental teaching levels should be equipped to improve the overall level of experimental teaching. A special experimental research fund should be set up for the reform of Public Relations experimental courses and experimental methods to mobilize the enthusiasm of laboratory personnel.
2. Optimization of curriculums
   Experimental teaching should not be looked as supplement of theoretical lessons. Independent experimental teaching curriculum should be set. Experiment type consists of verification experiment, comprehensive experiment and innovative experiment. Each part achieves certain aim. Verification experiment can train students' basic skills and understanding of theoretical knowledge. Comprehensive experiment mainly reflects students' knowledge, skills and understanding of knowledge. Such experiments are mainly designed and implemented by students independently.
3. Reform of teaching methods
   Using experimental teaching with different types, different levels and different methods can develop students' innovation awareness and practical ability. The teaching method of "demonstration experiment" can help students be familiar with experimental equipment and operation. It can use multimedia presentations. The teaching method of "simulation experiment" requires students to complete the whole process under the guidance of teachers and solve problems independently. The teaching method of "open experiment" lets students give full play to the imagination to design experimental program with unique style.

3.3 *Perfecting forms of organization and management*

1. Innovating teaching content
   A comprehensive simulation environment allows students to experience the confrontation and collaboration in the training process. In innovation of teaching content, frequent replacement of materials is not advocated, but it must be done with the times. New experimental projects must be explored and old experimental methods must be improved according to professional requirements and socio-economic development. Content of designing experiments and research experiments should be increased to give students ample space for displaying talents.
2. Innovation of experimental teaching management
   Experimental teaching management combines centralized management and decentralized management. Centralized management is mainly reflected in the laboratory building. It is implemented by unified planning, resource allocation and technical support. Decentralized management is mainly reflected in the implementation of the experimental teaching. In this mode, teaching faculty is responsible for confirmation of experimental curriculums and arrangement of experimental teaching programs and contents. At the same time, measure of setting office according to the profession or curriculums is changed and replaced by "team-based teacher organization". Full-time teachers from the experimental teaching center are responsible for development of comprehensive experiment programs. Professional teachers from teaching positions are responsible for the design and planning of the experimental teaching activities, and helping teachers in the experimental teaching center complete related work. Corporate officers or distinguished teachers from outside the school set up seminars or special lectures.
3. The use of advanced technology experimental teaching
   Experimental teaching environment can be constructed under modular architecture ideas based on information technology. The whole process from the experimental teaching design, implementation of experimental teaching to the experimental teaching evaluation is informationized. Each sub-module can be configured and selected flexibly. Teachers can develop sub-modules personally.

4 SETTING MODE OF PUBLIC RELATIONS LABORATORY

Setting mode of public relations laboratory is related to the effectiveness of the laboratory building directly. It needs to be combined with the actual situation of universities.

4.1 Setting mode of laboratory

Public Relations laboratory can be divided into two classes. One class is managed by the department, the other is co-managed by corporate and university. The first way is adopted by a comprehensive university. The second way is usually adopted by high-tech service industries and universities and still rare.

4.2 Selection of laboratory setting mode

The mode of co-management by enterprises and universities has many restrictive conditions. Not all universities can have it. Laboratories managed by departments lack capacity to mobilize resources directly in construction and management due to lower levels of leadership. That often leads to inadequate laboratory construction. If public relations laboratory can realize two-level management of university and department, it is easier to get the attention of leaders and is conducive to comprehensive laboratory work. Besides, it is beneficial to sharing of resources, standardized management, multidisciplinary analogy and use. Cross and fusion between multidisciplinary and more professions is easier to achieve teaching objectives.

5 PROBLEMS AND SOLUTIONS IN LABORATORY CONSTRUCTION

5.1 Focusing on software development

Public relations software generally includes two classes, one is software developed by professional companies; the other is software supporting experiments. No matter what kind of software, they have problems of deficient quantity and low quality. The reasons are as follows. Firstly, it is lack of investment. Secondly, professional teachers do not understand experiments and do not know what kind of software is more effective. Thirdly, experimental teachers' quality is low. Therefore, the substance of constructing public relations laboratory is software construction. The establishment of appropriate software resources is important to improve the level of laboratory. State education authorities should provide guidance on the use of relevant educational software to standardize software development and marketing, so as to solve the problem of lacking software.

5.2 Strengthening the continuous construction

Public Relations lab is composed of computers and network equipment. Its loss is not obvious as engineering laboratories. Therefore, maintenance of public relations laboratory will relax over time. Laboratory building is a long and ongoing process. Performance of lab hardware devices and environment of software applications will decline over time, thus affecting the normal operation of experiments. Meanwhile, content of experiments will change with the social development. Therefore, just by using without update, maintaining and innovative construction is bound to make the laboratory not play its due role.

To solve this problem, good laboratory construction plan should be formulated to ensure the laboratory's simultaneous development along with advances in technology and update of management philosophy.

5.3 Play function of laboratory

Public Relations laboratory building is generally considered to be things of laboratory management. Specialized teachers do not care about it. That results in laboratory being a simple experimental place, only a simple alternative to the classroom. And its function is hard to be played fully. Strengthening professional team building is an important way to improve the effective use of laboratory. First, improve professional teachers' level of experimental teaching and experimental design through training, demonstration, seminars etc. Second, improve the technological level of laboratory technicians continuously through projects research and professional training and controlling the technical structure. Through this way, it can meet the demand for all types of laboratories to optimize the effect of laboratory use.

6 CONCLUSIONS

Public Relations laboratory building is a systems engineering, and should adapt to science personnel training programs. It should take into account of the needs of students, teachers and managers. Scientific construction of experimental course system, reasonable set of highly effective experiments and training of high-quality experimental teachers are all needed to establish laboratories with compatibility, forward-looking and extension. That will provide good practical environment for students.

REFERENCES

[1] Shaohui Huang. Problems and countermeasures of university laboratory management. Guangdong Technical Teachers College, 2008 (6).

[2] ShuLing Zhang. Analysis and countermeasures of university laboratory management. Experimental Technology and Management, 2006 (1).

[3] Xinlong Gong. Exploring experimental teaching of arts under science and engineering environment. Research and exploration of laboratory, 2009, 2010 (4).

[4] Jin Li. Laboratory construction and management model. Research and exploration of laboratory, 2009, 26 (6).

[5] Zhijian Jiang, Weiliang He. Problems and countermeasures in constructing high-quality experimental teaching demonstration center. Experimental technology and management, 2009, 26 (2).

Education Management and Management Science – Zheng (Ed.)
© 2015 Taylor & Francis Group, London, ISBN 978-1-138-02663-6

Connotation, investigation, methods and evaluation of international economic and trade major higher education internationalization

Jing Xu

School of Economics, Northwest University for Nationalities, Lanzhou, China

ABSTRACT: Based on connotation of education internationalization of international trade major, this paper focuses on analysis of current status, implementation methods and effect evaluation of education internationalization of international trade major of the colleges and universities, the way to realize education internationalization of international trade major from the perspective of curriculum arrangement and teaching practices, and effect evaluation from such three aspects as teaching, teaching practices and teaching quality.

Keywords: international trade major; education internationalization; connotation; investigation of the current status; implementation methods and effect evaluation

1 INTRODUCTION

Internationalization of higher education is mainly reflected in more and more international cooperation ways, such as international flow of knowledge, international exchange of scientific personnel, and cooperation in education, research, development and training and so on.

2 CONNOTATION OF INTERNATIONALIZATION OF HIGHER EDUCATION OF INTERNATIONAL TRADE MAJOR

The new round of education internationalization since the 1990s is a process of mutual exchange and cooperation of different countries in education ideas, contents, methods and modes in the context of economic globalization, aiming at training international talents.[1] The connotation of internationalization of higher education of international trade major should include internationalization of clear education ideas, training objectives, teaching facilities, teaching environment and faculty.[2]

3 INVESTIGATION OF THE CURRENT STATUS BASED ON PRACTICE

Through investigation into students and graduate students of international trade major and their parents, 100 valid questionnaires have been obtained. The main findings are as follows.

3.1 *Information about international experience of the respondents*

Exchange experience, and the reasons mainly include no way of participation, high cost and language proficiency. Most of them like such exchange ways as short-term study abroad, overseas volunteer activities, exercitation abroad, and only a few wish long-term studying abroad and international conferences.

3.2 *Information about the learning of the respondent*

Nearly 90% of the respondents rarely read the foreign language books of their major. About 35% of the respondents cite foreign literature in their course and professional papers. More than 90% of the respondents consider themselves to be poor in reading and analyzing foreign literature. More than half of them are concerned about foreign affairs and are conscious of forming international culture thinking capability. 95% of them believe that the colleges rarely invite foreign scholars to give a speech or lecture and rarely provide the conditions for students to go abroad to participate in international exchange meetings. Over 70% of them feel that the library has enough professional foreign materials but it cannot be linked to famous foreign academic website that allows the students to download or it is uncertain whether this service is available. Most students are not sure whether there is content about analysis of "current foreign research status" or do not include such contents in their papers. 87% of them fail to pass CET-6.

Professional courses in bilingual languages or English (including courses in English) only accounts for about 0–25%. 87% of the respondents believe that bilingual or English teaching for international trade major will bring more benefit than harm, and believe that the main reason for more harm than benefit for education internationalization is the language barrier of the students.

3.3 Understanding of internationalization of higher education of international trade major

The respondents mainly understand information about internationalization of higher education of international trade major through network, presentations and press and so on, other than intermediaries and scholarly articles. The majority of the respondents believe that the domestic international schools (or international classes) are a place for transition of studying abroad and for experience of international education in China. Such schools can overcome some weaknesses of the education in secondary schools and universities. Economic condition permitting, they are willing to go to the international schools (classes).

In terms of comparison of higher education of international trade major in China and abroad, most respondents believe that foreign students have clearer learning objectives and enjoy more active classroom atmosphere, higher classroom participation degree, more human-orientated school management, more equal teacher-student relationship, more attention to the education process, fairer, more scientific and more flexible educational methods, richer extracurricular activities and more active in learning than domestic students. Most respondents believe that measures for internationalization of higher education of international trade major should include expanding the scale of studying abroad under public support, introducing a large number of overseas high-level talents, developing overseas training of backbone teachers, developing comprehensive reform of education, promoting education, regional collaboration, establishing bilingual courses, training foreign talents, introducing outstanding foreign teaching materials, strengthening international understanding education, promoting Sino-foreign cooperation in running schools, establishing international schools, fully implementing quality education and promoting scientific development of education. The respondents believe that the main reason for poor internationalization of higher education of international trade major stems from low availability of fund, less related projects and inadequate information publicity.

In addition, most respondents wish to study abroad in their university or graduate stage if there is opportunity and good family condition.

4 IMPLEMENTATION METHODS OF INTERNATIONALIZATION OF HIGHER EDUCATION OF INTERNATIONAL TRADE MAJOR

4.1 Internationalization of the education system

There is certain gap between China's current education system and the international level in academic setting, education, credit and other aspects, resulting in the international cooperation in education, student exchange, etc. and hindering the process of internationalization of higher education. International education system should be developed. The government should change its functions, release and encourage the universities to independently participate in international competition, support various social forces to participate in various forms of education, innovate in educational system, strengthen exchange of credits, schooling, education, etc., improving mutual awareness and adopting open school system.

4.2 Curriculum internationalization

4.2.1 Internationalization of curriculum structure
With China's rapid economic development, some courses of international trade major can no longer meet the needs of social development. As such, the internationalized curriculum structure construction should be developed: evaluating the original courses by major construction and social research, removing or improving the courses that are unsuited to international teaching requirements or those with problems, learning and promoting foreign high-level university curriculum, performing teaching reform by combining with the actual situation of the students, strengthening the construction of key courses, conducting education reform research by focusing on key courses as well as improving the scientific rationality of the course structure.

4.2.2 Internationalization of curriculum contents
Currently, most teaching materials of international trade major are prepared by domestic scholars, and are lacking in international perspective. Some of the textbooks just copy materials from abroad which are far from China's reality. Innovation should be made for internationalization of teaching contents: selecting perfect teaching contents, appropriately increasing proportion of foreign language teaching and bilingual education, trying to reverse the tendency of exam-oriented education, and strengthening and improving students' basic skills in foreign languages. Besides, teachers should better combine the internationalization contents of the curriculum with China's actual situation instead of just copying related foreign courses.

4.3 Internationalized teaching practice

4.3.1 Case teaching

Internationalization of teaching practice refers to full reference with foreign practice teaching methods to help students to better apply the theory by simulating trading practices, case teaching method and so on.[3] Case teaching method pays attention to social reality, focusing on creative thinking and the ability to solve practical problems of the students. In the case of teaching, teachers should pay attention to the teaching process design, and the college should establish a professional case base so that the students can truly understand and apply it to practice.

4.3.2 Application of lectures

International trade major course is highly practical. The occasional lectures given by the experts and working staff can make up for the weakness of classroom teaching. Lectures can show the practical experience and methods to the students to change the former single classroom teaching mode.

4.3.3 Training of cognitive practice

The colleges should carry out cognitive practice for the students through various forms: summer social practice can improve students' cognitive abilities by field research in some companies and institutions in order to lay the foundation for future practice; conditions should be created to encourage students to participate in a variety of production research projects to improve their theoretical analysis and social practical ability by participating in research projects.

4.3.4 Practice base construction

The students of international trade major are inseparable from social practice. Compared with other teaching modes, job training is the most direct and practical way, which can make the students understand the operation mechanism of the unit in a relatively short period so as to improve their job resilience, thereby improving their ability of working and analyzing and solving social problems. International trade major should focus on the development of teaching practice, reasonable construction of practice bases and increase of external cooperation.

4.4 Internationalized faculty

Teachers' leading role is the key for implementation of internationalization of higher education. Under the background of international education, the teacher's task is not only to impart knowledge but also to guide students to master learning methods and train their capability of cross-cultural communication and transaction. In addition to the professional knowledge, the teachers should also have international teaching awareness and teaching philosophy, master the latest educational technology, teaching methods and research methods so as to train to compound professionals meeting the international development.

5 EFFECT EVALUATION OF INTERNATIONALIZATION OF HIGHER EDUCATION OF INTERNATIONAL TRADE MAJOR

5.1 Effect evaluation of curriculum teaching

International teaching evaluation system is to learn from foreign advanced curriculum and examination system, train students to learn actively, improve learning efficiency and better achieve the overall evaluation of the course study.[4–5] Students' ability training is a systematic process, which can be achieved through a series of teaching links, including the explanation of curriculum outlines, preview, asking questions and discussion at class, exercises, stage examination, social practices, etc. Thus, it needs to build a comprehensive evaluation system of teaching for comprehensive assessment of the students' ability so as to ensure comprehensive scientific assessment, truly reflect the level of students' ability and avoid extensive teaching assessment. Specific assessment method is to decompose the course scores into all teaching aspects, such as attendance, exercises, class discussion, literature review, course papers, midterm investigation, final exams, etc. so as to realize reasonable allocation of teaching resources, train the comprehensive ability of the students in teaching process, reduce pressure on the final exam and ensure that the examination can reflect the true ability of the students.

5.2 Assessment of the effect of practice teaching

In foreign practice teaching evaluation mode, teachers should guide students to complete practice teaching and actively communicate effectively with students and practice units or agencies for comprehensive scientific assessment of students' practice effect. Practice teaching mode should be innovated and specialized personnel should be arranged for objective, comprehensive and scientific evaluation of practice teaching. The substance of assessment should be clarified. Assessment is just a means instead of the purpose. Efforts should be made to sum up experience and weakness of practice teaching mode in the assessment and to develop and improve practice schemes so as to continuously improve practice teaching.

5.3 Teaching quality control and evaluation system

International trade major requires a constantly revised and constantly improved evaluation system of teaching activities and the overall teaching effectiveness. Foreign teaching evaluation methods should be learnt for occasional evaluation of teaching activities of teachers and teaching system and adjustment according to the teaching effectiveness in order to continuously improve the quality of teaching. The specific approach is to revise the teaching programs and outlines each year based on the actual teaching situation and changes in demand of talents of the society. The teaching administration departments and teaching units should together demonstrate the teaching course assessment regulations, develop quantitative assessment criteria and summarize after course teaching and assessment in order to ensure the quality of international teaching mode.

6 CONCLUSION

The connotation of internationalization of higher education of international trade major should include internationalization of clear education ideas, training objectives, teaching facilities, teaching environment and faculty. The development of internationalization of higher education of international trade of our college should focus on: the respondents do not have overseas exchange experience; the students' language barriers limit the implementation level of internationalized education activities; and most respondents wish to study abroad in their university or graduate stage.

The specific implementation of internationalization of higher education of international trade major should focus on internationalized construction of education system, curriculum arrangement, teaching practice and teachers. Effect evaluation of internationalization of higher education of international trade major should focus on course teaching, practice teaching, teaching quality control and evaluation system.

ACKNOWLEDGMENT

This work was supported by the education reform research project of Northwest University for Nationalities (Grant No. 12JG-16706637).

REFERENCES

[1] Liu Zhengliang, Shi Huayu; Operating Performance of New Round of Internationalized Education and Educational Development in China;. Era Figures; 2008.09.
[2] Xie Wenwu, Wu Jie; Means, Method and Effect Evaluation of Internationalized Education of Business Major;. Researches in Higher Education of Engineering; 2010 Supplement.
[3] Huang Lijun; Discussion of Internationalized Practice Teaching Mode of Business Administration Major;. Chinese and Foreign Education Research; the 8th issue, 2009.
[4] Li Qin; Research of Construction of Postgraduate Education Internationalized Evaluation Index System. Nanjing Agricultural University; 2010.
[5] Zhang Yan; Construction of University Internationalized Evaluation Index System; China Higher Education Evaluation; the 01st issue, 2012.

Education Management and Management Science – Zheng (Ed.)
© 2015 Taylor & Francis Group, London, ISBN 978-1-138-02663-6

The cultivation of language communication skills of students majoring in design

Wenhui Li

Zhengzhou Huaxin University, Zhengzhou, China

ABSTRACT: Design was born to be connected with language. Good communication skill is a requisite in design industry. Education in this new period should adhere to the principles which are people centered, should firstly teach students moralities while emphasizing on ability so as to bring about students' all-round development. Schools educate students according to their own abilities, making every student a useful person to the society. However, there is a shortage of verbal communication ability in contemporary college students, especially those majoring in design. Having been aware of the serious communication problems that college students are now facing, many scholars and experts hold the point that effective methods should be taken to ameliorate current education situation so as to cultivate talents who can make greater contribution to the society.

Keywords: philosophy of education; communication skill; personnel training

1 INTRODUCTION

Professor Lin Jiayang, director of Teaching Guidance Committee of vocational education in Design of Ministry of Education, doctoral supervisor in Tongji University indicated in the Teaching Seminar of Art and Design held in Zhenzhou that students during their four-year college study should try to go to several cities or countries to experience different lives, to visit museums, to take part in some projects and social activities, to exchange courses so that when they graduate they would enjoy professional knowledge and social ability which enable them to create, speak, write and communicate.

Professor Lin Jiayang who had studied in Germany, is a pioneer of China's design education. The reform he started in the aspect of design education in Wuxi Light Industry College, now Jiang Nan University, in 1980s has influenced the design education all over the country. Professor Lin has been attaching great importance to college students' verbal communication skill which is crucial to their work and life. Hereby, I will focus on the education of communication ability of students majoring in design, which I hope will also benefit other students.

2 DESIGN AND LANGUAGE

Design in Chinese is "sheji". The radicals of those two characters are "yanzipang" which means language. Just like other characters with "yanzipang", such as "shuo, ping, jiang, tao, lun, yi", design also has had connection with language since the time it was built. As the name of a profession, design is a product of hands and brains, including a series of actions, planning, making stuff, etc. package design, which is very practical, is called "soundless salesmen", advertisement design "a key" which can impress people. The relationship between design and language is very complicated, contradictory and interdependent.

Design is a comprehensive interdisciplinary subject, integrating many subjects such as Painting, Humanities, Communication, Psychology, Marketing and so on. For example, Visual Communication Design, usually called Visual Communication, accurately and clearly conveying information to target audience through pictures or words, actually is Communication Design, in which language plays a crucial rule.

Design belongs to service industry. When design works, after being revised again and again, are put in front of clients, oral presentations are often a prerequisite. In professional advertising design companies of level 4 A, designers usually do not communicate with clients directly but through professional customer service staff who though having accepted professional training, may misinterpret design concepts and design philosophy in the process of information delivery. Bosses or designers of medium or small-sized design companies directly communicate with clients using PPT or videos to present their works, which if added with background music or dubbing will produce better results. Under the circumstance that clients demand that several design companies' design works should be compared together, language competence of presenters is obviously very important. It is very common in today's society when many people run their own business that bosses also undertake the mission as designers and customer service staff, which leads to the consequence that communication with clients is more than design works presentation.

3 DESIGN AND COMMUNICATION

As admission score of cultural courses of design majors in college enrollment is much lower than other majors, the comprehensive quality of students majoring in design is lower than that of other students. In addition, every student should accept extra tests of painting art. Painting tends to shape students into relative introvert, independent personalities which are not easily influenced by other people's thoughts. There are even some students who are stubborn, unsociable and eccentric. Designers who usually work under great pressure need to devote long time and energy to think deeply in the process of designing, which worsen their problems in communicating with others. The factors above lead to some issues of designers communicating with others which are:

1. with no team cooperation ability, tending to think by themselves and insist that their works are the best.
2. do not like to communicate with clients and place clients in their opposite, holding the view that clients have no aesthetic taste.
3. timid and nervous, not good at stating design ideas.
4. bad writing ability, logic and summarizing ability.
5. do not care about the society, not good at socializing with people out of design industry, isolated from the society, eccentric.

People often judge designers through tainted glasses, labeling them with "someone who deals with art" "individuality" "long hair" "beard" "non-mainstream" "abnormal clothes". Even before getting in touch with designers, people already have the impression that designers are not easy-going.

4 EDUCATION AND VERBAL COMMUNICATION

"CCP's Decision on Deepening the Reform of Some Major Issues" approved in the Third Plenary Session of the 18th Central Committee of the Communist Party of China stated that "government should make efforts to teach every student in accordance with their aptitude, making every child a useful person." Yuan Guiren, the Minister of Ministry of Education noted in his speech "Learning the spirit of the Third Plenary Session of the 18th CCCPC, deepening the reform in education" that "education should cling to the basic orientation which is to educate talents while shaping their morality. The essence of education is to cultivate talents. Moral education should come first and the training of ability should be paid more attention so that children can be developed in an all-round way. We should try our best to provide every student with the education that is suitable for them so that they all can become useful talents who can take over the responsibility of building our socialism with comprehensive development in moral, intelligence, sports, aesthetics and labor." He stated in expert forum that "now, there are many newly-graduated college students are not good at communicating with others. As for written communication, some students even can't write a notice correctly; and in the aspect of oral one, they also can't express their thoughts clearly."

Nowadays, with computers, mobile phones and social network app QQ, Wechat, Weibo becoming more and more widely used, the opportunities of verbal communication increasingly reduce, as well as the time of interpersonal interaction. Hidden behind those crisis such as "Poisoning case in Fu Dan University" is the problem that college students have in verbal communication. Vice director of students' office in Beijing University, associate professor Zha Jing noted that "in Beijing University, bad communication skills may cause interpersonal tensions which will bring about conflicts in dormitory, mostly undergraduates'."

Language communication can be divided into two categories: written and oral. Professor Lin Yonghe, vice director of China's college students psychological advisory committee and director of Social Science Department of Beijing Industry and Business University stated "communication, from the perspective of Psychology, is the process of two or more people sharing one piece of information, meeting each other's demand. In the aspect of interpersonal relation, communication requires art, tricks, humor, compromise, research, study, knowledge reserve, all of which are what college students lack and what they need to learn."

Written communication, with TV programs such as Chinese Characters Dictation Contest and Chinese Character Heroes becoming popular, has already drawn people's attention. Many people have been greatly inspired and make efforts in writing. However, oral communication is not so good.

After doing some research by visiting college students, we find that there are many problems in communication existing in campus, some of which even became jokes. Those problems mainly are:

6. do not know Phone Etiquette

The following is a true story. A student talked to his teacher on the telephone, "Teacher, are you in the office?" The teacher answered, "Today is weekend, so I am resting at home." "I want to hand in my homework," said the student. The teacher said, "I can't go to school because I have something to do." It is obvious that the teacher couldn't be in the office that day, but the student still tried to persuade the teacher to come to school, which gave the teacher bad impression. If he had expressed his demand in another way, for example, "Mr or Miss, how are you? I am so sorry to disturb you on the weekend. I have my homework to hand in today, which is very urgent. Is it possible that you come to the office or may I go to your home? I am very sorry to cause you such trouble". The teacher after hearing his words will feel that the student really has difficulty with his thesis, so even if the teacher has some important things to do, he or she will spare some time to meet the student while having the impression that the student is polite.

Besides, students make phone calls whenever they want without considering whether the people they call are sleeping or not; when calling others, they don't introduce themselves, not using such words as "please" "thank you" "goodbye" and so on; when talking on the phone, they often hum and haw without point; they do not use "if is it possible" "speak, please" and so on. These problems are very common.

7. use inappropriate names

Some students directly call their teachers' name in the office; in the canteen, they call workers "the old man" "that person" etc. Some students call their classmates "pig" "bitch" and so on.

In the campus, it is polite to call teachers "Lao Shi" which means teacher. It is better to add teachers' family name if students know the teachers well, such as Li Lao Shi or Wang Lao Shi. "Shi Fu" "Da Niang" can be used to call old logistic staff. Students can call each other some informal names. Nicknames may make students much more closer, but names with discriminating and insulting meanings should be avoided. Using inappropriate names is apt to cause students' resistance and irritation which will affect communication.

8. bad expressive skill

In the classes, students' oral expressive ability is bad as they often can't get to the point and their logic is disordered. Students' leaders may make mistakes or can't catch the main point when passing teachers' messages to students.

With the computers becoming more and more popular, students today mainly copy or duplicate something on the internet which leads to the result that their writing ability is becoming weaker and weaker. There are many problems in students written assignments such as illogical and inconsistent sentences, wrong format and misuse of punctuation.

What oral expressive ability shows is people's ability to analyze, think and create; more importantly, essential accomplishment and working attitude. People with good communication ability can correctly and proficiently use oral and written language.

5 THE CULTIVATION OF LANGUAGE COMMUNICATION ABILITY

To solve the problems mentioned above, the following measures can be taken:

1. encourage students to take part in discussions, seminars and presentations and some other interactive activities, encourage them to show their own style and get rid of nervousness.
2. reform the form of examinations, adding oral tests. For example, in design classes, test students' ability of describing their design works.
3. organize speech and debate contests on majors, including students as many as possible, providing students with professional expression material.
4. reform the teaching method of the course Practical Writing, attaching more importance to the training of writing ability to increase students' written communication skill.
5. add courses on communication, setting situations and letting students simulate real scenes, which can make them pay attention to their problems in communication and teach them the ways to deal with their problems.
6. organize activities such as prompt conversation, writing and characters dictation to improve students' language expressive ability in entertainment.
7. pay more attention to students' etiquette, delivering lectures on etiquette to change students' knowledge structure so as to improve students' social ability.
8. encourage students to join a club or volunteer service, increasing their opportunities to communicate with students from different departments, different professions and different industries.
9. encourage students to do part-time jobs. Direct contact with the society can train students'

adaptive ability to the society as well communication skills.

At school, there are more and more students who are able to create and design but are not good at expressing their ideas which is restricting students' future career development. In world-level design companies, designers' salaries are divided into three levels according to designers' abilities. Designers enjoying high level salary should have the ability to design, write and communicate; middle level salary, design and write, while people having low level salary are those who can only design. When educating students, we should focus on their whole life development, that is to teach them skills which can benefit their whole life. These skills can not only help students find jobs but also lay the solid foundation for their promotion.

We need to pay more attention to the training of communication ability and other comprehensive qualities.

REFERENCES

[1] Li Ning, The Training of College Students' Communication Skills Is Necessary. Technology Newspaper, 2007, 12, 27.
[2] Feng Ai jing, Study on the Cultivation of College Students' Communication Skills. Education Science, 2009, 1.
[3] Deng Hui, On the Cultivation of College Students' Communication Ability. Economy and Social Development, 2005, 6.
[4] Liu Yunlian, Research on the Problems of College Students' Communication Skills. The Journal of Liao Ning Educational Administration Institute, 2009, 10.

Education Management and Management Science – Zheng (Ed.)
© *2015 Taylor & Francis Group, London, ISBN 978-1-138-02663-6*

The relationship between Corporate Social Performance and Corporate Financial Performance: A literature review of twenty years (II)

Gang Fu
College of Economics and Management, Sichuan Agricultural University, Chendu City, Sichuan Province, China

Ping Zeng
Accounting Department, Sichuan Business Vocational College, Chendu City, Sichuan Province, China

ABSTRACT: To explore the real relationship between corporate social performance and financial performance, this research takes 63 studies from 1990s as samples to analyze synthetically the conception of CSP, basic theories, methodologies, industries and control variables, stakeholder groups, the measures of CSP and CFP and the correlation between CSP and CFP. We find that (1) dominant researchers assert the positive or neutral relationship of CSP and CFP, (2) basic theories are vague and unclear, (3) the measures of CSP and CFP is a "complex" phenomenon, (4) control variables play an important role in those studies, and (5) many researchers perform the across-industry studies, and view the industry as a key control variable.

Keywords: corporate social performance; corporate financial performance; control variable; stakeholder, research methodology

1 STAKEHOLDER GROUPS, MEASURES OF CSP AND CFP

1.1 *Stakeholder groups, behaviors and environment*

How to measure or reflect CSP relates to the internal elements of CSP in the process of examining the CSP-CFP relationship. As Surroca, Tribo and Waddock (2010) conceptualize CSP as the broad array of strategies and operating practices that a company develops in its efforts to deal with and create relationships with its numerous stakeholders and the natural environment (Waddock, 2004), the stakeholder related to CSP is not single. CSP should reflect the outcomes and levels of actions activated by stakeholders, who are stakeholder groups, behaviors and environment. After reviewing 63 studies, we divide them into the following categories. The first category is based on consumers, which mostly relate with product quality, diversity, product safety, quality of services, innovativeness, alcohol, tobacco, gambling and nuclear power (Preston, O'Bannon, 1997; Mahoney, Roberts, 2007; Moore, 2001, etc.). The second category is based on government, community or military, such as governments, community investment, community relations, community and society, defense/weapons, military contracting (Brik etc, 2011; Andersen & Dejoy, 2011; Surroca, Tribo,

2005; Fauzi, 2009; Baron, Harjoto & Hoje, 2011, etc.). The third category is based on employee, such as employees, human right, labor relations, ability to attract\develop and keep talented people (Fauzi, etc., 2007; Laan, etc. 2008; Waddock & Grave, 1997, etc.). The fourth type is based on shareholder or investor (Moore, 2001; Brik, etc., 2011; Laan, etc., 2008; May, Khare, 2008, etc.). The fifth type is based on internal management or governance, such as governance, quality of management, wise use of corporate assets, ownership in other companies, South Africa investments, financial soundness, non-U.S. operations (Peter, Sarah, 1998; Andersen & Dejoy, 2011; Fauzi, etc., 2007; May, Khare, 2008; Anderson, Olsen, 2011; Nelling, Webb, 2009, etc.). The sixth is based on society and environment, including responsibility to the community and the environment, natural environment, philanthropic donation, NGOs, contribution to society (Mishra, Suar, 2010; Clyde, etc., 2011; Surroca, Tribo & Waddock, 2010; Brammer & Millington, 2008, etc.). The seventh is based on suppliers (Mishra, Suar, 2010; Peters, Mullen, 2009; Brik, Rettab, Mellahi, 2011, etc.). The eighth type is based on women's and minority issues (Ruf, etc., 2001; Waddock & Graves, 1997; Nelling, Webb, 2009, etc.). The last angle is based on others, such as fairness (Choi, etc., 2010), other concerning compensation (Fauzi, etc., 2007), animal testing (Barnett &

Salomon, 2002). In nine groups, researchers often refer to these stakeholder groups, such as employees, communities and consumers, and environment. Of 63 studies, respectively, 33 studies, 30 studies, 28 studies and 35 studies refer to communities, consumers (products), employees and environment. Otherwise, few researchers just refer to single stakeholder group. For example, Simpson & Kohers (2002) just take community relations into account, Schuler & Cording (2006), Gromark & Melin (2011) just think of consumers, and Brammer & Millington (2008) are only concerned about philanthropic donation. In fact, stakeholders considered by numerous researchers are over two categories. Especially, Laan, Ees and Witteloostuijin (2008) divide stakeholders into two levels, which are primary stakeholders and secondary stakeholders. The former include employees, consumers and environment, and the latter includes communities, diversity, investors and human rights.

1.2 The measures of CSP and CFP

1.2.1 The measures of CSP

"CSP is a complex phenomenon" (Griffin, 2000). The measures of CSP have three categories, which are one-dimension single index, multidimensional comprehensive index, and multiple range indexes. In the samples, few studies use one-dimension single index and multiple range indexes to measure CSP. For example, Stephen Brammer and Andrew Millington (2008) use a special element of CSP, charitable donation, to investigate the link of CSP and CFP; Phillips (1999) applies two traditional variables, median household income and number of medicare patients days as a percent of patients days to research how social responsibility drives long-term financial performance of not-for-profit hospitals.

CSP is a multidimensional construct (Griffin, 2000), with behaviors ranging across a wide variety of inputs, internal behaviors or processes and outputs (Carroll, 1979; Waddock & Graves, 1997)), and each dimension has multiple variables and multiple operationalizations (Griffin, 2000). In the past twenty years, numerous researchers used multidimensional comprehensive index to measure CSP, such as the KLD index, the Fortune reputation rating, the Toxics Release Inventory (TRI), the composite CSP score, Dow Jones Sustainability Indexes (DJSI).

Among 41 empirical studies, the composite CSP score are used more frequently than others, and 24 studies apply the category to measure CSP (Tang etc., 2011; Gromark, Melin, 2011; Choi etc., 2010; Fauzi, 2009; May, Khare, 2008; Fauzi etc., 2007, et al.). Secondly, 14 studies have used the KLD index (McWilliams & Siegel, 2000; Johnson & Greening,

1994; Waddock & Graves, 1997; Harrison, Coombs, 2006; Callan, Thomas, 2009, etc.). The TRI (Griffin, Mahon, 1997), the Fortune reputation rating (Peter, Sarah, 1998) and the DJSI (Lee etc., 2009) is used, respectively, for one study.

1.2.2 The measures of CFP

Corporate financial performance is "a subjective measure of how well a firm can use assets from its primary mode of business and generate revenues". There are three broad subdivisions of CFP, which consist of market-based (investor returns), accounting-based (accounting returns), and perceptual (survey) measures (Orlitzky etc., 2003). First, market-based measures of CFP, such as price per share or share price appreciation, reflect the notion that shareholders are a primary stakeholder group whose satisfaction determines the company's fate (Cochran and Wood, 1984). Secondly, accounting-based indicators, such as ROA, ROE, or EPS, capture a firm's internal efficiency in some way (Cochran & Wood, 1984). Lastly, perceptual measures of CFP ask survey respondents to provide subjective estimates of, for instance, the firm's 'soundness of financial position', 'wise use of corporate assets', or 'financial goal achievement relative to competitors' (Conine & Madden, 1987; Reimann, 1975; Wartick, 1988).

The application of CFP measures used by researchers shows the characteristics of diversification. Some researchers use individual indicator, such as ROA (Mahoney etc., 2008; Peters, Mullen, 2009; Tang, 2011), accounting profit (McWilliams & Siegel, 2000), market value/book value (Andersen & Dejoy, 2011; Baron, Harjoto, and Hoje Jo, 2011), ROS (Peter, Sarah, 1998), debt-to-total asset ratio (Harrison, Coombs, 2006), Market Value-Added (MVA) (Surroca & Tribo, 2005), free cash flow (Phillips, 1999), Tobin's Q (Dugar etc., 2011; Surroca etc., 2010), EBITA (Gromark, Melin, 2011; Brammer & Millington, 2008), operating income (Andersen, 2010), ROE (May, Khare, 2008). However, most researchers adopt more than one indicator to reflect CFP. For example, Griffin & Mahon (1997) measure CFP with ROE, ROA, total asset and five year's ROS; Preston & O'Bannon (1997) use ROE, ROA and ROI; Ruf & Muralidhar, Brown, Janney, Paul (2001) have used the growth of ROS, ROE and ROS; Brik, Rettab and Mellahi (2011) have used the growth of ROS, the growth of profit, ROA and ROI etc. Of 41 empirical studies, there are 18 studies with individual indicator, 23 studies with two indicators or more. ROA, the most common index, is used by 21 studies. Accordingly, there are 15 studies with ROE, 7 studies with ROS, 5 studies with Tobin's Q, and 4 studies with MV/BV. Other indicators, such as P/E, EPS, annual return, total assets, turnover, Treynor Ratio, Sharpe measure, Jensen's alpha, are used less. Otherwise, most of researchers don't classify the measure of CFP except for Johnson & Greening (1994) et al.

2 THE RELATIONSHIP BETWEEN CSP AND CFP: POSITIVE, NEGATIVE OR NEUTRAL?

2.1 Positive relationship

Society—and many corporate managers—have embraced the idea that business organizations have a moral obligation to promote social welfare. Strategic philanthropy has been defined as "the synergistic use of an organization's core competencies and resources to address key stakeholders' interest to achieve both organizational and social benefits" (McAlister & Ferrell 2002). It appears that both corporations and society can benefit when businesses engage in behavior intended to do more than maximize shareholder wealth. This observation is supported by decades of empirical research. The majority of the articles support a positive, causal relationship where CSP is a determinant of corporate CFP (Margolis & Walsh, 2003; Pava & Krausz, 1996). Seen from the results of 63 studies in past twenty years, half of the studies, 31 studies, revealed the positive correlation. Not only with meta-analysis, but also with empirical analysis or normative analysis, the positive linkages are the conclusive recognitions. For example, in the studies of meta-analysis, Orlitzky, Schmidt, Rynes (2003) reviewed 52 studies and found out that "across studies, CSP is positively correlated with CFP". Marc Orlitzky (2001) asserts that the present study does not confirm firm size as a third factor which would confound the relationship between CSP and CFP, even if firm size is controlled, CSP and CFP remain positively correlated. Meng-Ling Wu (2006) analyzed the first set of 38 studies and revealed the positive relationship between CSP and CFP, and the further findings, those are "market-based measures are weaker predictors of CSP than other financial measure" and "perceptually based measures reported a stronger CSP-CFP relationship than performance-based measures".

Furthermore, some researchers (Orlitzky, 2005; Peloza, 2006; Becchetti, 2007; Gond etc., 2008; Beurden, Gossling, 2008; Chang; 2010; Neal, Cochran, 2010) argued with normative analysis that the relationship of CSP and CFP is positive, because they asserted that "firms practicing good ethics and good corporate governance are rewarded by the financial markets, while firms practicing poor ethics and poor corporate governance are punished"

(Neal, Cochran, 2008). Orlitzky (2005) pointed out that high social performance might be both a determinant and a consequence of high financial performance, and business can benefit financially from higher social and environmental performance. Gond and Palazzo (2008) view the positive relationship between CSP and FP as the 'Holy Grail' of the business and society field of research.

Of the 63 studies, 21 studies have the opinion of positive correlation between CSP and CFP. Preston and O'Bannon (1997) based on data for 67 large U.S. corporations for 1982–1992, and found overwhelming evidence of a positive relationship between social and financial performance indicators, that is, no significant negative social-financial performance relationships and strong positive correlations in both contemporaneous and lead-lag formulations.

2.2 Negative relationship

Within the economic and finance literatures there is a long history of commentators who have argued that a focus on CSR will reduce financial performance (Lee etc., 2009). For example, Friedman (1970) asserted that the only social responsibility of a company is to use its resources to engage in activities designed to increase profits. Many others, including Bragdon and Marlin (1972), Vance (1975) and Brammer, Brooks & Pavelin (2006), support this view. It has been argued that a negative relationship between CSP and CFP arises due to the additional costs incurred to improve social or environmental performance, which does not contribute to enhancing shareholder value. Also, profitable business and investment strategies that are rejected only because of CSR concerns must result in lowering economic performance (Langbein & Posner 1980; Aupperle, Carroll & Hatfield 1985; Knoll 2002).

Of 63 studies, there is only 1 study which finds a negative relationship between CSP and CFP. Makni, Francoeur and Bellavance (2009) performed the empirical analyses on a sample of 179 publicly held Canadian firms and used the measures of CSP provided by Canadian Social Investment Database for the years 2004 and 2005. Using the "Granger causality" approach, they found no significant relationship between a composite measure of a firm's CSP and FP, except for market returns. However, using individual measures of CSP, there was a robust significant negative impact of the environmental dimension of CSP and three measures of FP, namely return on assets, return on equity, and market returns. This latter finding is consistent, at least in the short run, with the trade-off hypothesis and, in part, with the negative synergy hypothesis which states that socially responsible firms experience lower profits and reduced shareholder wealth, which in turn limits the socially responsible investments.

2.3 *Neutral or no relationship*

For researchers, the relationship between CSR and financial performance is complex and nuanced (Pava and Krausz, 1996). Some researchers find the paradoxical results, and some find that there is non-linear relationship, or the curvilinear relationship. Therefore, Pava and Krausz (1996) exclaim that the association between CSP and CFP is the paradox of social cost.

Of 63 studies, the results of 30 studies are no relationship, or neutral relationship, or the curvilinear relationship. For example, Barnett and Salomon (2002, 2006) have researched SRI (social responsibility investment) funds, and found that the relationship between financial and social performance is neither strictly negative nor strictly positive. Rather, it is curvilinear, with the strongest financial returns to low and high levels of social responsibility, and significantly lowers financial returns to moderate levels of social responsibility. Phillips (1999) investigates that managerial efficiency and social responsibility drive long-term financial performance of not-for-profit hospitals, and results show minor links between social responsibility and long-term profitability. McWilliams and Siegel (2000) demonstrate a particular flaw in existing econometric studies of the relationship between social and financial performance. They find that some studies estimate the effect of CSR by regressing firm performance on corporate social performance, and several control variables. This model is misspecified because it does not control for investment in R&D, which has been shown to be an important determinant of firm performance. This misspecification results in upwardly biased estimates of the financial impact of CSR. They find CSR has a neutral impact on financial performance when the model is properly specified.

We usually argue that engaging in social activities pays to be good, but only in a limited context. It does not pay to be bad—engaging in illegal activities has a negative impact on profits (Johnson, 2003). Of course, some views of researchers are fuzzy and unclear. For example, Mahoney and Roberts (2007) find no significant relationship between the composite CFP measure and either ROA or ROE, but both the environment and international dimensions of CSP measure were significantly related to ROA. Moore and Robson (2002) draw blurred conclusions: (a) statistically significant support was found for the negative synergy hypothesis; (b) a statistically significant positive association was found between profitability (both lagged and contemporaneous) and community contributions. Moreover, Surroca & Tribo (2005), Surroca, Tribo and Waddock (2010) indicate the linkage of CSP and CFP, and find there is no direct relationship between corporate social and corporate financial performance and only an indirect connection that relies on their mutual connection with a firm's intangible resource.

3 FINDINGS AND ENLIGHTENMENT

3.1 *Findings*

The relation between a firm's financial performance and its Corporate Social Performance (CSP) has been a topic of interest and controversy for more than half a century (Dodd, 1932; Preston & O'Bannon, 1997), and serious empirical research on the association between financial and social performance indicators has been going on for several decades. Yet in spite of this long record of discussion and analysis, the connection between corporate social and financial performance has not been fully established (Preston, 1997), and the nature of the relationship remains unresolved (Andersen, Olsen, 2011). With the so-called 'vote counting' technique (Orlitzky, Schmidt, Rynes, 2003), we can find the following from reviewing 63 studies on the relationship between CSP (and CSR) and CFP.

Firstly, the relationship of CSP-CFP is still the key issue focused by many researchers. In the past twenty years, numerous researchers have investigated the link between CSP and CFP, but the studies are not saturated (Vishwanathan, 2010). Actually, it is worth noting that more researchers study their linkage from different angles. Secondly, the connotation of CSP is still uncertain. Whether outcome view and activity view, or comprehensive view, no definition has been approved highly. Thirdly, basic theories employed in studies are different. Although most of researchers fulfill the analysis based on stakeholder theory, but stakeholder groups taken into the analysis by researchers are very different. Fourthly, the measures of CSP and CFP are the complex phenomenon. The measures of CSP exist in two categories: the individual and the composite, but the composites lack the consistency. Fifthly, the positive relationship between CSP and CFP are the majority, and the neutral relationship are in less than half of the studies, and the negative relationship is less. A number of factors would influence their relation, and control variables play an important role in these studies. Lastly, dominant researchers have performed the across-industry studies, and they look at the industry as a key control variable to investigate the CSP-CFP relationship in many countries or a certain country.

3.2 Enlightenment

The current studies show that the relationship of CSP and CFP is same as the association of chicken and egg (Dean, 1998), and still is a vexing question. It is important to clarify the CSP-CFP relationship for stakeholders. For instance, the firm's top management decides whether they afford social responsibilities, the government improve the policies or not, the shareholders increase or decrease investments, the employees service for the company or not, the banks stop the loan or not, the suppliers go on providing the commerce or not, and the consumers search for new products or not. Therefore, we may draw lessons from the prior studies, and apply the appropriate methodology to research the CSP-CFP relationship.

Firstly, we have to rebuild the theoretical construction of CSP and CFP, which lead to identify the direction of the causality. These involve the definition of CSP, stakeholder groups, methodology et al. Secondly, the relationship of CSP and CFP is researched in a certain country and in different industries. In across-industry studies, the industry as control variable usually plays an key role (Waddock and Graves, 1997; Brik, Rettab, Mellahi, 2011; Surroca and Tribo, 2005; Brammer and Millington, 2008, etc.), which shows the CSP-CFP relationship is different in different industries. Thirdly, the measures of CSP and CFP selected are in accordance with reality. The requirements are (1) that stakeholder groups ought to be considered fully while measuring CSP, (2) that the measures of CSP are divided into individual index and composite index, which come from financial data and perceptual survey, (3) that the measures of CFP are based on both market and accounting, (4) that the factors, such as risk and firm size, are taken as control variables, and (4) to fully consider the impact of incremental changes, and to use both lagged and contemporaneous data. Finally, not only should we research the influence of CSP on CFP, but we should also study the influence of CFP on CSP. The scale of samples is large enough to examine the relationship.

It is always helpful to increase the general awareness about corporate social responsibility. The companies have to consider environmental, social, and ethical awareness in their business strategy and try to do philanthropy or charity and adopt a long-term sustainable strategy in the business to become ethical corporate citizen (Hameed, 2010).

ACKNOWLEDGMENT

This research was financially supported by the program (13XJC630005) from Ministry of Education of the People's Republic of China.

REFERENCES

[1] Moses L. Pava, Joshua Krausz, 1996, The Association Between Corporate Corporate-Responsibility and Financial Performance: The Paradox of Social Cost, Journal of Business Eethics, 15: 321–357.

[2] Marc Orlitzky, Frank L. Schmidt, Sara L. Rynes, 2003, Corporate Social and Financail Performance: A meta-analysis, Organization Studies, 24(3): 403–441.

[3] Meng-Ling Wu, 2006, Corporate Social Performance, Corporate Financial Performance, and Firm Size: A Meta-Analysis, The Journal of American Academy of Business, Carmbridge, 8(1): 163–171.

[4] John Peloza, 2006, Using Corporate Social Responsibility as Insurance fo Financial Performance, California Management Review, 48(2): 52–72.

[5] Douglas A. Schuler, Margaret Cording, 2006, A Corporate Social Pefformance-Corporate Financial Performance Behavioral Model For Consumers, Academy of Management Review, 31(3): 540–558.

[6] Michael L. Barnett and Robert M. Salomon, 2006, Beyond Dichotomy: The Curvilinear Relationship Between Social Responsibility And Financial Performance, Strategic Management Journal, 27: 1101–1122.

[7] Leonardo Becchetti, 2007, Corporate Social Responsibilit: Not Only Economic and Financial Performance, Finance & The Common Good/Bien Commun, 28–29: 152–158.

[8] Margaret L. Andersen and John S. Dejoy, 2011, Corporate social and financial performance: the role of size, industry, risk, R&D and advertising expenses as control variables, Business and Society Review, 116(2): 237–256.

[9] Jeffrey S. Harrison, Joseph E. Coombs, 2006, Financial leverage and social performance, Academy of Management, F1–F6.

[10] Gary Woller, 2007, Trade-offs between social & financial performance, ESR Review, 14–19.

[11] Gerwin Van der Laan, Haans Van Ees, Arjen Van Witteloostuijn, 2008, Corporate social and financial performance: an extended stakeholder theory, and empirical test with accounting measures, Journal of Business Ethics, 79: 299–310.

[12] John Peloza and Lisa Papania, 2008, The missing link between corporate social responsibility and financial performance: stakeholder salience and identification, Corporate Reputation Review, 11(2): 168–181.

[13] Jean-Pascal Gond, Guido Palazzo, 2008, The social construction of the positive link between corporate social and financial performance, Academy of Management, 1–6.

[14] Pieter van Beurden, Tobias Gossling, 2008, The worth of values-a literature review on the relation between corporate social and financial performance, Journal of Business Ethics, 82: 407–424.

[15] Jonas Nilsson, 2008, Investment with a conscience: examining the impact of pro-social attitudes and perceived financial performance on socially responsible investment behavior, Journal of Business Ethics, 83: 307–325.

[16] Stephen Brammer and Andrew millington, 2008, Does it pay to be different? An analysis of the relationship between corporate social and financial performance, Strategic Management Journal, 29: 1325–1343.

[17] Hasan Fauzi, Lois S. Mahoney, Azhar Abdul Rahman, 2007, The link between corporate social performance and financial performance: evidence from Indonesian companies, Issues in Social and Environmental Accounting, 1(1): 149–159.

[18] Peter May, Anshuman Khare, 2008, An exploratory view of emerging relationship between corporate social and financial performance in Canada, Journal of Environmental Assessment Policy and Management, 10(3): 239–264.

[19] Richard Peters, Michael R. Mullen, 2009, Some evidence of the cumulative effects of corporate social responsibility on financial performance, the Journal of Global Business Issues, 3(1): 1–14.

[20] Edward Nelling, Elizabeth Webb, 2009, Corporate social responsibility and financial performance: the "virtuous circle" revisited, Review of Quantitative Finance & Accounting, 32: 197–209.

[21] Scott J. Callan, Janet M. Thomas, 2009, Corporate financial performance and corporate social performance: an update and reinvestigation, corporate social responsibility and environmental management, 16: 61–78.

[22] Lois Mahoney, William LaGore, Joseph A. Scazzero, 2008, Corporate social performance, financial performance for firms that Restate earnings, Issues in Social and Environmental Accounting, 2(1): 104–130.

[23] Darren D. Lee, Robert W. Faff, Kim Langfield-Smith, 2009, Revisiting the Vexing Question: Does Superior corporate social performance lead to improved financial performance, Australian Journal of Management, 34(1): 21–49.

[24] Rim Makni, Claude Francoeur, Francois Bellavance, 2009, Causality between corporate social performance and financial performance: evidence from Canadian firms, Journal of Business Ethics, 89: 409–422.

[25] Hasan Fauzi, 2009, The determinants of the relationship of corporate social performance and financial performance: conceptual framework, Issues in Social and Environmental Accounting, 2(2): 233–250.

[26] Leonard A. Jackson, H.G. Parsa, 2009, Corporate social resonsibility and financial performance: a typology for service industries, International Journal of Business Insight & Transformation, 1: 13–21.

[27] Hasan Fauzi, 2009, Corporate social and financial performance: empirical evidence from American Companies, Globsyn Management Journal, 3(1): 35–34.

[28] Roberto Garcia-Gastro, Miguel A. Arino, Miguel A. Canela, 2010, Does social performance really lead to financial performance? accounting for endogeneity, Journal of Business Ethics, 92: 107–126.

[29] Jordi Surroca, Josep A. Tribo, and Sandra Waddock, 2010, Corporate Responsibility and Financial Performance: The Role of Intangible resources, Strategic Management Journal, 31: 463–490.

[30] Jong-Seo Choi, Young-Min Kwak, Chongwoo Choe, 2010, Corporate social responsibility and corporate financial peformance: evidence from Korea, Autralian Journal of Management, 35(3): 291–311.

[31] David J. Flanagan, K.C. O'Shaughnessy, Timothy B. Palmer, 2011, Re-assessing the relationship between the Fortune Reputation Data and financial Performance: overwhelming influence or Just a part of the puzzle?, Corporate Reputation Review, 14(1): 3–14.

[32] Margaret L. Andersen, 2010, Corporate social and financial performance: a canonical correlation analysis, the Academy of Accounting and Financial Studies, 15(1).

[33] William S. Chang, 2010, Social network and corporate financial performance: conceptual framework of board composition and corporate social responsibility, International Journal of Business and Management, 5(6): 92–97.

[34] Pushpika Vishwanathan, 2010, The Elusive Relationship Between Corporate Social and Financial Performance: Meta-Analyzing Four Decades Of Misguided Evidence, Academy of Management, 1–7.

[35] Kranti Dugar, Brian T. Engelland, Robert S. Moore, 2010, Summary brief: exploring corporate social responsibility, product recalls, customer satisfaction and financial performance in pulicly-traded firms, Society for Marketing Advances, 122–123.

[36] Johan Gromark, Frans Melin, 2011, The underlying dimensions of brand orientation and its impact on financial performance, Journal of Brand Management, 18: 394–410.

[37] Robert Neal, Philip L. Cochran, 2008, Corporate social responsibility, corporate governance, and financial performance: Lessons from finance, Business Horizons, 51: 535–540.

[38] Marc Orlitzky, 2011, Institutional Logics in the study of organizations: the socia construction of the relationship between corporate social and financial performance, Business Ethics Quarterly, 21(3): 409–444.

[39] Margaret L. Anderson, Lori Olsen, 2011, Corporate social and financial performance: a canonical correlation analysis, Academy of Accounting and Financial Studies Journal, 15(2): 17–37.

[40] Zhi Tang, Clyde Eirikur Hull, Sandra Rothenberg, 2011, How corporate social responsibility is pursued affects firm financial performance, Academy of Management.

[41] David P. Baron, Maretno Agus Harjoto, and Hoje Jo, 2011, The economics and politics of corporate social performance, Business and Politics, 13(2): 1–46.

[42] Clyde, Eirikur Hull and Sandra Rothenberg, 2008, Firm Performance: The Interactions of Corporate Social Performance with Innovation and Industry Differentiation, Strategic Management Journal, 29: 781–789.

[43] Supriti Mishra, Damodar Suar, 2010, Does corporate social responsibility influence firm performance of Indian companies?, Journal of Business Ethics, 95: 571–601.

[44] Marc Orlitzky, 2002, Book Reviews, The International Journal of Organizational Analysis, 10(2): 191–202.

[45] Grigoris Giannarakis, Ioannis Theotokas, 2011, The effect of financial crisis in corporate social responsibility performance, International Journal of Marketing Studies, 3(1): 2–10.

[46] Syed Kamran Hameed, 2010, Corporate Social Responsibility (CSR) theory and practice in Pakistan, Swedish University of Agricultural Sciences.

[47] Philip L. Cochran, Robert A. Wood, 1984, Corporate Social Responsibility and Financial Performance, Academy Of Management Journal, 27(1): 42–56.

[48] Kenneth E. Aupperle and Deane Van Pham, 1989, An Expanded Investigation into the Relationship of Corporate Social responsibility and Financial Performance, Employee Responsibilities and Right Journal, 2(4): 263–274.

[49] Jean B. Mcguire, Alison Sundgren, Thomas Schneeweis, 1988, Corporate Social Responsibility and Firm Financial Performance, Academy of Management Journal, 31(4): 854–872.

Based on the project of independent interaction open-ended engineering training teaching mode of reform and research

X.F. Wang, S.G. Wang & Y.J. Cai
School of Mechatronics Engineering, Qiqihar University, Qiqihar, China

ABSTRACT: Engineering training is an integral part of practical teaching to train the innovation spirit, engineering consciousness, and practical ability of students. By analyzing the problems of traditional engineering training and taking into consideration the engineering practical skills and invention ability of students into consideration, the necessity of project-based independent interaction open-ended engineering training teaching mode in the application of modern engineering training was proposed, with emphasis on project settings, project implementation, etc. Through the project pilot with exploration based on training, so that students can graduate into the professional fields to carry out continuous line of exploratory research, to cultivate students' engineering quality, innovation and practice ability; conducive to the students in the practice of autonomous learning, improve the learning initiative and self-consciousness; the ability to train all kinds of comprehensive application of the students knowledge to solve engineering problems.

Keywords: engineering training; innovation abilities; the independent and interactive; open-ended; teaching mode

1 INTRODUCTION

Engineering training is a practical teaching link, which is the comprehensive engineering college of engineering practice teaching essential technology basic course and modern manufacturing technology. It is laden with improving students engineering quality and engineering practice ability, train comprehensive, practical and creative, and with high quality, high level modern engineering and technical personnel, in cross century higher engineering education's role is irreplaceable by other courses [1]. To meet the needs of high-quality innovative talents, we must set up the training project scientifically, use the advanced training method and means, and pay attention to the advanced nature, openness, innovation, form a systemic and scientific and complete innovative talent training system, comprehensively training students in the scientific style, engineering consciousness and comprehensive analysis, finding and solving problem ability, which gives students innovative and entrepreneurial spirit and practice ability [2].

Based on the project teaching method that puts each project as the main line, makes the relevant knowledge into the project of each link through a project in the teaching process. Through the deepening of the problem or function expansion, it broadens the breadth and depth of knowledge, and gets a complete solution of the project, so as to achieve the purpose of learning knowledge and cultivating ability. This method has been used in the students' autonomous learning ability training [3], teaching model reform [4], practice teaching [5], teacher training, etc. In recent years, our school made some reforms and innovations in engineering training section. Our engineering training center conducts some innovations of teaching method based on the project of independent interactive engineering training teaching model, and achieves good teaching effect.

2 THE SHORTCOMINGS OF TRADITIONAL MODE OF ENGINEERING TRAINING TEACHING

The colleges and universities of engineering training center generally come from machining centers or electrical engineering practice bases. The curriculum setting of the project, mainly reflects the inheritance of the original curriculum, and also combines the school discipline settings and comprehensive innovations. Years of teaching practice proves that this model is suitable for the current level of development of science and technology in our country and has certain general teaching meanings. But compared with the training goals and requirements of the research university, there still exist certain gaps on the degree of interdisciplinary innovation need in ability training and the breadth of knowledge.

Traditional engineering training is according to the teaching plan and curriculum. Training programs, contents, time and so on are planned. It has not changed for many years. The training mode is based on work and the training content of each type of work is independent. There is no link between different professions. Teaching method is that the guide explains the work standards and demonstrates on the site. Students shall operate by the given drawings and process operations under the advice of the guide. The engineering training teaching mode has the following disadvantages: (1) The teaching process thinks the teacher as the center, training content, training methods are determined by teachers and instructors according to the syllabus in advance, students have no choice, and no creative space; (2) The training work is independent from each other, a component in the process of production might involve a variety of processing technology and equipment and connection between each type of work. The traditional engineering training methods do not pay attention to train students' comprehensive ability. Therefore, the training methods are unfavorable for the cultivation of the engineering consciousness and innovation ability of the students.

3 THE ESTABLISHMENT OF INDEPENDENT INTERACTIVE OPEN ENGINEERING TRAINING TEACHING MODEL BASED ON PROJECT

The independent interactive open training mode is based on the teaching mode. Its task is training to complete a project. And you can choose training program by the interest and ability. During the process, students can communicate with teachers and other students, they can learn from each other, interactively discuss questions. Students gain knowledge under the open premise of the teaching content, teaching methods, training time, and effective complete practice.

1. Reforming of traditional metalworking practice content and building a new mode of engineering training. In the traditional metalworking practice, the content is about a single type of work and gives priority to pure operation skill training. Training content is single, and technological process is simple. More importantly, the training content is fixed, monotonous and not rich, that makes training in a basic theory and skills of cognitive level of learning. The function of modern engineering training is based on the background of big engineering and manufacturing. It fully embodies the development direction of modern design technology, modern manufacturing technology, modern control technology and modern management technology. It gears to the needs of engineering students, provides a modern comprehensive industrial simulation environment, through that you can experience the process of product design, manufacture, integration of control management. So the students can get the procedural knowledge from the typical scene or device. So we need to build a new type of engineering training teaching model to try to meet the requirements of engineering training under the new situation.

2. Establishing a project training mode can expand the connotation of the engineering training and improve the students' autonomous learning ability. The training mode which gives priority to the project training refers to the actual situation of different professional and different grade students to select traditional training content. It expands the big project under the background of modern design, manufacture, control and management of new technology content, increase the content of the innovation practice. Binding energy that meets the market needs of practical engineering projects and products, we introduce scientific research items; optimize the integrating teaching resources for students with different levels, different types, and the typical training projects. The content is to meet the needs of different schools and different professionals and personnel training. Students can choose suitable training program about different projects and different credit according to his circumstance, such as interests.

4 THE SETTINGS OF INDEPENDENT INTERACTIVE PROJECT TRAINING MODE

According to the laws of the students' cognitive process, setting of training program is from simple to complex, from perceptual to rational, from foundation to comprehensive, from inside to outside, from junior to senior. It is multi-level and comprehensive. The content of the project which is throughout the whole process of teaching of undergraduate course highlights the cultivation of innovation ability.

The training program of engineering training center set is divided into four levels, as shown in Figure 1. Each level has different content and different credit training programs for students to choose.

1. Cognitive training program: The purpose is to train the students with the manufacturing system and manufacturing process of cognitive engineering. Through different professional presentations and some practical ways for different professional designs with different sets

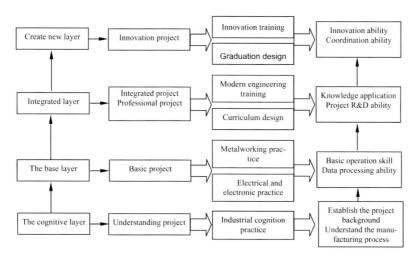

Figure 1.   System of engineering training structure based on project.

of content of the project, students lay a certain engineering background.

2. Base layer training program: Engineering skills based training to the students give priority to engineering skills training, use the combination of concentrated and dispersed training methods. According to the students' major in machine and the different requirements of different training content of the projects and credits, it makes students learn process knowledge, improve the ability of engineering practice, train engineering quality, and preliminary train the students' innovation consciousness and ability.

3. Composite layer training project: According to the request of different professional teaching plans and students' interests, the project gives priority to the training of comprehensive thinking ability and innovation ability to the students of engineering. It combines with the advanced engineering technology, provides comprehensive practice training which is not attached to the course of engineering projects. And it improves the students' ability to analyze and solve problems, enhances the comprehensive engineering quality and ability. It further develops students' innovative spirit and thinking ability.

4. The innovation training project mainly emphasizes the practice teaching mode of extending outside-school from inside-school. It will extend outside-school from the creativity of inside-school, establish students' innovative project activity groups, import scientific research projects and create subject project competition. The project combines engineering design and practice teaching. It will improve students' capacity for innovation and the ability of operating practical process.

5. In addition to the engineering training center training programs, students can also, completely by personal interests and abilities, apply for their own design develop training project. To review the center according to the actual situation, to the approved project approved credits, arrange training time, equipped with teachers training project.

5   THE IMPLEMENTATION OF INDEPENDENT INTERACTIVE OPEN TEACHING MODE BASED ON PROJECT

To set up independent interactive open teaching method based on the project, there must be a feasible teaching plan to support, guarantee and complete practice teaching and orderly, independent interactive open teaching is an open, orderly not disorderly open, is a student in the teaching content, teaching methods, teaching time in the open teaching environment, according to the engineering training center management system, choose according to their own time, interest and ability training, appointment training time, in order to achieve autonomous choice training content for the task of the project, for training. Engineering training center equipped with teachers guidance, management, supervision and examination, to ensure that each link carries on smoothly. Concrete implementation scheme has the following aspects:

1. Relying on the university campus network resources, establish the engineering training center of teaching management system. The Center will upload the teaching management system, safety management system, all levels of training project topic, project training content,

teaching plan, teaching courseware, teaching information related to internet, for the students as a select training project preparation, reference and network system set up the network entrance guard management, student login system by student number and password system.

2. Online booking application management way, in the engineering training center teaching management, online booking in advance to apply for training programs and training time, after approved by the center to accept, notify the students on the net, center according to choose the number and time condition, the rational allocation of hardware and software of teaching resources, arrangement of specific teaching work ahead of time, to ensure an open practice teaching smoothly.

3. Using network preview, combining the teaching of centralized and decentralized, field training and online learning, the combination of online interaction, interactive panel discussion on a variety of methods, such as the implementation of teaching. Training program to make an appointment after the application is successful, the student can be in the first online self preparation training content of the project, there is a problem can be put forward on the net, teachers' answering questions on the Internet, and can and teachers and classmates in time online booking for knowledge exchange interaction, at the beginning of the actual training teachers adopt the way of teaching or dispersed on training, teachers in a leading position. As the project progressed gradually transition to the students become the center of the project implementation completely independent training. The teachers can check the progress of the project training, timely guide a difficult problem, gradually guide the student to use a variety of information resources, solidarity and cooperation between classmates, and independently explore the method and ability to solve the problem.

4. Establishing the practice of assessment methods, assessment method combined with grades and evaluation results, operation and theory of examination, the combination of basic skills and comprehensive innovation evaluation way. The students' attendance, grades, credits, and so on through the center of teaching management system online inquiry.

5. Setting up and improving the engineering training teaching quality evaluation standard, establish teaching quality supervision system and quality assurance system, to ensure that for the implementation of the practice teaching concept, strengthen the student's work, practice report and the evaluation of the standardization of the operation, the training refers to the teachers teaching evaluation by the training center for teaching supervision constantly, in the form of regular examination and evaluation, to ensure each link of training is completed with good quality.

6 CONCLUSION

Project-based independent interactive open engineering training teaching mode can give full play to the engineering practice dominated by teachers, students as the main body the superiority of teaching methods. To strengthen engineering training focus on the "organic combination, training, mechanical and electrical engineering, open operation" the connotation of the construction, for the smooth realization of traditional metalworking practice to contemporary engineering training transformation provides the effective method, provides a real college students for a large project as the background, a comprehensive quality education in the classroom, to enhance the students' engineering practical ability, improve the students' comprehensive engineering quality, effectively cultivate students' innovation consciousness, innovation ability, to improve the practice teaching effect, more suitable for the modern requirements of training high-quality talents.

ACKNOWLEDGEMENTS

This work was supported by Higher Education Comprehensive Reform Pilot Projects in Heilongjiang Province (Grant No. GJZ201301036); the Education Scientific Research Project of Qiqihar University (Grant No. 2013073 and 2013083).

REFERENCES

[1] Yu Zhao-qin, Wu Fu-gen, Guo Zhong-ning. Project-driven Based Modern Engineering Training Methods. Research and Exploration in Laboratory. 2012, 31 (8): 131–133.

[2] Ma Lingling. An Empirical Study on Developing Students' Autonomy Learning Ability through Project-Based Learning. Journal of Shanxi Radio & TV University, 2010 (3): 54–55.

[3] Li Ze-hui. Application of Project-driven Mode in the Engineering Knowledge Course. Experiment Science and Technology. 2011, 9 (2): 133–134.

[4] Sha shu-jing. Based on project driven mechanical professional teaching model to explore and practice. Journal of Changchun University of Science and Technology, 2012, 7 (1): 207–208.

[5] Xiong fan, Li wei-bo. Project driven software engineering experiment teaching discuss. China Electric Power Education, 2012 (2): 77–79.

Education Management and Management Science – Zheng (Ed.)
© 2015 Taylor & Francis Group, London, ISBN 978-1-138-02663-6

Study on the modes of resource integration and utilization of foreign schools and enterprises

P.J. Chen
School of Automobile, Linyi University, Shandong, China

ABSTRACT: In the background of rapid development of higher education and vocational education, in order to train workers meeting the social needs, the modes of resource integration and utilization of school and enterprise abroad were studied. On the basis of analyzing the resource integration modes in major foreign developed countries, the main methods to integrate and use resources in foreign schools and enterprises were discussed, including human resource, financial resource, material resource and information resource. Those advanced experience were summarized. The research hopes to provide some insights for the cooperation between school and enterprise and resource integration in Chinese higher education.

Keywords: cooperation between school and enterprise; resource of school and enterprise; integration and utilization; integration mode

1 INTRODUCTION

At present, higher education and vocational education has developed very rapidly, the relationship between school and society has been increased annually, in which, the combination between school and enterprise is developing widely and deeply. The most direct form of expression is the cooperation between them. It is a talent training mode, whose goal is to improve the students' professional qualities and professional skills. It connects the schools and enterprises tightly and combines the classroom learning with practical work. The core is to enhance the employment ability of students and promote their all-round development (*Yu*, 2006).

The essence of the school-enterprise cooperation is the complementarity and correlation of resource utilization for school and enterprise; the core is the integration and utilization of resources of both sides (*Zhou*, 2011). The school resources mainly refers to human resource including teachers and students, financial resource including educational fund, internship fund, material resource including training base, experiment equipment, information resource including school brand and system rules. The enterprise resources are mainly staff, technical equipment, production plant, and so on.

The cooperation between school and enterprise can take full advantage of different environment and facilities, the integration and utilization of resources can effectively meet the interests of both schools and businesses, and it is a win-win for both parties (*Ren*, 2012). The cooperation and resources integration is an important way to develop high skilled talents for the new times, and it is the base of enhancing the educational strength and establishing competitive advantage (*Wang*, 2011).

The integration and utilization of school and enterprise resources is one of the experiences of achieving successful higher vocational education in developed countries. The system reflects the specific division and cooperation in academic education and skills training for schools and enterprises and close integration of theory and practice, especially the principle of resource sharing. Combined with the current development status of higher education and vocational education, the integration and utilization of resource of school and enterprise are studied in this project. The popular, effective and typical modes of resources integration and utilization between schools and enterprises are analyzed, and their advanced experience are summarized.

2 THE MAIN MODES OF RESOURCE INTEGRATION AND UTILIZATION OF FOREIGN SCHOOL AND ENTERPRISE

The main modes of resource integration and utilization of school and enterprise in other countries are analyzed. There are several modes as follows.

2.1 Cooperative education mode in America

The mode of cooperative education in America started in 1906, the University of Cincinnati launched an educational program requiring the students of some professional or education programs must have a certain time in one year to obtain the practical knowledge. It aims to reduce the burden on the equipment and facilities of the university, integrate the resource of school and enterprise, and allow students to gain employment skills and experience during the internship. Its main characteristics are as follows.

1. The cooperative education is school-based. According to the professional needs, the schools actively communicate with enterprises. They sign contract, specifying their rights and obligations.
2. For teaching time allocation, it is more or less evenly divided. That is, half in school, half in enterprise.

2.2 *Dual system mode in Germany*

German began to resume quickly after the war, which mainly depended on a large number of high-quality talents. Their training system is the dual system education. The system has been deeply rooted in the society, promoting the higher vocational education. The training of workers is carried out in schools and businesses. The main characteristics are as follows.

1. The main body of training is enterprise and the training time in business occupies a higher proportion, highlighting vocational skills training.
2. It has high pertinence and high enterprise engagement. The school sets up professional committee for every major, which member include business representatives and school teachers. The both sides undertake the formulation, inspection and adjustment of teaching programs, and complete the teaching task together.
3. The government supports and the cooperation is institutionalized. The schools and businesses have corresponding rights and obligations. The businesses share the achievements of education based on their funding amount to schools.

2.3 *Sandwich mode in England*

Sandwich, a flexible study system of combining with work and study, is a modern apprenticeship teaching model including field teaching and school education. The entire education stage usually lasts four to five years, the students are full time at school in the first year, and are trained in enterprises in the following years. Its main features are as following.

1. The system focuses on the individual difference of human. Besides the prescribed learning

contents, increasing learning contents and extending training time are allowed, and the government can give appropriate funding.
2. If the students complete the training program and pass the relevant examination, they can obtain the corresponding professional qualification certificates.

2.4 *Industry-academy cooperation mode in Japan*

This education mode is the cooperation of education and research activities between industry and school (*Shi*, 2006). The cooperation in university level includes private companies, public companies, public bodies, and so on. The main specific implementation methods are as follows.

1. Industry provides financial assistance. The universities use them to purchase teaching facilities to train employees for corporate.
2. Contract research. The universities study by the entrustment of enterprise and the research results are paid by the grantor.
3. Personnel dispatch and communication. The enterprises hire teachers to act as consultants or lecturers. The schools hire industry experts as teachers and enroll business employees as students.
4. Workshop practice. The schools commission enterprises to provide production site, making students receive practical training.

2.5 *Teaching factory mode in Singapore*

The mode of teaching factory is bought out by Nanyang Polytechnic. It is to enable graduates to adapt to the job demands as soon as possible. The advanced production equipment and real business environment are introduced into the schools, which are integrated with the school teaching to form trinity comprehensive teaching mode with schools and enterprise (*Zhang*, 2011). On the basis of existing educational facilities, the school-based teaching mode creates all around factory practice environment. Its main features are as follows.

1. Teaching facilities are in line with enterprise. The equipment of business are introduced into school labs to ensure that the teaching equipment are practical and advanced. Real teaching and working environment is created for students.
2. Teachers are in line with enterprise. They have not only qualified education degree, but also work experience of more than five years. The teachers can bring not only knowledge, experience, but also human relations and corporate projects.

3. Curriculum system is in line with enterprise. The curriculum content should have progressiveness and forward looking to meet the needs of employers. They can be improved at any time in accordance with technological developments.
4. School-enterprise cooperation for project development. The corporate engineers, teachers and students research projects cooperatively to integrating teaching and project studying.

2.6 Industry-led mode in Australia

After years of education reform, Australia built universities and colleges of technical and further education, and gradually formed an industry-led education mode. The characteristics are as follows.

1. The industry widely participates in education, especially in research and decision of major issues of education development, identifying training objectives and professional settings. They set industrial ability standards and operation specification, help schools to establish training bases. The schools and enterprises together determine the school-running orientation and quality assessment. The enterprises provide part-time teachers schools accept students learning.
2. The industry provides financial support for the education. It has become another important source of education funding besides the government which can ensure the healthy development of education (*Gong*, 2011).

3 THE MAIN WAYS OF RESOURCE INTEGRATION AND UTILIZATION OF FOREIGN SCHOOL AND ENTERPRISE

On the basis of analyzing the main modes of resource integration and utilization of school and enterprise in developed countries, the approaches for sharing resource are summarized as follows.

3.1 Human resources integration

The human resource sharing between schools and businesses is an important aspect of integration and utilization of resources. The sharing of human resources is primarily in terms of teachers, especially practical teaching staff. For example, the cooperation education in the United States, dual system in German, and industry academy cooperation in Japan attach great importance to the integration of training teachers, so as to nurture highly qualified workers.

3.2 Material resources integration

Material resources integration between schools and businesses are mainly in the areas of teaching facilities, especially internship training equipment. The dual system in German and industry-led mode in Australian are the models for the integration of material resources. In the aspect of practice teaching, the enterprises provide necessary equipment to achieve the sharing of advanced equipment, advanced technology and other resources.

3.3 Financial resources integration

The financial resources integration mainly means the education funding sharing. The enterprises provide funding for school development, whose purpose is to require schools to train qualified personnel for enterprises (*Kong*, 2007). The mode of industry-led education in Australia provides considerable funding, promoting the healthy development for schools.

3.4 Information resources integration

The integration of information resources between schools and businesses are primarily the sharing of social information and system security. With the advent of the information age, companies have urgent increasing demand for information. As the information resource center, on the basis of a wide range of investigating and analyzing the information needs, the university can provide effective information services for enterprises.

Many national governments, such as Australia, Britain, Japan, Germany, establish vocational qualification certificate system to promote the cooperation between school and enterprise. For example: the German government passes legislation to establish the education system, strictly regulates the rights and obligations for schools and enterprises (*Chen*, 2011).

4 EXPERIENCES SUMMARY OF RESOURCE INTEGRATION AND UTILIZATION OF FOREIGN SCHOOL AND ENTERPRISE

The developed countries have made obvious achievements in cooperation between schools and enterprises, and resources integration and utilization. Their experiences are summed up as follows.

4.1 Government attention and financial support

From the development history of foreign cooperation between school and enterprise, the governments of developed countries create conditions actively to facilitate the integration and utilization of resources smoothly. Financial support is the most effective education policy; the governments provide funds and preferential policies that the education and training need. They encourage different sectors of society, especially the enterprises, to actively participate in and support education. The dynamic mechanism carrying out school-enterprise cooperation and resource integration has been formed.

4.2 Legislative guarantee and legal basis

The developed countries have very complete system on law framework and keep to the legislation principle. They guarantee the smooth completion of school-enterprise cooperation. For example, the Orientation Law on Higher Education of France regards it as an important principle that the higher education is open to businesses. These legal measures make the school-enterprise cooperation obtain legal support, and strengthen efforts in support of government and the community on the school-enterprise cooperation, which fundamentally ensure the integration and utilization of resources for schools and enterprises smoothly.

4.3 Adequate system and personnel security

The developed countries establish a set of comprehensive personnel security mechanism for the education development. They fully implement the vocational qualification system. For instance, England has begun to implement the unified system of National Vocational Qualification. The mechanism is binding upon both parties for the cooperation between school and enterprise. The good social environment for school-enterprise cooperation and resources integration is built.

4.4 Industry guidance and enterprise participation

In the process of implementation of school-enterprise cooperation, the developed countries attach great importance to the initiative of enterprises, and the enterprises are involved with all aspects of running school. For instance, the dual system mode in Germany requires that the ability of graduates are uniformly formulated by industry associations. The enterprises not only develop training programs, strengthen skills, but also provide training funds to ensure adequate teaching material for school.

5 CONCLUSIONS

For the cooperation between school and enterprise, integration and utilization of resources,

the foreign researchers have made a lot of study, and got many achievements. Domestic academe has also carried on active exploration in different forms, at different levels and with different characteristics. But compared with developed countries, there is a certain gap in the depth, levels and models of the resources integration. It is hoped that the domestic research can learn from foreign advanced experience. Combined with the actual conditions of local economic development, stimulate the joint need and motive of the government, schools and enterprise, build a dual win mode of resources integration for schools and enterprises.

ACKNOWLEDGMENT

This work was supported by the teaching reform project of Linyi University, Shandong Province, P.R. China in 2014.

REFERENCES

[1] Chen, L. 2011. Exploring the Library and Information Sharing Mechanism of College and Enterprise Which Based on Deep Mix of College and Enterprise: The Investigation for the Requirement of Enterprises in Ningbo Beilun. *Journal of Ningbo Polytechnic* 5(3): 101–105.

[2] Gong, P. 2011. Analyzing the School-enterprise Cooperation Model of Higher Vocational College. *Weifang Higher Vocational Education* (7): 45–47.

[3] Kong, F. 2007. Research on Vocational Education Resources of the School-Enterprise Cooperation from Abroad. *Journal of Tianjin Professional College* 16(2): 89–92.

[4] Ren, G. 2012. Study on Human Resource Development under the Teaching Resources Platform of School and Enterprise. *Management & Technology of SME* (12): 229–230.

[5] Shi, L. 2006. Analysis and Research of Education Mode of School-enterprise Cooperation. *Higher Agricultural Education* (2): 81–84.

[6] Wang, J. 2011. Study on Bidirectional Flow Ways of School and Enterprise Resources Based on Deep Integration. *Education and Vocation* (8): 21–23.

[7] Yu, Q. 2006, Use Superior Resource to Innovate School-enterprise Cooperation Models. *Vocational and Technical Education* 27(13): 88–89.

[8] Zhang, J. 2011. The Revelation of Teaching Factory of Singapore to the Construction of Practical Training Base for Vocational Education *Guanli Xuejia* (4): 225–226.

[9] Zhou, M. 2011. Practical Teaching Model Experiments of Trinity Program-oriented Courses Based on Resource integration of School and Enterprise. *Occupation* (5): 137–138.

Education Management and Management Science – Zheng (Ed.)
© 2015 Taylor & Francis Group, London, ISBN 978-1-138-02663-6

Information overflow from futures and spot markets to underlying assets: An empirical analysis based on China's copper industry

Lin Wu
Nanjing Xiaozhuang University, Nanjing, China

ABSTRACT: The relationship between Shanghai Commodity Exchange, London Metal Exchange's copper futures index, the copper spot index in China and China copper stock price was examined based on vector autoregressive model, vector error correction model, and impulse response and variance decomposition method. The results show that there is a long-term equilibrium relationship between the price of copper futures, spot market and stock prices. The price information spillover effect is obvious in China's copper markets. London Metal Exchange's one day lagged futures price changes could be helpful to predict the changes of the stock price.

Keywords: futures price; stock price; spot price; information overflow

1 INTRODUCTION

Price discovery is an important function performed by futures markets. The development of futures markets makes it possible to predict spot prices efficiently. Whether the futures and spot prices could guide the underlying asset price has been the focus of researches and practices. Furthermore, this is also a key question to be solved by governments, institutions and investors, especially in developing countries.

Most of current studies focus on the relationship between futures market and the spot market, and achieved many important results. Garbade & Silber (1983) established a dynamic model for the relationship test between futures price and spot price. Bessler & Covey (1991), first applied the cointegration theory on the futures market analysis of the U.S. Brooks, Rew & Ritson (2001) analyzed the relationship of the FTSE 100 index futures and spot price using vector autoregressive model (VAR) and Error Correction Model (ECM), and found that the changes of futures prices is helpful in forecasting the changes in spot prices. Shawky, Marathe & Barrett (2003) studied the benefits relationship of futures and spot market by the Exponential Generalized Autoregressive Conditional Heteroskedasticity (EGARCH) model and the VAR model.

On information overflow from futures and spot market to underlying asset, scholars mainly research the impact on real estate. Okunev & Wilson (1997) tested monthly data of the U.S. and found there is nonlinear relationship between the real estate market and the stock market, but it is not significant. Titman & Quan (1998) conducted a quantitative analysis of real estate market and stock market price changes based on the 17 country and 14-year data and found that there are clear long-term positive correlations between real estate market and stock market price. Tse (2001) analyzed the changes in real estate prices in Hong Kong, and found that unexpected changes in real estate prices directly affect the stock price changes. Wu Lin & Ding Hao (2012) studied the information overflow from futures and spot markets to underlying assets based on the China sugar in industry.

These studies were mainly based on the long-term relationship between two markets. Although it is simple and intuitive, the models fail to catch the complex interactions effect and mechanism of the markets. Furthermore, the current scholars mainly focus on the prices' quantitative relationship between markets, while few on information overflow between futures, spot markets and underlying asset price fluctuations based on industrial sectors. In view of this, this paper studies the information overflow effect of the future and spot markets price to stock price changes using VAR and VEC models as well impulse response and variance decomposition analysis. An empirical analysis of China's copper industry is given based on the models above.

2 VARIABLE SELECTION AND SAMPLE DATA

2.1 *Variable selection*

As an important industrial raw material, copper's price changes are closely related to the world economic development. Copper price is becoming much important in China, which is the world's major copper producing and importing country.

Considering the important role of copper in China, we select it as the example for the analysis of the information overflow effect from the future and spot markets price to underlying assets.

2.2 Sample data

The data employed in this study comprises observations for all trading days in the period from January 1, 2009, to December 31, 2012. Copper futures prices are from the London Metal Exchange and Shanghai Futures Exchange (SHFE). Chinese copper spot price are from SHMET (www.shmet.com). To form China's copper company stock price index, we select a sample. The stock price index is calculated by weighted stock closing prices of Jiangxi Copper, Yunnan Copper stock. All sample data come from Wind Information database.

Spot price, stock price index, LEM and SHFE's futures price series recorded as $\{P_t\}$, $\{S_t\}$, $\{LF_t\}$ and$\{SF_t\}$. Accounting to the correspondence of two time series data, excluding the record which only the futures price and spot price is not, or only the spot price and futures price is not, then we got 1174 groups valid data. In order to avoid the effect of economic time series data's heteroskedasticity, all variables were removed to the natural logarithm form of the actual value. LnS_t is the variable as natural logarithm return of stock price index; LnP_t is the variable as natural logarithm return of Spot price; $LnSF_t$ is the variable as natural logarithm return of SHFE's futures price; $LnLF_t$ is the variable as natural logarithm return of LEM futures price. Then, this paper will be analyzed based on the vector autoregressive system of the variables.

2.3 Data stationarity test

For further analysis and model building, it is required to test the stationarity of time series. The unit root of time series is tested by Augmented Dickey-Fuller test statistic. The test result is shown in Table 1.

The test results of Table 1 show that the null hypothesis that each variable's time series of raw data has unit roots should be rejected at 1% significance level. That means at 99% confidence level, those

variable's time series is stationary. Therefore, the variable LnS_t, LnP_t, $LnSF_t$ and $LnLF_t$ are I (0) series.

3 MODELS AND METHOD

C.A. Sims (1980) introduced the VAR model into the economics research, and promoted the wide use of dynamic analysis in the economic system. VAR (p) model's mathematical expression is:

$$y_t = \Pi_1 y_1 + \Pi_2 y_2 + \ldots + \Pi_p y_{t-p} + \varepsilon_t \qquad (1)$$

where Π_i is the $k \times k$ order parameter matrix, ε_t is the $k \times 1$ column vector of random error.

Johansen (1988), Juselius (1990) used maximum likelihood estimation proposed test method (Johansen & Juselius, 1990) to analyze the multi-cointegration of the variables for the vector autoregressive model (VAR). The test method works in the following process. First, it is required to build VAR (p) model. Then, according to the model equation transformation, the following formula is established.

$$\Delta y_t = \sum_{i=1}^{p-1} \Gamma_i \Delta y_{t-i} + \Pi \Delta y_{t-p} + \varepsilon_t \qquad (2)$$

where Δ represents a first difference operator; Δy_t is expressed as a $k \times 1$ column vector; Γ and Π are coefficient matrix. When choosing p, it should be ensured that ε_t is a mean zero and has finite covariance matrix of multivariate normal white noise process. Then, the number of cointegration vectors is equal to the order of Π, cointegration could be tested from the test of Π order number.

Engle and Granger combined cointegration theory with the error correction model to establish the Vector Error Correction model (VEC). A VEC model without exogenous variables can be described as:

$$\Delta y_t = \sum_{i=1}^{p-1} \Gamma_i \Delta y_{t-i} + \alpha \beta' y_{t-1} + \varepsilon_t \qquad (3)$$

or

Table 1. Augmented Dickey-Fuller test statistic.

Variable	ADF t-Statistic	Test critical values			Prob.	Stationarity
		1% level	5% level	10% level		
LnS_t	−32.62	−3.44	−2.86	−2.57	0.00	*Stationarity*
LnP_t	−31.24	−3.45	−2.87	−2.57	0.00	*Stationarity*
$LnSF_t$	−34.70	−3.45	−2.87	−2.57	0.00	*Stationarity*
$LnLF_t$	−21.98	−3.45	−2.87	−2.57	0.00	*Stationarity*

$$\Delta y_t = \sum_{i=1}^{p-1} \Gamma_i \Delta y_{t-i} + \alpha\, ecm_{t-1} + \varepsilon_t \qquad (4)$$

where $ecm_{t-1} = \beta y_{t-1}$ is the error correction term, reflecting the long-term equilibrium relationship between variables. Coefficient vector α represents the weight of the equilibrium error correction terms, which is to describe the adjustment rate of model from any non-equilibrium to the long-term equilibrium.

C.A. Sims (1980) proposed variance decomposition method based on VMA (∞) description. This method encloses the quantitative relationships between the variables. The definition of relative variance contribution is:

$$RVC_{ij}(s) = \frac{\sum_{q=0}^{s-1}(\theta_{ij}^{(q)})^2 \sigma_{ij}}{\sum_{j=1}^{k}\left\{\sum_{q=0}^{s-1}(\theta_{ij}^{(q)})^2 \sigma_{ij}\right\}} \qquad (5)$$

where θ_{ij} is the impulse response function, σ_{ij} is the standard deviation of variable j, $RVC_{ij}(s)$ is the variance contribution of variable j to variable i.

4 ECONOMETRIC ANALYSIS AND RESULTS

4.1 Long-term correlation between stock and futures, spot markets

4.1.1 Vector Autoregressive Model
Before establishing VAR model, the lag period of VAR (p) model should be determined. This paper selects AIC and SC as indicators to determine the lag period of VAR (p) model. According to the AIC, SC minimum criteria, the VAR model established with LnS_t, LnP_t, $LnSF_t$ and $LnLF_t$ can be identified to be VAR (2) model. Thus, the VAR (2) model is built as follows:

$$
\begin{bmatrix} LnS_t \\ LnLF_t \\ LnP_t \\ LnSF_t \end{bmatrix}
=
\begin{bmatrix} 0.0000 \\ 0.0007 \\ 0.0001 \\ 0.0003 \end{bmatrix}
+
\begin{bmatrix}
0.0504 & 0.5627 & -0.2929 & -0.2593 \\
0.0481 & -0.1391 & -0.1064 & 0.1286 \\
0.0176 & 0.5001 & -0.4958 & 0.2224 \\
-0.0264 & 0.5046 & 0.0455 & -0.3811
\end{bmatrix}
$$

$$
\times
\begin{bmatrix} LnS_{t-1} \\ LnLF_{t-1} \\ LnP_{t-1} \\ LnSF_{t-1} \end{bmatrix}
+
\begin{bmatrix}
-0.0178 & 0.2358 & -0.0716 & -0.0269 \\
-0.0260 & -0.0237 & -0.0321 & 0.0820 \\
-0.0062 & 0.1899 & -0.1042 & 0.0995 \\
-0.0164 & 0.2307 & 0.0359 & -0.0164
\end{bmatrix}
$$

$$
\times
\begin{bmatrix} LnS_{t-2} \\ LnLF_{t-2} \\ LnP_{t-2} \\ LnSF_{t-2} \end{bmatrix}
+
\begin{bmatrix} \varepsilon_{1t} \\ \varepsilon_{2t} \\ \varepsilon_{3t} \\ \varepsilon_{4t} \end{bmatrix}
\qquad (6)
$$

Equation 6 shows that there is a long-run equilibrium relationship between stock prices change and copper spot prices, domestic and foreign futures price. In the long-term impaction, foreign copper futures have significant impact on the stock price changes. Stock price lag term change's impact on the current stock price is lower than the impact of copper futures, and copper spot price's impact affect on the stock price changes is not significant.

Figure 1 shows that the established VAR (2) model is obviously stable. Then, impulse response, variance decomposition could be carried on.

4.1.2 Vector Error Correction model
To further analyze the long-term equilibrium relationship between the stock price movements and changes of the spots and futures market's price, VEC model was built based on the existing VAR (2) model, as follows:

$$
\begin{bmatrix} dLnS_t \\ dLnLF_t \\ dLnP_t \\ dLnSF_t \end{bmatrix}
=
\begin{bmatrix}
-0.5914 & -0.1551 & -1.0162 & 1.1617 \\
0.0298 & -0.4169 & 0.2864 & -0.7652 \\
0.0301 & 0.3643 & -0.9651 & 0.2288 \\
0.0252 & -0.2243 & -0.7541 & 0.4108
\end{bmatrix}
$$

$$
\times
\begin{bmatrix} dLnS_{t-1} \\ dLnLF_{t-1} \\ dLnP_{t-1} \\ dLnSF_{t-1} \end{bmatrix}
+
\begin{bmatrix}
-0.3239 & -0.0090 & -0.6138 & 0.6368 \\
0.0077 & -0.1625 & -0.0384 & -0.3350 \\
0.0209 & 0.2495 & -0.3816 & 0.0993 \\
0.0141 & -0.0265 & -0.3739 & 0.1210
\end{bmatrix}
$$

$$
\times
\begin{bmatrix} dLnS_{t-2} \\ dLnLF_{t-2} \\ dLnP_{t-2} \\ dLnSF_{t-2} \end{bmatrix}
+
\begin{bmatrix} 0.7560 \\ -0.4835 \\ 0.0482 \\ 0.7838 \end{bmatrix}
vecm_{t-1}
\qquad (7)
$$

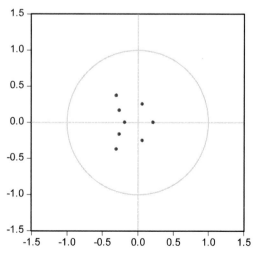

Figure 1. Inverse roots of AR characteristic polynomial.

271

$$vecm_{t-1} = LnLF_t(-1) + 1.3992 LnP_t(-1)$$
$$- 2.3561 LnSF_t(-1) - 0.0432 LnS_t(-1)$$
$$- 0.0001 \qquad (8)$$

Equations (7), (8) show that when the system deviates from the equilibrium state, the error correction have a positive effect adjustment on the spot price changes, while it plays a negative adjustments role on LEM futures price changes. As a main indicator of products future prices, the futures price change will directly affect the expected return on stock. So, there is a long-term equilibrium relationship between stock and futures and price fluctuations are inevitable and reasonable. In fact, the stock price may occur away from its equilibrium state in the short term, due to the certain impact of economic, institutional or human speculation factors in the market. And it would keep a long-term equilibrium relationship between stock price and futures price.

4.2 Information spillover effects to spot price

Since the variables LnS_t, LnP_t, $LnSF_t$ and $LnLF_t$ are all stationary, Granger Causality test could be directly used to analyze the spot and futures price effects on the stock prices. As Table 2 shows that the original hypothesis "$LnLF_t$ does not Granger Cause LnS_t" should be rejected at the 1% significance level. And the original hypothesis "LnP_t does not Granger Cause LnS_t" should be accepted. It means that London copper futures prices could effectively explain the early changes in the current Chinese copper company's stock price changes while the spot price changes fail to effectively explain stock price changes. There exist significant information spillover effects from international futures market to Chinese copper company's stock price. However, the domestic spot market has no significant information overflow effect to stock price. It is also found that the information spillover effect from London futures market to Shanghai futures market futures price is significant.

Table 2. VAR Granger Causality.

Hypothesis	Chi-sq	df	Prob
LnLF_t does not Granger Cause LnS_t	109.49	2	0.000
LnP_t does not Granger Cause LnS_t	5.74	2	0.057
LnSF_t does not Granger Cause LnS_t	6.69	2	0.035
LnLF_t does not Granger Cause LnSF_t	410.54	2	0.000

4.3 The impact of futures prices short-term fluctuations on spot prices

For further analysis of the short-term impact of variables to the changes of stock price, this paper uses generalized impulse method which does not depend on the order of VAR model variables to analyze variables impulse response based on VAR (2) model. The results are shown in Figures 2 and 3.

Figures 2 and 3 describe the impulse response function of copper futures price impact on stock price changes. After a positive impact on the London and Shanghai copper futures, the stock price changes reach the highest at the first period; the return increased by more the 1%. Then, it rapidly declines to the lowest at the 3rd period. It shows that the London copper futures price's short-term impact could transmission by the market information to the stock market, and bring the same direction impact to the stock price.

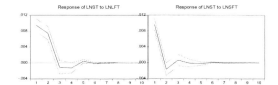

Figure 2. Response of LnS_t to $LnLF_t$ and $LnSF_t$ innovation.

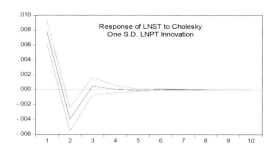

Figure 3. Response of LnS_t to Lnp_t innovation.

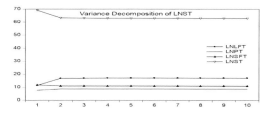

Figure 4. Variance decomposition of LnS_t.

The variance decomposition is to decompose the system mean square error into the impact of the contribution of each variable. Only consider the contribution of the spot market and futures market price changes on the stock price.

As shown in Figure 4, the initial Shanghai copper futures price's greatest contribution to the stock price is 12%, and the initial London copper futures price's greatest contribution to the stock price is 20%. Spot price's contribution rate to the stock price is not obvious.

Stock price changes are mainly affected by international and domestic futures. The effect of domestic copper spot price changes on stock price changes is not significant. In the short-term price impacts, the London Metal Exchange copper futures price impact is more effective on the stock price changes.

5 CONCLUSIONS

The research could get the following conclusions: Firstly, there is a long-term relationship between copper futures and stock market price changes. And there is no significant long-run equilibrium relationship between copper spot price and the stock price changes. The effect of future stock price movements mainly comes from the unexpected commodity price fluctuations. Secondly, there is an information spillover from the copper futures markets to the stock market. The stock markets have response to the information overflow from copper futures markets. Finally, stock price movements can be explained by the futures market price changes. Copper futures prices plays an important role in predicting future stock returns, and are important factors and indicators to analyze and predict future changes in stock prices.

ACKNOWLEDGMENTS

This work was financially supported by the Nanjing Xiaozhuang University Foundation (2012 NXY53).

REFERENCES

[1] Chris Brooks, Alistair G. Rew & Stuart Ritson. 2001. A trading strategy based on the lead–lag relationship between the spot index and futures contract for the FTSE 100. *International Journal of Forecastin.* 17(1):31–44.
[2] David A. Bessler & Ted Covey. 1991. Cointegration: Some results on U.S. cattle prices. *Journal of Futures Markets.* 11(4):461–474.
[3] Hany A. Shawky, Achla Marathe & Christopher L. Barrett. 2003. A first look at the empirical relation between spot and futures electricity prices in the United States. *Journal of Futures Markets.* 23(10):931–955.
[4] Kenneth D. Garbade & William L. Silber. 1983. Price movements and price discovery in futures and cash markets. *The Review of Economics and Statistics.* 65:289–297.
[5] Okunev & Wilson. 1997. Using nonlinear tests to examine integration between real estate and stock markets. *Real Estate Economics.* 25:487–503.
[6] Raymond Y.C. Tse. 2001. Impact of property prices on stock prices in Hong Kong. *Review of Picnic Basin Financial Markets and Policies.* 4(1):29–43.
[7] Sheridan Titman & Daniel C. Quan. 1998. Do real estate prices and stock prices move together? *Real Estate Economics.* 27:183–207.
[8] Wu Lin & Ding Hao. 2012. Impact of commodity price volatility on capital return—experimental analysis based on China's sugar industry. *Inquiry into Economic Issues.* 5:155–159.

Education Management and Management Science – Zheng (Ed.)
© 2015 Taylor & Francis Group, London, ISBN 978-1-138-02663-6

Study of the construction of personalized college English teaching system based on investigation of learning needs

X.X. Chen
China University of Petroleum, Qingdao, China

ABSTRACT: This article probes the construction of a personalized college English teaching system based on investigation of students' learning needs. A tentative study has been conducted by helping and guiding students to explore their own English learning needs on the basis of a full range of dynamic investigation, enriching the curriculum to satisfy their individual needs, conducting personalized teaching to promote their utmost development and facilitating autonomous learning by establishing the progressive autonomous learning mode, so as to solve the problems of personalized development and autonomous learning.

Keywords: learning needs; personalized teaching; autonomous learning

1 INTRODUCTION

The College English Teaching Requirements issued by the Ministry of Education stipulates that "The goal of college English teaching is to cultivate students' ability of using English, especially their listening and speaking ability, enhance their ability of independent learning, and improve their comprehensive cultural literacy." Therefore, many domestic colleges and universities are conducting college English teaching reform by taking advantage of modern information technology, hoping that the single teacher-centered traditional teaching mode can be transformed by implementing the "classroom instruction + Internet autonomous learning" mode. The ultimate goal is to make English teaching and learning not restricted by time and place to some extent and targeted towards a personalized and autonomous learning direction.

But because of the impact of exam-oriented education through the ages, the curriculum and teaching mode in our college English teaching generally focus on the teaching of basic knowledge and skills, which is obviously not conducive to the students' personalized learning and autonomous learning. The resulting problems such as insufficient ability in autonomous learning and lack of comprehensive ability in English application are still serious. Therefore, how to make full use of modern information technology to innovate the college English teaching system, so as to solve the problems of personalized learning and autonomous learning will determine the success or failure of our college English teaching reform.

2 PROBLEMS

2.1 *Lack of enthusiasm and motivation in English learning*

With the increasing importance attached to English teaching and reform in primary and middle schools, and under the pressure of college entrance examination, students spend a lot of time and energy in learning English. Their English levels are improving significantly and continuously. But after entering university, the non-English major students lose the goal in English learning. Students lack motive of English learning and only aim at passing the College English Test Band 4 (CET4) and Band 6 (CET6) only.

2.2 *Low efficiency in English learning*

Exam-oriented education leads to the emphasis on rote learning of basic knowledge and repeated drilling of examination skills. Moreover, because of the lack of scientific and reasonable learning methods and personalized instruction, students waste a lot of time and energy in rote learning, and cannot do as "using in learning, learning in using", which result in the low efficiency and insignificant effect.

2.3 *Insufficiency of ability in application, especially the ability in listening and speaking*

Profoundly affected by the exam-oriented education and the utilitarian, there is always a tendency of "emphasizing exams, ignoring abilities" in our college English teaching, which results in the fact that our teaching still centers around strengthening students' language knowledge and cultivating students' ability for tests, and ignores the cultivation of students' ability in using English. Influenced by this tendency, students regard passing CET4 and CET6 as their sole target, without paying attention to language application. The outcome is students are poor in comprehensive ability in English application, especially the ability in listening and speaking.

2.4 *Weak awareness and abilities in autonomous English learning*

Due to the great importance attached to classroom instruction, teaching of knowledge and the authority of teachers, the cultivation of students' autonomous learning ability has never been placed in the necessary important position. In universities, the contradiction between the demand for students' autonomous learning and their lack of autonomous learning abilities is becoming more and more prominent. On the one hand, the college English teaching is faced with the reality of reduction in credits and class periods, and the more and more limited classroom instruction cannot guarantee the fulfillment of the teaching target. On the other hand, students who have been accustomed to traditional classroom instruction lack awareness and ability in autonomous learning.

3 REASON ANALYSES

3.1 *Individual needs are not met*

Learning needs are people's desire for the unknown, a series of reaction driven by learning motivation for the sake of the needs in production, living and development. Benson (2001) defined learning needs as "including not only the actual needs for English in their jobs, but also the notion of their English abilities from the perspective of society". Needs are the driving force of all behaviors, so are they to foreign language learning. On the one hand, a considerable proportion of non-English major college students have little idea of their English learning needs (what should they learn in terms of their current level of English; what do they want to learn; what should they learn with regard to their future professional learning and vocational development). They are just satisfied with passing CET4 and CET6 for the sake of employment. On the other hand, the stereotyped curriculum and teaching mode in many universities cannot meet the individual needs of students. These have caused the problem of students' lack of English learning enthusiasm and motivation. Where students are not aware of their learning needs or their learning

needs are not satisfied, there will be no higher goal and lasting learning motivation.

3.2 *Personalized learning is not guaranteed*

The college English teaching in most universities at present focuses on the general English teaching in the elementary stage (freshmen and sophomores). Whether in terms of credit allocation or course arrangement, English teaching in the elementary period accounts for the vast majority, and is conducted mainly in compulsory curriculum (comprehensive lesson and audiovisual lesson). The college English teaching in the subsequent stage is mainly in the form of elective courses for the purpose of ability extension. Students cannot get many credits from these courses and just choose according to their own interests, which cannot draw enough attention from students. Although most universities have set the professional English as a compulsory course, not enough efforts have been spared on it, because on the one hand, the difficult content of this course is quite different from that of the general English, and on the other hand, credits and time are too limited to ensure students' enough investment of time and energy. Therefore, the teaching of professional English is not very efficient. The stereotyped curriculum and uniformed teaching mode cannot meet the needs of individual students.

3.3 *The ability of autonomous learning is not cultivated effectively*

The Ordinary High School English Curriculum Standard explicitly points out that the general goal of the senior high school English curriculum is "to further clarify the purpose of English learning, and develop the ability of autonomous learning and cooperative learning; to formulate effective English learning strategies; to cultivate students' comprehensive ability of using English on the basis of the English learning in the compulsory education stage" while T*he College English Teaching Requirements* stresses "to enhance students' ability to study autonomously". From "development of ability of autonomous learning and cooperative learning; formulation of effective English learning strategies" to "enhancement of students' ability to study autonomously", the teaching goals at the two stages reflect the requirement for gradual and sustainable cultivation of autonomous English learning abilities. But because of the profound influence of the exam-oriented education and the utilitarian, the cultivation of autonomous English learning abilities has always been neglected and students' abilities are generally not strong. "Learning is a lifelong process, and the ultimate goal of education is to cultivate the students' learning autonomy, and make them autonomous learners." (Zhu Yumei, 2007). But in the present case, the demand for students' autonomous learning is high, while we are not effective and successful in cultivating students' autonomous learning abilities which lead to the widespread phenomenon of "the alleged autonomous learning".

4 PERSONALIZED COLLEGE ENGLISH TEACHING SYSTEM

It is not difficult to find that the root causes of these problems are that the learning needs of students have not been excavated and mobilized, and our stereotyped and uniformed teaching mode cannot meet the individual needs of students and promote their personalized development and autonomous learning. Therefore, to solve these problems, the first and most important thing is to identify the students' English learning needs, whether they are the teachers or the students themselves. Furthermore, it is necessary to formulate personalized teaching plan and conduct personalized instruction according to students' diverse learning needs so as to solve the problems of personalized learning and autonomous learning.

4.1 *To fully meet the polyphyletic individual needs of students*

Wen Qiufang (2012) concludes from research that the college English curriculum objective should be formulated according to needs for development of students, of society and of discipline as well as changes in needs.

First of all, a full range of dynamic investigation of students' English learning needs should be carried out. Through the English proficiency test and questionnaire investigation and interview with students and English teachers, students' level needs (what should be strengthened in their English learning in terms of their current English level) and interest needs (what are students interested in for their English learning) are determined; Through the questionnaire investigation and interview with students, teachers in their majors and human resources department staff, students' professional learning needs (which aspect of English learning should be strengthened in terms of professional learning, communication and development) and occupational development needs (which aspect of English learning should be strengthened in terms of employment and occupational development after employment) are identified. In addition, at different stages, the learning needs of students will change. Therefore, a regular investigation of students' learning needs and a dynamic

understanding of the change in needs are crucial to the adjustment of the English teaching scheme.

Secondly, it is necessary to help and guide students to explore their own learning needs. Guide students to distinguish "wants" from "needs" rationally, and be clear about their learning objectives and development direction. The students' English learning needs include both subjective needs (level needs and interest needs), and objective needs (professional learning needs and occupational development needs). Therefore, only by helping and guiding students to explore their own learning needs, can their inner motives be aroused. Respecting students' choices and guiding them to make rational choices can help to maintain their interest and motivation in learning English.

Moreover, our college English curriculum has been further improved to be more conducive to students' personalized learning and language application abilities. With a richer curriculum, students' individual alternatives have been met and students are encouraged to choose from the courses according to their actual needs. The diversified curriculum promotes students' multidimensional personalized development, thus contributing to the cultivation of multi-specified and diverse talents. At the elementary stage, students are divided and taught according to their different English level needs while at the subsequent development stage, students are taught according to their interest needs, professional learning needs and occupational development needs. Besides, the teaching scheme is timely adjusted according to results of the dynamic investigation, thus manifesting the principle of "student-oriented" and maximizing the satisfaction of students' individual learning needs.

4.2 *To actually realize the multi-dimensional personalized development of students*

Firstly, hierarchical and classified teaching fully guarantees students' individual alternatives. Conduct hierarchical teaching at the elementary stage according to the different level needs of students. Set different teaching targets based on the general requirements, intermediate requirements and advanced requirements. A hierarchical teaching in conducted in three level groups: the basic level, the common level and the top level according to students' different English levels, with the principle of "different starting points, different targets, personalized cultivation, optimal development". At the subsequent development stage, conduct teaching in different directions to strengthen students in different aspects according to their interest needs, professional learning needs, and occupational development needs. Courses are offered in different categories: English for language skills, English for language culture, English for academic purpose, English for special purpose, and English for examination training. Students can choose their own courses according to the principle of "optional but guided" so as to fully meet their own needs.

Secondly, innovate the teaching mode, and effectively realize the students' personalized training. Innovate the 3 + 1 + X (3 periods of comprehensive lessons per week; 1 period of audiovisual lesson per week; X refers to the time for after-class autonomous learning) teaching mode based on computer/network and classroom and increase the intensity of autonomous learning. By strengthening the guidance and supervision of students' autonomous learning through the campus network language learning platform, the effect of autonomous learning has been guaranteed and personalized cultivation has been realized. Through the teaching mode reform in the audiovisual lesson, we construct our progressive autonomous learning mode gradually, and students' abilities in autonomous learning have been elevated.

Thirdly, reform teaching methods and promote students' abilities to the maximum. Adapting to the reform of curriculum and teaching mode, we are further enriching and improving our teaching methods. The traditional method, the task-based method, the inquiry method, and the project method have been taken, respectively, according to different levels of students. And students are encouraged to keep "using in learning, learning in using" and develop themselves in the direction of "basic knowledge → application ability → academic ability → professional ability" so as to fulfill themselves to the maximum. By changing the traditional method of teachers' giving lectures in class and trying the probing method, we have realized the student-centeredness.

4.3 To effectively strengthen the multilayer autonomous learning of students

Nunan (1997) believes "a completely autonomous learner does not exist. It is only an ideal, but not a reality. But a learner can achieve different levels of autonomous learning". We solve the problem of students' autonomous learning well through the construction of the progressive autonomous learning mode.

On the one hand, we focus on the autonomous learning strategy training besides strengthening language knowledge and language skills in the comprehensive lesson, so as to teach students how to learn; on the other hand, change the traditional teaching method of teachers' giving lectures in audiovisual lesson and adopt the students' semi-autonomous learning mode under the teacher's guidance. Teachers assign the task, guide the students to

control their own content, speed and schedule, and provide on-site guidance and help, help students master autonomous learning strategies and transit from passive learning to autonomous learning naturally. Similarly, for the students' after-class autonomous learning, teachers give full play to the role of guide, supervisor, cooperator and facilitator, ensuring a desirable effect. Under the inspection and monitoring of teachers, reactive autonomy is achieved. Although this is only a low level of autonomous learning, an incomplete autonomous learning, it is an effective transition to a higher level of autonomous learning. Finally, by taking advantage of all the Internet and natural learning environment such as the learning resources on the platform, other cyber sources, libraries, reading rooms and the foreign language corners, proactive autonomy, a much higher level of autonomy is fulfilled.

Through the construction of progressive autonomous learning mode (classroom teaching → semi-autonomous classroom learning → reactive autonomy → proactive autonomy), we realize the transference from teachers' "teaching" to students' "learning" and students change from being passive to being independent. Students' autonomous learning strategies have been cultivated systematically, with their autonomous learning consciousness increasing ceaselessly and the ability of autonomous learning improving continuously.

5 CONCLUSION

Being well aware of students' individual learning needs, we are exploring the optimal and dynamic personalized English teaching system through a series of reform of the curriculum, the teaching mode, the teaching methods and the learning resources to maximally meet the individual learning needs of the students, which will stimulate students' interest and motivation in learning, cultivate their awareness and ability of autonomous learning and improve their learning efficiency, thus contributing to the success of our college English teaching reform.

REFERENCES

[1] Benson, P. 2001. *Teaching and Researching Autonomy in Language learning*. London: Longman.
[2] Nunan, D. 1997. Designing and Adapting Materials to Encourage Learner Autonomy. In P. Benson & P. Voller (eds), *Autonomy and Independence in Language Learning*: 192–203. London: Longman.
[3] Wen Qiufang. 2012. Challenges and Strategies in College English. Foreign Language Teaching and Research (2): 283–292.
[4] Zhu Yemei. 2007. Autonomy in Language Learning. Foreign Language Journal (5): 137–139.

Education Management and Management Science – Zheng (Ed.)
© 2015 Taylor & Francis Group, London, ISBN 978-1-138-02663-6

Research on instrument and equipment information management platform of testing laboratory

Lixiong Zhang, Wei Hu, Jin Ding & Yanan Zhao
China Automotive Technology and Research Center, Tianjin, China

ABSTRACT: For the use of management of instrument and equipment of testing laboratory, the paper designed an information platform system using B/S and MCV system structure with Oracle database modular programming. The platform has the abilities of dynamic tracking with real-time monitoring and convenient querying of instrument and equipment information.

Keywords: management; instrument and equipment; platform

1 INTRODUCTION

Testing laboratory is an independent setup where scientific research, technology developement and product identification are undertaken. It is an important carrier where many advanced equipment are used. With the development and expansion of testing center's laboratory, the number of equipment is increasing, and also the equipment update is speeding up. As the hardware conditions and test environment have been improved, the laboratory testing capability has been enhanced, which indirectly promoted the improvement of the test level of the whole industry. However, the traditional equipment management methods lack effective management platform, dynamic monitoring and tracking means, and are also short of communication and sharing mechanism. The management of instruments and equipment usually has the problems of inefficiency, monitor traceability mismanagement, etc. Based on these problems, an equipment testing laboratory information management platform, using information technology tools instead of manual operation, achieves that laboratory's overall monitor and dynamic management of the equipment operational state on the internal network. The platform saved time, human and material resources, improved the laboratory's management capacity and service level.

2 DEVELOPMENT ENVIRONMENT

According to the purpose of the development application, in order to achieve the system functions of the equipmens information management platform efficiently, the service used Microsoft Windows 2008 Server 64-bit systems. The database is Oracle's latest version of 11 g. Oracle database. This database has the advantages of availability, good expansibility and high data security, strong stability. In the data management, Oracle can easily achieve data store technology. Software system structure is generally divided into C/S and B/S structure. Compared with C/S structure, B/S structure, as shown in Figure 1, which we have used has lower client coupling. Whatever operating system we have used, we can land on LAN as long as you install IE6.0 and up version. B/S which runs faster keeps the focus on the server, and what main computation and data operation are completed at. The programming language which the system used is Java, which has a series of characteristics such as efficient interpretation and execution, safety and neutral system construction, transplant and efficient, multi-thread and distribution. Web

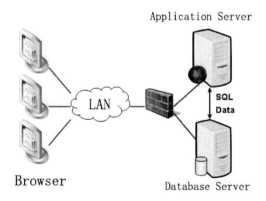

Figure 1. Diagram of B/S structure.

server used Tomcat 6.0 stable version, which has the advantages of open source, supporting the latest standard, faster and newer, cross-platform and good expansibility. The equipment information management platform chose Eclipse 8 as development tool and used Dreamweaver to develop the front pages.

3 SYSTEM ARCHITECTURE

The system architecture adopted the MVC (Model-View-Controller), as shown in Figure 2. It is a software design paradigm: using a method of datum show separate and business logic organizes code, gather business logic into a component. You don't need to rewrite the business logic while improving an individual customizing interface and user interaction. MVC is developed particularly for mapping the traditional input, processing and output in a structure of logic GUI (Graphical User Interface).

MVC has the advantages of low coupling, high reusability and low life cycle cost, fast deployment and high maintainability; it is propitious to the management of software engineering.

Because various layers fulfill their proper function, different applications in each layer have some same characteristics, which are good for managing program code through engineering and instrumentalization. The controller also provides a benefit that you can use the controller to connect different models and views for meeting users' needs. Then, the controller provides a powerful method for structural applications. If some reusable models and views are given, the controller can select models to process according to the users' needs, then select views to display results to the user.

The view layer uses the technology in accordance with Web 2.0 standard, of which main technologies including Jsp, Servlet, HTML, CSS, JavaScript, Jquery, Ajax, etc. JSP is a scripting environment on the server which can create and run the application of dynamic and interactive Web server. You can combine HTML page, JavaScript script commands and Active X components using JSP, in order to create interactive Web pages and powerful applications based on the Web. Ajax is a kind of asynchronous transmission technology and is characterized by using the document object template based on Web standard (Document Object Model) for dynamic display and interaction. What's more, it uses XML and XSLT to exchange data and related operations and uses XML HTTP Request for asynchronous data query and receiving.

The system controller is based on Struts 2.2 and Spring 3.0. Struts implementation of the MVC model, is a clear structure, thus developers only pay attention to the realization of the business logic. It is rich in tags which can greatly improve the efficiency of development. And the page navigation is one of the future directions of development, by which it can make the system clear. Through a configuration file, you can grasp the connection among each part of the whole system. This is a great advantage for the latter part of the maintenance. Especially when another group of developers take over the project, this advantage is reflected more clearly.

According to the requirements and the development cost of practical control of our system, we adopt MVC mode and the technology of every layer is of the lowest cost and the highest performance.

4 FUNCTION MODULES

According to the actual situation of testing laboratory and the purpose of development, the paper determined the function modules of the system, mainly including settings of platform, files management, equipment information, traceability management, purchase and development, as shown in Figure 3.

'Settings of platform' is used as the back-stage management of the system, including general settings, authority management and or so. 'Files management' is used for the loan management of equipment documents which can make the documents' state always clear and available. 'Equipment information' provides users with every detailed information of the instrument and equipment, including specifications, measurement range, precision, manufacturer, calibration record, date of purchase, with easy searching. 'Traceability management' is an important modular of the platform system. The calibration and verification of all equipment in the lab is of dynamic tracking. The platform can remind the validity of calibration

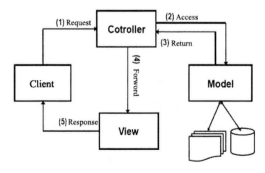

Figure 2. Diagram of MCV structure.

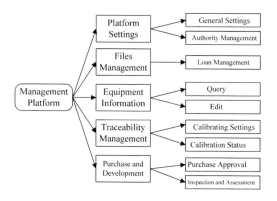

Figure 3.   Diagram of management platform modules.

can automatically and timely remind the equipment administrator to send the equipment to calibrate. Each traceability record including certificate of calibration and calibration results will be saved in the database of system, which can be easily queried by the users.

The platform system is flexible, with friendly human-machine interface, and can be conveniently queried. Through the modular design and the working mode of B/S, system can be expanded or upgraded conveniently. The platform webpage interactive window display is comfortable and can be conveniently operated, which makes it easy for the user to use. The fuzzy query system provides multiple conditions, multi field, automatic identification and intelligent matching.

to let the manager send equipment to calibrate in time, which ensured the traceability of equipments. 'Purchase and development' modular is used for development or procurement of equipment to do the process standardization management. The use of JBPM workflow engine strengthened the three examination and approval systems, which can effectively control the laboratory procurement joint.

5   PLATFORM FEATURES

The platform system owns a perfect authority management, by which each operator and administrator's role is clear. According to different users, the system's various functions module permissions are set, including editing, querying, approval, permission to remind etc. The users can only operate in their scope of work and duties, by which they can avoid unauthorized misoperation and realize the rights defined by hierarchical management mode.

The platform system has realized the real-time dynamic monitoring. By using the system, the equipment whole lifetime account, parameters, files, usage, transfer and calibration traceability can be dynamic tracking and information sharing. According to the calibration project, each instrument and equipment has set a calibration cycle, by querying the last calibration record, the platform

6   CONCLUSIONS

This paper built a set of equipment information management platform which is modular programming of process monitoring and dynamic management to meet the requirements of testing laboratory instrument management, using B/S mode, MCV architecture and Oracle database. The platform effectively improved the level of laboratory management and laid a solid foundation for the testing work.

REFERENCES

[1] Wangwei Zhong, Xiaoou Huang 2007. Library Management System Based on C/S and B/S Mixed Mode. *Modern Computer* (8): 124~126.
[2] Liu Yajing, Tian Guie 2012. Design and Implement of University Laboratory Equipment Network Management Platform Based on C/S and B/S Model. *2012 International Conference on Industrial Control and Electronics Engineering*: 1474~1476.
[3] Bingrong He 2009. Laboratory Equipment Management System Designed on B/S Model. *Journal of Mudanjiang University* 18(1): 104~106.
[4] Burton, Debra 1992. Teaching Computer Science With Closed Laboratories, *Computer Science Syllabus*, Number 3, pp. 8–11, September/Octorber.
[5] Stone, H.S. 1987. High-performance Computer Architecture, *Addison Wesley, Reading Massachusetts*.

Education Management and Management Science – Zheng (Ed.)
© 2015 Taylor & Francis Group, London, ISBN 978-1-138-02663-6

The investigation and analysis of college students' ESP autonomous learning

Xin-jian Fan
School of Energy and Power Engineering, Lanzhou University of Technology, Lanzhou, China

Xin-zheng Li
School of Foreign Languages, Lanzhou University of Technology, Lanzhou, China

ABSTRACT: This thesis reports a survey study of the ESP autonomous learning by 100 senior students via the instruments of a questionnaire and semi-structured interview after the questionnaire. The results of the data analysis reveal the following main findings in line with the research questions: Their gain in the average means of all the questionnaire items about the overall degree of the students' ESP autonomous English learning is 3.43. The means of determining the learning objectives and evaluating the efficacy are 3.29 and 3.08, respectively. The means of watching English movies to practice their English and listening to VOA and BBC to improve their listening comprehension is 3.19. Female students perform better in the category of determining their objectives of autonomous L2 learning than male students. This study will enlighten later researches about ESP.

Keywords: ESP; autonomous learning; gender

1 INTRODUCTION

With the development of economic globalization, international communication and cooperation are becoming more frequent. The demand for talents with continuous learning ability in the 21st century becomes more urgent. Autonomous learning is an effective way to solve this problem and is consistent with the principles of pedagogy. Autonomous learning is an important concept in English study in which learners take more responsibility for their own learning, including identifying learning targets, contents and speed, choosing learning strategies and skills, monitoring learning process and evaluating learning results. Autonomous learning plays a decisive role in language learning.

ESP (English for Specific Purpose) originated in the late 1960s. At present, most Chinese colleges set ESP courses for junior non-English major students. The importance of ESP has been recognized by the people, but the teaching of ESP is currently facing difficulties. The main problem is the limited teaching hours to complete the task of teaching and to achieve teaching objectives. Autonomous learning is an effective way to solve this. By improving the learner's own interest in learning and learning initiative and consciousness, autonomous learning can help students finish the task which cannot be completed in limited class time.

Theoretical researches on autonomy are abundant; however, empirical researches are fewer. These empirical researches are generally done in two aspects: focusing on individual differences of various factors related to autonomy and the relations among these factors or focusing on the relationship between autonomy competence and English achievements. The empirical research on autonomy attempted to explore individual differences on various factors of autonomy and the relations among these factors based on the questionnaires or interview (Zimmerman 1990; Pintrich and De Groot 1990; Garcia and Pintrich 1991; Wolters 1998; Xu Jinfen et al 2004; Zhoujie 2004). Pintrich and De Groot (1990) used a self-report measure of self-efficacy, intrinsic value, test anxiety, and self-regulation to examine relationships between motivational orientation, self-regulated learning and classroom academic performance.

In order to make a clear picture of the current situation of students' ESP autonomous learning, 100 questionnaires were distributed to students. According to related theories about autonomous learning, this paper sums up the current situation about students' ESP autonomous learning and the problems reflected in students ESP 'autonomous learning, and to explore how to improve students' ESP autonomous Learning Ability.

2 CATEGORIES OF THE QUESTIONNAIRES

Autonomous learning refers to learners' willingness and ability to move from teacher-dependence to teacher-independence, shoulder responsibility for their ESP learning and then to take corresponding actions, such as setting their goals, adjusting their learning strategies, checking their own learning process, evaluating their learning outcomes in their language learning. What is more, learners are able to avail of the rich resources of learning, for example, internet.

The focus of this study is two-fold. First, it aims to survey the extent of autonomous ESP learning by senior non-English major. Second, it is to find out what needs to be done to help them become more autonomous and independent in ESP learning. This thesis reports a survey study of the autonomous ESP learning by 100 junior non-English major via the instruments of a questionnaire and semi-structured interview after the questionnaire. It aims to address the following research questions:

1. To what extent do Chinese students conduct autonomous ESP learning?
   a. What are their beliefs in autonomous ESP learning?
   b. Can they determine their objectives of autonomous ESP learning?
   c. Can they implement the strategies of autonomous ESP learning?
   d. Can they monitor their process of autonomous ESP learning?
   e. Can they evaluate the efficacy of their autonomous ESP learning?
2. To what extent do junior students in ESP learning conduct the internet-based autonomous learning?
   a. What are their beliefs in internet-based autonomous learning?
   b. What are their behaviors in autonomous learning?
   c. Are there any statistically significant differences between male and female postgraduates on autonomous learning?
3. Are there any statistically significant differences between male and female postgraduates on autonomous L2 learning?

3 RESULTS ANALYSIS

3.1 *Non-English majors postgraduates' general autonomous L2 learning*

The average means of the subjects' response to these five questionnaire categories and the overall means are presented in Table 1.

The mean of believing in autonomous L2 learning is 3.43; the mean of determining the learning objectives is 3.29; the mean of implementing the strategies is 3.62; the mean of monitoring the process is 3.41; the mean of evaluating the efficacy is 3.08.

As shown in Table 1, the mean value of the responses to these five areas was below the medium level. The results listed evidently indicated that the overall beliefs of autonomous learning among the students. Among them, the average mean of the students' beliefs in ESP autonomous learning was the highest (Mean = 3.62). The average mean of the postgraduates' competence of evaluating the ESP autonomous learning was the lowest (Mean = 3.08).

3.2 *Non-English majors autonomous L2 learning postgraduates' internet-based*

The frequencies, means and standard deviations of are exhibited in the beliefs in internet-based autonomous L2 learning in Table 2. Based on Table 2, it was encouraging to note that most of postgraduates believe that internet-based autonomous learning is helpful to improve their comprehensive English.

In Table 2, a considerably large number of postgraduates are found to have the similar identity for the significance of internet in the autonomous L2 learning (Item 41. Mean = 3.91; Item 39, Mean = 3.79). A majority of postgraduates agree that the English learning resources in the campus net are rich (Item 40, Mean = 3.75).

Even if most of postgraduates stress the role of internet-based English in the autonomous L2, it is a must for the university and educators to create such websites and keep on updating the information in it.

As internet has been long claimed to be the most facilitative resource for ESP autonomous learning and the major field easy to conduct ESP autonomous learning, the subjects' internet-based learning activities are also investigated via the questionnaire with the focus on their general beliefs in and behaviors of their ESP autonomous learning. The findings are as follows (See Table 3).

From Table 3, it was found that many students are unable to practice English (Item 42, Mean = 3.32). In particular, most students fail to make the best of the English movies and VOA and BBC programs in the autonomous L2 learning (Item 43, Mean = 3.18) although they prefer English movies and songs. At the same time, many of them cannot ensure to chat with others in English in the chat room or through other internet media (Item 44, Mean = 3.43).

Why is it that so many students cannot make the most of the internet-based autonomous L2

Table 1. Average mean of all categories of overall picture of the postgraduates' AL.

Category	Content	N	Mean	S.D
1	Believing in autonomous learning	100	3.43	0.51
2	Determining the learning objectives	100	3.29	0.47
3	Implementing the strategies	100	3.62	0.36
4	Monitoring the process	100	3.41	0.42
5	Evaluating the efficacy	100	3.08	0.41
Average mean		100	3.35	3.37

Table 2. Average means of the beliefs in internet-based autonomous L2 learning.

Item	Content	Mean	S.D
39	Interact-based ESP learning is an ideal model	3.79	0.81
40	The English learning resources on the web are rich	3.75	0.79
41	Internet is helpful to improve my autonomous learning ability	3.91	0.85

Table 3. Average mean of their behaviors in internet-based autonomous L2 learning.

Item	Content	Mean	S.D
42	While learning on line, I avail the opportunities to practice English	3.32	0.80
43	I imitate pronunciation and intonation while watching English movies or listening to VOA or BBC	3.19	0.63
44	Sometimes, I chat with others in English in the chat room	3.43	0.79
45	BT downloading, QQ communication, BBS and SO on are beneficial to my autonomous learning in web-based situation	3.65	0.72

learning? There are some reasons accounting for this: in the first place, the learners in the survey get used to the traditional classroom learning model, and it is hard for them to change the learning concept and habits in a short time. Secondly, the web-based autonomous learning is novel to most of learners, and they need more time to adapt to it. Thirdly, due to the unfamiliarity with the web-based learning technology and fear of technology, it is difficult for most of them to judge whether the web-based learning is ideal or not. We instructors should make them realize the significance of

internet-based learning in the autonomous L2 learning.

3.3 The gender differences in autonomous L2 learning

Figure 1: Comparison of the gains in the ESP autonomous learning differences between Female and Male Students. The figure showed that there were no significant differences between male and female students in almost all the categories except Category 2 and Category 6. But the mean of female students are higher than male students in all the categories and the whole questionnaire. That means female students excel their counterpart a little bit in ESP autonomous learning, but only the differences in determining the objectives of ESP autonomous learning and the awareness of the internet-based autonomous learning are statistically significant. The further analysis of all the items demonstrates that female students do better than male students in Category 2 and Category 6. Further analysis demonstrates that:

1. In Category 2, female postgraduates are more mentally autonomous than male students. Compared with males, they are more willing to determine their objectives of English learning according to the instructors' demands because the female students were more obedient than male ones.
2. In Category 6, based on the chart, it was found that male students have much easier access to internet than female students.

The reasons for the above results maybe like this: Firstly, female students are educated to be modest and cooperative. They are more willing to follow what the instructors ask and more careful to complete the objectives the teachers offer. Secondly, the speech advantages of girls are also likely to arouse their interests in English learning. A strong interest in learning will stimulate girls' internal driving force, which provides a strong support for girls studying independently and to set practical learning planning, time and objectives. Thirdly, in our society, girls get more parental support for

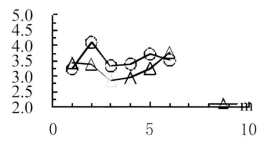

Figure 1. Comparison of the gains in the autonomous L2 learning differences between female and male students.

language learning. Whatever the rights and wrongs of the matter, some parents seem to regard learning languages as suitable for girls, while the boys are perhaps encouraged in the direction of subjects like science and technology. As a result, the male students are more competent in technology.

4 CONCLUSION

The results of the data analysis reveal the following main findings in line with the research question: (1) The overall degree of the students' ESP autonomous learning is not so satisfactory as expected. (2) They have no definite motivation and objectives of autonomous L2 ESP learning. What is more, they fail to be aware of the significance and the role of English in their major research. (3) The interaction does not play a real role in the ESP autonomous learning. In most cases, it is a tool to relax and

entertain themselves. (4) Female students perform better in the category of determining their objectives of ESP autonomous learning than male ones. However, the male students perform better in the category of the internet-based ESP autonomous learning than female ones. But the differences in the whole questionnaire between male and female students are not statistically significant.

REFERENCES

[1] Lirtlewood, w. (1996) "*Autonomy*": *an anatomy and a framework*.(4)427–43. System.

[2] McDevitt, B. (1 997) *Learner autonomy and the need for learner training*.34–39. Language Learning Journal.

[3] O'Malley, J.M. & Chamot, A.U. (1990). *Learning strategies in second language acquisition*. Cambridge: Cambridge University Press.

[4] Oxford, R.L. & Cohen, A.D. (1990) *Training for language learners: Six situational case studies and training model*(22): 197–216. Foreign Language Annals.

[5] Oxford, R.L. (1990) *Language Learning Strategies: What Every Teacher should Know.*

[6] Rowley, Mass: Newbury House. O'Malley, J.M., & Chamot, A.U. (1990).*Learning strategies in second language acquisition*. Cambridge: Cambridge University Press.

[7] Harey p. (1985) *A lesson to be learned: Chinese approaches to language learning*. ELT Journal, 39, 3: 183–186.

[8] Hedge T. (2000) *Teaching and Learning in the language Classroom*. Oxford: Oxford University Press.

[9] Little D.C (1991) *Learner Autonomy I: Issues and Problems*. Dublin: Authentik.

[10] Croft et al. (2004) *Cognitive Linguistics*. Cambridge: Cambridge University Press.

Education Management and Management Science – Zheng (Ed.)
© *2015 Taylor & Francis Group, London, ISBN 978-1-138-02663-6*

Application of AHP in graduate enrollment target allocation for instructors

De-Bo Fu & Fei Cheng

The School of Electronics and Information Engineering, Tongji University, Jiading, Shanghai, China

ABSTRACT: According to the research of the reasonable distribution for the graduate admissions index, deeply analyzing the factors of instructors in research funding, awards, research projects, academic papers, patents and other aspects. After the weight indexes are calculated based on the program level, the application of AHP enrollment index allocation model could be constructed. Upon checking the consistency, this method is fair, impartial, scientific, operable. It provides more scientific references for the allocation of graduate enrollment targets.

Keywords: index terms—AHP; graduate enrollment targets; allocation

1 INTRODUCTION

Great importance has always been attached to the problem of graduate enrollment target allocation for instructors all the time by the recruit units and graduate supervisor (mentioned as instructors below). The graduate enrollment target allocation for instructors influences the academic development, scientific research achievements of instructors and the quality of graduate training. With the variation in national graduate enrollment, the development of disciplines and the fluctuation in market demand, the contradiction between the enrollment index assigned to instructors and the demand of instructors for graduate students is very prominent. For instance, for the discipline of electrical engineering in a certain university, in 2008, the total enrollment indicator of 15 instructors was 39, i.e. every instructor had an enrollment index of 2.8. While in 2014, the total enrollment indicator for 25 instructors was 40, i.e. the quantity of graduates diminished to 1.6 per instructor. On one hand, allocation disparity of enrollment targets conduces to serious problems of unfair distribution; on the other hand, the strict average of target allocation will definitely hinder the development of instructors' research capacity. As a consequence, how to deal with the contradiction between the enormous quantity of instructors and the limited enrollment index is a significant issue. A reasonable allocation system plays an important role in promoting discipline development and improving the enthusiasm of instructors. Based on practical experience of allocating indicators in the electrical engineering discipline in university,

noting that enrollment targets should be oriented by academic development and instructors' capacity, this article considers research funding, awards, research projects, scholarly articles, patents and other factors of instructors and tries to allocate using the method of Analytic Hierarchy Process.

2 THE BASIC THEORY OF ANALYTICAL HIERARCHY PROCESS

Analytical Hierarchy Process was founded by T.L. Saaty, the famous American operational research expert, the professor of University of Pittsburgh in the 1970s, which is an effective method of non-quantitative analyses for quantitative events. AHP has been widely used in water allocation, evaluation of sustainable development, electrical safety assessments, telecommunications service quality evaluation, strategic process selection, security science and environmental science.

2.1 *Principle of Analytical Hierarchy Process*

The principle of analytical hierarchy process is that according to the law of the human mind, when in the face of complex selection problem, the problem is divided into its component elements, and then these factors are grouped by hierarchical structure formation on the basis of dominance relationship, to determine the relative importance of each factor level by pairwise comparison approach and then generalize the overall judgment of the decision makers and determine the order of relative importance of the decision-making overall program and to make choices and judgments. The key to the whole thought process is hierarchical division, the determining of weights and rules for merging arrays. That is, AHP builds the weight vector of n factor $W^T = [W_1 \, W_2 \dots W_n]$, gets the prioritization matrix $A = (a_{ij})$ by pairwise comparison approach and then obtains the equation $AW = nW$. If the consistency check of matrix A is met, the eigenvalue of matrix A is n and W is the eigenvector of matrix A.

2.2 Steps of Analytical Hierarchy Process

1. *Build the model of Analytical Hierarchy Process:* According to the understanding of the problem and preliminary analysis, determine the evaluation objectives, evaluation criteria for the program, and then construct hierarchical model with the goals, evaluation criteria and an action plan. In this model, goals, evaluation criteria and action programs are at different levels and the relationship between them are represented by each line segment. The model can be divided into three levels, as shown in Figure 1.
2. *Construct judgment matrix:* Use pairwise comparison approach and 1–9 scaling method, build judgment matrix starting from the second floor, to prepare for determining the weights of each group.
3. *Determine the weights:* It should be pointed out that, in this case, the weights refer to those of groups and levels.
4. *Check consistency:* Examine the consistency of prioritization matrixes and the overall ranking matrix, so as to determine whether the data is reasonable.
5. *Sort the weights and obtain the result:* Get the overall weights out of the weights of Scheme level and criteria level, as well as the comparison of importance.

2.3 Assumptions

1. All indicators truly and accurately reflect the ability of instructors for enrollment of graduate.

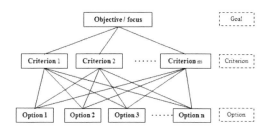

Figure 1. Generic hierarchy structure of the Analytical Hierarchy Process.

2. The weights of evaluation indicators are objective.
3. The weight of every indicator is only related to the weight calculated.
4. According to the criterion C, we use the 1–9 scaling method to determine the importance of the two factors u_i and u_j, referred as a_{ij}, $a_{ij} = 1/9$, 1/8 … 1/2, 1, 2, 3 … 8, 9. The larger a_{ij} is, the more important the u_i is to u_j.

2.4 Relevant definitions

Definition 1 Weight: In hierarchical model, assuming the factor C in upper level is a criterion, the importance of elements $u_1, u_2 \dots u_n$ dominated in the lower level to criterion C is the weight. In this case, n is the number of the elements.

Definition 2 Prioritization Matrix: Comprehensively consider the elements in certain level, make pairwise comparisons and then obtain the prioritization matrix. It is donated as:

$$A = (a_{ij})_{n \times n} = \begin{bmatrix} a_{11} & a_{12} & \dots & a_{1n} \\ a_{21} & a_{22} & \dots & a_{2n} \\ \dots & \dots & \dots & \dots \\ a_{n1} & a_{n2} & \dots & a_{nn} \end{bmatrix} \quad (1)$$

In this case, a_{ij} is referred to the importance of u_i to u_j, and it is determined by 1–9 scaling method. The prioritization matrix A has the following properties, i.e. $a_{ij} > 0$, $a_{ij} = 1/a_{ij}$, $a_{ij} = 1(i, j = 1, 2 \dots n)$.

Definition 3 Complete consistency: If the prioritization matrix A is satisfied with $a_{ij} = a_{ik} * a_{kj}$ $(i, j, k = 1, 2 \dots n)$, matrix A is completely consistent. That is, when experts are making pairwise comparison, the judgments remain consistent.

Definition 4 Satisfactory consistency: If the consistency overall sort satisfies the criterion $CR = CI/RI < 0.1$, the satisfactory consistency is met. In this case, $CI = (K_{max} - m)/(m - 1)$, CI refers to indicator of the consistency of overall sort. RI is the average random consistency of overall sort, which is shown in Table 1. K_{max} represents largest eigenvalue of pairwise comparison and m stands for the order of pairwise comparison A.

In the actual decision, the expert judgment can only be estimated, it is impossible to ensure that the prioritization matrix is given with complete consistency, satisfactory consistency is used to test the consistency of experts' judgment.

Table 1. RI of prioritization matrix from 1 to 9 order.

n	1	2	3	4	5	6	7	8	9	10
RI	0	0	0.58	0.9	1.12	1.24	1.32	1.41	1.45	1.49

3 APPLICATION OF MODEL

For system analysis with analytic hierarchy process, firstly, make the question hierarchical, according to the nature of the problem and the total target, decompose the problem into different factors, and then aggregate factors into the different hierarchical combination on basis of the mutual influence between these factors and the subordinate relationships, form a multi-level analysis structure model. Finally, according to the weight of bottom level (solution level), as well as the quota of the subject, determine the index allocated for the instructors participated.

3.1 Graduate enrollment target allocation system based on analytic hierarchy process

1. *Building the model of hierarchical structure*
   Considering that many factors influence the target allocation of Graduate enrollment, only a function is far from enough to express the complexity of the relationship between them, so these factors must be divided into multiple levels and assessed by level comprehensive evaluation. This article divides the Graduate enrollment target allocation system into three levels. The first level is object level A, i.e. allocating the Graduate enrollment target reasonably. The second level is criterion level B, which includes research funding, awards, research projects, academic articles, patents, works and student employment, excellent papers, etc., and reflects the comprehensive quality of the instructors. And the third level is solution level C, in which we select seven representative instructors in electrical engineering. After building the hierarchical structure, we determine the subordinate relationships and obtain the hierarchical model, as shown in Figure 2.
2. *Building the prioritization matrix and determining the weights*
   Factors that affect graduate enrollment target allocation are various and this analysis is based on the judgments of the relative importance by making pairwise comparisons. These pairwise comparisons are represented by the appropriate values, and written as the prioritization matrix, which is shown in Figure 2. We could use a_{ij} to show the comparison result of factor i to factor j, thus building the criterion matrix (only listed the case of prioritization matrix A-B and B1-C as shown in Table 2, Table 3).
3. *Checking consistency*
   We could obtain the weight of each factor by calculating the prioritization matrix.

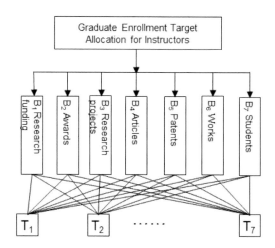

Figure 2. Hierarchical chart of graduate enrollment target allocation for instructors.

Table 2. Prioritization matrix A-B.

A	B1	B2	B3	B4	B5	B6	B7	B8
B1	1	3	4	2	5	6	8	9
B2	1/3	1	2	1/2	3	4	6	7
B3	1/4	1/2	1	1/3	2	3	5	6
B4	1/2	2	3	1	4	5	7	8
B5	1/5	1/3	1/2	1/4	1	2	4	5
B6	1/6	1/4	1/3	1/5	1/2	1	3	4
B7	1/8	1/6	1/5	1/7	1/4	1/3	1	2
B8	1/9	1/7	1/6	1/8	1/5	1/4	1/2	1

Table 3. Prioritization matrix B1-C.

B1	C1	C2	C3	C4	C5	C6	C7
C1	1	1/7	1/6	1/8	1/4	1/2	2
C2	7	1	2	1/2	4	6	8
C3	6	1/2	1	1/3	3	5	7
C4	8	2	3	1	5	7	9
C5	4	1/4	1/3	1/5	1	3	5
C6	2	1/6	1/5	1/7	1/3	1	3
C7	1/2	1/8	1/7	1/9	1/5	1/3	1

By calculating the prioritization matrix A-B, B1-C, B2-C, B3-C, B4-C, B5-C, B6-C, B7-C, B8-C, we gain the CR, 0.0363, 0.0392, 0.0376, 0.0247, 0.0389, 0.0366, 0.0387, 0.0247, 0.0315. The prioritization matrixes are all reasonable, as their CR are all less than 0.1.

4. *Obtaining the overall ranking matrix*
   We could get the graduate enrollment target allocation scheme for instructors after gaining the overall ranking matrix, as shown in Table 4.

Table 4. The overall ranking matrix.

Criterion	Research funding	Awards	Research projects	Articles	Patents	Works	Students	Outstanding students papers	Overall weight
Weight	0.3312	0.1586	0.1078	0.2313	0.0734	0.0509	0.0269	0.0200	
The weight of solution level									
In1	0.0328	0.0435	0.1036	0.0865	0.0242	0.0784	0.2399	0.2451	0.0660
In2	0.2562	0.0959	0.0676	0.1747	0.0337	0.0527	0.3543	0.0548	0.1635
In3	0.1789	0.1446	0.2399	0.0586	0.3732	0.2264	0.0312	0.3584	0.1685
In4	0.3679	0.2150	0.1587	0.3646	0.2617	0.4230	0.1036	0.1653	0.3042
In5	0.0932	0.0637	0.0448	0.0226	0.0493	0.0255	0.0448	0.0374	0.0579
In6	0.0473	0.0239	0.0312	0.0404	0.0727	0.0358	0.0676	0.0267	0.0417
In7	0.0237	0.4135	0.3543	0.2527	0.1851	0.1581	0.1587	0.1123	0.1982

3.2 Analysis of the application of AHP

1. Gaining the overall ranking matrix

   We call the matrix, which is donated as overall ranking matrix, by calculating the weights of all the factors in the same level to top level. This is proceeded from the highest level to the lowest level layer by layer. If the upper level A has m factors A_1, A_2 ... A_m and their weights are a_1, a_2 ... a_m, and the lower level B has n factors B_1, B_2 ... B_n and their weights for factor A_j are B_{1j}, B_{2j} ... B_{nj}, (if B_k is independent of A_k, B_k is 0), and also C_{ij} is the consistency index of the factors in level B to A_j, and the R_{ij} is average consistency index, then we can get that the CR for level B is $CR = \sum_{j=1}^{m} a_j C_j / \sum_{j=1}^{m} a_j RI_j$. If CR < 0.1, we could judge that prioritization matrixes are all consistent and reasonable. If not, we should adjust the factor in matrixes.

   The weights of each factor are shown in Table 4. By calculating, we get CR = 0.0363 < 0.1.

2. Confirming the scheme for graduate enrollment target allocation

   After adjusting the weight of each factor in level C, we could confirm the graduate enrollment target allocation.

   The enrollment targets of instructors for the current year:

   N = total index of subject * overall weights of Instructors.

   For example: a total of seven instructors of electrical engineering graduate instructors participate in index distribution in 2013 and the total index is 20, so the indicators of seven instructors are 1, 3, 4, 6, 1, 1, 4.

   Using the above criteria as the indicators in instructor assessment, the quantification process must be scientific, rational and objective to determine the amount of each index. By collecting information of instructors, testing in the model and comparing with the data in previous years, we can see that the model has high accuracy and maneuverability.

4 EPILOGUE

This article concentrates on the complexity and difficulty of the quantitative evaluation in graduate enrollment target allocation for instructors. Applying AHP, it obtains the initial result of the model after assessing comprehensively the factors of the instructors, and then determines a rational, scientific graduate enrollment target allocation scheme. The model, to some extent, overcomes the subjectivity and arbitrariness; in the process of target allocation planned distribution of subjective and arbitrary process, and achieves satisfactory results in the aspects of scientificity and accuracy. It is worth noting that the method establishes hierarchical model and prioritization matrix depending on the accuracy of the assumptions and input data. The selection preference of the decision makers will result in the error and inconsistency of the result, which requires further analysis and research.

Future work could further research in the aspects of the appropriate scale, considering the uncertainty of its input values, improving accuracy of pairwise comparisons to seek better algorithms.

REFERENCES

[1] Satty T.L. The analytic hierarchy process. New York: McCraw-Hill, 1980.
[2] Zhuang suo fa. Study of synthetic evaluation model based on Analytical Hierarchy Process. Journal of Hefei University of Technology (Natural Science), 2000, 04: 582–585+590.

[3] Wang Weiren, Tu Meizeng. Calculating of Proportion Based on Analytic Hierarchy Process in Basin Water Allocation. Journal of Tongji University, 2005, 33(8): 1133–1136.

[4] Wang Xiaopeng, Zeng Yongnian, Cao Guangchao. Evaluating Model and Application to Sustainable Development for Pastoral Areas on Qinghai Tibetan Plateau Based on Multivariate Statistical Analysis and AHP. Systems Engineering-theory & Practice. 2005, 25(6): 139–144.

[5] Jie Yuxin, Hu Tao, Li Qingyun. Application of analytical hierarchy process in the comprehensive safety assessment system of Yangtze River levee. Journal of Tsinghua University (Science and Technology), 2004, 12: 1634–1637.

[6] Chen Guohua, Tao Zhaoling. Using the Analytic Hierarchy Process to Prioritize Business Processes for Reengineering Based on Business Strategy. Operations Research and Management Science. 2003, 12(2): 101–105.

[7] Guo Jinyu, Zhang Zhongbin, Sun Qinyun. Applications of AHP method in safety science. Journal of Safety Science and Technology, 2008, 4(2): 69~73M. King, B. Zhu, and S. Tang, "Optimal path planning," Mobile Robots, vol. 8, no. 2, pp. 520–531, March 2001.

Education Management and Management Science – Zheng (Ed.)
© 2015 Taylor & Francis Group, London, ISBN 978-1-138-02663-6

Measuring impatience: Experiments of intertemporal choices

Y. Lu
School of Business and Administration, Northeastern University, Shenyang, Liaoning, P.R. China

C. Yang
School of Materials and Metallurgy, Northeastern University, Shenyang, Liaoning, P.R. China

ABSTRACT: Intertemporal decision making draws attention in econophysics and neuroeconomics. One of the main objectives of the researches in this field is to examine the degree of impatience. The process to measure impatience involves various types of experiments. In this paper, we illustrate the traditional experiment procedure and some variability factors of the procedure. Consequently, we show that current experiment settings cannot examine the effect of time-perception, and thus future studies shall reform the current experiment.

Keywords: experiments; intertemporal choices

1 INTRODUCTION

Studies in econophysics and neuroeconomics show that humans or non humans prefer sooner but smaller monetary rewards to larger but later rewards [1–6], i.e. the present value of a reward decreases as the delay to that reward increases. For example, one typically would choose to receive $100 now over $100 tomorrow, $100 tomorrow over $100 the day after tomorrow, and so on. The classical economic theory assumes that people make rational decisions, and thus have consistent time preference. According to Strotz [7], such time-consistent behavior can be only modeled by exponential discounting. However, empirical studies have provided us with ample evidence, which shows anomaly in intertemporal choices [10–15, 17–19]. People generally exhibit dynamic inconsistency when making intertemporal choices, namely that people tend to make patient plans in the distant future, but act impatiently in the near future. Thus, discount factor between adjacent periods close by is smaller than between similar periods further away [2, 3]. This preference reversal over time is referred as dynamic inconsistency and can be better described by hyperbolic discount models [9, 13].

2 THEORETICAL DISCOUNT MODELS

Intertemporal choices refer to decision making between options (typically, monetary rewards) delivered at different times. Individuals subjected to intertemporal choices face a trade-off between the utility (value) of an immediate but smaller reward and a delayed but larger one. Consider the examples: options between receiving $100 today or $105 in a month; options between spending $100 today or spending $105 in a month; deciding whether to smoke or not smoke a cigarette. The physical time duration between the present time and the time when the reward is delivered is referred as delay [9]. The non-discounted (real) value of a given reward is called objective value. The value to be received immediately, which is equivalent to the receipt of $V(D)$ on a specified delay, is referred as the subjective value of the reward or indifference point. $V(D)$ is a function of delay D. The shape of the discount curve is a decreasing monotonic function with null asymptotic value.

2.1 Exponential discounting

Classical economic theory assumes consistency temporal discounting, which follows exponential discounting function [10–13].

$$V(D) = A\exp(-r_e D) \qquad (1)$$

where $V(D)$ is the subjective value of a monetary reward, A is the objective amount of the reward, D is the physical time duration of delay of the reward, and r_e stands for a constant discount factor: $-(\mathrm{d}V(D)/\mathrm{d}D)/V(D) = r_e$

2.2 Simple hyperbolic discounting

Several studies have proposed simple hyperbolic discount model to capture the inconsistency of imtertemporal choice [10–13].

$$V(D) = A/(1 + r_s D) \qquad (2)$$

where $V(D)$, A, and D stay the same as in eq.(1), r_s is the simple hyperbolic discount parameter. For $r_s > 0$, the discount rate of simple hyperbolic discount model is a decreasing function of D: $-(dV(D)/dD)/V(D) = r_s/(1 + r_s D)$.

2.3 Quasi hyperbolic discounting

Quasi hyperbolic discount model is another hyperbola-like discount model, which is very popular in econophysics. It has been thoroughly tested and applied in retirement decisions or investment decisions, because a continuous change of its parameters can induce a jump in consumption and savings [3, 17–19].

$$V(D) = A\beta\delta^D \qquad (3)$$

where $V(D)$, A, and D are the same as in eq. (1), β and δ (both between 0 and 1) are discount parameters meant to capture the essence of hyperbolic discounting. Indeed, the discount factor between the first 2 periods is $\beta\delta$, and between any two adjacent periods later it is δ.

2.4 Q-exponential discounting

q-exponential discount function is a deformed algebra inspired in nonextensive thermodynamics [16]. It was firstly utilized to study intertemporal choices by Cajueiro (2006) [8], and then proposed by a number of researchers [10–13, 15].

$$V(D) = A/[1 + (1 - q)r_q D]^{1/(1-q)} \qquad (4)$$

where $V(D)$, A, and D stay the same, r_q and q are discount parameters of the model. For $q \to 1$, q-exponential discount recovers the classical exponential discount. For $q \to 0$, it yields the simple hyperbolic discount [8] [15]. Hence, with two free parameters, q-exponential discount model is a general form of exponential discount model and simple hyperbolic model, in which $1-q$ indicates the degree of inconsistency [10].

3 EXPERIMENTS

3.1 General procedure

In general, in an intertemporal choices experiment, the participants are asked to choose between two monetary rewards, an immediate but smaller reward and another delayed but larger one. The duration of the delay is usually from days to years. For each delay, the experiment begins with equal values for both rewards, so that a given participant chooses the immediate reward. The delayed reward value is kept constant while the immediate reward value is decreasing. This procedure is repeated until the delayed reward is preferred to the immediate one. The last immediate reward value chosen is described as the indifference point of the respective delay. To avoid a possible influence of the rewards presentation order in the experiments, the reverse procedure is also examined. The reversed experiment starts from the lowest value for the immediate reward, so that the delayed reward is preferred. The immediate reward is then increased until its first value is chosen. The average of the two indifference points is considered as the subjective value $V(D)$. In most experiments involving intertemporal choices, hypothetical rewards are used. Also, the delays are not experienced by the individual during the experiment. This type of procedure has the advantage of being cheap and time efficient.

For example, participants were seated individually in a quiet room, and faced the experimenter across a table. After that, participants received the simple instruction that the monetary reward in this experiment was hypothetical, but you need to think as though it were real money. Then, the participants were asked to choose between the card describing money delivered certainly and the card describing money delivered with a certain delay of receipt. The left card viewed from participants indicated the amounts of money that could be received immediately (a smaller but immediate reward), and the right card indicated 10,000 RMB (about $1,500) that could be received with a certain delay (a larger but delayed reward).

For the intertemporal choice task, monetary rewards and the delay were printed on cards. The 27 monetary reward amounts were 10,000 RMB, 9,900 RMB, 9,600 RMB, 9,200 RMB, 8,500 RMB, 8,000 RMB, 7,500 RMB, 7,000 RMB, 6,500 RMB, 6,000 RMB, 5,500 RMB, 5,000 RMB, 4,500 RMB, 4,000 RMB, 3,500 RMB, 3,000 RMB, 2,500 RMB, 2,000 RMB, 1,500 RMB, 1,000 RMB, 800 RMB, 600 RMB, 400 RMB, 200 RMB, 100 RMB, 50 RMB, and 10 RMB. The 8 delays of receipt were 1 week, 2 weeks, 1 month, 6 months, 1 year, 5 years, 25 years and 30 years. The experimenter turned the 27 immediate cards sequentially. The cards started with 10,000 RMB, down to 10 RMB and then back to 10,000 RMB. For each card, the participant pointed to either the immediate or delayed reward. The experimenter wrote down the last immediate reward chosen in descending order, and the first immediate reward chosen in ascending order, and the average of them were used as the subjective values in the following analysis. This procedure was repeated at each of the 8 delays [22].

Some researches used real-time and real money instead of hypothetical rewards. Participants experience the consequences associated with their choices (rewards and delays) while completing the experiment. Considering the time and expense associated with this type of procedure, researches usually examine comparatively shorter delays (<90s) and smaller rewards (<$0.50).

3.2 Variations of experiments

Besides the standard setting of intertemporal choices experiment, some variability factors can also be introduced in the experiment procedure [20].

3.2.1 The effect of presenting order of the rewards
The rewards are presented either in ascending order or descending order [10].

3.2.2 The effect of presenting order of the delays
The delays are presented in ascending order, descending order, or sometimes presented to the participants randomly [11].

3.2.3 The magnitude effect of the rewards
Participants' discount behavior may vary with the amount of the rewards [21].

3.2.4 The signal effect
Some experiments in econophysics discover that human discount differently for gains and losses [11].

3.2.5 The classification of participants
The education background, working experience, age, and gender of the participants also influence the results of the experiments. This implies experiment shall classify the participants accordingly [22].

4 FUTURE DIRECTION

Studies in behavioral economics have confirmed the logarithmic time-perception theory of irrational discounting, i.e. the irrational discounting behavior of participants is merely a consequence of their psychological time perception. Neither the general process nor the variations in Section 3 can analyze the independent influence of each factor in determining discount functions.

Takahashi (2012) [11] employed time-perception task originally developed by Zauberman et al. (2009) [21] to study intertemporal choices. However, the research still does not mention the independent influence of time-perception. The future studies should reform the procedure of the experiment to understand anomalies in intertemporal choice.

ACKNOWLEDGMENT

The research reported in this paper was supported by the Fundamental Research Funds for the Central Universities (No. N120306001) from Ministry of Education of P.R.China.

REFERENCES

[1] G.F. Loewenstein, D. Prelec, Anomalies in intertemporal choice: evidence and an interpretation, Quarterly Journal of Economics 57(1992) 573–598.

[2] D. Laibson, Golden eggs and hyperbolic discounting, Quarterly Journal of Economics 62(1997) 443–478.

[3] P. Diamond, B. Koszegi, Quasi-hyperbolic discounting and retirement, Journal of Public Economics 87(9)(2003) 1839–1872.

[4] G. Angeletos, D. Laibson, A. Repetto, J. Tobacman, S. Weinberg, The hyperbolic consumption model: calibration, simulation, and empirical evaluation, Journal of Economic Perspectives 15(3)(2001) 47–68.

[5] L.R. Keller, E. Strazzera, Examining predictive accuracy among discounting models, Journal of Risk and Uncertainty 24(2)(2002) 143–160.

[6] K.N. Kirby, N.N. Marakovic, Modeling myopic decisions: evidence for hyperbolic delay-discounting within subjects and amounts, Organizational Behavior and Human Decision Processes 64(1)(1995) 22–30.

[7] R.H. Strotz, Myopia and inconsistency in dynamic utility maximization, Review of Economic Studies 23(3)(1955) 165–180.

[8] D.O. Cajueiro, A note on the relevance of the q-exponential function in the context of intertemporal choices, Physica A 364(2006) 385–388.

[9] T. Takahashi, Theoretical frameworks for neuroeconomics of intertemporal choice, Journal of Neuroscience, Psychology, and Economics 2(2009) 75–90.

[10] T. Takahashi, H. Oono, M.H.B. Radford, Psychophysics of time perception and intertemporal choice models, Physica A 387(2008) 2066–2074.

[11] H. Ruokang, T. Takahashi, Psychophysics of time perception and valuation in temporal discounting of gain and loss, Physica A 391(2012) 6568–6576.

[12] A.L. Gustman, T.L. Steinmeier, Policy effects in hyperbolic vs. exponential models of consumption and retirement, Journal of Public Economics 96(5–6)(2012) 465–473.

[13] T. Takahashi, A. comparison of intertemporal choices for oneself versus someone else based on Tsallis' statistics, Physica A 385(2)(2007) 637–644.

[14] A. Kapteyn, F. Teppa, Hypothetical intertemporal consumption choices, Economic Journal 113(486)(2003)C140–C152.

[15] S.C. Rambaud, M.J. Torrecillas, A generalization of the q-exponential discounting function, Physica A 392(14)(2013) 3045–3050.

[16] C. Tsallis, What are the numbers that experiments provide? Química Nova 17(6)(1994) 468–471.

[17] J. Benhabib, A. Bisin, A. Schotter, Present-bias, quasi-hyperbolic discounting, and fixed costs, Games and Economic Behavior 69(2)(2010) 205–223.

[18] M.J. Salois, C.B. Moss, A direct test of hyperbolic discounting using market asset data, Economics Letters 112(3)(2011) 290–292.

[19] B. Wigniolle, Savings behavior with imperfect capital markets: when hyperbolic discounting leads to discontinuous strategies, Economics Letters 116(2) (2012) 186–189.

[20] N. Destefano, A.S. Martinez, The additive property of the inconsistency degree in intertemporal decision making through the generalization of psychophysical laws, Physica A 390(10)(2011) 1763–1772.

[21] K.B. Kim, G. Zauberman, Perception of anticipatory time in temporal discounting, Journal of Neuroscience, Psychology, and Economics, 2(2)(2009) 91–101.

[22] Y. Lu, X. Zhuang. The impact of gender and working experience on intertemporal choices, Physica A 409(2014) 146–153.

Education Management and Management Science – Zheng (Ed.)
© *2015 Taylor & Francis Group, London, ISBN 978-1-138-02663-6*

Research on the experimental teaching content and structure of business administration

Gangquan Xu, Lingzhi Liu & Shuili Yang
School of Economics and Management, Xi'an University of Technology, Xi'an, Shaanxi, China

ABSTRACT: Experimental teaching is an important part of college business administration education and an important procedure to foster the students' innovation ability. This paper discusses the existing problems and conditions of business administration experimental teaching and sets the experimental teaching goals based on the targets of business administration. Then we establish the experimental teaching content system of business administration, which is hierarchical, multi-module interrelated with each other and ability-oriented and give the quantitative analysis of experimental teaching ability level. Finally, optimization recommendation is given to experimental teaching of business administration.

Keywords: business administration; experimental teaching; innovation ability

1 INTRODUCTION

Experimental teaching is an important part of college education. It plays a non-substitutable role in cultivating the students' innovation ability[1]. The experimental teaching of business administration is an important approach to solve the disaccord problem between theory teaching and social practical of business administration education[2]. Therefore, deepening reform about experiment teaching of business administration has important role and realistic meaning to improve the quality of teaching and promote the innovative talents training of business administration. Dai Qiang's opinion about experimental teaching is representative: "we should focus on cultivating research talents in the overall design of the experimental teaching system and experimental curriculum structure, resolve the experiment teaching target into multi-level and multi-module to achieve". Jiang Linhua[3] considers that the practical teaching system of business administration should include three levels that are "professional basic skills training, professional skills training and comprehensive practical training and the three levels are the training process from the basic to comprehensive". Zhang Zhenji's[4] opinion is that the experimental teaching system of experimental center consists of three levels and two systems, the professional experimental teaching system consists of basic experiment, special experiment and comprehensive experiment.

2 ANALYSIS OF THE CURRENT SITUATION AND PROBLEMS OF THE EXPERIMENTAL TEACHING CONTENT OF BUSINESS ADMINISTRATION

In recent years, although the experimental teaching of business administration has been rapidly developed, it has been greatly valued than before. From the whole, the experimental teaching can't attract enough attention, the experimental teaching system is still not perfect, there is great disparity compared with the professional training objective demand of experiment teaching, the present situation of the experimental teaching cannot adapt to the innovative talents cultivation.

2.1 *The experimental teaching content and training goals are not coordinated*

Business administration aims to cultivate the creative talents that have the ability to analysis and solve problems. At present, the experimental teaching content of business administration is outdated and obsolete, the experimental class is not separately open but as a part of theoretical teaching, not taking fostering students' innovation into account. The experimental courses are mostly demonstrated by the teacher, students only need to follow the prescribed steps to complete the experiment. Students seldom have the opportunities to think carefully about the experiments all by themselves so that they can't meet the real demands in future work and even can't get a promotion on their capability of analyzing and solving problems without the combination of theory and reality.

2.2 The experimental teaching content lacks of top-level design

At present, the experimental teaching content of business administration lacks of systematic and experimental curriculum system has not been established. First of all, we don't have enough courses to meet the training objective demands of business administration. Second, the relationship between the experiment courses is indistinct and the arrangement of them is incoherent. There is no organic connection between the courses set up, which is resulting from cutting regularity of courses' relation factitiously, as some are repetitive, some are disjointed and even some have blank spot on knowledge. Experiment courses and theory courses disconnect each other. In other words, there are no corresponding courses as a theoretical guidance for experiment courses, andvise visa[5]. Finally, there are too many replication experiments, comprehensive experiments, and little innovative experiments in the experimental courses have been set, it couldn't cultivate the good students' ability of innovation.

2.3 The content of experiment teaching is not according with the rule of training

The experimental teaching content designed for students of business administration should gradually progressive form foundation to innovation, so the experimental class hours should be reasonable arranged to better cultivate students' innovative ability. By investigating many colleges in Shaanxi, we know that the average class hours of total experiments are 144 hours of business administration, the basic experiment class hours are 20, the professional basic experiment class hours are 84, and the professional experiment class hours are 40. We can see that there are too many professional basic experiments and few innovative experiments. So the proportion of different experiments that have been opened can't cultivate student's operational ability and innovative ability and can't fully accord with the rule of training based on students lacking of innovative ability.

3 BUILD THE EXPERIMENTAL TEACHING CONTENT SYSTEM OF BUSINESS ADMINISTRATION

3.1 The experimental teaching goal of business administration

In order to build experimental teaching content system of business administration, we should first determine the goals of experiment teaching, then build the system that coordinate with the target. The goal of experimental teaching is training the students' practical ability, cultivating the students' innovation ability. Through the combination of experimental teaching and theoretical teaching, they can train the students' attitude of integrating theory with practical, seeking truth from facts, enable students to understand and consolidate the knowledge, so as to improve the ability of finding problems, analyzing and solving the question, stimulate the independent thinking and innovation consciousness, more match the experimental teaching goal of business administration. The experimental teaching goal can be divided into four aspects: mastering subject knowledge, strengthening basic ability, improving comprehensive application ability and enhancing innovation ability.

3.2 Design the experimental teaching content module of business administration

When design the experimental teaching content system of business administration, we should focus on the experimental teaching goal, and decompose it into multi-level and multi-module to achieve, and optimize the experimental project with the principle that combine basic experiment with professional experiment, verification experiment with demonstration experiment, single experiment with comprehensive experiment. We determine the horizontal latitude of the experimental teaching content system of business administration which includes basic experiment module, basic skill experiment module, comprehensive experiment module and innovative experiment module based on the thesis of The Ministry of Education that is the scientific experimental teaching content system of combining multi-level and multi-module.

1. Basic experiment module. These module experiment courses are based on compulsory basic courses which include management, economics, statistics, operations research, economic law, accounting and other basic courses. These courses are designed for students to master the basic management knowledge and lay a good foundation for professional learning in future.
2. Basic skill experiment module. These module experiment courses are based on training plan and training goal of business administration which include computer basic operation experiment, office automation software experiment, database management system, and the experiment of e-commerce web site development experiment. These courses are designed for students to master the necessary professional basic skill and improve the professional level.

3. Comprehensive experiment module. These module experiment courses are based on professional courses of business administration which include project management software experiment, logistics management simulation experiment, human resources management simulation experiment, e-commerce simulation experiment, information resource management system experiment, quality management simulation experiment, production and operation management simulation experiment simulation test, and enterprise management simulation experiment. These courses are designed for students to improve students' comprehensive application ability based on mastering the knowledge of management and professional skills.
4. Innovative experiment module. These module experiment courses are based on the social's demands for talents of business administration which include the ERP sand table simulation experiment and ERP simulation experiment. These courses are designed for students to cultivate their innovative consciousness and innovative ability, proving their comprehensive quality to meet social's requirements based on improving their comprehensive application ability.

3.3 Design the experimental teaching ability training level of business administration

The longitudinal latitude of the experimental teaching content system of business administration has four different levels based on training innovation ability: confirmatory experiment, demonstrative experiment, comprehensive experiment, and innovative experiment.

1. Confirmatory experiment. The courses of basic experiment module all belong to the confirmatory experiment. Students learn and master the basic theory and knowledge through experiments. This kind of experiments often give details, demands, procedures, methods of experiments, students just follow the steps and draw a conclusion according to the guidebook.
2. Demonstrative experiment. The courses of basic skill experiment module all belong to demonstration experiment. This kind of experiments are mainly skilled experiments, it is the necessary knowledge to be a qualified student of business administration. Students can master basic skills through teachers' demonstrate, students themselves, teachers' guide and other means.
3. Comprehensive experiment. The courses of comprehensive experiment module all belong to comprehensive experiment. This kind of experiment requires students to apply all kinds of knowledge and accomplish the experiment under the guidance of teachers.

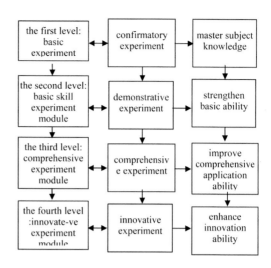

Figure 1. The structure chart of experimental teaching content system of business administration.

4. Innovative experiment. The courses of innovation experiment module all belong to innovative experiment. This kind of experiment requires students complete the whole things all by themselves; it is a kind of exercise for students' comprehensive application ability and innovation ability. The students must digest the knowledge before the experiments which need students have strong practical ability and creative spirit. The experimental teaching content system is designed as shown in Figure 1.

4 QUANTITATIVE ANALYSIS OF EXPERIMENTAL TEACHING ABILITY TO CULTIVATE HIERARCHY OF BUSINESS ADMINISTRATION

Number many domestic scholars have put forward different kinds of experimental teaching content system, but they only make the qualitative analysis, while seldom pay attention to the quantitative researches. In this paper, we use AHP to quantify the four experimental teaching levels based on experimental teaching of business administration, determine the proportion of each experimental level, and provide foundations for the experimental teaching of business administration. Construct hierarchical structure diagram of experimental teaching target.

1. Construct hierarchical structure diagram of experimental teaching target
   The experimental teaching goal of business administration is cultivating innovative talents, it can be divided into four specific aspects: master

the discipline basic knowledge, strengthen the basic skills, improve comprehensive application ability and enhance innovation ability. The experimental curriculums set by experimental goal are confirmatory experiment, demonstrative experiment, comprehensive experiment and innovative experiment. The structure of experimental teaching goal of business administration is shown in Figure 2.

2. Single sort for each level

Compare every two experimental level and get judgment matrix based on grading by specialist.

$$\begin{pmatrix} B1 & C1 & C2 & C3 & C4 \\ C1 & 1 & 2 & 2 & 2 \\ C2 & 1/2 & 1 & 1 & 1 \\ C3 & 1/2 & 1 & 1 & 1 \\ C4 & 1/2 & 1 & 1 & 1 \end{pmatrix} \begin{pmatrix} B2 & C1 & C2 & C3 & C4 \\ C1 & 1 & 1/3 & 1/2 & 1/2 \\ C2 & 3 & 1 & 2 & 2 \\ C3 & 2 & 1/2 & 1 & 1 \\ C3 & 2 & 1/2 & 1 & 1 \end{pmatrix}$$

$$\begin{pmatrix} B3 & C1 & C2 & C3 & C4 \\ C1 & 1 & 1/2 & 1/5 & 1/2 \\ C2 & 2 & 1 & 1/3 & 1/2 \\ C3 & 5 & 3 & 1 & 2 \\ C4 & 2 & 2 & 1/2 & 1 \end{pmatrix} \begin{pmatrix} B4 & C1 & C2 & C3 & C4 \\ C1 & 1 & 1/2 & 1/3 & 1/6 \\ C2 & 2 & 1 & 1 & 1/2 \\ C3 & 3 & 2 & 1 & 1/2 \\ C4 & 6 & 4 & 2 & 1 \end{pmatrix}$$

$$\begin{pmatrix} A & B1 & B2 & B3 & B4 \\ B2 & 3 & 1 & 1/2 & 1/5 \\ B3 & 5 & 2 & 1 & 1/2 \\ B4 & 7 & 5 & 2 & 1 \end{pmatrix}$$

we can calculate: CI(B1) = 0 < 0.1, CI(B2) = 0.0043 < 0.1, CI(B3) = 0.015 < 0.1, CI(B4) = 0.0048 < 0.1, CI(A) = 0.0247 < 0.1.

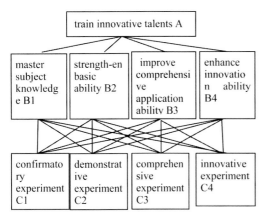

Figure 2. The structure chart of experimental teaching goal of business administration.

Table 1. The weight calculate w_i.

B	C				
	B1 (0.0561)	B2 (0.1573)	B3 (0.2845)	B4 (0.5021)	W_i
C11	0.4	0.1207	0.0976	0.0792	0.1089
C12	0.2	0.4138	0.1701	0.1485	0.1992
C21	0.2	0.2328	0.4882	0.2574	0.3159
C22	0.2	0.2328	0.2440	0.5149	0.3757

3. Total sort for each level, the weight calculate process and result are shown in Table 1.

According to Table 1, confirmatory experiment, demonstrative experiment, comprehensive experiment and innovative experiment account for 10%, 20%, 32%, and 38% respectively. We should reasonably arrange class hours of confirmatory experiment, demonstrative experiment, comprehensive experiment and innovative experiment based on the quantitative results of different experiments and the experimental teaching goal of business administration to coordinate with training goal. We also should make detailed experimental teaching plan that theory teaching and experimental teaching complement each other to better cultivate students' operational ability and innovative ability.

5 CONCLUSIONS

Consistency in this paper, we construct "four vertical and four horizontal" experimental teaching content system that on the standard of training ability and base on experimental teaching content, make a quantitative research on experimental levels based on training goal and experimental teaching goal of business administration, it has important significance to improve the experimental teaching quality of business administration. In order to make sure to carry out the experimental teaching smoothly, we should first change teachers' and students' concept in teaching process of business administration, emphasize the combination of experimental teaching and theory teaching, especially emphasize the development of students' practical ability and innovation ability. Second, we should reform the experimental teaching method, strengthen cultivating ability. We can take various methods based on the experimental teaching level. Finally, we should formulate experimental examine standard that measure the students of business administration to insure detailed and quantities examine way.

ACKNOWLEDGMENTS

This work was financially supported by the teaching research project of Xian University of Technology.

REFERENCES

[1] Gao Guanghua & Li Xinqiang (2009). The understanding and thinking of experimental teaching reform and talents training in college. Journal of Jiangnan University, 9(3):252–253.

[2] Dai Qiang (2010). Research on the application of germany "learning field" curriculum projects in higher vocational education. Higher education forum, (7):74–76.

[3] Jiang Linhua (2011). Research on reforming the experimental teaching system of business administration. Economic research guide, (18):264–265.

[4] Zhang Zhenji & Liu Shifeng & Chang Dan & Lin Zikui (2009). Research on the experiment teaching system of the economics and management. Experimental technology and management, (4):179–181.

[5] Zhang Qiao (2011). Research on the experimental teaching system of business administration major. Journal of Qiongzhou University, (1):27–29.

Education Management and Management Science – Zheng (Ed.)
© 2015 Taylor & Francis Group, London, ISBN 978-1-138-02663-6

A new management model of university scientific research funds based on the BSC

Y. Sun

Financial Department, Leshan Normal University, Leshan, Sichuan Province, China

ABSTRACT: The higher education is developing rapidly in China, while the corresponding management for it is not synchronized to improve, especially in the management of scientific research funds; it has significantly lagged behind the growth in research funding. In order to strengthen the management of college and university scientific research funding, improve the efficiency of funds to enhance the comprehensive strength of university research, BSC (Balanced Score Card) is proposed in this paper to evaluate the performance of these funds.

Keywords: BSC; research funds; performance management

1 INTRODUCTION

With increasing Chinese investment in higher education, college and university education is flourishing; the technological innovation capability is further enhancing and the research projects and funds are growing fast year after year, moreover, the proportion of research funds in the total revenue in college and university is increasing every year. However, the university authorities who should be responsible for the research funds don't pay great attention to the management, which leads to the poor efficiency, even waste of the research funds.

2 PROBLEMS IN THE MANAGEMENT OF RESEARCH FUNDS

2.1 *It is not clear that who should be responsible for the management of scientific research funds, research expenditures is random*

Currently, college and university adopt the management of responsibility for the research funds, which means that the research department, finance department and the head for the research project should take their role in the management for the funds respectively. The research department supervises the project primarily; the finance department is in charge of the financial management of funds and daily reimbursement; while the project leader is mainly responsible for the specific research, development and expenditures. However, in practice, in the one hand, the research department doesn't provide the financial department with the full range of project information and expenditure

of funds, doesn't track and evaluate the use of the research funds and wrongly considers the financial management of funds should be done by the financial department; on the other hand, the financial sector is unfamiliar with scientific research project, they mainly care about whether the reimbursement ticket is legal and the head for the project performs the appropriate funding approval procedures, rather than whether the expenditures meet the need of the research project. They have no rights and reasons to refuse reimbursement as long as related financial formalities completed; as for the leaders for the research projects, many of them wrongly believe that research funding is their own labor income, as long as the research project is completed and the project schedule knot, they can use the funds randomly and pay no attention to the efficiency of the funds. In this management style, on the surface, the financial sector is responsible for the management of scientific research funds, while the reality is there is no substantive administrative supervision for the funds. The head for the project can use the funds as long as he wants, which causes the serious loss of it.

2.2 *The accounting in the scientific research funds and education expenditure confused, it is difficult to account the scientific research costs*

It is common in the colleges and universities that the research funds are classified into vertical and horizontal funds according to their sources and channels obtained by the project leaders. While in the actual practice, the financial sector conducts accounting according to the project name instead of the horizontal or vertical research funds. It is useful to ensure that such accounts do not exceed the limits of project budget expenditure, but the scope of research funding and proportion of spending is unlimited, which is likely to cause some of the university expenditure items such as travel charge or entertainment expenses unrealistically high. In addition, it is difficult for the college and university to separate the teaching expenditures from the scientific research costs, the majority of them only extract a small amount of research management fees, the rest of the research funds can be used by the project leaders freely as long as they finished related financial formalities. As to the resource usage fee such as housing fee, utilities, laboratory equipment depreciation costs, it is not included in the research funds, which further crowding out spending for the education on a tight budget. Moreover, the college and university administrative control the total amount of research funding, the indirect costs such as resource usage fee arising from research activities are not calculated in the research expenditures, which makes the heads for the projects and researchers mistakenly believe that the project management fees have been extracted, the remaining project funds should belong to themselves, they even use false invoices to take research funding reimbursement until the project funding runs out. This management method by extracting management fees by a certain percentage to provide reimbursement, regardless of the nature, content and direction of the projects, is resulting in a distortion of scientific research costs, to make matters worse; it affects the efficiency analysis of scientific research funds.

2.3 The assets funded by the research projects suffered loss, the efficiency of it is low

As the project leaders have independent right to use their own funding, which provides for scientific funding institutional vacuum, they can purchase any of the assets in accordance with the project budget, although these assets are nominally state-owned, in fact, many of them have become the private property of project members and can be disposed freely, which is resulting in serious loss of the state-owned assets. In addition, college and university asset management department lack of effective asset management information platform, which may provide the project leaders with the detailed catalogue of assets. Thus some assets should be shared by the other projects. The project leaders buy assets according to the need of their respective projects regardless of whether they should be, which is resulting in purchasing assets repeatedly, some assets idled away even at the beginning, this leads to inefficient use of state-owned assets or even serious waste objectively.

As it can be observed from the above-mentioned analysis, the current traditional scientific research funds management has a serious shortcoming, which cannot promote the effective management of scientific research funds. Therefore, it is necessary to introduce a new management model that can implement the all-round management of scientific research funds.

3 SIGNIFICANCE OF THE SCIENTIFIC RESEARCH FUNDS PERFORMANCE MANAGEMENT BASED ON BSC

BSC (The Balanced Score Card) was proposed by Professor Robert Kaplan at Harvard Business School and Executive David Norton from Nolan Norton Institute, which is a new corporate comprehensive assessment system. It puts vision and strategy at the core, evaluates the corporate performance from the four dimensions such as finance, customer, internal business processes and learning and growth, and sets appropriate indicator systems to achieve the goal. BSC applies not only to companies, but also to the non-profit sectors such as colleges and universities. BSC, as a strategic management tool, was used to evaluate college and university scientific research funds from four dimensions such as funds management, service objects, project management and project staff, it will reform the current project control as the central management mode, and avoid many problems that exists in the current management of scientific research funds.

3.1 Help to improve the efficiency of scientific research funds

Traditionally, the management on scientific research funds is separated from the projects artificially, which has been resulted in the management vacuum for it. The BSC, emphasizing the goals and process management, will make target specific, process standardization, and enable researchers to clearly know the use of funds while undergoing scientific research at the same time. In this way, we can avoid the funds being idle or be used for other purposes, to ensure that scientific research funds were earmarked to improve efficiency further.

3.2 Help to enhance the strength of college and university scientific research

The key to enhance the level of college and university scientific research lies in the competitiveness of the inner core, which depends on the core competitiveness and creativity of teachers. How to motivate them to participate in scientific research with enthusiasm, a good set of essential funding evaluation criteria is vital. BSC that evaluates the performance of scientific research funds from four aspects of funds management, service objects, project management and project staff, can effectively avoid the short-term management of funds, for examples, pays attention to the total amount of scientific research funds, and ignores the quality of the project. It can create a valid evaluation system for the improvement of scientific research strength.

4 ESTABLISHMENT OF EVALUATION SYSTEM OF SCIENTIFIC RESEARCH FUNDS BASED ON BSC

We should take full consideration of various circumstances in designing performance indicators, because each college and university varies in scientific research objective, funds resource, spending structure and size. Let's take a teaching and research university for example to give the performance evaluation system of scientific research funds based on BSC, as shown in Table 1. Of course, this evaluation system will continue to adjust and improve according to the actual situation.

4.1 Scientific research funds management

It is well known that college and university which are funded by the state don't pursue the maximization of economic benefits as its ultimate goal in the management of scientific research funds. Due to the nature of the commonweal of public college and university in China, they can obtain scientific research funds from the government as well as the community; in this case, they care about whether they can obtain adequate scientific research funds from the government, enterprise and research

institution, they also pay a great attention to the efficiency in the use of these funds, which is an important part of the financial aspects. In order to improve scientific research work, in the one hand, it is necessary for the college and university to increase scientific research funds by all means; on the other hand, they should try their best to eliminate wasteful expenditure. The main evaluation indicators are as follows: scientific research funds income, scientific research funds sources and increase, expenditure details of scientific research funds and project achievements conversion rate. It should be noted that the increase in scientific research funds be calculated on the annual basis, moreover, vertical and horizontal funds analyzed separated; as to the expenditure detail, it should be classified by the individual project which is based on the accounting; it should not be difficult for the research department to calculate the conversion rate according to the total projects. By these indicators, the college and university can have a more objective understanding of the amount and efficiency of these funds.

4.2 Service object

The college and university are mainly serving the students, the fund providers and the society, which is different from the enterprises. The college and university are carrying out scientific research to satisfy the needs of the fund providers and the community. For the fund providers, they care about whether the college and university fulfill the scientific research requirements on schedule, of course, they pay special attention to the quality of the finished projects. For the society as a whole, they are concerned about whether the college and university combine the scientific research with the actual situation, cultivate a sense of innovation and pioneering consciousness to promote social progress,

Table 1. Evaluation indicators system of scientific research funds based on the BSC.

Scientific research funds management strategic objectives: increase the total funds; improve the efficiency of funds; enhance the level of college and university scientific research; promote scientific and technological progress and social development	Dimensions	Main evaluation indicators
	Funds management (finance)	Scientific research funds income; scientific research funds sources and increase; expenditure details of scientific research funds; project achievements conversion rate
	Service objects (clients)	Satisfaction with funds provider; satisfaction with higher authorities; satisfaction with local government
	Project management (internal business processes)	Scientific research progress control; level of the scientific research projects; the number and quality of scientific research papers; monographs and patents; awards for achievement in scientific research projects
	Project staff (learning and growth)	Satisfaction with project staff; academic exchanges; promotion in academic degree and professional title; team building and personnel training; satisfaction with the students; level of students in scientific research

increase qualified personnel and improve the level of local science and technology. The main evaluation indicators are as follows: satisfaction with fund providers, satisfaction with higher authorities and satisfaction with local government. Among them, the indicator for satisfaction with fund providers could be obtained through the questionnaire, then, for the satisfaction with higher authorities, it should be achieved by the research capacity evaluation of the college and university, as to the satisfaction with the local government, it may be a good method to evaluate how the technological development contributes to the local economy.

4.3 Project management

The third dimension of BSC, internal business processes, is playing an important part in the management of scientific research funds and the achievement of the strategic objectives in college and university. Scientific research funding management involves researchers and teaching administrative departments, the former can enhance the scientific research level of the college and university, while the latter can provide funding and logistical support for researchers. Whether the project management is reasonable and effective depends on the progress of the researchers and teaching and administrative departments have made. The specific indicators include: scientific research progress control, level of the scientific research projects, the number and quality of scientific research papers, monographs and patents, and awards for achievement in scientific research projects. Among them, the indicator for scientific research progress control can follow the schedule of the project declaration to supervise project staff to carry out the project as planned and use scientific research funds rationally; as to the indicators for level of the scientific research projects, the number and quality of scientific research papers, they can be obtained from the scientific research management system in the college and university, of course, data from which should be calculated and analyzed carefully; awards for achievement in scientific research should be classified according to prize level as to reflect the scientific research strength objectively.

4.4 Project staff

Learning and growth are the fourth dimension of the BSC. College and university are trying their best to strengthen the management of scientific research funds, with the ultimate aim to promote the construction of teaching staff, enhancement of innovative team and development of social services. To achieve these objectives, it is essential for the college and university to strengthen the training of teachers,

staff and students, encourage them to learn actively, strengthen exchanges and establish an aggressive team with cooperative spirit. The specific indicators are as follows: satisfaction with project staff, academic exchanges, promotion in academic degree and professional title, team building and personnel training, satisfaction with the students, and level of students in scientific research. Among them, the indicators for satisfaction with project staff should be obtained through qualitative analysis, while for the other indicators; the data in demand should be collected from the personnel and other relevant department and achieved through quantitative analysis.

5 CONCLUSIONS

Scientific research is one of the social functions of the college and university's commitment, in order to strengthen the management of scientific research funds to enhance the scientific research strength of college and university, this paper puts forward the performance evaluation management based on the BSC, which consists of four dimension such as funds management, service objects, project management and project staff. Every dimension was designed by specific indicators so that all aspects of the management of funds could be closely linked with the funds management objectives, which can make the target of scientific research specific. In this way, the efficiency in the use of scientific research funds could be greatly improved; scientific research capabilities in the college and university could be enhanced further.

ACKNOWLEDGMENT

This paper is a result of Sichuan Provincial philosophy and social science research base—Sichuan Tourism Development and Research Center research topic (LYC13-25).

REFERENCES

[1] Ministry of Education of the People's Republic of China & Ministry of finance of the People's Republic of China. 2005. Opinions on further strengthening the management of college and university scientific research funds (Jiao Cai [2005]11).
[2] Wang, Haihong. 2013. Discussion on the information management of scientific research funds in colleges and universities. *Friends of accounting* (10):119–121.
[3] Wu Zhigang & Jiang Junpeng. 2012. Thinking about improving the efficiency of the scientific research funds in colleges and universities. *Agricultural science and technology management* 31(5):41–43.
[4] Robert, S. Kaplan & David, P. Norton. 1996. *The Balanced Score Card: Translating Strategy into Action*. Boston: Harvard Business School Press.

Education Management and Management Science – Zheng (Ed.)
© 2015 Taylor & Francis Group, London, ISBN 978-1-138-02663-6

On risk evaluation index system of carbon finance in commercial banks in China

Wenjuan Pan
Netherlands School of Economics, Northwest University for Nationalities, Lanzhou, P.R. China

ABSTRACT: In this paper, the author analyzes four dimensional evaluation index systems for comprehensive evaluation on risk of carbon finance business in China. The analysis was conducted by means of an Analytical Hierarchy Process (AHP)-based approach that builds a system analysis technique for multiple criteria decision-making and by market risk, credit risk, operational risk and project risk. The result shows that evaluation index system based on AHP algorithm for comprehensive evaluation on risk of carbon finance in China has good reliability.

Keywords: carbon finance; commercial banks; risks; AHP

1 INTRODUCTION

In order to mitigate the impacts of the subprime mortgage crisis in 2007, the world must try to find new methods for recovering the economy in coming stage. The appearance of low-carbon economic just satisfies the need of these countries. It has already become the important chance to achieve the goal of sustainable development. After the promulgation of "Kyoto Protocol", the demand for carbon emission right fostered greatly and the carbon market enlarged considerable. The carbon emission right derivates to a financial asset which is worth to invest.

In China, carbon finance has promising developing perspective. With huge demand and potential profit, diverse financial institutions could take part in different intermediate business. Based on the statistics by the World Bank, during the years of 2006, 2007 and 2008, clean development mechanism in China has increased gradually to 54%, 73% and 84%, which are higher than the level of developing countries by a great extent. If enterprises in China are unable to follow the trend of international finance, change business strategy to develop carbon business, the commercial banks will lose the opportunity to become one member of the standardized international banking fields as well as binding up its competitive power towards rivals. China is one of the biggest countries which elisions greenhouse gases. It determined that the potential market of carbon business is huge. According to above-mentioned analysis, it is very necessary to prepare for launching carbon business strategy.

Since 2007, many countries started to intensify the supervision for financial institutions as well as the self-realizing of the importance of avoiding risks. However, commercial banks, the main source of fund supply, are difficult to avoid risks when be involved in green credit business of gaining benefit. As innovation, carbon business is relating to institutional risks with the unpredictable elements and unstable policies. When encountering policy and market risks, the priority of inner control is the theory that should be established at first. Therefore, many commercial banks are launching carbon business and paying great attention to risk controlling simultaneously. For the commercial banks in China, the shortage of related qualified personnel and stable policies, carbon business is still a business with high risk and uncertain return. Without effective managements and regulations, the economic losses which are bringing by the potential risk for commercial banks, even to the whole macro economic in China, is beyond imagination. The rest of this paper is organized as follows: section 2 describes the methodology and Analytical Hierarchy Process (AHP)-based model. Section 3 introduces basic principles and structure of index system on risk of carbon finance in China. Evaluation process and empirical analysis are discussed in section 4. Finally, section 5 concludes the paper.

2 METHODOLOGY AND MODEL

AHP is a system analysis technique for multiple criteria decision-making developed by Saaty in the early 1970s. The method was a multiple decision-making method that combined qualitative with quantitative analysis. It could effectively analyze the non-sequential relationship between levels of target criteria system and effectively measure the judgment and comparison of decision-makers. The procedure of the comprehensive evaluation index systems on risk of carbon finance business in China with AHP method is as follows.

2.1 Establishing the index system

2.2 Constructing the judgment matrix and single permutation of layer

The indexes of the same level are compared one by one to create judgment matrix. For example, if the elements of layer A have relation to the elements of lower layers $B_1, B_2, ..., B_n$, the judgment matrix is shown in Table 1.

In the above matrix, b_{ij} shows the significance extent of corresponding B_i and B_j, which takes 1, 2, 3 ..., 9 and their reciprocals as its value. When b_{ij} equals to 1, it shows that B_i and B_j are same significance; when b_{ij} equals to 3, it shows that B_i is a little more significant than B_j; when b_{ij} equals to 5, it shows that B_i is more significant than B_j; when b_{ij} equals to 7, it shows that B_i is much more significant than B_j; when b_{ij} equals to 9, it shows that B_i is extremely more significant than B_j; when b_{ij} is 2, 4, 6 or 8, it shows that the significance is between above adjacent values; and the corresponding reciprocal shows the insignificance extent.

2.3 Figuring out the eigenvector

Eigenvector (W_i) shows the significance extent of each element in the same layer of judgment matrix, and the biggest eigenvalue λ_{max} is given as follows:

$$\lambda_{max} = \sum_{i=1} (AW_i)/nW_i \tag{1}$$

2.4 Making statistical test about uniformity

The value of CI is an index evaluating the departure of judgment matrix from uniformity:

$$CI = (\lambda_{max} - n)/(n-1) \tag{2}$$

Table 1. Judgment matrix of the significant extent of the indexes in layer B.

A	B_1	B_2	...	B_n	W_i
B_1	B_{11}	B_{12}	...	B_{1n}	W_1
B_2	B_{21}	B_{22}	...	B_{2n}	W_2
⋮	⋮	⋮	⋮	⋮	⋮
B_n	B_{n1}	B_{n2}	...	B_{nn}	W_n

RI is the index of average random uniformity that judges uniformity (the value can be gotten in the statistics table). The proportion of the random uniformity is CR, which can be calculated by:

$$CR = CI/RI \tag{3}$$

If $CR < 0.1$, it shows that judgment matrix has an uniformity. Otherwise, the judgment matrix must be adjusted.

2.5 Sequencing the all layers

This step is to sort the values of relative importance of the layer C.

2.6 Grading of single element and comprehensive evaluation

Single element grading is to disposal single element as a standardized value through dividing each grade by its accumulation. The comprehensive evaluation (P_i) is to multiply the elevated value of single element multiplied by its weight, then to sum those values to get P_i, and to compare the results P_i, and finally to confirm their grades.

$$P_i = \sum_{j=1}^{n} C_{ij} \times E_{ij} \tag{4}$$

In the above expression, C_{ij} is the standardized data; E_{ij} is weight (W_i) in the aim layer j, n is the number of the aimed layer.

3 ESTABLISHMENT OF THE INDEX SYSTEM

3.1 Basic principles for an index system

In order to evaluate the index on risk of carbon finance in China, an appropriate index system is needed to achieve the complicated objection. The index system should be diversity and multi-dimensional. The indexes must also have mutual constraint and influence each other, and conform to the following principles: considering the completeness, independence and principal component of indicators comprehensively; selecting the number of indicators according to the evaluation purpose and accuracy; indicator screening methods should reflect the evaluation objects and evaluation purpose well; and indicator screening methods should be concise and simple and not complex algorithms.

3.2 Structure the evaluation index system

The evaluation of index system on risk of carbon finance should be established in the framework

Table 2. Evaluation index system (layer A–C).

Aim layer A	Guideline layer B	Sub-guideline layer C
The comprehensive evaluation index systems on risk of carbon finance business in China	B1: market risk	C1: interest rate risk C2: exchange rate risk C3: CERs delivered price risk
	B2: credit risk	C4: contractual capacity on buyer C5: whether CDM projects generates the CERs risk under the contract
	B3: operational risk	C6: risk of personnel quality C7: technology risk C8: system defects risk
	B4: project risk	C9: cyclical risk of CDM project development C10: domestic risk of examination and approval of CDM project C11: methodology risk for examination and approval of CDM project C12: verification risks C13: registration risks

Table 3. Weights and values of all evaluation indexes.

Layer C	Eigenvector Wi	Weight Ei	Value Vi
C1	0.733	0.031	4.1
C2	0.327	0.073	2.9
C3	0.432	0.081	2.9
C4	0.958	0.232	2.8
C5	0.417	0.005	4.1
C6	0.811	0.03	7.9
C7	0.547	0.032	5.3
C8	0.608	0.125	7.8
C9	0.085	0.231	4.6
C10	0.711	0.098	4.7
C11	0.242	0.012	7.4
C12	0.512	0.007	6.9
C13	0.435	0.004	4.3

of four-dimensional indicators. The market risk refers to the risk of assets loss in commercial bank influenced by market factors of carbon finance, mainly including interest rate risk, exchange rate risk, and CERs delivered price risk; the credit risk refers to the risks resulted from failure to fulfilling obligations or change of credit quality of counterparty in the process of investments on CDM projects in which the commercial banks engaged the carbon finance activities, including the risk of non-performing loans ratio of CDM project, contractual capacity on buyer, (real purchases of CERs/committed purchases of CERs), the risk of whether CDM projects put into operation on schedule and whether CDM projects generate the CERs risk under the contract; the operational risk: risk of personnel quality, internal process risk, system defects risk, technology risk, external

factor influence risk; the project risk refers to the risks despite the credit risk when the commercial banks invests in CDM projects, including cyclical risk of CDM project development, risk domestic of examination and approval of CDM project, methodology risk for examination and approval of CDM project, verification risks, registration risks, risks of international climate negotiations and force majeure risks, and so on. This paper selects 13 indicators as the preliminary evaluation index system, which is composed of the Aim layer (A), the Guideline layer (B), the field layer (C). The indexes in each layer are listed in Tables 2 and 3.

4 EVALUATION PROCESS AND EMPIRICAL ANALYSIS

4.1 Evaluation process

On the basis of the evaluation index system on risk of carbon finance in China, the evaluation indexes are divided into three layers according to AHP, the relations among layers and factors are lined out and comparing judgment matrixes, which are composed of the ratios of comparative importance of every pair of factors, and are determined by using experts method. For layer A, the matrix is as follows:

$$A = \begin{bmatrix} 1 & 5 & 6 & 1/4 \\ 1/5 & 1 & 3 & 1/7 \\ 1/6 & 1/3 & 1 & 1/8 \\ 4 & 7 & 8 & 1 \end{bmatrix}$$

By calculating the values of the relative weight, we get the eigenvector $W_A = (0.147, 0.072,$

0.083, 0.355), for the aim layer A, the maximum eigenvalue $\lambda_{max} = 4.2605$, CI = 0.087, RI = 0.89, and the relevant CR = 0.0976 < 0.1, thus high acceptability in the consistency of the judgment matrix. And for Guideline layer B, by using expert method, we can also get the relative matrix and the eigenvector. For layer B, $W_{B1} = (0.643, 0.074, 0.283)$, $W_{B2} = (0.750, 0.250)$, $W_{B3} = (0.667, 0.333)$, and comparing judgment matrixes are as follows:

$$B_1 = \begin{bmatrix} 1 & 1/3 & 1/6 \\ 3 & 1 & 1/5 \\ 6 & 5 & 1 \end{bmatrix} \quad B_2 = \begin{bmatrix} 1 & 6 \\ 1/6 & 1 \end{bmatrix}$$

$$B_3 = \begin{bmatrix} 1 & 7 & 6 \\ 1/7 & 1 & 1/2 \\ 1/6 & 2 & 1 \end{bmatrix} \quad B_4 = \begin{bmatrix} 1 & 1/3 & 1/6 & 1/8 & 1/7 \\ 3 & 1 & 1/4 & 1/7 & 1/6 \\ 6 & 4 & 1 & 1/6 & 1/5 \\ 8 & 7 & 6 & 1 & 2 \\ 7 & 6 & 5 & 1/2 & 1 \end{bmatrix}$$

The relevant indicators of the above matrices are $\lambda_{max} = 3.0940$, CI = 0.047, RI = 0.052, and the relevant CR = 0.0904 < 0.1; $\lambda_{max} = 2.0000$, CI = 0, RI = 0, and the relevant CR = 0 < 0.1; $\lambda_{max} = 3.0324$, CI = 0.0162, RI = 0.52, and the relevant CR = 0.0311 < 0.1; $\lambda_{max} = 5.4313$, CI = 0.0108, RI = 1.12, and the relevant CR = 0.0963 < 0.1, respectively, and suggest that the evaluation system has a satisfactory consistency.

4.2 Empirical analysis

According to the comparing judgment matrixes, the values of the relative weight are calculated, and the coefficients of values of the relative weight of all indexes are also computed layer by layer. After getting the relative weight of each layer, the relative weight of each evaluation index is obtained. Because the evaluation system is very complicated and some of the indexes are difficult or impossible to be quantified, expert method is used to decide the value of grade (V_i). Each index is divided into 5 grades from 1 to 5. The higher the value is, the better the situation of index is. Twenty experts are chosen to give the value of each index and weighted average is calculated to determine the value. The weight of each layer and the values of evaluated indexes are listed in Table 1.

The extent of evaluation indexes is calculated by the following equation:

$$G = \sum_{i=1}^{13} E_i \times V_i \quad (5)$$

The extent of integration is assessed: from 0 to 1, the integration degree is very bad; from 1 to 2, the degree is not good; from 2 to 3, the degree is normal; from 3 to 4, the degree is good; from 4 to 5, the degree is extremely good.

According to equation 5 and the data in Table 2, we get the following evaluation degree:

$$G = (E_1, E_2, E_3, ..., E_{13}) \times (V_1, V_2, V_3, ..., V_{13})^T$$
$$= 3.895$$

The result shows that the integration situation for comprehensive evaluation index systems on risk of carbon finance business in China is normal and in good reliability.

5 CONCLUSIONS

An index evaluation system for comprehensive evaluation on risk of carbon finance Business in China is constructed by AHP Algorithm. The ideas in this paper are unique and different from those in the literature. Although it should be noted that the evaluation model is mainly based on experts' numerical scoring through a series of mathematical methods with limited accuracy, the results are still helpful in constituting reasonable policy in the process of evaluation on risk of carbon finance business in China.

ACKNOWLEDGMENTS

This work was supported by the Fundamental Research Funds for the Central Universities of Northwest University for Nationalities (grant number 31920140026).

REFERENCES

[1] L.R. Yan, et al. 2009 "Boosting Low-Carbon Economic Development of the Green Financial Innovations," Fujian Finance, No. 12, 2009, pp. 4–8.
[2] Miao XiaoYu, Carbon finance risk analysis and the response of Chinese commercial Banks. Journal of rural financial research, 2010,(9):12–16.
[3] P. Krugman & J.A. Venables, 1995 "Globalization and the Inequality of Nations," The Quarterly Journal of Economics, Vol. 110, No. 4, 1995, pp. 857–880.
[4] Qin Xueping, "The Development of China's Carbon Finance and Carbon Finance Mechanism Innovation Strategy," Shanghai Finance, No. 10, 2010, pp. 23–29.
[5] R.G. King & R. Levine, 1993 "Finance and Growth: Schumpeter Might Be Right," The Quarterly Journal of Economics, Vol. 108, No. 3, 1993, pp. 717–737.
[6] V. Bencivenga & B. Smith, 1991 "Financial Intermediation and Endogenous Growth," Journal of Review of Economics Studies, Vol. 58, No. 2, 1991, pp. 195–209.
[7] W.F. Ren, "Low-Carbon Economy and Environment of Financial Innovation," Shanghai Economic Research, No. 3, 2008, pp. 38–42.

Education Management and Management Science – Zheng (Ed.)
© 2015 Taylor & Francis Group, London, ISBN 978-1-138-02663-6

On the translation teaching reform based on Project-Based Learning

Minggui Zou
College of Foreign Languages, North China Institute of Science and Technology, Sanhe, Hebei Province, China

ABSTRACT: Nowadays, the increasingly frequent international communication inspired by the economic globalization and cultural multiplication calls for more highly qualified translation professionals. However, the traditional translation teaching fails to accomplish the goal of training professional translators for too much emphasis is put on the translation theories and skills. Project-Based Learning (PBL), one of the important education theories in 20th century, is applicable to translation teaching for being project-based, practice-oriented and student-centered. This paper tends to explore the translation teaching reform in Chinese colleges and universities based on PBL.

Keywords: translation teaching; reform; PBL (Project-Based Learning)

1 INTRODUCTION

Nowadays, the increasingly frequent international communication inspired by the economic globalization and cultural multiplication calls for more highly qualified translation professionals. As college graduates of English major are presently the main force to do translation work, translation teaching for English majors in colleges and universities plays a rather important role in training translation personnel. However, the traditional translation teaching, as a teacher-oriented teaching model, fails to accomplish the goal of training professional translators for putting too much emphasis on the translation theories and skills and neglecting students' practical ability. As a result, to meet the demands of the information society, it is really necessary and urgent to reform the traditional translation teaching method.

To solve the problems in traditional translation teaching, researchers turn to new models of "imitating real situation" teaching and translation process teaching which will help to improve students' practical translation ability by combining translation teaching with translation markets. Under such circumstances, Project-based Learning (shorted for PBL), which, based on real projects, aims at training professional staff with practical ability, is applicable to translation teaching.

2 ABOUT PROJECT BASED LEARNING

2.1 *Theoretical bases of Project Based Learning*

PBL, one of the important education theories in 20th century and a hot topic for the study of international applied linguistics, is based on the following three education theories: Constructivism Learning, John Dewy's Pragmatic Theory of Education and Jerome S. Bruner's Discovery Learning.

According to Constructivism Learning theory, knowledge is acquired by means of meaning construction of the learner in a certain situation with the help of teacher and learning partners. It emphasizes real situation and cooperative learning.

Developed from American philosophers C.S. Peirce's and W. James' pragmatism, John Dewy's "Pragmatic Theory of Education" advocates learning from doing with experience, activity and problem-solving as the three cores of learning. It suggests that students acquire knowledge by taking part in exploring activities and making products.

Famous American educationist Jerome S. Bruner proposed Discovery Learning. He holds that learning is an activity of students' active acquiring knowledge, which emphasizes the process instead of results. During the process of learning, learners' intuitive thinking and inner motivation play a crucial role.

2.2 *Main features of Project Based Learning*

Developed from the above three theories, PBL is characterized of being project-based, practice-oriented and student-centered.

2.2.1 *Project-based*

First, the teaching content is based on real projects. By completing a real project, learners can accomplish new knowledge construction and professional training by applying learned knowledge into practical usage. The selection of projects should satisfy the following conditions: 1) the project should be closely and directly related to the actual working position and professional activities in enterprises; 2) it should provide students with the opportunity of organizing and arranging their learning behavior on their own; 3) it can help improve students' ability to solve certain problems by making use of newly learned knowledge and skills; 4) it should have specific products to be presented for appraisal.

2.2.2 *Practice-oriented*

Second, being practice-oriented is another characteristic of PBL. The ultimate goal of PBL is to help students improve their occupational competence including concerned professional knowledge, practical skills, and professional quality. Different from the traditional teaching model that neglects the importance of students' practical ability, the new model can guarantee more chances to students to practice the learnt knowledge and skills by accomplishing a real project. All the problems and obstacles occurred during the process of practice should be solved by their own efforts. By this means, students' ability of finding and solving problems is greatly promoted.

2.2.3 *Student-centered*

Finally, PBL is a student-centered teaching model. In the traditional teaching model, the teacher is the center of teaching activity with students' passively absorbing knowledge by means of the former's lecture in the classroom. While PBL teaching model requires the teacher to shift his role from an instructor to an organizer and assistant. Instead of indoctrinating knowledge and instructing students, he is required to organize activities for students and assist them accomplish their tasks. Under such circumstance, students will gain more access to practical operation and thus acquire skills more efficiently. In this sense, they become the subject and center of learning.

3 THE NECESSITY FOR THE APPLICATION OF PBL TO TRANSLATION TEACHING

3.1 *Demand from the translation market for professional translators*

With the unprecedented amounts of international communication, there appears an urgent need for a larger number of professional translation staff. According to the requirements of the translation market, in addition to a solid bilingual foundation translators should also have a good mastery of professional knowledge of other fields as well as

a sound interpersonal communication capability. First, the requirement from the market defines the discourse type of translation teaching. Nowadays, the translation work is not restricted to literary world, more and more translators are need in the fields of economy, law, science and technology. The traditional translation teaching, just focusing on literary translation, will certainly fail to satisfy the need of the market. Second, a professional translator also needs to possess good interpersonal communication capability that will help him to obtain useful translation resources, improve translation quality and work efficiency as well. However, the traditional translation class can not afford chances to students for communication with co-workers, customs, professionals and target readers. In a word, the application of PBL is the objective requirement of the translation market.

3.2 *The development tendency of studies on translation teaching*

With the introduction of advanced education theory and foreign language teaching ideology, studies on translation teaching have undergone tremendous changes in terms of education philosophy and psychology. The traditional translation teaching, a typical teacher-centered teaching model, follows the teaching steps of "translation exercise—presenting suggested translation—explaining translation skills and techniques", which will greatly eliminate students' learning enthusiasm and initiative with few chances for participation in the teaching activities, thus result in unsatisfactory teaching efficiency. To renew the teaching concepts, translation teaching researchers in Europe turn to "imitating real situation" teaching method and "translation process" teaching method, expecting integrate translation teaching with translation market requirements and students expectation to improve students' translation competence.

The teaching model of "imitating real situation" calls for the construction of real translation project in the class which covers all the factors in the translation situation—translation requester, reviser, target text user, translator, and so on. Such a situation can narrow the gap between classroom translation activity and professional translation practice, thus effectively inspire students' learning motivation, and strengthen their sense of participation and responsibility. Process teaching model stresses the description and explanation of students' translation process. In general, the following methods are adopted in this model: "brainstorming" and "journal writing", "discussion in groups", "communications between teacher and student". By encouraging students to find and solve problems by themselves, this model is helpful for students to improve their comprehensive competence including recognition, strategy-making, professional operation and psychology.

From the above-mentioned development tendencies of translation teaching in European colleges and universities, it is noticeable that the focus of translation teaching studies has shifted from the teacher to the student, from translation product to translation process, and from isolated classroom teaching to market-oriented professional education. These changes have laid theoretical foundation for translation teaching reform in colleges and universities in China.

3.3 *The renew of translation teaching technology*

The informative teaching based on modern information technology, has been a trend in college translation education. By applying Computer Assisted Translation Technology (CAT) including computer word processing, electronic resources and translation software, it can effectively improve the teaching quality.

Since 1980s, translation teaching based on CAT has aroused more and more interests and attention from researchers in translation field and education circle with the appearance and popularity of multimedia technology and the global network. By the year of 2010, about 115 kinds of relevant CAT software had been available. These advanced technologies are beneficial to improve students' skills of utilizing modern information technology to assist translation work, and thus to improve translation quality. The renew of translation teaching technology has provided technological support and assurance for the reform of translation teaching.

4 THE REFORM OF TRANSLATION TEACHING

4.1 *The reform of teaching goals*

The ultimate goal of translation teaching in colleges and universities is to develop qualified professional translators for the translation market. In Europe, it has been the mainstream of translation training and teaching, and in China, such a tendency is also becoming more and more obvious. Professional translation competence does not only refer to the translator's bi-linguistic competence, but also covers his (her) occupational behavior (occupational morality) and ability to utilize translation tools (e.g. dictionaries, internet resources, computer-assisted translation tools, and so on). However, such competence is difficult to be gained in the traditional translation classroom for the main goal of traditional translation teaching is to improve students' linguistic competence.

Therefore, in the context of PBL, the teaching goal needs to be adjusted to train students' comprehensive translation capability and professional quality.

4.2 *The reform of teaching contents*

One of the important factors affecting the efficiency of translation teaching is that the teaching content is far from being scientific and pragmatic. In the traditional translation class, the teaching content involves the introduction of translation theory and explanation of translation skills. Moreover, translation practice focuses on literary works with few pragmatic discourses. In other words, what the student can learn from the translation class can hardly satisfy the requirements of the society. To achieve the goal of training professional translators, the teaching content should also be reformed.

First, appropriate translation materials should be chosen for teaching based on the market demands. To be more specific, more pragmatic materials concerning economy and trade, science and technology, laws, tourism, education, and so on, should be adopted for translation practice because these materials are the major part of translation work in the market.

Second, what should call for more attention in translation teaching is the ways to search and utilize various translation resources including text resources, electronic resources as well as human resource, which can dramatically facilitate translation work.

Third, another important part which should be covered in translation teaching deserving teachers' attention is the professional behavior guidance. At present, students' deficiency in professional quality is mainly accounted for the fact that they have few opportunities scarcely to participate in real translation tasks. Therefore, appropriate instructions on professional behaviors will help them to get a better understanding of the requirements for a professional translator, and thus successfully adapt to the future professional work.

4.3 *The reform of teaching steps*

Traditional translation teaching generally follows the steps of presentation, practice and production (shorted for 3 Ps). To be more specific, the translation class often starts with teacher's presentation of some translation theories and skills, then some exercises are provide for students' translation practice after class, and the last step is the discussion of students' productions. Based on such a mistake-correcting teaching method, translation class is confined to knowledge dissemination without capability training. For neglecting the importance of students' creative thinking and comprehensive capability, it is conducive to the elimination of students' learning initiative and enthusiasm.

In practical translation companies, a completion of a project involves the following steps: accepting translation tasks—signing contracts—allocatingwork—collectingresources—translation—proofreading—delivering translation—translation evaluation. In order to train professional translators, the steps of translation teaching should also be readjusted in accordance with real translation process with real translation projects being introduced into the class. The specific translation steps are as follows.

Step one: identifying translation projects. For the lack of authorized and scientific translation textbooks, the teacher is responsible for assist students to investigate the local translation market to find appropriate real projects suitable for their needs and competence. Step two: pre-translation preparation. To make students achieve a qualified translation, the teacher is expected to help them get well-prepared by searching for relevant background information and translation resources, such as information about translation situation and parallel texts concerning the subject of the project. Step three: arranging and guiding translation activities. As the completion of a project is often based team-work. At the very beginning, students are to be divided into groups to accomplish certain specific translation work, with each individual responsible for a certain part of job according to his own features and needs. During the translation, the students are required to write diaries to describe the process of translation and record problems and solutions occurred in the process. After that, various forms of discussions (e.g. team discussion, class discussion, teacher-students discussion, and so on) are necessary for the proofreading and revision of the translation draft to achieve a quality translation. Step four: the evaluation of translation production. The evaluation of the translation production is to be conducted in the form of teacher's evaluation, students' mutual evaluation, their self-evaluation and customers' evaluation (if possible).

4.4 The reform of evaluation method

In the traditional translation class, the students' efficiency in translation learning can only be judged by means of teacher's evaluation of their exercises and final examination which focus on testing the degree of students' mastery of translation knowledge and skills, so the traditional evaluation can hardly be comprehensive and scientific. Moreover, the subjectivity of the teacher will also exert negative influence on the evaluation.

Fortunately translation teaching based on PBL can improve the situation by reforming the evaluation method. On the one hand, the evaluation will be made not only based on static exercises and tests, but also on the performance in the completion of the translation project, it tends to be more comprehensive for it can make test students' competence in a roundabout way by covering their professional translation skills, cooperative and communicative ability, as well as the capability to find and solve problems. On the other hand, as mentioned in the previous section, the evaluation is not made by the teacher himself, students and customers can also function as the subject of evaluation, which can certainly make the evaluation more objective and reasonable.

5 CONCLUSIONS

Based on the previous studies and the study made in this paper, main findings can be summarized as follows. First, with the rapid development of the global economy and modern information society, the present translation teaching fails to satisfy the requirements of the translation market. Second, being project-based, practice-oriented and student-centered, the new teaching model of PBL is applicable to translation teaching to achieve the goal of training professional translators. Third, based on PBL, the reform of the traditional teaching should be conducted from the aspects of teaching goal, teaching contents, teaching steps and evaluation method.

ACKNOWLEDGMENTS

This work was supported by the Educational Research Project of North China Institute of Science and Technology "A Study on the Application of Project-based Learning to College Translation Teaching" (number HKJYZD201334).

REFERENCES

[1] Dewy, J. 1994. The school and society. Beijing: People's Education Press.
[2] Liu, J.F. & Zhong, Z.X. 2002. A study on the model of project-based learning. Studies in Foreign Education 29(11): 18–22.
[3] Thomas, J.W. Mergendoller, J.R. & Michaelson, A. 1999. Project-based Learning: A handbook for middle and high school teachers. Novato, CA: The Buck Institute for Education.
[4] Yu, J.S. & Wang, H.S. 2010. A master program in computer aided translation. Chinese Translators Journal (3): 38–42.

Education Management and Management Science – Zheng (Ed.)
© 2015 Taylor & Francis Group, London, ISBN 978-1-138-02663-6

Research and practice on double-tutor system in undergraduate cultivation for Food Science & Engineering

G.R. Sun, F.G. Du, S. Liu, Y.L. Zhang, G.Q. Jiang, X.X. Gao, D.R. Mao, K. Chang & P. Xv
Forestry College of Beihua University, Jilin, China

ABSTRACT: The teaching reforms were taken for double-tutor system to increase abilities both in practice and theory. The pilot group was randomly selected from specialty of Food Science & Engineering and the reforms were focused on graduate thesis and courses practice. The results showed that the pilot group had given the students stronger abilities to invent new food products, have more opportunities to recognize problems then try to solve these problems and have more opportunities to further studies and employment.

Keywords: double-tutor system; undergraduate cultivation; food science & engineering; teaching reform

1 INTRODUCTION

1.1 Double-tutor system in undergraduate cultivation

The reasons that the current tutorial system does not produce significant effects include the budgets and policies of schools, the greater independence of students, and the infrequent interaction of tutors with students (Chiu, 2008). Tutor system is an effective teaching model undergraduate cultivation (D'Mello, 2012). Double-tutor system (Yang, 2010) in undergraduate cultivation refers to the exploitation for new teaching models by the effective combination with teaching systems of theory and practice to cultivate the high quality and high level undergraduate for the developing society. Traditional double-tutor system needed two tutors, one is responsible for daily life and the other for professional instruction.

1.2 Double-tutor system in undergraduate cultivation of Food Science & Engineering

Cultivation of students from Specialty of Food Science & Engineering (FSE) needs more practice abilities, the studies of teaching models of FSE had been complained for several years (Liu, 2009). "Double Tutorial System" (DTS) in Food Science & Engineering (DTS) refers to the combined tutor system with tutor group including college professional tutors and company practice tutors composed of responsible system. The tutor from college is the traditional sense of the professional basic course teachers and those from company grasp certain professional knowledge and have rich practical experience. The cores of DTS

are focused on training students by tutors from both college and company. In present situation, the tutor system should be further researched to meet the needs both in practice and theory for training undergraduate students for their multiple abilities. After many years of practice, the framework DTS has achieved good results, but also reflected some inevitable problems, mainly "double lack" means the lack of practice place for student and lack of practical experience for teachers. To solve this problem, the colleges have taken many steps including cooperation with professional company, building practice bases with food enterprises and has brought great convenience to practice teaching, but not fundamentally reverse the "double lack" situation.

1.3 Research goals and significance of double-tutor system in undergraduate cultivation

To solve this "double lack" problem, the DTS teaching model should be further researched. Study on the tutor system has positive role for promoting the integration of undergraduate tutor system and the credit system teaching mode. With the expansion of college enrollment, the tutor system has been unable to meet the teaching needs, innovation mentor system, adopt the DTS for the whole of the student counseling, help students to establish correct professional thinking, forming a full education atmosphere and promote the improvement of teaching quality. The reform of undergraduate tutor system will increase the enterprise joint efforts and the cooperation between college and enterprise collaborative innovation, contribution to training has certain innovation ability first applied talents for local economic construction.

1.4 Research contents of double-tutor system in undergraduate cultivation

To explore effects and problems of the double-tutor system in research practice of undergraduate students in Food Science & Engineering. Methods: We used double-tutor system in six-month research practice of four-year undergraduate students in Food Science & Engineering. The performance of graduation thesis writing and investigation by course practice was used to evaluate the effects.

2 THE TECHING REFORM OF DOUBLE TUTORIAL SYSTEM

2.1 Introduction to object of study

To research on the reform of double-tutor system, two classes of FSE specialty were selected as object. There were 75 students including 36 girls and 39 boys age between 21 and 24. The enrollment of these students in the fall of 2010 came from most area of China, graduating in spring 2014, they have studied in college for four years. Fourteen students were randomly select from above classes, including eight girls and eight boys as a pilot group with mean average of 22.4 years old, mean height of 1.69 meters. Tutors from college and enterprises instructed the pilot group. The tutors from college were responsible for instruction of professional research and enterprise tutors for instruction of practice.

2.2 Practice in undergraduate graduation thesis with DTS

The steps of undergraduate graduation thesis with DTS were separated three continuous periods. The first step was called preparing step, mainly including tutor double selection, choose research area; The second step was called opening step, mainly including project demonstration and opening; The final step was called practice step, mainly including practice in company, analysis on problem in practice and writing graduation thesis.

2.3 Practice in course study with DTS

Practice in course study with DTS included two fields, one is the class teaching by college tutors and the other is practice out of college by enterprise tutors.

3 THE RESULTS AND DISCUSSION

3.1 The results of practice in undergraduate graduation thesis with DTS

The results of practice in graduation thesis with DTS were listed in Table 1.

Table 1. The ratio of simulation and actual project of practice in undergraduate graduation thesis with DTS.

Projects	Pilot group		Control group	
	Results	Ratios	Results	Ratios
Simulation	0	0%	48	79%
Actual	14	100%	13	21%
Products	11	79%	0	0%
Solve problems	3	21%	0	0%

Table 2. The average scores between pilot group and control group.

Courses	Pilot group		Control group	
	Mean	Best	Mean	Best
Practice	89.2	86%	77.3	12%
Theory	81.1	75%	72.1	9%

In Table 1, the "Simulation" refers to the thesis researched on simulation title by college tutors and "Actual" refers to actual project from enterprise. Products means processed products by graduate thesis researches and solve the problems.

From Table 1, we can deduce that the pilot group had higher ratios of actual thesis with 100% than that of control group with 21%. That showed that pilot group had more opportunity to practice and increase their practice abilities.

The products number showed that the pilot group had given the students stronger abilities to invent new food products with 79% to zero. And in the same situation, they had more opportunities to recognize problems and try to solve them.

In sum, the reform in practice with DTS can increase students abilities of processing and designing food products, solving problems.

3.2 Effectives of reforms in course study with DTS

The average scores between pilot group and control group were listed in Table 2.

In Table 2, "Practice" means the courses contained practice and "Theory" means no practice and "mean" refers to average scores of these courses and "best" means best proportion in scores.

The results showed that students in pilot group not only had stronger abilities in practice courses with average scores of 89.2 higher than that of controlled group with 77.3, and more best proportion 86% than 12%, but also have abilities in theoretical courses with mean 81.1 higher than 72.1 best proportion 75% than 9%.

The results showed that students in the practice with DTS can increase their abilities both in practice and theory.

To this practice with DTS, the students with DTS had gained good scores in the courses with practice instructed by double tutors to increase abilities in practice. But to theoretical courses, the students with DTS had the same situation. Perhaps, these students with DTS have gained abilities to solve problems that gave them stronger abilities to study theoretical courses.

In sum, we should take every possible measure in practice to increase practice abilities to increase other abilities of analysis and solving problems. In practice, the instructions of students with DTS are carried out in practice sessions including operation, observation, demonstration and imitation and so on. The most important steps in practice is operation.

3.3 Total effectives of reforms in course study with DTS

The total effectives of reforms in course study with DTS were compared by employment rate and postgraduate rate were showed in Figures 1 & 2.

From Figure 1, the pilot group had more opportunities to further studies with postgraduate rate of 35.7% more than that in controlled group only with postgraduate rate of 8.2%. And from Figure 2, the pilot group had more opportunities of employment with employment rate of 100% more than that in controlled group with employment rate of 87.3%.

Students from pilot group with DYS had increased their abilities in practice to meet the need in accrual work and had more chance to communicate with experts or staff from enterprise to understand the market and enterprise and improve their abilities. As a result, these students had gained higher employment rate. These facts suggest us a

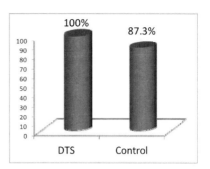

Figure 2. Compared effectives in employment rate.
Note: DTS stands for the group with double-tutor system; control stands for controlled group without double-tutor system.

simple way of improvement of employment rate by increasing abilities in practice and theory.

A question which claims precedence over all others is why the postgraduate rate with DTS is higher than that of controlled group.

First, the students in pilot group with DTS had more opportunities to contact experts from enterprise to discuss with tutors to widen greatly shot, realize intellectual value and try to lean more.

Second, these students needed to spend more time going through all related reference material to solve the problems encountered in practice. These students had to spend more time to study, and in this learning process, they could aware lack of knowledge and seize every chance to learn more, passing post graduate examination was one of their ideas.

Furthermore, these students had received guidance according to individual needs so as to enhance their self-confidence and performance.

Therefore these students in pilot group with DTS had more chance to improve their abilities and increased their self-confidence and performance to meet challenges in their study lives, as a result, they had gained higher postgraduate rate.

Postgraduate rate and employment rate are key index in evaluation of teaching in universities. The reform of double-tutor system can improve students abilities and gain higher postgraduate rate and higher employment rate to improve qualities of talents.

3.4 Reform of teaching models with DTS

Models of teaching with DTS are important measures in improving teaching qualities in college. From the above analysis, we can learn that measures in practice and classroom teaching with TDS can improve abilities, these measures include:

1. Practice with DTS in graduation design (paper) work. Graduation design (paper) work is one in which links and factors interact with each other.

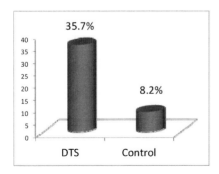

Figure 1. Compared effectives in postgraduate rate.
Note: DTS stands for the group with double-tutor system; control stands for controlled group without double-tutor system.

The teaching model in this step is that three steps were designed as tutor-student mutual selection, opening thesis and the thesis oral defense for undergraduate.

2. Practice with DTS in course practice. This practice was linked with building practice base. In these bases, students can carry out practice unit operation to improve their abilities.

3. Practice with DTS in practical activities. During practice in practice bases or other chance, these students had opportunities to attend practical activities by organized enterprise or college.

4. Practice with DTS in academic activities. To resolve the practical problems from enterprises, tutors from enterprise and college had to meet together to discuss and study measures, students had chance to take part in as members in the same team.

4 CONCLUSIONS

DTS was used in different steps in undergraduate training for improving abilities, these steps included graduation thesis, course practice and other activities.

As a strong practical discipline, the practical teaching occupies a very important role in food processing technology. Each course of the basic courses and professional courses contains the content of practice. Practice teaching plays a very key role to cultivate the student to analyze the question, solves the problems ability and practical ability. Combined with the teaching situation, this paper takes double-tutor model to food processing technology of professional practice teaching system, has obtained the good effect.

In practice of graduation thesis, abilities of students with DTS are improved as exploiting new products, solving practice problems because they had more opportunities to recognize problems and try to solve them. These practices improved their abilities of inventing food products, analyzing and solving problems, actual practice and adapting to work. The reform in practice with DTS can increase abilities of processing products, and solving problems.

In practice of course study with DTS, students with DYS had more opportunities to further studies with higher postgraduate rate and more enterprise to work with higher employment rate. The above facts suggest us a best way of improvement of employment rate by increasing abilities in practice and theory. These students with DTS had more chance to improve their abilities and increased their self-confidence and performance to meet challenges in their study lives, as a result, they had gained higher postgraduate rate.

To better promote the results of our research, models of teaching with DTS were built as four steps: step 1 is practice with DTS in graduation thesis; step 2 is practice with DTS in course practice; step 3 is practice with DTS in practical activities; step 4 is practice with DTS in academic activities.

In this paper, the double-tutor system was used in teaching reform for developing abilities of undergraduate students from Food Science & Engineering. The teaching reform can be spread in other Engineering specialties. As rules to spread, some regulations should be taken into account. First, DTS can be used in any undergraduate specialty for training students' abilities of inventing food products, analyzing and solving problems, actual practice and adapting to work; second, the DTS can be used in courses practice in engineering specialty for developing practice abilities according to the requirements of professional courses or professional basic courses; In addition, DTS requirements building teaching models before implementing in practice; finally, other aspects should also be considered, including mutual selection and mutual trust of teacher and student, design of practice projects, the requirements for abilities of students, and the requirements for students and teachers.

ACKNOWLEDGMENT

This project was financially supported by the Major educational project from Beihua University.

REFERENCES

[1] Arnott E. 2008. Research Methods Tutor: Evaluation of a dialogue-based tutoring system in the classroom. Behavior Research Methods 40(3): 694–698.

[2] Chiu J.M. & Liu W.L. 2008. A Study of the Feasibility of Network Tutorial System in Taiwan. Educational Technology & Society, 11(1): 208–225.

[3] D'Mello, S.; Olney A.; Williams C. & Hays, P. 2012. Gaze tutor: A gaze-reactive intelligent tutoring system. International Journal of Distance Education Technologies 8: 66–80.

[4] Liu, A.W. 2009. Implementing double-tutor system in research practice of undergraduate nurse students. Journal of Hunan University of Arts and Science (Social Science Edition) (4): 130–132.

[5] Rodrigues, J.J.P.C.; João, P.F.N. & Vaidya, B. 1996. Edututor: An intelligent tutor system. International Journal of Human-Computer Studies 70(5): 377–398.

[6] Roll I. Aleven V.; McLare B.M.; Kenneth R. & Koedinger K.R. 2007. Designing for metacognition—applying cognitive tutor principles to the tutoring of help seeking. Metacognition and Learning 2(2–3): 125–140.

[7] Yang, X.; Shao, W.L. & Yu N. 2010. Comparison of the Graduate Tutor System and the Postgraduate Tutor System. Chinese Nursing Management 10(2): 35–37.

Education Management and Management Science – Zheng (Ed.)
© 2015 Taylor & Francis Group, London, ISBN 978-1-138-02663-6

An empirical study on college students' mobile assisted English learning

Dongsheng Wang & Lijie Wang
English Departments, Jilin Medical College, Jilin, China

ABSTRACT: The subjects of the study are 218 Jilin Medical College sophomores. Before designing the experiment, a survey was conducted to investigate students' opinions of mobile assisted language learning. First, the researchers gave experimental group students audio materials, then send experimental group 1students corresponding key words and pronunciations via Fetion everyday. Students sent back their English summaries to the researchers by Fetion, after which the researchers sent listening materials to students. At the very beginning of the experiment, the experimental group 2 students were given paper materials that include key words, pronunciations and listening materials. The control group students learn on their own. After 40 days' trial, the researchers implemented another survey and interviews about the effects of learning, students learning habits, students' opinions of this experiment and so on. It has been shown that students of both experimental group 1 and experiment group 2 made significant progress, but the experimental group 1 made significantly progress in listening, and positive vocabulary. Most of the students had positive attitudes towards mobile learning, and meant to continue with it.

Keywords: college English; mobile assisted English learning; empirical study

1 INTRODUCTION

With the development of mobile technologies and the enhancement of calculation and storage of mobile devices, mobile phones and other handhelds are becoming an indispensable part of our lives. From the advent of the mobile devices, mobile learning has slowly changed our lives. Mobile learning refers to learning mediated via handheld devices and potentially available anytime, anywhere. Such learning may be formal or informal. (Kukulska Hulme, 2008) Nowadays mobile learning is extensively being applied in all subjects and fields and Mobile Assisted Language Learning (MALL) is becoming the attention of research.

2 LITERATURE REVIEW

Research of mobile devices at home and abroad mainly focused on creating mobile language learning software and system. It also concentrated on the application mode of mobile devices in language teaching. Uther et al. designed MAC to help Japanese students to distinguish /r/and /l/. Boticki et al. designed a software to support collaboration learning by sending different learning materials to different students via it and dividing them to dynamic collaboration groups. Li Xiaodong and Zhang Hong designed context-aware mobile English learning software. Stanford University set up a system that can help students to improve Spanish by using the functions of email and voice in mobile phones, which include vocabulary exercises, quizzes, translation and the real time conversations between teachers and students. Ogata et al. have set up LOCH system that can help overseas students learn Japanese in authentic situations. Cavus designed CAMLES system that can identify the location, learning time, learning style and the level of the learners. Chin and Chen collaborated to set up GLSS system that can provide learners ubiquitous learning environment, which combines the resources of real world with the digital world by using the functions of camera, bar code tag recognition and GPS in the mobile learning based on android.

Teacher also make use of the functions of mobile devices to promote English teaching level. Thornton and Houser have used the email function of mobile phones to send vocabulary to Experimental Group Students (EGS) and made a good result. In the UK, Braton University the TAMALLE program used digital interactive TV and mobile phones to help language learning. Zhang Jie explored how to use mobile technologies to help listening and speaking teaching. Waycott and Kukusha collaborated to study the difference between PDA reading and paper reading for students. Anna Coms-Quinn et al. have studied the blog's effects on writing.

These studies often proposed to improve students' single ability, while this paper tries to use mobile devices to improve students' multiple abilities.

3 THEORIES FOR MOBILE LEARNING

Learning theories went through behaviorism, constructivism and situated cognition theories. Behaviorism emphasized "stimulus—reaction", which proposed that knowledge can be transmitted to the learner's brain by repeated displays. Constructivism, derived from cognition processing theories, emphasized the students' subjectivity, and proposed that teaching is not a process of inputting knowledge but a process of stimulating students to construct knowledge actively. Therefore, teachers should be guides, tutors and material suppliers for the students. The researchers of this period not only supplied students with learning materials by using mobile devices, but also encouraged interactions between teachers and students and among students to solve real problems. Situated cognition theory emphasized the importance of outer learning environment to learning, and proposed that learners may have a meaning learning by applying the gained knowledge in the situation. Therefore, teachers should try to create authentic situations for students in education to make them learn how to use the gained knowledge in real life.

4 SUBJECTS AND METHODS

The subjects are 218 sophomores of Jilin Medical College who are taught by the same teacher. Students are pre-tested and divided into Experimental Group (EG) and Control Group (CG), 75 students in EG1, 63 students in EG2 and 80 in CG. The initial grades of EG and CG do not differ obviously. This paper applies the methods of documentary research, questionnaire, interview, comparative analysis to study the difference of mobile English learning and traditional English learning and analyzes the gained data by SPSS 11.0.

5 RESEARCH PROCESS AND ANALYSIS

At first this research did a survey of college students' ownership situation of mobile devices and their interest towards mobile learning. The experiment was designed on its basis. The experiment lasted 40 days and the learning result was surveyed and the data was analyzed and summarized.

5.1 *The questionnaire before the experiment*

We sent out questionnaire to 218 students and got 206 effective ones. The questionnaire consists of four parts: the ownership condition of mobile devices, the function of mobile devices, the application of mobile devices, and the interest toward mobile learning.

Ownership condition: mobile phone 100%; audio player 54.85%; PDA 4.61%

Function of mobile devices: smart phone 75.24%; function phone 24.76%

3G card 55.34%; others 44.60%

Application condition: <1 hour 19.42%; <2 hours 33.98%; >2 hours 46.60%

Mobile English learning: already 77.67%; never 22.33% as dictionary 73.30%; English songs 56.31%; video 22.30%; website browsing 11.17%; others 16.50%

Abilities to be improved: listening 77.67%, oral 52.43%, vocabulary 39.80%, reading 27.18%, translation 27.18%, writing 24.27%

According to the result of the survey we know that mobile phones are the most common devices among students. Smart phones are not so popularized yet. Students spend long time surfing the Internet by using the phones after class. Many students have used mobile devices to learn English, and electronic dictionary function is used most frequently, after that is listening to English songs, watching English videos and relatively few students browse the English website. Students are most eager to improve listening comprehension, next to it is oral English, vocabulary, reading, translation and writing.

5.2 *Experiment process*

As the subjects are sophomores who are facing the pressure of pass College English Test (CET), their learning motivation is relatively high. In college, students can use the English autonomous learning platform on campus to learn English, but during the winter vacation students can not get access to it because of the restriction of network. Therefore, we decide to do the experiment in the winter vacation. Given the fact that it will cost students money to download audio files by mobile phone network, we let EGs to copy the sequential audio files into the mobile devices. The audio files mainly involve VOA special English, modern English poetry, English songs, English stories, and English jokes. We gave EG2 paper materials and the content is the original text of the listening material and the important words of CET Band 4 and Band 6 which exist in the listening material, pronunciation is included. At the same time the mobile phone numbers of EG1 are added in Fetion. In order to encourage students to participate in it, we tell all the students that they will take a test after the vacation, and the result will be added to the next term's grades.

During the experiment we send EG1 the important words with the pronunciation of listening materials by Fetion every morning. After listening to the materials, the students make an English summary and send it to us by Fetion. The next day the original text of the listening materials is sent to the students by Fetion.

5.3 Questionnaire and interview after the experiment

After finishing the experiment, we make a questionnaire and interview for the 75 students of EG1. 71 effective answer papers are collected. The questionnaire paper mainly involves students' mobile learning habits, mobile learning effects and willingness of continue it. The interview mainly involves students' suggestions about mobile learning content and style.

Application of mobile devices: mobile phone 42.25%, mobile phone + computer 38.08% mobile phone + audio video broadcaster 19.72%

Amount of time: <10 minutes 5.63%, 10–20 minutes 15.49%, ≈30 minutes 36.62%, ≈60 minutes 16.90%, >60 minutes 11.27%

Times of listening: 1 time 4.23%, 2 times 28.17%, 3 times 50.70%, 4 times 11.27%, ≥ 5 times 6.33%

Times of modifying: 1 time 32.39%, 2 times 11.27%, several times 56.34

Whether memorizing words: yes 50.70%, no 49.30%

Helpful in English learning: a lot 29.58%, a little 23.94%, common 38.03%, little 8.45%, no 0%

Whether helpful in listening: yes 97.18%, no 2.82%

Whether helpful in writing: yes 78.87%, no 21.13%

Whether helpful in reading: yes 63.38%, no 36.62%

Whether continue mobile learning: yes 92.96%, no 4.23% both 2.81%

According to the above data, we know that the majority of students use mobile phones to study English, some students also combine the mobile phones, computers and other audio video broadcasters. The amount of time for listening varies and 30 minutes or so account for most. Most students listen to the audio files for 3 times and over. Most students modify the summary several times. Most students would memorize the received words, and others only use the words to help listening comprehension. Almost all the students think mobile learning is helpful for English learning and would continue it.

The interview shows that students think that audio files are difficult to understand and it is necessary to lower the difficulty level, and they hope the teacher may send them more questions and content related to CET Band 4 and Band 6. At the same time students hope the teacher send the questions related to the listening content and the styles of questions may be multiple choice, ask-and-answer and others. Most students think that teacher should test students about vocabulary periodically. Students hope that teacher can not only send them audio files and words, but also guide them on listening methods and strategies. Students think this experiment mainly improves their listening, reading, writing abilities, at the same time, their vocabulary is also enlarged. The major problems in the mobile English learning is that there are so many activities during the winter vacation, and sometimes it is hard for them to keep up. The stress symbols of pronunciation can not be displayed on mobile phones. Student can not stop and repeat the difficult parts. The duration of battery is also a problem.

5.4 Data analysis

To know the difference of learning results between EG1, EG2 and CG, we give all the students a test, which includes four parts: vacation listening, extracurricular listening, listening words and reading words. For the vacation listening, we choose parts of the materials in the vacation; for the extracurricular listening, we choose some new materials to know the students' improvement. For the words test, we choose parts of the important CET Band 4 and Band 6 words that exist in the vacation materials to test the students. The word test consists of listening words and reading words. Listening words here refer to the words students can recall the meaning when listening; reading words refer to the words students can recall the meaning when seeing. After the test, we get the analysis of unit factor variance of general grades, vacation listening grades, extracurricular listening grades, reading words grades and listening words grades, here is the result:

	F	P
General grades	34.883	0.000
Vacation listening grades	23.865	0.000
Extracurricular listening grades	3.261	0.040
Listening words grades	134.518	0.000
Reading words grades	10.686	0.000

The graph show that among the three classes that involve the general grades, vacation listening grades, extracurricular listening grades, reading words grades and listening words grades, at least two classes have differences. To further explore which two classes have differences for every aspect, we compare the three classes by every two ones, the result shows there is obvious difference among three groups in general grades ($P < 0.05$). From the specific data, we can observe that the average grade of EG1 is 50.63, of EG2 is 41.46, of CG is 31.90, so EG1 has the best result, next is EG2, and the grade of CG is low. There is obvious difference among the vacation listening grades in three groups ($P < 0.05$). The average grade of EG1 is 13.23, of EG2 is 10.95, of CG is 7.50, so EG1 has the best result, next is EG2, and the grade of CG is low. There is not obvious difference in the extracurricular listening grades between the EG1 and EG2 ($P > 0.05$), but the difference is obvious between the EG and CG ($P < 0.05$). According to the specific data, we can observe the average grade of EG1 is 13.72, of the EG2 is 13.37, of the CG is 10.59, so the average grade of EG1 is a little higher than that of EG2, but the difference is not obvious, and the grade of CG is low. There is obvious difference in grades for listening words ($P < 0.05$). The average grade of EG1 is 9.67, of EG2 is 3.89, of CG is 2.01, so EG1 has the best result, the grade of EG2 is low, and the worst is CG. As for the reading words, there is no obvious difference between EG1 and EG2 ($P > 0.05$), but there is great difference between EG and CG ($P < 0.05$). From the specific data, the average grade of EG1 is 14.01, of EG2 is 13.25, of CG is 11.80, so the grade of EG1 is a little higher than EG2, but the difference is not obvious, and the grade of CG is low.

6 CONCLUSION

The result of the questionnaire shows that the ownership of mobile devices for students is very high and they spend much time surfing the internet by the mobile devices in their daily life and study. Students have much interest in mobile English learning and actively took part in it and would like to continue it. In addition, students hope that the learning materials can closely relate to CET Band 4 and Band 6 and the teacher can send them learning strategies, methods when sending learning materials. They hope there will be more interactions between teacher to students and students to students.

The data show that 40-days mobile English learning has improved the general English level for EG1. Compared with EG2 and CG, the students of EG1 have obviously improved in vacation listening and listening words. Compared with CG, students who took part in mobile English learning have higher grades when testing them

new listening materials, which illustrates mobile learning improves students' listening comprehension, but compared with students who got the paper original materials, this improvement is not obvious. It may be caused by the short time of the experiment. At the same time, students who took part in the mobile learning and got the paper materials have no difference in grasping the reading words of the material. So it may be caused by the fact that students get used to memorizing words by paper, or the inconvenient reviewing by Fetion. The students of EG1 did self assessment for reading and writing, 78.87% of them think mobile learning is helpful for writing, 63.38% of them think reading is improved.

Considering the flow of the mobile phones, we did not have oral communication with the students, so we can not know the impact of mobile English learn on oral English. In the follow-up experiment, we will improve this aspect. We believe that with the improvement of mobile phone's function and the full coverage of 4G network, the mobile English learning will have a broader future.

ACKNOWLEDGMENT

This paper is of the "twelfth-five-year-plan" social science research project "An Empirical Study on College Students' Mobile Assisted English Learning" of the Jilin Province Education Bureau.

REFERENCES

[1] Alley, M., McGreal, R., Schafer, S., Tin, T. & Cheung, B. Use of Mobile Learning Technology to Train ESL Adults. Proceedings of the Sixth International Conference on Mobile Learning, Melbourne, 2007.

[2] Boticki, I., Wong, Lung Hsiang & Looi Chee-Kit. Designing Technology for Content-Independent Collaborative Mobile Learning. IEEE. Transactions on Learning Technologies, 2013: Vol. 6, No. 1.

[3] Chin, Kai-Yi & Chen, Yen-lin. A Mobile Learning Support System for Ubiquitous Learning Environments. The 2nd International Conference on Integrated Information, 2013.

[4] Kukulska-Hulme, A. An Overview of Mobile Assisted Language Learning: from Content Delivery to Supported Collaboration and Interaction. ReCALL, 2008 (3):271–289.

[5] Thornton, P. & Houser, C. Using Mobile Phones in English Education in Japan. Journal of Computer-Assisted Learning, 2005 (21):217.

Education Management and Management Science – Zheng (Ed.)
© 2015 Taylor & Francis Group, London, ISBN 978-1-138-02663-6

English teaching method reform seen from the perspective of students

Xingrong Xiang
Department of English, North China Electric Power University, Baoding, Hebei Province, China

ABSTRACT: With the development of globalization, the industrial demand for foreign high-end talent is increasing. The English, as an "Esperanto", the importance of which is obviously visible. The high quality of English major education helps to develop English talents, but now, the contradictory between the sophisticated demand of the cop-oration and the personnel training of the universities, which explains the English teaching methods are in a need of reform. This paper summarizes the survey results and personal interviews, analyzes and summarizes the current situation and problems of English teaching, and summarizes responses to these questions, trying to raise some effective English teaching reform proposals.

Keywords: students' angle; English; teaching reform

1 INTRODUCTION

In 2000 the Ministry of Education's official promulgation of the "Curricular for English Major" made it clear that the training objectives of English majors in colleges as: "The interdisplinary English talent should have a solid foundation and broad cultural knowledge and should be proficient in the use of English in foreign affairs, education, economy, trade, culture, science and technology, military and other departments engaged in translation, teaching, administration, research and other work." With the globalization of social development and the increasing industry demand for high-end foreign talent, English talents have been asked with higher requirements. In this background, college English teaching reform has been carried out in recent years, and made some achievements, but the contradictory English Majors Training divorced from social and economic development has not been solved. Such situations call for the emergence of new methods of English teaching reform. English teaching method can improve the overall quality of English Majors to lay a good foundation for the sustainable development of English majors. Therefore, the reform of English teaching methods, has a sense of entire education reform pioneer and forward-looking.

2 THE CURRENT SITUATION AND ANALYSIS OF ENGLISH MAJOR TEACHING

In order to understand the current college English teaching situation and to find out the problems and loopholes of the English teaching method, members of our group did a survey and personal one on one interviews to the English majors in North China Electric Power University (Baoding). 120 questionnaires were used with 111 copies returned, including students at school 13, 12, 11 and 10, a total number of eight classes.

Lectures for teacher satisfaction survey showed that 74% of students were on the part of the teacher satisfaction, 21% of students satisfied with all the teachers, 5% of the students were not satisfied with all the teachers. Overall, the way teachers teach got higher satisfaction. As for ideal classroom teaching of English, 32% of students chose the traditional way of the teachers' explanation with the combination of classroom activities, 47% of students chose networked multimedia self-learning and classroom activities combined. 21% of students chose a network independent study courses and teachers assisted answering. For existing teaching content and employment issues associated with the future, 26% of students said they had a close relationship and 68% of the students believed that little while five percent of the students thought that there was no association. The question that what is more focused on the teaching content, 79% of students thought we should focus on employment, 16% of students believed that more focused on academic, 5% of students believed to maintain the existing model. For the question whether the existing curriculum should be adjusted to, 65% of students thought it need a big adjustment, 20% of students believed that it need small changes, 5% of the students saw no need to change, 10% of students thought it did not matter.

From the above questionnaires and personal interviews, it can be seen that the current situation and problems of teaching English. They can be summarized as follows:

1. Teachers were overwhelmingly dominant in the teaching process, to a lesser extent the students to participate in class. In the English teaching classroom, many teachers still follow the teaching model, which is still based on the traditional teacher-centered teaching method, in 50-minute sessions to impart knowledge into the students. This method focuses on imparting knowledge, but ignores the students' initiative and dynamism that can only be used to cultivate English graduates with high exam scores, but poor oral interview.
2. Students' enthusiasm for learning English is not high, and many students just mechanically complete the learning task. As we all know, interest is the best teacher. Students have little interest in English, which will largely affect English learning. But the status quo is, in the usual English learning, that many students learn English with little interest in class, unwilling to spend extra time in learning English.
3. English curriculum evaluation mechanism is unreasonable, too much emphasis on the final exam in this single evaluation mechanism. In each discipline of study, attendance and the final exam are the only standards to measure whether to pass or not, resulting in a passive student attendance. And they focus only on the final exam questions and ignore the usual range of learning and accumulation.
4. Curriculum is not reasonable. In the foundation stage, curriculum does not involve translation. Translation course is not practical in junior stage and has less hours to study. Some of the specialized elective courses need to adjust in the curriculum, and fewer second foreign language class leads to poor learning outcomes.
5. Sophisticated professional direction is not divided sufficiently. Specialized courses to the junior stage of learning, linguistics, literary translation force uniform, no focus, students easily learn the pan without fine, broad and not deep. There is no clear and good professional direction.
6. Students lack of the opportunity to practice. Four years of undergraduate study at the university, most students do not have relevant practical experience and internship opportunities, likely to be restricted to mere campus environment, with less workplace exposure and related work opportunities at a disadvantage in the job competition in the workplace.

3 THE ADVICE PROPOSED FROM THE PERSPECTIVE OF STUDENTS

Based on the above situation and problems of English teaching method reform, it can be concluded that English teaching reform is to improve the quality of teaching and put forward an effective way to meet the social demands of English talent cultivation. As English majors, proposed from the perspective of students, to education, sociology and psychology as a starting point, we attempt to make a few suggestions for the reform of English teaching methods.

1. Put students in the dominant position of the classroom, to get rid of the traditional teacher-centered instruction mode.

    One of Cognitive Psychology representative Bruner believes that learning is to organize and reorganize cognitive structure, students gain knowledge not by teaching of teachers, but the students themselves take the initiative to explore and discover. The teacher's role is to guide and inspire students, in other words, the leading role of teachers is reflected in student-centered classroom teaching. Teaching methods should highlight on transforming the teacher's own teaching philosophy, focusing on students' ability to innovate and create, encouraging students to personality development through quality education, cultivating independent thinking, independent discovery and problem-solving skills, outstanding students teaching activities in the dominant position. The traditional teacher-centered teaching methods should be improved and reformed.
2. Teachers are expected to guide and train students' interest in learning. "Interest is the best teacher," especially for foreign languages. Once students like English, interested in it, resulting in a passion, you do not teach him, he will take the initiative to learn, to practice, to drill. Teachers should innovate teaching techniques, such as inspiration, discussion, role, and so on. Teachers could provide an individualized and timely way to teach students that "Teach students to learn how to learn".

    Teachers can organize a rich variety of extra-curricular activities in English, to mobilize the classroom atmosphere, such as, to conduct high-quality English Corner, to do English film screenings, open hearing classroom to help students do hearing training; to provide English journals in the student dormitory rooms to facilitate students' reading; to encourage students to do spoken English communication and English teaching broadcasting; to organize various activities such as the English original fiction contest, creating a strong atmosphere of learning English.

3. English courses evaluation mechanism needs to be reformed and a variety of test patterns should be implemented. Reducing the total score of the final exam in proportion, implementing a dynamic test mechanism, creating a comprehensive evaluation system are feasible and effective. Student test scores, classroom participation and performance are all included in the evaluation system to arrive at the student's final results. In order to achieve a comprehensive measure of all aspects of a student's level, an objective criteria is in need.

4. Adjust the curriculum. Increase hours of the Second Foreign Language and translation courses. Adjustments relating to professional elective courses are in need. To begin a comprehensive training from the basic learning stage, to improve the training intensity of professional quality.

5. Specialize small professional direction. English majors mainly have three directions: Anglo-American literature, linguistics and translation courses. If students are learning everything, it may lead to no good results. Students should be given the opportunity to choose a professional direction and acquire knowledge intensively, prompting the emergence of high-end English talent; increasing training in basic knowledge learning stage; adding a second foreign language of more choices, such as Korean, Spanish, Arabic and other minority languages settings.

6. To provide a rich resource of practical and internship opportunities. Schools can contact their local enterprises and institutions to provide summer holiday interns and other opportunities for students to ensure that students have the workplace sensitivity and adaptability in the future competition.

4 CONCLUSION

As English continues to develop, innovation and reform of English teaching methods will also be a process of continuous development and improvement. Our group made such a survey through questionnaires and interviews, from the perspective of students to solve this problem. College English teaching reform is a process full of difficulties and hopes, although the process may be long and complicated. But as long as students and teachers work together, follow the objective law of foreign language teaching and overcome difficulties, the firm will be able to move forward effectively. It can train for the industry with a number of interdisplinary English professionals, to promote the sustainable development of English majors.

REFERENCES

[1] Yao Limei improve the overall quality of college teaching methods innovation in English Social Sciences, Jiamusi University, 2012, 30 (1): 178–180.
[2] College Foreign Language Teaching English Group Steering Committee syllabus for English Majors Shanghai: Shanghai Foreign Language Education Press; Beijing: Foreign Language Teaching and Research Press, 2000.1.
[3] Lu Xiumei Foreign Language Training Mode Theory and Practice—Reflections on Teaching Ningxia University: Humanities and Social Sciences, 2004, 26 (3): 76–80.
[4] Wufeng Song Reform of university English teaching methods to explore English Place: academic research, 2013 (5): 76–77.

Education Management and Management Science – Zheng (Ed.)
© 2015 Taylor & Francis Group, London, ISBN 978-1-138-02663-6

The problems and countermeasures of "micro-class" from the point of view of cognitive psychology

Lihui Wang & Lijie Wang
Jilin Agricultural Science and Technology College, Jilin, China

ABSTRACT: The "micro-class" refers to the record of wonderful teaching and learning activities on the basis of video that emphasize a particular knowledge point or teaching process in the classroom. It has many advantages including short time, accurate content, small space requirements and rich resources. From the students' cognitive perspective, the "micro-class" has many problems that do not meet students' cognitive rules such as perception, attention, thinking and imagination. University teachers should change their ideas and improve the professional level to improve the quality of the micro-class. University administrators should provide good environment and materials for the micro-class.

Keywords: the micro-class; cognitive psychology; PPT

1 INTRODUCTION

At present, education informatization has become the important part of the education reform in colleges and universities. Then the key point of the education informatization is teaching informatization. How to use modern information technology such as multimedia and Internet to improve the teaching quality has become a topic of teachers in colleges and universities. As new type of online education, micro-class is booming. Experts predict that the micro-class will be one of the most promising education technologies.

In September 2012, the Ministry of Education issued the notice of the first "micro-class competition of China" which required the competitors be teachers in primary and secondary school. On November 20 2012, the National Higher Education Teacher Network Training Center of the Ministry of Education issued the notice of the first national university micro-class teaching competition and announced the program and rules. By the early March of 2013, more than 300 universities including Jilin University announced the competitors who would join in the competition, meanwhile 98 universities including Dalian University of Technology have finished the management accounts register of race website and Xiamen University, Beijing Science and Technology University and many other universities finished selecting the competitors. With the emphasis of the education authority and the active participation of the front-line teachers, the design and application of micro-class will become a hot topic. Then what is the micro-class? What are the characteristics? How can we make a micro-class that can be in accordance with students' cognitive characteristics?

2 DEFINITION AND CHARACTERISTICS OF MICRO-CLASS

2.1 Definition of micro-class

In other countries, the concept of "microlecture" has already emerged and it was first used by David Penrose of America in the fall of 2008. The term of micro courses do not refer to the micro content in order to develop the micro teaching, but refer to the actual teaching content which on the basis of constructivism for the purpose of online learning or mobile learning. David Penrose called the micro-lecture as "knowledge burst". In our country, "micro-class" was first proposed by Hu Tiesheng of Foshan Department of Education. From the perspective of Hu Tiesheng: the "micro-class" refers to the record of wonderful teaching and learning activities on the basis of video that emphasize a particular knowledge point or teaching process in the classroom. Micro-class has many functions. It provides learning environment, promotes the change of learning manners, and improves teachers' ability of developing issues. With the development of Internet and multimedia technology, it has new definition and characteristics.

2.2 Characteristics of micro-class

As a new teaching resources construction and application mode, the micro-class in colleges and universities has the following features.

2.2.1 Short time of teaching

The teaching video is a core content of the micro-lesson. According to the cognitive characteristics and learning rules of students, the micro-class in colleges and universities is generally 10 to 20 minutes (the length of micro-class of the primary and secondary is 5 to 8 minutes, and the longest is not more than 10 minutes). Compared with the traditional 45 minutes or even 90 minutes, "micro-class" can be called "lesson fragment" or "micro lesson".

2.2.2 Accuracy of teaching content

"Micro-class" is set up to grasp the key points, break through the difficulties, and seize key issues for students. The emphasis of theme is helpful to teachers' and students' need. Compared with the complexity of traditional class, the content of "micro-class" is accurate. Therefore it can be called as "micro-class".

2.2.3 Small space requirements of micro-class

From the size of micro-class, the total capacity of the video and supporting resources is generally tens of megabytes. Video format should support the online play. College teachers can use the campus network to observe the lessons and view lesson plans and some other supporting resources. Teachers also can download and save to mobile hard disk, U disk, MP4 or mobile phone flexible and easily. It is not only suitable to teachers' observation, evaluation, reflection and research, but also convenient to college students to use mobile phones and MP4.

2.2.4 Rich resources of teaching

"Micro-class" resources have the characteristics of video teaching cases. The teaching video is the main line on the basis of which teachers integrate teaching design, multimedia material and courseware, teaching reflection after class, students' opinions and specialists' comments. It constitutes a theme unit resource of which theme is clear, type is diverse and structure is compact. All of that make a real "micro-teaching resources and environment.

3 THE PROBLEMS OF MICRO-CLASS DESIGN AND PRODUCTION

The design of script and PPT is the main content of micro-class and teaching is the main clue for students' learning. From the micro-class we can see, there are some problems that do not adapt to students' cognitive psychology. The problems are the design and production of PPT and script and the combination of teaching activity and video.

3.1 The problems of the PPT design

3.1.1 Small difference of the object and the background

Small difference of the object and the background is not conducive to students' choice. A survey analysis shows that there is a significant positive correlation between the appearance of the courseware and teaching effect. Actually, the process of learning knowledge from PPT is also the process of perceived selection. The selectivity of perception is to distinguish the object from the background. The bigger difference between the object and background is, the more outstanding students can select from background. When the information that should be grasped by the students is new, different, beautiful, moving and strong, students are more likely to choose them. If the difference between the object and the background does not reach students' level, then the object is not easy to be perceived. The perception is the first step of cognition and also the basis of memory and thinking. When the teacher makes or uses PPT, there are some misunderstandings as follows, the color of background is too dark that cannot clearly contrast with the words; the size or color is not good enough for the students who sit a little away from the PPT; the photo is too small or not clear and so on.

3.1.2 Tedious content of PPT

The content of PPT is too much and the key points are not outstanding, all of which are not conducive for students to memorize and think. Psychology divided memory into two processes which are "remember" and "recall". "Remember" includes memorizing and maintain while "recall" includes recognition and memory. "Remember" is the basis and the premise of the "recall" while "recall" is the purpose of the "remember". In order to firmly hold, students should improve memory scientifically. Maintaining is the key point of the process of memory. The research of the psychologist Miller found that: the short-time memory of people is 7 around units of information. If there are too many words, students cannot store all of them in their short-time memory. However short-time memory is the basis of the long-time memory and long-time memory is the main type of information storage of the students. In addition, memory is the workshop of thinking and thinking is the production and abstraction of memory. If the content of PPT designed by teachers is clear, it is not only in accordance with memory rules, but also helpful to students' abstract thinking.

The content of some PPT is too random and some teachers put the content on the PPT randomly, so the PPT does not reflect the design and organization of the teachers. PPT will be changed into presentation. So it lacks system and generality. It is not conducive to the students' memory processing of knowledge and is not conducive to analysis and understanding. So it is contrary to the essence of the "micro-class".

3.1.3 *Speciosity of the screen*

Psychological research shows that attention is the prerequisite for the smooth psychological process. When the students feel, memorize, think and image, they need attention which is the guard of cognition. Attention can be divided into voluntary attention and involuntary attention. When the object of studying is new, different, beautiful, moving and strong, it is easy to arouse involuntary attention, which could help students to learn knowledge easily. If all above characteristics are not relevant to the content of studying, it tends to disperse students' attention.

In order to increase the pattern, some teachers will use audio, flash and some special effects, which does not only flood the theme, but also disperse students' attention. For example, if unsuitable flash is added to the micro-video, students' attention will be attracted by the flash not the content. So when the teachers produce PPT, they should pay attention to the content and avoid being flashy.

3.2 *Problems of video*

Micro-class video generally includes PPT, teacher's explanation, subtitles, sound, and students. How to organize all these elements into one video perfectly is the producer's consideration. The following problems are likely to exist: the image is not clear, sound is too small, screen is not harmonious and the content is monotonous. If the image is not clear or the sound is too small, the content cannot arouse the students' interests. If the sound is too large, students cannot listen to the teacher carefully which can affect students' learning. Similarly, if the organization is not harmony, it cannot arouse students' attention. Monotonous content is the biggest problem of micro-class. Psychological research shows that monotonous stimulus is easy to disperse students' attention. For example, some PPT only has teachers' explanation that is so monotonous. The PPT that is funny and knowledgeable is more likely to stimulate learners' motivation.

4 METHODS AND APPROACHES TO IMPROVE PRODUCTION LEVEL OF COLLEGE TEACHERS

With the attention of department of education and further development of education technology, Internet and online education, more and more teachers will take part in the design and application of the production of micro-class. So the administers and teachers should pay more attention to improving the production and design of micro-class.

4.1 *Changing concepts*

Changing concepts is the basis of improving the production of micro-class. As a new education technology and education reform type, micro-class does not attract some teachers' attention. Some teachers did not recognize the advantages of micro-class, so they will not research and use it. Especially older teachers' acceptance of micro-class needs a long psychological process, so they will ignore or even refuse it. The combination of traditional teaching methods and modern education technology will improve teaching effect. Changing concepts is the basis of improving the production of micro-class. Teachers can use Internet, consult colleagues or research students to explore the advantages of micro-class and the need of students. So the micro-class will be in accordance with students' cognitive psychology.

National educational administration and higher education managers should publicize the micro-class, so the teachers could change their concepts soon so as to satisfy the requirements of the information and Internet to college teachers.

4.2 *Strengthen business learning*

Strengthening business learning and improving level are the protection of knowledge and technology of high-level micro-class. First, the so-called learning includes the understanding of theories, pedagogy and psychology. Only grasping the three aspects of knowledge can the teachers make PPT from the perspective of students and make them study easily. Second, teachers should be familiar with the use the Internet and the resources. Some teachers do not design or use micro-class; maybe they are not familiar with it. For example, the speed of type is too slow or they could not use Internet to get their information. So the college managers should help these teachers use Internet. Teachers should learn hard and improve their theoretical and production level.

4.3 *Cooperation*

A good micro lesson requires teachers to design their class according to the teaching goal; and requires education teachers to record the teachers' design and explanation in order to upload, save and use video. This requires each teaching unit and the department of education technology or network center to cooperate. School managers should coordinate the relationship between the various departments, the education technical personnel's assistance. Obviously, the main teachers should obtain some knowledge of education technology, and then discuss with education technology personnel, and it is easier to produce high levels of micro video.

4.4 *Equipment is the material guarantee*

Micro course construction needs the security of hardware and software. It needs hardware device, needs a high-definition camera, the desktop or laptop computers, smart phones, and other learning equipment, Internet needs a router, network card and so on. And it needs software, such as a computer, mobile phone operating system, the installation package, network learning platform client and video player software, and so on. These equipments are in need of university logistics management department to purchase and installation, for teachers to record the classes; teachers and students can use them in micro class. What's more a quiet micro-standard classroom is also needed.

From what has been discussed above, micro lesson as "new-comer" of online education, it will increasingly cause people's attention, Micro lesson is to improve the quality of teaching service, bad micro lesson will waste resource, and it is no good to learners, good micro lesson should go with students' development of cognition. University administrators and teachers should strengthen the management and learning, in order to improve the quality of the production and use of micro course continuously.

REFERENCES

[1] http://weike.enetedu.com/news/html/2013-3-20/20133202039221.htm.
[2] Zhongke Guan. micro-class. China Information Technology E Education, 2011(17):14.
[3] Tieshe Hu. "micro-class": The new trend of development of education information resources. E Education Researc, 2011.10.
[4] Ye Dapeng. College of Educational Science of Anhui Normal University. Journal of Suihua College. 2013.10.
[5] Jia-wei Zhou. A Discussion on the "Micro Class" Teaching Resources Development Focused on "learning" Career Horizon. 2013.10.

Education Management and Management Science – Zheng (Ed.)
© 2015 Taylor & Francis Group, London, ISBN 978-1-138-02663-6

Research and practice of practical teaching mode of core example through teaching

J.C. Zhang
Hebei College of Industry and Technology, Shijiazhuang, Hebei, China

L.C. Wang
Hebei University of Science and Technology, Shijiazhuang, Hebei, China

ABSTRACT: The practical teaching projects of higher vocational colleges are usually divided by courses and the practical contents are independent and unrelated. Mechanical manufacturing and automation specialty in our college has explored a new practical teaching mode of core example through teaching. This mode selects a specific machinery product as the core example. According to the production tasks of the core example to construct the practical teaching system. The different practical projects are connected through a core example. Through the application of this mode, students can put distributed knowledge organically, and can experience the production characteristics of the enterprises.

Keywords: core example; practical teaching; machinery manufacturing

1 INTRODUCTION

In recent years, China's economy is integrating into the world economy in a fast speed. With the influx of foreign enterprises, China has become one of the important base of manufacturing industry. This requires a lot of mechanical manufacturing and automation professional talents, who can design mechanical products by computer and can manufacture products by numerical control machine. In order to better meet market demand, the practice teaching of mechanical manufacturing and automation specialty in Higher Vocational Education should be reformed. The new practice teaching should enable students to obtain the ability to use new digital tools for product design and to use new CNC equipment for manufacture.

2 THE MAIN PROBLEMS EXISTING IN PRESENT PRACTICAL TEACHING

At present, the higher vocational colleges have attached great importance to practice teaching, and have explored many new teaching mode to develop students' practical ability. But there are still many problems that hinder the cultivation of students' occupation quality. Mainly include:

– The practical teaching concept still lag behind. We focus on theory rather than practice.

– The practical teaching management system is not scientific or perfect and the operation mechanism lags behind.
– The practice content is outdated, and the test methods and steps are dogmatic. The validation and demonstration experiments are more than the practical training of production, process, design and innovation.
– The practical teaching projects are generally set up by single course and the projects are independent of each other. The training content is not coherent or systemic, so the knowledge and skills that students have learned are fractional and discrete. The students' comprehensive ability can not get effective training.
– The laboratories are independent of each other. It is not conducive to a coherent and comprehensive practical teaching. The utilization and efficiency of equipment and resources are low.
– The teachers' practice skills are not good. Many teachers are lack of the experience of designing and implementing a comprehensive project.
– Skills and ability training are not enough to enable students to obtain occupation accomplishment needed in work.
– The teaching material emphasizes theoretical knowledge and neglect practice guidance.

In order to solve the problem, we must change the traditional practical teaching system that organizes single practice for a course. We should break the course limit to set training project. The training process should be consistent with the production process of mechanical products. According to the process of design, manufacturing and assembly, the produce content is separated into the training projects. Students can experience the factory production process of mechanical products when they do the training projects in school.

3 THE CONCEPTION OF "CORE EXAMPLE THROUGH TEACHING" PRACTICAL MODE

3.1 Design ideas

A specific mechanical product is selected as carrier, and practice teaching system is built according to the production process of this carrier.

The production process of mechanical products is including product design, process design, tooling design, manufacturing, assembly, inspection and other links. We choose a specific and typical mechanical product as the carrier, which is the core example, and the practical contents are gained according to its production process. As shown in Figure 1.

Each project in this system should be gained from the production process of the core example, should have practice name, practice content and related courses. As shown in Table 1.

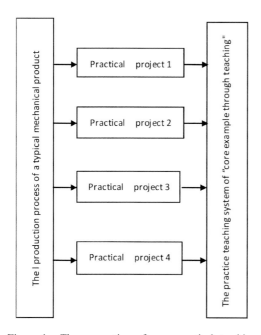

Figure 1. The conception of new practical teaching system.

Table 1. Practical teaching design.

Production process	Practice name	Practice content	Related courses
1	......	......	......
2	......	......	......
......	......	......	......
n	......	......	......

3.2 Key problems

Recombine the training rooms. Change the condition of the training room designed for a course, and recombine the training rooms according to the function and characteristics of the rooms. Set public training center, professional basic course training centers and professional training center. Realize the sharing of resources, reduce the duplication of investment and improve the efficiency of funds.

Carefully select the core examples. As the training example that runs through the whole process of practical teaching, the product should be the enterprise online product, and it has a typical structure, and its manufacturing difficulty is moderate.

Reform the traditional teaching mode. Put to use the open teaching mode. Training rooms are open, and students can go to the training rooms to do training projects at any time.

Provide instructors according to the job requirements. The instructors not only have rich theoretical knowledge but also have good practical skills.

4 THE APPLICATION AND RESULT OF NEW PRACTICE MODE

4.1 A survey of application

For the new practice teaching mode, our machinery manufacturing department conducted a pilot in the 2012 level students of mechanical manufacturing and automation specialty. We selected 5 products from Huabei diesel engine plant as examples to organize the practice teaching content. As shown in Table 2.

We divided the students into 5 groups, each group has 8~10 students. Each group carried out the practical activities according to the production process of the actual product from the enterprise.

4.2 One case for wind pushing mechanism

One group selected a wind pushing mechanism as the core example.

The enterprise product production process can generally be divided into four stages, they

Table 2. Practice teaching design.

Work process	Practice name	Practice content	Related courses
1	Scheme design	Design scheme	Drafting
2	Structure design	Overall design	Tolerance and measurement
		Structure design	Mechanical design
			The engineering materials
			Heat treatment
3	Machinery manufacturing	Process planning	Machinery manufacturing
		CNC operation	Numerical control machine
		Tool selection	Advanced manufacturing
			NC programming and machining
4	Assembly & debugging	Parts assembly	Fault diagnosis and repair
		Product debugging	Hydraulic and pneumatic
			Automaton and automatic line

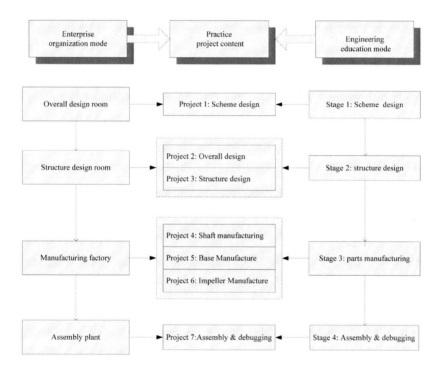

Figure 2. Design idea of practical projects for wind pushing mechanism.

are: scheme design, structure design, machinery manufacturing, and assembly & debugging. The whole project of the wind pushing mechanism was divided into four stages with seven sub projects. As shown in Figure 2.

The concrete teaching design is shown in Table 3.

4.3 The teaching achievements of new practice mode

After two years of practice, the new practice mode has made some achievements.

Through the "core case", the students have put the isolated, distributed knowledge linked, and the students' ability of comprehensive application of knowledge has improved.

Through the "core case", the teachers have improved the ability of combination of theory and practice, and have enhanced the ability of practical operation, and have strengthened the ability to communicate with the enterprise.

Training resources and equipment have been optimized, and the utilization ratio of the equipment has been improved.

Table 3. The concrete teaching design for wind pushing mechanism.

Core example name	Work process	Practice number	Practice name	Practice content
Wind	Scheme design	1	Scheme design	3D Schematic plot Program instructions
Pushing	Structure design	2	Overall structure design	General assembly drawing 3D simulation map
		3	Structural design of spare parts	Parts structure diagram (Impeller, Transmission Shaft, Base)
Mechanism	Parts manufacturing	4	NC programming and machining of the transmission axis (Lathe)	Process scheme Machining program Machine operation Parts inspection
		5	NC programming and machining of the case (Miller)	Process scheme Machining program Machine operation Parts inspection.
		6	NC programming and machining of the Impeller (Machining center)	Process scheme Machining program Machine operation Parts inspection
	Assembly & debugging	7	Assembly and debugging	Product description

The students have got more occupation qualification certificates. Students attending the pilot have all gained the Certified Solid Works Associate (CSWA) and the senior certificate of Numerical control lathe.

5 CONCLUSIONS

Recognizing the deficiencies to organize training project according to the curriculum, we have explored a new practical teaching system of "core example through practice". Training project is organized according to the production process of this core example and it connects all the training projects. After two years of practice, we have made some achievements and experience.

ACKNOWLEDGMENT

This work was supported by Hebei province occupation technology education association (Hebei province occupation technology education research institute) (ZJY13115).

REFERENCES

[1] Wang, W.B. & H. Li (2011). Teaching Reform of Object-oriented Programming Course with Instance. *Computer Education.* 01, 91–94.
[2] Zou, J.S. (2011). Construction of curriculum system based on working process. *Reform and opening.* 10, 158–159.
[3] Liu, Y.L. (2013). Based on the working process of "machine tool electrical installation and debugging" Curriculum Construction. *Course education research.* 6, 249–250.
[4] Yue, L.Y. (2014). Study on the curriculum development process and method oriented by work process. *China Power Education.* 08, 152–154.
[5] Yu, J.L. (2014). On the teaching reformation practice of curriculum model based on working process systematization. *Ship & Ocean Engineering.* 43, 73–74.

Education Management and Management Science – Zheng (Ed.)
© 2015 Taylor & Francis Group, London, ISBN 978-1-138-02663-6

From incremental innovations to really new innovations: On the development of modern Chinese ink paintings

Q. Zheng
Jinhua Polytechnic, Jinhua, China

Z.X. Chen
Zhejiang University, Hangzhou, China

ABSTRACT: This paper aims to explore the relationship between innovation and the development of modern Chinese ink painting by the methods of comparative with traditional ink paintings, and the results indicate that the development of modern Chinese ink painting is always accompanied with innovations. In the early stages of development, the modern ink paintings made breakthrough by "incremental innovations" in terms of composition, form of expression and style, which maintained connection with traditions and meanwhile kept distance from traditions, and produced the matured form of the modern ink paintings. In the 21st century, the modern ink paintings have a significant innovation in materials, techniques and concepts, and show a distinct development tendency. This paper argues that the innovations of the modern ink paintings shall be subject to the principle of "really new innovations" in order to unshackle the border of the ink wash painting and make use of its strength to spread the spirits of the Chinese ink paintings.

Keywords: incremental innovations; traditional ink painting; modern ink painting

1 INTRODUCTION

The modern Chinese ink paintings started from 1980s and became matured in 1990s. As a kind of new art distinct from classical ink paintings and realistic ink paintings, they play an important role in the current Chinese ink paintings. Despite that they were criticized as separation from traditional ink paintings and blind imitation of the modern western art at the time of emerging, they are quietly accepted by people at present. They are often displayed in ink painting exhibitions organized by national or local art organizations, receiving quite positive feedback. Their academic results have been cited by many painters. Experience has proved that reference to the modern western art, as reflected in a majority of modern ink paintings in a certain period, is way of breaking the bondage of rigidified, ossified and routinized conception of traditional ink paintings as well as seeking for a new pattern of artistic expression. During practical exploration, painters spared no efforts in critically assimilating, reconstructing and sinicizing the modern western art, and also in re-exploring the implicit modern elements in the traditional art with the modern concept, which significantly impacted and boosted the modern transformation of the ink paintings.

2 TRANSFORMATION FROM TRADITIONAL INK PAINTINGS TO MODERN INK PAINTINGS

Modern ink paintings are distinct from ancient ink paintings (such as those in Tang, Song, Ming and Qing dynasties) or paintings by Chen Shizeng, Qi Baishi and Huang Binhong or even ink paintings in 1980s. As time goes by, modern ink paintings have been added with something new, while have lost and vanished something.

2.1 Weakening of ink language

The core of ink painting techniques is well-known "brush & ink". The "brush & ink" rule basically requires painting in the way of calligraphy writing. Although the "brush & ink" concept has been extended for a long time, e.g. deemed as a kind of cultural spirit or aesthetic attitude (Wan 1997), it is undeniable that some art forms irrelevant to brush and ink also represent our cultural spirit and aesthetic attitude, such as ancient drawings, portrait bricks and portrait stones not created with brush & ink. It implies that the cultural spirit and aesthetic attitude, reflected by the brush & ink, are not representative of "brush & ink", but representative of Chinese culture in which the "brush & ink" is rooted. The history of Chinese paintings is not the history of brush & ink, while the brush & ink rule is a product of the traditional Chinese painting development at some phase (Liun 1997). In practice, similar to foreign paintings, traditional Chinese paintings embrace a variety of forms, techniques and rules, namely multiple aesthetic factors. After Yuan Dynasty, scholars advocated the dominance of the brush & ink rule instead of other skills and rules, which dramatically impacted the pattern of traditional ink paintings and made it simple.

The debate on "brush & ink equal to zero" and "no brush & ink equal to zero" as well as the question of "what is the value of ink paintings by Wu Guanzhong and Zhou Shaohua" suggested that: the brush & ink rule was unprecedentedly "disseminated" and "safeguarded" at the end of the 20th century because of unprecedented challenges it had. Under the background of great changes of writing by Chinese scholars, it removed the root of brush & ink development and turned it to be industrial skills, which completely contradicted the primary intent of elevating the status of brush & ink by ancient scholars.

2.2 *Disintegrating of "poetry, calligraphy, painting & engraving"*

Different from the "brush & ink", the whole set of "poetry", "calligraphy", "painting" and "engraving" have never turned out to be the rule in fact. We can find a lot of masterpieces without poetry or stamps by ancient masters. Since modern painters receive different educations compared with ancient painters, it is obviously an outrageous demand to have modern painters inscribe poems on their works. Frankly speaking, there were some excellent ink paintings in recent 20 years, but scarcely any high-level inscribed poem. The combined "poetry", "calligraphy" and "engraving" may be seen in future ink paintings just as the preservation of usual practice in the past (Shui 1999).

2.3 *Fading of spirits of Tai Ji, the eight diagrams, the book of changes, Chuang-tzu and Buddhism*

It is typical of Chinese art to set rules for ink paintings with the Book of Changes, the theory of Chuang-tzu and the theory of Buddhism which are even interpreted to be "oriental art". Through the ages, philosophers were always used to connecting philosophic theories with art phenomena, but historically philosophers and theologians were not art creators of their contemporaries. After Song and Yuan dynasties, some painters were addicted to philosophy and Buddhism, but they never regarded them as the criteria of calligraphy and paintings. Under certain conditions, philosophic theories and religious doctrines may indirectly impact artists' personality, character and generation of artistic ideas, but artists never take these as guidance for creation. An artist may claim that he suddenly gets inspiration from some mysterious ideas or theories (such as Tai Ji Diagram and the Book of Changes, or inferiorly Qigong and Chinese health study) and magically produce masterpieces. It is just the talk and means nothing. If you take it serious and praise it as the systematic painting theory, it appears to be too far-fetched. One worrying downside for the development of modern ink paintings is extremely constrained form and rambling art theories.

2.4 *Vanishing scholars' ink paintings*

From "scholars' paintings" by Su Shi in North Song Dynasty to "scholars' paintings" by Dong Qichang in Ming Dynasty to "scholars' paintings" by Chen Shizeng in the Republic of China, they all stressed that scholars had different moralities as professional painters had (Wan 1996). Chen Shizeng suggested "scholars' top qualities: first morality, second knowledge, third talent, and fourth thinking." He applied the standard of judging ancient outstanding persons as painting theories which actually was not the definition of rigorous "scholars' paintings". Literally he explained factors for "scholars' paintings", but he focused on setting up ideal "scholar painters" examples and standards of "moralities and characters for scholars" (Liu 1998). It is misinterpretation if taken as the basic standard of evaluating paintings today.

Although the concept of "scholars' paintings" leaves us freedom and space for interpretation, we can not deny the fact that traditional Chinese scholars no longer exist anymore and the environment for traditional scholars to grow up has vanished. Some (not all) ink painters at the beginning of the 20th century may be regarded as traditional scholars, since their basic living conditions and spiritual status are similar to those of ancient scholars. However, the social reform and ideological remoulding in the Mainland China have put an end to the possibility of "traditional" scholars. In fact, we can learn much from "scholars' paintings" and the most valuable is its rebellious spirit, its natural of amateur, and its artistic attitude of keeping away from utilitarian purposes. All of these values are eternal.

3 MATURED FORM OF MODERN INK PAINTINGS

The modern ink paintings are the logical development of traditional scholars' paintings compatible with the contemporary cultural environment after twists and turns for nearly one century. Not making a rebellious break with the tradition, they keep some connection with the tradition while some distance from it. Compared with traditional ink paintings, they demonstrate new features in respect of composition, form of expression and style. It is a kind of really new innovations according to modern innovation theories (Wu et al. 2007). In other words, by really new innovations, the modern ink paintings have transcended traditional ink paintings, made breakthrough and formed the almost matured "contemporary form" mainly reflected by expression style, abstraction style and margin style.

3.1 *Expression style ink paintings focusing on spiritual exploration*

The expression style ink paintings were one of the best products of the "modern ink paintings" in 1990s. On the one hand, they preserve the expression factor in traditional ink paintings; on the other hand, they benefit from the expressionism paintings in the west. "Expressionism" origins from the west, but "expression" in paintings is universal.

From the perspective of traditional Chinese paintings, "expression" is always a big potential factor, especially in ink paintings. The birth of ink paintings signifies the jump of traditional paintings from concrete to imaginary, from realistic to abstract, and from representational to expressional. The "Record of Famous Paintings in Tang Dynasty" records that Wang Qia was painting "by wiping, waving and sweeping" and achieved the results of "rosy clouds and natural wind and rain in paintings without sign of ink". It demonstrates that the ink paintings appeared to be expressional at the time of birth. However, as these paintings were historically used as a tool for emotional expression by scholars for a long time, they had to adapt to scholars' aesthetic taste. Scholars' thought of indifference to fame and seclusion from the society became the theme of ink paintings and its "wild" tone, slightly shown at the beginning, was gradually replaced by "literal" and "scholarly" tone. The improvised passionate expression by Wang Qia was criticized as beyond the rule of painting. Thus, passion was substituted by ration, and improvised expression with explosive force was constrained by stylized ink painting. In response to scholars' attitude of reclusion, ink paintings tended to be modest, gentle, and dispirited without visual tension.

In the 20th century, overwhelmed by the strong western culture, traditional paintings showed intensified lifelessness and decadency. To change this situation, one generation of masters made great efforts in reconstruction. Powerful and magnificent calligraphic works created by Wu Changshuo, the grandeur of paintings with new spirits like in North Song Dynasty reflected by Huang Binhong, the refreshing style integrated with folk life by Qi Baishi, aggressive expression by Pan Tianshou, reconstruction of realism by Xu Beihong, and foundation of modernism by Lin Fengmian all represent the reforms of traditional ink paintings with new spiritual orientation and painting forms instead of powerlessness.

After the new tradition of realism of ink paintings in 1930s, Shi Lu shifted from realism to expressionism as a pioneer at the end of 1970s. Before the "Cultural Revolution", his paintings were criticized as "wild, disordered, strange and black". After the "Cultural Revolution", he straightforwardly expressed internal anger and passion by enduring physical tortures. Nevertheless, his expression was still based on subjects commonly seen in scholars' paintings and had little direct connection with the western expressionism. From 1980s, young and middle-aged painters further explored the new expression way between the tradition and the modern as well as between the east and the west on the foundation of their predecessors. Zhou Sicong apparently drew on many elements from Germany expressionism (mainly Katie Kollwitz) and applied these expressional elements in the "Miner Picture". In 1990s, the expressionism of ink paintings was gradually matured, reflected by a number of works with certain quality and influence. Painter Tian Liming enlightened us through a series of his work: learning from the tradition should take expression of modern people's feelings into account, and learning from the modern western art should take pursuit for traditional aesthetics into account. If not, painters might be misled to blindly imitate the tradition or the western art. The expressionism of ink paintings overturns the traditional elegant and gentle art style of ink paintings and brings in images with visual impact as pioneering "modern status" of the modern ink paintings.

3.2 *Abstraction style ink paintings targeting language experiments*

It is an inevitable trend for ink paintings to be abstract. Just as mentioned above, ink paintings appeared to be quietly abstract at the beginning of development. "Shaping as it is" by ink splashing, as grasped by Wang Qia by chance, was actually an imaginary creation behavior with abstractionism skills. These vanward skills were not accepted by modern painters, just as Zhang Yanyuan said "ink splashing is not painting". However, such skills were so attractive that it drew great attention as the ink properties developed. In the modern times, when the modern abstractionism was flourishing in the west, this potential factor was further enhanced and developed from ink splashing to color splashing as Zhang Daqian and Liu Haili did. However, in their paintings, "abstract" was a kind of skills learned from others, but not real abstractionism. Abstractionism in ink paintings started from the modern art movement in Taiwan in 1960s. A group of painters, represented by Liu Guosong, initiated the new style of abstract ink paintings. In 1980s, abstract ink paintings emerged in the Mainland China. Though abstractionism was not popular, it absolutely was a new form of evolving ink painting patterns. The language form of abstract ink paintings is mainly caused by three factors:

First, evolved from painting skills. Just as Huang Binhong suggested, it broke images with skills, which meant that the charm and spirit of brush & ink were highlighted by deconstructing the "form" so as to replace the value of image with the value of brush & ink.

Second, evolved from ink. It develops from ink breaking, ink splashing, color splashing to creation of texture and quality. Just as Liu Guosong suggested, it completely made use of properties of water, ink and paper to produce abstract images by natural diffusion of water and ink.

Third, evolved from structure. Just as Pan Tianshou suggested, it accentuated compositional division and overall visual power, and applied "blocks" to get rid of elegance of brush & ink (here, brush & ink were not the visual center, but only basic elements of paintings and their structural relations dominated. "Structure" displayed its internal significance with its own tension).

In addition, the avant-garde in the calligraphy circle abandoned "character pronunciation", "character meaning" and deconstruction of "character shape" to be close to abstract ink paintings.

However, currently the abstractionism in most ink paintings is not so pure since painters cannot give up their pursuit for image and artistic conception. Even initiators such as Liu Guosong and Wu Guanzhong are still aimed at achieving ideal images and artistic conceptions. The difference between the imaginary ink paintings and the abstract ink paintings: the former seeks for images and produces artistic conceptions in abstract ink paintings, which it targets; the latter may display some kind of image, but not seeks for images and artistic conceptions as its target (Lu 1999). Even though abstract ink painters get inspiration from concrete objects, they have to process them before painting.

3.3 Margin style ink paintings with modern concepts

Classic artistic presentation of ink paintings materially relies on highly developed medium materials.

The elasticity of writing brushes, the delicate layers of ink, and the gentle texture and diffusion property of Xuan paper matter a lot for excellent expression of ink paintings. The enriched brush & ink language is dependent on further exploitation and improvement of old materials as well as introduction of new materials. A group of artists exploring on the margins are emerging in the ink painting field. They are not satisfied with painting with water, ink and Xuan paper, and try to apply other materials. They may break through limitations of ink and create a completely new world.

As a material, ink is neither the patent owned by ink painters, nor the fetters for them. During artistic development, language and style transformation are mostly caused by material changes. Extension of materials is accompanied by promotion of languages. Any new material introduced may be threatening for paintings as they are alien, but for artistic development, it means the start of reform. When some artists develop ink paintings from two-dimensional to three-dimensional, it not only signifies introduction of new materials, but also changes how ink paintings exist.

4 NEW DEVELOPMENT OF THE MODERN WATER INK PAINTINGS

Started from 1980s, the modern ink paintings showed a distinct development tendency of conventional breakthrough as a trend in the 21st century, which is easily observed. Based on applying traditional skills, modern ink painters subvert traditional concepts and stylized models, make innovations, focus on learning skills of western paintings, boldly explore new techniques and expressions, and endow modern Chinese ink paintings with new life, new image, new language and new structure (Mo 2010). By preserving national features, they arouse the vitality of ancient Chinese paintings and rebuild the aesthetic space, rhythm and artistic conception of ink paintings.

4.1 Innovation of materials

The development of ink paintings is highly interrelated to the development of painting materials. The concept of Chinese ink paintings is produced as paper, brush, ink and other materials are developed and creatively applied. As a number of new materials are created and applied in the changing times, painters keep experimenting and exploring new medium materials as well as their properties and effects, such as mineral pigment, acrylic paint, and plant pigment. It also includes application of some special materials in ink paintings such as cloth, gold and silver foil, wood chips, hemp ropes, printing materials, vitriol, gel and salt. Since the materials of ink paintings are greatly extended, the ink paintings have more contemporary features.

4.2 *Innovations of techniques*

Any ink painting is created by painters by applying concrete medium materials such as brush, ink, paper and pigment. Technically, unfamiliarity with material properties or improper application of materials is an obstacle for internal and external expression of artworks. However, different painting materials have different properties, aesthetic attributes and skills, which require in-depth research and analysis of properties, features and aesthetic attributes of materials, as well as adjustment of painting skills to adapt to new demands in creation of ink paintings. Therefore, modern ink painters shall not only get familiar with mineral pigments, acrylic paint, and plant pigments as well as production of painting materials, but also know the principle and skills of inscription rubbing, splashing, piling, collaging and spraying. They shall break the shackles of the texture mode of traditional ink paintings and create the physical quality of integrated materials to enhance the visual perception.

4.3 *Innovations of concepts*

Innovations of materials and techniques of ink paintings play an important role in improving the expression of images and conveying the spirits of works. Different materials, expression skills or applications enrich and diversify expressions of ink paintings, and also greatly change the painting style and even the aesthetic principle. Thus, the evaluation standard of traditional ink paintings based on aesthetic orientation of scholars is not suitable for the development of modern ink paintings. Some new concepts and aesthetic values of ink paintings are emerging. These conceptual innovations give freedom of expression for painters during painting, and maximize their artistic creativity. Meanwhile, they provide development opportunities for ink paintings as a kind of traditional Chinese paintings in respect of expression form, expression language and expression method, which gradually boosts the diversified pattern of modern ink paintings.

Material, technical and conceptual innovations make it feasible to innovate the modern ink paintings. However, with such innovations, the modern ink painters are confronted with choice between innovation and tradition as well as between nation and world, which leads to subversive test for traditional ink paintings and even extreme variation of Chinese paintings. It is well known that ink paintings are closely related to traditional Chinese culture and embedded in Chinese spirits. Therefore, any innovation of modern ink paintings concerning materials, concepts and techniques shall be based on advocacy of Chinese artistic spirits (Liu 2011). In other words, the modern ink paintings shall be innovated by holding the principle of "really new innovations". Such innovation shall neither be bound by traditions, nor be blind pursuit for changes. Painters shall learn from other arts and paintings by removing the boundary, make use of features of ink paintings, and disseminate the spirits of Chinese art. Any form of innovation of ink paintings shall be in line with the above basic principle. Otherwise, "ink paintings will be unlike ink paintings", or even worse "ink paintings will be not ink paintings any more."

5 CONCLUSION

The modern Chinese ink paintings are the logical development of the traditional ink painting adaptive to the modern cultural environment. Its' development is always accompanied with innovations. Started from 1980s, the modern ink paintings made breakthrough by "incremental innovations" in terms of composition, form of expression and style, which maintained connection with traditions and meanwhile kept distance from traditions. It produced the matured form of the modern ink paintings. In the 21st century, the modern ink paintings are remarkably reformed in respect of material, technique and concept, and show a distinct development tendency. However, the innovations of the modern ink paintings shall be subject to the principle of "really new innovations", based on which it is able to unshackle the border of the ink painting and make use of its strength to spread the spirits of the Chinese ink paintings.

ACKNOWLEDGMENTS

The authors wish to thank the foundation of education department of Zhejiang province of China for contract Y201327073, under which the present work was possible.

REFERENCES

[1] Wan. Q.L. 1997. Painter and painting history. Beijing: Publishing House of China Academy of Art.
[2] Liun. X.C. 1997. Ink paintings: Huang Binhong and Lin fengmen. Hangzhou: Zhejiang People's Fine Arts Press.
[3] Shui. T.Z. 1999. Ink paintings in the new century. Art Researches 12(94): 37–40.

[4] Wan. Q.L. 1996. Literati painting and literati painting tradition. Literature & Art Studies 17(1): 140–144.

[5] Liu. X.L. 1998. Mr. D.A. Cun and Mr. Chen Shizeng: modern revival of literati painting by two exotic struggling. Quarterly Academic Journal of the Imperial Palace 15(3): 45–50.

[6] Wu. X.B., Hu. S.C., Zhang. W. 2007. The review of innovation classification. Journal of Chongqing University (Social Sciences Edition) 13(5):35–41.

[7] Lu. H. 1999. Modern Ink two decades. Changsha: Hunan Fine Arts Press.

[8] Mo. Z.X. 2010. The art of innovation: on the Innovation of Modern Chinese ink painting Journal of Guangxi Normal University (Social Science Edition) 31(2): 16–19.

[9] Liu.C.S. 2011. The use of new materials and innovative of contemporary ink painting. Art Panorama 23(7): 48–56.

Education Management and Management Science – Zheng (Ed.)
© 2015 Taylor & Francis Group, London, ISBN 978-1-138-02663-6

Agricultural economics case teaching and effectiveness evaluation based on Markov chain

S. Yu & G. Li

Agronomy College of Anshun University, Anshun, China

ABSTRACT: Teaching evaluation is a basic and indispensable link in the teaching activities. It has become an important urgent task of all kinds of school to establish and improve the teaching evaluation system. In this paper, limit distribution and penalty factor method of homogeneous Markov chain are the evaluation of teaching effectiveness about case teaching mode in "agricultural economics" course teaching process, the evaluation results are objective and accurate. Results showed that the implementation of case teaching students good teaching effect in the implementation of case teaching is not students, illustrate the case teaching in the teaching of mathematics is an effective teaching method.

Keywords: case teaching; agricultural economics; Markov chain; teaching evaluation

1 INTRODUCTION

Case teaching originates from Harvard University, which has become one of the important methods in modern education and teaching after a century. It has many advantages such as enhancing students' understanding of views and principles, mobilizing their enthusiasm, and improving their ability to solve problems, and so on. Agricultural economics has a strong application, as a study on the movement discipline of agricultural activities in productive forces and production relation. Agricultural economics occupy a pivotal position in Chinese economics, for such a large agricultural country. It is necessary to accomplish the teaching mission on agricultural economics in order to solve development problems of agricultural economics and train more agricultural economic talents of socialist agricultural industry. Thus, to use case teaching model in agricultural economics has important significance in theory and practice.

Teaching evaluation as the final part of teaching process, whose purpose is to provide feedback for the first few sectors of teaching activities. For teachers, the effective implementation of teaching evaluations can help teachers to obtain dynamic information about the quality of teaching, get the success and main problems in teaching process, make reasonable adjustments, and improve their business standards and teaching effects. For students, the effective implementation of teaching evaluations can help students to understand their own learning effects, find the problems, and adjust their working direction [1]. Currently, many educators have researched on how to conduct the case teaching mode in agricultural economics [2–4]. But most of their teaching effectiveness evaluation is qualitative research, lack of empirical analysis and specific discussion, which needs for further improvement. In addition, the teaching evaluation is a multi-factor, multi-variable, fuzzy nonlinear process, this paper try to see a process of case teaching as a homogeneous Markov chain, using the limit distribution of homogeneous Markov chain, excluding the differences in learning foundation, finishing objective and fair assessment of the teachers' effectiveness in the teaching process.

2 PRINCIPLES AND METHODS OF EVALUATION

Markov chain has limit state which determined only by changing reason, that is only related with current state. In this paper, the student' academic performance is divided into several finite states; the entire teaching process is divided into several time periods. Changes of the student' learning state from one time period to next time period, which can reflect teaching effects of teacher's teaching activities. This effect is a transformation from one state to the next state, which has nothing to do with the previous state, so the entire teaching process can be seen as a homogeneous Markov chain. Using the limit distribution of Markov chain can be used to evaluate teaching effects of teacher's teaching activities, and compare the teaching effects with several teachers [5].

2.1 Data sorting and grading student achievement

The student achievement is graded from low to high with standardized exam papers. Considering less academic performance is lower than 60 point, this study divided student achievement into five levels, such as 0~59, 60~69, 70~79, 80~89, and 90~100. Then the ratio (state vector A) of each level to the total number of students was calculated, which is calculated as $A = (n_1/n, n_2/n \ldots \ldots n_i/n)$, where n_i is the number of students in i level.

2.2 Calculation of the one-step transition probability matrix

According to Markov's no aftereffect, the next state of the system is only concerned with the current state, regardless of the state even earlier, so transition probability matrix P has become an important decision factor. During teaching evaluation, transition probability matrix P is epitomized by many factors such as the quality of teaching [6]. Transition probability matrix as $P_{ij} = n_{ij}/n_i$, where n_{ij} represents the number of students whose achievement transform from i level to j level. Because this paper divided student achievement into five levels, the formula of transition probability matrix P is calculated as $P = (p_{ij})_{5\times5} = (n_{ij}/n_i)_{5\times5}$.

2.3 Evaluation of teaching effectiveness

For a long time, evaluation of teaching effectiveness is a difficult issue in teaching research. The choice of evaluation method determines the accuracy of teaching effectiveness evaluation. This paper used limit distribution and penalty factor of Markov chain to evaluate teaching effectiveness and quality of agricultural economics case teaching. The reason is that the aperiodic normal returns irreducible Markov chain exist a unique stationary distribution [7]. Under the conditions of a considerable level of education, the proportion of students at all levels will be possible to achieve stable, this stability is only concerned with the level of teaching has nothing to do with the students on the basis. So limit distribution of Markov chain can be used to evaluate teachers' teaching effects, for it meeting the requirement of teaching effectiveness evaluation. In addition, for teaching is a interact process of teachers and students; it is one-side to evaluate teaching effectiveness only from teachers' teaching effects. This paper introduced penalty factor indicators to examine changes in student achievement, analyze achievement in learning process, and verify the merits of teaching effectiveness.

2.3.1 Evaluation of teaching effectiveness based on limit distribution

Solution of the limit as $\prod = (\pi_1, \pi_2, \pi_3, \pi_4, \pi_5)^T$, according to smooth equation $\prod P = \prod$ and

$\sum_{j=1}^{s} \pi_j = 1$, the final equation of teaching effectiveness as $Y = \sum_{j=1}^{s} \pi_j x_j$, where x_j represents the student achievement.

2.3.2 Evaluation of teaching quality based on penalty factor

Each factor in transition probability matrix is multiplied by the penalty factor $2(j-i)$ to get the improve efficiency matrix as $P^* = (2(j-i)\, n_{ij}/n_i)_{5\times5}$, and $E = \sum_{i=1}^{5} \sum_{j=1}^{5} (2(j-i)\, n_{ij}/n_i)$ is the extent of the cumulative change in student achievement, E greater, indicating that the quality of teaching, the better.

3 EFFECTIVENESS EVALUATION OF AGRICULTURAL ECONOMICS CASE TEACHING

There are two agricultural economics exam results (terminal and midterm) from 30 students of Agricultural and Forestry Economy Management (Class One) and Rural Regional Development (Class Two), see Table 1.

According to the Table 1, the initial state vector was:

$$A_1\left(0 \quad \frac{1}{15} \quad \frac{1}{3} \quad \frac{1}{2} \quad \frac{1}{10}\right), \quad A_2\left(\frac{1}{30} \quad \frac{1}{10} \quad \frac{1}{3} \quad \frac{2}{5} \quad \frac{2}{15}\right),$$

After the first test, Class Two carried out case teaching model in teaching activities, while Class One continues using conventional teaching method. After the second half of semester, the number of metastasis for each grade in the second test, see Table 2 and Table 3.

Table 1. First results and statistics.

Number/ score grades	0~59	60~69	70~79	80~89	90~100
Class One	0	2	10	15	3
Class Two	1	3	10	12	4

Table 2. Second results and statistics of Class One.

	Second results				
	90~100	80~89	70~79	60~69	0~59
First results					
90~100	2	1	0	0	0
80~89	1	6	3	2	3
70~79	2	1	4	2	1
60~69	0	1	0	1	0
0~59	0	0	0	0	0

Table 3. Second results and statistics of Class Two.

	Second results				
	90~100	80~89	70~79	60~69	0~59
First results					
90~100	3	1	0	0	0
80~89	3	5	3	1	0
70~79	0	4	3	3	0
60~69	0	0	1	1	1
0~59	0	0	1	0	0

The one-step transition matrix of student achievement from Class One and Class Two was:

$$P_1 = \begin{pmatrix} \frac{2}{3} & \frac{1}{3} & 0 & 0 & 0 \\ \frac{1}{15} & \frac{2}{5} & \frac{1}{5} & \frac{2}{15} & \frac{1}{5} \\ \frac{1}{5} & \frac{1}{10} & \frac{2}{5} & \frac{1}{5} & \frac{1}{10} \\ 0 & \frac{1}{2} & 0 & \frac{1}{2} & 0 \\ 0 & 0 & 0 & 0 & 0 \end{pmatrix},$$

$$P_2 = \begin{pmatrix} \frac{3}{4} & \frac{1}{4} & 0 & 0 & 0 \\ \frac{1}{4} & \frac{5}{12} & \frac{1}{4} & \frac{1}{12} & 0 \\ 0 & \frac{2}{5} & \frac{3}{10} & \frac{3}{10} & 0 \\ 0 & 0 & \frac{1}{3} & \frac{1}{3} & \frac{1}{3} \\ 0 & 0 & 1 & 0 & 0 \end{pmatrix}$$

The smooth equation of Class One from transition matrix P was:

$$\Pi = (\pi_1, \pi_2, ..., \pi_5) = (\pi_1, \pi_2, ..., \pi_5) P_1$$

$$= \begin{pmatrix} \frac{2}{3} & \frac{1}{3} & 0 & 0 & 0 \\ \frac{1}{15} & \frac{2}{5} & \frac{1}{5} & \frac{2}{15} & \frac{1}{5} \\ \frac{1}{5} & \frac{1}{10} & \frac{2}{5} & \frac{1}{5} & \frac{1}{10} \\ 0 & \frac{1}{2} & 0 & \frac{1}{2} & 0 \\ 0 & 0 & 0 & 0 & 0 \end{pmatrix}, \pi_j > 0, \sum_{j=1}^{s} \pi_j = 1$$

The smooth equation of Class Two from transition matrix P was:

$$\Pi = (\pi_1, \pi_2, ..., \pi_5) = (\pi_1, \pi_2, ..., \pi_5) P_2$$

$$= \begin{pmatrix} \frac{3}{4} & \frac{1}{4} & 0 & 0 & 0 \\ \frac{1}{4} & \frac{5}{12} & \frac{1}{4} & \frac{1}{12} & 0 \\ 0 & \frac{2}{5} & \frac{3}{10} & \frac{3}{10} & 0 \\ 0 & 0 & \frac{1}{3} & \frac{1}{3} & \frac{1}{3} \\ 0 & 0 & 1 & 0 & 0 \end{pmatrix}, \pi_j > 0, \sum_{j=1}^{s} \pi_j = 1$$

The limit vectors of Class One and Class Two from smooth equations were:

$$\Pi_1 = (0.169 \quad 0.423 \quad 0.141 \quad 0.169 \quad 0.098),$$

$$\Pi_2 = (0.390 \quad 0.390 \quad 0.173 \quad 0.035 \quad 0.012)$$

Finally, the five grades were given point as 95, 84.5, 74.5, 64.5, 29.5, the comprehensive assessment of each teacher from Class j (j = one or two) was given value as $Y_j = \pi_j(x_1, x_2, x_3, x_4, x_5)^T$, so $Y_1 = 76.0945$, $Y_2 = 85.505$. Since $Y_1 < Y_2$, the teaching effectiveness of teachers from Class Two was considered superior to Class One. The superior was in consonance with the fact that Class Two used case teaching model to improve teaching effects.

Each factor was multiplied by penalty factor 2 (j-i), reveals a new transition matrix was:

$$S_1 = \begin{pmatrix} 0 & \frac{2}{3} & 0 & 0 & 0 \\ -\frac{2}{15} & 0 & \frac{2}{5} & \frac{8}{15} & \frac{6}{5} \\ -\frac{4}{5} & -\frac{1}{5} & 0 & \frac{2}{5} & \frac{2}{5} \\ 0 & -2 & 0 & 0 & 0 \\ 0 & 0 & 0 & 0 & 0 \end{pmatrix},$$

$$S_2 = \begin{pmatrix} 0 & \frac{1}{2} & 0 & 0 & 0 \\ -\frac{1}{2} & 0 & \frac{1}{2} & \frac{1}{3} & 0 \\ 0 & -\frac{4}{5} & 0 & \frac{3}{5} & 0 \\ 0 & 0 & -\frac{2}{3} & 0 & \frac{2}{3} \\ 0 & 0 & -4 & 0 & 0 \end{pmatrix}$$

The efficiency of the two classes was: $E_1 = 7/15$, $E_2 = 91/30$.

343

Since $E_1 < E_2$, the teaching efficiency of Class Two was considered superior to Class One, students whose test scores progressed from Class Two were more than Class One. While the teaching efficiency of Class One was less, indicating that the use of case teaching model made students whose test scores progressed from Class Two become much more, no use of case teaching model caused students' progress transformed not significant from Class One.

4 CONCLUSION AND IMPLICATION

Agricultural economics as a professional discipline with strong practical and application, if the related theories were introduce by force lack of actual cases, students will fall into rote learning, cannot learn the way of thinking and advance concepts in agricultural economics. The case teaching has good application prospects, for its obvious advantages such as students' ability to think independently, comprehensive analysis capabilities, innovation ability, and so on. Class Two with case teaching is better than Class One regardless of performance metrics or students' progress indicators from paper above, indicating case teaching in agricultural economics has a good effect, should be vigorously promoted.

It's often more difficult to choose evaluation criteria in teaching evaluation process. The usual practice is to evaluate teaching effectiveness by exam pass rates and average scores, which has a big limitation with no consideration of teaching effectiveness impact by students' raw scores. Because Markov has no aftereffect, this paper introduced Markov chain to evaluate teaching effectiveness and relative improvement of students' achievement after using case teaching model, which can really reflect teachers' teaching effectiveness and students' progress. The paper analyzed application effect of case teaching in agricultural economics by limit distribution and penalty factor. Both of the two methods confirm that case teaching promoted students' performance in agricultural economics.

It has more scientific and practical compared to former evaluation with criterion as average and passing rate, for the evaluation system considering with student achievement and teaching effectiveness [6]. Therefore, effectiveness evaluation based on Markov chain has good maneuverability, can be widely used in teaching evaluation of various subjects.

ACKNOWLEDGMENTS

This paper is supported by Guizhou Province Education Science Project named Application of case teaching method in the teaching of agricultural economics (2013C038), Guizhou Academy of Agricultural Science and Technology Project [NY (2010)3014], and Natural Science Research Project from the Education Department of Guizhou Province [KY(2013)148]. G. LI is the corresponding author.

REFERENCES

[1] Ji Yq. Curriculum and instruction theory. Nanjing: Nanjing University Press, 2009:170–171. (in Chinese)
[2] Geng W.C. Application of case teaching method in agricultural economics. China's Collective Economy, 2009, 30:187–188. (in Chinese)
[3] Gu L.L. & Jiang H.M. Study on the application of case teaching in agricultural economics. Modern Communication, 2012, 2:252–253. (in Chinese)
[4] Zhao Y.K. & Zhao C.P. Analysis on the application and effect of case teaching in agricultural economics. Journal of Henna Institute of Science and Technology, 2013, 12:95–97. (in Chinese)
[5] Zhang H. An application of Markov chain. Journal of Changchun Institute of Optics and Fine Mechanics, 1994, 3:44–49. (in Chinese)
[6] Liu L.W., Chen X.R., He T. The teaching effect evaluation method based on Markov chain. Statistics and Decision, 2014, 3:93–94. (in Chinese)
[7] Wan L.Y. The application of Markov chain in teaching quality evaluation. Nantong Vocational College Journal, 2011, 2:67–69. (in Chinese)

Education Management and Management Science – Zheng (Ed.)
© 2015 Taylor & Francis Group, London, ISBN 978-1-138-02663-6

Thinking about the transformation of application-oriented undergraduate colleges into applied technology colleges

X.G. Song

Economic Management School, Huaiyin Institute of Technology, Huaian, Jiangsu Province, China

ABSTRACT: Recently, the State Council's decision to accelerate the development of the modern occupation education, opened a new round of higher education reform. According to the spirit of the document, the Ministry of Education decided to put more than 700 local colleges gradually change in applied technology colleges of modern higher occupation education. Based on elaborating the development confusion of application colleges, this article analyzed the difference of application-oriented colleges and applied technology colleges, and pointed out the regression of modern higher occupation education is the realia development path of application-oriented colleges. Finally, this article proposed application-oriented university changes to the application technology in preliminary measures.

Keywords: vocational education; higher education; reform

1 INTRODUCTION

1.1 *Type area*

In May 2, 2014, the State Council announced the decision to accelerate the development of the modern occupation education. This decision puts forward our country modern occupation education the medium-term development objectives and tasks. An important aspect to achieve the goals and tasks is to explore the development of undergraduate education in the higher occupation to consolidate the foundation of development for medium vocational education and higher occupation college education. To establish occupation demand as the guidance, the practice ability as the key point, combining production and study as a way of professional degree graduate training mode. The decision will undoubtedly lead China higher education a revolutionary adjustment getting started.

2 ANALYSIS OF THE BACKGROUND OF A NEW ROUND OF REFORM IN HIGHER EDUCATION

From 1999, with the enrollment expansion, China's higher education has made great achievements, but also produced many new problems. The most prominent problem is the serious departure between the personnel training and social needs, caused the difficult employment situation. With the progressive realization of higher education transition from "elite education" to "public education" in China, the number of college graduates is facing overcapacity and excessive pressure on employment issues. At the same time, a large number of manufacturing enterprises are suffering from a dilemma not to hire qualified workers. In March 7, 2013, a member of the CPPCC National Committee, proposed the ACFTU vice chairman Duan Dunhou made a speech on the second plenary meeting of the CPPCC National Committee of the twelve session: national technical supply and demand gap in between 22 million to 33 million peoples, strengthen occupation skill training, improving the quality of workers is a pressing matter of the moment. Our senior technicians account for only about 5% of the total number of workers, and a far cry from the developed countries the proportion of 40% senior technicians. College graduates employment difficulty has continued for many years. Therefore, the higher education of our country is facing a major reform to adapt to the transition of social economic structure. Compressing general higher scale and expanding the scale of higher vocational education are the main direction of this reform.

3 THE EXPLORATION OF THE APPLICATION-ORIENTED COLLEGES TRANSITION TO APPLIED TECHNOLOGY INSTITUTIONS

3.1 *The development of China's higher education classification and application-oriented colleges confusion*

At present, full-time higher education institutions are divided into three categories: the first category is the high level academic research universities, including 985,211 institutions and some provincial key university; second is the application type undergraduate colleges and universities, mainly local colleges. Third category is numerous in higher vocational colleges and a small amount of normal, medical and the public security colleges.

Application type undergraduate colleges refer to the provincial colleges of application for the school running orientation, rather than the key universities to take the scientific research as the orientation. Local application oriented undergraduate education has played a positive role to meet China's economic and social development of the high-level application-oriented talents need and promote the popularization of higher education in China.

Higher vocational education emphasis on applied technology training personnel, currently in China has basically refers to levels of specialist qualifications in vocational education, and higher vocational education qualifications abroad have risen to PhD level.

China's current application-oriented colleges training objectives are positioning the application of talent, but not the vocational education system running. In response to recent college graduates difficult employment situation, application-oriented colleges gradually opened up some new professional-oriented social demand for professionals, and has also implemented a program of applied education reform. But these partial reform measures can not solve the fundamental problem. From personnel training targeting perspective, there is not much difference in application-oriented colleges and vocational colleges. But in recent years, as two types of schools graduate employment situation, the employment rate of graduates of vocational colleges actually much higher than ordinary colleges. Application-oriented colleges are in an awkward position that they basically did not develop into a high-level academic research universities, but in terms of applied talents of view, they are down to compete vocational colleges. In fact, from the rest of the world view of the history of higher education, there are differences between general education and vocational higher education. General higher education engaged in training academic personnel and there is no such a role positioning applications oriented colleges. Therefore, the return of higher vocational education is the rational development path of application-oriented undergraduate college.

3.2 The distinction between application-oriented colleges and the applied technology colleges

From the name, the two seem very different, but essentially there is a fundamental difference.

The former attributable to ordinary education series, which belong to vocational education series. There are continuing education or adult education, in addition to general education and vocational education. Our general education series includes basic education (including preschool and compulsory), secondary education and higher education. Vocational education in China mainly includes primary vocational education, secondary vocational education and higher vocational education. Being lack of vocational education at undergraduate and postgraduate levels of education, vocational education in our country is a broken road, which is not conducive to advanced applications training technical personnel.

Above, the application-oriented colleges are engaged in ordinary education series undergraduate and graduate degree levels of education in provincial colleges, because of its location-based applications running talents distinguished academic and research universities. And the application of technology-based colleges engaged in a series of vocational education refers to undergraduate and graduate degree-level vocational colleges. Obviously the latter is to improve the appearance of the system of vocational education, to help to form a complete chain.

3.3 The difficult and countermeasures of the applied undergraduate college transformation for the application of technology colleges

3.3.1 The whole community should abandon prejudice against vocational education

Application-oriented colleges transformation into the application technology colleges, means the transformation of higher education series. And in our society as a whole has long formed a prejudice against vocational education that is lower than general education vocational education, only poor school performance students and poor economic conditions of families will choose to receive vocational education. Confucius words make no social distinctions in teaching, education itself is not of high and low points. Including the departments of education, educational institutions and the family, the whole society should eliminate obsolete concepts of occupation education inferior. In fact in developed countries, to accept the occupation education students scale is far beyond the general education students scale. The German occupation education development is in a leading position in the world. According to the authoritative organization, graduates universities in Germany for peer ratio of only 20%, nearly 80% of the young people are to accept the occupation education, and then go to work. It is through the occupation education successfully, providing a large number of outstanding industrial workers as the "made in

Germany", has also become an important source of the German national competitiveness.

From vocational education in developed countries experience, vocational and technical education for families, schools and society are very meaningful. To set application technology college of undergraduate and graduate degree levels occupation education is in order to compensate for the shortcomings in our country over the years of higher vocational education. In the long run, this will raise the level of vocational education for the community to help reverse the bad impression.

3.3.2 *Application-oriented colleges should lay down the burden of thinking, deepen reform, restructuring initiative*

A. The higher occupation education in colleges and universities is the legal responsibility and obligation. Article XIII of People's Republic of China Vocational Education Law stipulates clearly that occupation school education is divided into primary, secondary, and higher occupation education. According to the needs and conditions the higher occupation school education is carried out by the higher occupation colleges, or by regular institutions of higher education. Therefore, the application type undergraduate colleges should put aside ideological baggage, actively participate in the cause of higher occupation education.

In recent years, the scale of higher education enrollment in China maintains the basic stability. According to our country population birth rules, with the scale expanding and abroad student studying abroad in young students, the future of ordinary higher education enrollment will be relatively reduced. Ten years ago, almost every applied undergraduate college was blind to expand school scale for college enrollment and the Ministry of education undergraduate teaching assessment. Now, most colleges have not yet repaid bank loans. Facing the trend of ordinary higher education enrollment relatively narrow, application-oriented in the crevice will appear the difficulty of enrollment, and the survival of the crisis if they can not take the initiative to change. Engaging in the higher occupation education, may resolve the plight of applied undergraduate colleges.

B. Application-oriented colleges should deepen reform inside to lay a good foundation for the transformation.

After putting aside ideological baggage to transit to the higher occupation education, application-oriented colleges should deepen the internal reform, and create a good internal condition for transformation.

First, they should be carried out quickly the internal mechanism reform to remove administration. Currently, the colleges and universities have strong color of administration. The general provincial leadership to configure multiple job number in universities (deputy departmental level and above) is more than 10 people, the number of deputy department level cadres over nearly 100 people. Deputy level cadres accounted for more faculty ratio is not less than 10%. Obviously, our current university management practices are following the management mode of the executive in planned economy era. The huge bureaucratic administration management team will take up too much of educational resources, and lead to a decline in efficiency of university management, and even administrative bureaucratic corruption. Therefore, universities should increase efforts to remove administration reform, reduce the administrative cadres and administrative institutions, attract businesses and social forces to participate in the school, and establish a true achieve professors management and service model by school board. Colleges and universities should retain the authority competent, change idea, and take as the major functions to provide services for teachers and students.

Second, colleges should change the mode of talent training, take the professional development as the central, curriculum construction as the key point. Application-oriented colleges should suppress inner impulse to upgraded to comprehensive academic research university. Colleges should truly realize that the application of technology-based training senior personnel are based on fundamental, abandon the discipline construction as the focus, center on professional development, and take curriculum development as the focus. Training objectives should be to stand humane quality and professional quality. To increase the set curriculum system reform, colleges should change the unreasonable phenomenon that such as college English, ideological and political courses takes up too much time.

Third, we must form a professional teaching team with double-position teachers to meet the needs of higher vocational education.

The teacher demonstration is playing a very important role in occupation education. As qualified professional teachers of higher vocational schools, they should have classroom teaching competence, understand the business and market, have a higher skill level, and effectively carry out the practice of teaching. Colleges should actively explore vocational teacher training model, vigorously develop the double-position teachers which have solid theoretical knowledge and strong professional practice, and can integrate theory and practice teaching task. Therefore, we should set up specialized training system that course teachers regularly go to enterprises to practice, or start the part-time job. Colleges should employ double-position part-time teachers from the business with relatively high technical expertise.

3.3.3 *State should provide a good external environment for the development of higher vocational education*

Leading application-oriented colleges to transit to technology-based colleges and even promoting the entire vocational education development needs of a favorable external environment. This aspect, the state assumes the bounden duty.

A. The state should accelerate the reform of the college entrance examination system. Since the general higher education and vocational education training objectives differ, they should not choose the same personnel selection methods. The college entrance examination to choose the academic talent and skill talent should separate. High school education should also be appropriate to adjust.
B. To take the ways of pilot to promote and demonstrate, the state should lead and guide a group of ordinary undergraduate colleges to transit to apply technical colleges, and focus on holding undergraduate vocational education. Not to engage in campaign-style vocational education reforms, the state should as soon as possible promote colleges to remove administrative reform.
C. To build the rational classification management criteria of institutions of higher learning and to take the reform of vocational education teachers titles assessment approach. In 2003, the Ministry of Education formulated the project to evaluate the undergraduate teaching work level of university or colleges and established a five-year assessment system. The valuation to take the same criteria for different types of colleges and universities is obviously unfair. If you have to implement a similar system for the evaluation of colleges after the transition, it is very detrimental to their development. Similarly, for the quality requirements of higher vocational education teachers and the general education teachers are not the same, its title appraisal approach should be different. The current title appraisal system is not conducive to higher vocational education teachers.
D. The state must as soon as possible to improve the system of modern occupation education, increase the higher occupation education investment. At the same time also give full play to the role of the market, with market forces to school. To gradually change the higher occupation education state ownership situation, can adopt the share-holding system, mixed ownership and other flexible educational system.

REFERENCES

[1] Department of higher education, teaching staff report higher occupation education.
[2] The state council, on accelerating the development of the modern occupation education decision.
[3] The people's republic of China occupation education law.

Education Management and Management Science – Zheng (Ed.)
© 2015 Taylor & Francis Group, London, ISBN 978-1-138-02663-6

Research on international business talents cultivation of innovative thinking

Jianling Li

Business College, Beijing Union University, Beijing, China

ABSTRACT: The creative thinking is a new genre of thinking and philosophy category at the beginning of the 21st century. In both personnel training and scientific research, it showed the unique function. In today's society, the world economic integration is trend to develop in the direction of depth, increasing of international trade. International economic exchanges have become more frequent, our demand for international business talents become more and more urgent. In the era of knowledge economy, with the development of economic globalization and information technology, international business talents cultivation of innovative thinking is very important. In this paper, from the aspects of concepts, elements, feature and mode, we discussed the innovative thinking. In the end, we discussed the international business talents cultivation of innovative thinking. Through the international business talent cultivation of innovative thinking, we enhance the innovative ability of the international business talent. We produce high quality international business innovative talents for the country.

Keywords: international business talents; innovative thinking; innovation ability; culture; quality education

1 INTRODUCTION

In the process of economic globalization, in order to adapt the rapid development of international trade, developing international business talents with international business knowledge and skills is very important. With the constant improvement of the economic globalization and internationalization of national, under the impetus of the development of economic globalization and information technology, the form and content of international business activities are profound changes have taken place. The 21st century is the age of knowledge economy. Knowledge economy is based on knowledge as the main body of economy. Talents are the carrier of knowledge. So international business talents with innovative thinking and innovative ability cultivation are the need of our research topic, it's also one of the countries needed to raise internationalization.

2 THE KEY OF PERSONNEL TRAINING

The knowledge economy era is an era of constant innovation. Innovation is the foundation of a country's economic sustainable development. The contemporary international competition is innovation talented person's competition. In today's social and economic globalization, the development of high and new technology quickly, the international competition is fierce. Science and technology determines the competitiveness of the country. The cultivation of high-quality talent is the key. The training of innovative talents cultivation is mainly. A nation, a people want to have the innovation ability, we must have a lot of creative thinking, innovation consciousness and innovation ability of talents. How to adapt to a knowledge economy develop innovative thinking, innovation spirit and innovation ability of high quality people, this is an important task to colleges and universities and the top priority.

In the evaluation of education of China, Dr Yang Zhenning has said that: "China education has two advantages and two disadvantages. Two advantages are more solid basis knowledge of students' and test ability. Two disadvantages are the weak of students' operating ability and practice ability. Students are lack of innovative spirit and ability." It said that our higher education only pay attention to imparting cultural and scientific knowledge, but ignored the cultivation of students' creative thinking ability. Educators think that the students' brains are stored knowledge warehouse. They attach importance to imparting scientific knowledge. Students accustomed to the teacher speak of memory book knowledge and existing solutions. Examination is given priority to with existing knowledge memory. Examination is the lack of students' creative thinking and the ability to deal with the problem of content. The cultivation of innovative thinking ability is the core and soul of cultivating creative personnel. Innovative thinking is a multi-angle, multifaceted, multi-level and the thinking of the structure. It is not restricted by the existing knowledge and the bondage of the traditional method. It puts forward new ideas, builds new theory, and discovers new things and new laws. It makes people to move to a higher and newer. Creative thinking is the sole and creative, spiritual activity, critical, agility, fluency, profundity, syndrome differentiation, sudden and comprehensive qualities. Master the creative thinking strategies and consciously forging innovative thinking is preparing a new talent key ring section. [1–4]

Talent training is need innovation. It basically is the cultivation of innovative talents. Because thinking determines consciousness and behavior, so the key to innovative talent training is the cultivation of the innovative thinking ability. This is also the key to talent training.

3 THE CONNOTATION AND MODE OF INNOVATIVE THINKING

3.1 *The concept of innovative thinking*

At present, the innovative thinking is not a clear and unified definition. And innovation thinking is the most close to the creative thinking. It can produce new ideas of thinking activity. It can break through the conventional and traditional. It is cramped by existing conclusions. It solves the new problems in a new unique way. It has openness, pioneering, flexibility, uniqueness, effectiveness and the basic characteristics of unconventional, it is shown in Figure 1. It is not confined to the pursuit of thinking results, but pays more attention to the thought process. Creative thinking can not only reveal the nature of the objective things and the internal relations of things, but also can produce novel, unique insights and ideas. It can come up with creative ideas. Therefore, innovative thinking is higher than general thinking form. It is creativity to find new problems. It is a innovative thinking process to find new questions and solve them. [5]

3.2 *The elements of innovative thinking*

Innovative thinking is the important foundation of international business talent innovation ability. Innovative thinking includes talents innovation and comprehensive development of their own quality. It has inspired, interests, and to predict three elements. Inspiration is a sudden burst of creativity. It is the foundation of creative thinking. The interest is formed in the brain of excitement and tendency. It can promote the development of creative thinking. Prediction is an important factor in creative thinking. It is an important means to control and grasp the opportunities in the talent. Inspiration, interests, and prediction is one of the important elements of innovative thinking, as shown in Figure 2. In the three elements of innovative thinking, inspiration and interest are the life of the innovative thinking and motivation, prediction is to better promote the development of innovative thinking. By analyzing the elements of innovative thinking, we can stimulate the talents further understanding the creative thinking, to deepen the understanding of the innovative thinking. In this way, we can greatly to mobilize talent innovation

Figure 1.   Innovative thinking.

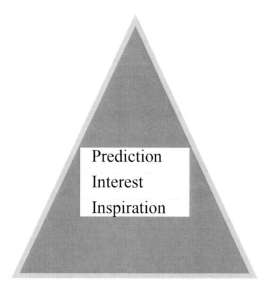

Figure 2.   The three elements of innovative thinking.

consciousness, cultivate talents of innovative thinking, and promote the birth and development of innovative thinking. So we can improve international business talents' innovative thinking ability.

3.3 The main feature of innovative thinking

The essence of innovative thinking is questioning mind. It is pursuit of innovative, unique and best. That is unknown to choose, have the courage to do things unconventional or unorthodox. We develop a new process, in order to obtain the initiative thinking of social value. Innovative thinking is on the way of thinking that characterized by the combination of divergent thinking and concentrated thinking. But it is more emphasis on divergent thinking. [6]

3.3.1 Active aggressive

Innovative thinking has the outstanding dynamic aggressive. Because it's thinking is the thinking process of strong creative desire and thirst for new knowledge, rather than a relaxed state of mind. Innovative thinking is prepared, and it is a must have.

3.3.2 Flexible variability

Multi-dimensional open way of thinking is put objects into the thinking of the three-dimensional network. We are from the perspective of forward, reverse and lateral multidimensional survey analysis. Such innovative thinking is access to new discoveries.

3.3.3 Strong challenge

Traditional thinking is admitted to the premise of existing things. It derived from deduce new knowledge. And innovative thinking is under certain conditions, we have been in the negative things. So innovative thinking has a strong challenge. It is a more powerful weapon of recognizing things.

3.3.4 High efficiency

According to conventional thinking, on the basis of existing knowledge, we have quantity degeneration of innovation. But break this kind of pattern of thinking, we can obtain breakthrough degenerative innovation. Visible, compared with traditional thinking, innovative thinking is a kind of quality and efficient thinking system.

3.4 The mode of innovative thinking

The traditional thinking is the thinking that divided into logical thinking, rational thinking, image thinking and dialectical thinking dimension. With the development of epistemology and the initiation

of the innovative thinking. Someone divides into the innovative thinking, divergent thinking, reverse thinking, lateral thinking, dimensional thinking and convergent thinking, inspiration thinking, positive thinking and negative thinking and identity thinking, and so on. There are mainly four kinds of patterns such as divergent thinking and lateral thinking, these constitute the main body of innovative thinking. [7–9]

3.4.1 Divergent thinking

It is also known as radiation spread dimension, thinking. It comes from an observation point, multi-angle thinking method of an object. It is characterized by dimensional vision, to a state of divergence. It is one of the major modes of creative thinking. It is also a form of multi-dimensional thinking.

3.4.2 Lateral thinking

It is put forward by the British scholar Edward DE bono. It advocates to flank broadening of thinking. It makes thinking on the two-dimensional spatial extension. But in depth, it is a certain limitation.

3.4.3 Reverse thinking

It is also known as questioning mind. It is of uncertain things or ideas to the way of thinking. With suitable to opposite thinking, it is from the results back to the known conditions. So that we can obtain a new discovery. Compared with the traditional thinking, it is to change the direction, but there is no increase dimension.

3.4.4 Multidimensional thinking

It is open to multi-dimensional thinking. It has a systematic, holistic, much to the characteristic of the thinking. It is close to comprehensive thinking, the overall thinking of. Thus multi-dimensional thinking can be basic covers the advantages of the above three kinds of thinking mode. It is the core concept of innovative thinking and the dominant mode of thinking.

Figure 3. The mode of innovative thinking.

Innovative thinking ability includes scientific imagination and sharp hole after forces, quick intuition, accurate judgment and skilled symbol processing ability and so on. Innovative thinking ability to focus on is related with the innovative thinking ability. Innovative thinking is a system. This system can make a comprehensive and in-depth thinking activities, orderly and rapid. The mode of innovative thinking is shown as in Figure 3. [10]

4 INNOVATIVE THINKING TRAINING OF INTERNATIONAL BUSINESS TALENTS

With the development of society and economy, service industry rapidly raised as the core of the third industry. Our economy is developing rapidly. Great changes have taken place in the industrial structure. Relative weight of the third industry is the fastest growth. Among them, international trade industry is developing rapidly with the world economic integration trend. International trade has been occupied the important strategic position in national economy of our country. Second thing is the demand for international business talents become more and more urgent. Creative thinking requires us when faced with a problem, we can quickly give a creative solution. In this way, we can timely find problems and correct understanding the problem, quick put forward the reasonable solution to solve the problem, effective implementation of the scheme to solve the problem. Creative talents with character are the thinking logic of originality, thinking choice of comprehensive, scientific create imaginary and thinking flexibility. [11–17]

4.1 Update the ideas

We support the excellent talents to break through the bondage of thought. We adhere to the scientific attitude of seeking truth from facts. We eliminate the original knowledge system in the old thinking mode of thinking obstacles. We are looking for the most suitable method of thinking in all existing way of thinking. We should firmly establish a dialectical materialism and historical only socialist view of the world. Exert oneself aspirant academic atmosphere, we offer innovative thinking type talented person growth loose external environment.

4.2 Grasp the dialectical thinking method

Creative thinking is the highest realm of human mind. Creative thinking ability is the concentration of dialectical thinking ability. Only the flexible use of dialectical thinking form and thinking method to analyze everything, remove the routine of the traditional way of thinking at the same time, we can seize the essence through the phenomena. We can find the inherent law of development and changes of things. So we can get a new conclusion, form a new understanding.

4.3 Innovation quality-oriented education

Man is the basic power to create value. We must start from the source of knowledge economy. We attach importance to human capital, to improve the quality of the people. The cultivation of the innovative thinking ability is not exclusive learning and mastering of knowledge. But we must take person's full scale development as the fundamental. We pay attention to a pluripotent, then form a reasonable knowledge structure. Therefore, we must break through the closed structure of education system. We carry out quality education seriously. We strengthen the personal special skill training. We enhance the ability of research questions, so as to develop independent thinking and to use new theory, new method to solve the problem of habit.

4.4 Foster development with innovation

Creative thinking is the first into the cultural accumulation and formed a kind of cultural form. Especially the first culture contained in the democracy and enterprising can prompt the development of innovative thinking dimension. Democracy is the social culture foundation of creative thinking. Creative thinking is the cultural force. Open is a creative thinking activity will be conditions. Only pushing into the advanced social culture construction, get rid of traditional culture hamper innovation thinking, extensively absorb the culture of mankind nutrition, as well as the valuable achievements of innovative thinking to give social return, we can increase the innovative thinking of new energy, become a strong driving force to promote the development of knowledge economy.

4.5 Develop a healthy psychological personality

The object of creative thinking is the practice of the potential has yet to recognize the elephant. It is in do not break change from potential to reality. The subject is that object of existing conditions of innovative thinking, though an understanding of an object but not completely. So the object is to have the potential. This creative thinking process is a process of exploring the unknown. Due to the limitation of various subjective and objective factors and influence of innovative thinking sometimes make the mistake of invalid or conclusion, even it is hampered by tradition and authority. Only with the mental health of people, we can maintain normal mentality in creative thinking activities. We insist on continuous improvement, and thus to achieve the ideal of innovative thinking.

4.6 *Build the perfect system of creative talents*

Age in progress, career development, national demand for creative talents is more and more big. Set up perfect national innovation talent system is inevitable. We will accelerate the establishment of a conducive to retain talent and use talent mechanism. Income distribution, we should institutionally ensure that all kinds of talents to get adapted to their work and contributions. Through various efforts and all the work, we strive to cultivate innovative talents.

5 SUMMARY

Today's knowledge economy era, innovation is the key to the training of thinking. With the development of economic globalization and information technology, international business talents cultivation of innovative thinking ability is very important. We discussed the concept, elements and characteristics of creative thinking and mode, and discussed the international business talents cultivation of innovative thinking ability. Through the international business talent cultivation of innovative thinking, we enhance the innovative ability of the international business talent. Our international business innovative talents for the country to create high quality, it is the need of China towards the internationalization.

ACKNOWLEDGMENT

This research was financially supported by the Natural Science Foundation of Beijing, China (Grant number 9132001).

REFERENCES

[1] Guo Zhenjun. Applied talents innovation ability training. The Chinese and foreign education research. 2011.10 (in Chinese).
[2] Wang Chongtao, Zhao Wuyi. International business personnel training mode of exploration research. Science and technology innovation herald, 2013.16 (in Chinese).
[3] Huang Tao, Fan Yanping. Theory of innovative talent training. Science and technology talent. Market. 2004.1 (in Chinese).
[4] Hayfaa A Wahabi and Lubna A Al-Ansary. Innovative teaching methods for capacity building in knowledge translation. Wahabi and Al-Ansary BMC Medical Education 2011, 11:85.
[5] Zhou Xiangjun. The research and practice of cultivating creative thinking ability. Science and technology in western China, 2008.1 (in Chinese).
[6] Yu Hongjuan. The cultivation of the innovative thinking ability. Technical secondary school physics teaching, 2001.6 (9) (in Chinese).
[7] Sun Guangyou. Innovative thinking and high quality talents. Journal of Harbin institute, 2012.6 (33) (in Chinese).
[8] James L Wofford and Christopher A Ohl. Teaching appropriate interactions with pharmaceutical company representatives: The impact of an innovative workshop on student attitudes. BMC Medical Education 2005, 5:5.
[9] Philipp Koellinger. Why are some entrepreneurs more innovative than others?. Small Bus Econ (2008) 31:21–37.
[10] Wang Xueli. Innovative thinking is how tempered. Invention and innovation, 2013.8 (in Chinese).
[11] Gao Dengcheng. Innovative thinking type talent cultivation. Enterprise reform and management. 2001.9 (in Chinese).
[12] Zhang Hongyan. I see a new person to bring up. Wu han financial college journal, 2002.6 (69) (in Chinese).
[13] Xun Ge. University of international business practice teaching research of talent training. Intelligence development of science and technology and economy, 2007.12 (17) (in Chinese).
[14] Li Xiaohong, Mo Xikun. WTO and international business talents. Journal of shanxi university of finance and economics (higher education), 2002.4 (56) (in Chinese).
[15] Wang Xiuqin. Train of thought to explore innovative thinking training in knowledge economy era. The China science and technology information, 2006.1 (in Chinese).
[16] Alphonse Uworwabayeho. Teachers' innovative change within countrywide reform: a case study in Rwanda. J Math Teacher Educ (2009) 12:315–324.
[17] Connie M. Lee, Siyuan Gong, Chao Tang and Wendell A. Lim. Bridging cross-cultural gaps in scientific exchange through innovative team challenge workshops. Quantitative Biology 2013, 1(1): 3–8.

Education Management and Management Science – Zheng (Ed.)
© 2015 Taylor & Francis Group, London, ISBN 978-1-138-02663-6

On the multimodal evaluation in Business Spoken English Course

Jian Zhang
Ningbo Institute of Education, Zhejiang, China

ABSTRACT: This paper describes the significance of application of multimodal evaluation in Business Spoken English Course, proposes three evaluation methods which are presentation in class, multimedia technology application and group capability training program, and explores the practical application of multimodal evaluation in Business Spoken English.

Keywords: multimodal; evaluation; Business Spoken English

1 INTRODUCTION

Nowadays, the single mode in communication field such as written text has gradually given way to multiple-mode integrating various media containing text, voice and image, and the multimodal means of communication is creeping into our life and study unconsciously[1]. The way of multimodal evaluation was also generated from multimodal discourse. China is still at the initial stage of applying multimodal evaluation to learning, and only in recent years did we realize the importance and start research. However, how to effectively combine the both of them? This has always been a problem which scholars and professionals concern and want to solve, because it's the inevitable development trend of language teaching.

2 MULTIMODAL EVALUATION AND THE THEORETICAL FRAMEWORK

2.1 Multimodal evaluation

Multimodal discourse analysis means applying various sensory modality in human communications to analysis and communication and has been widely used in each field of education circles. Multimodal evaluation, as the new means of language education and learning, mainly means the communication, interaction and learning in multiple aspects of image, text, color, video, voice and other elements with the participation of a variety of senses of vision, audition, touch, and so on, which shall be taken as the method to judge and evaluate a person[2]. To apply multimodal evaluation to Business Spoken English Course, it means the way to comprehensively use various evaluation methods, integrate the opinions of evaluation and then reflect the evaluation result on teaching performance.

2.2 Theoretical framework of multimodal evaluation

The theoretical basis most suitable for multimodal evaluation is constructivism, and both are inseparable. Constructivism is a learning view that regards learning as a kind of constructive activity.

According to constructivism, learning should not be thought as the passive receiving of knowledge taught by teachers, instead, it should be the learners' constructive activity of taking an active role in receiving new knowledge on the basis of the knowledge and experience they already have with the combination of their own real conditions. And of course in this process, the communication with others is unavoidable. So constructivism lays stress on positivity, initiative and innovation in learning. In this sense, multimodal evaluation cannot do without the guidance of constructivism, because multiple-mode is not only the integration of voice, text, image and multimedia equipment, but also the organic combination of students and teachers. In the process of multimodal evaluation, the students could show their knowledge and skills with diverse accomplishments and by various means and the teachers could take multimodal evaluation as the evaluation and test method for constructivism learning.

The theoretical framework of constructivism is composed of four parts of environment, exchange, cooperation and goal, and in a similar way, multimodal evaluation could refer to this theoretical framework. The environment must be in favor of students' learning and the cooperation should be in the whole process of learning and evaluation, even the evaluation on students should be embodied in the way of cooperation. Interchange of students in this process, which is the most basic way of cooperation, is a must, and the ultimate goal of multimodal evaluation is to improve students' learning outcome and promote their all-round development. Therefore, environment, exchange, cooperation and goal constitute the theoretical framework of multimodal evaluation. When teachers assess the students, they shall follow this theoretical framework and go deeper and deeper gradually, so as to reach the ultimate aim of improving the quality of teaching and cultivating well-rounded talents.

3 THE APPLICATION OF MULTIMODAL EVALUATION TO BUSINESS SPOKEN ENGLISH COURSE

Business Spoken English Course aims to train students' practical ability in business communication and oral English, boasts particular linguistic characteristics, and manifests the strong features of pragmatism and operability. The conventional way of single mode teaching evaluation concerning more about language or memory capacity has been far from meeting the demands for applied talents in business English teaching, nor can it satisfy the needs of interchange in business English teaching. In order to reflect business English interchange activities more truly and comprehensively investigate students' ability of effective business English interchange during study at ordinary time, the conventional single mode evaluation could be innovated in three aspects.

3.1 *Presentation in class*

Presentation offers students a platform for study and interchange, and is the key way of experiencing learning results, as well as a very effective means of teaching evaluation. It is absolutely not the simple repetition of teaching contents, but the "presentation, revelation, evaluation and enhancement" of the learning outcomes. Main ways of presentation include written presentation, oral presentation and dynamic presentation. Written presentation (e.g. design the schedule of business activities) requires written and methodized contents to be presented and is one of the methods to evaluate students' ability of abstracting main points of questions and ability of logic thinking. Oral presentation (e.g. introduce the company to customers) requires the students to speak fluent and concise languages in explanation, discussion and communication and to properly state a point, a problem or research conclusions. In the course of Business Spoken English, the high standard requirement for oral expression is no doubt the priority among priorities. Thus it can

be seen that oral presentation plays a vital role in exposing students' problems in pronunciation, linguistic organization and verbal communication. Dynamic presentation mainly refers to the teaching method of role play and situational teaching. Dynamic performance requires students to play a part in business activities under specific simulation situations, such as business meeting, business and price negotiation with foreign trade businessmen. The evaluation on presentation in class could be the combination of teacher's comments on site and students' anonymous comments, and students could become aware of the essential problems when hearing teacher's comments and making comments on others, so they could further enhance language use ability and foreign trade business ability.

3.2 *Multimedia technology application*

Multimedia technology fully mobilizes student's visual and auditory sense, as well as associative and operational ability through such teaching approaches of projector, slide and film, and could be taken as an evaluation means with forcible basis of feasibility. First of all, multimedia technology is an important guaranty for success of presentation evaluation and group task-based teaching evaluation and greatly boosts the enjoyment and feasibility of these two evaluation means[3]. Besides, Business Spoken English attaches more importance on language communication ability and business operation ability; however, the conventional single mode evaluation can hardly realize the investigation on students in multiple aspects, for which multimedia technology provides a channel. For example, with remote operation technology, students could directly communicate and exchange with foreign businessmen to evaluate their ability of intercultural communication and adaptability; ask the students to work in groups and shoot video about simulation business situations, let other groups point out the mistakes, and then evaluate their knowledge of language and business operation under such situations. Besides breaking through the time and space limit of conventional teaching evaluation, multimedia technology also makes the evaluation more vivid and variable. However, due to the restriction of all kinds of objective conditions, the application of multimedia technology to evaluation still needs further study.

3.3 *Group capability training program*

Group capability training program is the combination of group teaching and task-based teaching. Based on the entire business activity, the teacher could divide all links into four capability, training programs of old customers retaining, developing new customers, business negotiation and business services, and design several tasks in each program, so that students could have comprehensive exercise through the practice simulation. The tasks require students' collaborative work in groups, while each student is different from one another in terms of leaning ability, strong points and subjective initiative. Group work enables students to make the best of them, which not only promotes the participation rate, but also strengthens their team spirit.

In group capability training program, evaluation of the group inside and outside could be implemented with the group as a unit. Evaluation outside of the group refers to the objective evaluation made by the teacher and other students according to accomplishment of the entire tasks, and that inside the group is carried out by group members according to participation in the tasks. In this way, all group members could air their own opinions and cooperate with others actively to fulfill each task, thus their dominant roles are promoted potentially and each student's potential is explored.

4 EFFECTS OF MULTIMODAL EVALUATION

4.1 *Change the subject of evaluation*

In classroom teaching, we often emphasize the dominant role of students and support of teachers. However, when it comes to evaluation on students, they become passive, and results given by teachers become the only standard to measure their ability. Obviously, this single mode evaluation imposes restrictions on students' innovation and practical ability and their initiative and flexibility in solving problems[4]. Multimodal evaluation, on the other hand, abandons the conventional single mode mechanism of evaluation by teachers and replaces it by the mechanism of combination of students' self-evaluation, mutual evaluation and teachers' evaluation, which is to make comprehensive evaluation on students' learning attitude, strategy and accomplishments, thus changing students' role of passive participation into positive evaluation, and enhancing their participation and confidence. Apart from this, multimodal evaluation has many means to comprehensively investigate a student's language communication ability, business communication ability, professional quality and innovation ability. For students, multimodal evaluation provides sufficient time and space to show themselves during evaluation, and teachers could learn about students' abilities in various aspects, thus teaching them in accordance with their aptitude.

4.2 *Make up the insufficiencies of single mode evaluation*

Since long, in Business Spoken English Course, the single mode final evaluation means of sole dialogue between two people has given rise to the problem that teachers cannot make comprehensive evaluation on students' actual application ability of Business Spoken English due to the single purpose of evaluation and ignorance of differences and value of personality development during evaluation. We have hardly noticed that the purpose of evaluation is to find out students' deficiencies and facilitate their further development. Single mode evaluation leads to students' thinking set, improves ability in exams and weakens ability in adaptability[5]. Such particular characteristics of flexibility and practical operability of Business English Conversation have been ignored. Comparatively speaking, the means of multimodal evaluation runs through the entire teaching process, flexibly and reasonably carries out evaluations with pictures, texts and even videos based on content of course, for example, presentation of scenario, correction of errors in videos, and group tasks, and turns final evaluation into formative evaluation, so that the result of one oral test will not be the only criteria to measure the students.

4.3 *Increase learning interests and strengthen teaching effect*

A thing can be very attractive only if a person has deep interest in it. The means of multimodal evaluation adopts various ways of evaluation and changes the conventional single mode evaluation, for example, presentation in class, group tasks, and so on. Of all the ways, there's always one students like, so their interests in Oral English for Foreign Trade will be deepened, their attention be attracted and learning efficiency enhanced[6]; on the other hand, these ways of evaluation are varied, some are supported by text and some need the students to make images or video for evaluation, so they would get rid of the rigid method of memory. In the meanwhile, students could make use of all time to make sufficient preparation and practice and then strengthen what has been taught by the course Foreign Business English Conversation.

5 CONCLUSION

In conclusion, multimodal evaluation is a brand new choice for the evaluation in Business Spoken English Course; it not only has far-reaching significance on our evaluation models reform, but also brings invisible challenges. Under the condition of deep-rooted conventional single mode evaluation, at first we should recognize the importance and feasibility of multimodal teaching evaluation, positively introduce in multimodal evaluation ways, and make constant improvements. And undoubtedly, during the implementation, there're problems to be solved, so we need to make continuous explorations and researches in practice, thus improving the evaluation model adaptive to development of times and needs of courses.

REFERENCES

[1] Dai Shulan. Focusing on Multimodal Teaching to Improve the Students' Communicative Competence. Shandong Foreign Language Teaching Journal, 2011(3):48–53.

[2] Li Baohong, Yin Pi-an. An Empirical Study on New Model of College English Teaching under the Context of Multimodality. CAFLE, 2012(148):72–75.

[3] Li Chunguang. Study on the Application of Multimode Learning Environment Based on Internet to the College English Teaching for Music, Sports and Arts Majors. CAFLE, 2012(153):71–75.

[4] Yuan Chuanyou. Construction of Multimodal Information and Cognition Model from Teaching and Learning Practice. Research in Teaching, 2010(4):50–55.

[5] Li Jian. On Multi-modal teaching of College English. Chongqing and the World, 2012(5):59–62.

[6] Wang Lifei, Wen Yan. Using Multi-modal Approach in Applied Linguistics Research. CAFLEC, 2008(5):8–12.

Education Management and Management Science – Zheng (Ed.)
© 2015 Taylor & Francis Group, London, ISBN 978-1-138-02663-6

Study on personnel training modes for engineering majors based on credit system

X.H. Jiang & P.J. Chen
School of Automobile, Linyi University, Shandong, China

ABSTRACT: Credit system is not a teaching management tool, but a mode for innovative talents training. Practicing the credit system becomes the development direction of the teaching management system reform in colleges and universities in China. On the basis of analyzing the status of credit system education at home and abroad, the meaning of credit system is discussed. Combined with the characteristics of engineering professionals, the reform ideas and safeguard measures of personnel training modes meeting the credit system in colleges and universities are discussed. This study can provide advice for the realization of credit system in China.

Keywords: credit system; engineering major; personnel training mode; development strategy

1 INTRODUCTION

The reform of credit system in colleges and universities aims at individualized education. In the elastic study period, the students who get or exceed the specified credits ruled by the related undergraduate training program can apply for bachelors' degree and can be approved for graduation. Then, the reform of credit system is to encourage students to choose courses, instructors and school timetable according to their own circumstances. They can independently arrange their learning process, and this freedom makes learning arrangement more personalized (*Zhou*, 2009). For students, it is the need of personality development, democratization education and lifelong education. In addition, it can benefit the deepen reform of education and the promotion of quality education.

2 STATUS OF CREDIT SYSTEM EDUCATION AT HOME AND ABROAD

2.1 *Status of credit system education in China*

The universities in China began to implement the reform of credit system comprehensively from the late 1990s. Currently, the credit system is the adjustment of the conventional scholastic year system. In essence, the so-called credit system in this country has not got rid of the framework of the scholastic year system. Basically, it is an incomplete credit system. There are several problems in the teaching system.

2.1.1 *Strong sense of scholastic year system*

Many colleges do not provide green channel of early or delayed graduation for the students in the fundamental system of students' status management. In the traditional training model, the students can only learn the courses what school arranges (*Zhang*, 2013). Due to the limited proportion of elective courses, both the learning time and place and the course contents are planned by the school.

2.1.2 *Elaborate division of specialties*

The elaborate division of specialties constrains the selection courses for students. In recent years, some universities attempt to weaken the professional for lower students, which provides the broad background for the implementation of the credit system relatively. But, the matched college teaching management system has not been established, which lags the implementation of the credit system.

2.1.3 *Serious position thought in teaching management*

When implementing the credit system, many colleges and universities consider the conveniences of teaching management, even look it as the modern management tools (*Xia*, 2013). They do not think about the benefits of learning from the point view of students, then, the credit system cannot play its role.

2.1.4 *Cannot develop initiative and flexibility*

The credit system does not start from the perspective of students. As the learning body, students do not take initiative. Lacking the power to choose, the students cannot arrange their learning time and space actively and reasonably.

2.2 *Status of credit system education abroad*

The credit system began quite early in universities of western countries. Those colleges have formed a relatively mature credit system. The foreign education is to carry out the credit system in modern higher education stage. The credit system has special place in the teaching process and plays an important role. The main representative types of foreign credit systems are as follows.

2.2.1 *Flexible teaching modes in America*
Almost all universities in America adopt the full credit system. But, there are differences in different colleges, such as, the ratio of required courses, the semester arrangement, credit measurement standards, required total credits, and so on.

The American credit systems are found to have many concordances (*Chang*, 2010). First, they pay more attention to guiding the learning process for the students. Second, the systems focus on basic education, broaden their knowledge, and emphasize that the low-grade students should choose foundation courses from different professionals, which aims to broaden students' horizon and to receive training in different thinking ways. Third, the credit can be converted, that is to say, the credits in different universities can be recognized with each other. Fourth, the students have large degree of freedom to select courses and professionals. Finally, the teaching programs are so flexible that the students can arrange personal learning plans according to actual situation.

2.2.2 *Balanced credit system in Japan*
Japanese universities use scholastic year credit system. It is meticulous and balanced and can ensure that the students receive appropriate knowledge within the stipulated time. The country rules that the undergraduates must learn in college for four years or more to obtain the minimum credits for graduation (*Zheng*, 2008).

The flexibility of the system is expressed in optional courses. The students have extensive and broad contents to select in addition to the second language. Japan has also implemented the convenient credit transfer system. They have great autonomy in practice and in space, thereby facilitating the exchange of learning between schools.

2.2.3 *Diverse models and parallel development in Europe*
The credit system in Europe is relatively lack of freedom and flexibility. Many schools adopt the idea of parallel development with a variety of teaching modes.

In France, the comprehensive universities use credit system. It is divided into two phases. The first two years are mainly basic education, and the students study the basis and common courses systematically. In the second stage, the students learn the specialized knowledge, some schools abandon credit system and some schools convert to certificate system.

In England, some colleges take the parallel education management methods of scholastic year system and credit system. The country gives more emphasis on the learning of basics knowledge, and focuses on students' learning abilities and learning methods. The universities turn to develop refined and specialized talents.

In addition, the credit transfer system has been recognized by the majority of European countries, which promotes the mobility and exchange of student between countries.

3 CHARACTERISTICS OF CREDIT SYSTEM

By studying the successful personnel training modes of credit system at home and abroad, it can be seen that the essence of credit system is in the teaching process. Converting from the traditional teacher-centered into student-centered, the student-oriented can be realized and the purpose of students selecting learning field on their own initiative can be gotten. The key of the credit system is elective system. Compared with the scholastic year system, the credit system has the following several features.

3.1 *Flexibility of learning time*

Referring to the learning years required by the academic education, the credit system does not strictly limit the learning times. Usually the general scholastic year system rules that the learning time is four years, but the credit system is flexible, from three to six years (*Li*, 2013). Students may graduate early by getting full credits, or graduate late. The system allows students allocate education time according to their arrangement, helping to develop their personality.

3.2 *Selectivity of learning contents*

The credit system should establish autonomous elective system, the students can choose professional and courses what they think necessary and interesting. This can optimize the overall course structure and knowledge structure, and be conducive to improving the overall quality of students and to meet the learning knowledge needs of students at different levels. What's more, it can help students to realize self-design and selection of talent direction, as well as the permeation of humanities and science and the implementation of quality education. The system without elective rights cannot be considered as a real credit system.

3.3 Flexibility of course assessment

If a student fails the exam of a course, he can retake it until pass the exam, or he can take another course to obtain equivalent credits. This allows students regulate themselves within certain limits according to their own development. It not only reflects the importance of credit, but also helps students develop the ability of adapting to the society needs. Some universities assess and award the students for their innovations, designs and papers.

3.4 Guidance of training process

The credit system creates the necessary conditions for the independent learning, playing specialty and training overall quality. But, the students, especially the low-grade students do not find out their own target mode meeting the social demands. Therefore, in order to guide the students, it is necessary to implement the tutorial system to deal with problems in the studying process.

3.5 Share of teaching resources

The mutual recognition of credits benefits the teaching resource sharing among colleges and universities, improving the operating efficiency, mobilizing the enthusiasm of teachers and students, and promoting the improvement of teaching quality. For teachers, it can enhance the sense of competition, mobilize the teaching enthusiasm, improve their teaching methods, and improve teaching effectiveness. For students, it can stimulate learning enthusiasm, initiative and independence, and effectively develop intelligence.

4 DEVELOPMENT STRATEGY OF CREDIT SYSTEM REFORM OF ENGINEERING MAJOR

From the current development status of universities in China, the implementation of credit system is to change the mode of personnel training program, teaching arrangements, and student status arrangement. It emphasizes that the students can select course freely (Cheng, 2008). If the student receives full credits in the elastic period, he can

graduate and receive the degree. In order to convert from the old scholastic year system to the credit system smoothly, according to the nature of the credit system, the several aspects should be focused on.

4.1 Design personnel training mode of credit system scientifically

According to the regulations of credit system teaching management, the teaching contents and curriculum system in the scholastic year system are updated, integrated, optimized and reorganized to form the personnel training program under the credit system. It can achieve a certain degree of freely choosing course and mutual recognition of credits and promote personality development of students through the model change from professional education-oriented to student-centered. It should realize the optimal allocation of educational resources, carry out quality educational activities, and create environment of diversified personnel training to train comprehensive talents.

4.2 Optimize curriculum system adapting to credit system

The overall curriculum system and teaching contents should be optimized to break the professional boundaries in the past rigid scholastic year system. The profession should be weakened in the curriculum design. It can transform independent course system to the integrated curriculum system. The closed practice teaching should be converted to the open practice teaching. It can strengthen the integrity and scientificity of curriculum structure, increase the set of general education curriculum, practical courses, innovation and entrepreneurship courses, and establish the curriculum adapting to individual differences. The curriculum modularization is necessary that makes students can choose different courses in different course modules freely. It must highlight the systematicness of the curriculum modules to form complete and optimized curriculum system.

4.3 Establish new model emphasizing theory and practice teaching

Engineering majors emphasize both theory and practice. In order to train strong practical ability for students, it is necessary for practice curriculum to build the operating mechanism with task-driven, project-oriented, integration of the teaching, learning and doing. Professional probation and practice must run through each semester, so that the students have enough time and effort to understand the actual work situation of the specific major, thereby it can effectively play the function of the probation and practice. It should arrange practical courses, take the form of from simple to complex and from easy to difficult to ensure that the technical training gradually deepens and increases in teaching contents.

4.4 Construct teachers compliance with credit system

It has high requirements for the quantity and quality of teachers to implement credit system personnel training program. First, teachers must change teaching ideas, establish service consciousness for students, respect the students' personality, esteem student rights of learning, choice and speak. Second, it requires adequate teachers to give sufficient number of high quality courses. Third, it encourages teachers to innovate classroom teaching to stimulate student interest, mobilize learning potential, creativity and innovative potential.

4.5 Establish management system adapting to credit system

The reform of credit system is a systematic project, every department of universities should fink out its responsibilities, coordinate, collaborate, re-enact the relevant management system according to the relevant requirements. It should reform systematically enrollment and employment guidance mode, daily education and management, comprehensive evaluation methods, accounting and payment methods of student tuition, teaching funding channels, service delivery, teacher workload assessment, positions identification, and so on, so as to ensure the smooth implementation of the credit system.

4.6 Establish advanced and practical education management platform

It is need to build and improve flexible academic system, credit examination system, major and minor professional system, innovative training quality control and assurance system, so that teaching management is conductive to innovative talents training monitoring and security system. The education management system platform should be established to realize resource sharing, achieve automation of course selection, and so on. These methods can provide strong guarantee for the implementation of personnel training program of the credit system.

5 CONCLUSIONS

In short, the fundamental purpose of the credit system reform is to give students the learning autonomy, help them arrange learning process by themselves and build their own knowledge framework adapting to personal development. It can truly reflect the dominant position of students in learning, thus improve the quality of education and teaching and train high-qualified innovative talents. For this purpose, it is very important to scientifically build engineering innovative personnel training mode in the credit system for the innovation and practice ability of students.

ACKNOWLEDGMENT

This work was supported by the teaching reform project of Linyi University, Shandong Province, P.R. China in 2014.

REFERENCES

[1] Chang, S.L. 2010. The development history and features of credit system in American universities. *Theory and Practice of Contemporary Education* 2(5): 24–26.
[2] Cheng, R.F. 2008. Analyzing of training approach for applied talents based on credit system. *China Higher Education Research* (3): 90–91.
[3] Li, Y.H. 2013. Study on reform of innovative talents cultivation mechanism in universities, Taking as credit system an example. *Journal of Shanxi Normal University* (*Social Science Edition*) 40(5): 154–156.
[4] Xia, J.P. 2013. Study on Project Design of Credit System in University. *Time Education* (6): 96–97.
[5] Zhang, X.L. 2013. Research on the System of Innovative Talent Production for College Students in the Innovation Credit System. *Journal of Higher Education Management* 7(1): 95–99.
[6] Zheng, Y.C. 2008. Comparison and analyzing of credit system model in universities of America, Japan and China. *Heilongjiang Researches on Higher Education* (1): 73–75.
[7] Zhou, G.J. 2009. Credit System and the Cultivation of Innovative Talents at College. *Journal of Guizhou University for Ethnic Minorities* (*Philosophy and social science*) (4): 175–178.

Education Management and Management Science – Zheng (Ed.)
© *2015 Taylor & Francis Group, London, ISBN 978-1-138-02663-6*

Service links in the global value Chains and the Sino-US trade imbalance

Lin Kong
School of Management, Capital Normal University, Beijing, China

ABSTRACT: The distinct mounting up of the trade volume based on the global value chains breaks the country limit and final product limit in the traditional international division of labor. And multinational corporations also supply service links among sections in the global value chains, when they transfer the assembly segment of final goods to China. Thus, there is also this kind of connection in the Sino-US trade. Through structural analysis of the Sino-US trade data, this paper finds that the Sino-US trade relationship is much more complex than the data of final goods trade shows. Service links and other multi-crossing border trade relations among the segments in the global value chains are also important reasons for the expansion of the US trade deficit with China.

Keywords: Sino-US trade imbalance; global value chains; service links; producer service investment

1   INTRODUCTION

As multinational corporations, which dominate the global value chains in the world, optimize, allocate and utilize resources, the free flow and agglomeration of factors realize in a larger range. Thus this affects the Sino-US trade flows in the goods exchange relationship. The expansion of the US investment in China, especially the increase in the service links investment, enhances the export capacity of the US corporations, and further enlarges the Sino-US trade imbalance.

The required inputs of parts production are not only labor factor, but also service links such as efficient communication, transportation, marketing, insurance and so on, which reflect a great amount of producer service demand in the production process. Meanwhile, the information technology changes the ways of service provision, promotes the development of modern service industry and the transformation of its own structure. This causes the proportion of the knowledge and technology intensive service, which is mainly for producers and transaction process, rising higher and higher. With the acceleration of factors international flow and the evolvement of modern service industry, global service trade is expanding rapidly. At the same time, the liberalization process of service trade also creates a strong system guarantee for the development of global producer service trade. Global direct investment gathers quickly towards the service sectors. And the service sectors' share in the global Foreign Direct Investment (FDI) stock grows from one-fourth in the early 1970s to about two-thirds at present. In addition, most

producer service trade can be offered only through direct investment or establishing branches without property in the foreign market, so the liberalization process of the FDI in service sectors has effectively promoted the growth of service trade. According to the World Trade Organization (WTO) data, the total amount of world service exports is only $71 billion in 1970, and the figure reaches $1.44 trillion in 2000 and then jumps to $4.383 trillion in 2012, in which the US service exports accounts for 14.38%. (It is the largest and the most competitive service exporter around the world.)

Because having abundant advanced factors such as technology, capital, information, human resources etc., developed economies represented by the United States occupy the producer service segments with high added value in the global value chains and dominate the development of the world service trade. Their share in the world service trade exceeds 75%, and most of them have rising trade surplus year after year. In contrast, developing countries accounts for less than one-fourth in the global service trade. Moreover, these countries, especially their technology-intensive producer service sectors, are mainly in the subordinate status and have trade deficit. Contemporary international division of labor has shifted from the situation that developing countries provides primary products and developed countries provide manufacturing goods to the labor division between low-end segments and high-end segments in the global value chains. And it strengthens progressively because of path-dependence. Therefore, when studying the current Sino-US trade issues, we need to consider these economic realities.

Although most China and the US trade statistics do not cover trade in services, but in fact, the US multinational corporations not only move their parts production and assembly overseas, but also provide service links between various fragments in the global value chains as well as coordination and management. In other words, as a part of the global value chains evolvement, production-related services, especially that activated by knowledge and information technologies, have become more and more feasible in trade, which is different from the traditional concept of international trade in services. Therefore, this paper studies the Sino-US trade derived imbalance based on the global value chains from the perspective of the US service factor output and export, including two aspects of substitution for trade, one is the US producer service investment and the other is its sales networks in China.

2 THE US PRODUCER SERVICE INVESTMENT IN CHINA

At present, service industry is the dominate industry, and is the main industry of direct investment abroad in the US. In 2009, the US Outward Direct Investment (ODI) in service industry accounts for more than 70% of its total ODI. As China implementing its WTO accession commitments, its domestic service sectors also gradually open to the foreign direct investment. There are American investments in all of the one hundred service divisions in China now, which are promised to open at the time of accession to the WTO. Producer services (including financial service, communication service, insurance service, computer and information service, copyright and licensing service, construction service and other business service) have also become major sectors of the US investment in China. Statistics of the United States Bureau of Economic Analysis (BEA) shows that the US producer service investment in China reaches $9.897 billion in 2012. This data is only $ 88 million in 1992, with an average annual growth rate of more than 25%. Figure 1 shows the US producer service investment changes in China from 1992 to 2012.

From the perspective of subsectors, the US producer service investment in China is still in its starting stage before 2002, mainly in warehousing, transportation, leasing and so on. After 2002, the US direct investment in China's producer service sectors grows steadily, especially the investment in business, professional and technical services rising fastest, with an above 30% average annual growth rate. By 2008, the US direct investment in these service subsectors in China reaches $2.353 billion. With the increase of US multinationals' investment

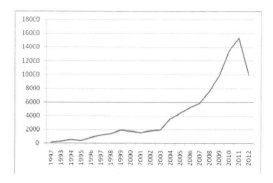

Figure 1. The US producer service investment in China, 1992–2012 (in millions of dollars).
Source: Author's calculations based on US Bureau of Economic Analysis (BEA) statistics.

in China, producer services provided by them, as a form of service exports, play an increasingly important role in China. Many US multinational companies in manufacturing industry also invest in sectors such as finance, insurance, marketing, business and so on. Thus, the products exported by these multinationals contain the value-added in both production and services segments. For example, automobile manufacturing enterprises provide consumer credit services; electronic communication corporations transfer their Research and Development centre to China, etc.

The services provided by the US multinationals in China are not reflected in the related items on the balance of payments, and the statistics on the Sino-US trade in services and the Sino-US trade balance do not cover this part. However, when considering the issue of Sino-US trade imbalance, we could not ignore this producer service export to China, which is offered by the US multinational corporations based on the global value chains.

3 THE SUBSTITUTION OF THE US MULTINATIONALS' SALES NETWORKS IN CHINA FOR THE US EXPORTS

Production and trade globalization characterized by multinational corporation system and value chains coordination and management have significant effects on China's exports and foreign investment sectors. Although exports are the main functions of the foreign companies in China, it is often overlooked that many products made by foreign enterprises are sold in China's market. Moreover, the US long term restrictions on exports of high-technology products to China have also contributed to its capital and service factors direct output to China. Over a long period of time, the US government has strengthened its export controls to China on many fields such as military and civilian technologies, which make the US companies, lose many direct export opportunities to China based on comparative advantages. In order to bypass policy restrictions and compensate for the losses, American companies begin to increase direct output to China in producer service segments and higher-tech factors, build sales networks in China and export lots of high-tech products back to the US. That is to say, the US multinationals in China changes two trade figures: One is exaggerating China's export value to the United States. The US multinationals allocate their low-end manufacturing segments based on the global value chains in China, making lots of their manufacture and service value-added segments transfer to China directly. When the manufactured goods are re-exported from China to the United States, the total value-added contained is fully brought into China's exports by the current trade statistics system. The other is the factors that the US outputs to China replacing the Sino-US import. The purpose of some US multinational corporations operating in China is to bypass the US government technology export control and build its manufacturing and marketing

segments there. As a result, these multinationals' sales in China have an export nature.

In addition, the US multinationals also adjust their production and marketing strategies in China constantly based on their overall coordination and management of the global value chains, transiting from simple processing and assembling sections to production base and sales market. And the US companies in China are mostly built in the eastern coastal areas where the retail price and the per capita income are relatively higher, which reflects the investment consideration of the US multinationals on the potential size of the local market as well. The United Nations Conference on Trade and Development (UNCTAD) statistics shows that the $3.8 trillion production and sales value made by the overseas branches of the US multinational corporations is not included in the US export statistics from 2002 to 2006. Merrill Lynch's report in 2005 also illustrates that a portion of the Sino-US trade surplus is caused by the US companies in China re-exporting to their home country in recent years. Research data provided by Bergesten et al. (2006) also indicates that about 70% of the US multinationals processing and assembling products in China are sold in the local market. Yuefen Li (2006) thinks that the US multinational

Table 1. Sino-US trade balance and the related business activities, 2000–2013 (In millions of dollars).

Year	US trade deficit with China by US Census Bureau	US trade deficit with China by UN comtrade	Business activities of the US affiliates not reflected on the trade balance			
			US related parties' imports from China		US affiliates sales in China*	Total
			Value	Proportion of total US import from China		
2000	83833	29782	18061	18.1%	29914	47975
2001	83096	28138	18487	18.1%	36547	55034
2002	103065	42789	25538	20.5%	46718	72256
2003	124068	58682	34839	23.0%	56695	91534
2004	162254	80401	53172	27.1%	73006	126178
2005	202278	114439	62716	25.8%	92642	155358
2006	234101	144487	70701	24.6%	115478	186179
2007	258506	163621	82404	25.5%	139407	221811
2008	268040	171258	89339	26.5%	168109	257448
2009	226877	143540	84829	28.7%	238857	323686
2010	273042	181046	107038	29.4%	303281	410319
2011	295250	201887	111599	28.0%	294947	406546
2012	315111	218672	121813	28.7%	N.A.**	N.A.
2013	318711	215669	124490	28.4%	N.A.	N.A.

*Data up to 2008 is gathered from All Nonbank Foreign Affiliates, and data for 2009 and forward is gathered from All Foreign Affiliates.
**N.A. indicates that the data in the cell is not available.
Source: US Census Bureau Statistics; US Bureau of Economic Analysis (BEA) Statistics; UN Comtrade; Author's calculations.

corporations use their manufacturing and marketing networks in China to replace partial bilateral trade. In 2013, the Organization for Economic Cooperation and Development (OECD) and the World Trade Organization (WTO) also release a new trade statistics method tracking added value changes. The US trade deficit with China would be significantly reduced if calculated according to this new trade statistics standard with added value. Table 1 illustrates the differences between two kinds of Sino-US trade balance statistics and the replacement of trade by the production and marketing networks of the US multinational corporations in China in detail through data.

Furthermore, the production and international trade of China generally only involve processing and assembling segments in the global value chains, which are very small parts in the multinationals' value chains. The leading corporations in the global value chains could resolve finance, marketing, transportation and other matters through transnational dealers, channels, brands, international financial centers and so on. So processing and assembling segments allocated by the dominant firms in China can be connected conveniently with their global marketing networks. On one hand, making full use of its sales networks in China in order to achieve some final products sold in China, substituting for trade; On the other hand, after processing and assembling, exporting the final products to Europe and other developed countries.

It shows that the establishment of the US sales networks in China reduces the US export to China in statistics. However, in fact, it does not decrease, but increase this export value.

4 CONCLUSIONS

Parts Production requires not only capital and labor inputs, but also marketing, transportation, insurance, efficient communication service and so on. And the US multinational corporations also supply service links among segments in the global value chains as well as coordination and management, when they transfer parts production and final goods assembly abroad. Thus, the US continued investment in various sectors and its management in service links in China, which are because of the US' higher position in the global value chains, have become important factors in widening the US trade deficit with China. In addition, as the US government controls the exports of high-tech products to China in the long term, multinational companies combine the processing and assembling segments in China with their powerful anti-trading sales networks in the global value chains and achieve replacement of some US exports by final goods sales in China. In the meantime, processing and assembling sections allocated in China by leading firms could be connected conveniently with their global marketing networks, which has also become one of the critical elements to expand the US trade deficit with China. This is the trade performance that the dominant power in the global value chains coordinates and governs production and business activities dispersed in space as well.

ACKNOWLEDGMENTS

The author gratefully acknowledges the support from the Youth Talent Program in Beijing Colleges and Universities for the project titled Research on the Sino-US Trade Imbalance: the Latest Progress Based on the Global Value Chains.

REFERENCES

[1] Deardorff, A.V. 2001. International Provision of Trade Services, Trade, and Fragmentation. *Review of International Economics,* 9(2):262–286.
[2] Francois, J.F. 1990. Trade in Producer Services and Returns Due to Specialization Under Monopolistic Competition. *Canadian Journal of Economics*, 23(1):109–124.
[3] Gereffi, G., Humphrey J. & Sturgeon T. 2005. The Governance of Global Value Chains. *Review of International Political Economy*, 12(1):78–104.
[4] Golub, S.S., Jones R.W. & Kierzkowski H. 2007. Globalization and Country-Specific Service Links. *Journal of Economic Policy Reforms*, 10(2):63–88.
[5] Helleiner, G.K. 1973. Manufactured Exports from Less-Developed Countries and Multinational Firms. *Economic Journal*, 83(329):21–47.
[6] Helpman, E. 2006. Trade, FDI, and the Organization of Firms. *Journal of Economic Literature*, 44(3):589–630.
[7] Rodrik, D. 2006. What's so special about China's exports?, *China and the World Economy*, 14(5):1–19.

Education Management and Management Science – Zheng (Ed.)
© 2015 Taylor & Francis Group, London, ISBN 978-1-138-02663-6

Study on factors which influence CEO turnover in SOEs—the moderating effect of CEO power

Kai Wang
China Academy of Corporate Governance, Nankai University, Tianjin, China
Business School, Nankai University, Tianjin, China

Yaqi Liu
Business School, Nankai University, Tianjin, China

ABSTRACT: The existing studies have identified some factors which influence turnover of senior managers from different perspectives, such as enterprise efficiency, political process of senior managers' power. Based on these studies, this paper focuses on the specific context of Chinese local SOEs and extracts factors which influence CEO turnover. By empirically examining our hypotheses, we find that, for the CEO who has more power, the impact of investment of the local SOE which he/she sits in on his/her turnover is significantly negative. However, this kind of impact is little for the CEO who has less power. Our conclusion not only enriches the studies of CEO turnover, but also has some practical implications for Chinese SOEs' reform and completing the evaluation system of local government.

Keywords: CEO turnover; investment; CEO power

1 INTRODUCTION

The decision announced on the 3rd Plenary Session of the 18th Central Committee of CPC points out that, SOEs should "establish the system of professional managers". To achieve this goal, we need to know which factors influence CEO turnover in SOEs significantly at present, so that we can shoot the arrow at the target. The existing studies have identified some factors (Giambatista et al. 2005). Based on these studies, this paper focuses on the specific context of Chinese local SOEs and explores whether there are other specific factors which influence CEO turnover in local SOEs. In addition, we analyze whether factors influencing CEO turnover vary with the degree of CEO power.

The structure of this paper is as follows: literature about factors of senior managers' turnover is summarized in the next section; the third section analyses how investment influences CEO turnover in a local SOE and proposes related hypotheses; the fourth section empirically examines hypotheses proposed in the third section using data about local state-owned listed companies; and the final section concludes our finding and discusses it.

2 LITERATURE REVIEW

Scholars mainly identify factors of CEO turnover from two perspectives, that is, enterprise performance and political process of senior managers' power. This section reviews these studies.

2.1 *Enterprise efficiency*

The performance of an enterprise which has a capable CEO is often better. If the performance of an enterprise is poor, its CEO may be dismissed. Some empirical studies have offered evidence. For example, Warner et al. (1988) found that if the stock price is very low, the possibility of CEO turnover will increase. Other scholars examine whether the expected earnings of securities analysts can influence CEO turnover. Farrell & Whidbee (2003) found that if the real performance doesn't reach analysts' expectation, CEO turnover will be more likely to happen.

Whether incapable CEO can be identified and dismissed depends on the arrangement of corporate governance to a great extent. So it's necessary to examine the effectiveness of governance mechanisms. For instance, in the aspect of internal governance mechanisms, Weisbach (1988) found that if the board of a company is dominated by outside directors, the relationship between performance and CEO resignation will be stronger. Huson et al. (2004) found the shareholding of institutional investors can increase the impact of CEO turnover on performance. In the aspect of external governance mechanisms, Defond & Hung (2004) found a strong law enforcement system can significantly increase the sensibility between CEO turnover and performance.

2.2 Politics

Some scholars analyze factors of senior managers' turnover from the perspective of politics, such as Boeker (1992). He found that if CEO has less power, CEO turnover will be more sensitive to performance. In another study, Boeker and his colleague found the proportion of insiders in a board and shareholding of insiders can restrict an external successor (Boeker & Goodstein 1993). Shen & Cannella (2002) found that if a CEO is an outsider, his/her tenure is short, the percentage of non-CEO inside directors is high and the ownership of non-CEO senior managers is high, the probability of CEO dismissal will increase.

In a word, scholars have identified some influence factors of CEO turnover. However, because of the fiscal decentralization system in China, there are other specific factors which influence CEO turnover in local SOEs. We will analyze this in next section.

3 RESEARCH HYPOTHESES

3.1 Factors of CEO turnover in local SOEs

Resource dependence theory indicates that, to eliminate uncertainty, an organization often absorbs the elements in its environment which it depends on. However, the dependence between organizations sometimes is mutual. For example, to achieve development and maintain stability at the same time, local government relies on SOEs for tax, investment, employment and so on. Also, the senior managers of SOEs rely on government for providing political identity and promotion (Hung et al. 2012). Since the existence of this kind of interdependence, if SOEs do not satisfy the investment demand of local government, the government who serves as an ultimate shareholder will tend to replace the CEO. Based on this, we propose:

H1: The lower investment level of a local SOE, the higher probability its CEO turnover happens.

3.2 The moderating effect of CEO power

We further analyze whether the influence of factors of CEO turnover varies with the degree of CEO power. According to modern structure of corporate governance, the core of corporate governance is board of directors, and it's responsible for great decisions of its company (Li 2008). So the investment decision, which belongs to great decisions, is in the charge of board. In this light, for the CEOs who have significant impact on board decisions, government is more likely to determine whether to change them according to the investment level of SOEs. Based on this, we propose:

H2: The more power a CEO has, the more significant is the impact of investment level of his/ her local SOE on his/her turnover.

4 EMPIRICAL ANALYSIS

4.1 Data and sample

The local state-owned listed companies in A share from 2010 to 2012 form the starting sample. Then, we excluded the companies that were lack of key variables, had abnormal values of the variables (such as debt ratio was greater than 1). Also, the companies whose CEO tenure was less than one year are not included. The resulting sample contains 1757 observed values. We collected the data from the China Stock Market and Accounting Research (CSMAR) database.

4.2 Model and variables

To test the relationship between the investment of local SOEs and CEO turnover, we construct the regression model:

$$\text{Logit} \frac{p(\text{CEO\_turnover})}{1-p(\text{CEO\_turnover})}$$

$$= \alpha + \beta_1 \cdot \text{invest} + \beta_2 \cdot \text{roa} + \beta_3 \cdot \text{size} + \beta_4 \cdot \text{id}$$

$$+ \beta_5 \cdot \text{dual} + \beta_6 \cdot \sum_{i=1}^{11} \text{ind}_i + \beta_7 \cdot \sum_{i=1}^{2} \text{year}_i + \varepsilon$$

where the dependent variable (CEO_turnover) is a dummy variable that takes the value of one if the CEO turns over; the independent variable (invest) is the level of investment, which is the proportion of the cash used to acquire fixed assets, intangible assets and other long-term assets to total assets; referring to Wiersema & Zhang (2013), we control the profitability (roa, net profit divided by total

Table 1. Descriptive statistics of variables.

Variables	Mean	SD	Min	Max
Turnover	0.15	0.36	0	1
Invest	0.06	0.05	0.00	0.27
Roa	0.04	0.06	−0.21	0.20
Size	22.25	1.21	19.44	25.15
Id	0.37	0.05	0.30	0.57
Dual	0.10	0.30	0	1

Table 2. Results of Logit regression.

Variables	Model 1	Model 2 dual = 1	Model 3 dual = 0
Invest	−0.03 (1.29)	−14.00** (5.65)	−0.36 (1.33)
Roa	−3.28** (1.32)	−8.04 (5.52)	−3.00** (1.38)
Size	0.02 (0.06)	0.54** (0.22)	−0.03 (0.07)
Id	1.30 (1.23)	17.90*** (5.52)	0.38 (1.33)
Dual	0.05 (0.24)		
Ind	Yes	Yes	Yes
Year	Yes	Yes	Yes
Constant term	−2.01 (1.42)	−17.31*** (5.65)	−0.53 (1.53)
Wald chi^2	58.64***	33.59***	47.93***
Pseudo R^2	0.04	0.23	0.03
N	1757	146	1584

Notes: Numbers in the brackets are the clustering-robust standard errors at the firm level; *, **, and *** denote statistical significance at the 10%, 5%, and 1% level, respectively.

assets), size of the company (natural logarithm of total assets), proportion of independent directors (id, number of independent directors divided by the size of the board), CEO duality (dual, dummy variable that takes the value of 1 if CEO is also chairman of the board), and industry (ind) and year (year). Among the variables, invest and roa are lagged by one year, and all the continuous variables have been processed by 1% winsorize to eliminate the influence of abnormal value.

4.3 Descriptive statistics

Table 1 shows descriptive statistics for the variables. The table shows that CEO turnover happens in 15% of the sample companies. The level of investment varies from nearly zero to 27%. And in 10% of the sample, CEO is also the chairman of the board.

4.4 Logit regression analysis

To test the hypotheses, we use Logit regression for the regression analysis. Table 2 shows the results. In the table, model 1 is to test the effects investment on CEO turnover in hypothesis 1, and it shows that the coefficient is negative as expected but statistically insignificant. We further divide the sample into two groups according to CEO duality, and then test the relationship between investment and CEO turnover. Finkelstein (2002) indicated that CEO would have more power when CEO is also chairman of the board. Model 2 shows the regression results of the subsample in which CEOs have more power. And the coefficient is negative and statistically significant at the level of 5%. This indicates that investment is an important factor of

CEO turnover to CEOs with more power. What's more, the coefficient of roa is not statistically significant, which implies that performance has little effects on the turnover of CEOs in this subsample. Model 3 shows the results of CEOs with less power, and the coefficient is not significant any more, but the coefficient of roa becomes negative and significant. We can see that performance has more effects on CEO turnover to CEOs with less power. From the chi-square test and Pseudo R^2, the goodness of fit of the three models is well.

5 CONCLUSIONS AND DISCUSSIONS

This paper focuses on the specific context of Chinese local SOEs. By analyzing the exchange relationship between government and senior managers, we extract the factors that influence CEO turnover and examines whether the effects of the factors differ because of characteristics of CEO. Our empirical analysis finds out that, when its CEO has more power, the investment level of the company is higher, and turnover is less likely to happen; when CEO has less power, the effects of investment is not significant any more.

By identifying the specific factor of CEO turnover in local SOEs and examining the moderating effects of CEO power, this paper enriches the studies of turnover of senior managers. What's more, the conclusions have important practical implications, such as, the improvement of the evaluation system of local government is good to lower its intervention in SOEs. Without doubt, this paper has some limitations. For example, SOEs has to assume other tasks in spite of the investment requirement of local government, which can be explored in future studies.

369

ACKNOWLEDGMENT

This paper is supported by the Ph.D. Candidate Research Innovation Fund of Nankai University.

REFERENCES

[1] Boeker W. 1992. Power and managerial dismissal: Scapegoating at the top. *Administrative Science Quarterly* 37(3): 400–421.

[2] Boeker W. & Goodstein J. 1993. Performance and successor choice: The moderating effects of governance and ownership. *The Academy of Management Journal* 36(1): 172–186.

[3] Defond M.L. & Hung M. 2004. Investor protection and corporate governance: Evidence from worldwide CEO turnover. *Journal of Accounting Research* 42(2): 269–312.

[4] Farrell K.A. & Whidbee D.A. 2003. Impact of firm performance expectations on CEO turnover and replacement decisions. *Journal of Accounting and Economics* 36(1–3): 165–196.

[5] Finkelstein S. 1992. Power in top management teams: Dimensions, measurement and validation. *Academy of Management Journal* 35: 505–538.

[6] Giambatista, R.C., Rowe W.G. & Riaz S. 2005. Nothing succeeds like succession: A critical review of leader succession literature since 1994. *The Leadership Quarterly* 16(6): 963–991.

[7] Hung M., Wong T.J. & Zhang T. 2012. Political considerations in the decision of Chinese SOEs to list in Hong Kong. *Journal of Accounting and Economics* 53(1–2): 435–449.

[8] Huson M.R., Parrino R. & Starks L.T. 2001. Internal monitoring mechanisms and CEO turnover: A long-term perspective. *The Journal of Finance* 56(6): 2265–2297.

[9] Li W. 2008. *Corporate governance in China: Research and evaluation.* John Wiley & Sons (Asia) Ltd.

[10] Shen W. & Cannella A.A. 2002. Power dynamics within top management and their impacts on CEO dismissal followed by inside succession. *Academy of Management Journal* 45(6): 1195–1206.

[11] Warner J.B., Watts R.L. & Wruck K.H. 1988. Stock prices and top management changes. *Journal of Financial Economics* 20(1–3): 461–492.

[12] Weisbach M.S. 1988. Outside directors and CEO turnover. *Journal of Financial Economics* 20(1–3): 431–460.

[13] Wiersema M.F. & Zhang Y. 2013. Executive turnover in the stock option backdating wave: The impact of social context. *Strategic Management Journal* 34(5): 590–609.

Education Management and Management Science – Zheng (Ed.)
© 2015 Taylor & Francis Group, London, ISBN 978-1-138-02663-6

Survey on rural middle school English education in Shaanxi—a case study of Shanyang No. 2 Middle School

Qiangfu Yu & Zhen Chen

Faculty of Humanities and Foreign Languages, Xi'an University of Technology, Shaanxi, China

ABSTRACT: This paper is based on a survey conducted in Shanyang No. 2 Middle School (SN2MS), using questionnaires and interviews, to analyze the present situation in this school and to point out the reasons for the existence of problems. Finally, on the basis of analyses, some suggestions and ideas are provided in the hope of improving the quality of English education in rural junior middle schools in China.

Keywords: rural English education; current situation and problems; reasons; suggestions

1 INTRODUCTION

Rural English education has been making progress during recent years. However, due to the limitation of rural economic situation, traditional education philosophy, structure of teaching faculty, students' learning attitude and other factors, many problems and risks still exist in rural school English teaching, exerting some negative influences on improving the quality of rural English education. So we need to study the problems and risks in rural junior school English teaching and come up with effective solutions. This paper takes SN2MS, a rural middle school in Shangluo, Shaanxi, China as an example, aiming at describing the present situation in this school, analyzing the problems and the challenges in this rural junior school's English teaching, giving some effective suggestions as well, and hopefully shedding some light on rural middle school English education in China.

2 METHODOLOGY

2.1 *Subjects*

SN2MS is one of the public junior schools in Shanyang, located in the east of Shanyang county seat, near the countryside, so it can reflect the real condition of rural English teaching. There are 1,020 students in this middle school and all students are divided into three grades: Grade 1, Grade 2, and Grade 3. There are 81 teachers and only 6 English teachers in this rural school.

2.2 *Instrument*

This investigation includes two questionnaires designed by the author, together with interviews with some students and teachers randomly chosen from the above-mentioned school. The questionnaire for student consists of 10 questions, which aims at finding out three aspects as follows: students' attitude, students' interest and students' learning methods. The questionnaire for teacher includes 5 questions for collecting teachers' thinking about English teaching and their suggestions. The two questionnaires are written in Chinese so as to avoid misapprehension on the survey. The interview consists of 7 questions which are mainly focused on the present situation, the problems and the challenges of rural English teaching.

2.3 *Data collection*

The questionnaires for students were collected back 55 of 58, in other words, the return rate is about 94.83%. The questionnaires for English teachers in SN2MS were returned 6, that is to say, the return rate is 100%.

3 RESULTS AND DISCUSSION

3.1 *Results of the investigation*

3.1.1 *The situation of the school*
SN2MS is one of the five rural junior schools in Shanyang county. Because of the location in the countryside, approximately 2/3 of the students are from rural primary schools close to the school, and the rest 1/3 of the students come from far regions with their parents working in Shanyang.

As Table 1 reveals, the present situation of English teaching equipments in the school under investigation is not optimistic. There are only two electronic reading rooms, without any language lab

Table 1. Modern equipments in SN2MS.

Item	Multimedia classroom	Electronic reading room
Number	1	2

Table 2. Number of teachers in SN2MS.

Item	English teachers	Other teachers
Number	6	75
Percent (%)	7.41%	92.59%

Table 3. Genders of teachers in SN2MS.

Item	Number	Percent
Male	25	30.86%
Female	56	69.14%

Table 4. Genders of English teachers in SN2MS.

Item	Number	Percent
Male	2	33.33%
Female	4	66.67%

Table 5. Age and length of teaching of English teachers in SN2MS.

Item	Age		Years of teaching		
Figure	20–35	>35	1–5	6–10	>10
Number	2	4	1	2	3
Percent	33.33%	66.67%	16.67%	33.33%	50%

Table 6. Education background of English teachers in SN2MS.

Education background	Number	Percent
Junior college degree	4	66.67%
Bachelor degree	1	16.67%
Master degree	1	16.67%

or audio-visual classroom. With the development of science and technology, there is just one multimedia classroom in this school. But it is seldom used for teaching because of the cost of the facilities maintenance. The multimedia classroom is just used for having important meetings, organizing teachers' training or having pubic classes. In short, the utilization rate of the modern equipment is very low. What English teachers often use in teaching are textbooks, blackboard, chalk and mouth.

The problems facing SN2MS are mainly lack of teaching equipment and teaching resources and outdated mode of teaching.

3.1.2 *The situation of English teachers*
The total number of teachers in SN2MS is 81. And there are 6 English teachers in this school, about 7.41% of the whole teachers. The rest 75 teachers are major in Chinese, Maths, Geography, History and other subjects.

Table 3 shows the proportion of male and female teacher in SN2MS. Table 4 shows the distribution of male and female English teacher in SN2MS. There are two male English teachers and four female English teachers. From these two tables, we can conclude that the distribution of male and female teachers is not balanced in SN2MS.

A total of 33.33% of the English teachers are between 20 to 30 years old, as is shown in Table 5, and 66.67% of the English teachers are above 35 years old, which shows the middle-aged teachers take up the most among the English teachers in SN2MS. Concerning the length of teaching, 16.67% of the English teachers have been teaching for 1–5 years, 33.33% for 5–10 years, and 50% for more than 10 years.

Table 6 shows that about 66.67% of the English teachers have obtained junior college degrees, about 16.67% with bachelor degree and the rest 16.67% with master degree.

Problems concerning teachers in SN2MS are that the proportion of male and female English teacher in SN2MS is not balanced, that the age structure and teaching experiences are not fit for

English teaching and that the level of teachers' quality is not high on the whole.

3.1.3 *The situation of students*
The author collected data by using questionnaire among 58 students randomly in SN2MS and got 55 returned. The author chooses some data from the questionnaire in the survey and states some results are shown in Table 7.

From Table 7, 47.27% of the students like English, they gave the reasons such as widening their view, getting higher score in the examination or they just like it for they have to. And no one shows that they will use English for communication or they like English because they are interested in English. A total of 27.27% of the students do not show their attitude clearly. Maybe they learn English as a subject. 16.36% of the students like English sometimes, that is to say, the number of this kind of students is not stable for some reasons like the degree of their attitude, their interests, their confidence, their learning methods, the teaching methods and other reasons. And 9.09% of the students show that they really do not like English, for they can not understand what is taught in the

Table 7. Question 1: Do you like English? And give reasons.

Item	Number	Percent (%)
A. Yes	26	47.27%
B. No	5	9.09%
C. Sometimes	9	16.36%
D. Do no hate	15	27.27%

Table 8. Question 2: What's your method of learning English words?

Item	Number	Percent (%)
A. By rote	19	34.55%
B. By sentence	11	20%
C. By phonetic	21	38.18%
D. Never	4	7.27%

Table 9. Question 3: Do you learn English after class?

Item	Number	Percent (%)
A. Often	7	12.73%
B. Seldom	25	45.45%
C. Never	10	18.18%
D. demand	13	23.64%

English class or they think that however hard they learn, they can not catch up with English talents in their class.

There are about 34.55% of the students who recite English words with traditional way called rote. 38.18% of the students are likely to remember English words by phonetic that is a new way to remember English words by the word's pronunciation. 20% of the students put words that need to be remembered into sentences or passages. 7.27% of the students demonstrate clearly that they recite English words seldom or they never remember English words unless to be asked by teacher. Some of them declare that they can deduce and guess the meaning of words in phrases, sentences, and passages.

From Table 9, there are about 12.73% of the students often do some English practices by themselves after English class among the 55 students in this survey. The ways that they often use to continue to learn English after class are doing more and more practices by homework, listening to the English tape and reading English newspapers such as China Daily, 21st Century and other English books. 45.45% of the students show that they seldom study English after class. And 23.64% of the students declare that they just do homework about English that should be handed in next English class. The rest 18.18% of the students announce that they never study English after class by any way whether teacher asks or not because they do not have enough time to do the extra homework.

Problems concerning students are mainly focused on lack of motivation and interest of English learning and not possessing scientific and efficient methods of learning English.

3.2 Reasons of problems in SN2MS

3.2.1 Low economy status of Shanyang

Located in the southeast of Shaanxi Province is Shanyang, a mountainous county, where 83% of the region is mountain, 8% water and 9% farming land. Its economy is underdeveloped and there is no hi-tech, no modern building and even no convenient transportation. Because of its low development of economy, there is no enough capital for education development. Advanced English teaching equipments and good English teaching environment cannot get guaranteed. For example, the only one multimedia classroom is seldom used for English teaching, because maintenance cost is too high for them to afford; there is no enough money to build English equipment and employ foreign English teachers for English teaching; young English teachers and teachers with higher abilities are eager to go to the big city for higher salary.

3.2.2 Misunderstandings of motivation of English

Motivation was considered as the main factor in language learning. An expert said that learning motivation was the internal motivation directly promoting the learning activities (Liu, 1993: 43). At present, for rural junior English teaching, many students, teachers and schools failed to recognize its value and meaning. Though the teachers and school could recognize its importance, they usually ignore its fundamental purpose because of teaching goals. Some teachers consider English teaching as a task, a pressure and homework, passive to cope with, and waste a lot of time to improve students' score rather than to cultivate students' ability. So does school. And most students regard English as one subject so as to get high score in the examination, neglecting English listening and speaking, which makes English become dumb or deaf.

3.2.3 Lack of career training for rural junior English teachers

Further education for English teachers is a kind of training to push forward the professional growth and life-long development of rural English teachers. Usually, the trainings for rural junior English teachers are short-term training. There exists the limitation of trainer's number, training time and training expenditure. From the interview, we find that teachers support the idea of teachers' training, but such problems as the faculty level, education views, operating conditions and so on are taken into their consideration. And many teachers hold the idea that the meaning of teachers' training is to get through the training exam and to get certificate, which makes teachers and trainers neglect the value of training.

3.3 Suggestions to improve rural English education

3.3.1 Suggestions for the Department of Education

The Department of Education is the department that organizes, leads and manages education business in accordance with educational guideline, policies and laws put forward by the state. So the decision made by the Department of Education is important for the development of English teaching in rural junior schools. For the present situation of English teaching in SN2MS, the Department of Education needs to insist the guide of policies, laws and guideline by the state, to provide sufficient funds for the development of education, encouraging rural English teaching innovation.

3.3.2 Suggestions for rural middle schools

First, rural junior middle schools should provide more funds to improve teaching facilities and learning environment by, for example, building more multimedia classrooms and audio-visual classrooms, using computer to see English films and broadcast the English news to attract students' attention, setting up an English corner to show some excellent English compositions, and sticking slogans such as "No Smoking", "Be Quite", "Washroom" and "Classroom" to create immersive English learning atmosphere. What's more, middle schools in rural area should strengthen communication and cooperation with other rural schools and with schools in urban area. Activities such as organizing cultural festivals, arranging meetings to discuss teaching affairs and share teaching experiences and resources can narrow the gap between schools and make contribution to the development of rural schools. Finally, schools should encourage teachers to participate in career training to improve their teaching abilities by offering them funds and opportunities to study in better middle schools at home and abroad.

3.3.3 Suggestions for teachers

First, teachers should learn to employ scientific and efficient way to teach instead of the traditional teaching mode with one textbook, one chalk and his/her mouth. Second, teachers should organize some interesting class activities such as making daily dialogues, singing English songs, reading English poems, making up stories for songs, and role-playing some famous dramas to arouse students' curiosity and interest of English learning. As a saying goes, give a man a fish and you feed him for a day while teach him how to fish and you feed him for a lifetime. It's of more importance for rural English teachers to teach students scientific methods to learn English. Finally, English teachers in rural area should actively participate in career training to improve their teaching abilities.

3.3.4 Suggestion for students

First and foremost, students need to cultivate internal motivation of English learning. Learning motivation can be divided into surface motivation and deep motivation. Students with surface motivation are lack of learning enthusiasm. And learners with deep motivation can learn English well. Second, students need to study consciously, especially after class. The junior students have ability to study by themselves with the help of computer and guidance from partners. According to the questionnaire for students, there are 23.64% of the students who continue to learn English after class with teacher's demand. Third, students can learn more from group study and peer work.

4 CONCLUSION

In this investigation, the author adopts the method of questionnaire and interview to analyze the present situation of rural junior English teaching in Shanyang No. 2 Middle School, finds out some problems such as shortage of English teaching equipment, unbalanced quality and number of English teachers, misunderstanding of the value of English learning and deficient views of teachers' training, and provides some suggestions to prompt the development of rural English teaching and learning.

REFERENCES

[1] Borko, H. and Putnam, R. Professional development and reform-based teaching. Introduction to theme issue. Teaching and Teacher Education, 14(1998), 1–3.
[2] Brown, H.D. Principle of Language Learning and Teaching. Shanghai: Foreign Language Teaching and Research Press, 2008.
[3] Butler, D.L. and Schnellert, L. Collaborative inquiry in teacher professional development. Teaching and Teacher Education, 28(2012), 1206–1220.
[4] Carroll, David. Psychology of Language (5th ed). Shanghai: Foreign Language Teaching and Research Press, 2008.
[5] Ellis Rod. The Study of Second Language Acquisition. Shanghai: Shanghai Foreign Language Education Press, 2003.
[6] Ono, Y. and Ferreira, J. A case study of continuing teacher professional development through lesson study in South Africa. South African Journal of Education, 30(2010), 59–74.

Education Management and Management Science – Zheng (Ed.)
© 2015 Taylor & Francis Group, London, ISBN 978-1-138-02663-6

Research on the teaching reform of the SEE training

W.J. Luo, L.P. Chen & X.D. Tian
School of Computer Science, Hebei University, Baoding, China

ABSTRACT: With the rapid development of information technology, the demand for software engineers is increasing, how to cultivate Software Excellent Engineer has become a key issue in current computer education reform research. This paper discusses reform teaching of the improvement of teaching methods, teaching contents and establishing two-way teaching evaluation mechanism, elaborates the method of applying the theory of Personal Software Process and Team Software Process of excellent engineer software training, a comprehensive teaching computer application talents supporting environment and facilitate the development of engineering and innovation.

Keywords: Software Excellent Engineer (SEE); teaching reform; Personal Software Process (PSP); Team Software Process (TSP); teaching supporting environment

1 INTRODUCTION

1.1 *Problems*

In order to meet the need of the social development of personnel training, training system as the main training application, clear compound, skilled personnel, the education mode, innovation mechanism, optimize the structure of professional, perfect quality assurance system, improve the students' practical ability and innovation ability, has been the core and focus on teaching reform research.

However, in the current computer undergraduate teaching still exist the following problems: (1) the basic teaching and professional practice is not close enough; (2) on cultivating students engineering ability not to combine the teaching of specialized courses set up step by step level; (3) the lack of teaching, opening teaching practice content, influence the cultivation of students' innovative ability; (4) curriculum is closely related with the lack of practice application industry; (5) cultivating the ability of students' lack of scientific system and evaluation mechanism.

1.2 *Research background*

China's Ministry of education and established the "Excellent Engineer education program in 2010 June" (referred to as the "excellent plan"), the plan is to implement the "national long-term education reform and development plan (2010–2020)" and "national long-term talent development planning outline (2010–2020)" of major reform projects, and an important move to promote the engineering

education power toward engineering education in China. The implementation of the plan is subordinate categories of colleges and universities in the implementation level; engineering undergraduate, graduate and doctoral students; implementation of all walks of life goal is to cultivate high quality innovative engineering and technical personnel.

At present, Chinese has 194 universities to join the "excellent plan". The urgent problem is how to expand the curriculum organization and practice matched in each specialty.

The teaching reform of Watt S. Humphrey on PSP and TSP theory is applied in computer application specialty, reform from teaching methods, teaching content and teaching evaluation of the three aspects, establishing comprehensive teaching supporting environment, in order to meet the training objectives of the requirements of SEE, improve the university graduates in engineering and innovation ability.

2 TEACHING REFORM

2.1 *PSP and TSP*

PSP focuses on the training of software development person. Its objective is to improve estimate, plan and quality management capabilities of software engineer. The PSP process consists of a series of methods, forms, scripts, how to guide software developers to ensure their quality of work, how to estimate and plan their work, how to measure and track individual performance, how to improve the software process and the quality of their own.

The basic measurement data of PSP include: each stage of software scale, the time required for each stage, implanted into the defect and the various stages of the elimination of the defects. Need to collect data and the actual data in these data items.[1]

TSP is a best practice has been defined and shown that support construction and management team, by definition, the team development process measurement and improvement, to guide developers in the shortest possible time to a predetermined cost to develop high quality software products.[2]

Introduced to the professional course teaching theory and practice in the practice teaching reform will be PSP and TSP, to enable students to accumulate individual software data, understand the individual software ability, grasp the development team, and team development ability.

2.2 The teaching framework for training SEE

Teaching framework is shown in Figure 1 for SEE training.

The teaching framework has the following characteristics.

1. The teaching activity of teachers and students learning activities are driven by the teaching plan, teaching plan input is theory courses,

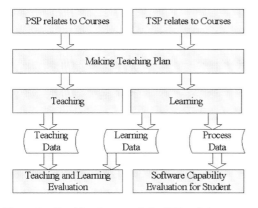

Figure 1. Teaching framework for SEE training.

practice course or other activities (such as curriculum design, simulation projects and graduation design).

2. In the training of the SEE, focus on the cultivation of students' engineering ability and innovation ability, the software development course, course experiment, course design and graduation design, simulation project respectively in accordance with the personal software process related courses and team software process related courses are divided, as shown in Table 1.

3. The teacher completes each link of teaching according to the teaching plan, to obtain the corresponding evaluation.

4. Students according to the teaching plan, complete the fixed learning task and non-fixed learning task, process data of personal or team accumulation, and obtain the corresponding evaluation.

5. The teacher can check the students to finish the teaching plan data, also can process data have been accumulated for students.

6. Every student can view their learning task completion data and process data and related evaluation.

2.3 The reform of teaching methods

The teaching plan driven teaching method, namely teaching leader develop overall teaching plan, unified management of teaching schedule, control course chapter, knowledge point and test difficulty; teachers' teaching plan specific content, including the teaching unit, exercises and unit test settings, and solve the students encounter in learning. The students to finish the teaching plan can communicate with the teacher according to the teaching unit.

With the methods of teaching the following improvements in the implementation process of the teaching framework, so as to promote the improvement of teaching quality, achieve the desired teaching objectives.[3]

1. Classroom teaching: teaching method combined with the use of multimedia teaching and

Table1. The classification of teaching plan input.

Classification	Contents	Examples
PSP relates to the contents and courses	Exercises, experiment, course design, graduation design, some individual entry title	Program design and experiment, data structure and experiment, operating system and experiment
TSP relates to the contents and courses	Course design, simulate project, some team entry title	Software engineering curriculum design, database curriculum design

traditional teaching, according to the specific content of the essence about it, coarse, clear hierarchy, prominence to the key points, grasp the situation of the knowledge of students to understand and grasp.

2. Exercise: exercise is an important part of supplementary teaching. By exercise students to consolidate the classroom learning content, extended master the knowledge point of flexibility; through exercise the teachers to understand students, the course focuses on the difficulties and various knowledge points used to grasp the situation, thus can be targeted teaching. In the teaching exercises of the course, the companion study guide books and teaching website, the specified exercises, exercises and simulation exercises three exercises, including objective and subjective questions, covering all knowledge points of course, and the important points of knowledge by various forms of exercises.

3. Course discussion: teachers through seminars and courses such as network forum exchange method make the students to master the content of classroom teaching quickly, strengthen the focus on learning, the difficulties in understanding, answer as soon as possible problems in learning, plays an important role.

4. Application practice: application practice is to cultivate SEE is essential, in teaching through the experimental curriculum, graduation design, simulation project three methods respectively from the basic, comprehensive and innovation in three levels of practice and training.

5. Guide self-study: guide the students to self-study learning by teaching websites and teaching support environment, help students master correct learning methods, learning materials, extended learning content, learning interest, and answering questions for students.

6. Assessment: the assessment methods, not only to help students understand their own level, but also teachers can grasp the learning of students, adjusting the teaching contents. Such as teaching the use of regular assessment combined with random examination; unit assessment combined with integrated assessment, appraisal and assessment theory practice combination evaluation method.

2.4 *The reform of teaching contents*

In order to train SEE adjust the teaching content of theory teaching and practice.

1. In the construction of theory course, correctly handle the relationship between the course content of the basic and advanced, classical and modern, this course and related courses. Remove prove cumbersome theory, abundant examples, exercises and typical case analysis, focusing on breakthrough and application of knowledge. And the relevant teaching outline, teaching plan, thinking questions, exercises, experimental guidance, reference catalog released to the teaching website, realize the sharing of high quality teaching resources, to promote the curriculum construction.

2. In the construction practice of the course, in order to stimulate the students' personality and creative ability, and to meet the needs of cultivation of students at different levels, the redesign of the experimental subject curriculum practice, set the requirements of different experiments for each experimental subject. The curriculum is divided into three levels: basic experiment, comprehensive experiment, and innovative experiment. Basic experiment, also known as the verification experiments, mainly used to verify the use of the typical data structure in the application, and master the design specification and basic method; comprehensive experiment mainly inspects the student to analyze the question and the comprehensive application of data structure design ability, students are required to complete the comprehensive use of the knowledge innovation; the experimental requirements students into the new elements in the experiment, completed the design of practical significance.

2.5 *The reform of teaching evaluation*

To change the teaching evaluation at the end of the semester will usually students evaluation of teachers to two-way evaluation between the teachers and the students, the following principles. (1) Teaching evaluation conforms from two aspects of evaluation of teachers and evaluation of students; (2) the evaluation of teachers and evaluation of students are composed of subjective evaluation and objective evaluation; (3) a subjective evaluation of subjective evaluation of each teaching unit is divided by the teaching plan; (4) objective evaluation of objective evaluation form each teaching unit is divided by the teaching plan.

The evaluation mechanism has the following characteristics. (1) It will be combined with the subjective evaluation and objective evaluation, weaken the influence of subjective factors, which makes the evaluation more objective; (2) the smallest unit of evaluation are each teaching unit, the evaluation of the assessment of teachers and students more specific and timely; (3) urged teachers and students find problems.

3 A COMPREHENSIVE TEACHING SUPPORTING ENVIRONMENT (CTSE)

According to the requirements of training SEE, build the CTSE that includes some main module below.[4]

1. User management: users of the system include senior administrator, administrator, teaching leader, teachers and students. The senior manager responsible for import administrator user name, password. The administrator will be responsible for all the teachers and students information into the system database, and manage the course information for each department designated teaching leader. Teaching leader directly into the teachers' information, teachers can directly into the students'.
2. Test management: teachers can add test and its properties. Teachers and teaching leader can modify the question or item attributes, and can then query the students completed the relevant teaching plan times, various questions and error rate.
3. Making and implement teaching plan: students will be on the course, to choose class they belong to, the purpose is to accept the teacher's teaching plan. Add the teachers teach the curriculum, class information, semester teaching leader, automatically generating system of teaching plan number to determine the only teaching plan. The teacher taught the class to add the corresponding teaching plan, teaching content of each unit, the teaching plan the teaching plan content or existing added to their teaching plan. Teaching leader and teacher can inquiry teaching plan contents and situation.
4. Knowledge point and chapter management: add the corresponding teaching leader course sections, teachers to add section add curriculum knowledge, knowledge points according to the attribute structure of input that can clearly reflect the relationship between the various levels of knowledge points. And teachers can choose to delete and modify the chapter and knowledge point.
5. Teaching evaluation management: teachers and students in CTSE can be in each teaching unit at the end of each subjective evaluation, and at the same time according to the teachers and students to participate in the teaching unit, the objective evaluation system is given, both for the teachers to improve teaching, students can detect early learning problems, strengthen the communication.
6. Data management of PSP: the accumulated data of the PSP support students to complete homework, daily exercise and the unit test process, including the number of lines of code, number of errors, the coding time, debugging time, and provides query and data analysis functions related to. Students can through the accumulation of data, to understand their software development of error rate and coding efficiency, to improve the estimation of software development cost and schedule accuracy.
7. Data management of TSP: supporters of the new software development team, support estimation in team work schedule and cost the team members used the PSP data, and provide records, search and analysis team data function.
8. Discussion area: discussion area includes user questions, answer, users to search the user, administrator, the administrator audit arrangement module, teaching support environment will discuss the content automatically saved to the database to store information and make optimization, to help students better understand the knowledge in the discussion.
9. Integral management: integral rule adopts "the double integral" which is used to determine the user level and will be used for consumption integral are distinguished. By this way users can be encouraged to share more quality knowledge. While the single integral system is a user level changes and consumer behavior are from the same integral system, encourage students learning interest.
10. Sending mail: teaching environment through the judgments of students' learning situation, automatic query the database and send mail to inform students to complete their tasks, also can send mail regularly to encourage the students.
11. Answer area: Q & A area mainly to solve the problems encountered in students' learning settings, students can post questions, questions will notify the teacher to answer, improves the efficiency of teacher-student interaction.
12. Questionnaire investigation: teachers can release evaluation or questionnaire by the function, quick to understand the status of students, interest in learning and mastery of knowledge and other aspects of the information, to strengthen the communication, adjust the teaching schedule and plan.

4 SUMMARY AND FUTURE WORK

Need to cultivate SEE, found the current problems in teaching, reform teaching methods, improve teaching efficiency and interaction quality, the professional courses and related practice in accordance with the personal software process and team software process division, software development related data acquisition and accumulation of students, with data enables students to see their own software and through the CTSE to lead students to promote individual software ability and team software capability. The need to further improve the teaching system in the future, and constantly improve the comprehensive teaching supporting environment, making it more suitable for the cultivation of SEE.

ACKNOWLEDGMENTS

This work was supported by National Natural Science Foundation of China (61375075), Soft Science Program (13455317D), the Doctor Foundation of Hebei University (2010–190) and the Seventh Teaching Reform Project of Hebei University (JX07-Y-28).

REFERENCES

[1] Watts S. Humphrey, PSP A self-Improvement Process for Software Engineers, Addison-Wesley, 2005.
[2] Watts S. Humphrey, Introduction to the Team Software Process, Addison-Wesley, 2000.
[3] Shen Liyan, Research on the domestic and foreign network curriculum development case, Information Education and Research Weekly, 2008, vol.10 p.140–141.
[4] He Kekang, Teaching System Design. Higher Education Publishing House, 2006.

Education Management and Management Science – Zheng (Ed.)
© 2015 Taylor & Francis Group, London, ISBN 978-1-138-02663-6

Analysis of road network accessibility demand based on different service objects

L. Wang, X.Y. Wei & C.N. Gou

School of Economics and Management, Chang'an University, Xi'an, China

ABSTRACT: According to the development philosophy of the transportation industry that "easy traveling of people and smooth flow of goods", this paper analyzes demands for transportation supply quality for different service objects such as passengers and goods based on the smooth characteristics and the meaning of the road service object. Passenger transport demand level and goods transport demand level were proposed from the perspective of demand, and then the accessibility of meeting the demand of both passengers and goods transport was discussed. Research results can provide beneficial reference for better overall efficiency of the system performance optimization of the road network.

Keywords: transport demand; road network; smooth characteristics; service objects

1 INTRODUCTION

The idea that "easy traveling of people and smooth flow of goods" is the basic characteristics to judge whether the transportation industry of the country is able to adapt to the needs of the national economy and social development. With the development of national economy and society, the country continues to increase investment in infrastructure, especially for the road infrastructure, the purpose of which is to better serve transportation. However, the phenomenon of "disconnect" between road construction and transportation is relatively common in practice. However, the number of cars has been saturated or even overload when the road was built soon. Meanwhile, the road traffic flow and passenger flow are little although these roads were built long, which did not play its proper economic and social benefits. This made the superiority of the road network system function and the advantages to the development of transport not evident. Therefore, it is necessary to analyze the demand of different service objects for road smooth characteristics.

2 THE MEANING OF ROAD SERVICE OBJECT

The road service object refers to the object entity that accomplishes displacement by running on the road, which can be divided into the direct service objects of road and the most basic service object of road.

The purpose of building roads is to ensure vehicles complete transportation service on the road conveniently, safely and comfortably. Therefore, from angle of the most direct benefits, the direct service objects of road are various passable vehicles.

The most basic service objects of road are passengers and shippers, which use the road to realize passenger and goods displacement to meet the demand of transportation. Road and vehicles are traffic infrastructure in the service of transport activities, and both construct road traffic system, thus serving passengers and goods displacement. Traffic infrastructure is just the means used to serve passengers and goods displacement, and successfully completing the displacement is the real purpose. That is to say, road is built for the exchange of passengers and goods (Qin 2005 & Amy 1998). In order to evaluate the smooth general characteristic of road more directly, the demand analysis of passenger and shipper is selected to research the evaluation in this paper.

3 THE DEMAND ANALYSIS OF THE SERVICE OBJECT FOR THE SMOOTH CHARACTERISTIC OF ROAD

The demand of the passengers and goods displacement which is good for the national economy and society, whatever it's aim can achieved should belong to the category of transport demand. Because different passengers and goods face the different environments, so do the demand time, demand location, demand quality, demand scope and demand quantity for transportation, the transport demand is diversified. Based on this, this paper analyzes the service objects in the respect of the basic needs and demand level, and so on.

3.1 The basic needs of the service object

People's demand for transportation is not determined by the number of transport supply. However, it focuses on what transportation supply quality the market can provide (mostly known as the "quality" characteristics). The transportation supply quality should be for "quality" rather than "quantity", and the "quality" here measures by the implementation of transport supply or the level which has been achieved actually, which is the expectations of the transportation subject (passenger and shipper) for the transportation services. The standards of transportation supply quality can be summarized as accessibility, timeliness, security technology, economy and comfort (Wu et al. 2007 & Yan et al. 2003). The different requirements of passengers or goods on transport service leading to passengers and goods transport have different determinants on the smooth characteristics of road. To be specific, there are two differences. On the one hand, the demands of transportation supply quality for passengers travel, including timeliness, comfort and economy. On the other hand, the demands of transportation supply quality for different kinds of goods, including security, timeliness, space value and efficiency.

Generally speaking, in order to achieve the desired goals, both passengers and shippers hope to realize people or goods space displacement through transport activity. From the point of view of passengers or shippers, it is the people or goods displacement that the actual value or potential value are promoted; utility is played better and emotion is expressed or met, and so on; from the point of view of country, people or goods space displacement can promote the development of national economy and the long-term development of the society (Wang 2010).

3.2 The hierarchy analysis of the service object

1. The analysis of passenger transport demand level.
   With the increase of people's living standard, the requirements of passenger travel quality become increasingly higher, which not only requires traveling smoothly, but also successfully. Thus transport demand is proposed for all levels based on such smooth characteristics as feasible, safe, convenient and comfortable. In this paper, passenger transport demand is

divided into four levels: feasibility demand, security demand, convenient demand, and comfort demand[4], as is shown in Figure 1.

The change of passenger transport demand level reflects the people's travel choice changes along with the change of their own conditions to a certain extent. The most fundamental purpose of passenger transport is to achieve passenger displacement; however, the smooth degree of road determines whether passenger displacement can be realized. Therefore, the study of the smooth characteristics can help road passenger transportation to meet the transport demand.

2. The analysis of goods transport demand level.
   From the above-mentioned, it can be seen that the determinants of goods transport demand are different with passenger. That is because the passenger transport demand is affected by people's subjective desire, but the direct factors affecting the goods demand are the structure of the industry and the types of goods. With the development of economy and layout of production, the transport demand of small batch and high value-added products was increasing day by day, and the transport demand of the primary products and raw materials gradually reduced at the same time, and the types of goods transported increased dramatically. Goods transport demand had diversified development and flow of commodities changed randomly. From these phenomena we can see that the hierarchy of goods transport demand does not change with a variable, but from one demand level to another, and the change of the hierarchy is mainly manifested in different transport demand which produced by the change of the goods. According to that, this paper constructs a figure to show the goods transport demand level, as is shown in Figure 2.

3. The accessibility of meeting the demand of both passengers and goods transport.
   Security is the most basic requirements of transportation, and also the main starting point of the transport demand.

Figure 1. The passenger transport demand level in China.

382

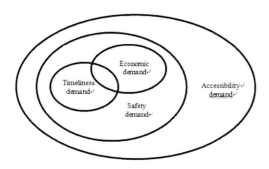

Figure 2. The goods transport demand level in China.

Convenience is the popular demand for transportation quality, whether it is a convenient traveling is the prior factor in the case of security being ensured.

Timeliness refers to reducing the traveling time as much as possible, and ensuring that time is accurate for passengers and shippers so that they can arrange work easily. Timeliness has two aspects: high speed, punctual and arrive on time. Timeliness is the important aspect of transport quality and the most basic forms of the smoothness of road, and it is related to the time value of passengers and goods.

Comfort is primarily for passenger transport, which is the comprehensive feeling of passengers when they travel. It includes the environment of the car, vehicles travel, the quality of the road, the evaluation and impression of the car service, and so on. Comfort is a higher level of transport demand, which is established on the basis of the first three. Transportation quality is difficult to be quantified because it lays particular stress on the feelings of the passenger or the shipper in quality.

4 CONCLUSIONS

Different passengers or goods have different demand satisfaction when they measure the demand of the smooth characteristics of road. From the view of traffic, the smooth characteristics of the same road does not change because of passenger transport and goods transport; from the view of the service object of the road (or transport demand), different passenger transport and goods transport will have different demand of the smooth characteristics of road. From the perspective of demand, this study analyzed the demand of the smooth characteristics of road based on different service objects, and the research results can provide beneficial reference for better overall efficiency of the system performance optimization of the road network.

ACKNOWLEDGMENT

This work was financially supported by The Central University Special Fund for Basic Scientific Research (2013G6231001).

REFERENCES

[1] Qin, H.R. 2005. The configuration research of highway service area based on the demand analysis of service object. Xi'an: Chang'an University.
[2] Amy, H. 1998. Changing intra-metropolitan accessibility in the USA: evident from Atlanta. *Transportation Research Part B*, 49(2): 55–107.
[3] Wu, Q.Q. & Yang, X. & Wang, Z. 2007. The quality analysis and demand forecast of road transport supply and demand. *Journal of Chang'an University (JCR Social Science Edition)*, 2007(12): 6–19.
[4] Yan, Z.R. & Zhang, R. 2003. Transport Economics. Beijing: China Communications Press.
[5] Wang, L. 2010. The Technical and Economic Assessment of the Road Network Flow Based on The Transport Demand. Xi'an: Chang'an University.

Education Management and Management Science – Zheng (Ed.)
© 2015 Taylor & Francis Group, London, ISBN 978-1-138-02663-6

A comparison of second language teaching methodology content-based instruction vs task-based instruction

Yu Zhou
School of Qiu Zhen, Huzhou University, Zhejiang, China

Sheng Zhou
Department of Finance, Military Economics Academy, Wuhan, China

ABSTRACT: In second language acquisition, methodology is regarded as a set of systematic teaching practice, which links language theories to learning practice. To facilitate second language acquisition in a classroom context, a lot of research has been carried out to explore more effective pedagogical methods. This article draws an examination to content-based instruction and task-based instruction, two mainstream approaches widely used in second language teaching. From different perspectives, a comparison is made to demonstrate their interpretation and application in the classroom setting.

Keywords: second language acquisition; content-based instruction; task-based learning

1   INTRODUCTION

Cognitive science has confirmed that high levels of content knowledge can promote the development of human's intellectual competence. Some linguists believe that second language acquisition can be most successfully acquired when the language conditions are similar to those present in first language acquisition, which is characterized by meaningful content. Therefore, it is suggested that the focus of second language should be academic content and the modification of the target language facilitates language acquisition and makes academic content accessible to second language learners (Crandall, JoAnn, 1994).

The learning principle underlying the task-based approach is that learners will learn a language best if they engage in activities that have international authenticity (Bachman 1990). The philosophy and rational of Task-Based Learning originate from Communicative Language Teaching based on the theories of British functional linguistic. As a more interactive instruction approach, it suggests that language is a tool or a resource for communicating, rather than academic knowledge to be studied. "Task" is identified as a central component part in designing language teaching syllabus.

2   FOCUS OF INSTRUCTION

2.1   *Meaning*

Content-Based Instruction is also called integrated language and content instruction (Rebecca Oxford, 2001). The focus of a CBI lesson is on the topic or subject matter. During the lesson students' learning is concentrated on "meaning" rather than "form". Something meaningful refers to anything that could arouse learners' interests. The subjects could be mathematics, science, social studies, and other academic subjects. Appropriate teaching materials, learning tasks, and classroom techniques are selected to develop learners' intellectual abilities according to academic content.

"In a content-based approach, students simultaneously acquire subject matter expertise and greater proficiency in English, the medium of instruction. Additionally, they learn to master skills necessary for academic success." (*D. Raphan & J. Moser: 1994*)

In the process of learning about the specific subject, learners are encouraged to use and practice the language they are trying to learn. The target language is served as a tool for developing not only content knowledge in some specific field but also learners' linguistic ability. Many linguists believe that the second language can be successfully acquired in this natural way.

2.2   *Practice*

In Task-Based Methodology, target task (Nunan 2004: 4) is an important concept underlying the instruction. When tasks are turned into learning practice, they become pedagogical tasks (Nunan 2004: 4). Here are two definitions of a pedagogical task.

A task is a workplan that requires learners to process language pragmatically in order to achieve an outcome that can be evaluated in terms of whether the correct or appropriate prepositional content has been conveyed. To this end, it requires them to give primary attention to meaning and to make use of their own linguistic resources. (Ellis 2003: 16)

A task is a piece of classroom work that involves learners in comprehending, manipulating, producing or interacting in the target language while their attention is focused on mobilizing their grammatical knowledge in order to express meaning... The task should also have a sense of completeness, being able to stand alone... with a beginning, a middle and an end. (Nunan 2004: 4)

The focus of the teaching is on the completion of a task which could arouse learners' interests. What is studied in a second language class depends on what happens in the process of activity. Learners use the target language to complete the task; there is little correction of errors and meaning is much more important than form. The learning context reflects real life. For learners, playing a game, discussing a problem in a small group or sharing information or experience, can all be considered as meaningful and authentic tasks.

3 INSTRUCTION MODEL

3.1 *Content-based teaching models*

The Sheltered Model, The Adjunct Model and The Theme Based Model are three major models for CBI (Stephen Davies, 2003).

The aim of the Sheltered Model is to enable second language learners to study the same content material as regular first language learners. Special assistance is given to help second language learners understand regular classes. Two teachers, a content specialist and an ESL specialist, can work together to give instruction to student in the specific field.

The Adjunct Model is to prepare students to join the classes with as regular first language learners. Emphasis of Adjunct Model is placed on acquiring language proficiency as well as study skills, such as note taking and skimming and scanning texts. Adjunct classes are usually taught by ESL teachers.

Theme based CBI is usually found in EFL contexts. Theme based CBI can be taught by an EFL teacher. The content can be chosen from an enormous number of diverse topics. Syllabus is tailored to build on learners' interests.

3.2 *Task-based teaching models*

In the model of task-based learning, students start with the task. When they have completed it, the teacher draws attention to the language used, making corrections to the students' performance. In A Framework for Task-Based Learning (Jane Willis, Longman), Jane Willis presents a three-stage process: pre-task—Introduction to the topic and task; task cycle—Task planning and report; language focus—Analysis and practice.

In the pre-task stage, teacher introduces the topic to the class, points out useful words and phrases, and helps learners understand task instructions and gets them prepared. Learners may read part of a text as a lead in to a task.

In the task cycle, students do the task in pairs or small groups. The teacher monitors at a far distance and can give some help if he is asked to so. Students feel free to fulfill the task with the target language, because there is no correction and mistakes are allowed. Students are required to present what happens during their task in an oral or written form. Before that, they can practice what they are going to tell in their group. The teacher is available to help them clear up any language questions they may have. Some groups present their reports to the class, or exchange written reports, and compare results. Teacher gives comments on the content of the reports.

In language focus and feedback stage, the teacher then highlights relevant parts from the text for the students to analyze or also highlight the language that the students used during the report phase for analysis. Practice of new words, phrases, and patterns occurring in the data is conducted for learners to consolidate what they learned during the class.

4 ADVANTAGES

4.1 *Authentic*

Second language learners can be easily motivated in content-based model, which can make second language learning more interesting. Real purpose is fulfilled in an authentic context. Both content and linguistic learning can make students feel more independent and confident. A much wider knowledge of the world, got through CBI, would benefit learners' general educational need.

In content-based model, some valuable study skills are also cultivated, such as note taking, summarizing and extracting key information from texts. With information from different sources, students are encouraged to re-evaluate and re-construct that information. The whole process can help students to develop their reading, writing and critical thinking competence. In addition, a group work element within the framework gives students the opportunities to cooperate with each other, which can have great social value.

4.2 *Communicative*

In task-based learning, second language learners are required to use their language resources to solve the problem rather than just practicing one selected linguistic item. Students are provided a relevant or authentic context to practice the target language and the "task" makes learning a personalized experience.

Through situational activities, the target language is naturally explored for the communicative needs from language learners, instead of from the course book or a premeditated teaching plan. Second language learners are exposed to a whole range of lexical phrases, collocations and patterns as well as language forms and are driven by a strong desire to communicate in order to complete the task.

5 POTENTIAL ADVENTURE

5.1 *Different levels of proficiency*

Although it has been believed that language skill could be learned through a focus on meaning, there is increasing evidence indicates that focus on meaning is insufficient to ensure completeness of learning itself and grammatical perfection. Since content-based instruction isn't explicitly focused on language learning, some students may feel confused or may even feel that they aren't improving their language skills.

To some extent, the nature of the content might differ by proficiency level. For beginners, the content often involves basic social and interpersonal communication skills, but past the beginning level, the content can become increasingly academic and complex. An effective and natural integration of language and content highly demands Sound Teaching Practices, Methods for Promoting the Acquisition of Content and Language, Techniques for Incorporating Levels of "Positive Complexity" into Instruction and Approaches for Building Curricular Coherence (Fredericka L. Stoller, TESOL 2002), which means a huge task for second language teachers.

5.2 *Conversational forms of a task*

Task-based classes focus on the completion of a task, which involves lots of activities to engage students, and majority of the responses takes conversational form from both teachers and students. As for second language starters, they could encounter much more difficulties during the task when they listen to the instruction, try to express their ideas with target language or make themselves understood by their teachers and peers. It is argued that tasked-based instruction does not suit lower-level second language learners.

The content in tasked-based instruction can be predetermined, but not the "task". In practice, there is often significant disparity between what is supposed to happen and what actually happens. The task could be affected by many factors. For example, learners can simply disengage from the intended pedagogical focus in group works when the discussion becomes heated. Sometimes the communication confusions also arise due to teachers' failure to set up their intended pedagogical tasks or student's misinterpretation for teachers' real intention.

6 CONCLUSION

In content-based instruction, students practice the language in a highly communicative fashion by means of learning content such as science, mathematics, and social studies. In task-based instruction, students participate in communicative tasks in language learning. Both of these methods can be applied in second language acquisition to generate an authentic and meaningful communication context and promote the integration of all kinds of language skills of second language learners.

ACKNOWLEDGMENT

This article is sponsored by Zhejiang Higher Education Teaching Reform Research Project (No.: jg20132460).

REFERENCES

[1] Crandall, JoAnn, 1994, Content-Centered Language Learning, ERIC Clearinghouse on Languages and Linguistics Washington DC.
[2] Bachman 1990, Fundamental consideration in Language testing, Oxford: Oxford University Press.
[3] Rebecca Oxford, 2001, Integrated Skills in the ESL/EFL classroom http://www.cal.org/topics/ilc.html.
[4] Raphan, D. & J. Moser. (1993/94). "Linking Language and Content: ESL and Art History." TESOL Journal. 3, 2 17–21.
[5] Ellis, R. 2003. 'Task-based research and language pedagogy'. Language Teaching Research 4/3: 193–220.
[6] Nunan 2004: 4, Task-based language teaching, Oxford: Oxford University Press.
[7] Stephen Davies, 2003, Content Based Instruction in EFL Contexts, The Internet TESL Journal.
[8] Jane Willis, Framework for Task-based Learning, Longman Express.
[9] Fredericka L. Stoller, TESOL 2002, Content-based Instruction: A Shell for Language Teaching or a Framework for Strategic Language and Content learning?

Education Management and Management Science – Zheng (Ed.)
© 2015 Taylor & Francis Group, London, ISBN 978-1-138-02663-6

The exploratory examination of the teaching methods of business communication

Bei Ju

International Business Faculty, Beijing Normal University, Zhuhai Campus, Tangjia Bay, Zhuhai, P.R. China

ABSTRACT: This paper attempts to review the related studies on the instruction of business communication course and it has found that the teaching methods of business communication are closely related to course content and the studies are of great variety. Therefore, the teaching approaches are discussed on the basis of three essential topics in business communication including international communication, ethics in communication and communication technologies, whose significance is not yet emphasized at present instruction but definitely worth developing in the future. Finally, the appropriate teaching methods of each topic are suggested to be found in the future by experimental analysis.

Keywords: business communication; course content; teaching methods

1 INTRODUCTION

Setting against the new era of technology, globalization and pluralism, the effective approach of communication serves as the decisive role in the business activities, which contributes to the characteristics of business communication course offered in undergraduate business schools as practical, up-to-date, communicative and intercultural. It is noticeable that business communication course has made strides forward but it still has many aspects unchanged. Therefore, this paper conducts the exploratory examination of the teaching method of business communication course and provides an innovative pattern of its teaching for instructors to help students develop their knowledge and skills in the modern workplace.

2 RELEVANT STUDIES OF THE INSTRUCTION OF BUSINESS COMMUNICATION

2.1 Studies of the business communication course in China

In China, the relevant studies of this course are mainly overlapped in two domains, namely ESP (English for Specific Purpose) and business field. The new world of business communication in China is introduced by Baolin Zong and H.W. Hildbrandt (1983) through the explanation of how the three courses in business communication are integrated, organized and instructed.

In terms of instruction, they found that the translation exercises and case studies are much stressed. Ying Li (1996) generally introduced its teaching methods such as group discussion, case study, role-play, oral presentation, and so on. Furthermore, Xiumei Fu in 2007 explored the in-class teaching methods of business communication course from the communicative approach, simulation approach, collaboration approach and heuristic approach under the teaching philosophy by integrating English language communication ability, expertise of international business and cross cultural difference. To be more specifically, Yourong Guo (2010) analyzed the implication of interactive teaching method in business communication course by illustrating four specific ways of instruction and emphasizes the significance of teacher's role as organizer and guider. In Addition, the experiential teaching method is discussed by Juxiang Zhang (2010) from three approaches as simulation, role-play and transposition with the suggestion of its proportion in the overall teaching procedure. Similarly, Fengjuan Ji (2013) examined the implication of such teaching methods as case analysis, cosplay, task driving and interaction and group teaching in business communication. It can be seen from the previous studies on the teaching methods of business communication that most Chinese researchers limit their studies within the discussion of instruction ways, but not combine them with the development of new communication channels or the needs of learners in their workplace.

Table 1. The content of *Communicating in Business* based on topics.

Topics	Chapter	Content summary
International communication	1	Language differences create a verbal barrier to communication.
	2	International business would not be successful without international communication.
	3	Facial expressions, voice qualities, the meaning given to time and the importance of touch behavior.
	4	Cultural norms for email vary.
	5	Slang is particularly challenging to nonnative English speakers.
	6	Goodwill messages that may be appropriate in one country may be improper in another.
	7	The direct and indirect organizational plan chosen by culture.
	8	People in high-context cultures may prefer the indirect organizational plan for bad-news messages.
	10	Different countries have different laws regarding plagiarism.
	11	Presenters need to prepare and practice carefully to avoid miscommunication.
	12	International jobs may require a curriculum vitae instead of a résumé
Ethics in communication	1	Companies expect employees to communicate ethically, beyond legal requirements.
	2	Team members have an ethical responsibility to respect each other's integrity and emotional needs.
	4	Using BCC when copying people on email messages may be considered an ethical issue.
	5	Companies need to present information accurately and language should express honest evaluations.
	6	Goodwill messages are sent out of a sense of kindness.
	7	Persuasive communication is prone to questionable ethics; customers may react negatively to questionable sales letters; specific, objective, truthful evidence.
	8	Ethical communicators use a buffer to help the reader accept disappointing news.
	9	Data in tables and charts; everyone involved in a reporting has a responsibility to act ethically.
	10	Emphasis and subordination; falsifications; distortion by omission is unethical.
	11	Negative information should not be ignored; humor in presentations is risky.
	12	Lying on a résumé is one of the worst mistakes; an appropriate résumé.
Communication technologies	1	Common communications use technology: email, phone, voice mail; instant and text messaging; and social media.
	2	Technology such as wikis and Google Docs can help a team manage documents.
	3	Building meaningful relationships through social media; voice technologies and text messaging; conference calls, videoconferences.
	4	Email and web writing.
	5	The "Track Changes" feature.
	6	Instant messaging; social media sites.
	7	Negative customer feedback on social media sites is common.
	8	Email has advantages for communicating bad news.
	9	Internet resources.
	10	Reports in programs such as PowerPoint and Keynote are standard; saving a report as a PDF file; the standard for documenting sources in business reports.
	11	Online presentations; tools for creating presentations; video is useful in presentations.
	12	Social and video résumés; saving a résumé as a PDF file; video interviews; the importance of managing your online reputation.

Not only by looking into the above-mentioned teaching pedagogy, the relevant studies in foreign countries seem more inspiring and innovative. Jean (1999) described AL-L, academic service-learning, that is an innovative business communication pedagogy and it has been found that the greatest strength of using it is students' trying theory into actual application of principles, student passion for the work, faculty enthusiasm for the results and community satisfaction. In Addition, Kathleen (2002) exploited library resources to accomplish four assignments which are respectively related to the business report, professional journal, style manual format and international business communication. Along with the social development, Michael (2012) discussed why business communication scholars should focus on social media as an important part and how to integrate it in its MBA program. Focused on teaching written business communication, Preeja in 2013 examined the conceptual framework of Bloom's taxonomy and its objective in current education, especially in course design and delivery, which gives a brand-new aspect of teaching.

The previous studies not only introduced the teaching methods in a general way, but also integrated the teaching pedagogy with vivid description of why and how the course should be delivered which cannot be separated from course content. Therefore, this paper will propose an integrated teaching method varying with the business communication course content.

3 THE APPROPRIATE TEACHING METHOD OF BUSINESS COMMUNICATION

Based on the recent study conducted by Matthew and Eva (2013), course content in business communication seems mainly similar to that in 1999, with no much change. Actually, the world is under the trend of innovation and development, how to equip students with new and practical communication skills to meet their needs in the current workplace, the more feasible course delivery should be developed on the basis of course content. By examining the Top 50 Undergraduate Business Schools, these topics, found to be mentioned in line with their frequency in course description by Matthew and Eva (2013) are teamwork, intercultural, teamwork, job-seeking skills, interpersonal, visual, ethics and so on. Therefore, they suggest a possible trend of increased attention to topics as intercultural communication, technology, visual communication and

ethics which are believed to be essential for business communication students. Along with the latest and popular textbook of teaching business communication in China, *Communicating in Business* (Scot & Amy, 2013), the exploratory teaching approach is put forward for the traditional and new course content which is essential to teaching business communication in China. There are five parts (labeled as twelve chapters) in *Communicating in Business* (China Student Edition), namely Communicating in business: the foundations, developing your business writing skills, written communication, writing reports, oral and employment communication. Compared with the original textbook, two chapters as Team and Intercultural Communication and Interpersonal Communication Skills are omitted which are believed to be of great importance. Consequently, these two parts will also be included when discussing the teaching methods. Sorted by essential topics such as international communication, ethics in communication and communication technologies, the content of textbook could be organized as followings.

Although different teaching methods could be used for the different tasks, the previous studies have not classified them in accordance with the topics in a systematic way (with reference to Table 1) which will provide the instructors with the more specific and accurate guidance. In terms of international communication, students should be encouraged to establish their cross-cultural awareness and apply it into the daily communication. Under this circumstance, it is suggested to use the method of case analysis by digging into the cultural differences. Meanwhile, the video materials about international communication could be used to illustrate the key point by providing the vivid images. As for the ethics in communication, students will be encouraged to do the role-play by experiencing the various scenarios from which they could identify the ethics on their own and find out what the ethics are demanded in communication. When it comes to the communication technologies, different ways of technology could be introduced by task-based teaching along with the usage of library resources. Currently, this is a new perspective into the instruction of business communication based on the grouping of different topics.

4 CONCLUSION

The studies on the teaching methods of business communication course have gained much progress at present. Moreover, the approach of finding the appropriate methods according to the essential topics of business communication gives a new idea to the relevant researches. To find out the most effective teaching method for each topic, the questionnaire and experimental teaching are suggested to administrate in the future by examining the students' preference and the efficiency of different teaching methods.

REFERENCES

[1] Baolin Zong & H.W. Hildebrandt (1983). Business Communication in the People's Republic of China. *The Journal of Business Communication 20*, pp. 25–32.

[2] Fengjuan Ji (2013). Implication of Multiple Teaching Methods in Business Communication Course Based on Skills Training. *Journal of Changchun University of Science and Technology*, Vol. 8 No. 3, pp. 191–192.

[3] Jean L. Bush-Bacells (1999). Innovative Pedagogy: Academic Service-Learning for Business Communication. *Business Communication Quarterly*, Vol. 61 No. 3, pp. 20–34.

[4] Juxiang Zhang (2010). The Implication of Experiential Teaching in Business Communication Course. *Journal of Jilin Business and Technology College*, Vol. 26 No. 4, pp. 100–102.

[5] Kathleen M. Hlemstra (2002). Library-based Assignments That Enrich the Business Communication Course. *Business Communication Quarterly*, Vol. 65 No. 3, pp. 55–63.

[6] Matthew R. Sharp & Eva R. Brumberger (2013). Business Communication Curricular Today: Revisiting the Top 50 Undergraduate Business Schools. *Business Communication Quarterly*, Vol. 76 No. 1, pp. 5–27.

[7] Michael J. Meredith (2012). Strategic Communication and Social Media: An MBA Course from a Business Communication Perspective. *Business Communication Quarterly*, Vol. 75 No. 1, pp. 89–95.

[8] Preeja Sreedhar (2013). Using Bloom's Taxonomy as a Pedagogical Tool for Teaching Written Business Communication. *The IUP Journal of Soft Skills*, Vol. 7 No. 3, pp. 51–55.

[9] Scot Ober & Amy Newman (2013). *Communicating in Business* (Eighth Edition), China Student Edition, Beijing: Tsinghua University Press.

[10] Xiumei Fu (2007). On the Teaching Design and Teaching Method of Intercultural Business Communication. *Heilongjiang Researchers on Higher Education*, 2007, No. 2, pp. 144–146.

[11] Ying Li (1996). Business Communication—A New Course of Effectiveness, Usefulness and Attraction. *Journal of Beijing International Studies University*, Vol. 3, pp. 120–127.

[12] Yourong Guo (2010). An Analysis of Interactive Teaching Method in Business Communication Course. *Journal of Chifeng University (Natural Science Edition)*, Vol. 26 No. 6, pp. 200–201.

Education Management and Management Science – Zheng (Ed.)
© 2015 Taylor & Francis Group, London, ISBN 978-1-138-02663-6

Training way of ethnic education characteristic on border areas in Yunnan, China

Y. Yang
School of Tourism and Geographical Sciences, Yunnan Normal University, Kunming, China
School of Information, Yunnan Normal University, Kunming, China

Y.L. Sun
School of Education Sciences and Management, Yunnan Normal University, Kunming, China

L. Xu
School of Tourism and Geographical Sciences, Yunnan Normal University, Kunming, China

ABSTRACT: Special geographical environment gives international characteristic for border education; ethnic education on border areas has mutual influences to international education. Geography education regards that geographic environment influence education content significantly. But it is a low level for combination degrees of education content and geographical environment. Focusing on the situation that lack of local related curriculums, as well as aiming to develop a ethnic education within a local and ethnic's characters, this paper proposes that ethnic education must fit the needs of local requirement. In order to protect national security, ethnic education needs to strengthen the patriotism education. Set up international language, culture curriculum and ethnic resource curriculums is necessary for enhancing the understanding of international and local ethnic.

Keywords: ethnic education; regional characteristics; curriculums; border areas

1 INTRODUCTION

The concept of border in the dictionary means "a certain width area for administrative management inside the border line" (Zheng, X.R 2012), and is key sites where the 'inside versus outside' (Sprke, 2009, p. 52). It means the political divide on border line. This expression is mostly from the perspectives of geography and politics, but with the development of the scientific, technological and economic globalization, concept of border is changing from geographical border to be the "Soft Border" which concludes nation's politics, economic, culture, society, education, and information. National security changes from the military confrontation to the traditional and nontraditional threats intertwined. In view of these conditions, the education security especially on the border areas, becoming important to the national security.

2 PARTICULARITY OF BORDER ETHNIC EDUCATION

2.1 The special geographical environment

Yunnan borders Vietnam, Laos, and Burma. The border length in the province is 4060 km, 20% of Chinese land border line. As adjacent to Southeast Asia, South Asia, it has unique advantage to communicating with Pacific Ocean and India Ocean. In Yunnan region, countries are separated by a mountain, or by a river, some places just have border markers. Be linked with mountains, rivers and roads, creates advance geographical condition for economic, cultural and educational exchange among countries.

2.2 International characteristic for border education

National education on border areas includes the characteristics of ethnic education in other kinds of regions, but it has special characteristic itself. Among those characteristics on the border education, internationalism is the most significant feature of border education.

Border areas are mostly multi-ethnic regions, some minorities are live across the border areas. Yunnan province has 25 minorities, and 16 of them live across the border areas. These cross-border ethnic groups are resident in complex geopolitical environment, and, of course, they are the special culture groups. Although a minority residents in different countries, they has the same religious belief, culture, living habits, and they can not be divided easily. Once the cross-border minority's identity of the outside border's part is exceed the national identity, the national security will become under threat, especially when face the hostile forces' temptation (Su 2012).

2.3 The mutual influence of national education

Because the special geographical environment, the border areas are more susceptible to the other countries. From 1990s, Vietnam, Laos, and Burma were formulated a number of policies on their border areas, carried out education reform to develop boundary education, and some reform measures have more strengthen and more influence range than China. For example, Vietnam government regulates that: 1) student's cost of book and study things for following third grade are free; 2) from grade four to university, student's basic necessities of life were given from government; 3) gives most 50 RMB living subsidy to ethnic school students and secondary specialized school. Those regulations also suitable for Chinese students on border areas (Ying 2003).

Those policies create great effect to the marginal people on Guangxi and Yunnan provinces, some students (mostly ethnic students) went to foreign school. Since 2000, within the implement of "Free for book, study things and other cost" projects and "Vitalize Border Areas and enrich the people living there", students go outside school are decline and becoming rare. On the contrary, some foreign students come to the Chinese schools for study.

3 THE CURRENT SITUATION OF BORDER EDUCATION

3.1 Significant progresses

Geography of education regard that geographic environment can influence the education content (Luo 2003, p. 68). The features of producing and living have mutual relation with geographic environment, especially in the lower level of productive forces, isolated societies. Geographic environment can influence the curriculum design and the arrangement of textbook.

If we envisage the influence of geographic environment, adjust measures to local conditions, combining region economic feature, education also can influence geographic environment. For example, environment training people, improves people's cultural quality (Luo 2003, p. 91). Through delivery man's nature and social experience, people can cognize, utilize, transform geographic environment in macro scales and deep levels. Through people's cultivation, they use the knowledge that studied in school to utilize and transform natural.

What we need to do is make benign and harmonious relationship between education and geographical environment. But, there was low level for combination degree between education content and their geographical environment.

3.2 The lack of local related curriculum on border education

In China, schools, including rural schools and ethnic schools are followed standard and uniform curriculums. The purpose of curriculum system, values, learning objectives are enter to higher school, leave the rural village. Much of content in textbooks takes from city life, the rural and ethnic students driven by it, and as lacking necessary materials related to these contents; as a result, they can't understand the content, of course, they don't like learning. On the border area, as the education level was backward, many rural and ethnic students can't go to high school or university, and because they leave home to study, they have few chances to work with family, they become disgust labor on rural, what they want is get away from home and go to big cities. Elementary education can't teach student the technology and skill for producing, but fosters many "young emperors" in poor family. One investigation shows that the practical skill is the most scarcest knowledge for students (Jin, 2008, p. 98).

The complex mountain environment makes the modern agriculture technology is difficult to develop in these areas. As a result, the economic advance on border areas is very slow, some places still in primitive or traditional agriculture, this lead some students who graduated from middle school or secondary school can't use their knowledge in work. It causes villagers think learning is useless.

Zhang (2012) concluded that education activity has no relationship with local nature geography, social culture, and economic development. Especially in rural area, schools are independent kingdoms, isolated local nature, economic, and culture; only accept the educational administration's lead and management. In a word, schools are "enclaves", "isolated islands".

In terms of interview by author, 2 elementary schools on border of Jancheng are just study national curriculum and their county curriculum. One of them has some courses on adjacent countries culture, but it was a national research project, as the end of project, course had been over.

3.3 The lack of curriculum for communication with adjacent country

According to Ying (2003, p.), besides the reason of poverty that some parents send their children to Burma, "a part of parents in China consider that compared with Chinese schools just teach Chinese, children go in Burma's schools can learn not only Burmese and English in Burmese school, but also Chinese in ethnic China's schools, thus their children will study more knowledge and language." Language is the media to communication, and communication brings understanding. On the border areas, people need communicate with each other in many situations, especially for trade. Within the 12 national ports and 8 provincial ports, Yunnan is building into China's southwest gateway. If we want to achieve this goal, we need more talents who can use adjacent countries languages and knows adjacent country's politics, economy and culture. Border areas have advantaged condition for cultivating these talents.

In author's interview, mostly Lao people can use Lao, English, Chinese, even French in their daily life. They can communicate well with different peoples. But Chinese people just only use Chinese and little simple English. It will cause an obstacle to communication.

In Yunnan province, some universities enrolled major for adjacent languages, but just some of students doing the work related with the adjacent countries when they graduated; of course, they will forget these language, unless they use them in their daily life. But on border area, people really need this knowledge in their life and communicating. Now it is necessary to establish an international curriculum for them.

4 THE METHODS FOR CULTIVATE NATIONAL EDUCATION CHARACTERISTIC ON BORDER AREAS IN YUNNAN

National education on boundaries is different with the mode of developed regions. Education must fit the ethnic societies' appear and potential needs. Curriculum resources concentrated reflection the border education's advantages and characteristics. Boundaries' schools can create school-based curriculum or curriculum resource that full of region characteristics, and it can realize the border education characteristic development.

4.1 Strengthen the patriotism education curriculum development

In order to protect stability of border areas, it is need to strengthen the patriotism education on border areas. The primary tasks for boundaries education are enhancing national identity education, and ethnic group's national consciousness.

In the classrooms, teachers can use local conflict (between countries) histories in some historical

content, and extend class outside. Using the geo-advantages, establish cooperative relationship with the local frontier forces. Soldiers can be invited to schools at regular time to teach national defense in their perspective as well as drugs education. Integrating with local tourist attractions gives students opportunities to experience the history. Through these methods, students will realize the patriotism is not only the school request for them to study and remember, but also is the realistic needs.

Schools can use the geographical advantage to doing patriotism education, giving students some investigative study tasks, let students to visit local villagers to collect historical events, and to compare the changes on education, economic, transportation with past. These methods will foster students to love their country and hometown.

4.2 Set up the international curriculum

With the construction for China and ASEAN Free Trade Area (CAFTA) and National Gateway Project of Yunnan, the international cooperation becomes frequent to China and adjacent countries on economic, culture. People who can use adjacent countries languages and knows their culture is valuable for society. Adapting to this situation, the boundaries schools in Yunnan can use the geo-political advantages to form a special education space on border areas. Setting up different languages in different adjacent countries areas, such as Vietnamese, Lao, Burmese, even Hindi curriculum, and border countries' cultural curriculums. It will offer students many chances to contact adjacent countries' condition earlier, forming fundamentality knowledge about other countries, and helpful for their future work. Such curriculums content are mostly related with student's future, and students will interest in.

4.3 Strengthen ethnic cultural curriculum development

Yunnan border areas are mostly minority concentrated areas, as well as cross-border ethnic groups areas. National education on border need to undertake the mission to inherit and protect the ethnic culture's, and also need to enhance the national unity idea.

Christine (1999) concluded that elementary Maori students who participated in kapa haka showed increased levels of self-esteem and motivation which were linked to increased academic performance (Paul, 2010).

Ethnic cultural curriculums will teach ethnic languages, life styles and life circumstances. Schools can use ethnic festivals to do activities curriculums, or adopt social investigation to combine the task that already mentioned. The ethnic curriculums are not only enhances the understanding for groups, preserves their ethnic cultural in economic modernization process, but also enhances the attraction for school, benefit for national identity (Gerand 1999).

5 CONCLUSION

Ethnic education on boundaries is special for it international feature. Education can help protecting the national security. Education must fit the boundary minority societies' needs. Under the advantage for geographic, schools on border areas can develop special curriculums. We can strengthen patriotism education and ethnic culture education to enhance national and ethnic identity. Set up international curriculum to let communication more conveniently. There are the boundary education's characteristics. The boundary education can construct these characteristics and need to do it.

REFERENCES

[1] Gerand, A.P. 1999. *China's national minority education: culture, schooling, and development*:17. Basing stoke, England: Falmer Press.

[2] Jin, Z.Y. 2008, *The Inheritance of Ethnic Culture and Curriculum Reform of Ethnic Elementary Education*: 98. Beijing, China: Minority press. (in Chinese).

[3] Luo, M.D. 2003. *Gography of Education*: 69,91. Kunming, China: Yunnan University press. (in Chinese).

[4] Paul, W. 2010. Indigenous-based inclusive pedagogy: The art of Kapa Haka to improve educational outcomes for Maori students in mainstream secondary schools in Aotearoa, New Zealand. *International Journal of Pedagogies & Learning*, 6(1): 3–22.

[5] Sparke M. 2009. Border. In: Gregory D, Johnston R, Pratt G et al (eds.). *The Dictionary of Human Geography*. 5th edition. Chichester, England: Blackwell.

[6] Su, D., Wang. Y.B. 2012. National Identity Education: Strategic Choice of Educational Development in Yunnan Border Areas. *Journal of Research on Education for Ethnic Minorities* 23(5): 5–9. (in Chinese).

[7] Ying. H.W. 2003. The Competition in Education on Border Line. *Southern Window* 6: 50–52. (in Chinese).

[8] Zhang. Z.J. 2012. Educational Geography and Geography of China's Education. *Journal of Yangteze Normal University* 28(8): 10–15. (in Chinese).

[9] Zheng, X.R., Ouyang, C.Q., Shi, M. 2012. A Window to Understanding Ethnic Education in Border Areas of China—: A Review of On Ethnic Education in Border Areas of China. *Journal of Minzu University Of China (Philosophy and Social Sciences Edition)* 39(6): 157–157. (in Chinese).

Education Management and Management Science – Zheng (Ed.)
© 2015 Taylor & Francis Group, London, ISBN 978-1-138-02663-6

Reform and practice of teaching system of VC++ practice course based on CDIO

Pingle Yang, Qinge Zhang & Yong Wang
Jiangsu University of Science and Technology, Zhangjiagang, China

ABSTRACT: Nowadays engineering education in colleges neglects the enterprise reality and college students lack practice experience. The CDIO model makes theory and practice with good that represents the advanced engineering education methods and lays emphasis on the development of personal abilities and qualities, complying with the principle of cultivation. The teaching of VC++ practice course can be based on CDIO mode emphasizing on the development of personal abilities and qualities. This paper discusses how to conduct teaching reform from the aspects of teaching program, teaching methods, teaching management and testing methods.

Keywords: CDIO engineering education; teaching reform; project teaching

1 INTRODUCTION

CDIO represents Conceive, Design, Implement and Operate, with R & D operation life cycle as the carrier, the organic connection between practice and curriculum for students to learn and acquire the ability of engineering practice. CDIO training syllabus divides ability training into four levels of engineering basic knowledge, personal abilities, interpersonal and team skills, systems engineering capability, prompting students to achieve the intended target in the four aspects based on the integrated cultivation mode.

Domestic and international experiences show that the idea and method of CDIO "learning by doing" is advanced and feasible, fitting reforms for all links in the teaching process of engineering education. From 2005 on wards, led by Shantou University, the domestic CDIO pilot colleges and universities have been playing the positive role in the further promotion in China. The Ministry of Education has been promoting it by organizing the CDIO forum since 2007, and the seminars are held each year to promote the CDIO in the professional training.

This model accords with the engineering personnel training pattern and combines theory and practice well, representing the advanced engineering education methods. From the point of view of the CDIO engineering education concept, it requires not only knowledge but also pays more attention to the cultivation of personal ability and quality.

2 ANALYSIS OF THE PRACTICE TEACHING OF VC++ COURSE

Over the years, the reform of computer basic course teaching has not changed conventional form that teachers teach the theoretical knowledge. Students don't have access to enterprise working environment, not having the opportunity to study in the actual project.

"VC++" is one of public basic courses for science and engineering students. It is not only a programming course but also an important basis for these specialized follow-up professional courses, it is an important tool of expressions and development to carry out the research and application of computer. Therefore, this course is an important basic course.

At present, the VC++ teaching mode is still the traditional mode of examination oriented education to impart knowledge as the main teaching objective. Although it has abandoned the traditional theory—training apart method and begin to use of the teaching method of theory practice integration, but this kind of teaching method is not reflected in the ability, quality training, can not fully meet the target of application oriented talents cultivation, and can not reflect the advanced concept of CDIO engineering education; teaching process is mainly reflected in the classroom teaching, is not conducive to the cultivation of students' autonomous learning ability and ability to get information.

Because of China's current education system, students have developed the passive learning habits.

Most university teachers and students lack the engineering background, the system engineering concept and practice of education are deficient. The "computer program design in VC++" course teaching it is difficult to reflect the capability and quality training effect, and it takes a long time for students to do the job well after entering the enterprise.

At present, the students can not fully meet the requirements of employers, so university should train students to adapt to the society as soon as possible. To solve the problem of how to integrate knowledge, ability, quality in the course of "computer program design in VC++", CDIO engineering education model provides a good solution. In view of this, our school took CDIO engineering education model as a guide from the beginning of last year to reform the course "computer program design in VC++", in order to achieve better results in training students the basic programming skills and engineering ability.

3 TEACHING REFORM MEASURES ON FUNDAMENTALS OF COMPUTER APPLICATION

3.1 *CDIO teaching objectives and curriculum outline*

In order to meet the requirements of the major construction with the concept of CDIO engineering education, the reform of teaching contains mainly the following two aspects.

The computer programming course based on CDIO teaching model not only emphasizes basic concepts and programming syntax, but also emphasizes more on application design methods that achieve the ability to use computers to solve the problem, in particular focuses on improving students' computer skills and the training of the engineering practice ability. According to CDIO requirements.

3.2 *The reform of teaching methods*

With the goal to develop students's knowledge, ability and quality, which focuses on students' ability of self-learning, information obtain, software application and teamwork, the traditional teaching methods can not meet this demand. The reform of teaching methods contains mainly the following two aspects.

3.2.1 *The implementation of "project assistant" system, creating a new learning model of team project*

To improve the effect of student's teamwork when carrying out cooperation projects, since 2008, in the process of curriculum reform, the "project assistant" is gradually introduced into the teaching process, and the new learning relationship linked by "project assistant" is initially established, forming a "Teacher→Project→Student→Project Assistant" new learning mode. Through guidance, the students' learning effect has been improved in the project implementation process. The "project assistant" not only guide and help project team members to improve their own capacity, but also became a dominant force in college innovative practice activities, gradually forming a "second class" competition echelon.

Practice is carried out in accordance with the project management and project development model, focusing on the needs analysis, program development, testing, and so on, and it also developes students' ability to adapt to requirement changes to simulate real program development process. It stresses the normalization of documents and the necessity of review stage, emphasizing communication and collaboration within the team. Throughout the course of the program training, each member of each team is provided at least one opportunity to explain, report and reply, and the preparatory work should be done in advance to make sure of effective participation. Combined with instructors' review, students can master the programming system development process, methods and tools, to lay the foundation for students'future professional development.

CDIO capacity building methods are listed in the following table (Table 1).

3.2.2 *The establishment of a new scientific teaching feedback system and diverse collaborative learning methods*

The achievement of curriculum module project is the main line of the reform, which strengthens team communication and presentation. We strengthen the assessment stage of the curriculum module project, and guide the students to present the project. With respect to the theoretical knowledge of the course modules, periodic reexamination is made to make timely instructional feedback. The theoretical teaching must be reformed from a teacher-centered method into a project-based student reporting approach to report their relevant knowledge, supervised through questions, project evaluation, and so on. Through the adjustment of assessment methods, and the method to link the quantity and quality of student participation to the students'assessment to encourage students to actively participate.

By establishing curriculum learning website to support the students' project training, the multi-faceted communication platform is created for diverse collaboration. By "project assistant", project-based counseling is reinforced

Table 1. Methods the cultured CDIO capacity.

Training ability	Methods	Training venues
Autonomous learning	Combined with the project, the teaching method guided, heuristic	Classroom
Software application	The module of project implementation	Classroom, laboratory
Access to information	Do some comprehensive project implementation	Laboratory
The team cooperation	Realization method of using group learning and discussion	Classroom

to achieve enhanced effectiveness and quality of the practice.

Research group has developed a course learning sites to serve as teaching assistance, which includes curriculum learning platform, homework platform and assessment platform. Rich related curriculum resources can be uploaded, including course introductions, curriculum outline, electronic textbooks, electronic lesson plans, multimedia courseware, exercises, laboratory guidance, and so on. Module assessment also takes advantage of this platform to greatly stimulate students' independent learning and innovation.

3.3 The reform of performance assessment and grading methods

Performance assessment methods are very important, it determines the whole effect of the teaching method. In order to fully demonstrate the teaching objective, reflecting the actual student learning situation and direct students' learning orientation, the traditional assessment methods are reformed in the teaching process of VC++ programming language.

According to teaching requirements and characteristics of VC++, assessment contents focus on both the team performance and individual contributions, balancing the process quality and the final project work. Therefore, the assessment method is made up of the various stages' assessment of the teachers and student project managers. By the weighted gain of the progress of the project, the results of quality, documentation quality and extent of regulation and other aspects of teamwork, the overall team score is given by the teacher. Overall score accounts for 50% of each team member's personal grade, the remaining 50% of which is composed of the following parts: bonus award that is made up of classroom question performance (including the certain questions and discussive questions) and extra-curricular teaching question scores; module scores that refers to the students' work scores after the study of a software module, and results of this module is constituted by multiple software modules learning achievements. As shown in Figures 1 and 2.

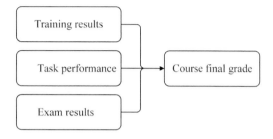

Figure 1. The examination mode before the reform.

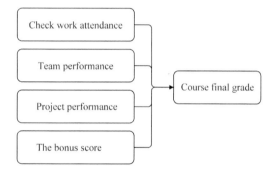

Figure 2. The examination mode after the reform.

4 IMPLEMENTATION RESULTS AND CONCLUSION

The VC++ practice teaching based on CDIO patterns can stimulate students' interest in learning. This paper redesigns the CDIO syllabus according to the international engineering education of CDIO concept. And it also designs and reforms the teaching methods, teaching management and performance assessment.

Under the guidance of the concept CDIO, and with the research group's efforts, "computer program design in VC++" was assessed as a campus "key construction" course in 2011 April. In the trials of electrical 2010 in Zhangjiagang campus, it not only made the student to experience the interest of learning at the beginning, but also strengthened the relevant theoretical knowledge, ability and trained students to observe, analyze and solve problems, enhanced students' teamwork skills, helped students to build up the confidence of learning, greatly improved the students' learning initiative, laid a solid foundation for the subsequent programming ability of professional learning. And in computer grade examination, Jiangsu Province, the pass rate is 98.7% (only one did not pass), and students also achieved excellent results in "Blue Bridge Cup" design contest.

In short, "computer program design in VC++" belongs to the public basic courses, having the characteristics of public basic courses, which cannot be completely in accordance with the CDIO engineering education mode to carry out the teaching reform of the course. However, we can grasp the cultivation of quality and ability which CDIO advocates and project mode. Specific reform measures can be determined by school situation, but must revolve around the concept of CDIO.

REFERENCES

[1] Gu Peihua, Li Yiping, Shen Minfen. EIP_CDIO model of cultivating innovative engineering talents to design oriented. China higher education, 2009 (3): 47–49.

[2] Gu Xueyong. Linking theory and practice CDIO. Research on higher engineering education, 2009 (1): 11–23.

[3] Edward F. Crawley. The CDIO Syllabus: A Statement of Goalsfor Undergraduate Engineering Education [2008-7-2]. http://Cdio.org/cdio-syllabus-rept/index.html.

[4] Cha Jianzhong. Engineering education reform strategy of "CDIO" and university industry cooperation and international. Chinese university teaching, 2008 (5): 16–19.

[5] Kang Quanli, Lu Xiaohua, Xiong Guangpin. CDIO syllabus and training innovative talents. Journal of higher education research, 2010 (4).

[6] Chen Chunlin, Zhu Zhangqing. Engineering course reform in education and practice of based on the concept of CDIO Education. Education and modernization, 2010 (1).

[7] Zhang Shuyun, Jiang Shuju. The teaching method of programming language course. Computer education, 2005 (5).

Education Management and Management Science – Zheng (Ed.)
© 2015 Taylor & Francis Group, London, ISBN 978-1-138-02663-6

Construction experience of teacher team of mining engineering major

S.W. Sun & H. Ding
*School of Resource and Safety Engineering, China University of Mining and Technology (Beijing),
Beijing, China*

ABSTRACT: Belonging to geological and mining majors, mining engineering major mainly analyzes theories and methods to exploit ore deposit and develops new technologies concerning mining industry. China University of Mining and Technology (Beijing) is the first university that established the major in China. The mining engineering major of China University of Mining and Technology (Beijing) ranked first in 2013 undergraduate teaching appraisal conducted by Ministry of Education. This paper introduces our university's experience in constructing teacher team, such as professional teacher team scale, knowledge structure, professional background, academic level, education reform, course construction, as well as guiding strategy in students' major, technological innovation, and social practice. In addition, it also explains measures to improve teaching quality as well as appraisal methods.

Keywords: mining engineer; teacher team; teaching reform; course construction

1 INTRODUCTION

Belonging to geological and mining majors, mining engineering major mainly analyzes theories and methods to exploit ore deposit and develops new technologies concerning mining industry [1].

The mining engineering major appeared early in western countries [2]. During Industrial Revolution, it had developed in certain scale. In China, China University of Mining and Technology (Beijing) first established the major in 1909. At that time, the name of the university is Jiaozuo Railway and Mine School. Our school's mining engineering is national key discipline, which enjoys high reputation in China and foreign countries. Through learning mining engineering, students can acquire systematic basic scientific theories and broad professional technological methods, have the abilities to conduct drawing, calculation, experiment, test expression, and proficient in basic technical skills and strong computer application skills. The construction of teacher team is the basis of major construction. This paper mainly introduces the experience of constructing teacher team and aims at providing China's colleges with references.

2 TEACHER QUANTITY AND STRUCTURE

Currently, the mining engineering major of China University of Mining and Technology (Beijing) has 32 teaching staffs, as shown in Table 1, including 10 professors, and 8 associate professors (senior engineer). The teachers with senior title account for 56%. The teachers with doctor's degree are 27, accounting for 84% of all teachers. In addition, the major includes 12 part-time teachers.

Table 1. Summary table of general situation of teacher team.

Title	Below 35 year-old	36–45 year-old	46–60 year-old	Above 60 year-old	Total of the left side	Doctor	Master	The major	Similar major	Other major
Senior professor	0	0	7	3	10	10	0	10	0	0
Associate professor	2	5	1	0	8	7	0	8	0	0
Middle level	10	2	2	0	14	10	2	14	0	0
Others	0	0	0	0	0	0	0	0	0	0
Total	12	7	10	3	32	27	2	32	0	0

At present, mining engineering major has formed the team which has reasonable structure, attaches importance to teamwork, and is dedicated to academic ideas. In addition, it has good professional quality, high teaching level, professional dedication and cohesive force, which lay the solid foundation for cultivating high-quality talents. Currently, the team has replaced old cadres with younger ones, forming the combination of old and young teachers, of whom young teachers and associate professors are subjects. Moreover, in practice programs such as cognition practice, production practice and graduation practice, the school invites mine engineering technicians to give special report or guide practice. Besides, through various activities the school will also employ retired teachers or part-time experts to guide students or make special report. For example, carry out special event of mining life sometimes.

3 TEACHING LEVEL

The major always attaches importance to construction of teacher team. It has established complete incentive mechanism in terms of bringing in, fostering and giving full play to talents. While the university scale is gradually expanding, the ratio of students to teachers is in reasonable level, which effectively meets the requirements of talent cultivation.

The major guarantees new teachers to get tile of lecturer and further improves the quality and teaching level by taking the following effective measures: bringing in teachers, intensifying pre-job training, attending other teachers' lectures, giving lectures, implementing tutorial system of young teachers, and encouraging new teachers to engage in coal enterprise production and scientific research.

The university defines teacher in strict accordance with the qualification standard of lecturer. Currently, the lecturers have complied with the interim provisions concerning the training of relevant teachers. The new teachers shall participate in pre-job training. After they pass trial lecture, they can give lectures.

The teachers shall have enough teaching ability and professional level and actively participate in engineering practice. They shall carry out scientific research in combination with practice problems of mining engineering, and promote wide domestic and overseas academic exchange, so as to meet the requirements of teaching [3].

4 TEACHING REFORM

In the major, professor and associate professor participate in teaching reform, which is an important measure and a conventional system to improve teaching quality of undergraduate education. In addition, participating in teaching reform program, guiding graduation design, and guiding practice are necessary conditions for job promotion and employment. If a teacher applies for senior professional technical position, the teacher must participate in and pass teaching appraisal organized by the school. The above measures have intensified professor and associate professor's sense of duty, which promotes more professor and associate professors to participate in teaching reform. Currently, all teachers undertake teaching reform. In aspects of talent training mode, teaching management, education philosophy, etc., the major comprehensively carries out exploration and practice, which fully guarantees the teaching level of lecturers.

In aspect of teaching reform, according to notice concerning finishing project approval of course construction and teaching reform issued by China University of Mining and Technology, the major offers less than 10,000 Yuan fund for the course which is less than 40 credit hours; the course which is 44~56 credit hours will get 12,000 Yuan fund; the course which is equal to or more than 65 credit hours will get 15,000 Yuan fund. For the general education, discipline basis or basic required course in which several teachers and classes are involved, the fund shall properly increase [4]. In addition, the major offers certain subsidy for project approval of teaching reform. The major always abides by the goal of improving teaching level, intensifies teaching abilities, promotes "research before teaching", and encourages teachers to carry out teaching research, so as to improve teaching level. During recent years, the major has continuously increased fund of teaching research program through multiple channels, which has achieved preliminary teaching achievement.

5 STUDENT CULTIVATION

The major requires professor and associate professor to guide students and encourages teachers to carry out students' innovation (entrepreneurship) training in combination with scientific research training. It uses the following conditions as the necessary conditions of job promotion and employment: guide students' practice or social investigation, guide course design, graduation design (paper), and innovation (entrepreneurship) training, take charge of a class, and hold the post of assistant. The teachers of the major shall guide students' professional study, technological innovation and social practice.

In aspect of professional guidance, academic leaders shall explain to freshmen the discipline characteristic, major prospect, and requirements on students at the beginning of each term, so that freshmen can define their development goal early; the major also employs domestic and overseas experts and scholars to give professional academic report, cultivates students' learning interest, and organizes students to visit modern mines, so that students know the coal production develops toward high technology and are more interested in it. In addition, the major arranges one teacher to take charge of one class, who will be selected from young doctors. They will guide students to finish professional task and course selection, periodically hold class meeting, introduce the courses to students at the beginning of each term, and guide students to finish review and examination preparation at the end of each term. For the clever and diligent students, teachers may guide them to enter research of professional stage in advance. In addition, carry out a series of enrollment education activities and comprehensively standardize students' daily routine; through a series of strict regulations, make students develop good habits of morning reading and self-study at night; the self-study at night implements attendance checking system, so that students may form a healthy learning environment. According to self-characteristics, the school praises advanced group and individual in learning atmosphere construction, which promotes the construction of good learning atmosphere and achieves significant effect. Select good students as exchange student to study in foreign first-class universities, which stimulates students' enthusiasm [5–10].

In aspect of technological innovation guidance, the school strictly implements Management Method of Students' Innovation Training Program of China University of Mining and Technology (Beijing), Evaluation and Performance Appraisal Method of Students' Innovation Training Program of China University of Mining and Technology (Beijing). Through innovative research practice teaching of combining teachers' guidance and students' self-study, reach the goal of fostering innovative thinking, training innovation ability, and improving innovative quality, which greatly promote students' enthusiasm to participate in technological innovation activities. In addition, actively carry out student technological innovation festivals; while selecting members to participate in various technological innovation activities organized by school, carry out diversified technological innovation activities in combination with practical situations of the major; improve students' qualities of technological innovation. At the same time, positively carry out "Challenge Cup" activity. Motivate and organize students to participate in Challenge Cup and energy-saving competition

activities each year. The quantity of participators increases gradually, and they have got awards in competitions of school level, provincial level and national level. Besides, vigorously carry out mathematics contest, mechanics contest, English contest, etc. and get good results. Positively carry out discipline innovation ability competition according to discipline characteristics each year. The competition consists of two parts: namely professional knowledge competition and research design competition. Professional knowledge competition mainly investigates contents of main courses; the research design competition requires students to put forward solutions according to on-site practical situations and use PPT to report. The innovation ability competition fully combines practical situations of the major, greatly promotes students' innovation ability, and brings about significant influences among students.

In aspect of social practice guidance, the discipline attaches much importance to students' social practice work and carries out social practice activities in multi-level and multi-aspect in summer. The social practice activities are conducted in three levels: the school uniformly organizes a team; the school uniformly organizes a team; students go back to hometown to carry out social practices. Students are required to participate in social practice activities and write social practice report. Participating in social practice activity is one necessary condition of annual general appraisal. Social practice includes mobilization, organization, communication, appraisal and prize.

6 MAIN MEASURES TO IMPROVE TEACHING QUALITY

In order to improve teaching quality of undergraduate students and intensify central position of undergraduate teaching among teachers, the teaching administrators of the school actively and carefully finish their due work. The teaching dean, teaching secretary, and academic dean define their duties clearly, perform their due obligations and closely cooperate with each other. They effectively monitor teaching quality through comprehensively implementing supervision mechanism of undergraduate teaching, including teaching supervision group's appraisal on teaching quality, school leaders and teachers' appraisal on teaching quality, etc. The objective of the major is clear; the major ranked first in 2013 undergraduate teaching appraisal conducted by Ministry of Education. During teaching, teacher team has strong sense of responsibility, continuously solves problems in teaching process, and guarantees implementation of major objective.

While conducting daily appraisal on teaching quality, the school takes comprehensive measures to improve teaching level of young teachers in order to enhance teaching quality. In addition, organize young teachers to view and emulate, discuss in seminar, assist class teaching, and learn from senior teachers, etc. In order to help young teachers improve teaching quality, periodically organize lecture competitions among young teachers, and stipulate that the teachers under 35-year-old must participate in lecture competition, which effectively promotes the improvement of lecture quality of young teachers.

ACKNOWLEDGMENTS

This work is supported by the National Natural Science Fundation of China (Grant number 41002090 and 51034005) and the Fundamental Research Funds for the Central Universities (Grant number 2011QZ05). The authors would like to express our gratitude to the editors and reviewers for their constructive and helpful review comments.

REFERENCES

[1] Peng Guojun (2010): Discussion on Practice Teaching Reform of Civil Engineering Major of Local Undergraduate Universities. Education Theory and Practice, 33 (36): 15–17 (in Chinese).

[2] Zhou Feifei, Mo Weifeng (2010): Application of Multimedia Teaching Method in the Courses of Science and Engineering. Education Theory and Practice, 33(15): 44–46 (in Chinese).

[3] Sun S.W., Zhu B.Z. and Zheng J. (2010): Design method of micropile group for soil slope stabilization based ultimate resistance of micropile, Chinese Journal of Geotechnical Engineering, 32 (11), 1671–1677 (in Chinese).

[4] Sun S.W., Zhu B.Z. and Ma H.M. (2009): Model tests on anti-sliding mechanism of micropile groups and anti-sliding piles, Chinese Journal of Geotechnical Engineering, 31 (10), 1564–1570 (in Chinese).

[5] China National Standards GB50010-2002 (2002): Code for design of concrete structures, Ministry of Construction of China, China Architecture & Building Press, Beijing (in Chinese).

[6] China National Standards GB/T50123-1999 (1999): Standard for Soil Test Method, the Standardization Administration of China, the Ministry of Construction of China, and the Ministry of Water Resources of China, China Planning Press, Beijing (in Chinese).

[7] China National Standards GB50330-2002 (2002): Technical code for building slope engineering, Ministry of Construction of China, China Architecture & Building Press, Beijing (in Chinese).

[8] Lizzi F. (1982): The pali radice (root piles), Symposium on Soil and Rock Im-provement Techniques including Geotextiles, Reinforced Earth and Modern Piling Methods, Bangkok, Paper D-3.

[9] Plumelle C. (1984): Improvement of the bearing capacity of soil by inserts of group and reticulated micro piles. International Symposium on in-situ reinforcement of Soils and Rocks, Paris, ENPC Press, 83–89.

[10] Palmerton J.B. (1984): Stability of moving land masses by cast-in-place piles, Fed-eral Highway Administration, Washington, DC, Final Report, Miscellaneous Paper GL-84-4.

The influence of traditional culture on the safety of production

Meiyun Zhang & Yin Guo
North China Institute of Science and Technology, Hebei, China

ABSTRACT: Nowadays, safety culture which is called "fast food culture" in China is based on the safety culture of western countries. It hasn't stood the test of time. So, only when we study the safety culture and criticize it, we can build the safety culture that actually fits the situation of China, and help the safety production management.

Keywords: the safety of production; the traditional culture; the security management; national characteristics

1 INTRODUCTION

Reason, as the essence of human, was made the basement of the western rational civilization in modern times by ancient Greece. The western safety culture which played an important role in enterprise security management was based on the Christian culture. On the other hand, Chinese safety culture wasn't built on the traditional culture. Nowadays, safety culture which is called "fast food culture" in China is based on the safety culture of western countries. It hasn't stood the test of time. So, only when we study the safety culture and criticize it, we can build the safety culture that actually fits the situation of China, and help the safety production management.

2 THE ABUNDANT SAFETY PHILOSOPHY IN TRADITIONAL CULTURE

In spite of the dregs, there are lots of deep thoughts and sayings about the concerning, rescuing and prevention for the disaster. For example, we can find Kongzi cared much more about the slaves' safety rather than the horses' when the stable was on the fire in Lunyu, although the slaves were much cheaper than horses. We need to find and collect all these resources for the Chinese safety culture.

3 THE REALISTIC MEANING OF THE CULTURE INHERITANCE

"Any new thoughts won't last long in the history river, if it has nothing to do with the old culture. It should be an old fashion." Said Master Lin He. Chinese broad and profound traditional culture makes the profound impact on Chinese people. The modern safety culture has just been born, which makes us think it over again. The building of modern safety culture won't be smooth because of the differences between modern safety culture and traditional culture. We should analyze the similarities and differences, and make the building easier.

Conditions, which contains eternal substance suits all of the Ages and humanity, such as the space-time, the universe, the secret of lives and the meaning of living. These questions are argued since people appeared on the world.

The intension of culture just captures the essence of objects, which is above the age and nationality expressed as the reason, the wisdom and the truth, so it is powerful. In the aspect of the existing form of traditional culture, it exists on its national thinkers, values, moral principles, personalities, behaviors and customs, whose sediment is a kind of inheritable gene of Chinese culture infiltrates through all the fields of the modern society of Chinese political, economic and spiritual lives. The traditional culture also comes as the appearance of people's livelihood and usual behavior, produces an effect on the national member's psychology and characters. The traditional culture is an undeniable existing factor, playing the part of the bridge of history and the present.

4 PROMOTING AND DEVELOPING THE NATIONAL CULTURE, AND ENHANCING THE EFFECTS OF SECURITY MANAGEMENT

The government takes the issue seriously, it supervises the reformation of the safety of production management institution, requires establishing a system to get rid of the hidden troubles during the producing process, and keeps the serious accidents within limits. The issues of security management are an important part in the reformation of the government, whose new concepts of safety will monitor the safety of production all-around. It will raise the consideration of the source, environment, the benefits, the wasted energy, the technology creation, the safety of production and the debts, which based on the economical increase. The whole trend for the safety of production is not optimistic, so much as severe. Removing and preventing the troubles, is a significant part of the safety of production. The safety factors in customs can be internalized in the staff's safety conscious, which is the mental foundation in the safety preventing and controlling system. Consequently, the traditional culture is significant for the study of the safety of production.

Besides, the pattern of the security management originates abroad, which cannot suits the situation in China as the cultural and spiritual differences. We should reveal our problems in the security management, and combine it with the spirits in the custom as thinking highly of the results, chasing brief, attaching importance to harmonious, and taking the experience seriously. On the other hand, we also share the characteristics of lacking of the conscious of democracy and legality, attaching importance to the families, bearing huge burdens, expecting realizing self-fulfillment and lacking of caring. As a result, we should abstract the specialty and the trend of development for Chinese security management, based on the situation of China, which is vital.

5 CREATING THE SAFETY METHODS DEPEND ON THE IDEA OF TRADITIONAL CULTURE

5.1 Enhancing the researching of traditional theory

The presupposition of the theoretical study is explaining that the "safety of production" accurately, knowing that the production promoting the development of human society. The safety is the basic need for human beings, the safety of production is the premise of the sustainable development of our society.

According to the history, the research of the traditional theory could be separated to three stages: the first stage is the exploration of the safety of production in ancient times, searching the production management experiences of stoneware and woodenware from the time immemorial which can be traced back to Banpo clan, before 6000–7000 years such as Emperor Yu Tames the Flood and Dujiangyan irrigation system. The second stage is summing up the cultural specialty of the shifting of the human production from animal husbandry age to the mining age with machinery tools. During this stage, the accidents appeared in the production of human beings, as the arise of the machines. The safety problems come out along with the development of the manufacture, the safety technology is improving, including the safety of construction, bronze smelting, mining and the well salt mining. The third stage is discussing the safety of production in modern times in China. We should pay close attention to the Chinese idea about safety of mining during 1840 to 1919, such as "preserving lives and coal pit, paying attention to prevention" and "the law of mining in Qing dynasty" before 100 years.

We should discuss the moral principles in the safety of production, including samsara idea which grown from the samsara in Chou Yi to "people are my brothers, animals are my friends" by Zhangzai, Wei dynasty; the traditional concept of morality and justice, and the conscious of the responsibility for safety; the perception of credit in tradition and the credit mechanism; the doctrine of knowing and doing and execution; not to repeat previous mistakes and handle the accidents.

5.2 Searching the security methods based on the traditional culture

Nowadays, the safety history researching is not valued enough. It is very important to establish a new concept of safety history, reorganize the material of history, compare all the results of study in all aspects, judge the phenomenon appeared in the development of safety, and find out the reasons bound to the human security in society, institution, technology and culture, to provide a foundation to make the right safety policy.

First of all, exploring the values of using the experiences of the safety methods and concepts that flowing in the historical river for reference, contains that "Preparedness ensures success", concept of viewing the situation as a whole, and supervisory institution. However, we should take care of the negative effects in traditional culture, such as ignoring the individuality, the patriarch system, and "Domestic shame should not be made public". Moreover, we should put the conflicts and compromising between the traditional culture and the western modern civilization in the area of safety of production.

According to the investigate, many problems have arisen in the effect on the security management in many companies. The safe producing is more than institution and equipments input, which should be lead by values and culture.

Surpassing the bound of innate limits in security management and arousing the staff's born ability of "concerning safety, loving lives" in a deeper cultural layer, we can contribute a long-acting mechanism in safety of production. Therefore, exploring a long-acting mechanism in safety of production with culture inside, institution external, moral and law combined and the culture from the Orient and the Occident mixed together, which is depended on the study of theory, could be very vital.

ACKNOWLEDGMENT

Supported by The Fundamental Research Funds for the Central Universities Project (3142014016).

REFERENCES

[1] LingShi. Brilliant ancient safety culture. Modern occupational safety. 2009(9).
[2] Shiying, Wu Chao, Yang fuqiang. Research of safety historiography methodology. Journal of China Safety Science. 2010(5).
[3] GeRong jin. Mind control theory—The modern interpretation of the traditiona Confucian concept of morality and justice. Journal of F the party school of the centralcommitte. 2010(3).

Education Management and Management Science – Zheng (Ed.)
© 2015 Taylor & Francis Group, London, ISBN 978-1-138-02663-6

Present situation and causes analysis of male early childhood teachers' professional identity in dilemma

Fu Chen
Shandong Women's University, Jinan, Shandong, China

ABSTRACT: Professional identity of early childhood teachers directly affects the quality of early childhood education, the development of preschool children, and professional development of early childhood teachers. As a tiny yet quite important part of early childhood teachers, male preschool teachers' professional identity is directly related to the mental and physical development of children. This paper explores the related concepts of professional identity of male early childhood teachers, and uses case interview and investigations to systematically analyze the present dilemma of male early childhood teachers in professional identity and further discusses and digs out the causes behind the dilemma.

Keywords: male early childhood teachers; professional identity; dilemma

1   INTRODUCTION

Early childhood teachers' professional identity is a profession-related and positive attitude, including cognition, emotions and behavioral tendencies. It is the overall view of preschool teachers towards their profession, and can be summarized as self-image recognition, professional preparations, professional emotions, professional abilities, professional wills, and so on. The professional identity changes all the time, and preschool teachers can have better understanding of professional identity by self reflection and evaluation.

Professional identity is at the core of professional development of male early childhood teachers, referring to their understanding and experiences of preschool teachers defined by the society, their conformity with professional ethics, attainments, and standards as teachers and appreciation of early childhood teachers' value and status as real and specific "human beings". The concept emphases preschool teachers' values, life experiences, cognitive features, and the importance caused by the intrinsic value of individual preschool teacher, that is the organic unity of "social self" and "individual self".

However, it is customary for people to think that early childhood teachers are females, that taking care of children is female's work, therefore, professional male teachers in early childhood education are often questioned, leading to the huge gap in professional identity between male and female, which means that male early childhood teachers have lower professional identity than their female counterparts, and male teachers are easy to be in a dilemma in their professional identity.

2   PRESENT DILEMMA OF MALE EARLY CHILDHOOD TEACHERS IN PROFESSIONAL IDENTITY

2.1   *Marginalization of male early childhood teachers' professional identity*

Due to the particularities of kindergartens, such as paying equal attention to education and child care, principal leadership, single gender teachers, teachers in charge of classes, kindergartens' rules, regulations and responsibilities all cater to female teachers, and it seems the cultural environment is more suitable for female teachers. When male teachers enter this kind of female dominated cultural and educational place, they are not only faced with conflicts between their gender features, psychophysical characteristics and administrative supervisions of kindergarten, but drowned by their inconformity with the universal voice of the community. They are also tasked with extra administrative work and other non-teaching duties thanks to their gender. All these lead to their professional identity marginalization to some extent, and affect their positive professional identity.

2.2   *Low specialization of male early childhood teachers*

Low specialization is one of the main dilemmas of male teachers. Its main performances are as follows:

2.2.1 Male teachers are not interested in early childhood education

Influenced by social customs and present education conditions in China, male early childhood teachers are inclined to lack interest or less interested in their work. First, very few male students choose early childhood education as their major in college; second, few male graduates of the major in normal universities are willing to work in the early childhood education field; finally, male early childhood teachers who work as an early childhood educator seldom show interest in daily educational activities, and avoid responsibilities in their work.

2.2.2 Male teachers lack sense of achievement

It is clearly shown in that they cannot fully realize their values in their teaching. Comparing with female early childhood teachers, male teachers are more easily lacking in sense of security. They cannot find focuses in their teaching, have no clue of children's daily life and necessary teaching activities and do not know what to do, so they are more likely to feel frustrated and defeated. This leads to their inability to realize their dreams thus triggers their frustration towards their work and themselves.

2.2.3 Male early childhood teachers cannot adjust to working environment

As is generally known that stuff at a kindergarten from principal, directors to teachers responsible for daily teachings, child-care workers, cleaners, health service stuff are almost female. Male teachers would inevitably feel constrained in the all-female working environment, so they often leave their job behind or neglect duty. At the same time, they will be dissatisfied with their work and worried about the future of their jobs. They will feel unbalanced if they are described with expressions such as "feminization" and "low salary".

2.2.4 Low social identity of male early childhood teachers

This is caused by traditional occupation concept in our country. First, males are the "heaven" according to traditional ideas, and they are the backbone of the small family and the country, the main driving force for social development. Therefore, their jobs are considered "powerful" according to traditional concepts, and the public often despises and scorns the males those take up feminine jobs. Second, the traditional professional views believe that males enjoy higher social status than females, mainly based on males' economy level. We tend to think that males' salaries are the main source or the only source of income for a family. While salaries of those who work in early childhood education are relatively low. Finally, whether male early childhood teachers will be respected by the society is another question. Male teachers are called "male aunt". Though few children and parents call them that out of sarcasm or bad intentions, the name inevitably shows male teachers embarrassment and imbalanced status in society.

3 CAUSES ANALYSIS OF PRESENT DILEMMA OF MALE EARLY CHILDHOOD TEACHERS IN PROFESSIONAL IDENTITY

3.1 Constrained by social concepts and economic development

The number of employed people and gender ratio in different industries are related to social development and social concept that is cultural quality. Social and culture development especially the development of gender concept mainly contribute to the fact that few males devote themselves in early childhood education. It is shown in two aspects: on the one hand, the number of male normal university students is quite small due to social gender concept. Traditional ideas believe that females are kind and attentive, so they are fit for caring for children, while males are rough and wild, so they are not perfect for taking care of children. Early childhood education therefore becomes a female-only profession, and few males choose to work in this field. On the other hand, salaries for kindergarten stuff are relatively small, since females often don't have to support a family, they are ok with that. However, males are burdened with the whole family income according to traditional beliefs, so male teachers are not satisfied with their income. That is to say there is a huge gap between their salaries and their expected values. Salaries are the direct cause for few male teachers.

3.2 Male teachers' mentality, professional qualities and difficulties in real work

First, their mentality: it is the subjective reason of whether he can do well. Mentality here not only means how they see their profession, but also their attitudes and performances in the work. If male normal university students regard early childhood education as utter most important, noble and sacred, then he will engage himself in the career, while if they scorn the profession and even want to avoid it, then he will not put his heart and soul in the work and perform negatively.

Second, professional qualities and level of male teachers: their professional qualities are important factors affecting their attitudes towards their future work and career development. During the training of male preschool teachers, if they focus on attaining professional knowledge and skills, they will be confident and capable of handling their work. On the contrary, if they do not fully understand professional knowledge and skills such as psychological and physical characteristics of preschool children at different stages, playing piano, they will be incompetent, feel negative and gradually lose confidence.

Third, their ability to solve troubles in real work: after male teachers enter early childhood education with enthusiasm despite traditional social ideas and gender differences, they will face a lot of troubles. How to correctly tackle with and whether they can well handle difficulties in their real work are their challenges. If they do not handle them properly, they will feel more than tired and want to quit the job. The difficulties may be: relations with colleagues. Colleagues are important companies in social interactions, and in kindergartens, female teachers still dominate. Therefore, it requires male teachers to know how to cope with relations with colleagues, including relations with managing stuff, exchanging teaching experience with experienced teachers, and daily interactions with fellow female teachers. Relations with parents: just as we know many parents believe that new teachers don't know how to take care of children, and they are particularly worried about male teachers doing the job. They think male teachers are not kind enough and lack patience, and don't know if children are happy with their food. This requires male teachers to improve themselves on these matters and learn how to communicate with parents. They cannot avoid these activities nor let their work and confidence be affected simply out of concerns; relations with kids: besides professional qualities, male teachers should be patient and properly handle problems caused by children, for example children often tell on each other, at this point, whether male teachers can properly handle this indicates whether he can cope with his primary work, whether he is capable of doing this, and is also related to his confidence and professional identity.

After all, if male teachers are willing to engaging in early childhood education, they will be faced with more pressure and worries than their female counterparts. This requires them to have to strong tenacity and psychological enduring capacity, but also properly handle difficulties in the work, have a positive attitude and tackle difficulties head on.

REFERENCES

[1] B.A. Vasily Sukhomlinskii. Zhou Zao, translated, etc. One Hundred Pieces of Advice to Teachers. Tianjin: Tianjin People's Publishing Company, 1981.

[2] Gu Minyuan. Comprehensive Dictionary of Education. Shanghai: Shanghai Educational Publishing, 1990.

[3] Zhao Changmu. The Growth of Teachers. Lanzhou: Gansu Educational Publishing, 2004.

[4] Ju Yucui. Entering Teachers' World—Reflections of Personal Practice as A Teacher. Shanghai: Fudan University Press, 2004.

[5] Shi Jinghuan, editor. Entering World of Gender of Textbooks and Education. Beijing: Educational Science Publishing House, 2004.

[6] Gan Ruyi. East of Water. Shanghai Yuandong Publishing House, 2005.

[7] Yu Shengquan & Ma Ning. Modern Educational Technology and Teachers' Professional Development: New Technology, New Concepts. Beijing: Peking University Press, 2009.

[8] Yuan Lan. Research on Current Trainings of New Teachers in Private Kindergartens. Thesis of Hunan Normal University, 2010.

[9] Lin Peisheng. About Promoting Occupational Maturity of Young Teacher. Journal of Tianjin Normal University. 1993 (1).

[10] Zhang Yanting. My Views on Male Teachers in Kindergartens. Journal of Educational Development, 1997 (5).

[11] Liu Jianxia. Analysis of Adverse Effects of Single-gender Teachers. Journal of Luoyang Normal University. 1996 (0).

[12] Farquhar, Sarah (1997). Are Male Teachers Really Necessary New Zealand.

[13] Santiago, Anthony (1999). Male Early Childhood Montessori Teachers: Why they Chose To Teacher. U.S.: New York.

[14] Kind-R (2004). The (Im) possibility of Gay Teachers for Young Children. Theory into Practice.

Education Management and Management Science – Zheng (Ed.)
© 2015 Taylor & Francis Group, London, ISBN 978-1-138-02663-6

The trade effects of the US producer services investment in China

Lin Kong
School of Management, Capital Normal University, Beijing, China

ABSTRACT: The development of the production and trade based on the global value chains blurs the boundaries of trade in goods and services. The continuous increase of the US investment, especially its producer services investment in China, changes the Sino-US trade pattern and trade flows of the final products to a large extent. This paper analyzes the trade effects of the US producer services investment in China through establishing an empirical model in order to show the present Sino-US trade relations in the global value chains more clearly.

Keywords: producer services investment; Sino-US trade relations; global value chains; trade effects

1 INTRODUCTION

The disintegration of production and the integration of trade have become the symbols of economic globalization, and the factor intensity of the final products and the factor used intensity in the production process also become two different concepts. Under this trade pattern, when China exports goods to the United States (US) and imports services from the US, China will inevitably have goods trade surplus and services trade deficit. At the same time, the US will also have goods trade deficit and services trade surplus. In fact, the Sino-US trade surplus in goods is just a mirroring of the US trade surplus in services with China and the comparative advantage of the US service sectors.

China and the US have complementary factor endowment, the US multinationals combine their advanced factors with the labor resources in China, promote Sino-US exports and expand Sino-US trade surplus in goods through final products processing and assembling. The US department of commerce statistics also shows that there is a large amount of continuous trade surplus in the US cross border trade with China in services, including the trade in producer services. The phenomenon that multinationals transfer downstream processing segments and export capacity in the global value chains in order to achieve value-added and serve target markets better lays a good foundation for us to discuss the present Sino-US trade relations.

Many scholars conduct relative research on producer services and their impact on trade structure and trade relations. Markusen (1989) discusses the complementarity between domestic and foreign specialized inputs and that between trade in final goods and producer services. Deardorff (2001) points out that trade in producer services deepens the international division of labor in goods and services and promotes the development of trade. Jones and Kierzkowski (1990, 2005) analyze the effects of producer services on deepening the division of labor through production fragmentation and service links. Francois and Woerz (2008) discuss the role of service in manufacturing as well as the indirect exports of services through merchandise exports. Lu (2007) explains the internationalization of the present service outsourcing from the intra-product specialization perspective. Zeng and Li (2006) also make econometric analysis on trade in producer services by using the gravity model and panel data.

In order to indicate the effects of the US producer services investment in China on its processing and assembling exports to the US in the global value chains more clearly, this paper intends to analyze the trade effects of the US investment in producer services sectors by establishing an empirical model.

2 THE DATA

This paper intends to test the trade effects of the US producer services investment on the Sino-US bilateral trade, including export effects and import effects, so in the regression, three dependent variables are chosen respectively, which are the volume of the Sino-US trade surplus, the volume of the China exports to the US and the volume of the China imports from the US. Trade flow data is obtained from the United Nations Commodity Trade Statistics Database (Comtrade) from 1993 to 2012. In the independent variables, the US producer services investment flows and stocks in China are mainly the sum of five subsectors investment, including financial, insurance, banking, information, business services, and professional scientific and technical services. Flow data comes from the United States Bureau of Economic Analysis from 1993 to 2012; Stock data is the sum of the flow data with a lag of one year and before, sorted and calculated from the United States Bureau of Economic Analysis from 1992 to 2011. To eliminate heteroskedasticity in the data, all the variables are in the natural logarithm form.

3 THE MODEL

According to the theories of international direct investment and international economics, an empirical model established in this paper is:

$$\ln TRD_t = \alpha + \beta_1 \ln PFDIR_t + \beta_2 \ln PFDIE_{t-1} + \varepsilon$$

The dependent variable TRD_t denotes the value of the Sino-US trade surplus (The value of the Sino-US exports; The value of the Sino-US imports) in period t.

The independent variables on the right hand side of the above equation are as follows:

$PFDIR_t$: The US producer services investment flows in China in period t.

$PFDIE_{t-1}$: The US producer services investment stocks in China in period t−1.

α: Intercept.

ε: Random error.

4 EMPIRICAL ANALYSIS

4.1 *Test of stationarity*

In order to prevent the occurrence of spurious regression, we first use ADF method to carry out unit root test for the time series of the variables. The results are shown in Table 1.

The results in Table 1 illustrate that all of the original time series ADF statistics of the $\ln TRD$, $\ln PFDIR_t$ and $\ln PFDIR_{t-1}$ are greater than the critical values of 10% significance level, which are nonstationary. After first differences, the ADF statistics of $d\ln TRD$, $d\ln PFDIR_t$ and $d\ln PFDIR_{t-1}$ meet the stationarity requirements. Thus, the series of the Sino-US trade surplus value (the Sino-US exports value, the Sino-US imports value), the US producer services investment flows in China and the US producer services investment stocks in China are I (1) process.

4.2 *The cointegration analysis*

Through Engle-Granger two-step procedure, we get the export and import effects of the US producer services investment in China on the Sino-US trade, which are shown in Table 2.

As shown in the regression results, the coefficients of the US producer services investment flows and stocks in China are not zero and very significant, so the US producer services investment flows and stocks in China have positive effects on the Sino-US trade surplus (exports, imports). The fitted degree of the regression is good, and there is no autocorrelation.

Then we need to do stationarity tests on the residual series of the regression equations, using ADF method. The results are shown in Table 3. The residual series of the cointegration equations are I (0) process. This proves that there is a stable long-run cointegration relationship between the US producer services investment stocks in China and the Sino-US trade surplus (exports, imports).

The regression results demonstrate that both of the US producer services investment stocks in China with a lag of one year ($PFDIE_{t-1}$) and the US producer services investment flows in China ($PFDIR_t$) have strong positive correlations with

Table 1. The results of the data stationarity test.

Variables	ADF statistics	Critical values	Stationary series or not
$\ln TRD_t$ (Sino-US trade surplus)	−1.2804	−2.6745*	No
$d\ln TRD_t$ (Sino-US trade surplus)	−3.3029	−3.0818**	Yes
$\ln TRD_t$ (Sino-US exports)	−0.9870	−2.6745*	No
$d\ln TRD_t$ (Sino-US exports)	−3.9316	−3.0818**	Yes
$\ln TRD_t$ (Sino-US imports)	0.4467	−2.6745*	No
$d\ln TRD_t$ (Sino-US imports)	−3.8204	−3.0818**	Yes
$\ln PFDIR_t$	−1.5521	−2.6745*	No
$d\ln PFDIR_t$	−8.2514	−3.9635***	Yes
$\ln PFDIE_{t-1}$	−1.9597	−2.6745*	No
$d\ln PFDIE_{t-1}$	−3.1979	−3.0818**	Yes

Notes: ***, ** and * indicate significance level at 1%, 5% and 10% respectively.

Table 2. Regression results.

Variables	Trade effects equation (dependent variable is the Sino-US trade surplus)	Export creation effects equation (dependent variable is the Sino-US exports)	Import creation effects equation (dependent variable is the Sino-US imports)
$\ln PFDIR_t$	0.1101*** (0.0339)	0.1083*** (0.0346)	0.1002** (0.0384)
$\ln PFDIE_{t-1}$	0.6607*** (0.0449)	0.6092*** (0.0434)	0.4658*** (0.0481)
Intercept	9.1307*** (0.5581)	10.5292*** (0.8106)	12.2696*** (0.8962)
R^2	0.9633	0.9603	0.9243
Adj R^2	0.9580	0.9547	0.9135
F	183.91	169.67	85.49

Notes: ***, ** and * indicate significance level at 1%, 5% and 10% respectively. Numbers in parentheses are standard errors.

Table 3. Stationarity tests on the residual series.

Dependent variables	ADF statistics of the residual series	Critical values	Residuals are stationary or not
Sino-US trade surplus	−4.1975	−3.9228***	Yes
Sino-US exports	−4.0893	−3.9228***	Yes
Sino-US imports	−3.9967	−3.9228***	Yes

Notes: ***, ** and * indicate significance level at 1%, 5% and 10% respectively.

the Sino-US trade surplus, which are significant at the 1% level. So the marketing, transportation, insurance, efficient communication service and other producer services provided by the US multinational corporations for their OEM firms in China in components production and final goods assembly play an important role in the US trade deficit with China. Because investment tends to have lagged effects on trade flows, the US producer services investment stocks in China with a lag of one year $PFDIE_{t-1}$ have more significant effects on the increase of the US trade deficit with China.

Moreover, in view of the export effects equation and import effects equation of the US producer services investment in China, the Sino-US exports and imports increase (or decrease) by 0.61% and 0.47% respectively along with the US producer services investment stocks in China with a lag of one year ($PFDIE_{t-1}$) goes up (or drops) by 1%. It shows that the US producer services investment in China has a higher export creation effects compared with its import effects. And the first regression equation of trade effects also corroborates that the US producer services investment in China has creation effects on the Sino-US trade surplus.

5 SUMMARY AND CONCLUSIONS

It shows that the Sino-US bilateral trade is of the full range, including production segments and

service segments in the global value chains, due to the realization of trade feasibility in the production related services. And the development of production and trade based on the global value chains even blur the boundaries of trade in goods and services.

The services links and the corresponding factors output and export among the fragments in the global value chains offered by the US multinational corporations for their components production and final products assembly (the US producer services investment in China) have more obvious export effects on the Sino-US trade than import effects. Thus it further exacerbates the trend of the Sino-US trade imbalance.

ACKNOWLEDGMENTS

The author gratefully acknowledges the support from the Youth Talent Program in Beijing Colleges and Universities in China for the project titled Research on the Sino-US Trade Imbalance: the Latest Progress Based on the Global Value Chains.

REFERENCES

[1] Deardorff, A.V. 2001. International Provision of Trade Services, Trade, and Fragmentation. *Review of International Economics* 9(2):233–248.

[2] Francois, J. & Reinert, K. 1996. The Role of Services in the Structure of production and Trade: Stylized Facts from a Cross-Country Analysis. *Asia-Pacific Economic Review* 2(1):1–9.

[3] Francois, J. & Woerz, J. 2008. Producer Services, Manufacturing Linkages, and Trade. *Journal of Industry, Competition and Trade* 8(3): 199–229.

[4] Guerrieri, P. & Meliciani, V. 2005. Technology and international competitiveness: The interdependence between manufacturing and producer services. *Structural Change and Economic Dynamics* 16(4): 489–502.

[5] Javorcik, B. 2004. Does Foreign Direct Investment Increase the Productivity of Domestic Firms? In Search of Spillovers Through Backward Linkages. *The American Economic Review* 94(3): 605–627.

[6] Jones, R.W. & Kierzkowski, H. 1990. The Role of Services in Production and International Trade: a Theoretical Framework. In R.W. Jones & A.O. Krueger (ed), *The Political Economy of International Trade*: 31–48. Oxford: Basil Blackwell.

[7] Jones, R.W. & Kierzkowski, H. 2005. International Fragmentation and the New Economic Geography. *The North American Journal of Economics and Finance* 16(1): 1–10.

[8] Lu, F. 2007. *Economic Analysis on Service Outsourcing: From the Intra-product Perspective.* Beijing: Peking University Press.

[9] Markusen, J. 1989. Trade in Producer Services and Other Specialized Intermediate Inputs. *The American Economic Review* 79(1): 85–95.

[10] Markusen, J. Rutherford, T. & Tarr, D. 2005. Trade and Direct Investment in Producer Services and the Domestic Market for Expertise. *Canadian Journal of Economics* 38(3): 758–777.

[11] Park, S.H. and Chan, K.S. 1989. A Cross-Country Input-Output Analysis of Intersectoral Relationships between Manufacturing and Services and their Employment Implications. *World Development* 17(2): 199–212.

[12] Yeaple, S.R. 2006. Offshoring, Foreign Direct Investment, and the Structure of U.S. Trade. *Journal of the European Economic Association* 4(2–3): 602–611.

[13] Zeng, Y. & Li, J. 2006. The Analysis of Trade Model of Producers Service Trade: the Analysis Based on Panel Data. *Statistical Research* 12:48–54.

Education Management and Management Science – Zheng (Ed.)
© 2015 Taylor & Francis Group, London, ISBN 978-1-138-02663-6

IAT application in technical translation

D.F. Ge
North China Institute of Science and Technology, Hebei, China

N. He
University of International Business and Economics, Beijing, China

ABSTRACT: Along with the coming of information age, the emergence of computer and the popularization of the internet provide new equipments for translators. In the field of Internet-Assisted Translation (IAT), such research projects as web corpus and search queries have been carried out increasingly. However, it is noticeable that few researches have been made as to the application of IAT in English-Chinese technical translation. This paper, on the basis of introduction of IAT, technical texts and technical translation, analyzes IAT application in technical translation from the aspects of lexicon, phrase, and sentence. This study shows us the applicability of IAT and the negative effects of IAT in technical translation. It's hoped that this study will facilitate the future development of IAT application in technical translation.

Keywords: Internet-Assisted Translation (IAT); technical translation; negative effects

1   INTRODUCTION

The past century has witnessed great developments in science and technology innovation, which have exerted a revolutionary impact on translation work by profoundly changing the environment and the way of it (Li, 2004). Translators in today's market face challenges such as higher requirements on efficiency and accuracy. Meanwhile, it is estimated that technical translation accounts for some 90% of the world's total translation output each year (Kingscott, 2002). In such a situation, the application of IAT in English-Chinese technical translation attracts people's attention.

2   INTERNET-ASSISTED TRANSLATION (IAT)

2.1   *Current situation and importance of IAT*

In the early time, internet was limited to specific groups and it did not have the search function. Not until the occurrence of Hypertext Transfer Protocol (HTTP) and World Wide Web, through which can people surf the Internet conveniently. When the search engine was invented, one needs only to open the search engine and type in the keywords to get access to any information needed.

Through Yahoo, Google, Baidu, AltaVista and some other search engines, translators can also get a lot of useful information to facilitate translation. The Internet has been widely applied to commerce, education and some other fields as well as the development of the linguistics field. There are free resources on internet and the information is updated quickly, so the translators can take full advantage of modern translation tools to improve efficiency and quality of translation.

2.2   *Main functions of IAT*

2.2.1   *Resource sharing*
Internet provides a variety of resources including giant shared database, where one can find rich encyclopedia, dictionary and thematic resources. In addition, cyber source can be used conveniently. Test shows looking up a word with the traditional paper dictionary takes on average 32.5 seconds, but it takes on average only 16.8 seconds with the help of online dictionary, both not including the reading time.

2.2.2   *Search query*
Internet is a giant free database, and a search engine is used to make efficient use of the database. The search engine can help us find the required resources promptly. The search query is taken as the core function of internet. Presently, the world's most famous powerful search engine is Google, while in China Baidu is frequently used.

2.2.3   *Information exchange*
Internet also provides us a perfect platform across time and space. Currently, internet offers services like e-mail, forums (BBS), blog, chat rooms and web conferencing.

3 TECHNICAL TEXTS

3.1 *The nature of technical texts*

According to Reiss, based on the informative function, expressive function and operative function, text types can be divided into content-focused text, form-focused text, and appeal-focused text, and technical texts are one of the principal content-focused texts (Reiss, 2000).

3.2 *The linguistic features of technical texts*

3.2.1 *Lexical feature of English and Chinese technical texts*

Generally, English technical texts feature frequent use of nominalization. Many nominalized structures are used in the technical texts to represent the semantic meaning of a verb and a noun or of an adjective and a noun simultaneously. Besides, prepositions are also frequently used out of grammatical or semantic consideration.

While in Chinese technical texts, the syntactic structure is verb dominated, and verbs are more frequently used to fit in with the demands of Chinese syntax. Prepositions are not as frequently used as in English technical texts.

3.2.2 *Syntactic feature of English and Chinese technical texts*

The syntactic feature of English technical texts can be generalized as: frequent use of passive voice, nominal structures, long and complicated sentences, as well as frequent use of ellipsis, inversions and separations. By contrast, Chinese are paratactic in nature and sentences are generally short and simple. Chinese technical texts also feature in syntax a different frequency in the use of passive voice.

4 TECHNICAL TRANSLATION

4.1 *Emphasis on terminology*

Terminology is regarded as the systematic designation of defined concepts within a specific field (Bowker, 2002). In technical translation, terminology cannot be over-emphasized. Newmark (2002) claimed that "the central difficulty in technical translation is usually new terminology" and "technical translation is primarily distinguished from other forms of translation by terminology", although terminology accounts for 5-10% of the total content of technical texts. Besides, consistency of the terminology translation should be guaranteed in technical translation.

4.2 *Assessment of technical translation*

Based on the features of technical texts and the functions of technical translation, the assessment of technical translation can be listed as accuracy, fluency, and conciseness.

5 A CASE STUDY OF THE APPLICATION OF IAT IN TECHNICAL TRANSLATION

In this part, the translation of a technical text of "How the Heart Works" is used as example to display the application of IAT. The source text is presented in appendix.

5.1 *From the aspect of word*

5.1.1 *Translation of polysemy*

Polysemy is a very common phenomenon in English, which poses a great challenge to online translation. Four examples were selected in this part to discuss about the translation quality of polysemy by Google online translation. Table 1 lists the polysemy and their meanings in the two versions of translation.

Table 1 shows that sometimes Google online translation cannot give the correct answer to some polysemy, as indicated in the case of "beat", "about" and "contract". But these problems can be easily solved by taking the context into consideration or by using search query.

5.1.2 *Translation of terminology*

The typical terminology in the source texts are displayed in Table 2.

Table 2 shows that Google online translation made a good work in the translation of terminology. Seventeen words are translated totally correct.

Table 1. Translation of polysemy.

Polysemy	Google online translation	Feng Qinghua's translation
beat		
beats (about 72 times)	to pulsate	to pulsate
(in a) beat	the tempo	to pulsate
about		
about (the size)	preposition	almost
about (... times)	almost	almost
right		
(to the) right	of the side	of the side
right (auricle)	of the side	of the side
right (ventricle)	of the side	of the side
contract		
contracts (in a beat)	binding agreements	shrink
contracting (muscle)	document	shrinking

Table 2. Translation of terminology.

Terminology	Google online translation	Feng Qinghua's translation
pump	a device	a device
contraction	action of contracting	action of contracting
relaxation	lengthening that characterizes inactive muscle	lengthening that characterizes inactive muscle
lungs	organs for breathing air	organs for breathing air
partition	segment	segment
sections	segments	segments
blood	liquid circulating the body	liquid circulating the body
auricle	*pinna*	atriums
ventricle	chamber of heart	chamber of heart
chamber	enclosed space within the body	enclosed space within the body
oxygen	chemical element	chemical element
cell	the smallest structural unit of living matter	the structural unit of living matter
carbon dioxide	one kind of gas	one kind of gas
vein	blood vessel	blood vessel
valve	*mechanical device*	structure in heart
aorta	the large artery	the large artery
artery	tubular branching muscular-walled vessels	tubular branching muscular-walled vessels
quarts	unit of liquid measurement	unit of liquid measurement
capillaries	small tubes carrying blood within the body	small tubes carrying blood within the body

5.2 From the aspect of phrase

Nineteen phrases were taken from the text to discuss phrase translation by Google online translation.

As we can see from Table 3, Google online translation offered wrong translation of four phrases of "the dark, used blood", "in a beat", "push… through" and "without faltering". By analyzing we know that the wrong translation of "the dark, used blood" is caused by wrong parsing, the incorrect translation of "in a beat" is caused by misunderstanding of the polysemy "beat", the wrong translation of "without faltering" results from mechanical transference, and the translation of "72 times a minute" ignores the expressing habit in the target language.

5.3 From the aspect of sentence

5.3.1 Translation of word order

There is a big difference in word order between English and Chinese. The order of each clause is generally consistent with the original order in Internet-assisted translation, which sometimes will result in inaccurate translation of the text.

Example: A healthy heart keeps this up for a lifetime without faltering.

In this sentence, "for a lifetime" and "without faltering" are two adverbial phrases modifying the action of "keep this up". In English, the adverbial phrases are often put after the verb they modify. But in Chinese, the adverbial phrases are usually put before the verb. So in English-Chinese translation the order of verbs and adverbial phrases should be changed. However, Google online translation of this sentence is just a mechanical transference of the source sentence, violating the expressing standards in the target language.

5.3.2 Translation of voice

Passive voice is more frequently used in English than in Chinese, so in English-Chinese translation, sometimes passive voice should be transformed into active voice to meet Chinese expression standard. However, it does not mean that all passive sentences in English should be translated into active sentences in Chinese. Sometimes, to better highlight the objects or to achieve objectivity, it's better to translate some English passive sentences into Chinese passive ones.

Example 1: "…is placed snugly between the lungs…"

Google online translation transfers this sentence into a passive Chinese sentence, which here does not conform to Chinese expression standard. And it is preferable to translates it into a Chinese active sentence.

Example 2: "From there it is forced by the contracting muscle through a valve into the aorta." The same is true to this example. This passive English sentence should be translated into a Chinese active sentence where "contracting muscle" serves the subject, and "force" the predicate.

Example 3: "The blood is pumped through all four chambers in turn in the course of being circulated through all parts of the body".

Here, to translate it into one Chinese passive sentence by following the original subject and predicate is better than to transform it into one active sentence as that made by Google online translation.

Table 3. Translation of phrases.

Phrases	Google online translation	Feng Qinghua's translation
72 times a minute	*a minute 72 times*	72 times a minute
divide ... into ...	separate ... into ...	separate ... into ...
at the same time	in the meantime	in the meantime
deal with	cope with	cope with
in turn	no translation	by turns
in the course of	in the process of	in the process of
dissolved foods	dematerialized food	dematerialized food
carry away	take away	take away
pour ... into	cause to flow into	cause to flow into
the dark, used blood	*the used blood in the darkness*	the used blood of dark color
in a beat	*in a tempo*	in a strike
push ... into	move ... into	move ... into
be replaced with	be substituted with	be substituted with
be forced into	be forced to enter	be forced to flow into
all over the body	within the body	throughout the body
push ... through	*shove by means of* ...	cause ... to flow by
keep up	maintain	maintain
for a lifetime	all one's life	all one's life
without faltering	*with no stagger*	unceasingly

5.3.3 *Translation of long and complicated sentences*

Long and complicated sentence translation is a headache for online translation. Since there is a great disparity between Chinese and English sentence structures, it is difficult for IAT to discern the main idea and the layout of the structure and cut it into separate sections and then translate them one by one.

Example: "The muscle surrounding this part contracts in a beat that pushes the blood into the lungs where the carbon dioxide is removed and replaced with vital oxygen".

In this sentence, the subject is "The muscle", and the predicate "contracts". The present participle of "surrounding this part" is the post-modifier of "the muscle"; the attributive clause of "that pushes ..." is modifying the noun "beat"; "where the carbon dioxide is removed and replaced with vital oxygen" is the adverbial clause. To correctly translate this sentence into Chinese, one should, based on correct parsing of the English sentence, adopt methods of reversing or recasting to make the Chinese translation fluent and expressive. The Google online translation of this sentence is confused due to its unchanged order of the original sentence.

6 CONCLUSION

The analysis of the application of IAT in technical translation in this thesis contributes to the following conclusions.

First, IAT can assist translators to work faster and better in their translation activities, and it can help keep the consistency of terminology translation.

Second, IAT also has negative effects. Internet itself does not have the ability to understand natural language, the translation support it provides is mechanical, so IAT can only play an auxiliary role.

Third, translators must play their initiative in translation activities, they should read the text carefully, analyze the intention of the author combined with linguistics, stylistics and analysis of the source text, then express the original idea accurately and concisely and ensure that the translation in compliance with the expressing criteria in the target language.

REFERENCES

[1] Bowker, L. 2002. *Computer-aided translation technology: A practical introduction.* Ottawa: University of Ottawa Press.
[2] Feng Qinghua. 2002. *A Practical Coursebook on Translation.* Shanghai: Shanghai Foreign Language Education Press.
[3] Kingscott, G. 2002. *Technical Translation and Related Disciplines. Perpectives: Studies in Translatology* 10(4), 247–255.
[4] Li, Changshuan. 2004. *Theory and Practice of Non-Literary Translation*: 103. Beijing: CTPC.
[5] Newmark, P. 2002. *A Textbook of Translation.* Shanghai: Shanghai Foreign Language Education Press.
[6] Reiss, K. 2000. *Translation Criticism: The Potentials and Limitations.* Shanghai: Shanghai Foreign Language Education Press.

The source technical text:

How the Heart Works

The heart, which is a muscular pump, beats about 72 times a minute through a continuous and automatic process of muscular contraction and relaxation. It is about the size of a fist, weighs about 9–11 ounces and is placed snugly between the lungs, a little more to the left than to the right. A partition runs down the centre of the heart, dividing it into left and right sections which work at the same time but deal with two different types of blood. Each section is again divided into upper and lower parts, the auricles and ventricles. The blood is pumped through all four chambers in turn in the course of being circulated through all parts of the body.

The heart's first purpose is to supply a steady flow of oxygen to all the body cells and to return carbon dioxide to the lungs. On its journey the blood distributes dissolved foods and carries away wastes.

Two large veins pour the dark, used blood into the first chamber, the right auricle, which passes it into the chamber below, the right ventricle. The muscle surrounding this part contracts in a beat that pushes the blood into the lungs where the carbon dioxide is removed and replaced with vital oxygen. Meanwhile, fresh scarlet blood from the lungs enters the left auricle to be transferred to the left ventricle. From there it is forced by the contracting muscle through a valve into the aorta, the body's largest artery which distributes it all over the body.

The heart beats about 100,000 times every 24 hours and pushes several quarts of blood through miles of arteries, veins and capillaries. A healthy heart keeps this up for a lifetime without faltering.

Education Management and Management Science – Zheng (Ed.)
© 2015 Taylor & Francis Group, London, ISBN 978-1-138-02663-6

Research on moral education curriculum of the elementary schools in China

Xuan Chu
Research Institute of Higher Education, Shenyang Aerospace University, Shenyang, China

ABSTRACT: Moral education is developed through many different kinds of moral curriculum in China. In this paper, content analysis is used as its principal research method. Analyzing the contents of selected textbooks, it realizes the features and dimensions of moral education curriculum, also discusses the teaching and learning problems of moral education curriculum currently faced in China.

Keywords: China; moral education; curriculum; elementary schools

1 INTRODUCTION

In Chinese elementary schools, moral education is carried through moral education textbooks which named "Morality and Society". "Morality and Society" is an important curriculum which aims to guide children to learn how to live, to become integrated into the community and to become good citizens. In a new round of curriculum reform since 2010, the moral education curriculum has received a great deal of attention as it is a platform for students to learn how to sort out their lives, a window on society, and a door to enable pupils to move into society.

The moral education in elementary schools not only teaches students knowledge, but also enables the students to master the essential survival skills for modern society, and leads them in gaining a better understanding of the world. One of the most important responsibilities of education is to guide children to "learn to live", especially to guide children to have a civilized and healthy life, to develop a high level of civic virtue, to develop the courage to explore and to develop the innovative spirit of science.

2 DATA AND METHOD OF RESEARCH

The data is an analysis of moral education textbooks published by the Educational Science Press, 2002–2008 editions (hereafter referred as "ESP Edition"). Two key textbooks were analyzed: the textbooks named "Moral Character and Life" (QiWanXue 2002) for elementary grades 1–2 and others named "Morality and Social Studies" QiWanXue (2002–2005) for elementary grades 3–6. Each grade is divided into two levels so that there are 12 levels in the textbooks altogether.

This paper uses content analysis as its principal method. This method makes objective and systematic descriptions of the selected contents. We have used this method to analyze the "ESP Editions" textbooks. In this paper, the textbooks are divided into many categories according to the Moral and social curriculum standards which published by Ministry of Education of the People's Republic of China.

The textbooks are systematically subdivided. Each chapter is divided into a number of units, each unit contains many sections and each section may contain a number of different elements. So, for the purposes of this research we have set the "section" as the basic unit of analysis. Sections are more useful as an analysis unit than "units" or "chapters".

3 DIMENSIONS OF MORAL EDUCATION

3.1 Knowledge standards for moral education curriculum in the "ESP Edition"

The framework of textbook will be analyzed against the content of the Morality and Society Standards Ministry of Education of the People's Republic of China, 2004. The social circumstances include time, space, the human environment and the natural environment. Social activities include daily life and cultural, political and economic activities. Social relations consist of human relationships, social norms, rules and the legal system. Based on above contents, we have modified these headings to construct a number of dimensions for analysis. These are: individual dimension; social relationships; social knowledge; and skills. Through these dimensions individuals are able to address the three areas of learning addressed by the textbooks named Morality and Society, which are: "emotion and attitude", "action and habits", and "knowledge and skills". Furthermore, the curriculum is able to develop children who have a good moral character, good behavior habits, would explore and can live loving lives.

3.2 *Dimensions of moral curriculum*

All three aspects (moral and values education, social relationships and social knowledge) are represented in the content of the textbooks. Moral and values education is included, as well as individual physical and psychological development. In particular, in the textbook for Grade 2 and Grade 5, there are some sections whose contents are about the development of the body to help children learn about their own development. For example: what height and weight I am, how I can keep healthy, and so on.

Social relationships are included in the contents of the textbooks. For example, the relationship between family and school are covered in the 1st Grade and 3rd Grade. The relationship between community and hometown are addressed in the 2nd and 4th Grade. The relationship between community and hometown are also addressed in the 5th Grade. In the 6th Grade, there are more contents about the motherland and the world. This arrangement seems likely follow the student's life experience. When they are in grade one, they should learn more about school. At the same time, they should know more things about their family and hometown. These are issues which are familiar to children so that they can relate their own experiences to the materials covered in the textbooks.

Social knowledge is the second main aspect which is addressed in the textbooks. Children learn about knowledge of everyday life, science, four seasons, geography and so on in the lower grades. In Grade 3, they learn some cultural knowledge, such as festivals and New Year. But when they get up to Grade 5, they learn more culture and history about motherland and world. They learn about the long history and the fast development of China. They will also learn about other countries and the changes happening globally.

Social skills are the third aspect addressed by the textbooks. Children in the early grades will learn more life and learning skills than children in higher grades. For example, the chapter "the flood coming—Tingting's adventure", "When the earthquake occurred-how to survive", "Our handbook of survival" contain many examples of life skills, which develop children's own survive skills so that they are able to avoid risk and disaster.

There are little contents about learning skills in the lower grades. For example, in the first year, the chapter "beautiful little interrogation mark" encourages students to ask more questions, so that we can develop their creative spirit. The chapter "where to find the answer" teaches students how to resolve problems. It tells students that they can find the answers through asking others, looking them up in a book, looking up online, or using first-hand experience. The chapter "we have found" tells students to consider the relevance of science, encourages them to think and do the research by themselves and to display their results. One chapter in Grade 3 asks students to draw a map of their school, so that they can learn to distinguish orientation. One chapter in Grade 4 asks students to make an investigation plan. They must design, collect materials and publish a class handbook collaboratively. Some chapters ask the children to observe the numbers of stores around their homes and draw a map based on observation, interviewing and investigation. Students are asked to keep accounts of the family expenses.

3.3 *Moral values and significance level*

We can divide the moral value implied in the moral education curriculum into several categories. These include: loving the motherland, loving one's hometown, cherishing life, loving life, loving learning, loving school, loving collectivity, loving science, loving civilization, kindness to others, national unity, sense of democracy, sense of responsibility, awareness of rights, a sense of participation, being law-abiding, sanitation and healthiness, consciousness of safety, fidelity and trustworthiness, traditional virtues, respect for teachers, respect for older and younger individuals, cooperation, protection of the environment, consciousness of equality, justice and sense of humor, consciousness of creation, family virtues, occupational satisfaction, consciousness of peace, mental health, consciousness of competition and global consciousness.

We also have calculated the contents of textbooks and got the proportion of each moral value. Loving the motherland (13.5%), loving one's hometown (3.8%), cherishing life (2.3%), loving life (0.7%), loving learning (0.7%), loving school (1.4%), loving collectivity (1.4%), loving tradition (1.4%), loving science (13.1%), loving civilization (2%), kindness to others (3%), national unity (1.4%), sense of democracy (0.9%), sense of responsibility (0.2%), awareness of rights (0.45%), sense of participation (0.2%), being law-abiding (3.3%), sanitation and healthiness (5%), consciousness of safety (2%), fidelity and trustworthiness (2.4%), traditional virtues (5.6%), respect for teachers (1.4%), respect for older and younger individuals (1.8%), cooperation (2.3%), protection of the environment (6.8%), consciousness of equality (0.2%), justice and humorousness (-%), consciousness of creation (2.9%), family virtues (3.8%), occupational satisfaction (2.5%), consciousness of peace (2.3%), mental health (5.2%), consciousness of competition (-%), global consciousness (3.8%).

4 THE FEATURES OF MORAL EDUCATION IN CHINA

4.1 *Focus on children's lives*

The contents of these moral education textbooks surround the social lives of children, which helps children to learn common sense about social life, to identify the best sense about social life, to understand the norms of social life, to master and use their own knowledge and capabilities to participate in social activities. These elements, especially in the textbooks for the lower grades, are mainly based on the reality of children's lives and living needs, rather than abstract scientific concepts. The contents of textbooks and curriculum content are close to the lives of children, designed and arranged from the perspective of children.

4.2 *Comprehensive knowledge*

There is comprehensive knowledge in moral education textbooks. In the field of living, the knowledge refers to six dimensions: individual, family, school, hometown (community), motherland, and the world. In the field of knowledge, it refers to historical knowledge, cultural knowledge, economic knowledge, norms of life and so on.

4.3 *Developing civil and moral education throughout life*

Moral education tells the students how to be a good citizen and to cultivate their values. Through studying the facts of children's own lives, the students are taught about the ideological and moral, civil and moral, traditional values. The education guides the students to correct moral orientation, so that students have correct moral values. These facts are close to the children, so that students need to master the basic knowledge and skills for their lives and to develop their own learning and self-innovation skills.

4.4 *Include variety moral values*

In the moral and social standards of the new curriculum in elementary schools (pilot version), moral education is to encourage the students to develop good moral character and social development.

Moral education aims to lay the foundation for students to know about society, participate in society, adapt to society and become a qualified citizen of socialism, develop a caring character, responsibility and high standards of behavior and personality. Therefore, moral education for the schools must include moral values for elementary school students; there is a need to analyze the moral values implied in moral education curriculum of elementary schools.

5 THE PROBLEMS OF MORAL EDUCATION IN CHINA

5.1 *Content*

Currently moral education provides comprehensive knowledge, and it must cover a wide range of content, including social life, geography, history, morality, legislation, politics, economic, culture, and environment. However, a number of problems can be identified. The first problem is whether the contents of moral education textbooks address the principles of students' physical and psychology development. They should consider the goals for different grades and what content is contained in the different textbooks. For example, how many contents are about studying in schools and how many contents are about family relationships? In addition, it is necessary to consider whether the contents are too difficult or too easy for students. In particular, the contents of textbooks should be arranged based on the characteristics of students in different grades. Students do not want to learn things they already know, nor do they want to struggle with things they can't understand.

The second problem is about the ratios of different values in the contents of these textbooks. We have identified the ratio of different values appeared in textbooks. We found that the proportions of contents about sense of responsibility, sense of participation, sense of democracy and consciousness of equality is less than the proportions of other values. In addition, creativity and innovation are very important values, necessary to cultivate students' creative consciousnesses and their motivation. However, our analysis of moral education textbooks shows there are little contents about awareness of creativity.

The third problems identified about the contents of textbooks are relevance to all Chinese students. The contents of textbooks are close to the experience of children in cities, but there are so little examples and contents that relate to the rural experience of many children. There is a great deal of geographical and historical knowledge, which should help the students learn about the history of their nation and inspire love of China, but the contents also needs to include modern society. Students should objectively know the development characteristics and the disadvantages of China compared with OECD countries.

5.2 Teaching methods

The new curriculum standards claim that moral and social programs can be taught by discussion, data survey, site investigation, simulation and role-playing scenarios, operational and practical activities, teaching games, visits, enjoyment, practice, storytelling, lectures and so on. Therefore, a variety of teaching methods should be used in moral education curriculum teaching. Furthermore, the moral curriculum is a synthesis subject including aspects of culture, economics, and politics and so on. This means there are increased requirements for teachers' own subject knowledge and ability. It is a particular challenge for the rural teachers.

The moral education curriculum encourages children to participate in social activity by themselves and in their own way. So lots of time will be spent encouraging students using a variety of teaching methods. When there is not sufficient time, the learning may be compromised.

5.3 Curriculum resources

The new curriculum standards for moral education states that the resources for moral education curriculum in elementary school are divided into tangible and intangible resources. The first type of resources are commonly used in schools, such as textbooks, guide books for teachers, audio-visual materials, internet and other subjects, classes or school activities. The second type of resources includes children's experiences, interests within the community; teachers, peers, children's families, communities of all kinds of physical facilities, cultural and educational institutions. The third type of resource is cultural activities, festivals and people who engage in a variety of career-related activities etc. The fourth type of resource is nature world: animals and plants, mountains and rivers and other natural phenomena. Because of our relatively large gap between urban and rural education, there are also significant differences for moral (or called society) curriculum in elementary schools. Cities will be rich sources of tangible and intangible resources, but in most rural areas there is no library, no museum nearby, no education base, no memorial and no cultural centers, and so on, as well there are no major social organizations or government agencies. The lack of curriculum resources will affect the quality of moral education in rural areas.

5.4 Curriculum assessment

The main object of moral education curriculum is to affect the life of students and offer moral education for students. In particular, the curriculum helps children feel, experience, understand and develop in all aspects of the real world. The internal moral education of children is abstract. So, teaching and evaluating is difficult and complex.

In some rural areas, especially those regions where the quality of teaching, teaching conditions and teachers is not high, the assessment of the moral education curriculum is much more difficult. Some students are asked to remember the contents of textbooks and the knowledge of moral education curriculum. The factor which has not been considered in the teaching process or the ethical behavior of students is getting rid of evaluation. Process and approach are not considered in teaching activities. When teaching and learning methods find and resolve problems, developing creative and experience based capabilities is not considered in the process of understanding, experiencing and exploring the problems. The key issues of whether the children change their attitudes, form good habits, have a good range of emotions, access knowledge, train their capacity and solve problems are not paid sufficient attention or examined in enough details. In some school districts, all the students are ranked by scores on the examination conducted at the end of a course of social studies classes, and the scores are linked to the salaries of the teachers.

6 DEADLINE

In modern society, moral education is a very important route to cultivate citizenship. There are many good ways to develop and improve moral education, which are improving the content of social curriculum, strengthen professional development of teachers, sharing of curriculum resources and making better the evaluation system.

REFERENCES

[1] Ministry of Education of the People's Republic of China (2004) "Morality and Social Curriculum Standards," Beijing: Beijing Normal University press, 1–23.
[2] QiWanXue (2002–2005)."Morality and Society". Beijing: Education Science press.
[3] QiWanXue (2002). "Moral Character and Life". Beijing: Education Science press.

Education Management and Management Science – Zheng (Ed.)
© 2015 Taylor & Francis Group, London, ISBN 978-1-138-02663-6

Genre awareness in L2 reading and writing at the college level: A case study

W. Huang

Tianhua College, Shanghai Normal University, Shanghai, China

ABSTRACT: Despite considerable attention for the genre pedagogy across disciplinary areas, the impact of explicit genre instruction and detailed procedure has not been sufficiently discussed. This paper carries out a tentative case study on cultivating students' genre awareness in EFL reading and writing and show how it influences the students' ability to analyze and compose texts in English.

Keywords: genre; genre-based pedagogy; EFL writing

1 INTRODUCTION

Traditional English textbooks are mostly topic based, divided into different units, and students in use of these textbooks are particularly familiarized with bottom-up reading strategies, building meanings from vocabulary to larger units. However, students often encounter difficulties in real life communication and in academic reading and writing.

According to data released by ETS, the average TOEFL score achieved by the students from mainland China last year is 77, lower than the global average 79. Its UK counterpart IELTS (Academic) presents an even more gloomy picture with Chinese students getting 5.6 in average and China ranked the 8th lowest in the average score among the 41 countries taking IELTS test in 2013.

Hence the questions are: How can our students transfer input from the classrooms to the relatively more authentic academic context? How can we redesign our syllabus by connecting the mutually insulated units in the textbook and making it more contextualized? My attempt in this article is not to provide an absolute answer to these questions, but a discussion of a tentative research on explicit genre instruction by which novice academic students are scaffold and encouraged to participate in understanding and negotiating the meaningful academic or socio-cultural context.

2 GENRE-BASED PEDAGOGY

Genre has been widely discussed in recent years and has become "one of the most important and influential concepts in language education" (Hyland, 2004, p. 5). The word "genre" means "kind" or "type" in French and the study of genres can be traced back to Aristotle, and have extended from literature into linguistics.

Current genre theories and pedagogies have been well recognized in the literature with three established traditions: (1) Australian systemic functional linguistics (SFL), (2) North American New Rhetoric (NR), and (3) English for Specific Purposes (Hyon, 1996; Johns, 2002).

In this article, the notion of genre is more complementary than contradictory, drawing much on Johns' (2008) conceptualization of the nature of genres:

1. genres are both social and cognitive.
2. genres are purposeful, or responsive.
3. genres are named by those in power.

It goes far beyond the simple definition of "text types" and can be broadly understood as what Hyland (2007) referred to as "abstract, socially recognized ways of using language".

Despite numerous influential researches on genre pedagogies from three traditions, most of them fit into their own socio-cultural and academic context. For example, SFL school promotes genre acquisition for immigrant students in Australia (e.g. Bhatia, 2004); NR's contributions are mainly on native speakers of English with little discussion for novice students (e.g. Bawarshi 2003); ESP scholars design their research and pedagogy for professional or advanced academics in North America (e.g. Samraj 2004).

Therefore, as a teacher teaching novice academic students at a private college in China, I need to explore how genre-based pedagogy can be adapted in local EFL context, and how explicit genre-based instructions can be beneficial for EFL learners at a lower level, and how genre awareness can be enhanced to help students to access and negotiate genres to a situation?

According to Johns (2008), genre awareness helps students develop "rhetorical flexibility necessary for adapting their socio-cognitive genre knowledge to ever changing contexts" (p. 238), so I need to analyze the students' learning context first.

3 EFL STUDENTS' LEARNING CONTEXT

Traditional English classes are the English instructor's solo performance based on the only script—the textbook, of which the texts are adapted from English articles of different topics and various genres. The students are often required to translate or recite paragraphs or passages as their homework. In high schools, the majority of the English classes are exam-oriented. With the ultimate goal to pass the college entrance exams, students struggle in "an ocean of exercise papers". When they enter the college, they are still faced with similar textbooks with longer passages and a series of exams from course tests to national or even international English tests.

The Integrated English Course is designed for the English major students, with its first-year focus on the comprehensive improvement of basic language skills. Most students resort to their textbook as their major source of exposure to English reading and writing, and many of them choose to participate in a nation-wide proficiency test called College English Test Band 4 (CET 4) by the end of the first year at college.

As most part of the test paper is designed as multiple choices, students find it particularly difficult in the writing section. Traditional English learning strategies such as memorizing new words, making grammatical choices, and reciting textbook passages provide very limited facilitation in the improvement of their writing.

In this paper, I attempt to show the extent to which the students develop genre awareness enables them to better understand the text and thus transfer the knowledge to their writing.

4 METHOD

4.1 *Key concepts in the research procedure*

The following procedure is employed in my study:

1. course redesign based on genre-analysis (Table 1);
2. explicit genre awareness instructions;

Table 1. Course overview.

Week	Genre analysis	Writing assignment
1–2	Narrative	Story map
3–4	Memoir	Memoirs reading and writing
5–6	Memorial speech	Describing a person
6	Students presentation 1	
6–8	Exposition 1	Structure map
9–10	Exposition 2	Introduction writing
11–12	Exposition 3	Body and conclusion
13	Students presentation 2	
13–14	Persuasion 1	Comparison and contrast
15–16	Persuasion 2	Cause and effect
17	Students presentation 3	

3. students' analyses of the text based on generic features;
4. students' presentation and writing.

As mentioned above, the course textbook is topic-based, and there is a little connection between each unit. I re-group the units in the textbook according to different genres to relate them with common communicative purposes or rhetorical moves (Swales 1990).

The explicit genre awareness instructions are necessary as the students need to be exposed to learning experiences in which they work their way through explicit analysis of generic features, critical appraisal of the social function of the genre, and then writing in that genre. (Cope & Kalantzis, 1993, p. 85)

According to Vygotsky' (1978) scaffolding approach, students receive more guidance from the instructor at the beginning stage of each genre analysis, and then learn collectively on the group presentation, while work independently on the writing assignment at a later stage.

4.2 *Data collection and analysis*

This is a case study with data collection through multiple sources: classroom observation, students' writing and reflections, interview with some students, and a survey on the course.

The data were analyzed interactively from informal discussions to structured writings.

In the informal discussions, the students were divided into different learning groups and volunteered to meet the researcher and talk about what they had learned about genre and what were their difficulties in reading and writing. This stage aimed at finding out whether the students were developing their genre awareness after explicit instructions in the class.

Then they worked collaboratively on the feature analysis of a particular genre and made group presentation on their knowledge of the genre. The presentations were compared and discussed to see the procedural development of their awareness of different genre features and their knowledge of how to recognize the features in reading and apply them in writing.

The data analysis also entailed students' writing assignments and a simple survey, which indicated how students possibly transfer genre awareness into their writing.

5 FINDINGS

After one term of the course instruction, some remarkable changes of the students can be seen from the students writing assignment and their reflections. Students in the following cases appear under pseudonyms.

Generic structure comparison deepens the students understanding of different genre and improves their organization of writings.

Explicit genre instructions do not simply provide an answer to the text features, but offer opportunities for the students to be aware of more features of a specific discourse. In teaching the persuasive text, the students and I collaboratively figure out the generic structure of two texts in the textbook (Table 2), and both texts show that the author's opinion, or the thesis statement appear in the introduction part. But in the discussion time, Katherine and her group mates disagree with this structure by quoting their knowledge of Chinese writing pattern and their experiences in CET-4 preparation exercises. Katherine took out a CET-4 test paper and showed me the writing section (Table 3) and argued that all her high school English teachers have taught her that thesis statement should appear

Table 2. The generic structure of a persuasive text.

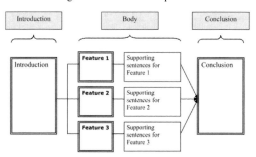

Table 3. Writing directions in CET-4 2006.

Directions: For this part, you are allowed thirty minutes to write a comosition on the topic Spring Festival Gala. Your should write at least 120 words, and base your composition on the outline given below:
1. Many people like watching Spring Festival Gala on Chinese lunar new year's eve.
2. Some people propose to cancel Spring Festival Gala.
3. My opinion.

Table 4. The generic structure of an expository text.

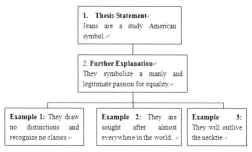

in the end of an essay, and she had been trained to write in this structure for many years.

Chinese students often put the thesis statement in the conclusion part, partly due to the negative transfer of the generic features in Chinese writing. In addition, years of exam-oriented teaching and learning repeatedly emphasize this writing template and leaves great impact on students' writing. Therefore, we revisited the three texts the compared the generic structures. The two authentic English texts formulate similar structure as follows:

1. Introduction (establish a topic area ^ public opinion ^ the author's opinion)
2. Body (sub-topic areas ^ supporting details)
3. Conclusion (restate the thesis ^ call for actions).

Students found in amazement that the outline provided in the CET-4 writing overlapped with the introduction part of the two English texts. " I guess the CET-4 writing was too short to develop into a full persuasive article, and only introduction part is required here." (Katherine).

By comparing similar texts, students' genre awareness seemed to show some procedural progress and more confidence in writing.

For instance, one unit was an exposition about a symbol of American culture-jeans. Before the genre analysis, the students in the outlining stage of their own writing would only come up with very colloquial expressions such as "I think", "you know", and the structure was incomplete or even

Table 5. Ann's outlining structure of her writing.

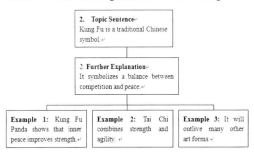

illogical. In contrast, after the genre analysis of the introductory part of the article (Table 4), the students related to Chinese culture and developed outlining structure map (Table 5).

6 DISCUSSION

In the limited scope of study, students' responses and writings suggests that genre awareness do transfer into students' writing decisions. For example, by the end of the term, students tend to follow the typical genre structure of the texts as well as apply different generic features in their own composition. As Cheng (2007) defines the process as "recontextualization", "learners ability not only to use a certain generic feature in a new writing task, but to use it with a keen awareness of the rhetorical context that facilitates its appropriate use" (p. 303).

7 CONCLUSION

This study is only an attempt to shed some light in the situated teaching and learning context which invites further research.

Bakhtin (1986) argues that writers must be able to control the genres they use before they can exploit them. My study is only situated on the particular textbook and the typical generic features of the texts, and my students can only partly translate

the knowledge of genre into effective reading and writing. However, further studies must be done on a more detailed and verified generic framework to a better instruction of genre awareness for the EFL students in a novice academic classroom.

REFERENCES

[1] Bakhtin, M. (1986). *Speech genres and other late essays.* Austin: University of Texas Press.
[2] Bhatia, V.K. (2004). W*orlds of written discourse: A genre-based view.* London: Continuum.
[3] Bawarshi, A. (2003). *Genre and the invention of the writer.* Logan, UT: Utah State Universit Press.
[4] Cheng. A. (2007). Transferring generic features and recontextualizing genre awareness: Understanding writing performance in the ESP genre-based literacy framework. *English for Specific Purposes* 26: 287–307.
[5] Cope, B., & Kalantzis, M. (1993). The power of literacy and the literacy of power. In B. Cope and M. Kalantzis (Eds.), The powers of literacy: A genre approach to teaching writing (pp. 63–89). Pittsburgh, PA: University of Pittsburgh Press.
[6] Hyland, K. (2004). *Genre and second language writing.* Ann Arbor, MI: The University of Michigan Press.
[7] Hyland, K. (2007). Genre pedagogy: Language, literacy and L2 writing instruction. *Journal of Second Language Writing* 16:148–164.
[8] Hyon, S. (1996). Genre in three traditions: Implications for ESL. *TESOL Quarterly* 30(4): 693–722.
[9] Johns, A.M. (1997). *Text, role, and context: Developing academic literacies.* NY: Cambridge University Press.
[10] Johns, A.M. (2002). Introduction: Genre in the classroom. In A.M. Johns (Ed.), *Genre in the classroom: Multiple perspectives* (pp. 3–13). Mahwah, NJ: Lawrence Erlbaum Associates.
[11] Johns, A M. (2008) Genre awareness for the novice academic student: An ongoing quest. *Language Teaching* 41(2): 237–252
[12] Samraj, B. (2004). Discourse features of student-produced academic papers: Variations across disciplinary courses. *Journal of English for Academic Purposes* 3: 5–22.
[13] Swales, J.M. (1990). *Genre analysis: English in academic and research settings.* Cambridge: Cambridge University Press.

Ideas on China's foreign trade talent demand and teaching reform of international trade specialty

F.H. Meng & M. Zheng
School of Business, North China Institute of Science and Technology, Sanhe, Hebei, China

ABSTRACT: Talent is the key factor in the survival and development of enterprises. This paper studied what China's foreign trade enterprises need should be high-quality, inter-disciplinary foreign trade talents. But the truth is that there are still some shortages of current foreign trade talents, such as, lack of analytical ability in economical situation and the environment of policy, marketing negotiation skills are insufficient, knowledge of foreign trade is relatively simple, the lack of trade and legal dual talents, etc, which cannot meet the changes and needs of the current international trade environment. Finally, we point out that the talent training should meet the needs of the market. Therefore, the reform of international trade major should be implemented in terms of curriculum provision, methods of teaching, teacher's quality, ways of running school and some other fields.

Keywords: demands for talents; problems; major reform

1 INTRODUCTION

At present, our country is facing a severe situation in exportation and its problems are mainly manifested in fields below. The trading friction is becoming more and more normal on a macro level and it's harder to exploit the overseas market. On a microcosmic level, the cutthroat competition between foreign trade enterprises is becoming more and more intense, leading to a lower commodity price in export. In the meantime, with the deepening reform of exchange rate of RMB, the risk of exchange rate is increasing. Not considering the enterprises themselves, and overlooking the trade protectionism in the international market, the quality of people working in the foreign trade industry cannot meet the changes and demands of the current international trade environment, which is the main reason of this issue. This article is to analyze the types of talents required in the foreign trade industry and discuss the deficiencies of people working in the foreign trade industry. Furthermore we will come up with some solutions to the improvement of the teaching methods in this major in colleges.

2 ENTERPRISES' CURRENT DEMAND FOR TALENTS IN TERMS OF FOREIGN TRADE

Along with the changes of international trade and the transformation of the foreign trade policy, the requirements for foreign trade talents from enterprises are higher. In general, a high-qualified talent should

be someone who not only has a good knowledge of marketing management, but also knows negotiating skills and laws well, besides, rules and knowledge of international trade should be memorized in heart.

The foreign trade talents who are equipped with the knowledge of marketing management can not only help with the product positioning and the development plan of the company, but also coordinate the different links of international trade effectively, in the purpose of exploring overseas markets effectively for the company in the highly competitive market environment. Those talents who own a good knowledge of international business negotiation can not only be familiar with the product of their company, but also thoroughly understand the knowledge in the area of trade, marketing, law, psychology and habits of negotiating clients. They can strive for the maximum profit for their company. A foreign trade talent who are aware of the law and rules of international trade can put their professional knowledge into practice and thus avoid law conflicts effectively. In addition to this, when there are conflicts concerning customs, commodity inspection, currency, governments, they can handle these issues with rules of trade.

High-qualified foreign trade talents are knowledgeable, practical, ideological and innovative. Knowledge is the basis of quality, impeccable system of knowledge plays an important role in tackling foreign trade problems. High-qualified foreign trade talents cannot be cultivated without knowledge. Applicability is the core ability of foreign trade talents. International trade involves marketing management ability, organizational and coordinating ability, communicating skills and so on. Applicability cannot be trained simply by theoretical learning, practical problem solving ability should be cultivated intensely. Logical thinking is the premise of other features. Ideology is the guidance of action. International perspective, competitive awareness, dedicating and aggressive conceptions and tough willpower should be grasped by heart. Creativity is the advantage of every industry. Enterprises cannot be highly competitive and own strong vitality unless they adjust to the changing situation in time, possess the awareness of high creativity and continuous innovation.[1]

3 THE PROBLEMS THAT FOREIGN TRADE TALENTS THEMSELVES OWN

As an important part of factors promoting the international trade, foreign trade talents are required with some new standards in terms of specialty, professional skills and personal abilities by the international environment. However, the cultivating programme of international trade major is not following the trend, which is not revised along with the improvement of international trade. What's more, the cultivating programme in different colleges are not in the same level, resulting in some foreign trade talents not qualified to meet the demands of international trade environment. The shortcomings of the foreign trade talents can be divided into the following parts.

3.1 Lack of analytical ability in economical situation and the environment of policy

Higher ability of analytical skills in international trade is of great value. In the activity of foreign trade, economic situation and the environment of policy have great influence in business decisions of enterprises. It is very hard for our country to develop the overseas market, part of the reason is that foreign trade workers cannot grasp the international economic situation correctly and comprehensively. Internationally, the prospect of international economy is not very clear after the financial crisis. Developed countries find every means to develop their manufacturing industry in order to rise the employment rate. The crisis in the Euro zone is continuing and its outlook is not optimistic. The recovery of American economy is not smooth, quantitative easing measures are constantly introduced. Many companies in our country still regard Europe and other developed markets as their main target, thus losing sight of the needs of emerging markets. Domestically, starting from 2010, our government has already come up with some relevant measures to promote importation.

The subsidies for export products and the import licensing system are gradually canceled. For the moment, import is not only easy to be done but its

cost far lower than domestic products, meanwhile, the margin of importing is fat, which is not seen by the employees, leading to the enterprises continue to export their products at low prices.

3.2 Lack of marketing ability to negotiate

A foreign trade professional with wealthy marketing knowledge can deeper investigate the overseas market and furthermore choose the potential targeting market for the company and their products and then make effective strategy of the export marketing as a whole. Besides, they can establish solid partnerships with international buyers and increase the rate of return on investment. They decide the direction of development, which is pretty much important for foreign trade enterprises. At present, many foreign employees still remain the thoughts of the last century which was the old-fashioned trade. They still search for potential clients on the Internet, then text them massively beginning with English, in other words, passively waiting for their clients, lacking of active-advanced awareness. Therefore, they can't adapt to the changing overseas market. International trade practitioners blindly exploit overseas market due to the lack of experience and negotiating skills. A number of domestic enterprises have no idea of how to make use of good marketing channel and marketing tools to attract the attention of clients of overseas market. They are not aware of using the negotiating skills in the process of negotiating with clients rather than simply the low price to attract the customers. This can not only affect the original market shares and the exploitation of the overseas market, but also have a certain influence on the maintenance of number of clients. Thus the situation of our country's foreign trade enterprises having a hard time surviving and their malignant low price competition are repeatedly occurred.

3.3 Relatively simple knowledge of foreign trade

International trade is more complicated than domestic trade. The nationalities, languages, cultural backgrounds, living habits, ways of thinking are all different between the two parts of negotiating, and the process involves politics, economics, laws and some other factors. Currently, our foreign trade practitioners are likely to be skillful in practical operating, but not very good at other aspects. Therefore, they are not familiar with the cultural conventions and behavioral habits of the clients during their communication. This phenomenon leads to the difficulty of communicating and exploiting the overseas market, thus the two parts cannot reach an agreement eventually. Meanwhile, the foreign trade practitioners of our country are not active enough to learn new knowledge relevant to foreign trade. They are not good enough to grasp the international trade policy, and conflicts are occurred during the process of dealing with foreign trade. Besides, they are not clear enough of the effective measures when it comes to this kind of issue. In a word, they can't solve these problems efficiently in time. Some emerging markets are making great progress. In addition, a number of foreign trade practitioners neither pay attention to the marketing trend, nor open up new markets. A majority of enterprises are participating in the malignant low price competition in occident.

3.4 Lack of talents in the field of both trade and law

Besides involving laws, regulations, agreements and routines, International trade also involves many complicated procedures and agencies, except from the importers and exporters, transportation departments, insurance companies, customs, customs clearance agencies, banks, wharfs, commodity inspection agencies, governments and some other official departments, it also involves problems like language, currency, jet-lag, spatial difference and usual trade practices, even all kinds of international trade documents as well.

Currently, those enterprises participating in the import and export trade are mainly small and medium-sized foreign trade enterprises. Plenty of foreign trade enterprises especially the small ones don't have full-time lawyers for legal affairs, they don't have a clear understand of the laws and regulations of the trade partners, which leads to some conflicts during the time. Therefore, this requires the foreign trade practitioners to master a certain degree of laws and regulations. On the other hand, along with the improvement of our foreign trade and economy, international and national laws and regulations, the trade frictions of our country is becoming more and more obvious. In the face of the reality that our country is being repeatedly investigated by anti-dumping and anti-dumping lawsuit, there are nothing they can do but passively accept the unfair deal and suffer economic loss because of the anti-dumping duty, in the long run lose the overseas market. In order to actively respond to anti-dumping, high-level foreign trade talents who are familiar with the international practices and standards of the WTO are needed of course.

4 TEACHING REFORM MEASURES IN INTERNATIONAL TRADETEACHING REFORM MEASURES IN INTERNATIONAL TRADE

The foreign trade talents can not satisfy the current demand from enterprises, which fully demonstrate that there are many problems in the training mode

in the existing international trade major in Colleges and universities of our country. This requires that we should carry out teaching reform of international trade major according to the market demand for talents in international trade.

4.1 Adjust the curriculum

With the rapid development of our society, subjects are subdivided, some new subjects are emerging, and some subjects are developing with international trade. The courses to be set should be based on the understanding of the current development of China's international trade and overseas market. And then decide the courses needed to be phased out and reduced. We can learn from the training mode and curriculum from internationally renowned business schools. We should pay attention to the following aspects in the area of curriculum: 1) Increasing the percentage of the practical courses is good for the cultivation of students' practical abilities. 2) Attach more importance to courses of international laws and regulations, pay attention to the training of professionals who are familiar with the trade rules in WTO and international economic laws. 3) Enhance the internationality of the curriculum concerned about marketing, accounting and business management. We can furthermore introduce some original materials from the first-class foreign universities and update the contents of the curriculum in time. We should also train the marketing awareness and management awareness of the foreign trade talents. 4) Pay more attention to bilingual teaching in the purpose of improving the English level of the students and cultivate the ability of making direct consultations and signing deals with foreign dealers. 5) Expand the scope of international trade. We should not only establish some courses concerned trade in goods, but also relevant to trade in service. 6) Implement "double certificates" system. Students are required to obtain some certificates about foreign trade. Curriculum and research should be combined for the convenience of students' obtaining certificates.[2] 7) Increase the proportion of international cultural knowledge and increase cross-cultural understanding and comparison of cultural knowledge and learning in the purpose of broadening the students' horizons. 8) Create some minority language courses for students to learn and enrich the variety of languages that foreign trade talents use.

4.2 Reform the teaching methods

Good, appropriate teaching methods contribute to the improvement of teaching. There are two different types of knowledge in general knowledge. One is declarative knowledge; the other is the procedural knowledge. Declarative knowledge is acquired by the memorization of knowledge, dealing with the issue starting with "what". While procedural knowledge is acquired by practical operating, dealing with issues starting with "how". In conclusion, general knowledge not only includes the acquisition of knowledge, but also the ability of applying the knowledge. Traditional teaching methods ignore the differences between the two kinds of knowledge. They mainly convey some declarative knowledge to students rather than practical skills of applying knowledge. Thus, in order to compensate for the lack of procedural knowledge, on the one hand, universities should improve the conditions of international business laboratories. Laboratories should not only include the simulation of the entire procedure, but also include the simulation of marketing analysis and international business secretarial operations. On the other hand, it is truly different between real international trade and the simulation. Therefore, we should strengthen the link with relevant foreign trade enterprises and make connections with the construction of the workers. And then establish bases in the true sense. This on the one hand can train the practical ability of the students, on the other hand make the enterprises culture the students with a purpose and it's more convenient for the companies to select talents. It's a true "win-win" situation for both the enterprises and schools. [3]

4.3 Improve the quality of teachers

Teachers are the initiator of knowledge, the quality of students depends largely on them. Therefore, a high-qualified team of teachers is the vital point. Only with the help of the creative activities which are organized by high-qualified teachers with modern educational thinking and advanced educational conceptions can the amendments of educational programme be implemented smoothly. On the one hand, we should actively train teachers in service and frequently assign teachers especially young teachers to domestic and foreign key institutions and universities in the purpose of broadening their horizons, getting to know the advanced academic situations and updating their knowledge. Teachers are required to add the international advanced conceptions to their lessons and renew their knowledge all the time. Only when the students are aware of the latest and advanced knowledge can they better adapt to the current international situation. On the other hand, invite some foreign experts to give lectures or teach in a short term. Teachers from different cultural backgrounds and academic backgrounds can have different understanding and angles of investigations of the same knowledge. Colleges and universities should actively introduce talents who have come back home. They not only have international perspective, but also have a deeper understanding of its profession academically, so that advanced knowledge can be conveyed to the students.[4]

4.4 Improve the way of running school

In order to broaden students' horizon, schools should encourage students to step out of the campus and experience by themselves, only in which way can they get a substantive improvement. At present, many colleges and universities have companionship with each other on the issue of running schools. Domestic and foreign universities have various of activities like exchanging teachers, students and books. They set up summer camp, have mutual recognition of credits and jointly grant degrees. But for now, these activities can only involve a limited number of students and the time is comparatively short, the degree in teaching, research, student training and international integration is not enough. Universities increase the fiscal subsidies for students and the form of studying abroad is diversified. They send exchange students majoring in international trade abroad to study. For instance, they can promote joint training methods like "2 + 2" or "3 + 1".

4.5 Attach importance to humanistic education

With the widening range of people's activities, international trade connects people of different cultures, different religions and different moral standards together; therefore, ambiguities and conflicts are easily occurred during the process of communicating. Thus, moral education of students majoring in international trade should be enhanced in teaching and the differences of eastern and western knowledge should be realized. When in contact with customers, the concept of seeking common grounds should be reminded by heart, learn its merits and throw away its demerits. Students should be told to obey the common behavioral norms of human, be honest, be justice and fair in the process of trade, oppose bribery, never sell unqualified products and abide by the order of world economy. In the world of globalization, we should enhance education of students in patriotism and safeguard our national interests and informational security in economical activities.

REFERENCES

[1] Mingjie, Zhang Hongyan, The analysis on the types of demand and characteristics of our country's high-qualified talents, Foreign Trade Practice, February 2012.
[2] Shi Yan, Yu Shuang, Nie Dingchang, The analysis of status quo of China's foreign trade development and personnel needs. Business Times, October 2011.
[3] Weixia, The analysis of China's situation of needs and forstering for foreign trade talents, Collective economy. December 2012.
[4] Liu Yusong, Zhang Sisi, Xie Chaoyang, Countermeasures of the changing needs of higher education personnel in the time of post international financial crisis, Marketing Modernization, The second half of September 2010.

Education Management and Management Science – Zheng (Ed.)
© 2015 Taylor & Francis Group, London, ISBN 978-1-138-02663-6

Study on the capacity of the quantity of agricultural labor force in urban-rural area in China

Lifang Zhang, Ying Li & Chunyang Jiang
College of Quartermaster Technology, Jilin University, Changchun, Jilin, China

ABSTRACT: There is a great significance to the theory and practice by estimating rationally both the capacity of rural labor force in urban area and quantity of agricultural labor force in rural need to transfer. This paper discusses and improves a number of the calculate method related to the quantity of labor force. By using the improved method, the capacity of the quantity of labor force in urban area from 2004 to 2011 and the quantity of labor force the rural need to transfer from 1978 to 2011 are calculated, getting the properly rational results. When the comparison between the two results is completed, the conclusion shows that the quantity of the rural labor force the urban economic develop could hold comes close to the quantity of the agricultural labor force which would transfer in the situation of full employment in agriculture from 2004 to 2011. Considering that the information flow, energy flow and material flow in the whole transferring system run without any obstacle, it is feasible to shift the agricultural labor force in the situation of full employment in agriculture into urban, which is significant to the harmony in China's urban-rural development.

Keywords: quantity of agricultural labor force; urban area; rural area; China

1   INTRODUCTION

With the development of the reform and opening-up policy in these 30 years, there are great changes in both the urban and the rural. Therefore, the urban area can contain a large number of the agriculture surplus labor force with economic development in the urban zone. In this background, the paper uses many kinds of improved methods to estimate the quantity of agriculture labor force, which is very significant to the theory and practice about the estimation on the quantity of the labor force the city could contain and the rural need to transfer. Based on the scientific definition on the urban and rural area of the Temporary Provisions on Delimiting Town or City by Statistic, the paper takes the authentic data of statistical yearbook of China over the years to estimate. It makes sure that the estimate results are scientific and reliable. In addition to this, the conception of the agriculture labor force the paper uses is broad and is equal to the population in rural area.

2   THE ESTIMATION OF THE AGRICULTURE LABOR FORCE THE URBAN AREA COULD HOLD

Urbanization refers to the process that the rural populations assemble in the urban area. In fact, the urbanization degree shows the absorptive capacity and carrying capacity the urban area could hold of the rural population. It has been a common perspective in academia that the level of economic development gets behind the economic development. In other words, the paper could estimate the agriculture labor force the urban area in China can contain according to the gap between the urbanization degree the economic development should get and the actual urbanization degree in China. The formula is:

$$L_p = (U_y - U_s) \times P_t \qquad (1)$$

In this formula, L_p refers to the population the urban could be able to hold.

U_y and U_s respectively refers to the urbanization degree in the particular economic development and the urbanization degree in fact. P_t means the total nationwide population or the total regional population. Obviously, the key in this formula is to fix the urbanization degree in the particular economic development.

This study, based on the standard structure the American famous economist Chenery put forward to reflect the relation between the economic develop and the urbanization level, estimates the urbanization level China should get over the years. Table 1 is the data based on Chenery's calculating

Table 1. Chenery's correspondence table between the urbanization level and economic development.

GDP	<100	100	200	300	400
Rate	12.8	22	36.2	43.9	49
GDP	500	800	1000	<1000	
Rate	52.7	60.1	63.4	68.5	

Table 2. 2004–2011 China's per capita GDP by dollars in 1964 discounted data.

Year	GDP	Deflator	Discounted GDP
1964	83.93	100	83.93
2004	1490	494.3	301.44
2005	1732	510.72	339.13
2006	2070	527.24	392.61
2007	2652	542.53	488.82
2008	3414	554.62	615.56
2009	3749	559.42	670.16
2010	4430	566.88	781.47
2011	5432	578.96	938.23

Table 3. 2004–2011 the potential population capacity (unit: ten thousand, %).

Year	U_y	U_s	U_y-U_s	P_t	L_p
2004	43.9	41.76	2.14	129988	2782
2005	45.94	42.99	2.95	130756	3857
2006	48.59	44.34	4.25	131448	5585
2007	52.26	45.89	6.37	132129	8412
2008	55.54	46.99	8.55	132802	11351
2009	56.89	48.34	8.55	133450	11412
2010	59.63	49.95	9.68	134091	12982
2011	62.38	51.27	11.11	134735	14965

Table 4. 2004–2011 the quantity of the agriculture labor force the urban can accommodate (unit: ten thousand, %).

Year	L_p	R1	L_r
2004	2782	58.24	1620
2005	3857	57.01	2199
2006	5585	55.66	3109
2007	8412	54.11	4552
2008	11351	53.01	6017
2009	11412	51.66	5895
2010	12982	50.05	6498
2011	14965	48.73	7292

urban can still accommodate (L_r). The estimating formula is:

$$L_r = (U_y - U_s) \times P_t \times R_1 \qquad (2)$$

In the formula, R_1 refers to the proportion of the rural people in the total population. Through the calculation, the capacity of the urban area can hold is showed in Table 4.

Although this is a kind of estimation, the results of the calculation explain something, i.e., there are still 400 million to 500 million population who stay in the countryside, even though the China's urbanization level reaches above 70% in the future. Just like the Chairman Xi said, the development of the urban and rural must synchronize which makes them improve together. What's more, the countryside must not be the wasted area, must not be the rural zone full of the kids and the aged, must not be the district which we can only remember in dreams.

3 THE ESTIMATION ON THE QUANTITY OF THE AGRICULTURE LABOR FORCE THE RURAL AREA CAN ACCOMMODATE

In China, the China's situation of more people and less farmland decides that the cultivated land determine the demand of the agriculture labor force. Hence, the quantity of the agriculture labor force the rural area can accommodate can be calculated according to the existing cultivated land in the rural area.

This paper will firstly introduce the ChenXianyun's formula which calculates the rural surplus labor based on the cultivated land. The formula is:

$$\begin{cases} LS_t = L_t - A_t / E_t \\ E_t = 0.4966 (1 + \beta)^{(t-1952)} \end{cases} \qquad (3)$$

on the relation between the economic develop and the urbanization level among more than 100 countries in at least 20 years.

As the GDP per capita in Table 1 is the year of 1964. According to the comparability principle, this research converts China's GDP per capita in recent years into the dollars of the year of 1964, showed in Table 2.

According to the data in Table 1 and Table 2, it can be calculated the urbanization rate under the China's economic level over the years. Furthermore, the potential population capacity over the years will be calculated according to the formula 1, showed in Table 3.

This research takes a further study on the calculating the number of the agriculture labor force

This formula 3 takes the fixed term in 1952 as the fully utilized agriculture labor force. LS_t refers to the quantity of the rural surplus labor force in the T-th year. L_t means the actual quantity of the agriculture labor force in the T-th year. A_t refers to the actual cultivated area in the T-th year. E_t means the ideal per capita cultivated area ignoring the rural surplus labor in the T-th year. The average of per capita cultivated area is 0.4966 hectares from 1949 to 1957. β is the business cultivated area rate of change (described the influence the agriculture knowledge improvement have on agricultural productivity), through the calculating, β is 0.0018.

In the Chen's formula, A_t is the actual cultivated area. However, in the practice of the agricultural production, the quantity of the agriculture labor force the cultivated land resource could accommodate is related not only to the cultivated area, but also to the multiple cropping coefficient of the cultivated land. With the same size of cultivated area, different the multiple cropping coefficient demands different agriculture labor force obviously. Furthermore, the multiple cropping coefficient of the cultivated land changes with the improvement of the technology, which has influence on the quantity of employment capacity. Hence, GaiMei and

TianShicheng take the planting acreage of cropper instead of the actual cultivated area [1]. The revised formula is:

$$\begin{cases} LS_t = L_t - Af_t / E_t \\ E_t = 0.8143(1+\beta)^{(t-1957)} \end{cases} \qquad (4)$$

In the formula, Af_t is the planting acreage of cropper. The average (unit: hectare) of planting acreage of cropper is 0.8143 in the year from 1952 to 1957. The value forms on the assumption that the agriculture labor force before 1957 is all productive, i.e., assuming that there is no surplus labor force. This assumption made is based on the survey written by John Buck in 1920s. The survey explained that there were no surplus labor force in the agriculture in the first half of the 20th century in China. This assumption is also based on the situation, i.e., the reason why the rural surplus labor force exists is that the rapid agricultural population growth happened after the foundation of the People's Republic of China. However, it influenced little on the quantity of the agricultural labor force in 1957.

So the formula to calculate the quantity of the agriculture labor force the agriculture recourses

Table 5. The quantity of the agriculture labor force in the situation of full agricultural full employment.

Year	Af_t	E_t	LY_t
1978	150104	0.85	17750
1980	146380	0.85	17248
1985	143626	0.86	16772
1990	148362	0.86	17170
1991	149586	0.87	17280
1992	149007	0.87	17182
1993	147741	0.87	17006
1994	148241	0.87	17033
1995	149879	0.87	17190
1996	152381	0.87	17446
1997	153969	0.88	17596
1998	155706	0.88	17762
1999	156373	0.88	17806
2000	156300	0.88	17766
2001	155708	0.88	17667
2002	154636	0.88	17514
2003	152415	0.88	17231
2004	153553	0.89	17329
2005	155488	0.89	17516
2006	152149	0.89	17109
2007	153464	0.89	17225
2008	156266	0.89	17508
2009	158614	0.89	17740
2010	160675	0.9	17938
2011	162283	0.9	18085

Table 6. 1978–2011 the computation sheet of the quantity of the labor force the Chinese agriculture need to transfer (ten thousand).

Year	Agricultural workers	LY_t	Transfer quantity
1978	28318	17750	10568
1980	29122	17248	11874
1985	31130	16772	14358
1990	38914	17170	21744
1991	39098	17280	21818
1992	38699	17182	21517
1993	37680	17006	20674
1994	36628	17033	19595
1995	35530	17190	18340
1996	34820	17446	17374
1997	34840	17596	17244
1998	35177	17762	17415
1999	35768	17806	17962
2000	36043	17766	18276
2001	36399	17667	18732
2002	36640	17514	19126
2003	36204	17231	18973
2004	34830	17329	17501
2005	33442	17516	15926
2006	31941	17109	14832
2007	30731	17225	13506
2008	29923	17508	12415
2009	28890	17740	11151
2010	27931	17938	9993
2011	26594	18085	8509

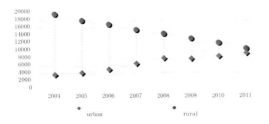

Figure 1.

(mainly refers to the cultivated land resources) could accommodate in the t-th year can be got according to the formula followed.

$$\begin{cases} LY_t = Af_t / E_t \\ E_t = 0.8143(1 + \beta)^{(t-1957)} \end{cases} \quad (5)$$

Table 5 shows the calculating results. According to it, the quantity of the agriculture labor force the agriculture needs to transfer can be calculated, as Table 6 indicates.

4 CONCLUSION

Table 4 and Table 6 are compared. It shows in the chart (Fig. 1).

In the chart, the under line mean the quantity of the agricultural surplus labor force the urban can still accommodate (ten thousand). The upper line mean the transferring quantity of the agricultural surplus labor force in the situation of the agricultural full employment (ten thousand).

The paper implies the conclusions by the comparison. As the urban economy develops rapidly after 2004, it creates much more jobs. Meanwhile, the higher salary in cities than the income of the agricultural production, the advanced urban public service facilities and more personal development opportunities attract much agricultural surplus labor force leaving the agriculture and countryside. The quantity of these years' transferring labor force results in the situation that in 2011 the quantity of the agriculture labor force the urban can accommodate comes close to the transferring quantity of the agriculture labor force in the agricultural full employment. Considering that the information flow, energy flow and material flow in the whole transferring system run without any obstacle, it is feasible to shift the agricultural labor force in the situation of full employment in agriculture into the urban, which is significant to the harmony in China's urban-rural development.

However, it is appropriate to take the vision of development to judge the final the transferring effect, as the per capita crop acreage when the labor force are fully utilized is based on the data of calculating the level of Chinese agricultural modernization in those years. With the continuously acceleration of the Chinese agricultural modernization process, the trend of the per capita crop acreage on condition that the labor force are fully utilized will expand and the quantity of the agricultural surplus labor force will be corresponding to increase. According to the HeChuanqi [2] and the other's research, with the continuously improvement of the Chinese agricultural modernization level in the next 40 years, more than 200 million peasants will be transferred from the agriculture. And the total quantity of the agriculture labor force will decrease to 31 million by that time. It will be a long and tortuous course to transfer the agricultural labor force from the rural area to the urban zone.

REFERENCES

[1] Mei Gai & Chengshi Tian 2010. The calculation on the quantity of the effective employment in the Chinese rural area. Statistics & Decision. 2:4–6.
[2] Chuanqi He. 2012. The report of Chinese modernization 2012, the study on the agricultural modernization. Beijing: Peking University Press.

Education Management and Management Science – Zheng (Ed.)
© 2015 Taylor & Francis Group, London, ISBN 978-1-138-02663-6

Empirical study on the operating performance of Chinese military quoted companies

Lifang Zhang & Zhichao Liu
College of Quartermaster Technology, Jilin University, Changchun, Jilin, China

ABSTRACT: This paper focuses on the operating performance of Chinese military quoted companies by using company financial data of 2012.The research uses factor, cluster analysis methods to take an in-depth analysis and discussion from 30 military quoted companies. Finally, conclusions are obtained as follows: the overall performance of Chinese military quoted companies is good, but six main factors' ranking of 30 military quoted companies are very unstable, including income factor, debt paying factor, equity factor, main business profit factor, growth factor, expansion factor, and defects are particularly obvious in the asset structure, profitability and expansion. Military quoted companies not only have characters of securitization and marketization, but also have feature of servicing defense and enjoying together as public goods. Therefore, the paper could provide advices from perfecting main business layout of military quoted companies, meeting demands of army and people, improving profitability, strengthening consciousness of market expansion and optimizing corporation management level. This will give a reference for improving performance of military quoted companies.

Keywords: military quoted company; operated performance; factor analysis; cluster analysis

1 INTRODUCTION

As an important component of national defense industry, the traditional military industry groups have not been able to cope with increasingly serious international economic and political environment. Therefore, China promulgates a series of policies to consolidate and improve the competitiveness of industry groups, which gradually integrate into military quoted companies which will promote collaborative development of defense industry and the securities market.

Military quoted companies are organic combination of military defense industry and securities market, also inevitable result of the development of military enterprises. Consequently, relevant research about military quoted companies has very great significance for national defense, economy. Existing research is mainly manifested in two aspects about qualitative and quantitative analysis, especially focusing on capital structure, evaluates management of military quoted companies, but lacks of overall performance evaluation. The transformation of joint-stock Chinese military quoted companies has achieved certain results. But because of characteristic of military quoted companies, there is still a series of problems in the operation and management that lead to poor in some military listing Corporation's management

performance. Therefore, it is great theoretical and practical significant for promoting harmonious and healthy development of national defense, economy to solve problems of operating Chinese military quoted companies performance. Consequently, this paper chooses 30 military quoted companies which take military products for the main business, belonging to 10 military groups, as the research object and chooses its operating performance as core, uses factor analysis, clustering econometric methods to analyze 2012 financial data which represent management of military quoted companies at this stage.

2 FACTOR ANALYSIS ON MILITARY QUOTED COMPANIES BUSINESS PERFORMANCE

Business performance of military quoted companies is reflected by financial statement data, but relevant financial index is too many and diverse to carry out effective comments. Therefore, choosing appropriate econometric methods will directly determine military quoted companies' performance evaluation. Factor analysis, which changes perplexing variables into a few principal factors, evaluates performance of military quoted companies effectively.

2.1 Samples and index selection

Considering data's comparability, validity, consistency, and some related factors of Chinese enterprise performance evaluation, the paper chooses 15 financial indicators of military quoted companies as samples by factor analysis. These are main business profit rate (X_1), Rate of Return on Total Assets (X_2), rate of return on net assets (X_3), per share earnings ratio (X_4), asset-liability ratio (X_5), current ratio (X_6), quick ratio (X_7), turnover of total capital (X_8), inventory turnover rate (X_9), average accounts receivable turnover ratio (X_{10}), total asset growth rate (X_{11}), growth rate of main business income (X_{12}), net profit growth rate (X_{13}), net assets per share (X_{14}), Per share fund (X_{15}). At the same time, 30 military quoted companies are named as 1 to 30 to facilitate analysis of consolidation.

2.2 Factor analysis

2.1.1 Normalization and descriptive statistics

It can be seen that the 30 military quoted companies X_1 average is about 0.2341, standard deviation is about 0.2540, profit of main business don't exist large difference; X_1, X_2, X_3, X_4 index minimum value is negative, profit of several military quoted companies is weak; mean value of X_5 is about 0.4495, X_6 value is about 2.0918, X_7 average is about 1.4297, paying ability of military quoted companies is well; the mean value of X_8 is about 0.7097, turnover is slow and needs to be strengthened; X_{11}, X_{12}, X_{13} minimum value is negative, X_{13} was negative, military quoted companies present negative growth in 2012.

2.1.2 Applicability test

KMO test value is 0.6220 and could carry out factor analysis; Bartlett ball test value is 0.0500 and suitable for factor analysis.

2.1.3 Extracting common factors

The paper uses principal component analysis method to extract 6 main factors. Cumulative contribution rate is 89.34%, so, these main factors could reflect data features well.

2.1.4 Naming factors

Factor 1 has a larger load in X_8, X_{10}, X_4, X_2, X_3, X_{12} variables, named for assets income factor (F_1); factor 2 has a larger load in X_7, X_6, X_5 variables, named for debt paying factor (F_2); factor 3 has a larger load in X_{15}, X_{14} variables, named for equity factor (F_3); factor 4 has a larger load in X_1 variable, named for main business profit factor (F_4); factor 5 has a larger load in X_9, X_{11} variables, named for growth factor (F_5); factor 6 has a large load in X_{13} variables, named for expansion factor (F_6).

2.1.5 Calculation of factor scores and explanation results

It is necessary to calculate factor scores and make a comprehensive evaluation of military quoted companies' performance after determining 6 common factors. According to the corresponding contribution rate of 6 common factors, comprehensive performance score is as follow.

$$F_1 = -0.061X_1 + 0.131X_2 + \cdots - 0.038X_{15} \quad (1)$$

$$F_2 = -0.079X_1 + 0.045X_2 + \cdots - 0.065X_{15} \quad (2)$$

$$F_3 = -0.062X_1 - 0.018X_2 + \cdots + 0.544X_{15} \quad (3)$$

$$F_4 = 0.654X_1 + 0.148X_2 + \cdots - 0.055X_{15} \quad (4)$$

$$F_5 = -0.045X_1 + 0.042X_2 + \cdots - 0.099X_{15} \quad (5)$$

$$F_6 = -0.175X_1 - 0.044X_2 + \cdots + 0.016X_{15} \quad (6)$$

$$F_A = 0.2846F_1 + 0.2302F_2 + \cdots + 0.0952F_6 \quad (7)$$

Factor scores of 30 military quoted companies, based on these equations, are obtained.

For factor index, comprehensive performances of 30 companies are good, and just only comprehensive score of Baoding Tianwei Baobian Electric Co. is negative. There are also differences in the factor index. Asset return ability is strong, and especially China Aerospace Science & Industry Corp is best in 30 military quoted companies. Solvency is more balanced and only factor score of Baoding Tianwei Baobian Electric Co. is negative. Benefit of military quoted companies is good and just only factor score of 3 companies is under zero. Factor score of main business profit is low and only Baoding Tianwei Baobian Electric Co. is positive. The overall performances of growth ability are ordinary: the best company is Taiji Computer Corp, factor score of 28.105, the worst company is Baoding Tianwei Baobian Electric Co., score of −2.0632. Expansion ability is poor and only expansion factor scores of HT-SAAE and Yunnan Xiyi Industry Limited by Share Ltd is positive.

For military quoted companies, various indicators are unbalanced. China Aerospace Science & Industry Corp, Baotou Beifang Chuangye Co., Ltd. and Taiji Computer Corp are the top three in the comprehensive performance scores. China Aerospace Science & Industry Corp is the first one in F1; Baotou Beifang Chuangye Co., Ltd. Is the first one in F_3; Taiji Computer Corp is the first one in F_2 and F_5. But in the meantime, Taiji Computer Corp is also the last one in F_3, F_4, the bottom second in F_6 and the third from the last in F_1. China Aerospace Science & Industry Corp, which faces the same dilemma, is the bottom second in F_4 and the third from the last in F_3, F_6. This will restrict the further growth of military quoted companies.

Table 1. Comprehensive evaluation.

C	F_1	F_2	F_3	F_4	F_5	F_6	F_A
1	1.87	0.11	3.05	−1.27	−0.59	−1.15	0.67
2	0.80	1.22	5.02	−1.42	−0.33	−1.35	0.90
3	2.45	0.66	2.21	−2.32	−0.10	−1.62	0.72
4	1.30	0.92	2.20	−1.35	0.65	−1.44	0.68
5	1.72	2.06	4.76	−2.55	0.62	−2.47	1.19
6	1.83	1.44	0.23	−3.13	3.32	−2.29	0.68
7	−0.20	0.58	1.46	−3.97	0.47	7.77	0.60
8	2.81	0.09	2.24	−1.04	−2.06	−1.14	0.67
9	11.46	2.85	−0.80	−11.4	2.78	−6.99	2.07
10	1.39	2.52	2.76	−1.66	0.03	−1.59	1.03
11	1.58	0.68	0.77	−1.68	1.05	−2.20	0.43
12	1.46	0.49	2.66	−1.31	1.18	−3.77	0.54
13	2.26	0.33	4.31	−1.74	−0.76	−1.24	0.93
14	3.37	1.11	2.75	−3.73	0.86	−2.36	1.04
15	3.19	1.09	6.03	−4.24	1.45	−3.02	1.41
16	1.90	0.80	0.80	−0.48	−1.12	−2.20	0.44
17	2.93	0.87	1.67	−2.98	0.45	−2.28	0.75
18	1.10	0.93	0.38	−1.98	−0.11	1.07	0.43
19	0.73	1.65	2.12	−2.53	4.53	−5.27	0.63
20	1.58	−0.07	4.60	4.05	3.33	−25.2	−0.39
21	1.37	1.58	0.27	−2.97	2.68	−1.27	0.63
22	4.07	0.32	−0.84	−4.18	1.96	−3.64	0.49
23	1.29	1.27	3.39	−2.60	2.51	−2.25	0.93
24	2.26	3.05	1.30	−2.79	1.27	−1.91	1.17
25	0.70	4.42	−4.24	−13.6	28.02	−10.6	1.29
26	1.11	0.60	1.93	−1.30	1.16	−1.57	0.57
27	0.39	2.49	1.31	−0.99	0.98	−2.25	0.66
28	5.33	0.39	0.83	−4.31	−0.89	−2.47	0.86
29	1.05	0.89	1.45	−0.99	1.32	−4.39	0.34
30	1.08	1.21	2.76	−1.24	0.15	−1.34	0.73

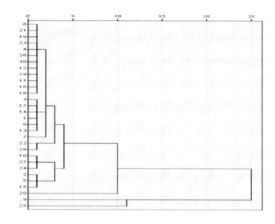

Picture 1. Dendrogram.

3 CLUSTER ANALYSIS OF MAIN FACTORS

The basic idea of cluster analysis about military quoted companies is converging, based on correlating degrees of 6 main factors. The total number of samples is 30 and valid samples are also 30, efficiency sample 100%. When 30 companies are divided into two parts, Taiji Computer Corp And China Aerospace Science & Industry Corp are in a group and the 28 companies are in the other one. When 30 companies are divided into three parts, Taiji Computer Corp is in a group; China Aerospace Science & Industry Corp is in a group and the 28 companies are in the last one. When 30 companies are divided into four parts, Taiji Computer Corp is in a group; China Aerospace Science & Industry Corp is in a group; Baoding Tianwei Baobian Electric Co. is in a group and the 27 companies are in the last one. Cluster analysis results are consistent with the results of factor analysis,

and they also have the analogous tendency in comprehensive factor scores.

4 EMERGENCING PROBLEMS

4.1 Main business profitability is poor

At present, main business profitability of Chinese military quoted companies is poor, but this disadvantage has no effect on overall business performance of military quoted companies. Owner's equity of 30 companies is positive and net profit of 24 companies is positive. This indicates that non-main business profits of military quoted companies make up for deficit of military quoted companies. But this situation does more harm than good. First of all, military quoted companies are the foundation of Chinese national defense industry, which is significant for national security. If these companies could not promote benign development of the defense industry, management of military quoted companies will also be affected seriously. Secondly, military quoted companies are also the combination of national defense industry and stock market. Although our country has absolute control, management of military quoted companies is influenced by CSRC and market. Because of poor operating performance, several military quoted companies have been stopped sign or reorganized of assets. It is not difficult to see that profitability of main business plays a vital role in the management of military quoted companies.

4.2 Expansion ability is poor

Our country holding controlling interest is a double-edged sword. When country promotes military quoted companies develop at the same time, it also makes management of military quoted companies more passive. Companies could not take the initiative to meet the market challenges. In the end, the expanding capacity of military quoted companies is poorer and poorer in the evaluation of performance. The key reason is development time of military quoted companies is rather short, and comparing with quoted companies of another industry, ability of financing is weak, management is not strong.

5 COUNTERMEASURES AND SUGGESTIONS

5.1 Optimizing main business layout and improving profitability

Main business of military quoted companies is military products and civilian industry products are supplementary. But domestic supplying channels for military products are extremely limited and a small amount of exportation has been restricted by national security, core technology security and other factors. These reasons restrict the development of military quoted companies, which directly lead to poor performance. Consequently, strengthening Civil-Military Integration, promoting technology transformation between military industry and civil industry and motivating coordinated development of overall industry is very significant for optimizing main business layout and improving profitability. First of all, we should plan civil military integration strategy, encourage and support to optimize industrial layout of military quoted companies, proceeding in an orderly way and step by step. Secondly, strengthen legal construction, promote civil-military integration strategy, support civil-military technology transfer and intensify industrial layout of military quoted companies. Lastly, military quoted companies also should establish a suitable management mechanism for military and civil industry and improve profitability of military quoted companies effectively.

5.2 Enhancing the expansion of consciousness and improving the level of market

In order to enhance expansion of consciousness and improving market level, we had better improve policies and strengthen management of military quoted companies. Because of the specificity of military quoted companies, policy support is very significant and necessary. But the national policy should be put in effect reasonably and moderately under the premise of safeguarding national security and following market economy. First, policy support is not food prepared in a large canteen

cauldron. We should choose the effective and competitive military quoted companies to support with the hand. Secondly, support should adhere to the principle of compensation and indirection and encourage military quoted companies to develop in the long term. Military quoted companies undertake the task of guarantee national security, increase economic benefit. But at this stage, managers of state-owned enterprise are too conservative to participate in management activities in military industry listing Corporation actively. In order to realize reunification of national defense, economic benefits, introducing marketing management ideas and breaking the shackles is of great urgency. Establish performance-related pay and stock-compensation system and motivate potential management to create economic benefits. At the same time, improve the management system of responsibility with accountability.

6 CONCLUSIONS

Development time of Chinese military quoted companies is not long and the foreign useful precedent are also very few. So, managers face many problems in the management process. If we could solve these problems properly, these troubles will constrain the development of military quoted companies and military industry. The overall performance of Chinese military quoted companies is good, but six main factors' ranking of 30 military quoted companies are very unstable, including Income Factor, Debt Paying Factor, Equity Factor, Main Business Profit Factor, Growth Factor, Expansion Factor, and defects are particularly obvious in the asset structure, profitability and expansion. Consequently, optimizing military quoted companies, improving profitability, enhancing expansion of consciousness and improving the level of market of industry is the key way to achieve the goal of enhancing performance, realize the organic integration of industry and securities market and national defense construction and economic construction ultimately.

REFERENCES

[1] Denis, D.J. & T.A. Kruse. 2000. Managerial Discipline and Corporate Restructuring Following Performance Decline. Journal of Financial Economics. 33.
[2] Harold Demsets & Belen Villalonga. 2001. Ownership structure and eorporate Performance. Joumal of Corporate Finance. 209–233.
[3] Brown L.D & Caylor M.L. 2006. Corporate govemance and firm valuation. Journal of Accounting and Public Policy. 25(4):409–434.

Education Management and Management Science – Zheng (Ed.)
© *2015 Taylor & Francis Group, London, ISBN 978-1-138-02663-6*

Financial performance evaluation on medicine-listed companies based on factor analysis method

Qing-Dong Li & Jia-Hui Lin
School of Economics and Management, Liaoning Shihua University, Liaoning, P.R. China

ABSTRACT: The paper chooses thirty-one indexes from corporate profits quality. The data is taken from balance sheet, income statement and cash flow statement in 2012. Data analysis utilizes factor analysis method. At the same time, the thesis evaluates the medicine-listed industry in China in 2012 from factors and corporations sides. The result of research finds that the factor of asset utilization has much effect on the medicine-listed industry of China.

Keywords: financial performance; factor analysis method; medicine-listed industry

1 INTRODUCTION

Physiological need is the basic demand in the Maslow's need hierarchy theory. Health is one of the contents of physiological need. Health care is carried out by the corporation of medicine industry, hospital and government. The industry of medicine has close links with the health of people. The operating situation of the corporation of medicine industry of China is closely connected to our people's health. How does the corporation of medicine in China operate in the current situation? This paper evaluates the listed corporation of medicine in China in 2012.

2 DEFINITION OF FINANCIAL PERFORMANCE

In the early 20th century, Frederick Taylor began to do his research on performance in the Principles of Scientific Management. The word of performance was definitely raised by Burmbrach in 1988. There are four viewpoints in the present document. The first viewpoint is action. Murphy believes that performance is action that reflects the aim of person and organization. Kane argues that performance is actions that are different from results, because results are influenced by system factors[1]. The second viewpoint is result. Abel considers that performance records result on special time, specific job function, activity and behavior[2]. The third viewpoint is process adding to result. In this view, performance is a mirror that responds to the completion of project. It reflects economy, efficiency, effectiveness and comfort level of society demand[3]. The forth viewpoint is a kind of ability. Spencer thinks that ability balance potential

personality traits. The article is positioned in the third viewpoint. Financial performance is one of performances, which shows by the indexes of finance. It includes effect, efficiency and benefit. Effect sums the quantity and quality of useful results. Efficiency is a scale of inputs and outputs. Benefit determines the completion of the intended target and economic benefits[4].

3 BASIC THEORY OF FINANCIAL PERFORMANCE EVALUATION

3.1 *Index choice*

The paper selects index from viewpoint of corporate profits quality. Profit can measure management results of enterprises in a certain accounting period. Quality of profit not only embodies the forming of profit of enterprise, but also measures compliance, profitability and fairness of profit results. It's easy to see that profit is different from quality of profit[5].

3.2 *Selection of evaluation methods*

Nowadays, there are all kinds of methods on enterprise performance evaluation, such as EVA evaluation method, Balanced score card and Fuzzy Comprehensive Evaluation. But these methods are affected by subjective factors. It will be to comparison in different enterprises in the same industry disadvantage. Meanwhile a number of samples are hard to deal with. However, we use the factor method to overcome the above disadvantages. The factor method that applies the idea of dimension reduction is one of multivariate statistical analysis methods. It is able to turn relationship between complex variables into a handful of comprehensive factors[6].

4 PRACTICAL ANALYSIS

4.1 Data sources

The paper excludes some companies that corporation of medicine industry were special treatment. In addition, the listed companies with missing data are ruled out. Finally, the paper chooses 120 corporations of medicine industry. The practical analysis is based on the financial performance evaluation indexes, which are put forward in the above paper, on the basis of the Balance sheet, Income statement and Cash flow statement in 2012. Data analysis of the article applies SPSS21.0 Statistical Analysis Software.

4.2 Factor analysis application

4.2.1 Sample data processing

Main business cost rate, liabilities and owners' equity ratio, equity ratio and asset-liability ratio are contrary index. The rest of 27 indexes are forward index. It turns contrary index to forward index through the efficacy coefficient method in order to ensure the accuracy of the results.

4.2.2 Factor analysis rightness testing

KMO of sample data is 0.725. Commonly used KMO measure standard is 0.7. Because 0.725 is greater than 0.7, it can apply the factor method[7].

4.2.3 Extracting the public factors

The article extracts nine factors to describe financial performance evaluation on the listed corporation of medicine, which is based on two standards-Characteristic Root values and the cumulative variance contribution rate.

4.2.4 The factor named

These factors are named after indexes that are high loading in rotating component matrix.

4.2.5 Comprehensive evaluation model

The paper constructs the comprehensive evaluation model of financial performance evaluation on the listed corporation of medicine.

$$F = 0.17826F_1 + 0.13623F_2 + 0.12266F_3 \\ + 0.11312F_4 + 0.06511F_5 + 0.06313F_6 \\ + 0.05624F_7 + 0.04997F_8 + 0.04272F_9$$

4.2.6 Results and analysis

4.2.6.1 Data analysis principle

Financial performance of corporation performs well when score of financial performance is greater than zero. If the score of financial performances equal or close to zero, financial performance of corporation is in the general level. Financial performance of corporation is relatively poor if the score of financial is less than zero.

4.2.6.2 General results observation

According to component score coefficient matrix and comprehensive evaluation model, the paper arrived at the ranking of every factor and the general observation on financial performance of the public corporation of medicine industry.

Double-aigrettes pharmacy, Staidson BioPharmaceuticals, Squire pharmaceutical, Salubris, The constant group, Dong'e E-Jiao Group, Yunnan Baiyao, Watson, Qianhong Bio-pharma and Hualan Biological rank in the top ten corporations with the financial performance. In general, there are fifty-six corporations whose score of financial performance comprehensive is greater than zero, accounting for 47.11% of the total. The rest of the financial performance comprehensive score is less than zero, accounting for 52.89% of the total. The highest score is 0.9 and the lowest score is −0.70269.

Table 1. Factor named.

Factor	Indexes that is high loading	Factor name
F_1	Return on total assets, rate of return on total assets, net assets income rate	Asset utilization effect factor
F_2	Main business cost rate, main business profitability	Main business profit factor
F_3	Operating cash flow to sales ratio, operating cash flow and debt ratio, cash flow ratio	Cash flow factor
F_4	Liabilities and owners' equity ratio, equity ratio, asset-liability ratio	Debt levels factor
F_5	Shareholders' equity and fixed assets ratio	Owners of capital factor of fixed assets
F_6	Growth rate of gross operating income, net profit growth rate	Growth ability factor
F_7	Rate of net value of fixed assets	Operating conditions factor
F_8	Growth rate of net asset	Asset value factor
F_9	Multiple interest payments	Debt pressure factor

On the whole, development on the financial performance on listed corporation of medicine is not balanced and the level of performance is not high.

4.2.6.3 Analysis from the factor viewpoint

The asset utilization effect factor reflects the enterprise asset utilization ability. In this factor, fifty-eight enterprises achieve the score greater than zero, and the rest achieve less than zero. The highest score is 2.46389 and the lowest score is −2.32874. The figures show that the enterprise asset utilization ability of corporation is poor. There are fifty-six enterprises' comprehensive score of financial performance is greater than zero. It plays very important role to the financial performance on the listed corporation of medicine industry of China.

Main business profit factor measures main business profit ability. Meanwhile it reflects the main source of corporate profits. If a large of profit is achieved by main business, the profit source of corporation is steady. In this factor, there are seventy-one corporations get the score larger than zero. From the data analysis, the main profit was achieved by main business to most corporations of China. Watson, Squire pharmaceutical, whose score of the factor is larger than 2, are at the top of comprehensive score on financial performance.

From cash flow factor, we can judge the ability to collect cash, while we can learn the profit quality of corporation. If a corporation achieves the number of profit, however, no cash flowing, it faces the risk of bad debts. From the perspective of audit, the truth of corporation profits was explored. In this factor, there are forty-four corporations achieve the score greater than zero, the rest of the score is less than zero. It illustrates that the ability of collecting cash has yet to be improved.

Debt levels factor indicates a business loan. There are seventy-three corporations' score of the factor is greater than zero. Most companies funded by borrowing are more than 50% of the enterprises. Owners of capital factor of fixed assets value describes what percentage of the source of the purchase of fixed assets would come from owner's equity. In the factor, there are fifty enterprises' score is greater than zero, the rest is less than zero. It explains the capital in purchasing fixed assets mainly comes from borrowing. Growth ability factor weighs the growth ability of an enterprise. Only a corporation's score is greater than eight. The lowest score of the factor is −1.72569. It shows that the corporations of medicine stand in different stages of development. Operating conditions factor measure the operating situation of an enterprise. There are fifty-four corporations' score greater than zero, which implies that half of the enterprises operation conditions are better. The asset value factor weighs

value of an enterprise asset. The score of fifty-three enterprises are greater than zero. It shows that our country enterprise in the different development levels. And it turns out to be the correctness of the above factor analysis. Debt pressure factor balances that an enterprise will face debt pressure. In the factor, there are fifty-two enterprises' score greater than zero. Facing different debt pressure, listed corporation locates in various stages.

4.2.6.4 Analysis from enterprises viewpoint

Double-aigrettes pharmacy, Salubris, The constant group and Dong'e E-Jiao Group whose comprehensive score rank in the top ten with the financial performance locate in the top ten corporations with the asset utilization effect factor, which demonstrates that the comprehensive utilization of assets have a significant effect on financial performance of corporation. But debt levels factor distributes middle or back in 120 corporations. In the other factors, distributions of these enterprises are unbalanced. In the asset utilization effect factor, Qianhong Bio-pharma locates in 116th, which shows that the comprehensive utilization of assets has yet to be improved. In cash flow factor, Staidson BioPharmaceuticals stands in 111th. The enterprise should strengthen collection of cash. The constant group is in 116th in owners of capital factor of fixed assets value, which indicates most capital is from loan. Hualan Biological is in 116th in growth ability factor, which suggests that the enterprises adjust business strategy in order to improve enterprises development potential.

Most enterprises are in the middle or lower levels in the ranking of asset utilization effect factor, main business profit factor and debt levels factor. Less than or equal to 50% corporations are at the rear in owners of capital factor of fixed assets value, growth ability factor, operating conditions factor and debt pressure factor. But some factors' score is at the top of 120 corporations. In the cash flow factor, Shanghai Furen is in 7th, which suggests collection of cash the corporation is well. In owners of capital factor of fixed assets value, Sea King biomedical is in the 9th, which shows that most capital purchasing fixed assets are from owner's equity. So it doesn't face huge debt pressure. In asset value factor, North China Pharmaceutical is in the first place, which shows that the corporation has great potential for development.

5 CONCLUSION

The paper conducts exploratory factor analysis from thirty-one indexes. Finally, it extracts the nine factors, including Asset utilization effect factor, Main business profit factor, Cash flow factor, Debt levels factor, Owners of capital factor of fixed assets value, Growth ability factor, conditions factor, Asset value factor and Debt pressure factor. The level of financial performance of on listed corporation of medicine industry is uniform from the data analysis results. Therefore, the listed enterprises of medicine industry learn from each other. At the same time, learn the experience of foreign pharmaceutical enterprises. In the end, the listed corporations of medicine industry enhance the overall competitiveness and serve the people better.

AUTHOR INTRODUCTION

1. LI-Qingdong, Doctoral Candidate, Associate professor. School of Economics and Management, Liaoning shihua University. P.R. China, 113001.
2. LIN-Jiahui, Master Candidate, School of Economics and Management, Liaoning shihua University. P.R. China, 113001.

REFERENCES

[1] Guo Yulian. Index System Research on Local Public Health Expenditure Project Performance Evaluation [Master Thesis]. Wuhan University of Technology. 2008.
[2] Lou Jianshe. Financial Performance Evaluation Reform Research of Vocational College in Innovation of Financial Management. Market of China, 2013, (9).
[3] Zhang Hongjing, Hu Xiaomeng. Financial Performance Evaluation Research of Listed Companies GEM based on Factor Analysis. Times Finance, 2013, (8).
[4] Li Qingdong. Financial Performance Evaluation on Listed Corporation based on Principal Component Analysis. Journal of Liaoning University of Petroleum Chemical Industry, 2006, (9).
[5] Meng Tongyu. Financial Performance Evaluation of Virtual Enterprises. International Business Accounting, 2013, (1).
[6] Wei Wei. Financial Performance Research Summarize on Enterprise Social Responsibility. Charming China, 2010, (6).
[7] Liu Xueqing. Financial Performance Evaluation System Construction on Social Responsibility of Listed Corporation. Special Zone Economy, 2010, (4).

Education Management and Management Science – Zheng (Ed.)
© 2015 Taylor & Francis Group, London, ISBN 978-1-138-02663-6

Discussion on physics teaching innovation: Taking Coulomb's law as an example

W. Shao, B. Jiang & J.K. Lv
Aviation University of Air Force, Changchun, China

ABSTRACT: Taking Coulomb's law as an example, the physics teaching shall be explained, during which the forming process of physical laws shall be highlighted, and students' scientific accomplishment and learning ability can be further improved.

Keyword: electric charge physical laws Coulomb's law

1 INTRODUCTION

In traditional physics teaching, a special attention is paid to physical conception, and a summary of physical laws, and problem solving by extrapolation with the use of laws; less attention is paid to the connotation laws of physical conception in the teaching process, causing that most students superficially understand physical laws, namely content and expression of physical laws. They use the formula to settle problems but ignore the study and understanding on connotation and essence of physical laws[1]. And today physics teaching will be talked about, taking Coulomb's law as an example.

Coulomb's law: acting force between two stationary point charges in vacuum is directly proportional to the absolute value of the product of electric quantity brought by the two point charges, and inversely proportional to the square of the distance between them, and the direction of acting force is along the two point charges.

2 OBSERVING PHENOMENA AND PUTTING FORWARD THE PROPOSITION PREPARING THE NEW FILE WITH THE CORRECT TEMPLATE

As for knowledge about electrics, people know about electric charge at the earliest. When talking about the electric charge, students all know: there is positive and negative charge; charge is quantized; charge has conservativeness; charge is relativistic invariant and so on. As the most fundamental and important particle, electric charge naturally becomes the starting point for us to beginning researching electrics. The first physical law of electromagnetism section-Coulomb's law is put forward in the process of studying interaction of electric charge.

3 GUESSING THE ANSWER AND THEORETICAL PREDICTIONS

The famous electrician Franklin (B. Franklin, 1706–1790) once observed that cork ball in the metal cup was completely unaffected by the electric charge on the metal cup. He wrote a letter for Priestley and told his observations, expecting him to redo this experiment, and confirm this fact. In 1766, Priestley did the experiment, and he electrified the metal cavity vessel and found there was no electric charge on its inner surface, and obviously there was no acting force of metal vessel on electric charge inside. He immediately thought that this phenomenon was very similar to universal gravitation situation, namely the matter within the homogeneous material spherical shell will not be subject to the acting force of the shell material. Thus, he guessed that electric power and universal gravitation are applicable to the same rules, and acting force between two electric charges shall be inversely proportional to the square of the distance between them. This is a very important analogy speculation, but the speculation at that time did not attract the attention of scientists. At the same time, Priestley himself had no confidence in strictly proving this speculation, so this finding was shelved.

4 DESIGNING EXPERIMENT FOR MEASUREMENT, MODIFYING THE THEORY, FINDING THE RULE AND FORMING THE LAW

In 1769, Robinson from Edinburgh (J. Robinson, 1739–1805) first confirmed law of electric power in the manner of direct measurement and he got repulsive force of two homo-charges, which is inversely proportional with 2.06 power of distance. The attraction of two hetero-charges is a little bit smaller than the inverse square power, so he concluded that the correct law of electric power is the inverse square law. His research results were published in 1801 and got known by people.

In 1772, the famous British physicist Cavendish (H. Cavendish, 1731–1810) followed Priestley's idea to verify electric power inverse square law by experiment. If the inner surface of charged cavity conductor is confirmed to be really free from electric charge by experiment, electric power complying with inverse square law can be determined. Quantitative results obtained from Cavendish's experiment are equal to the results directly measured with torsion balance by Coulomb (A. Coulomb, 1736–1806) 13 years later (1785). However, Cavendish's experiment is groundbreaking work, which exactly verifies electric power inverse square law. In the following 200 years, since the experiment is constantly repeated and improved, accuracy is greatly improved. Cavendish is an unsociable and eccentric scientist, who is dedicated to academic research and disregards scholarly honor or official rank, and his many research results haven't been published. 100 years later, Maxwell (C1Maxwell, 1831–1879) arranged a large number of his manuscripts and made the above mentioned results known to the world.

The most famous is the work of French physicist Coulomb. Coulomb has once been engaged in research of hair and wire torsion elasticity, which contributed to his invention of the torsion balance later called the Coulomb balance in 1777. In 1784, Coulomb published thesis and introduced the relationship between twisting force and wire diameter, length, twisting angle and other constants related to the wire physical properties he found as well as the method of measuring weak force with torsion balance. In the same year, in response to the rewarded solicitation of researching marine compass by French Academy of Sciences, Coulomb began to shift his scientific life from engineering and construction into electric and magnetic studies. In 1785, Coulomb designed and made a precise torsion balance, and proved that repulsive force of homo-charge complied with the inverse square law by torsion balance experiment and repulsive force of hetero-charge also complied with the inverse square law by vibration experiment His experimental squared error deviation is 4×10^{-2}. Coulomb verified the inverse square relationship between electric attraction and repulsion by experiment, and his work had been widely recognized, and he was named after the electric power law[2]- Coulomb's law.

5 INVESTIGATING ESTABLISHMENT CONDITIONS OF LAW

The establishment conditions of Coulomb's law consist of three points: point charge, vacuum and static, which are not all necessary conditions.

5.1 Point charge is an ideal model, which is an establishment condition but it is not limiting condition

Only the distance between the two charges has exact meaning but Coulomb's law can also be applicable to point charge group or continuously charged body with the use of superposition principle. And at this time they must be broken down into a point charge for consideration. Electrostatic force between two stationary charged bodies relies on the size and shape of charged body and distribution of electric charge besides quantity and relative position. When the dimension of the two charged bodies is much smaller than the distance between them, the problem will be greatly simplified, namely regarding the charged body as point charge. However, there is no absolute standard for how much smaller the dimension of charged body is than the distance for treating the charged body as point charge. It depends on the accuracy required for discussing the issue.

5.2 Vacuum condition is neither establishment condition nor limiting condition

It aims to ignore the impact of other electric charges, making the two point charges only interacted by each other. However, if a conductor or a dielectric exists in space, they will also have an effect on the two charges. At this time, the superposition interaction among all electric charges, such as electric charge in vacuum, inductive charge of conductor and polarization charge of dielectric medium shall be taken into consideration, so the total force borne by two point charges will be rather complicated. However, it does not mean that Coulomb's law is wrong when a conductor or dielectric exists.

5.3 Static condition is a fundamental limiting condition

Static refers to be stationary relative to inertial system, and two point charges are relatively static, which are stationary relative to observers. Static condition can be extended to a stationary source electric charge's effect on a moving electric charge, but cannot be generalized to moving source electric charge's effect on a stationary or moving electric charge. Spatial distribution of electric field generated by the stationary electric charge will not change with time, and moving electric charge bearing electric field force generated by the stationary electric charge is only relevant with the relative position of the two electric charges and their electric quantity, independent of the movement of mechanical electric charge, namely complying with Coulomb's law. But conversely, stationary electric charge bearing electric field force generated by moving charge will be different; according to the special theory of relativity, this force is not only relevant with the electric quantity and relative position of two charges, but also with the speed of a moving electric charge and accelerated speed, namely not complying with Coulomb's law. Establishment condition is an important basis for setting up a law. For a certain law, its establishment condition and existing background shall be grasped. Separated from the existing condition and background of law and formula, it is meaningless to indulge in empty talk about the relation among physical quantities.

6 APPLICABLE SCOPE AND ACCURACY OF COULOMB'S LAW

Applicable scope of Coulomb's law refers to r scale applicable to electrical power inverse square. Many verification experiments in the history indicate that Coulomb's law is applicable to the laboratory scale, but whether it is applicable to any scale range is still unknown. This problem is very easy to be ignored, and it must be stressed here which scale range Coulomb's law is applicable to shall be determined after being verified by experiment. In 1921, Rutherford (E. Rutherford, 1871–1937) proved that electric power inverse square is applicable to atomic nucleus scale 10^{-15} m by scattering experiment. So far the high-energy electron-positron scattering experiments have detected the interaction condition between 10^{-18} m, proving the electric power inverse square law still applies[3]. In addition, geophysical experiments show that electric power inverse square law still applies to the maximum of 10^7 m.

Whether the electric power inverse square law is applicable to the scale range of $<10^{-18}$ m and $>10^7$ m shall be further proved by experiments.

7 THEORETICAL STATUS AND MODERN SIGNIFICANCE

From Coulomb's law and universal gravitation, we can see that these tow formulas are inverse square relation; they are both field forces, of which interaction of electric charge and gravitation shall be transmitted through field; they are both conservation forces, of which working is independent of path, with the corresponding potential energy. Universal gravitation is so similar to Coulomb's law that we have to admire scientist Priestley's so sharp physical thought since he associated universal gravitation with Coulomb's law at the very beginning of research. And this similar comparison research is just the most common thinking mode of researching physical problems.

Coulomb's law is the basis of electrostatics, and it is also one of the most accurate physical experimental laws. Due to limit of teaching hours, teachers cannot explain each section in detail during class but they should guide students to pay attention to the connotation and essential understanding of physical laws, cultivate students to master scientific research method and establish physical thought and serve as a modest spur to induce students to come during teaching, which will make a great contribution when students study other physical laws or research questions on their own.

REFERENCES

[1] Chen Bingqian, Shu Yousheng Special Study on Electromagnetism Beijing High Education Press, 2001.
[2] Establishment, Verification and Its Theoretical Status of Coulomb's Law Wang Xiaolin Academic Journal of Northeast Normal University (Natural Science Version) No. 4 vol. 24.
[3] Xu Zaixin, Qian Zhenhua Accuracy of Coulomb's Law. Physics Teaching, 2010(4): 4–6.

Education Management and Management Science – Zheng (Ed.)
© *2015 Taylor & Francis Group, London, ISBN 978-1-138-02663-6*

From flipped classroom to physics teaching

J.K. Lv, W. Shao & B. Jiang
Aviation University of Air Force, Changchun, China

ABSTRACT: Flipped classroom is a new teaching mode, in which students are watching videos at home instead of listening to the teachers in the classroom, while doing exercises and communicating with their teachers and classmates in the classroom. The practice and research about how to use the new teaching mode—flipped classroom in physics teaching has become one of the popular research topics in the present college physics teaching.

Keywords: flipped classroom; Coulomb's law; conservation of angular momentum of rigid bodies

1 INTRODUCTION

How to make the teaching reform of college physics suitable for era's requirement has become one of the popular research topics in the present college physics teaching. The research in this field comprises of teaching forms, methods, means and content, etc. There are various related research topics, such as multimedia auxiliary teaching, network teaching, computer simulation experiment, application of teaching videos, etc. Especially we are in the face of the teaching mode of flipped classroom, which has attracted attention from the international educational circles. How to make it localized to become a teaching mode suitable for Chinese education is worth being thought about and studied by us teachers.

Flipped classroom is a new teaching mode, in which students are watching videos at home instead of listening to the teachers in the classroom, while doing exercises and communicating with their teachers and classmates in the classroom. It reverses the traditional teaching order—listening to the teachers in the classroom and finishing homework after class. About the flipped classroom mode, our school has had a trial study for one year and got some experience.

2 RISE OF FLIPPED CLASSROOM

The practice and research about "Flipped Class Model" in the early stage was carried on in American colleges and universities, characterized by emphasizing the interaction between teachers and students. For example, in the 1990s, Harvard's physics professor Eric Mazur started the research about "flipped study", and integrated the flipped study method and "peer instruction": before class, students watch videos, read articles, think by using the pre-learned knowledge, and then review what they have learned and ask questions; while their teachers make the teaching arrangement and develop the studying materials for the class on the questions asked by their students, and trigger discussions among students to solve the problems together in the class[1]. In 2007, two chemistry teachers in "Woodland Park" High School practiced flipped classroom model first. The flipped classroom model carried on in Khan Academy is regarded as the first glimmer of officially starting the future education[2]. Nowadays, the flipped classroom model has been developed into a new type of teaching mode which is very popular in the whole North America or even in the whole world.

In America, there are a number of schools who are flinging themselves into the teaching reform of flipped classroom model, such as the flipped classroom of mathematics in Stone Bridge Elementary School, "the Starbucks classroom" in Highland Village Primary School, the flipped model of the whole school in Clinton Dell High School, the Ap calculus class in Bullis School, the flipped class of English in San Aquinas School, the selective flipped class in Prairie South High School, the digital interaction textbooks in California Riverside Unified School District, etc. In China, we witnessed the initial development of flipped classroom practice, such as the flipped classroom procedures Chongqing Jukui Middle School—"three practical operations in the flipped classroom", "four sections before class", "five steps in class" and "six advantages", "teachers' teaching and students' studying on their own for half a day" respectively in Shanxi Xinjiang Middle School, the after "teahouse" teaching exploration in Shanghai

Yucai Middle School, etc. Moreover, in East China Normal University, they established the moocs (massive open online courses) centre, organized the relevant elementary and secondary schools all over the country to carry out the study of "flipped classroom", and built the C20 moocs alliance ④, focusing on the development of the teaching micro videos of all disciplines at the basic education stage. In addition to that, many other schools, like Shenzhen Nanshan Experimental School, Shanghai Qibao Middle School, Qingdao Second Middle School, Hangzhou Xuejun Middle School, Wenzhou Second Middle School, etc., have carried out the experimental project of flipped classroom, and achieved a lot. To a certain extent, the flipped classroom model has helped improve students' marks, meet the personal studying requirement of students, improve the ability of independent and self-learning study and make the relationships between teachers and students closer[3].

3 PRACTICE OF FLIPPED CLASSROOM MODEL

As a new type of teaching model based on information technology, "flipped classroom" is generally boiled down to the integration of digital campus construction and information technology and another major attempt on new curriculum reform by the domestic scholars. However, the objectives from different pilot units and different majors are not so identical with various forms. For example, some parts in the college physics textbooks are combined with practical application more closely, such as mechanics; some parts are of high theory, such as fundamental laws and theorems of physics. The physics teaching research office of our school has made an attempt to the practice of flipped classroom teaching according to different lectures and from different perspectives.

3.1 *Of theoretical and the heuristic*

This kind of lectures mainly introduce the basic theoretical law, such as Newton's law, the law of conservation of energy, Coulomb's law, etc., which are too theoretical. College students have learned these laws and theorems in middle schools and high schools. Thus they should be very familiar with them, but do not always understand the physical conception contained in them. The teachers may make the students understand the profound connotation and denotation of the physics laws from their formation and development. Take the Coulomb's law as example. The teacher may offer the related reading materials about the physics history to the students before class, making them to learn about the process of building a physics law. Coulomb's law originated from the very beginning when people observed that there is an interaction force between charged bodies: the same charges repel each other, while different charges attract each other. In the process of research on the charge force of the electric charge, Priestley was inspired by the gravitational field and guessed the inverse square relations between the charge forces by using the method of analogy. In the numerous experimental verifications, Coulomb's torsion balance experiment (electric repulsion) and electric pendulum experiment (electric attraction) were finally recognized. The physical law which describes the acting forces between two stationary point charges in the vacuum was finally called Coulomb's law to commemorate Coulomb's contributions. As one of the most basic laws of electromagnetism, Coulomb's law is the cornerstone of the whole electromagnetism. Gauss theorem and Maxwell's equations were put forward on the basis of its inverse square relationship. In the development of science, scientists have continuously verified the inverse square relations through experiments, and so its precision has been improving. In 1921, the scattering experiment by Rutherford (E. Rutherford, 1871–1937) proved that power inverse square law applies when the nucleus scale is in the range of 10 to 15 m. Up to now, the high-energy—positron scattering experiment has detected the interactions when the nucleus scale is in the range of 10 to 18 m and proved that power inverse square law also applies[4]. The teachers should guide the students to summarize the process of a general physics law being built: observing phenomenon—putting forward the proposition—guessing the answers—predicting the theory—designing the experiment—fixing the theory and finding the rule—forming the law and the theorem—studying the establishing conditions, scope of application, precision, theory status and the modern connotation. Through that, it will make the students learn physics in the way of a physicist thinking and establish the physics thinking, and make them understand that the purpose of studying physics is not to do physics exercises but to obtain the thinking ability.

3.2 *Application and discussion*

As the aim of our school is to train pilots, we introduce the theorems and laws of physics which are connected closely with flying practice to our students, such as the rocket flight principle of rigid bodies, the theorem of angular momentum, the law of conservation of angular momentum, electrostatic shielding, etc. The students didn't know much about these theorems and laws in middle school and high school, but they will use them in their future flight practice. So when introducing these theorems and laws to the students, the teachers should connect the flight practice closely with the theorems and laws.

Take the law of conservation of angular momentum as an example. After learning the angular momentum of particles system and rigid bodies, the students will study the theory in this section without so much difficulty. The teaching way we adopt in this flipped classroom is: the students study the theory and read the books on their own, while the teachers make conclusions in class and ask the students questions about this section to check their ability to integrate theories with practice. According to the teaching content, the teachers ask four questions in total: a. why do people swing their arms when walking? Why are they uncomfortable when walking with their arms and legs on the same direction moving at the same time? b. Why does the rotation rate of a person on the Zhukov swivel chair change with his different postures? c. Explain the relations between the speed and postures of the people who are diving or skating in the videos. d. Explain why the helicopter has two propellers. Obviously those questions above are all about the law of conservation of angular momentum. The students are divided into 4 groups and each representative from each of the 4 groups explains their answers after separate discussions. The students are very active in the warm classroom atmosphere. And the homework is not the traditional one which is doing exercises, but to discuss together after class: why doesn't a top topple and fall easily after it is spinning? The answer is that the ship's navigation principle works. Through those questions, the students are guided to integrate theory with practice. It enhances the students' interest in learning physics on one hand, and on the other hand the college physics with flight characteristics is formed, which increases the pilots' physical knowledge and is favorably commented by the leaders.

4 THE SIGNIFICANCE AND PROSPECT OF THE FLIPPED CLASSROOM MODEL

Harvard's physics professor Baker suggested that the teachers should be "a guide to the students" instead of "the saint on the platform", which becomes the slogan of the flipped movement of the university classrooms. Compared with the traditional class, the flipped classroom has broken through the limitations of traditional teaching and changed the teachers' role from stars to directors. Flipped classroom doesn't really lie in the form of videos but is to highlight the teachers' guiding position, making the students learn how to think instead of accepting knowledge passively, improving their ability in analyzing and summarizing problems, and the ability in learning in collaboration and interaction. The education department pointed out in the ten years of development planning of the educational informationization: flipped classroom is a creative teaching model, which has turned the traditional teaching mode, and is a significant breakthrough in the teaching reform under the conditions of the development of modern information technology. As educators developing synchronously with the times, we shouldn't just stay at the theoretical stage to research it, but should apply it more in the teaching practice, so as to make due contributions to the modern educational reform of our country.

REFERENCES

[1] November A. Mull B. Flipped Learning: A Response To Five Common Criticisms [2012-03-27]. http://www.Eschoolnew.com/2012/03/26/flipped-learning-a-response-to-five-common-criti cisms/.
[2] Review on Study and Practice of Flipped Classroom both at Home and Abroad Liu Jianzhi, Wang Dan, Theory and Practice of Contemporary Education.
[3] Teaching Design and Applied Research Based on Flipped Classroom Model Chen Yi, Zhao Chengling XDIYIS.
[4] Accuracy of Coulomb's Law. Physics Teaching, 2010(4): 4–6. Xu Zaixin, Qian Zhenhua.

Education Management and Management Science – Zheng (Ed.)
© 2015 Taylor & Francis Group, London, ISBN 978-1-138-02663-6

Research on community-based family education guidance for mobile children school of education

Fu Chen

Shandong Women's University, Jinan, Shandong, China

ABSTRACT: Family education occupies a pivotal position in early childhood growing up, and the mobile problem in early childhood family education is extremely unfavorable on children's development. According to the theory of social ecology, community is an important support system for family education on mobile children. Community-based education on parents of mobile children is an important way to improve the quality of early childhood education, thus contributing to the fairness of education. Therefore, this study, based in Jinan City, Shandong Province, attempts to do research on community-based family education of mobile children. The objectives are to understand the problems and needs faced by parents about education, to enhance the relevance of family education on one hand, to summarize the inadequacy, success and difficulties through specific practices, and to provide suggestions and recommendations accordingly for family education guidance activities in Jinan City.

Keywords: community; mobile children; family education

1 INTRODUCTION

Community is the basic unit of society, and learning community is the foundation of a learning society. Community, as an important support system for family education on mobile early children, is manifestation and extension of community education, as well as an important way to enhance the quality of early childhood education, to promote equity in education, to implement lifelong education and to build a learning society.

2 GUIDING IDEOLOGY: SOCIAL ECOLOGICAL THEORY

According to the social ecological theory by Bronfenbrenner, at the micro level, children come into contact with family, schools, kindergartens (nurseries), hospitals, and neighborhood playgrounds. For the mobile children in this study, their families, kindergartens, neighborhood playgrounds are no doubt in the community. As the community plays an important role in early childhood growth and development, it is feasible to help and guide parents of mobile children for better education.

The parents of mobile children form a special group, who are usually in self-employed occupations, such as: selling vegetables, cleaning, or driving a motorcycle taxi, etc., with no fixed units. The communities where they live are relatively more stable, thus it is more important to carry out such community-based education.

3 MAJOR WAYS OF CARRYING OUT ACTIVITIES

3.1 *Cooperate with medical institutions*

For young children, a healthy physical condition enjoys paramount importance. However, according to the results of the previous research, the parents of mobile young children lack the sense of health care, and because their parents are generally not so much literate, usually busy with their work, and the channels and their time for knowledge in this area is limited, it is therefore very necessary and important to cooperate with medical institutions and let them give parents professional guidance.

Hospitals, especially public hospitals with rich material and human resources, should take the lead to carry out various activities in the community so as to make a contribution to the healthy development of young children, the same goes for the community hospitals which receive the most attention in national support. Our government is now making great efforts to develop community health service system so that some small medical problems can be solved without going far, which greatly eases the pressure on medical treatment. However, there is no denying that community hospitals, on the way of development, are experiencing various difficulties, and they themselves are also facing many problems. Despite such a fact, the elderly and young children, as a group weak in immunity, should be taken into consideration with priority. Therefore, there also should be a professional pediatrician even it is a comprehensive department. In our country's great efforts to develop, regulatory mechanisms should also keep up with it and strengthen its social functions. With the cooperation of community hospitals and community workers, not only the stress of hospitals but also that of families can be eased. Besides, parents' increasing awareness of health care will benefit their children's growth.

3.2 *Attract university volunteers actively*

It has been pointed in Opinions on Further Improving community volunteer service work under the new situation which was jointly issued by the Ministry of Civil Affairs (2005) (hereinafter referred to as "Opinions") that: improving community volunteer service work under the new situation will facilitate the integration of community resources, improve the social service system, solve social problems relying on social forces, and satisfy residents' growing material and cultural needs; it will help to improve the relationship between people, thus contributing to the shape of a harmonious interpersonal environment of fraternity, respect for the elderly and care for the young. It will also melt many social conflicts, maintain social stability and promote social harmony. The reason why we should attract university volunteers is that university scholars have the following advantages: Firstly, they have strong expertise. Secondly, they have a more flexible schedule. Thirdly, there are no conflicts of interest between them. All these advantages can not only ensure a smooth development of high quality activities, but also increase their social practice, killing two birds with just one stone. Therefore, for communities who lack funds, attracting university volunteers becomes one of the important ways to carry out activities.

3.3 *Make use of the existing educational resources, and try to improve the current situation of communities*

3.3.1 *Make use of the existing educational resources*

Community education resources are the whole combination of the material and spiritual resources, which have educational functions on all the members within the community. They include: human resources, environmental resources, material resources, and policy resources. Community educational resources are the foundation of community resources, and the good use and development of them will help to create a harmonious and stable social environment, and promote educational equity. Besides, it is also conducive to a rational allocation of social resources, the harmony of different social classes, and community stability and sustainable development.

3.3.2 *To improve the current situation of communities, we can start from these three aspects*

The efforts of community workers themselves, the government's efforts and community funds. In their service work, community workers will meet people of various kinds, so they should have a sense of service and always be very patient. They need to patiently explain and answer all kinds of questions raised by families, and turn to experts for advice once they have come across some real problems, so that parents will feel the community can really eliminate their confusions. Because community workers bear heavy burdens of administrative work, especially demographics and family planning management, some residents especially migrants or those who have more than two children, are strongly against community workers, which is something unavoidable. Government should have impacts on directing public opinions, so that residents can realize that community workers are working for them. As early as in 2012, the country pointed out that the investment in education should reach 4% of GDP, and this investment should pay more attention to preschool education. Through the analysis of relevant policies, I found that the investment into pre-school is mainly for kindergarten, such as support for inclusive kindergarten, admission grants for children from disadvantaged families, etc. But I believe that investment should be multi-channel, and community is not a place to be overlooked, especially for mobile families with young children at the age of 0–3. These people need scientific guidance, because their children have not yet gone to kindergarten and early childhood institutions are too expensive for them. If guidance is not given on the basis of communities, education will be totally in a vacuum. Therefore, in addition to the government's procurement of services, communities should have relevant education funding, especially for mobile children and economically disadvantaged families.

3.4 *Cooperate with early learning institutions*

There are two advantages for early learning institutions: professional teachers, and venues and equipment, which are what the neighborhood communities are unable to provide, especially venues and equipment, making the cooperation with early learning institutions very important. This can save a great number of financial and material resources for neighborhood communities. Taking into consideration the importance of the age from 0 to 3 for the development of children and the fact that children over 3 years old have gone to kindergarten and therefore their parents can get, to some extent, some guiding education from kindergartens, I try to start from the cooperation with early learning institutions.

4 CARRY OUT EDUCATIONAL GUIDING STRATEGIES FOR MOBILE CHILDREN FAMILIES

4.1 *Make attempts during the operation process of government's procurement of services*

4.1.1 *Identify the problems and needs of parents*
Based on the fact that Municipal Women's Federation, the Municipal Board of Education, and City PTA are the departments responsible for family education, and the District Women's Federation, the District Board of Education for the management of district-level family education, staff from the District Women's Federation and the District Board of Education firstly should be sent to identify and then solve the problems that parents are facing in family education.

4.1.2 *Decide the guiding units for parents*
Community workers at first should inspect the resources within communities, sift out the enterprises meeting the standards, and make a list. After the inspection for these units, these 3 parties will discuss and decide the units, which will carry out guiding activities for parents.

4.1.3 *Carry out guiding activities*
After deciding the units undertaking the guiding activities, community workers will discuss with them on the specific details of how to carry out, in what form, and how much money we need, etc. The details of the activities are for informing parents to participate, and the confirming of the fund for reporting the Board of Education. After the check of the Board of Education, it will report the Ministry of Finance for funding. After that, the work of management and organization, and the specific details of the carrying out of these activities parents participating will be undertaken by the guiding units.

4.1.4 *Evaluate the guiding activities*
There is no doubt that the evaluation work is the focus of government's procurement of services. Whether parents are satisfied, whether the units

are responsible, and whether the activities are effective can all be reflected in this part, which is the basis for the government's next purchase. In this session, special staff of the Board of Education will list the detailed assessment rules. Community personnel operate and organize parents for the evaluation of the guiding units. The evaluation results will be sent to the guiding units, the Board of Education and the Women's Federation, which will decide whether these units will continue to provide guidance for parents, and if they will, then what are the deficiencies that can be improved in their guiding work.

4.1.5 *Mechanism of encouragement*
There should be mechanism of encouragement for units performing well in the guiding activities. It can start from these 2 aspects: honorary awards and material awards so as to encourage these units to provide better guiding services for parents.

4.2 *The main forms of guidance*

According to the mole theory in the ecological theory, the guidance for mobile young children family education should be put into effect with specific activities. According to the previous literature review, we can draw such a conclusion that the guiding activities for parents are mainly of the following kinds: lectures, salons, one-to-one, family activities, consulting rooms, and brochures.

4.2.1 *Parent-child activities have always been parents' favorite*
Activities in this form can not only give parents guidance, but also promote the intimacy between parents and children. And it is especially much more beneficial to mobile children who lack the company of their parents. However, the purpose of these activities should be fully understood by parents when carried out. Only in this way can these activities achieve salient effects.

4.2.2 *Salons are undoubtedly the most effective method when it comes to solving the problems parents encountering in children's education*
I have visited some parents in several activities, and asked them whether they liked the form of one-to-one or consulting rooms to get more specific answers. They just laughed without giving a direct answer; only saying that it was better when they gather together and just shot the breeze.
  The way of one-to-one or going to the consulting room may make parents feel uneasy and uncomfortable, so they are more willing to exchange their ideas with each other and solve these problems.

457

4.2.3 Lectures are the best way to give a systematic explanation to parents who lack certain knowledge

As has been mentioned in choosing toys, when parents do not have a systematic understanding about toys, such as their importance, precautions, how to choose, how to use, etc. giving them a systematic explanation in the form of lectures is beneficial.

4.2.4 Other aspects

In this study, forms like training, leaflets, and billboards and so on are also adopted. For these training methods, parents feel it is easy to learn and remember, having reaped a lot. But because the time is too long, parents' participation cannot be guaranteed. Leaflets are there mainly to strengthen the activities, which have already been carried out. However, the effects of guidance with sole leaflets can be too small to be noticed, and even though there are a lot of pictures, parents are not ready to take even a look. And what are mainly playing a subtle role are billboards whose contents will be updated every quarter by community workers. In other words, the contents of billboards will last three months and can be seen by parents passing by. Therefore, if the contents parents are most concerned about and want to know most are put on these billboards, the knowledge will gradually be improved by parents.

4.3 Strengthen the social service functions of kindergartens, hospitals, etc

Early learning institutions, kindergartens, hospitals, etc. go hand in hand with children's growth. Therefore, for these units, especially the ones that our country is vigorously supporting or pouring funds into, the government should impose policies on them and direct the public's opinions towards in order to strengthen their social functions.

In short, the family education of mobile children is disturbing, and their parents need guidance, which is anything but the business of just one person. To make a difference, the higher levels of government need to attach importance and give support in all aspects. In particular, the government needs to work out corresponding policies and put them into effects from specific operational level, such as how much money is to be invested, how many people are to be arranged, and how to follow one guideline instead of merely increasing investment just to reach the standard. What's more?

The relevant long-term mechanism should keep up to give awards to units who are responsible for specific guiding activities and have good performance, so as to encourage them to go on and provide guidance services of high quality for parents.

REFERENCES

[1] C. Edwards. A Hundred Languages of Children. Beijing: Psychology Press, 1998.

[2] Chen Xiangming. Qualitative Research Methods and Social Science Research. Beijing: Education Science Press, 2006.

[3] Ding Lianxin. Family Education on Pre-school Children. Beijing: Science Press, 2007.

[4] Huang Yunlong. Community Education Management and Evaluation. Shanghai: Shanghai University Press, 2000.

[5] Wang Qi. Chinese Migrant Children Education Survey and Research. Beijing: Economic Science Press, 2005.

[6] Ye Liqun. Family Education. Fuzhou: Fujian Education Press, 2000.

[7] Yang Yinsong. Introduction to States Community Education. Shanghai: Shanghai University Press, 2000.

[8] Zhang Yan. Fourth Ring Playgroup Story: Exploration on Non-formal Pre-school Education of Migrant Children [M]. Beijing: Beijing Normal University Press, 2009.

[9] Cao Nengxiu. Distinctive Thai Community Preschool. Preschool Education Research, 1994(2).

[10] Chi Jin. Related research and education concept on psychological characteristics of children. Educational Research and Experiment, 2003(2).

[11] Ding Yuanzhu. Nature and community building. China Development Observation, 2006(6).

[12] Dong Li, Chen Shangbao, Wo Jianzhong. Factors affecting 4–6 year-old children in the concept of parental education. Preschool Education Research, 2006(12).

[13] He Linxia. Thoughts on mobile parent-child relationship. Educational Development (Early Childhood), 2008(11).

[14] Ackerman B.P., Brown E.D. et al. Maternal Relationship Instability and the school behavior of children from disadvantaged families. Developmental Psychology, 2002.

[15] Branz-Spall, A.M., Roger Rosenthal, Al Wright. Children of the Road: Migrant Students. Our Nation's Most Mobile Population. The Journal of Negro Education. Washington: Winter, 2003.

[16] Branz-Spall A, Wright A. A History of Advocacy for Migrant Children and Their Families: More than 30 Years in the Fields, In Scholars in the Field: The Challenges of Migrant Education.

Education Management and Management Science – Zheng (Ed.)
© 2015 Taylor & Francis Group, London, ISBN 978-1-138-02663-6

Flipped classrooms and bilingual courses

Rui Wang
Tianjin Key Laboratory for Civil Aircraft Airworthiness and Maintenance, Department of Electrical Engineering, College of Aeronautical Automation, Civil Aviation University of China, Tianjin, China

Kathy Gause
Information Technology and Business Division, Tidewater Community College, Virginia Beach, Virginia, USA

Hui Sun
Department of Automation, College of Aeronautical Automation, Civil Aviation University of China, Tianjin, China

ABSTRACT: Innovation and leadership in science and business will depend on the ability to communicate in many languages. Unique teaching methods such as flipped classrooms, through the use of technology and more engaged instructors, promise to effectively engage students in bilingual classes such as engineering, business and science. Some advantages, disadvantages, challenges and solutions are explored.

Keywords: flipped classroom; bilingual courses; evaluations

1 INTRODUCTION

With the development of bilingual major courses in global higher education, more researchers have been attracted to developing new teaching methods to adapt to the new teaching environments [1]. In order to adapt to this trend, teaching methods are correspondingly emerging such as case teaching methodology, cooperative teaching, and differentiated teaching. Among these methods, flipped classrooms are becoming popular. This method turns the traditional teaching into a totally new model. By 'flipping' theory outside the classroom through use of personalized videos, the problem regarding inactive learning phenomena may be resolved.

This paper mainly considers and analyzes an approach to use flipped classroom teaching methods in Chinese-English bilingual major courses.

2 FLIPPED CLASSROOMS

Several years ago, two high school chemistry teachers Jonathan Bergmann and Aaron Sams, working in Woodland Park High School in Colorado's Pike's Peak, stumbled onto an idea [2]. In order to solve the problem of struggling to find the time to teach absent students living in the remote areas, they recorded and annotated lessons and posted them online. Therefore, absent students could make up what they missed. It was also a great opportunity for students who had not missed classes to review classroom lessons. This naturally evolved into Bergmann and Sams realizing that they had the opportunity to rethink how they used time in the classroom. Correspondingly, the "Flipped" classroom has occurred which means the common instructional classroom approach has been flipped. Instead of instructions being presented and lectured in class, the flipped classroom method provides teacher-created videos and other types of instructional materials virtually. Students can view them at home at any time ahead of class. In class, students are then able to engage in collaborative learning and maximize the learning resources by rethinking the material and gaining insight from their classmates. Therefore, traditional instructional knowledge shown in lectures has been finished in advance of the class, and concept engagement takes place in the class with the guidance of instructors. Students become the center of the class in flipped classrooms. This method overturns traditional teaching methods.

Compared to the traditional teaching methods, the flipped classroom method has its own advantages and disadvantages [3]. First, advantages include: 1) Learning becomes flexible. Students can learn at a pace suitable for themselves because of the lectures available online. They can learn at any time anywhere and may stop and rewind the video depending on their needs. 2) Motivating students learning. Since learning instruction is arranged in multiple forms, students remain motivated through different learning processes and diversification in roles. Students switch from inactive participants to active participants. There is more accountability present for individual learning and more possibilities for critical thinking. 3) Encourages teamwork spirit. Throughout the collaborative activities in class, students and teachers build on the diversity of experiences and perspectives in the classroom discussions. This may be risky for instructors at times as theory is challenged and questions raised may lead to complimentary topics not planned to be covered. 4) Accommodates differentiated learning for diverse learners. This type of instructional approach addresses learners who differ not only culturally and linguistically but also in their cognitive abilities, background knowledge, and learning preferences [4]. Classrooms are for learning and for enticing students to continue to learn. Teaching methods are transforming beyond brick and mortar buildings and territorial boundaries. Will this stimulate the human mind to move beyond a specific chapter of a book or a specific topic of the course to a more holistic approach? 5) Learning through questioning is more impactful. The days of instructors lecturing pure theory are being replaced by classroom interaction between faculty and students. At the core of student engagement is how questions are asked of students and how questions are answered [5]. In this flipped setting, students have a unique opportunity to capitalize on the instructor's wealth of information and experience in addition to any theoretical material that may accompany a class. The flipped approach connects both teacher and student in a curiosity-learning environment within the learning goals of the class.

Second, disadvantages include 1) Not all students are self-disciplined to learn at home online which makes them not keep the pace with others and limits their progress, participation and learning. Having an engaging, nurturing classroom atmosphere where instructors are trained in dealing with motivating students and cultivating a bit of friendly peer pressure may lessen the likelihood of this occurring. The flipped approach will give an opportunity for many students to become familiar with the hybrid approach of using technology in face-to-face classroom settings. Currently, in the U.S., many students are still not comfortable with using web-based software in the classroom setting and prefer face-to-face instruction. A combination of both introduces more students earlier to this future standard educational approach to learning. 2) Students may not have the Internet capabilities to view online in remote areas, although this is becoming less of an issue. This could be resolved on an individual basis or through technology, e.g., there must be an app for that. 3) More instructor time will be necessary to prepare and design engaging and creative learning activities. The instructor or team of instructors will definitely have to spend more time in keeping the class up-to-date, researching current events and changes in the field and providing rich real-life examples. There are numerous resources already in place. In fact, many zero textbook pilot classes are being experimented with in the U.S. Publishers are being forced to re-invent their businesses to keep a competitive advantage. Progressive instructors are creating videos and researching free resources to design courses to avoid students paying such high costs for textbooks [6].

Flipped classrooms bring several challenges for both instructors and students. First, this method requires instructors to master more complete and recent knowledge, especially being sensitive to state of the art technology or current events. Due to the format of the learning, some students will enjoy the challenge of challenging while other students will begin thinking "out of the box" and in new and innovative ways. This type of innovative thinking is learned from an early age and all of us have memorable experiences with our respected teachers who motivated us to learn more with real-life examples. An example would be in the international business field or the science field, where we all know that elements of international studies, sociology, economics, geography, government, finance, and culture all interplay together and are connected. Flipped courses enable instructors to customize learning and apply the topic of a discipline's theory with real-time world events and experiences. These courses might be customized for special projects based on the instructor's comprehensive knowledge or language. Project Based Learning is a concept that demonstrates this, using real-world problems to capture student interest and to build their problem solving skills [7]. That is how the real world of business or science works, not one topic at a time, not one chapter at a time. The virtual world has increased the speed of learning and space so that more is expected of both instructors and students.

Second, how to evaluate students and design an objective evaluation system is the third challenge [8]. Regardless of the educational approach, evaluation is an important metric in determining whether learning has occurred. It is an elusive metric. Assessment of learning is the process of collecting information about student learning and performance to improve education. Assessments should reflect the goals and values of the discipline, and prepare students for an evolving workplace. Best practice rubrics abound from major colleges and universities. Like any course, learning objectives and instructional strategies should be clearly outlined. Formative assessments are especially important in the Flipped approach, i.e., monitoring student learning and providing ongoing feedback that can be used to improve one's teaching and by students to improve their learning. A particularly effective assessment is the concept map to represent their understanding of a topic. It promotes non-linear thinking and could be used with assignments completed at any time in the classroom forum. Summative assessments or evaluating student learning at the end of an instructional unit would be done by comparing against a rubric or benchmark, i.e., an exam, a learning project, a paper or presentation [9]. A grading rubric could incorporate participation, effort, assessments and peer ratings on engagement. The Flipped approach of teaching would involve a self-assessment from the student. Other evaluation techniques that motivate might allow the "best" grade of a student out of a number of attempts or offering options when giving assignments. Depending on one's learning style, one student may choose to write for a team project and one student may choose to present. If projects are assigned to teams, private journaling with the instructor periodically should be encouraged. As with all grading, it is all about expectations. Clearly communicating expectations in advance will encourage a learning environment.

3 BILINGUAL TEACHING RESEARCH

The concept of bilingual teaching means using two languages to teach a major course with the exception of Language or English courses. In China, in order to adapt to the global era of the economy, technology and other areas, Chinese higher education must integrate with the world. This requires universities to cultivate students to meet the international competition requirements. Therefore, students should not only learn the solid theoretical techniques of a discipline, but the ability to communicate conversationally in other languages about that discipline. Correspondingly, the Ministry of Education of China encourages that a bachelor degree education should create conditions to promote bilingual teaching using English and other languages and these courses are to account for 5% to 10% of all courses offered [10].

Bilingual teaching will ideally improve the fundamental English of the student while learning professional capabilities. First, students have more opportunities to practice and listen to English. Basic Interpersonal Communication skills can develop in as little as two years while cognitive and academic language proficiency, such as in a scientific field, is estimated to take 5 to 9 years [11]. In general, this type of course uses the books or materials in English, but if students can also hear the authentic English language standards and styles, it is much more beneficial. Second, bilingual teaching could improve the student's professional terminology in English. English is the most popular business and communication language in the world. Therefore, if students could master it well, it will positively affect their ability to communicate with other countries in a technology setting. This is an important issue that is addressed by flipped teaching.

Unfortunately, there are some problems in a bilingual course. First, in practice, employing a uniform teaching method may not balance all of the student requests for knowledge because of the different English levels. Students who are weaker in English might lose interest in the major course. Another encountered problem for bilingual courses is that it takes a resilient instructor to judge and promote the class climate. Students will need to be engaged and encouraged to participate in the class, which may not be the style of all instructors. Learning and teaching both English and a technology topic at the same time will be challenging.

4 APPLICATION OF FLIPPED CLASSROOM IN BILINGUAL COURSES

Faced with the problems stated above in teaching current bilingual courses, the paper considers using the flipped classroom method.

First, in order to solve the problem with the English level difference, the lectures may be categorized into different levels. The materials are diverse. For example, simple chapters will be designed for students weaker in English. As for the important chapters, every student has to learn the materials according to the difficulty level of the class. In class, students will be divided into several discussion groups depending on their English level. In this way, everyone can learn from the course, not only the professional knowledge but also English.

With development of new education media such as education big data, cloud education and MOOCs (Massive Online Open Courses), they provide multiple and full teaching resources for bilingual courses. In 2001, MIT first launched Open Courseware, and opened about 1,800 courses to the world for free [12]. In general, these source courses always provide captions, and then students can catch up with the materials. Instructors could find the original English materials for these resources as well for students. Learning becomes more flexible.

Second, in order to promote student's learning interests and to cultivate the spirit of teamwork, more interaction activities are added in class. In class, a representative of each group may give a presentation in English to introduce their learning. Other students put forward questions to the other students in their group and enable instructors to evaluate each student. During the discussion process, students can learn how to communicate with others, as well as to find the salient learning points from each other. Instructors could give instructions for students according to their learning capabilities. This avoids spending too much time on learning vocabularies in class by the traditional teaching method for bilingual courses and improves the efficiency. In the meantime, students are more active in class and enjoy the learning process, and then independent thinking and cooperation bring them success.

5 EVALUATION OF STUDENTS IN BILINGUAL COURSES

The diverse teaching materials in bilingual courses also afford an opportunity to experiment with various evaluation criteria instead of the traditional final exam and midterm exam deciding the final grades. A combination of multiple evaluation criteria such as engaging in activities in class, language, quizzes, and group performance may be considered in evaluations. Even though only one or two representatives gives a presentation from the group, other group members will be asked questions and be knowledgeable in a particular area. Therefore, instructors could still evaluate the work of all students. Simultaneously, students from other groups may peer evaluate the presented group members. Students from the same group will also peer evaluate each other. This evaluation would use standard evaluation criteria and include effort, research, English level and professionalism.

6 CONCLUSIONS

Global science and business leadership will demand communicating and connecting in different languages. The Flipped classroom method has realistic possibilities for how learning and teaching is changing to meet what technology and modern thinking has to offer while serving as an effective tool to learn and teach in bilingual discipline classes such as engineering, business and science.

ACKNOWLEDGEMENT

This Research is supported by NSF of Tianjin Grant #13 JCYBJC39000, Tianjin Key Laboratory of Civil Aircraft Airworthiness and Maintenance in CAUC and Scholars of Civil Aviation University of China (2012QD21x).

REFERENCES

[1] Current Situations, Problems and Trends of International Higher Education Open Courses http://www.zlunwen.com/education/higher/48460.htm.
[2] B. Tucker, The Flipped Classroom Online Instruction at Home Free Class Time for Learning, Education Next, 82–83, 2012.
[3] Flipped Classroom-Instructional Module http://www.kokuamai.com/test/flipped/.
[4] Huebner, Tracy A., What Research Says About Differentiated Learning, Association for Supervision and Curriculum Development (ACSD). Educational Leadership, vol. 67, no. 5, Feb. 2012. Retrieved from: http://www.ascd.org/publications/educational-leadership/feb10/vol67/num05/Differentiated-Learning.aspx.
[5] Berkeley University of CA. Center for Teaching and Learning, Asking and Answering Questions, 2014. Retrieved from: http://teaching.berkeley.edu/asking-and-answering-questions-0.
[6] Perry, Mark J., The college textbook bubble and how the "open educational resources" movement is going up against the textbook cartel, American Enterprise Institute Ideas, Dec. 2012. Retrieved from: http://www.aei-ideas.org/2012/12/the-college-textbook-bubble-and-how-the-open-educational-resources-movement-is-going-up-against-the-textbook-cartel/.
[7] J. David, What Research Says about Project Based Learning, ACSD, Educational Leadership, vol. 65. no. 5, Feb. 2008. Retrieved from: http://www.ascd.org/publications/educational_leadership/feb08/vol65/num05/Project-Based_Learning.aspx.
[8] J.L. Zhang, Y. Wang, B.H. Zhang, Flipped Classroom Teaching Module Research, Remote Education Magazine, (4): 6–51, 2012.
[9] Sheridan, Harriet W. Center. Brown University Assessment Overview & Best Practices, 2014. Carnegie Mellon University Enhancing Education Site. Retrieved from: http://www.cmu.edu/teaching/assessment/index.html.
[10] Education Information A Decade of Developing Plan (2011–2020), Ministry of Education of China http://www.edu.cn/zong_he_870/20120330/t20120330_760603_3.shtml,2013-03-3.
[11] S.A. Reyes and T. Kleyn, Teaching in Two Languages. CA: Sage Company, 2010.
[12] Massive Open Online Courses http://baike.baidu.com/view/10187188.htm?from_id=8301540&type=search&fromti.

Education Management and Management Science – Zheng (Ed.)
© 2015 Taylor & Francis Group, London, ISBN 978-1-138-02663-6

The inspiration of the American moral education in university

Shasha Yin
College of Pharmacy, Harbin University of Commerce, Harbin, Heilongjiang, China

ABSTRACT: By reference to the moral education development history in the US, we find from learning the advantages of American moral education will efficiently carry out university moral education in China. In this paper, we tried to clarify the American moral education historical trail and character that we aimed to search for the source material which could be used for the construction of our college moral education. Then we positively studied the inspiration drawn from American college moral education: adhere to one-factor domination, innovate moral education ideas; Stress combined radiation, expand moral education approaches; Stress organization security, improve the contingent of moral education. As a specimen to look for inspirations on moral education for our own.

Keywords: moral education; American; inspiration

1 THE TRANSFORMATION HISTORY OF AMERICAN MORAL EDUCATION

1.1 *American moral education in 19th century*

However, with the rapid development of science and technology as well as the formation of a multicultural, the U.S. gradually lowered its attention to moral education, and the emphasis on humanity and moral disciplines gave way to a stress on science and practical disciplines: teachers became increasingly absorbed in a narrow scope of professional activities and refused to solve the broad moral issue any longer; school moral education witnessed a popularity of the educational methods of behaviorism, that is to educate students to abide by code of ethics, so as to maintain the social stability. Instead of taking into account the fact whether students will internalize the teaching, schools began to establish practice-based course system. At the end of the 19th century, moral philosophy courses in universities started to give way to the disciplines such as ethics, psychology, sociology and politics. And American moral education transformed from classroom instilling to broad penetrating education characterized by a trinity of family, school and society. At that time, American moral education stressed on individual growth and combat compulsory education.

1.2 *American moral education after the 20th century*

Lack of moral education system of unified standard resulted in the disorder in the U.S. realm of ideology. With the development of globalization and information, America witnessed an increasingly prominent social moral problem, a corruption of social morality as well as a social moral degeneration. In the U.S. campuses, there emerged endless insoluble social problems such as shooting, violence and drug taking. Until the end of 1990s, the Columbia school shooting case brought the summit of people's requirements for student moral character cultivation, with an increasingly louder cry for moral education promotion from all social circles.[1] With the development and maturing of moral cognition development theory, social cognitive theory and values clarification model, moral education study began to bottom out and revive. The U.S. government also issued a series of intervening measures concerning citizen moral education, organized special agencies to make major decisions, carried out such education systems as Character Counts, thus redirecting American moral education back to school education from one-sided emphasis on no-instilling education.

2 THE CHARACTERISTICS OF AMERICAN MORAL EDUCATION

2.1 *Moral education penetrability*

During the development course, schools, the society and families of the U.S. has day by day merged into a trinity of penetrating education network. Apart from merging moral education into the whole education progress, university moral education also attaches importance to conduct student moral education by ways of extracurricular activities, campus culture establishment and guidance upon consultation. The society is a significant carrier of American moral education, which, as a return, should be contained in the demonstration of social material, cultural and ideological civilization. Moreover, government houses, museums, memorial halls and historic relics are supposed to be built into moral education basement for students. Activities and education should be combined through activities of politics, religions and community service, with the moral levels of teachers being lifted as well. Since family is the origin of moral cultivation, American schools lay great emphasis on building ties with families, which mean that school will explore student-training programs together with students' parents by way of establishing parent committees and communication platforms. In the implementation of moral education, concerted efforts have been made by a coordination and interaction of schools, the society and families to carry out student moral education. While instilling education is combated, it is stressed that moral education should be penetrated into the learning and living of students. Moral education should be combined with practical student development through various educational forms. The use of heuristic and discussion methods of education should be emphasized to guide the individual development of students, so as to help them finally develop independent values and ethics after gradual groping and improving during the course of their learning and practice.

2.2 *Moral education subjectivity*

The values clarification theory, which rose in response to the requirements of the U.S. society, exerted an extremely profound impact on American moral education. The values clarification school believes that, instead of directly passing their values to students, teachers are supposed to guide their students to develop value systems appropriate for themselves in the learning and living by way of analysis evaluation, etc. And the fundamental idea of American education exactly embodies this theoretical viewpoint that is to give full play to the individuality and potential of every student by respecting and equally treating every one of them.

As for moral education, the U.S. pays full attention to the display of students' subjectivity. With reference to the issue of moral education, the government has not laid down unified standard and content, allowing each state and university to set its own distinct and characteristic moral education goal according to its own understanding of education. All sorts of living courses are offered by universities, while ethics groups are established to help develop value systems of students' own, thus allowing students to make their own choices. In terms

of teaching methods, stimulation and guidance are highlighted to secure students full autonomy, keeping teachers from rash intervention. Teachers should conduct student moral education by taking different measures and selecting different contents based on their characteristics, so as to lead students to take part in teaching activities. Since universities, teachers and students can all choose what to teach and learn according to their own needs; both the enthusiasm and initiative of teaching and learning can be maximized, so as to accomplish moral education goals to the greatest extent.

2.3 *Moral education practicality*

Education practicality is also stressed in American moral education. Because the formation of good virtue cannot be solely dependent on instilling of knowledge, instead, it is supposed to be educated through practice, as students will not understand their own behaviors until the knowledge is internalized as their own values through practice. In order to strengthen students' perceptual cognition and avoid the singleness and inflexibility of classroom instilling, practical activities such as extracurricular activities, community activities and campus culture establishment are carried out by universities to encourage student participation in university administration, which practices ethical education by action. Besides, American schools have long been encouraging students to join in social services, which play an important role in the cultivation of student character, citizen consciousness as well as social responsibility. And it has been a tradition that students take part in community voluntary services, since this is one of the targets of university undergraduate education and students must get involved as well as create practice profiles. Moreover, social services are regarded as significant assessment criteria of students. By advocating the practicality of moral education, we can not only avoid students' aversion towards moral education resulting from mere instilling, but also enhance their recognition of the moral education content in the course of university activities and social services, thus accomplishing the final goal of university moral education by implanting in their own value systems the state moral education ideas such as patriotism and courage to shoulder social responsibilities.

3 ENLIGHTENMENT ON CHINA'S MORAL EDUCATION

3.1 *Adhere to one-factor domination, innovate moral education ideas*

Indeed, mere theory instilling will lead to the inflexibility of education forms, individuality development being restricted and poor education effects. However, as revealed by the transformation history of American moral education, a total repudiation of concept recognition and one-sided emphasis on individual moral judgment will also direct moral education to turmoil. Yet China has been implementing the One-factor Domination method in university moral education. By sticking to the Marxist stand, opinions and approaches as well as a specific goal, we are able to conduct systematic ideological theory education with a unified standard and within classrooms, which will not only help students to set up a political belief but also form a unified value idea in ideology.

However, according to the moral education ideas of One-factor Domination, we should not only conduct a single cramming education, but also attaches importance to students' subjectivity during the teaching and educate them with abundant moral contents. With reference to that of the U.S., universities offer citizen education, so that students can get a basic knowledge of government form and its principle, cultivate sense of participation as well as understand citizen responsibilities; universities offer history education, so as to stimulate students as well as strengthen their cohesive and centripetal force by exploring traditional culture and enhancing national spirit; universities offer living education, by emphasis on cultivating the good virtue of students in learning, living and daily communication, try to blend in daily life the cultivation of values, ideals and faiths, and code of ethics. Through popularizing moral education in all the courses, we will be able to fully display course characteristics and the initiative of various teachers. Meanwhile, by use of living education, behavior guidance and consultation, we can also have all the problems confronting students in their learning and living solved. In this way, we will not only be able to combine patriotism and national spirits with moral character and belief of individuals, but also efficiently avoid students' potential aversion towards moral education resulting from mere instilling, thus greatly benefiting our universities to conduct moral education with Chinese features.

3.2 Stress combined radiation, expand moral education approaches

The U.S. moral education is integral and systematic, with a trinity of the society, schools and families sharing the same goal and fully displaying the role of each other, which forms huge join forces in moral education. Considering this, Chinese moral education should also develop an integral pattern with such a trinity, give full play to group power,

and establish a sound education atmosphere of everyone conducting moral education at anytime.

The society is the main practice field for students to internalize moral ideas instilled by universities into individual value and externalize it to actions. Therefore, our universities should pay enough attention to the connection with it, and try to make students understand more of the history as well as enhance the sense of national honor by visiting constructions of material, cultural and ethical culture and upholding national spirits. We should carry out social activities combined with students' majors, help them to locate themselves in the society and strengthen sense of participation in social undertakings. We are supposed to carry out community services and various social voluntary activities, so as to reinforce their sense of social responsibilities. With the help of the popularity of the media, we should first get a tight hold of public opinion, and then direct students to break away with ill influences and build correct values by way of multiple mass media forms. Universities ought to, by use of social forces during the social activities, direct students to display their subjectivity and establish correct individual value as well as good moral quality. Universities are the main fields of moral education, that's why we should make full use of such carriers as campus culture and university network media to expand more education methods, as well as help to improve moral education. Families are home to moral education and individual value start here. However, after students enter universities, their parents are in loose contact with the universities and unable to keep up with their on-campus situation. Therefore, families fail to further play their role in student moral education. Today, we should borrow experiences from the U.S., taking modern network and communication science and technologies as carrier; establish an enduring and timely communication platform between universities and families. By keeping in constant touch with universities, parents will be able to keep up with the on-campus situation of their children and make corresponding reactions. That's how families can be involved in student moral education and facilitate universities to achieve the goal of moral education.

3.3 Stress organization security, improve the contingent of moral education

To carry out university moral education, it is necessary to build a hierarchical professional team of moral education with high morality, business skills and decent style. In the U.S., student work has always been an integral part of school administration. Apart from requisite moral teachers, the U.S. universities also recruit professional talents of high degree as psychological consultants, student advisors and guides. These talents will give all-round guidance to students in their learning, living and work, and direct them to develop values. While in Chinese universities, moral education contingent consists of two specialized contingents: one is the teachers of Marxist Theory, and the other is the team which is responsible for the university's work of the Party, League and student. As to the former contingent, it is supposed to be equipped with profound Marxist Theory theoretical attainment, high morality and sound academic background; it is supposed to practice individual capabilities in classroom instilling and enhance the effects of student moral education. As to the latter one, it is made up of the student workers who are in the most frequent contact with students. Therefore, for the purpose of improving the effects of moral education, it is necessary to improve this contingent's quality first by introducing high-degree specialists, strengthening systematic learning of Marxist basic theories in the work, cultivating strong scientific research abilities and promoting a change towards specialists and scholars-based. Through this, our university contingents will be able to combine theoretical research and practice in the process of student moral education, which will not only promote the academic levels and work competence of student workers, but also help to explore the establishment of an updated moral education theoretical system with Chinese characteristics, thus laying a foundation of theory and experience for better moral education development of universities.

REFERENCE

[1] Rong Zeng, Li Hong, The characteristics of American moral education and It's inspiration, Education Exploration, 2012.

Education Management and Management Science – Zheng (Ed.)
© 2015 Taylor & Francis Group, London, ISBN 978-1-138-02663-6

Framing global brands: Frame alignment strategies on legitimacy in host country markets

Jing Huang
Economics and Management School, Wuhan University, Guangzhou, P.R. China
Economics and Management School, Jinan University, Guangzhou, P.R. China

ABSTRACT: Marketing scholars have focused increasing attention on legitimacy and markets development. However, little is known about the rhetorical strategies used to legitimate the global brands in host country markets. Drawing from fame theory and institutional theory in sociology, this article integrates the research findings of social movement and legitimacy, and conceptually proposes frame alignment strategies, which are hypothesized to have positive impacts on global brand legitimacy and increase the evaluation of global brands in host country markets.

Keywords: global brands; frame alignment strategies; legitimacy

1 INTRODUCTION

Developing and managing brand image is an important part of a company's global marketing program. Communicating a clearly defined brand image enables consumers to identify the needs satisfied by the brand (Park et al. 1986).

Many firms face challenges of legitimating in foreign markets. Developing a legitimacy-based image strategies provide the foundation for creating new markets in host countries. Firms are successful to the degree to which they can successfully navigate the environment.

Scholars have shown how impression management is a central part of legitimation (Ashforth & Gibbs 1990; Zimmerman & Zeitz 2002). According to this perspective, the management of legitimacy often involves targeted and even manipulative rhetoric aimed at presenting issues in a way that promotes the interests and protects the power position of specific actors (Elsbach & Sutton 1992; Elsbach 1994; Brown & Jones 2000). One main finding is that a successful framing requires that the audience can link the message to other discourses and identify with the key concepts and arguments. Scholars have then singled out specific elements in rhetorical justification and identified rhetorical legitimation strategies (Suddaby & Greenwood 2005).

Although firm legitimate strategies have been discussed in the global context, the relationship between global brand rhetorical strategies and legitimation has not been well understood. This article introduces global brand frame alignment strategy as an important brand communication approach of the legitimate process in host country markets.

2 REVIEW OF LEGITIMACY AND LEGITIMACY STRATEGIES

2.1 Legitimation and legitimacy

Legitimation is the process through which a product, idea or industry becomes commonly accepted (Dowling & Pfeffer 1975). Legitimation is the process of making a practice or institution socially, culturally, and politically acceptable within a particular context (Johnson et al. 2006; Suchman 1995). Legitimacy is solidified by a network of norms and beliefs—"the legitimate order"—that make some forms of power legitimate and some forms of power illegitimate (Weber 1978). Theory of legitimacy focuses on the way this network of norms and beliefs is constructed and maintained for a particular entity. Since Weber's initial theorization, legitimacy has been refined into a multidimensional construct.

2.2 Types of legitimacy and legitimacy-building strategies

Previous research has examined three types of legitimacy: cultural cognitive, pragmatic, and normative (Scott 1995; Suchman 1995). Cultural cognitive legitimacy refers to the degree to which an industry can be classified, understood, and integrated within existing congnitive schemas and cultural frameworks (Aldrich & Fiol 1994; Scott 1995; Shepherd & Zacharakis 2003; Suchman 1995). Pragmatic legitimacy rests on the self-interested calculations of an organization's most immediate audiences. Often, this immediacy involves direct exchanges between organization and audience; however, it also can involve broader political, economic, or social interdependencies, in which organizational action nonetheless visibly affects the audience's well-being. (Suchman 1995). Normative legitimacy refers to the degree to which an industry is viewed as being socially acceptable according to dominant norms and values (Ruef & Scott 1998;Scott 1995).

Legitimacy building is generally a proactive enterprise, because managers have advance knowledge of their plans and of the need for legitimation. Suchman (1995) proposed three clusters of legitimacy-building strategies: (a) efforts to conform to the dictates of preexisting audiences within the organization's current environment, (b) efforts to select among multiple environments in pursuit of an audience that will support current practices, and (c) efforts to manipulate environmental structure by creating new audiences and new legitimating beliefs. All three clusters involve complex mixtures of concrete organizational change and persuasive organizational communication (cf. Dowling & Pfeffer, 1975). However, these legitimacy strategies clearly fall along a continuum from relatively passive conformity to relatively active manipulation (cf. Oliver 1991).

2.3 Marketing and legitimacy

Previous research on legitimacy in marketing has studied the acceptance of brands (Fournier 1998; Holt 2002; Kates 2004), subcultures (Kozinets 2001), and business practices (Deighton & Grayson 1995), pointing to mechanisms that range from explicit manipulation of legitimacy through social cues and actions (Kates 2004; Kozinets 2001) to implicit manipulation of affective attachment through integration into daily life (Fournier 1998), the use of cultural scripts (Holt 2002), discourses (Thompson 2004), the transformation of foreign brand (Rui Guo et al. 2010), and the country of origin (Ling Zhou & Tao Wang 2012).

This previous research has tackled a number of disparate phenomena under the topic of legitimacy, a clearly articulated theory of the legitimation process and its relationship to market-oriented behaviors has yet to be posed and developed by marketing scholars; and little is known about the rhetorical strategies used to legitimate the global brands in host country markets.

3 A FRAMING PERSPECTIVE ON LEGITIMACY IN HOST COUNTRY MARKETS

Frame analysis has been used extensively in sociology to study changes in political and cultural discourses over time (Ferree & Merrill 2000; Gamson & Modigliani 1989; Johnston & Baumann 2007). From a framing perspective, this article theorizes the process of international marketing as an institutional process, provides frame alignment strategies for global brands to gain the legitimacy in host markets, and tries to explain the mechanism and boundaries.

4 THEORY AND RESEARCH PROPOSITION

The role of language in legitimation has been discussed by some scholars (Creed et al. 2002; Phillips et al. 2004; Vaara et al. 2006). Scholars have shown how issues can be framed in specific ways to advance or resist the legitimation of particular phenomena, decisions, practices, or changes in them (Martin et al. 1990; Creed et al. 2002). Different accounts can provide radically different understandings of issues, and thus lead to legitimation or delegitimation (Creed et al. 2002). Closely related, scholars have shown how impression management is a central part of legitimation (Ashforth & Gibbs 1990; Arndt & Bigelow 2000; Elsbach & Sutton 1992; Elsbach 1994; Staw et al. 1983; Zimmerman & Zeitz 2002). Lately, scholars have then singled out specific elements in rhetorical justification and identified rhetorical legitimation strategies (Green 2004; Suddaby & Greenwood 2005).

Frames are "individual cognitive structures … that orient and guide interpretation of individual experience" (Oliver & Johnston 2000). They enable a person to "selectively punctuate and encode objects, situations, events, and experience within one's present and past environment" (Snow & Benford 1988). To accomplish this punctuation and encoding, they "draw from the supporting ideas and norms of ideologies, but are understood as more specific cognitive structures advanced by social actors to shape interpretation and understanding of specific issues" (Johnston & Baumann 2007). Frames are the linguistic tools by which social actors attempt to manipulate legitimacy over time.

In sociology, not only has the framing concept been applied most to cognitive psychology, linguistics, discourse analysis, communication, media and political science, but also interest in framing processes in relation to the operation of marketing has animated an increasing amount of conceptual and empirical scholarship. Scholars have suggested that the framing of an object or concept in the media can have a profound effect on its legitimacy (Gamson & Modigliani 1989; Lakoff & Ferguson 2006).

Humphreys (2010) offered a new perspective on the creation of markets by viewing it as a process of legitimation (Dowling & Pfeffer 1975;

Handelman & Arnold 1999; Suchman, 1995), and assessed the deployment of framing strategies by managers, and examined company press releases and interviews with executives and reanalyzed quotations from industry executives in the newspaper data set. The research found that casino proponents employ the framing strategies of amplification, extension, and bridging at different points in the legitimation process. Stakeholders use specific frames to shape the perceived legitimacy of an industry and these frames are effective in markets.

What are the best strategies for becoming the accepted, dominant industry in a crowded competitive landscape. Research on social movements (Snow & Benford 1988) might be helpful to conceptualize the ways industries and firms compete for legitimacy through cultural frames (Humphreys 2010).

Such research can be applied here to assess the ways that framing strategies taken by executives to legitimate the international firms in host countries. The author extends this logic to argue that any internationalization process in a new market requires the same strategic efforts. International firms can benefit from its framing strategic implications for gaining foreign consumers' frame resonance in different legitimacy process in the host country.

Four rhetorical strategies have been outlined by previous research: amplification, extension, bridging, and transformation. Amplification is "the idealization, embellishment, clarification, or invigoration of existing values or beliefs" (Benford & Snow 2000).

In the case of the global brand marketing, amplification occurs when the company representatives emphasize the advantages of the products and service, the history and culture of the companies. This strategy idealizes the practice, making it congruent with the customers' value in host countries. Extension is a framing strategy in which proponents enlarge the initial concept, "extending it beyond its primary interests to include issues and concerns that are presumed to be of importance to potential adherents" (Benford & Snow 2000). Extension in the case of global brands takes the form of enlarging the initial concept of a product or service. The strategy of extension can be used to appeal to multiple stakeholders, like consumers, investors, and local partners. Bridging is the "linking of two or more ideologically congruent but structurally unconnected frames regarding a particular issue or problem" (Benford & Snow 2000). Bridging in the case of global brands takes the form of linking two cultures or concepts. Transformation refers to "changing old understandings and meanings or generating new ones" (Benford & Snow 2000). For example,

changing consuming habits in the host country markets, like coffee consumption and life styles.

The deployment of frame alignment strategies by the executives can be found from company press release and interviews with managers and reanalyzed quotations in the newspaper and on main comprehensive financial websites. Based on the literature and discussion, this article conceptually proposes the following:

Proposition 1: Amplification framing strategy will have a positive impact on the global brand legitimacy.

Proposition 2: Extension framing strategy will have a positive impact on the global brand legitimacy.

Proposition 3: Bridging framing strategy will have a positive impact on the global brand legitimacy.

Proposition 4: Amplification framing strategy will have a positive impact on the consumers' evaluation of global brands.

Proposition 5: Extension framing strategy will have a positive impact on the consumers' evaluation of global brands.

Proposition 6: Bridging framing strategy will have a positive impact on the consumers' evaluation of global brands.

Proposition 7: The global brand legitimacy will mediate between the frame alignment strategies and evaluation of global brands.

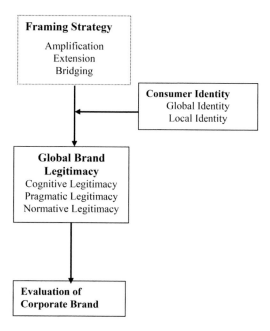

Figure 1. Frame alignment strategies, global brand legitimacy and evaluation of corporate brand.

Studies of social judgment found the way individuals respond to cultural cues depends on their cultural identity structure. Individuals with global identity tend to assimilate to cultural cues, yet individuals with local identity contrast to these cues. Therefore, the consumers' global/local identity will influence the effect of frame alignment strategies.

Proposition 8: The consumer global/local identity will moderate between the frame alignment strategies and evaluation of global brands.

In this article, alignment strategies for global brands have been conceptually proposed and they need to be empirically examined. Frame alignment strategy is analyzed as a brand communication strategy for global brands to enter new foreign markets and gain legitimacy. A grasp of legitimacy dynamics in new markets give managers methods to navigate complex host environments. This study also increases theoretical understanding of the role of frame and rhetoric in brand legitimacy among the host country markets.

REFERENCES

[1] Ashforth, Blake E., & Gibbs, Barrie W. 1990. The double-edge of organizational legitimation. *Organization Science* 1: 177–194.

[2] Benford, Robert D. & David A. Snow. 2000. Framing Processes and Social Movements: An Overview and Assessment. *Annual Review of Sociology* 26 (1): 611–40.

[3] Creed, Douglas, Maureen Scully, & John Austin 2002. Clothes make the person? The tailoring of legitimating accounts and the social construction of identity. *Organization Science* 13/5: 475–496.

[4] Dowling, John & Jeffrey Pfeffer. 1975. Organizational Legitimacy, *Pacific Sociological Review* 18 (1): 122–36.

[5] Giesler, Markus. 2012. How Doppelganger Brand Images Influence the Market Creation Process: Longitudinal Insights from the Rise of Botox Cosmetic. *Journal of Marketing* 76 (6): 55–68.

[6] Goffman, Erving. 1974. *Frame Analysis: An Essay on the Organization of Experience*, Cambridge, MA: Harvard University Press.

[7] Humphreys, Ashlee. 2010. Megamarketing: The Creation of Markets as a Social Process. *Journal of Marketing* 74 (2): 1–19.

[8] Martin, Joan, Maureen Scully, & Barbara Levitt. 1990. Injustice and the legitimation of revolution: Damning the past, excusing the present, and neglecting the future. *Journal of Personality and Social Psychology* 2: 281–290.

[9] Suchman, Mark C. 1995. Managing Legitimacy: Strategic and Institutional Approaches. *Academy of Management Review* 20 (3): 571–611.

[10] Vaara, Eero, & Janne Tienari. 2008. A discursive perspective on legitimation strategies in MNCs. *Academy of Management Review* 33/4: 985–993.

[11] Zimmerman, Monica A., & Zeitz, Gerald J. 2002. Beyond survival: Achieving new venture growth by building legitimacy. *Academy of Management Review* 27: 414–43.

Education Management and Management Science – Zheng (Ed.)
© 2015 Taylor & Francis Group, London, ISBN 978-1-138-02663-6

The relationship between the primary school enrollment and economic growth speed in Hubei province

Jin Luo

Wuhan Textile University, Hubei, Wuhan, China

ABSTRACT: Migration in China is mainly due to economic factors. The rapid growth of the economy will attract population migration, and population growth will increase the regional children's enrollment. We will research the relationship between speed of regional economic growth and primary school enrollment.

Keywords: Hubei province; educated population; enrollment; GDP

1 INTRODUCTION

With the acceleration of the process of socio-economic development and urbanization, migration and mobility of the population have become a very common phenomenon. Compared with the common international definition, as the individual Chinese household registration system, our definition to population migration and population mobility has Chinese characteristics. The mobility population is provisionally left their domicile, eventually to return; the migration population is usually to change the domicile, usually no longer return to the original resident ground.

Many reasons are there for population migration, but the main reason for the migration of China's current population is undoubtedly because of economic factors. When the economic development of a region, as Hubei province, is significantly higher than the surrounding areas, a large number of labor migrations will enter, on the anther hand, migration will accelerate the development of the regional economy; this will accelerate the development of the local Gross Domestic Products (GDP). Population migration is a slow and long process, so the impact is a slow and long-term.

The migration, including children, as the school-age children migrate, education authorities must pay attention to prepare for making arrangements to ensure this part of the school-age children should be compulsory. Even has a comprehensive system of household registration in China, predict migration is still very difficult. It is more difficult to forecast due to population migration brings school-age children, because with Chinese habit, there is some migration, although have a new domicile, still placed the children in the original place of residence (native place) to receive an education.

Since the affect each other due to the migration and GDP, while migration changes the number of school-age children, so we consider that there will be a link between the changes of children enrolled number in GDP. Taking into account the stage of compulsory education in China, especially in primary school enrollment rate has remained at more than 99%, this article will be in primary school enrollment changes.

2 ANALYSIS THE POPULATION DATA OF BASIC EDUCATION IN HUBEI PROVINCE

2.1 *The relationship between the enrollment change and economic growth in Hubei province*

We make Hubei province as the research objective, and take into account the migration from the scope of the whole of China. Firstly, let us research the primary school enrollment in Hubei province. As described previously, compulsory education in China, the enrollment rate of school-age children have been maintained at more than 99%, it is not possible to research the problem from the enrollment rates, so we need to research the relationship between the absolute number of children enrolled and the number of children born.

The age of children go into primary is 6-year-old, so we compared the enrollment (**Enr**) of children and the number of children born in corresponding year, the 6 year before. As shown in Table 1, taking into account the natural death of the child, that the enrollment is greater than the

Table 1. The enrollment and birth number in Hubei province.

Year	Enr (1000 person)	Bir (1,000 person)	Na
1995	1,289	1101.3	−0.171
1996	1,250	1144	−0.093
1997	1,160	1133.4	−0.024
1998	1,073	1056	−0.017
1999	997	1125.5	0.114
2000	905	1033.1	0.124
2001	804	929.6	0.135
2002	712	932.4	0.236
2003	652	866.2	0.274
2004	600	740.9	0.191
2005	567	685.2	0.172
2006	570	577.6	0.013
2007	611	508.4	−0.201
2008	626	501	−0.25
2009	642	495	−0.297
2010	680	506.6	−0.343
2011	692	526	−0.316

Table 2. The Hubei and Chinese GDP.

Year	Hubei	Country	Ggr
1979	0.156	0.076	−0.08
1980	0.064	0.078	0.014
1981	0.065	0.053	−0.012
1982	0.119	0.09	−0.029
1983	0.059	0.109	0.06
1984	0.209	0.152	−0.057
1985	0.162	0.135	−0.027
1986	0.055	0.089	0.034
1987	0.084	0.116	0.032
1988	0.078	0.113	0.035
1989	0.045	0.041	−0.004
1990	0.05	0.038	−0.012
1991	0.066	0.092	0.026
1992	0.141	0.142	0.001
1993	0.13	0.14	0.01
1994	0.137	0.131	−0.006
1995	0.132	0.109	−0.023
1996	0.116	0.1	−0.016
1997	0.119	0.093	−0.026
1998	0.086	0.078	−0.008
1999	0.078	0.076	−0.002
2000	0.086	0.084	−0.002
2001	0.089	0.083	−0.006
2002	0.092	0.091	−0.001
2003	0.097	0.1	−0.003
2004	0.112	0.101	−0.011
2005	0.121	0.102	−0.019
2006	0.132	0.116	−0.016
2007	0.146	0.119	−0.027
2008	0.134	0.09	−0.044
2009	0.135	0.091	−0.044
2010	0.148	0.104	−0.044
2011	0.138	0.092	−0.046

number of births (**Bir**) in the same year, it certainly due to the impact of immigration. We calculate the difference between the two numbers and come to the proportion as **Na** is (Bir-Enr)/Bir.

On the other hand, as shown in Table 2, we research Hubei province and the country's GDP, found that the Hubei province GDP and the pace of development in China, precisely because of the existence of such differences, which leads to national wide population mobility existence, clearly that the economic development in Hubei province higher than the national, then there will be the flow of population from outside to Hubei province. We calculate the difference between Hubei province and the national GDP growth rate, as **Ggr** is the country's GDP minus Hubei's.

Two of the rate of change in the contrast, we found that make the year of children enrolled as a standard, 13 years in advance, there are obvious links between the Na and Ggr, if ignored the tiny differences in individual years, they have the same positive and negative, shown as Figure 3. The same sign can be interpreted as in Hubei province, the place where economic development was achieved more quickly than the whole country, can attract population migration come in. The school-age children accompanied with migration will increase the enrollment. The time gap between the two rates, 13 year, can be interpreted as the time required for this trend.

The more rapid economic growth will attract immigrants to come in and the immigrants will increase enrollment population, and this effect will be appeared after a longer period of time (maybe 13 years). As the economic continue rapid growth in Hubei province, there will be more population immigrant entering, the compulsory education enrollment growth will be sustained for a long time, the education authorities need to plan for it in advance.

2.2 Analysis the population data of basic education in Hubei province

As take seriously in education, further consolidating and improving the level of compulsory education, Hubei province achieves a better result in basic education. In 2010, Hubei province has 3.65 million pupils, primary school enrollment net rate over

99.96%; have 2,184 junior schools and 2.18 million students in the junior school, net enrollment ratio is 99.63%; the general senior middle school students is 1.23 million; the new enrollment of secondary vocational school is 0.34 million, have students 1.12 million, increase over 51.5% as 2005; the secondary gross enrollment rate of 87.2%.

Since 1995, compulsory school-age child's enrollment rate of the Hubei province stage has remained above 99%, the gross enrollment rate of junior school is more than 97%. So, it is not necessary to analyze the basic education enrollment, then we analyze the relationship between the number of children enrolled in the absolute number of children born in the corresponding year. To facilitate statistical calculations, we will unify the children enter primary school age as 6 years old, and in order to correspond with the statistical data, we are ignoring the data entry errors stipulated time education sector brings.

As shown in Table 1, the corresponding school-age children enrolled in comparison with the corresponding born number of six years before, taking into account factors that naturally died, it is clear that in a closed environment, the enrollment number should be less than the number of enrollments children. According to the analysis in Figure 1, from 1999, in Hubei province, the enrollment greater than the born number in corresponding years, this happens may only be affected due to the inward migration.

To for this study, make the year when children enrolled as the standard, research the differences in the rate, we found that 13 years earlier, the same change in the relationship between the ratio and the difference in the rate of enrollment of children. Ignoring the small differences in individual years, the rate of change between two identical negative relationships, this result can be interpreted as: Is relevant that the speed of economic development in Hubei province with after 13 years the number of children enrolled, the economic development

in Hubei province faster than the speed of economic development in the nation, it can attract migration from other provinces. This is a long-term impact, because the differences in the rate of elastic strength of Hubei province showed economic growth within the region rather than a direct reflection of the strength of economic growth. Accompanied by migration brought into the school population changes will lag in migration change, so the difference of 13 years better reflects the long-term nature of this change, while indicating that the period ahead will be to maintain this trend.

2.3 Forecast the basic education population in Hubei province

The number of educated population will continue to maintain a slight increase in the number of the existing basis. With the national economic development and population policy implementation, the number of basic education population change significantly, primary school enrollment from 1995 to 1.29 million a diminishing year by year to 0.567 million in 2005, reducing the rate reached 56%, and primary school enrollment began to rise slowly.

Analysis and comparison of the number of people born in recent years, and since the economic development in Hubei province faster than the national average, we have reason to believe that in the next few years of primary and secondary school population will continue to grow. Make a linear regression based on the birth rate of population and economic development in Hubei province since 2005 we predict primary school enrollment will reach 0.756 million in 2014 and will maintain the annual growth in the next five years as 18,000 human scale and the peak will appear in 2022, the number reached 0.85 million. If we take into account the impact of 'separate second child' population policy factors, this number will be higher. Primary school population changes will directly affect the management of basic education.

Since 1995, accompanied with the enrollment sharp decrease, the primary school in Hubei province the number and the scale in rural, at the same time the urban elementary number unchanged. In 2009, Hubei province total primary school (including teaching points) is 976, the number of classes is 17,475. If the number of primary and secondary school enrollment in the coming years continue to grow, then the education authorities must be prepared, including increased education funding, adjust the layout of the schools, and to increase the number of primary school teachers.

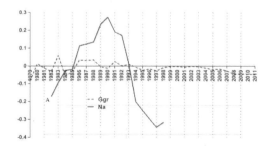

Figure 1. The relation between **Ggr** and **Na**. A is the year began statistics enrollment, 1995.

3 ANALYSIS THE POPULATION DATA OF BASIC EDUCATION IN HUBEI PROVINCE

The government departments in Hubei province are willing to invest in education, strict regulation of education, while a good education conditions and educational environment conducive to attracting talent and stimulate economic development. With the advent of peak into the population (Natural factors and policy implications) and foreign population, the population increases every year. The basic education situation will occur within the scope of Hubei province, primary school population will steadily improve over the next 10 years, response to increasing population trends in education, relevant education sector must be arranged in advance. The research of the specific quantitative impact of changes in the pace of economic growth and enrollment will be a later stage, the impact of specific factors are also worth exploring.

REFERENCES

[1] Che Wei-ping. 2008. The Effect of Changes of the Scale of Fiscal Expenditure on China's Education Allocation on Economic Growth and International Comparison. *Journal of Hebei University.* *4:29–34.*

[2] Chen YeLing. 2011. Analysis of the Return Model of Educational Investment Decision. *Journal of Educational Studies.* *7:103–107.*

[3] Ding XiaoHao. 2003. Analysis of Higher Education Enrolment Expansion on economic growth and increase employment. *Exploring Education Development.* *9:110–121.*

[4] Jin Fang. 2009. Correlation Analysis of Diversified investment on Higher Education and economic growth. *ShanDong Economy.* *4:153–157.*

[5] Liang Yanling. 2000. The Education of the population forecast. *Journal of Guangxi Normal University.* *36:48–52.*

[6] Shen Baifu. 2012. China's rural education investment analysis. *Rural Education.* *12:47–51.*

[7] Zhang Xiang-qian. 2002. Analyses of Human Resource and the Regional Economy Development. *Journal of Operations research and management science.* *5:117–122.*

[8] Zhuo Zuo-jie. 2011. The Study of Nonlinear cointegration reltaionship between Educational Investment and Economic Growth. *Journal of Guangxi University of Finance and Economic.* *8(24):59–65.*

[9] 2012 HuBei statistical yearbook. *China statistics Press.* (2012. 6).

Education Management and Management Science – Zheng (Ed.)
© 2015 Taylor & Francis Group, London, ISBN 978-1-138-02663-6

Competitive advantage and dynamic adjustment of capital structure

Dongqin Zhu

Dongwu School of Business, Soochow University, Suzhou, China

ABSTRACT: Using Tobit regression model based on Chinese listed companies, this paper examines the characteristic of capital structure for Firms with Competitive Advantage (FCAs) and its adjustment under the dynamic environments. The results show that FCAs' debt levels are lower than others. Further, FCAs will significantly increase their debt levels if the environmental dynamics increases. The findings of this paper indicate that FCAs display a tendency to combine conservative with flexibility in their capital structure policy.

Keywords: capital structure; competitive advantage; environmental dynamics

1 INTRODUCTION

It is a hot topic for a long time in the corporate financial field about how to choose a reasonable capital structure. Modigliani and Miller put forward the famous MM theorem in 1958. Since then, scholars relax the assumption of MM theorem step by step, and present various capital structure theories, such as pecking order theory, signal transmitting theory, agency theory and trade-off theory. With the development of economic globalization, firms have to face more and more changing environment. As a result, some financial scholars begin to pay much attention to the influence of the changes of environment on firm's capital structure. Important findings in the literature show that competition in the product markets and macro economic changes will bring different capital structure (Brander and Lewis, 1986; Robert and Levy, 2003; Su and Zeng, 2009).

Sustainable competitive advantage for a firm is an ability to maintain its leading position in the long-term competition in the industry. This ability can make a firm to create new competitive advantages before rivals successfully imitate or surpass it. However, a firm is more easily influenced by Schumpeter shock in the uncertainty of environment, which means that a firm will be probably destroyed or lose its competitiveness in a very short time. Experts in the strategy management areas point out that the sustainability of competitive advantage can be fully presented only in the dynamic environment (Eisenhardt and Martin, 2001). Therefore, with regard to financial field, it is interesting to study the following questions: Whether is the capital structure of FCAs different from other common firms? Whether will

FCAs adjust their capital structure on time with the changes of environment? This paper will build a tobit regression model to empirically test the financial feature for FCAs and answer the above questions. This paper will make up the lacks of existing literature research in this area.

2 HYPOTHESES

Sustained competitive advantage can keep a firm earn and grow over its industry rivals in the market competition. It often arises from firm's unique resource and ability. A strategy-oriented competitive advantage always emphasizes the sustainability of competition. As an important financial decision for a firm, the decision of capital structure must adapt to the firm's strategic management. Barton and Gordon (1987) ever emphasized that any capital structure is defensive. Besides capital structure, financial policies should follow firm's strategy. Therefore, it is obvious that the decision of capital structure should be put in the organizational objective, and can be well understood on the point of the strategic management. Andrews (1980) also pointed out that financing policy, as one of firm's basic financial policies, should support firm's long-term policies and be consistent with them. In practice, capital structure policy itself is in one part of corporate strategy. As to FCAs, they usually want to keep sufficient and stable internal cash in order to avoid financial risk, which helps their strategy to be enforced successfully. As a result, FCAs' bankruptcy costs are usually very low even if they encounter economic recession. Bettis and Prahalad (1983) thought that keeping low capital structure can increase firm's flexibility in operating activities. Based on the above analysis, this paper points out the following hypothesis:

H1: FCAs' debt level is lower than that of common firms.

In order to acquire sustained competitive advantage, the higher the environment dynamics, the more likely that FCAs are inclined to invest in special resource, technology and talent. Through this way can FCAs prevent industry rivals from simple imitation and thus improve their defensive capability. Given this objective, FCAs is likely to increase their cash requirements in dynamic environment. On the other hand, from a perspective of debt governance role, debt can increase restrictions on managers and align interest of managers with that of outside shareholders (Jensen and Meckling, 1976; Robert and Levy, 2003). Harris and Raviv's model showed that the debt information of a firm can be transmitted to investors through which investors can monitor managers' behaviors. They concluded that debt is a good monitoring way. Narayanan (1988) further pointed out that the probability of debt being overvalued is lower than that of equity. If a project is financed by debt, it will accept a positive NPV with no doubt. This view indicates that debt can reduce overinvestment by constraining the behavior of managers in a firm. Given the perspective of debt governance role, increased cash requirements in FACs can be satisfied by increasing debt levels, which helps to align the managers with shareholders. Moreover, with their high profitability in the past they seldom encounter financial constraints. As a result, it is not difficulty for FCAs to acquire loans from commercial banks if they require them. Based on the above analysis, this paper put out the following hypothesis:

H2: FCAs will increase their debt levels if environment becomes more dynamic, while common firms' debt levels probably do not change even if environment becomes more dynamic.

3 SAMPLE AND MODEL

3.1 Sample

This paper collected China's A-share listed firms in the manufacturing industry from 2007–2009 as our initial sample. It was further classified into 10 sub industries according to the classification standard of "Industrial Classification Guidance for Listed Firms" promulgated in 2001 by China's Securities Regulatory Commission. Those firms with missing, incomplete and abnormal data or negative owners' equity in the initial sample were deleted. The paper finally got 710 firms for three consecutive years in the sample which had 2130

panel data. The financial and other data in this study came from Database on Chinese Security Market provided by the Centre of China's Economic Research. Some missing data were acquired from firm's annual reports by hand.

3.2 Identification of FCAs

In the theory of strategic management, how to accurately define the competitive advantage of enterprises is considered to be very difficult. In this paper, we used principal component analysis to comprehensively calculate the index of competitive advantage for each firm. The analysis includes three indicators which are profitability, growth and market share. This index method used here can better overcome the limitations of single indicator. The profitability indicator uses firm's excess returns as proxy. The excess return for each firm for every observed year was measured by each firm's return of equity minus its sub manufacturing industry median. The growth for each firm for every observed year was measured by the growth of the operating income. The market share for each firm for every observed year was measured by the ratio of its operating revenue to the total of the sub manufacturing industry. After getting the index of competitive advantage for each firm for each observed year calculated by the principal component analysis, this paper further screened out FCAs in the sample. If a firm's comprehensive indexes of competitive advantage for three consecutive years were higher than the median of its sub industry, this firm was identified as FCA. The rest firms in the sample were considered as the common ones which index of competitive advantage was lower than the median of sub industry in at least one year. The method used in this study to obtain FCAs in the sample can effectively delete those firms due to the occasional success in the market opportunities in a short time, and better reflect the capability for FCAs that cannot be easily imitated or substituted.

3.3 Model

This paper builds up the following model to test the hypotheses:

$$CS_{i,t} = \beta_0 + \beta_1 Comp_i + \beta_2 DYN_{i,t} + \beta_3 Comp_i$$
$$\times DYN_{i,t} + \beta_4 Size_{i,t-1} + \beta_5 FDAssets_{i,t-1}$$
$$+ \beta_6 SOE_{i,t-1} + \beta_7 MNGshare_{i,t} + \beta_8 BAL_{i,t}$$
$$+ \sum_{i=9}^{17} \beta_i IND_i + \varepsilon_{i,t}$$

In the above model, Capital Structure (CS) is the explained variable. As high accounts payable and accounts receivable in advance can cause high ratio of liability to asset, using the ratio of liability to asset to measure capital structure has limitations. Reference to Xie and Chen (2009), this paper uses the financing debt level to measure capital structure. Financing debt level comes from financing activities which includes short-term loans, notes payable, long-term debt due within one year, long-term loans, bonds payable and long-term payable. Compared to accounts payable and accounts payable in advance, a firm often wants to pay interest charges for financing debt to commercial banks or other financial institutions.

This paper uses Tobit model to regress the above model because the explained variable, the financing debt level, is an interval one in the model, and the financing debt level of a few firms in the sample is zero.

The explanatory variable in the model is *Comp*, *DYN* and their interaction variable. *Comp* is a dummy variable. This paper uses common firms in the sample as a basis, if an observed firm is considered as a FCA, Comp equals 1, and otherwise it equals 0. *DYN* is represented by coefficient of variation for return of equity for each firm's industry from 2007 to 2009. The interaction variable in the model is used to test whether the choice of capital structure is different in the dynamic environment. Its coefficient and direction should be paid an attention in testing hypothesis 2.

By reference to previous studies, this paper use firm size (*Size*) and the proportion of fixed assets (*FDAsset*) as two control variables to control the effects of firm size and fixed asset size on capital structure. Large firms generally like to have high financial leverage. The firm size variable is measured by the logarithm of total assets for the observed firm in the observed year. Fixed assets are usually regarded as mortgage in financing. Previous literature shows that their sizes often have a positive relation with the debt level. The proportion of fixed assets for the observed firm equals the net fixed assets divided by its total assets in the observed year. Given that the debt levels for observed firms at the end of t year were probably influenced by firm size and net fixed assets at the beginning of t year, the data of total assets and net fixed asset came from the observed firms' last year. Previous studies show that firm's property has an important impact on the debt level. The Father Effect that governments bring to State-Owned Enterprises (*SOEs*) causes that these firms don't want to care much about financial risk (Xie and Chen, 2009). As a result, SOEs usually have higher debt levels than those in private ones. *SOE* in the model is a dummy variable which equals 1 if the observed firm's ultimate ownership is state-owned, and otherwise it equals 0. Besides firm's characteristic and property

factors, corporate governance mechanism also has important influence on capital structure. Given the existence of information asymmetry, if a firm is lack of incentives for its executives, its managers are likely to use their power to seek their private benefits and improve their control rights by enlarging firm size through debt expansion. Firms, which are lack of balance between shareholders, can easily induce that block shareholders expropriate or transfer corporate resource, thus increasing firm's agency costs and debt level. Therefore, *MNGShare* (the proportion of managers' shares) and *BAL* (ownership balance) are included in the model to control the effects of corporate governance on capital structure. In addition, industry dummies are also included in the model. According to the classification standard of manufacturing industry propagated by China's Securities Regulatory Commission in 2001, this paper classified the manufacturing industry into 10 sub ones.

4 EMPIRICAL RESULTS

4.1 Summary statistics

The results of summary statistics are shown in Table 1. The mean of financing debt level for total firms, FCAs and common firms in the sample is 0.26, 0.24 and 0.26 respectively. The results of t test indicate that the financing debt level for FCAs is 0.02 lower than that of common firms at 5% significant level. This result is consistent with the hypothesis 1, which shows that FCAs chose a lower financing debt level. With regards to the statistics of environmental dynamics, the mean for total firms, FCAs and common firms in the sample is 3.42, 3.20 and 3.44 respectively. It is noted that no difference is found about environmental dynamics between FCAs and common firms in the result of t test, which shows that the changes in the environment for two sub samples are almost the same. In the control variables, the results of summary

Table 1. Summary statistics.

Variables	FCAs		Common firms	
	Mean	S.D.	Mean	S.D.
CS	0.24	0.15	0.26	0.16
DYN	3.20	3.17	3.44	3.42
Size	9.38	0.40	9.28	0.46
FDAsset	0.31	0.17	0.31	0.15
SOE	0.41	0.49	0.38	0.49
MNGshare	0.01	0.03	0.01	0.04
BAL	0.59	064	0.52	0.51
Obs.	2130		2130	

Table 2. Results of tobit regression.

Variable	(1)	(2)	(3)
Comp	−0.04*** (−2.96)	−0.02* (−1.77)	−0.03*** (−2.80)
DYN	0.00 (0.18)	0.00 (0.54)	0.00 (0.07)
Comp × DYN	0.002** (1.96)	0.002* (1.80)	0.002** (1.99)
Size	0.09*** (11.94)		0.08*** (11.82)
FDAsset	0.19*** (8.53)		0.20*** (8.82)
SOE		−0.00 (1.08)	−0.02*** (2.70)
MNGShare		−0.10 (1.08)	−0.18* (1.95)
BAL		−0.04*** (−6.64)	−0.09*** (−6.27)
Intercept	−0.59*** (−8.47)	−0.29*** (−13.4)	−0.58*** (−8.09)
Industry	Yes	Yes	Yes
Obs.	2130	2130	2130
Log Likelihood	933.29	842.13	954.94

Note: The values shown in parentheses is the Wald ones, ***, ** and * represent significance at 1%, 5%, 10% level respectively.

statistics show that the means of firm size and ownership balance for FCAs are higher than those for common firms. As to the means of the proportion of fixed assets, whether ultimate shareholders are state-owned and the proportion of managers' shares, no significant differences are found.

4.2 Results of tobit regression

The results of tobit regression based on capital structure as the explained variable are shown in Table 2. The coefficient of *Comp* is above 1% significantly negative in the three regression equations. This result indicates that the financing debt level of FCAs is significantly lower than that of the common firms in the sample, which further supports hypothesis 1. The coefficient of environmental dynamics shows that environmental dynamics does not have real impact on capital structure. However, the coefficients of the interaction variable are significantly positive at above 1% level in the three equations, which indicates that environmental dynamics has important influence on capital structure. Specifically, the debt level of FCAs will be increased when environmental dynamics is increased. The regression results give significant evidence on hypothesis 2.

5 CONCLUSION

Based on the strategic management theory, pecking order theory and agency theory, this paper calculates the comprehensive index of competitive advantage by using principal component analysis, and examines the characteristics of capital structure and its dynamic adjustment in a dynamic environment for FCAs. The empirical results show that FCAs usually choose lower debt levels than that of common firms, which reflects the conservative financial policy in FCAs. If environmental dynamics is obvious, FCAs will adjust their capital structure. The more dynamic the environment is, the higher the financing debt level is, which reflects the flexibility of financial policy in FCAs.

The above conclusions indicate that FCAs can combine conservative with flexibility in capital structure decision. This conclusion has important meaning in firm's financial policies in operational procedure. To maintain a competitive position in the industry, low debt level helps avoid financial risk brought by the uncertainty in the enforcement of firm's strategy. It also helps prevent the vicious price and marketing competition from industry rivals which can easily cause shortage of cash in a short time for FCAs. Therefore, low debt level can support FCAs' sustainable and stable development. On the other hand, if the outside environment changes greatly, a firm needs to increase its capital structure to adjust to the changes of environment and cannot simply keep conservative capital structure.

REFERENCES

[1] Barton S.L. & Gordon, J.P. 1987. Corporate strategy: Useful perspective for the study of capital structure?. *Academy of Management Review*, 12(1): 67–75.
[2] Brander, J.A. & Lewis, T.R. 1986. Oligopoly and financial structure: the limited liability effect. *American Economic Review*, 76(5): 956–970.
[3] Eisenhardt, K. & Martin, J. 2000. Dynamic capabilities: What are they?. *Strategic Management Journal*, 21(10): 1105–1121.
[4] Jensen, M.C. & Meckling, W.H. Theory of the firm: Managerial behavior, agency costs and ownership structure. *Journal of Financial Economics*, 1976, 3(4): 305–360.
[5] Modigliani, F. & Miller, M.H. The cost of capital, corporate finance and the theory of investment. *American Economic Review*, 1958, (48): 261–297.
[6] Narayanan, M.P. 1988. Debt versus equity under a symmetric information. *Journal of Financial and Quantitative Analysis*, 23(1): 39–51.
[7] Robert, A.K. & Levy, A. 2003. Capital structure choice: Macroeconomic conditions and financial constraints. *Journal of Financial Economics*, 68(2): 75–109.
[8] Su, D. & Ceng, H. 2009. Macroeconomic factors and changes of capital structure. *Journal of Economic Research*, 6(12): 52–64.

Education Management and Management Science – Zheng (Ed.)
© 2015 Taylor & Francis Group, London, ISBN 978-1-138-02663-6

Implications of Paired Reading Schema for China's foreign language teaching

Binbin Wu

University of the Pacific, California, USA
Tianhua College, Shanghai Normal University, Shanghai, P.R. China

ABSTRACT: Paired Reading is a student-centered instruction schema with the support of peer tutoring learning strategy. More recently, constructivist learning theory is exerting an impact on foreign language teaching reform in China and leads to some attempts to apply student-centered instructional approaches to the classroom teaching environment. Paired Reading Schema can be a complement to the traditional teaching methods concerning its capability to foster student-centered learning by means of engaging student actively in knowledge construction. Hence, the purpose of this study is to inform foreign language teachers in China of a student-centered instructional approach and present its implications for China's foreign language teaching.

Keywords: Paired Reading Schema; student-centered learning; China's foreign language teaching

1 INTRODUCTION

Paired Reading is a student-centered instruction schema with the support of peer tutoring learning strategy. With its foundation rooted in Vygotsky's Zone of Proximal Development (ZDP), peer tutoring is a student-centered learning strategy engaging students in the learning process by means of a complementary relationship between student pairs. Vygotsky defined the zone of proximal development as the distance between the actual development level and the level of potential development (Vygotsky, 1978). He suggested that independent problem solving determine the actual development level and problem solving with the help from an adult or more skillful peers determine the level of potential development. Accordingly, Vygotsky encourages the cooperative learning activities where more capable peers offer assistance to less competent companions. Vygotsky's zone of proximal development provides a structured theoretical implications for peer tutoring.

However, in China, the dominant grammar-translation and audio-lingual methods of teaching English as the foreign language provide sufficient target language input; they do not foster student-centered learning environment as peer tutoring learning strategy would. Instead, they emphasize correct grammar provided by teachers. The classrooms of the grammar-translation and audio-lingual methods are teacher-centered so that teachers rarely manage to promote learners' active participation in the learning process. Although the theoretical underpinnings and the possible implementation of student-centered instructional approaches in Chinese foreign language teaching classrooms have been discussed for many years there are only a few experimental studies aiming to design an applicable student-centered instructional approach available to classroom teachers in China. The purpose of this study is to inform foreign language teachers in China of a student-centered instructional approach and address its implications for China's foreign language teaching.

2 PAIRED READING SCHEMA

Researchers, within the framework of peer-tutored learning strategy, have developed Paired Reading Schema to improve reading skills most, and also social skills and thinking skills, which is the most frequently adopted peer-tutored schema in the peer tutoring studies. Topping and Ehly (1998) presents Paired Reading from two aspects, Method and Organization. Method addresses the techniques students are expected to engage with in Paired Reading, such as correction, praise, reading together and so on. Organization involves how to organize this schema, such as pairing students. Tutor and tutee begin with reading together. If there is no error tutor praises tutee. They are encouraged to talk about the book and share their understanding of the book. If there is an error tutor allows four seconds for self-correction. If tutee can not self-correct the error tutor will say the word correctly. Tutee repeats the word behind tutor. If tutee is confident of his reading he can signal to tutor to read the book alone. As shown in Figure 1, this is a cycle.

Paired Reading Schema was reported to be effective to children at the age of eight through eighteen. The format in which Paired Reading Schema occurred was either cross-ability or same-ability and reciprocal. However, the cross-ability format in which pair students have ability gap was investigated to be more successful in promoting children's literacy development (Miller, Topping & Thurston, 2010; Topping, Miller, Thurston, McGavock & Conlin, 2011).

2.1 *Effectiveness of the schema*

Paired Reading Schema has positive effects on participants' reading skills, thinking skills and even social skills (Mckinstery & Topping, 2003; Topping, Campbell, Douglas, & Smith, 2003). The studies that evidenced gains on students' English reading skills range from kindergartens to middle schools. Those studies in relation to kindergartens and elementary schools were mostly associated with early literacy development. For example, in a classwide Paired Reading study, the phonemic awareness and letter sound recognition of kindergarten English learners in the experimental condition developed better than those of control English learners and similarly to those of kindergarten non-English learners with the same Paired Reading intervention (Mcmaster, Kung, Han, & Cao, 2008). In another study involving first graders, children in a classwide Paired Reading intervention demonstrated their improvement in phoneme segmentation fluency, nonsense word fluency, and oral reading fluency (Calhoon, Otaiba, Cihak, King, & Avalos, 2007). For adolescent students, the researchers were more interested in the effectiveness of Paired Reading on

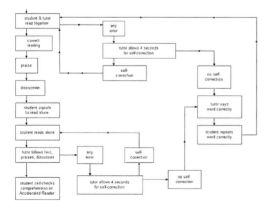

Figure 1. Paired Reading Schema (Topping & Ehly, 1998).

adolescent readers' reading comprehension. Like the findings of two studies above, it was adolescent readers in the Paired Reading group whose reading comprehension outperformed that of their peers in the control group (Fisher, 2001). Additionally, two studies witnessed the significant gains of social and thinking skills respectively among the participants in the Paired Reading context (Gumpel & Frank, 1999; Mckinstery & Topping, 2003).

2.2 *Paired reading studies in China*

There are only two experimental Paired Reading Schema studies aiming to improve English learning in China's Knowledge Resource Integrated Database (CNKI), the Chinese authority academic database (Gu, 2010; Wang, 2007). One study in the Shandong Province investigated Chinese university students' participation in Paired Reading Schema. Seventy-six freshmen from two classes participated in this study with 38 in the control group and 38 in the intervention group, who participated in same-age, reciprocal, classwide Paired Reading four times a week for 15 weeks. The instrument used to measure college students' reading comprehension was ten intact passages selected from Test of English as a Foreign Language (TOEFL). The questionnaire used to examine students' attitudes toward peer tutoring consists of close-ended 5-point Likert scale questions and an open-ended question (Fuchs, Fuchs, Mathes, & Simmons, 1997). For the post-test scores in terms of English reading comprehension, the intervention students overall significantly outperformed the control students though no significant differences between them existed at pre-test (Gu, 2010). For the pre- and posttest scores, average- and in particular low-performing students in the intervention condition demonstrated statistically dramatic reading progress, whereas high-performing students did not advance significantly.

Another study in Hunan Province investigated Chinese second-grade middle school (equal to 8th-grade in the United States) students' participation in Paired Reading Schema (Wang, 2007). Forty students selected from two natural classes participated in this study with 20 in the control group and 20 in the intervention group, who engaged in the reciprocal Paired Reading for three 45-minute sessions a week for 8 weeks. Wang (2007) used the mid-term and the final-term exam papers respectively to assess students pre-test and post-test English achievement. In this study, Illinois Foreign Language Attitude Questionnaire and Attitude and Aptitude Tests for Language Learning were used to measure students' attitudes towards and aptitude for English language learning. For Illinois Foreign Language Attitude Questionnaire and Attitude and Aptitude Tests for Language Learning, the posttest scores of students in the intervention condition were statistically higher than their pretest scores, which indicates that students in the intervention condition became more interested in and more capable of English learning after this peer tutoring project. However, in this study students in both the intervention and control conditions demonstrate statistically significant lower post-test English achievement scores than their pre-test English achievement scores. Wang (2007) explained that lack of cooperation between student pairs and failure to identify a reliable instrument to measure students' English achievement were probably two limitations of this peer tutoring study.

3 CONCLUSION

China's dominant foreign language pedagogy relies on traditional methodologies that focus on correct grammar: the grammar-translation and audio-lingual methods. (Chang, 2006). The grammar-translation method focuses on developing language learners' reading and writing skills, emphasizing learning about the grammar rules of the target language (Chastain, 1988). It has been over one hundred years since Chinese foreign language educators first adopted this teaching method at the end of 19th century (Peng, 1999). The audio-lingual method is an oral-based approach that drills students in the usage of grammatical sentence patterns. In the 1960s, the audio-lingual methods began to arouse the interest of some Chinese foreign language educators who believed that listening and speaking should be given the priority over reading and writing. More recently, constructivist learning theory is exerting an impact on theoretical discussions of foreign language teaching reform in China and leads to some attempts to apply student-centered instructional approaches to the classroom teaching environment. Paired Reading is a promising instruction schema for teaching English reading in Chinese middle schools. With its foundation rooted in peer tutoring learning strategy, Paired Reading is a student-centered instruction schema, characteristic of students' specific role-taking as tutor or tutee (Topping, 2005). Hence, Paired Reading Schema can be a complement to the traditional teaching methods concerning its capability to foster student-centered learning by means of engaging student actively in knowledge construction.

REFERENCES

[1] Calhoon, M., Otaiba, S., Cihak, D., King, A., & Avalos, A. (2007). Effects of a Peer-Mediated Program on Reading Skill Acquisition for Two-Way Bilingual First-Grade Classrooms. Learning Disability Quarterly, (3), 169. doi: 10.2307/30035562.

[2] Chang, J.Y. (2006). wai yu jiao xue fa de fa zhan ji qi dui wo men cong shi wai yu jiao xue de qi shi [The implications of the development of Foreign language pedagogy for Chinese foreign language teaching]. Foreign Language Teaching Abroad, 2006(4), 1–5.

[3] Chastain, K. (1988). Developing second language skills (3rd ed.). San Diego, CA: Harcourt Brace Jovanovich.

[4] Fisher, D. (2001). Cross Age Tutoring: Alternatives to the Reading Resources Room for Struggling Adolescent Readers. Journal of Instructional Psychology, 28(4), 234.

[5] Fuchs, D., Fuchs, L.S., Mathes, P.G., & Simmons, D.C. (1997). Peer-assisted learning strategies: Making lassrooms more responsive to diversity. American Educational Research Journal, 34(1), 174–206.

[6] Gumpel, T.P., & Frank, R. (1999). An expansion of the peer-tutoring paradigm: cross-age peer tutoring of social skills among socially rejected boys. Journal of Applied Behavior Analysis, 32(1), 115.

[7] Gu. H.Y. (2010). Tong ban hu zhu xue xi ce lue zai da xue ying yu yue du jiao xue zhong de ying yong [Application of Peer-Assisted Learning Strategies in College English Reading Teaching]. (Master thesis, Shandong University, Jinan, PRC). Available from China Knowledge Resource Integrated Database.

[8] McKinstery, J., & Topping, K.J. (2003). Cross-age peer tutoring of thinking skills in the high school. Educational Psychology in Practice, 19(3), 199–217.

[9] McMaster, K.L., Kung, S.H., Han, I., & Cao, M. (2008). Peer-assisted learning strategies: A 'tier 1' approach to promoting English learners' response to intervention. Exceptional Children, 74(2), 194–214.

[10] Miller, D., Topping, K., & Thurston, A. (2010). Peer tutoring in reading: The effects of role and organization on two dimensions of self-esteem. British Journal of Educational Psychology, 80(3), 417–433.

[11] Peng, W.Q. (1999). mei guo jiao xue fa gai ge yu zhong guo wai yu jiao xue fa [Pedagogical reform in the United States and foreign language pedagogy in China]. Educational Guide, 1999(8–9), 55–57.

[12] Topping, K.J. (2005). Trends in peer learning. Educational Psychology, 25(6), 631–645. doi: 10.1080/01443410500345172.

[13] Topping, K.J., Campbell, J., Douglas, W., & Smith, A. (2003). Cross-age peer tutoring in mathematics with seven-and 11-year-olds: Influence on mathematical vocabulary, strategic dialogue and self-concept. Educational Research, 45(3), 287–308.

[14] Topping, K.J., & Ehly, S.W. (Ed.). (1998). Peer-Assisted Learning. Mahwah, NJ: Lawrence Erlbaum Associates, Inc., Publishers.

[15] Topping, K.J., Miller, D., Thurston, A., McGavock, K., & Conlin, N. (2011). Peer tutoring in reading in Scotland: thinking big. Literacy, 45(1), 3–9. doi:10.1111/j.1741-4369.2011.00577.x.

[16] Vygotsky, L.S. (1978). Mind in Society: The development of higher psychological processes. Cambridge, MA: MIT Press.

[17] Wang, E.J. (2007). Zhong xue tong ban jiao xue de li lun yu shi jian yan jiu [Research on the theory and practice of peer tutoring in middle school] (Master's thesis, Qufu Normal University, Qufu, PRC). Available from China Knowledge Resource Integrated Database.

Education Management and Management Science – Zheng (Ed.)
© 2015 Taylor & Francis Group, London, ISBN 978-1-138-02663-6

Environmental regulation and the evolution of industrial structure

M.Q. Hu & P. Wang
Harbin Engineering University, Harbin, China

ABSTRACT: Supported by the date of our country from 2000 to 2011, we research on the effect of environmental regulation to industrial structure. And the conclusion shows that environmental regulation has positive effects on industrial structure adjustment, and these are consistent with the theoretical analysis results. In addition, human capital, science and technology innovation, market-oriented and financial development also have positive promoting effects on industrial structure adjustment, and FDI to the adjustment of industrial structure is relatively stable enough. Under the dynamic environment, at the same time, the inertia, the adjustment of industrial structure has obvious environmental regulation and government behavior for a longer duration of the influence of industrial structure, there is a lag effect.

Keywords: environmental regulation; the government behavior; industrial structure; pollution haven

1 INTRODUCTION

Aggravate of fog and haze, discussions have begun to break through the readjustment of the industrial structure of economic category, into the social life. In order to control environmental pollution, the government to use administrative, legal, economic, technical and other public policy tools to control environment. But it is a pity that environmental pollution is still severe, and its negative effects have been spread. Focus at this stage environmental problems erupted, fundamentally speaking, is the inevitable result of extensive industrial production development of high and low consumption industry. To reverse the environmental problems fundamentally, the optimization and upgrading of industrial structure is the inevitable path.

The process of industrial structure adjustment requires the synergistic action of diversified factors, like: Higher education human capital, R & D and innovation, financial factors, FDI, industrial integration, international trade and international industry transfer. In addition to the above reasons, the influence factors of environmental regulation are also important. In theory, the ascension of environmental regulation will be eliminated some high pollution industries, promote enterprises to improve production technology, transition to a clean. We will use the data to verify this.

2 STUDY DESIGN

2.1 *Theoretical framework*

Lucas R E B, Wheeler D, Hettige H (1993) think environmental regulation is the inevitable outcome of the economic development. Comprehensive Palmer K (1997), Wells P, Varma A's points of view, the influence of environmental regulation on the optimization and upgrading of industrial structure of the path as shown in Figure 1.

2.2 *Model set*

In order to empirically examine the impact of environmental regulation on the industrial structure optimization, in this paper, panel data regression equation is as follows:

$$Y_{it} = \beta_1 ER_{it} + \beta_2 OPEN_{it} + \beta_3 HC_{it} + \beta_4 MARK_{it} + \beta_5 FD_{it} + \beta_6 ST_{it} + c + u_i + \varepsilon_{it}$$

In order to further investigate the explained variable effects on industrial structure in the dynamic environment, item will be explained variable lag introduced into the model:

$$Y_{it} = \beta_1 Y_{it-1} + \beta_2 ER_{it} + \beta_3 OPEN_{it} + \beta_4 HC_{it} + \beta_5 MARK_{it} + \beta_6 FD_{it} + \beta_7 ST_{it} + c + u_i + \varepsilon_{it}$$

Among them, Y represents the industrial structure, I represents the provincial section, T represents the year. ER stands for environmental regulation, open represents the degree of opening to the outside world, HC stands for human capital, mark represents the degree of marketization, FD stands for financial development, ST represents the science and technology innovation, C represents the constant term, ui is used for fixing effect on control of different provinces on industrial structure, ε_{it} stands for the random disturbance term, β_i Parameters to be estimated.

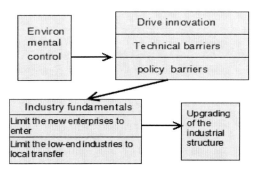

Figure 1.

Table 1. Statistical variable description.

Variable	Obs	Mean	Std. Dev	Min	Max
Y	372	225.80	11.83	202.80	275.3
ER	372	1.57	0.82	0.46	4.52
OPEN	372	546.41	898.80	3.31	5728.51
HC	372	8.25	1.11	4.05	11.7
MARK	372	6.39	2.33	0	12.34
FD	372	2.68	1.45	0.53	10.27
ST	372	1.06	0.98	0.01	5.82

This study on Chinese provincial panel data from 2000 to 2011 as sample, including 12 years time series data of 31 section unit. A total of 372 sample observations. Data sources: "Chinese Statistical Yearbook", "Chinese statistical yearbook on science and technology", "China Financial Statistics Yearbook" and "Chinese Energy Statistics Yearbook".

3 THE EMPIRICAL RESEARCH

3.1 Unit root test

In order to keep the panel data of assumptions, we conduct unit root test, results as shown in Table 2, all the variables through the unit root test.

3.2 The estimation results of static panel

Table 3 gives the results of static panel data estimation, in order to analyze the various explanatory variables on the impact of the explanatory variables, using hierarchical regression. Prior empirical research, Fixed Effect panel data (FE) and Random Effects (RE) recognition, according to Hausman test shows that the value of Hausman through the 1% level of significance test, using fixed effects estimation method.

Model 1 the regression results shows: Human capital, science and technology innovation and

opening to the outside world has a positive promoting effect on industrial structure upgrade. Opening to the outside world means that the entry of foreign capital, followed by the competition effect, demonstration effect, the overflow of capital and technology, through different paths affects the industry development direction. At the same time, international trade, impact on product demand structure and the influence of domestic and regional consumption structure, will promote the optimization and upgrading of industrial structure. Model 2 and Model 3 show that to raise the level of marketization and the financial development have played a significant role in the industrial structure upgrade. Marketization level is a big economic environment, a good market environment for industry development will produce positive role; financial development for industrial structure provides capital support, for industry to the advanced stage of evolution. Model 4, environmental regulation is a reverse index, so, although the return to symbol is negative, but its actual meaning is still the environmental regulation standard promotion plays a positive role in promoting the upgrading of industrial structure. This is consistent with the basic hypothesis, the present environmental regulation policy of China "reversed transmission" business transformation, industrial upgrading successfully.

3.3 Dynamic panel estimation results

When using dynamic panel data to consider the endogeneity problem model, tn this, we can use GMM (SY-GMM) estimation method to overcome the endogeneity problem. GMM set up conditions, there are two: Random perturbation terms ε_{it} nonexistent autocorrelation and all tools variables are effective, therefore, need to GMM estimation method for Autoregressive (AR) identification of the inspection and excessive Sargan test, in order to ensure the scientific nature and effectiveness of the model estimation results. The estimation results in Table 4, Model used in Table 4 system GMM method for parameter estimation is effective.

Model 1,The explanatory variables lagged one year Y, due to the change of industry structure has some inertia, from the empirical results, after one year lag, the industrial structure has a positive effect on the next stage of industry structure. Model 2 shows after the lag phase 2, then no significant influence on the current industrial structure. 3 ~ 4 model show that After environmental regulation lag issue, the adjustment of industrial structure still has a positive influence. Environmental regulation is more show the lag of public policy, environmental regulation in addition to the current valid, also has a strong influence on the development of the follow-up industry, and have a more stable and long-term role in promoting.

Table 2. Unit root test results.

Variable	Y	ER	HC	OPEN	MARK	FD	ST
Result	−7.651***	−5.764***	−10.44***	−5.674***	−14.141***	−6.225***	−51.097***

Table 3. The estimation results of static panel.

Model	(1)	(2)	(3)	(4)
Explained variable	Y	Y	Y	Y
Estimation method	FE	FE	FE	FE
HC	3.750*** (7.95)	2.317*** (4.85)	1.505** (3.64)	0.767*** (3.51)
ST	1.415* (2.50)	0.665*** (3.10)	0.403*** (3.05)	0.357* (2.53)
OPEN	0.00149** (2.66)	0.0000921 (0.17)	−0.000314 (−0.59)	0.000117 (0.22)
MARK		1.503*** (7.49)	1.704*** (8.71)	1.392*** (6.67)
FD			3.348*** (5.63)	3.177*** (6.23)
ER				−0.833*** (−3.79)
_cons	192.5*** (52.22)	197.0*** (56.76)	194.1*** (57.74)	205.8*** (45.57)
Hausman	117.94***	129.37***	138.29***	126.02***
Within R²	0.4173	0.4235	0.4439	0.5630
F	46.75***	48.29***	56.05***	34.93***
N	372	372	372	372

Note: Numbers in brackets for the t value, p < 0.05, **p < 0.01, ***p < 0.001. The same below.

Table 4. Consider endogenous dynamic panel estimation results.

Model	(1)	(2)	(3)	(4)
Explained variable	Y	Y	Y	Y
Estimation method	SY–GM	SY–GMM	SY–GMM	SY–GMM
L.Y	0.331*** (7.13)	0.337*** (6.79)	0.312*** (6.60)	0.322*** (6.66)
L2.Y		−0.0236 (−0.58)		
HC	−0.579* (−2.10)	−0.714* (−2.36)	−0.510 (−1.83)	−0.606* (−2.18)
ST	2.160* (2.43)	3.351** (3.25)	2.123* (2.41)	2.106* (2.37)
OPEN	−0.000495 (−0.88)	−0.000909 (−1.47)	−0.000605 (−1.09)	−0.000470 (−0.83)
MARK	0.962*** (5.32)	1.051*** (5.21)	0.947*** (5.29)	0.989*** (5.33)
FD	1.837*** (5.30)	1.839*** (5.01)	1.859*** (5.40)	1.859*** (5.34)
ER	−0.891 (−1.59)	−0.811 (−1.34)	0.150 (0.17)	−0.888 (−1.60)
L.ER			−1.682** (−2.66)	
_cons	139.6*** (15.14)	142.3*** (12.07)	144.6*** (15.23)	141.3*** (14.77)
N	341	341	341	341
AR(1)	0.033	0.021	0.039	0.034
AR(2)	0.434	0.249	0.311	0.278
Sargan test	0.682	0.729	0.499	0.531

4 DISCUSSION AND CONCLUSION

A feature of this study is the first mechanism analysis, on the basis of theoretical study validated by static panel model and dynamic panel, get the following conclusion: Through the theoretical framework and the design of path analysis showed that, environmental regulation by limiting the low-end enterprises to enter, "Reversed transmission" existing enterprise innovation actively, influence of the industrial structure and industrial chain; Through empirical research data found that, in a static environment, environmental regulation has significant effects to the optimization and upgrading of industrial structure, factors of human capital, technological innovation, the level of market and financial development also promotes the optimization of industrial structure through different paths. And FDI represents growingly opened to upgrade industrial structure is not stable; In a dynamic environment ,Upgrading of industrial structure has certain "inertia effect", environmental regulation of the stability of the optimization and upgrading of industrial structure is stronger, positive effect also exists in the lag period. This from is necessary for the effects of environmental regulation should be from the perspective of long-term and development, not the pursuit of short-term effect.

The significance of this study is to explore the effect of the adjustment of the industrial structure of the new point, environmental regulation as a passive reversed transmission of industrial structure adjustment policy tools, need to policy makers to locate again. From the opposite perspective of environmental regulation, we also must to control environment "refuge", this requires a regional environmental policy standard there should be a balance, as far as possible to avoid the policy factors and the occurrence of "pollution transfer" and "repeated pollution" phenomenon. From the perspective of other factors, industrial structure is a big shift in economic system, needs capital, technology, human capital and market to support, this provides a way for the adjustment of industrial structure. On the one hand, through the "reversed transmission" adjustment, on the other hand, provide basic shift requirements for industrial structure adjustment, guide the industry to the fundamentals, rationalization, "Push" and "traction" jointly play a role, can play a maximum effect of policy.

REFERENCES

[1] Firm organization, industrial structure & technological innovation. Journal of Economic Behavior & Organization (Teece D.J. 1996): 193–224.

[2] The co-evolution of technology, industrial structure & supporting institutions. Industrial & corporate change (Nelson R.R. 1994): 47–63.

[3] Finance, innovation & industrial change. Journal of Economic Behavior & Organization, (Dosi G.1990): 299–319.

[4] Trade, market size & industrial structure: revisiting the home-market effect. Canadian Journal of Economics/Revue canadienne d'économique, (Yu Z. 2005): 255–272.

[5] Economic Development, Environmental Regulation & the International Migration of Toxic Industrial Pollution, 1960–1988 [M]. World Bank Publications, (Lucas R.E.B, Wheeler D, Hettige H.1993).

[6] Environmental regulation & the competitiveness of US manufacturing: what does the evidence tell us? Journal of Economic literature, (Jaffe A B, Peterson S.R, Portney P.R, et al.1995): 132–163.

[7] Environmental regulation & innovation: a panel data study. Review of Economics & Statistics, (Palmer K, Jaffe A.B.1997): 610–619.

[8] Governmental regulation impact on producers & consumers: A longitudinal analysis of the European automotive market. Transportation Research Part A: Policy & Practice, (Wells P, Varma A, Newman D, et al. 2013): 28–41.

[9] Government size & economic growth: A new framework & some evidence from cross-section & time-series data. The American Economic Review, (Ram R. 1986): 191–203.

[10] Government size & macroeconomic stability. European Economic Review, (Gali J. 1994): 117–132.

[11] Government spending & budget deficits in the industrial economies. Economic Policy, (Roubini N, Sachs J.D. 1989):159–168.

[12] Financial systems, industrial structure & growth. Oxford review of economic Policy, (Rajan R.G, Zingales L. 2001): 467–482.

[13] Firm size & industry structure under human capital intensity: insights from the evolution of the global advertising industry. Organization Science, (Von Nordenflycht A. 2011): 141–157.

Education Management and Management Science – Zheng (Ed.)
© 2015 Taylor & Francis Group, London, ISBN 978-1-138-02663-6

Incorporating soft skills in software curriculum construction

N.K. Jiang, H.Y. Mao & J.Y. Zhang
Software Engineering Institute, East China Normal University, Shanghai, China

A. Wang
Microsoft Customer Service and Support APGC Region, Shanghai, China

ABSTRACT: Universities and colleges are undertaking the cultivation of talents for the country and society that the students not only have solid technology knowledge or hard skills, but also have good soft skills and comprehensive quality. In software engineering disciplines, usually the learning of technology is more emphasized, but the soft skills capability is ignored. In fact soft skills are very essential ability for students to apply technology and adapt enterprises developments. The study of technology and soft skills is complementary and coordinated; to some extent soft skills can enhance the students' study interests and learning capacity. The paper analyzes what are soft skills and the key elements, and depicts the soft skills cooperation curriculum system including courses architecture and teaching methods. It describes the necessary and importance of the soft skills in disciplines education and how to carry out in practical teaching. The soft skills learning model should be established to improve soft skills which would play the fundamental role in software engineering education.

Keywords: soft skills; technology training; software education; curriculum construction

1 INTRODUCTION

The "Soft skills" are the power and strength firstly coming from the relationship between the international competition and national countries' strategy, and later continually used in higher education and personnel cultivation, including the responsibility, communication skills, analysis and organization ability, and teamwork spirits. A survey study of Standford research institutions shows that success factors, 75% depending on soft skills ability and 25% depending on the professional and technical skills [1, 2]. Technology is also called "hard skills", as well as soft skills, which are the core competencies in working career [3]. The technical skills are taught by knowledge learning, assignments and projects training in classroom. However the soft skills are taught by means of practical cases, projects, scenarios exercising to improve students' self-confidence, responsibility, communication, presentation and teamwork capabilities. Soft skills are a kind of comprehensive quality and ability to adapting social and business development, which has a multi-disciplinary nature, universality and mobility [4, 5]. In software engineering education, we cannot cultivate software engineers with only professional knowledge but without accomplishments and thoughts. They have integrated soft skills training in the regular academic schedule ensuring individual attention for undergraduate engineering students [7]. In [8], the paper introduces the soft skills course and describes the soft concepts of computer science such as abstraction and readability, and soft skills should be learned gradually based on students' engagement and active learning. The paper shows the extensive set of soft skills was incorporated into a service learning courses in the Mathematical, Information and Computer Sciences disciplines [9].

2 THE IMPORTANCE OF SOFT SKILLS

With the continuous development of software technology, universities and enterprises stress that graduates should have solid professional skills, and at the same time make a full range of talent diverse requirements. In the recruitment stage, companies pay more attention to people's overall quality and comprehensive capabilities. Professional knowledge is build up through the work process, however it's hard to achieve good learning and working attitude, cooperation ability, communication skills and other soft skills in short-term training.

As we know, "Running Fan" graduating from Peking University abandoned students in the Sichuan earthquake, which makes people thinking of ethics and quality talent personality problems.

Zhang Mengsu whose college entrance examination scores only reached the third rank of Hubei province, was admitted by Singapore Polytechnic, because he helped a female teacher splitting awning in an admission counseling of Southwest University [4]. Through the event the overall quality of students began to be re-examined, so soft skills is becoming the important employing elements and the core competence.

The interview of the famous SAP enterprise is divided into three stages. The first stage focuses on the group ability of team discussion and self-summary; the second stage is the investigation of soft skills; the third part is the "hard skills" of technology. This shows that the overall quality of a person is important, and not restricted to a particular technical aspect. Only passing the overall quality and soft skills investigation, there is a chance to the third stage for technology examination.

Technical and soft skills are the key factors constituting human life, competitiveness of work, and they are mutually dependent on each other. For students, technology corresponds to the specific knowledge and professional skills studying in the classroom. Soft skills are about how to quickly and efficiently get knowledge and how to apply in practice and projects flexibly. Soft skills serve for the technical expertise, but beyond the technology itself. Soft skills need to rely on daily accumulation, show yourself, your enthusiasm and consideration with self-confidence, and seize the opportunities to show you.

3 WHAT IS SOFT SKILLS

Technology mainly includes specialized knowledge and professional skills which are measured and evaluated, and compared each other objectively. However soft skills content are quite broad, which has invisibility, emotional, uncertainty, variability and other features. It is difficult to be measured with hard targets. Many colleges and universities have achieved some results in cultivating students' professional and technical capabilities, however neglected the soft skills which are paid attention by the society and companies.

3.1 *Strong responsibility*

The responsibility is the most basic qualities for students. It's also a self-conscious. In projects developing, individual students participating in the software development suddenly left the project do not saying a word to classmates or the leader, and not transferring his work, while the project hasn't been finished. Many students have no group concept and project responsibilities. Their responsibility sense is weak,

and they consider the issue only from the personal views, regardless of other's feelings and projects benefits. We should cultivate students' strong responsibility from some practices and projects. Once he took part in projects, and he is part of the team which should undertake all and work together.

3.2 *Good communication ability*

Communication skills are the ability to clearly and accurately, fluently express them. It is also the capability to properly understand the meaning of others; whether through negotiation the two sides reached the consistency purposes. A man with team spirit, responsibility, helpfulness, and strong communication skills will be living and working in handy. Communication skills is the basis of good relationships with which the people can adapt to the environments and interpersonal and make timely adjustments [4].

3.3 *Good business writing and presentation capability*

The communication of information age is more dependent on Email, electronic report and other writing forms. The writing is different from business communication, which is non-face to face communication. Your idea and thoughts are expressed by means of the email or report, maybe the expression of a word or a sentence can make you lose the contracts and opportunities. So sometimes the use of language and content is more important than the interview. Good business writing skill is one important factor of soft skills.

3.4 *Powerful teamwork spirit*

Teamwork is built on the basis of team work together, and powerful teamwork and mutual complementarities can achieve the maximum productivity. In a group working, we should continue to share the advantages of our team members, draw on the good characteristics of the other members, and problems are promptly resolved. Teamwork can inspire incredible potential of team members, so that everyone can contribute the strongest capability.

Soft skills include self-management, leadership, analytical and problem-solving ability, and other capability. In the process of professional practice course, several students as a team, which allows students to manage their own team and project division. This way certainly encounters technical and workload problems, which require students to have team spirit, take consideration of the project and the group interests. Timely communication and resource sharing are required once problem comes. They must learn to accommodate members, be modest, and studious.

4 INCORPORATING SOFT SKILLS IN CURRICULUM CONSTRUCTION

4.1 Soft skills teaching

In order to strengthen international exchanges and promote multifaceted cooperation, enterprises and projects are now in line with the international standards, which require employees to be familiar with the different areas and cultures, and professionals in different areas to be cooperation with different departments. So employees must have good communication and teamwork soft skills.

As for curriculum reform, we should focus on soft skills cultivation objectives. In curriculum system setting, we should combine with the professional knowledge to select the appropriate use case, paying more attention to the importance degree of soft skills courses.

Universities tend to focus on teaching students professional knowledge, while ignoring the soft skills training, which make students integrate into enterprise environments slowly and hinder the development of students late for improvement. Especially in the field of software, a complex system or software cannot be finished by a single person, the teamwork is indispensable, so good communication and collaboration is essential. In some specialized courses and projects, it's necessary to adopt soft skills related cases and scenarios to train and enhance students' mastery of soft skills.

Soft skills training formally is incorporated into the teaching reform plan. According to professional requirements an enterprise needs, we open the self-learning and self-confident professional courses in low grades based on the actual situation of students. With the learning of technology, gradually teamwork courses are opened. In higher grades, we set up business communication and innovation capability curriculum. In the software practice course, students demonstrate mastery of soft skills and knowledge combining with actual cases.

In the courses construction process, on the one hand we reform course content and outline, on the other hand it's need to focus on training students the right ethics and values, We take staged and targeted teaching methods to enhance and improve students' soft skills, which lay a good foundation for future career and way of life of students.

University or college provides a four-year curriculum for students to learn basic knowledge and expertise, more knowledge is required to learn after graduation in practice and work. As we know soft skills continue to support how to efficiently learn new technical knowledge. With

a strong self-learning ability, responsibility and teamwork, technical and operational capacity will be improved faster. In the application of specialty and the assistance of soft skills, the comprehensive quality will be strengthening such as the project analysis and project organization ability, innovative capability, teamwork collaboration spirits as well as leadership and management skills.

4.2 Soft skills curriculum system

Soft skills can promote the production of the business and knowledge potentiality. And adapting and engaging in different levels of work will promote the accumulation of soft skills. Soft skills and the mastery of technology are complementary and mutually superposed relationship.

It's very important to strengthen the students 'soft skills education. While soft skills cultivation is a systematic project, our Software College and Microsoft Asia Pacific Global Technical Support Center jointly develop soft skills training courses and establish courses goals that we jointly develop students' creativity, responsibility, communication, presentation and collaboration capabilities, and enhance the comprehensive quality of students. Combining with the development of software technology and the actual business scenarios, students realize the importance of soft skills and deepen gradually from the heart.

Soft skills courses system is shown in Figure 1 involved with the foundation courses, professional compulsory and optional courses corresponding to specific soft skills. In teaching process, teaching methods of cases, heuristic, and seminar-style are adopted to improve teaching effectiveness. Political theory and arts humanities courses are mainly to give students the correct worldview and ethics. International Software Services focuses on developing awareness of service and international cultural ideas. In optional courses, business communication and writing, modern management, project management emphasize on the ability to analyze problems and solve problems, as well as team cooperation spirit.

Figure 1. Soft skills courses system.

In software development courses, we set systematic and comprehensive projects, which enable students to combine and select group leader freely, discuss the clear division, and work together to complete the project. In the form of project and team teaching model, it is easy to train students' organizational skills and team spirit, cultivate the students' self-learning and communication skills. While examining the project, it allows students to inspect and evaluate each other, learn and share their experiences. Project reports are periodically submitted to strengthen expression ability and report normative. Through completing the project, students not only study technology knowledge, also fully develop the students' comprehensive ability. Depending on the application scenario, the soft skills are trained by means of team cooperation, group discussions, practices, and mutually commenting form.

4.3 Soft skills in practice

In object-oriented programming courses, they just need to learn a language such as Java, and master design ideas of classes and objects consisting of the grammar, structure, interface, inner classes and user-defined classes. In the future study and work, they can completely use C# to develop projects, because the object-oriented principles are the same, and only the syntax is slightly different. The study motivation is changed from the forced to initiative learning. The self-confidence and self-expression ability are built to enjoy each homework and projects. The sense of joy and completion of the project would stimulate students' active thinking.

Students of software engineering disciplines are mainly male or female engineering IT, generally be considered as technical personnel, introverted, not good at communication. As long as given a little guidance, they also have strong communication, cooperation, innovation capacity and become management talent with solid technology expertise. In this way, we need to improve students' soft skills which also help to strengthen the knowledge study. Only the comprehensive ability is achieved, it's possible to become a remarkable and excellent software engineer.

While two interviewers who have the balanced technical capabilities, in the interview, the soft skills play a decisive role. And in some cases the lack of technology can be made up. Hence we must have a conscious exercise to strengthen the soft skills in the teaching process.

5 CONCLUSIONS

In the era of knowledge economy and the rapid development of information, we should take consideration of their own development and future employment from students, not only staying in the books of knowledge. In curriculum construction reform, some soft skills courses are gradually opened to corporate with enterprise and industry standards. Not only students' technical abilities are improved, but also the soft skills are progressively enhanced. In the meanwhile, the students can be deeply aware that they must work hard to learn professional knowledge, and they can have a chance to stand out in the fierce competition for talent. Thus they keep a sense of cultivating their comprehensive quality and training soft skills.

Soft skills cultivating is a long-term project, we should recognize the importance of soft skills from the heart. Through reforming software engineering curriculum and teaching mode, we integrate soft skills in the courses system and practice in teaching.

ACKNOWLEDGMENTS

We are grateful for the support of Shanghai Great Course Construction "Specification and Practice for Software Development ", and Specialty Reform Construction of Ministry of Education and IBM "Software Life Cycle and Quality Management".

REFERENCES

[1] G. Morales, Daniel, etc. 2011. Teaching "soft" skills in Software Engineering. 2011 IEEE Global Engineering Education Conference: 630–637.
[2] Ahmed, Faheem. 2012. Evaluating the demand for soft skills in software development, IT Professional, v 14, n 1: 44–49.
[3] S. Pressman. Roger. 2011. Software Engineering: A Practitioner's Approach [M], China Machine Press.
[4] T.S. Du. 2013. Review and Strategy: Soft Skills' Training for International Service Talents, China Higher Education Research.
[5] Y. Dai. 2011. Soft Skills Training of Vocational College Students, Heilongjiang Researches on Higher Education.
[6] Zhang, Aimao. 2012. Peer assessment of soft skills and hard skills, Journal of Information Technology Education: Research, v 11, n 1: 155–168.
[7] Panthalookaran, Varghese. 2010. Cultivation of engineering soft skills within the constraints of a prescribed curriculum, ASME International Mechanical Engineering Congress and Exposition, Proceedings, v 7: 241–247.
[8] O. Hazzan, H.S. Gadi, 2013. Teaching computer science soft skills as soft concepts, SIGCSE 2013 Proceedings of the 44th ACM Technical Symposium on Computer Science Education: 59–64.
[9] C. Lori. 2011. Ideas for adding soft skills education to service learning and capstone courses for computer science students. Proceedings of the 42nd ACM Technical Symposium on Computer Science Education: 517–521.

Education Management and Management Science – Zheng (Ed.)
© 2015 Taylor & Francis Group, London, ISBN 978-1-138-02663-6

Cultivating local university mechanical specialty undergraduates' innovative ability relied on discipline competitions

Y.B. Zhou, H.D. Wang, D.J. Zhu & H. Cheng
School of Mechanical and Automotive Engineering, Hubei University of Arts and Science, Xiangyang, China

ABSTRACT: With the development of advanced equipment manufacturing technology, there is an urgent need for local university mechanical specialty to cultivate a large number of professional engineering technical talents with innovative spirit and innovative ability. There are many approaches of cultivation of innovative talents; discipline competitions can play an important role in innovative talent cultivation. According to the characteristics of local university mechanical specialty and the existing problems of discipline competitions, this paper discusses some specific methods to develop the innovative ability of students relied on discipline competitions. By optimizing talent cultivation plan, improving discipline competition system, strengthening the guidance team construction, building practical innovation platforms, establishing incentive mechanisms, constructing campus innovative atmospheres and other measures, can effectively train the local university mechanical specialty undergraduates' innovative ability, and improve the comprehensive quality of students.

Keywords: discipline competitions; innovative ability; cultivation; mechanical specialty

1 INTRODUCTION

Cultivating innovative talents is the development of the society demand, is also the primary task of current higher education (Wu 2014). The national medium and long-term education reform and development plan outline (2010–2020) explicitly pointed out that higher education should focus on improving students' initiative spirit and the practical ability to solve the problem. With the "popularization" transformation of higher education, the university graduates quantity increases annually. The continuous development of advanced equipment manufacturing technology is an urgent need to cultivate a large number of engineering and technical talents with innovative spirit and innovative ability. How to cultivate mechanical college students' practical ability and innovative ability, enhance the employment competitiveness, is the urgent problem needed to be solved for local university mechanical engineering talents cultivation (Zhang 2013).

Carrying out discipline competitions in the mechanical college students, integrating classroom and extracurricular practical educational resources, not only can make college students practice the theoretical knowledge, but also play an important role to improve their comprehensive quality such as innovative ability, practical ability, cooperation consciousness, scientific spirit, and organization ability.

2 INNOVATIVE ABILITY CULTIVATING METHODS RELIED ON DISCIPLINE COMPETITIONS

In recent years, the local university especially attaches great importance to discipline competitions, cultivate students' practical ability and innovative ability through the discipline competitions, promote the implementation of innovative education, inspire students' individual participation, and have obtained certain achievements in the students' innovative ability practice ability training. But there still exist some problems such as poor active participation consciousness of students, insufficient coordination of learning professional knowledge and the competition, lack of competition teachers and excellent students (Wang & Wang 2012). Aiming at these problems, combining with the characteristics of mechanical engineering in local university, this paper discusses some effective measures and specific methods to develop the innovative ability of students relied on the discipline competitions.

2.1 *Optimizing talent cultivation plan*

Cultivate the innovative ability of college students need to step by step, different levels of students, innovation requirements and forms are also different. Local university shall, according to mechanical professional training target, carefully study the relation of discipline competition and innovative talent cultivation, adopt the cultivation of innovative ability in university talent training plan, to form a system of training plan (Duan et al. 2013). In curriculum and training scheme, appropriately increase the proportion of elective courses, students can choose different courses according to their own foundation, interests, abilities and needs. In the experiment, course design and other practice teaching links, students are encouraged to adopt, compare and discuss different scheme.

Local university should gradually improve, promote the discipline competition course teaching reform, revises training plan, course outlines and experiment outline, establishes independent elective courses for discipline competitions, and combine with the practical teaching such as curriculum design, graduation design, research and design new competitive events, enrich and perfect the discipline competition system. In addition, local university should actively support the student scientific research project and practice innovation activities, encourages the student to deeply investigate and study on hot issues, difficult problem and key technological problems in the process of social development, and puts forward solutions or ideas. Guiding the students to independent choice and independent design, finishing related works and writing summary report, these can mobilize and inspire students' learning initiative and enthusiasm, improve the students' independent innovative ability and practice ability.

2.2 *Improving discipline competition system*

The establishment and improvement of rules and regulations is one of the premises to complete discipline competitions. The discipline competitions for mechanical specialty students need to obtain the policy guarantees, school support of software and hardware facilities, and so on, which can maximize the play to the enthusiasm and initiative of students to participate in the competition. In order to ensure the smooth operation of the discipline competitions, local university should establish the corresponding management institution, formulate the perfect rules and regulations, and carry on the effective management of discipline competitions (Liang 2014). Also, local university should clearly define the discipline competition programs, the specific duties of management departments at different levels, funds sources and management, reward measures, etc., so as to strengthen the discipline competitions as an important part of the school routine teaching work, and make discipline competition management standardized and institutionalized.

As to the student-oriented innovation projects, the school need to provide the funds safeguard and patent protection on study results, in order to further stimulate students' enthusiasm to actively participate in the innovation, to improve and strengthen discipline competition education results and improve the students' innovation consciousness and innovative ability. Competition projects that have been good carried out, the school should encourage them to develop higher and deeper level. The disciplines and specialties that have not carried out competitions should find suitable competitions as soon as possible, keep trying, so that the students can get exercise in the competition, achieve the goal of cultivating students' innovative ability.

2.3 *Strengthening the guidance team construction*

High level competition guidance team is the most effective guarantee to combine the discipline competitions and the innovative talents training. Local university needs to set up a specialized discipline competition steering group, provides strong guidance and technical support for students engaging in discipline competitions (He 2013). Instructors should open the corresponding series of elective courses according to the needs of discipline competitions, strengthen the basic knowledge, introduce the latest trends of the scientific developments, broaden the students' knowledge, cultivate students' ability to analyze and solve problems, and improve students' understanding and competitiveness of discipline competitions.

At the same time of giving full play to the teacher's guidance, local university should strengthen the teachers' professional ethics education, and pay more attention to the improvement of guidance teachers' ability. Only access to the scientific frontier, new design ideas, new thinking in scientific research, teacher's academic level, business level and comprehensive quality can be improved, and the students can be more innovative and pioneering.

For teachers participating in the activities of discipline competitions, local university should provide a series of incentive policies in such aspects as education, employment and assessments. At the same time, local university can introduce teachers with expertise in competition, send teachers to participate in the competition training, and strengthen the experience exchange between competition teachers, improve teachers' guidance ability of competition, construct an excellent guidance teacher's team for discipline competition.

2.4 *Building practical innovation platforms*

Mechanical specialty undergraduates' practice teaching needs to adapt to the need of discipline competitions, and discipline competitions also effectively enriched the content of the practice teaching. Laboratories are the foundational teaching environment of cultivating students' innovative spirit and practice ability, establishing open innovation laboratories is an effective measure. The innovation laboratory should be equipped with corresponding equipments, invites teachers with professional, strong scientific research ability as instructors, and provides favorable conditions and environment for students training. The research subjects can be put forward by teachers or students, and the design scheme can be completed independently or in collaboration. The establishment of the innovation laboratory cultivates students' interest of scientific discovery and invention, stimulates students' creative thinking, and improves their innovative ability in practice (Zhu & Li 2013).

Related training plan and course system in local university need to be revised and adjusted to adapt the practical teaching forms and methods of innovative talents cultivating. Properly adjusting classroom and extracurricular hours can increase the time and space of students' independent thinking and innovative design. By increasing the innovative activities, carrying out the innovative methods of exploration, and choosing the discipline competitions subjects as all kinds of design topics, local university can implement the organic combination of classroom teaching and practice innovation.

According to the needs of discipline competitions, local university can lead machinery industry outstanding enterprises and scientific research institutes to participate in, employ technical staff with rich practical experience as competition guidance teachers, and establish discipline competition and innovation practice bases for mechanical specialty undergraduates. These enterprises and scientific research institutes can not only provide students with the usual study, experimental opportunity, also assist the local university in competition organization, competition training, and other important works.

2.5 Establishing incentive mechanisms

To ensure the discipline competitions to develop healthy and orderly, and encourage students participating enthusiastically, local university is necessary to make special incentive mechanism for discipline competitions, and according to the actual circumstances of the discipline competitions, give full play to the role of incentive mechanism (Shen & Han 2013). Local university can set up special competition awards funds, provide award-winning students and performance prominent certain material rewards; at the same time, take the competition award and related papers,

patents as the qualified condition for scholarship and recommended graduates.

The guidance teachers have played an important role in the course of competition, and have paid a lot of energy and labor, the discipline competitions incentive mechanism should also include them. Local university can offer teachers of certain material rewards, on the other hand, can link up the competition results with professional title evaluation, position employment and annual assessment. These measures are fully affirmation to the guidance teachers' work, and can fully arouse the enthusiasm of teachers to participate in the guidance of discipline competitions.

2.6 Constructing campus innovative atmospheres

Local university should attach great importance to the construction of campus innovation atmosphere, strengthen the importance propaganda of application and innovative ability training; at the same time, organize discipline competitions well, enlarge the students' participation, and encourage students to participate in the discipline competition, combining with interdisciplinary or multidisciplinary mode.

Discipline competition's website is an important window of discipline competitions, and plays an important role to promote the discipline competitions. The website can make positive propaganda for the competition types, competition contents, competition schedules, team situations, works show, and so on; so that the students can all-round understanding of the discipline competition events. The competition communication contents become more plentiful and the competition information publishing become more timely and accurate.

Local university must make a detailed training plan, arrange the training content reasonably, and being targeted to set up the training courses. Students should not only have certain professional knowledge quality and practice ability, but also need to capable of information processing technique, thesis writing ability, team cooperation consciousness, and oral communication ability, etc. This requires strengthening the skills training during daily training sessions aimed for competition key links.

Setting up related different type associations plays a powerful support and promotion of discipline competitions. Through publicity, recruit members, and activities, the associations can expand the influence of discipline competitions, make the students quickly understand the competition events and entry requirements, and increase their enthusiasm of participation in the discipline competitions. After the teacher's professional guidance and training, some outstanding members can be selected to participate in the discipline competitions.

Participating in the discipline competitions must attach much importance to the participating students' team echelon construction. Consciously absorb the first-year students take part in the discipline competitions, and let them participate in the actual training program as soon as possible. Forming a formal team should pay attention to optimize the team structure. The upperclassmen lead the underclassmen, and the underclassmen also promote the upperclassmen, these can ensure the team member have a reasonable distribution between the seniors and the juniors, and at the same time, ensure the stability and long-term effectiveness of the competition performance.

3 CONCLUSION

Carrying out the discipline competitions is the need of cultivating innovative talents, and the discipline competitions also play active role in students' innovative ability training. Local university mechanical specialty needs to organically combine the daily teaching, practice links and course construction with discipline competitions, promote the mechanical engineering curriculum content updates and teaching methods reform, rely on the discipline competition, motivate the students' scientific research enthusiasm and learning interest, cultivate the students' competitive consciousness, team spirit and innovative ability, gradually achieve the goal of training high quality innovative talents in the field of mechanical engineering.

ACKNOWLEDGMENT

This work is supported by the Hubei Province Educational Science "Twelfth Five-year" Plan Project (2013B204), and the Hubei University of Arts and Science Teaching and Research Project (JY2013033).

REFERENCES

[1] Wu, X.Q. 2014. The theory of colleges and universities innovation education and talents cultivation. *Journal of Educational Institute of Jilin Province* 30(6): 33–34.

[2] Zhang, H.W. 2013. Research and practice of cultivating applied personnel of mechanical engineering in local colleges. *Journal of Jilin Institute of Chemical Technology* 30(12): 85–87.

[3] Wang, H.S. & Wang, S.X. 2012. Taking subject competition as a carrier to cultivate college students' innovative ability. *Journal of Jiamusi Education Institute* (3): 135–136.

[4] Duan, J.F. et al. 2013. Mechanical outstanding engineers talent training scheme research. *China Electric Power Education* (20): 15–16.

[5] Liang, N. 2014. Discussion of colleges and universities discipline competition management and operation mechanism. *Education Teaching Forum* (6): 23–25.

[6] He, J.J. 2013. On subject contest of economy and management for higher vocational colleges based on tri-intelligence model. *Journal of Zhejiang Business Technology Institute* 12(1): 62–64.

[7] Zhu, X. & Li, J.J. 2013. Exploration of building innovative open laboratory management mode. *Journal of Education Institute of Taiyuan University*, 31(4): 121–122.

[8] Shen, H. & Han, X.M. 2013. Research on the significance and safeguard mechanism of university academic competition. *Journal of Zhejiang Shuren University (Acta Scientiarum Naturalium)* 13(3): 52–55.

Education Management and Management Science – Zheng (Ed.)
© 2015 Taylor & Francis Group, London, ISBN 978-1-138-02663-6

Differences between Chinese and Western ethics based on *The Joy Luck Club*

X.L. Chen

College of Foreign Studies, Guilin University of Technology, Guilin, Guangxi, P.R. China

ABSTRACT: According to the value system of ethics, there are four basic differences between Chinese and Western ethics: the basis of ethics; the goal of ethics; the orientation of ethics; the realizing mechanism of ethics. These differences can be found in the works of literature, especially Chinese American literature. Take The Joy Luck Club as an example. It is written by Amy Tan, a Chinese American writer. In the book, there are four pairs of Chinese mothers and American-born daughters. The conflict of Chinese and Western ethics is harsh, and it is obvious to discover the differences between Chinese and Western ethics in their lives.

Keywords: Chinese ethics; Western ethics; differences; *The Joy Luck Club*

1 INTRODUCTION

Ethics is a code that points to the standards put forward by the society or the cultural group. Ethics are more a code of conduct set forward by the community. The conflict of Chinese and Western ethics is a major difference between Chinese and Western culture. According to the value system of ethics, there are four basic differences between Chinese and Western ethics: the basis of ethics; the goal of ethics; the orientation of ethics; the realizing mechanism of ethics.

The Joy Luck Club is one of the internationally bestselling novels, which has got a lot of praise from the media like *The Washington Post Book World, The New York Times Book Review, Los Angeles Times, The New Yorker*, etc.. In *The Joy Luck Club*, vignettes alternate back and forth between the lives of four Chinese women in pre-1949 China and the lives of their American-born daughters in California. Thus, the differences between Chinese and Western ethics can be found in *The Joy Luck Club*. This paper tries to analyze the differences between Chinese and Western ethics based on *The Joy Luck Club*.

2 THE BASIS OF ETHICS—MORALITY AND NATURALNESS

The Chinese traditional ethics uses the moral eyes to view the human nature. It is looking for the ethical foundation and ethical standards from human morality. However, the Western ethics originates from ancient Greek view of nature. Naturalism ethics is the archetype of Western ethics. The human basis of Western ethics is one's naturalness. How to judge human morality depends on how to realize human naturalness.

In *The Joy Luck Club*, the marriages of mothers and daughters expose the different bases of Chinese and American ethics. Lindo Jong was betrothed to Huangs' son, a baby who was one year younger than her when she was just two years old. Because she was promised to the Huangs'son for marriage, her own family began treating her as if she belonged to somebady else. Her mother always reminded her of that. In her eyes, her mother did not treat her that way because she didn't love her. She knew that she would have a bad husband, but she had no choice because that was how backward families in the country were. They followed the old-fashioned customs that mothers chose their daughters-in-law, ones who would raise proper sons, care for the old people, and faithfully sweep the family burial grounds long after the old ladies had gone to their graves.

Another story is about An-mei Hsu's mother's marriage. It was even more miserable. An-mei's mother had been sent her way by her own mother because she had dishonored her widowhood by becoming the third concubine to a rich man, Wu Tsing. Her brother slapped her for calling him Brother; his wife cursed her. They accused her, "She has thrown her face into the eastward-flowing stream. Her ancestral spirit is lost forever. The person you see is just decayed flesh, evil, rotted to the bone." It seemed that she was a licentious woman.

Actually, the Second Wife of Wu Tsing conspired with him to lure her to his bed. And Second Wife complained to many people about the shameless widow who had enchanted Wu Tsing into bed. How could a worthless widow accuse a rich woman of lying? So she had to be his third concubine. Her brother kicked her, and her own mother banned her from the family house forever, but she came back to cut a piece of meat from her arm to make medicine for her mother when her mother was dying. She followed the Chinese traditional ethics—filial duty, obedience, endurance and ritual. The most miserable was that she gave birth to a son, which Second Wife claimed as her own. Eventually, she committed suicide. Hence, her daughter, An-mei, learned from her story to desire nothing, to swallow other people's misery, to eat your own bitterness. But she taught Rose, her daughter, the opposite because she learned to shout on the day her mother died.

But An-mei still told Rose to save her marriage when Rose decided to divorce. What Rose learned is that Chinese people had Chinese opinions; American people had American opinions. And in almost every case, the American version was much better. Rose finally made the decision as an American did that she tried to catch what belonged to her, not just followed her husband like her grandmother.

Lena, the daughter of Ying-ying, had to share everything with her husband. They made it very clear who would pay the bill. Facing mother's confusing, Lena explained with the words she and her husband had used with each other: "So we can eliminate false dependencies... be equals ... love without obligation ..." Actually, in her deep heart, Lena wanted to protest. She didn't like this way about money. When the marble end table which refers to the weird family rule collapsed, Lena would stop it as her mother told her to do.

In mothers' marriages, they had to obey the old-fashioned morality no matter how stupid it was. So Lindo escaped her marriage and An-mei's mother committed suicide. Filial duty, obedience, endurance and ritual are Chinese traditional ethics, but they were used wrongly. On the other hand, daughers could deal with their lives when they were not satisfied with their marriages.

3 THE GOAL OF ETHICS—
RIGHTEOUSNESS AND HAPPINESS

The general goal of Western ethics is looking for happiness on the basis of naturalism human nature. Nevertheless, the Chinese traditional ethics is pursuing the supreme morality. In Chinese tradition, the counterparts of Western morality and happiness are the righteousness and benefit. The

traditional Chinese ethics places the emphasis on the righteousness.

Mother wanted to teach daughter about Chinese character. How to obey parents and listen to your mother's mind. How not to show your own thoughts, to put your feelings behind your face so you can take advantage of hidden opportunities. Why easy things are not worth pursuing. How to know your own worth and polish it, never flashing it around like a cheap ring. Why Chinese thinking is the best. But this kind of thinking didn't stick to daughter. She thought mother was old-fashioned. Consider the case of Jing-mei, for having started half a degree in biology, then half a degree in art, and then finishing neither when she went off to work for a small ad agency as a secretary, later becoming a copywriter. But her mother was always having the argument with her about her being a failure, a college drop-off, about her going back to finish. Her mother even told others that she was going back to school to finish the degree.

What daughter learned is about American circumstance. If you are born poor here, it's no lasting shame. You are first in line for a scholarship. If the roof crashes on your head, no need to cry over this bad luck. You can sue anybody, make the landlord fix it. In America, nobody says you have to keep the circumstances somebody else gives you.

When they planned to get married, Waverly, the daugher of Lindo, had been worried whether her mother permitted her to marry Rich, an American man. To Rich, it was very simple, just to say "Mom, Dad, I'm getting married". But it was a big problem for Waverly, because she thought her mother "never thinks anyone is good enough for anything." Her mother always criticized everything. In fact, her mother wanted her children to have the best combination: American circumstances and Chinese character.

4 THE ORIENTATION OF ETHICS—
COLLECTIVISM AND INDIVIDUALISM

The orientation of Chinese traditional ethics is collectivism. The social group is the principal part of value in Chinese traditional ethics. Conversely, the individual value depends on the group. One cannot gain one's value until the group's need is met and the group's benefit is achieved. The orientation of Western ethics is individualism. Individualism is the basis of Western ethics, and the happiness for Westerners refers to the individual happiness. Ego is the core of value in Western culture.

In Chinese mothers'eyes, children are not only children themselves, but the part of the family, the honour of the family, even part of mothers themselves. Children's success means mothers' success, children's failure means mothers' failure. Lindo and Suyuan were both best friends and arch enemies who spent a lifetime comparing their children. Jing-mei, Suyuan's daughter, was one month older than Waverly, Lindo's daughter. From the time they were babies, their mothers compared the creases in their belly buttons, how shapely their earlobes were, how fast they healed when they scraped their knees, how thick and dark their hair, how many shoes they wore out in one year, and later, how smart Waverly was at playing chess, how many trophies she had won, how many newspapers had printed her name, how many cities she had visited. Therefore, Jing-mei's mother tried to cultivate some hidden genius in Jing-mei. She did housework for an old retired piano teacher who gave Jing-mei lessons and free use of a piano to practice on in exchange.

Lindo loved to show off her daughter, like one of the trophies she polished. She would show off her daughter's victory to the family friends who visited. In Saturday market, she even said to whoever looked her way, "This my daughter Wave-ly Jong." Waverly hated the way her mother tried to take all the credit. She told her she didn't know anything, so she shouldn't show off. She should shut up. Thereafter, her mother changed. She no longer hovered over her as Waverly practiced different chess games. She did not polish her daughter's trophies every day. She did not cut out the small newspaper item that mentioned Waverly's name. It was as if she had erected an invisible wall. Then, Waverly lost the games. When she refused her mother, she lost the gift of chess and turned into someone quite ordinary.

That's why Waverly needed her mother's permission for her marriage. Her American friend couldn't undestand why she was so nervous. It was just getting married. The friend advised her to tell her mother to shout up. Waverly said, you can't ever tell a Chinese mother to shut up. You could be charged as an accessory to your own murder.

Almost all the mothers hadn't had a satisfactory marriage, and their lives had not been happy, so they sailed to America. They wished they would have a daughter. Nobody would say her worth was measured by the loudness of her husband's blech. Nobody would look down on her. She would always be too full to swallow any sorrow. Anyway, their daughter would have a totally different life.

5   THE REALIZING MECHANISM OF ETHICS—EMOTIONALISM AND RATIONALISM

The basic social element in China is family. Chinese ethics has been given the family emotions. Western society is based on individual. The Western culture believes in a materialistic approach, and has a pragmatic attitude. In the realizing mechanism of morality, Western ethics resort to "intellect" while Chinese ethics resort to "emotions".

Mothers were raised the Chinese way. They were taught to be quiet, to listen and watch, as if your life were a dream. You could close your eyes when you no longer wanted to watch. Waverly's mother taught her the art of invisible strength. It was a strategy for winning arguments, respect from others. However, the daughters didn't think so. To Jing-mei, her mother always criticized. It seemed that her mother was always displeased with all her friends, with her, and even with her father. When she was taking Introdution to Psychology, she tried to tell her mother why she shouldn't criticize so much, why it didn't lead to a healthy learning environment. She told her mother that parents shouldn't criticize children. They should encourage instead. People rise to other people's expectation. And when you criticize, it just means you're expecting failure. But her mother continued to criticize.

When she got trouble in her marrigae, Rose would rather talk to her psychiatrist than to her mother. Her mother didn't understand why she could talk with a psychiatrist but not with mother. She thought a mother was the best. A mother knew what was inside a daughter. Mothers felt that there was a river between them and their daughters. They were watching their daughters from another shore, and accepted their American ways. They saw daughters who grew impatient when their mothers talked in Chinese, who thought they were stupid when they explained things in fractured English. They saw that joy and luck did not mean the same to their daughters, that to these closed American-born minds, "joy luck" was not a word and it didn't exist. They wished their daughters would not swallow sorrow, while the daughters swallowed Coca-cola. They visited their daughters by unannounced but daughters suggested they should call ahead of time.

Chinese mothers and American-born daughters lived different lives. They followed their rules. Mothers wished their daughters to be real Americans, to speak perfect English. But they still taught them in Chinese way, and instilled in them Chinese thoughts. The daughters were raised in America, they looked like Chinese, but they have never really known what it means to be Chinese. Yet, as Jing-mei's mother told her, "Someday you will see, it is in your blood, waiting to be let go." When Jing-mei meets her sisters in China, she feels what part of her is Chinese. It is so obvious. It is her family. It is in her blood. After all these years, it can finally be let go.

6 CONCLUSION

There are differences between Chinese and Western ethics, but neither superior nor inferior to each other. The world is a global village. Different peoples live in it with different cultures. It will help people to communicate to learn different cultures. If a country wants to build up a good relationship with others, people should respect each other's culture. *The Joy Luck Club* is such a book that opens a door to both Chinese people and American people. In the book, readers can see the conflicts between mother generation and daughter generation, Chinese culture and American culture, Chinese ethics and American ethics. At the end of the novel, when Jing-mei, the American-born daughter hugs her Chinese sisters, the two different cultures accept each other. They do not colliside but exist in the world together.

REFERENCES

[1] Zeying Wang, Ethics Beijing Normal University Publishing Group 2012, 4.
[2] http://en.wikipedia.org/wiki/Ethics
[3] http://www.differencebetween.net/business/difference-between-ethics-and-morals/
[4] Amy Tan, The Joy Luck Club Penguin Books 1989.
[5] Zhanyou Chen, On Main Differences between Chinese and Western Ethics Value Tianjin Social Science 2006, 1 P26–29.
[6] Chao Chen, A Study of Differences and Integration between Chinese and Western Ethics New View 2007, 3 P52–54.
[7] Zailin Zhang, Lianfu Yan, From Intellectual Ethics to Family-Based Ethics—A Comparison of Chinese and Western Traditional Ethics Journal of Hangzhou Normal University [Social Scinces Edition] 2009, 9 P68–74.

Education Management and Management Science – Zheng (Ed.)
© 2015 Taylor & Francis Group, London, ISBN 978-1-138-02663-6

Regulation of Internet Investor Relations: A comparison between USA and China

Yan Jie Feng, Min Chen & Ying Fan
Shanghai University of International Business and Economics, Shanghai, China

ABSTRACT: More and more listed companies use internet as a main investor relations activity channel. How to make the internet platform become an efficient and fair communication method while at the same time to avoid unfair and other illegal activity which have become the major concerns of regulators and listed companies. This paper compares relevant laws and regulation rules between the US and China and suggests that China should formulate regulations matching listed companies' internet investor relation activity needs and put in place some regulation rules to regulate and promote the development of internet investor relations.

Keywords: investor relations; financial regulation; internet investor relations

1 INTRODUCTION

With the fast development of the internet technology, more and more listed companies are engaged in Internet Investor Relations (IIRs) via designated websites, their own official websites, social media or other internet platforms. IIRs offer enhanced communications between companies and investors. However, the Internet itself is a double-edged sword. For instance, with the widespread information on the Internet, it is hard for supervision agencies to monitor all the online information, making misleading or inappropriate information appear. Therefore, internet investor relations need to comply with certain regulation rules.

Regulation concerning IIRs of listed companies in China has made some progress so far, but it is far from being perfect. By comparing regulation rules of internet investor relations between China and the USA, the essay shows history and status quo of this area of the two nations and hopes to promote laws and regulations development concerning IIRs in China.

2 LITERATURE REVIEW

Global research of IIRs began in 1996. According to Lymer (1997), listed companies have paid more and more attention to the development of Internet information from 1992. In 2000, the American Association of Investor Relations released a report stating that about 74% listed companies have set up investor relations column on the company website. Michael & Geodes (2005) thought the website investor relations can help more people share the

company's information. James (2006) found that the various functions of listed companies' websites could promote investor relations.

Some scholars (Feng & Wan 2013) conducted comparison studies of IIRs between different countries and areas, and stated that regulation difference is one of the main causes of different levels of the internet investor relations. However, research in detailed differences of laws and regulations related to IIRs can be rarely seen. This paper seeks to explore this area by comparison between the USA and China.

3 COMPARISON OF REGULATION ON IIRS BETWEEN USA AND CHINA

3.1 *Investor Relations (IRs) related laws and rules in the two nations*

Regulation of IRs is closely related to regulation of securities. Regulation includes at least two categories: regulation by purely governmental regulatory agencies in accordance with laws and regulatory rules, and listing requirements of exchanges. Sometimes rules of self-regulatory organizations are also encompassed.

In USA, the Securities Act of 1933 (the "Securities Act"), the Securities Exchange Act of 1934 (the "Exchange Act") are the most important laws related to Investor Relations. Some other laws involved include: Trust Indenture Act of 1939, Investment Company Act of 1940, Investment Advisers Act of 1940, Securities Investor Protection Act of 1970, Sarbanes-Oxley Act of 2002, Dodd-Frank Act of 2010, Jumpstart Our Business Startups (JOBS) Act. As the administrative department of securities activities, the USA

Securities and Exchange Commission (SEC) made a lot of rules and regulations by rulemaking in implementing these statutes. Besides statutes, rules and regulations, SEC also makes SEC Concept Releases, SEC Interpretive Releases and SEC Staff Interpretations. The federal statutes and the SEC rules and regulations have the force of law. Other SEC-issued documents vary in the degree to which they carry the force of law.

In China, The administrative department of securities is the China Securities Regulatory Commission (CSRC). Laws and regulations related to IR in China can be divided into 5 categories: State Laws, Administrative Laws, Judicial Interpretation, Department Rules and Self-regulatory Rules. Law of the People's Republic of China on Securities (the "Securities Law"), Companies Law of the People's Republic of China (the "Companies Law") and Criminal Law of the People's Republic of China (the "Criminal Law") are the most important laws related to IRs. The Administrative Measures for the Disclosure of Information of Listed Companies (Administrative Measures for Disclosure) promulgated in 2007 by CSRC is the most important department rule governing IR. CSRC released Working Guidelines for the Relationship Between Listed Companies and Investors (IRs Guidelines) in 2005, which is especially about IRs, but it's of very low law force.

As a special form of IRs, IIRs are also regulated by the regulations above. We focus on regulations of using of website for IIRs in the next parts, and comparing the history and provisions of related regulations in the two nations.

3.2 *Regulation on the use of website in USA*

SEC has kept promoting and regulating using of web sites as IRs and information bridges between listed companies and investors, mainly in two ways, EDGAR and listed companies websites.

EDGAR, the Electronic Data Gathering, Analysis, and Retrieval system, performs automated collection, validation, indexing, acceptance, and forwarding of submissions by companies and others who are required by law to file forms with the USA Securities and Exchange Commission (SEC). All USA listed companies, foreign and domestic, are required filing registration statements, periodic reports, and other forms electronically through EDGAR. Anyone can access and download this information for free through the SEC website.

Besides EDGAR, SEC has issued a series of interpretive releases and rules that promote and regulate the use of listed company web sites as a means for companies to communicate and provide information to investors.

On April 28, 2000, SEC issued interpretive release on the use of electronic media (the 2000 Electronics Release). The release addresses the use of electronic media in three areas: (1) electronic delivery of corporate communications; (2) web site content; and (3) online offerings. The release reminds listed companies that the federal securities laws apply in the same manner to the content of their web sites as to any other statements made by or attributable to them.

In August 2000, Regulation FD was promulgated by the USA Securities and Exchange Commission (SEC). The regulation FD sought to stamp out selective disclosure, in which some investors (often large institutional investors) received market moving information before others (often smaller, individual investors). As a whole, the regulation requires that when an issuer makes an intentional disclosure of material nonpublic information to a person covered by the regulation, it must do so in a way that provides general public disclosure, rather than through a selective disclosure. The regulation states that one of generally acceptable methods of public disclosure is making announcements through open conference calls "that the public may attend or listen to either in person, by telephone, or by other electronic transmission (including the Internet)". Internet was recognized as one of the legitimate (under certain circumstances) vehicles of making public disclosure.

Since 2000 Electronics Release and the adoption of Regulation FD, there emerged a dramatic increase in the use of company web sites. In 2008, the SEC has reached a point where the availability of information in electronic form—whether on EDGAR or a company web site—is the superior method of providing company information to most investors, as compared to other methods. They believed that to encourage the continued development of company web sites as a significant vehicle for the dissemination to investors of important company information, it was an appropriate time to provide additional Commission guidance specifically addressing company web sites. Issued in the form of an interpretive release, the SEC guidance provides helpful information for companies considering providing investors with interactive content on their Web sites, as well as summary information and links to third-party information.

In this guidance, SEC explained that its rules and interpretations that promote the use of web sites generally work in two different respects. First, when delivery of documents is required under the federal securities laws, SEC has encouraged the delivery in electronic format or recognized that electronic access can satisfy delivery—hence, prospectuses and proxy materials can be delivered or otherwise made available using electronic communications and the Internet in certain circumstances. Second, where disclosure of information is required under the Exchange Act, SEC allowed companies to make such information available to investors on their web sites with their web sites serving, depending on the circumstance, as a supplement to EDGAR, as an alternative to EDGAR, or as a stand-alone method of providing information to investors independent of EDGAR.

A fundamental principle underlying these interpretations and rules is that, where access is freely available to all, use of electronic media is at least equal to other methods of delivering information or making it available to investors and the market. Further, SEC have recognized that, in some cases, allowing companies to provide information on their web sites has advantages for investors over mandating that EDGAR serve as the exclusive venue and format for company disclosures.

After the releasing of these regulations and rules and releases, IIRs have made great progress in USA. Some companies announced to rely on the use of its corporate website as the channel for making public disclosures of information, such as e-Bay and Emulex.

3.3 *Chinese regulations on the use of website*

Compared to USA, Chinese securities market started much later. Yet CSRC obviously has noticed the positive meaning that Internet may bring to IRs. CSRC has been promoting using of designated websites. Relevant policies can be pursuant to Securities Law, of which Article 70 states that: The information which must be disclosed pursuant to law shall be released through the media designated by the securities regulatory authority under the State Council, and shall be placed simultaneously at the domicile of the company and stock exchange for public information. According to CSRC official hotline, for main board listed companies, designated media include six designated newspapers and three designated websites. The designated websites are the Shanghai Stock Exchange website, the Shenzhen Stock Exchange website and cninfo.com.cn. CSRC repeated and explained the designated media rule in every information disclosure related regulation.

In the July 2005 IRs Guidelines, CSRC stipulated that the information, which shall be disclosed under the relevant laws, regulations, provisions of the securities regulatory department and bourses, shall first be published on the designated newspapers and websites for company information disclosure; A company shall not let any other public medium precede the designated newspapers and websites in disclosing its information, nor may it replace the company's announcements by way of press conferences or answering questions of the reporters (Article 8 and 9). In the guidelines, CSRC showed an intention to promote development of list companies' websites by stating that listed companies may set up investor relations column on their websites and shall enrich and timely update contents of their websites continually (Article 10 and 11)

and that the company may regularly or irregularly carry out communication activities to improve its relationship with investors through the internet or other modern communication tools (Article 12).

In the 2007 Administrative Measures for Disclosure, CSRC mandated that listed companies and other information disclosure obligors shall submit the draft public announcement and the relevant documents for inspection to the stock exchange for registration when making information disclosure pursuant to law, and make the announcement on media designated by CSRC. The timing of the information announced on the corporate website and other media by an information disclosure obligor shall not precede the announcement of such information in the designated media; the reporting and public announcement duties may not be in the form of a press conference or reply to questions posed by reporters or in any other form (Article 6). In Article 71, it's explained that designated media are CSRC designated newspapers and periodicals and websites. It's not strange that in the administrative measures, articles about using of designated websites are repeated within the 2005 IRs Guidelines, partly because that these measures have much bigger law force than the guidelines.

In the 2012 and 2013 issued Criterion of Information Disclosure Content and Format for Public Securities Offering Company, it's stated that listed companies must publish the whole text of periodical reports on designated websites and at the same time publish abstract in at least one designated newspaper. The abstract contents shall not more than one quarter of a newspaper layout. Periodical reports can also be published in other media, but the timing shall not precede designated media. In the criterion, we can see that CSRC is stressing using of websites while limiting using of newspaper, as a sign for reducing cost of information disclosure, since newspaper disclosure is costly.

September 2009, CSRC approved 5 designated information disclosure websites for Growth Enterprise Market (GEM) listed companies. They are the cninfo.com.cn, cs.com.cn, cnstock.com, stcn.com and ccstock.cn. Four of them are CSRC designated disclosure newspapers' official websites, and one is shareholding by Shenzhen Stock Exchange. These websites was asked to build specialized platform for disclosure, and publish GEM company prospectus, IPO announce, temporary announce, periodical report and related policies and they do these in no fee charge to listed companies. Investors can get GEM companies published information for free on these websites.

3.4 *Comparison between USA and China*

After studying USA and Chinese laws and regulations about IIRs, we can find some differences between the two countries. The differences are reflected not only in the use of websites, but also in the use of social media.

In USA, in certain circumstances, SEC permits listed companies to make nonpublic material information public on their own websites or through their social media accounts. But in China, listed companies shall make information public through CSRC designated media, listed company website and social media releasing of those information shall not precede CSRC designated media.

Electronic formats of information disclosure are equal to traditional formats of information in USA, while in China, newspaper information disclosure in paper dissemination is still irreplaceable.

Press conference and some other forms of information releasing through Internet can be used for making public disclosure in USA. This can diversify IIRs activities and make IIRs activities more useful and interactive. In China, making public information disclosure shall through announcements in designated media and shall not through press conference and other forms. This means that in IRs activities, information disclosed shall not more than the company has announced in designated media, even if the activity is taken in a way that provides general public disclosure.

SEC is actively encouraging and regulating using of Internet for IRs, by releasing of Commission Guidance on the Use of Company Web Site, investigation report of Netflix, and some other related releases. For CSRC, sticking to the designated media mechanism, it's not strange that it hasn't released any systematic rules to promote IIRs actively.

4 CONCLUSION AND SUGGESTIONS

In order to study laws and regulations related to the internet investor relations, this paper made an in-depth comparison of the regulatory history and provisions on the use of website and social media between USA and China. Through the comparison, we can see that IIRs regulation has a longer history in USA than in China, SEC takes an open and encouraging attitude towards new internet technologies application in IIRs, and keeps taking steps to actively regulate and promote using of internet in IRs, including using of company website and social media. CSRC basically is sticking to designated media policy and holds a more conservative attitude towards IIRs.

With the fast development of Internet, many investors are depending on it to get information and communicate with listed companies more and more. Company website and other internet technologies can decrease disclosure cost and proliferate disclosure and IRs activities, which will benefit both listed companies and investors and hence promote development of the capital market. Development of IIRs is an irreversible trend all over the world. To welcome this trend and actively promote its development, we have the following suggestions for Chinese administrative agencies of securities and related entities:

Take active measures to cultivate using of company websites for both listed companies and investors under the current designated media mechanism. For example, encourage listed company disclose its official website in periodical reports, recognizing company websites as fulfillment method of Article 70 of the Securities Law that information disclosed shall be placed simultaneously at the domicile of the company if the company website can fulfill certain requirements, such as accessibility and safety. Some systematic policies shall be made in this area.

Eventually make company website and other internet communication method to become public information disclosure channel, depending on the circumstance, as a supplement to designated media, as an alternative to designated media, or as a stand-alone method of providing information to investors independent of designated media. SEC regulations of company website using and EDGAR can provide beneficial reference.

This research made some investigation into regulations about IIRs and compared them between USA and China, but it is far from complete in the subject. For example, requirements of exchanges about IIRs are not discussed and suggestions are preliminary. Studies in this area still need more exploration.

ACKNOWLEDGMENT

This research is sponsored by "Shanghai Colleges and Universities 085 Engineering Project".

REFERENCES

[1] Feng, Y. & Wan, T. (2013), Website-based Investor Relations: A Comparison between Developed and Developing Economies, Online Information Review, 37(6), 946–968.
[2] Lymer, A. (1997), Corporate reporting and the Internet—a survey and commentary on the use of the WWW in corporate reporting in the UK and Finland, Paper presented at EAA '97, Graz.
[3] MacGregor, J., & Campbell, I. (2006). What every director should know about investor relations. International Journal of Disclosure and Governance, 3(1), 59–69.
[4] Michael, E. & Jr., John G. 2005, Timelines of Investor Relations Data at Corporate Web Sites, Communications of the ACM, 48(1), 95–100.
[5] Tobias, K. & Andreas K. (2006), Investor relations for start-ups: an analysis of venture capital investors' communicative needs, International Journal of Technology management, 34(1/2), 47–62.

Education Management and Management Science – Zheng (Ed.)
© 2015 Taylor & Francis Group, London, ISBN 978-1-138-02663-6

Analysis of 'pyramid-shaped' undergraduate training system: Based on cultivation of excellence accountants (financial managers)

Lingna Mo
Department of Finance and Economy, Guangxi University of Technology, Guangxi, China

Jiadi Li
Lushan College Guangxi University of Science and Technology, Liuzhou, Guangxi Province, China

Juan Sun
Department of Finance and Economy, Guangxi University of Technology, Guangxi, China

ABSTRACT: As the 'Excellence Engineers Cultivation Program' have been started and implemented, the cultivation of excellence accountants (financial managers) is imperative. This paper systematically discuss the issues of 'pyramid-shaped' undergraduate training system which bases on cultivation of excellence accountants (financial managers) in terms of importance, foundation, as well as supporting measures and frame of the training system, and aim to offer theoretical and practical direction for the cultivation of excellence accountants (financial managers).

Keyword: 'Pyramid-shaped' undergraduate training system; cultivation; excellence accountants (financial managers)

1 THE IMPORTANCE OF SETTING UP 'PYRAMID-SHAPED' UNDERGRADUATE TRAINING SYSTEM WHICH BASES ON CULTIVATION OF EXCELLENCE ACCOUNTANTS (FINANCIAL MANAGERS)

The original meaning of 'Excellence' is a talent or quality which is unusually good and so surpasses ordinary standards. 'Excellence Engineers Cultivation Program' (abbreviate to 'Excellence Program') is to carry out major innovation issue of *<National Planning Outline of Education Innovation and Development for Medium and Long term (2010–2020)>* and *<National Planning Outline of Talent Development for Medium and Long term (2010–2020)>*. On 23rd June 2010, Ministry of Education convened Kickoff Meeting of 'Excellence Engineers Cultivation Program' in Tianjin, and has associated with relevant departments and industry associations to implement the program jointly. After that, Zhejiang University drew up an 'Excellence Engineers Cultivation Program' implement plan (trial implementation) of Information and Communication Engineering Major, and has turned out to be a model of others' universities in China. This also means that the cultivation of 'Excellence Engineer' has been developed in China gradually.

Under these circumstances, some experts said 'all types of engineers can pursue Excellence', likewise, all types of accountants (financial managers) can pursue excellence, moreover, excellence accountants' training schools have been established already. Specifically, the traditional model of undergraduate education has been showed the issue that value theory and despise practice which has lead to some of the graduates cannot master their job quickly; even some of them cannot work out financial statements. Therefore, it is essential to cultivate the 'excellence accountants (financial managers)' who are adapting to the requirements of socioeconomic development.

2 THE BASIS OF THE 'PYRAMID-SHAPED' UNDERGRADUATE TRAINING SYSTEM WHICH BASES ON CULTIVATION OF EXCELLENCE ACCOUNTANTS (FINANCIAL MANAGERS)

2.1 Design the training system bases on cultivation model of excellence accountants (financial managers)

The cultivation model of excellence accountants (financial managers) is a kind of coalition of schools and enterprises cultivation model, so, the training system should design an Inner-practice program includes single course training system, course group training system, comprehensive training system for specialties, and on the other hand, also should be comprehensively arranged, which means both of them are essential contents of cultivation program of excellence accountants (financial managers).

2.2 Design the training system bases on the social requirements of excellence accountants (financial managers)

After 1999's enrollment expansion in institutions for higher learning, the scale up of undergraduates and the embarrassment and complexity of our economic reform, therefore; requirements of education model of accounting major (financial management major) have been raised. It asks for the education close to the actual situation of economic reform; shorten the distance of theory and practice, strength the combination of theory and practice has became the objective demands of undergraduates' education. Therefore, the design of undergraduates training system must base on social requirements on excellence accountants (financial managers), and advances the learning and practical ability of the undergraduates.

2.3 Design the training system bases on the requirements of excellence accountants (financial managers)

The training system should be varied based on the training target, some of them emphasis on research talent, and some of them emphasis on advanced applied talent. For example, Guangxi University of Technology defined the talent cultivation targets based on the requirements of applied talent on accounting major (financial management major) which refers to under the premise that insist on the basis which includes broad range, strong knowledge basis, high quality and value application, cultivate the students who have strong and wider range of knowledge, have strong ability and high quality, possess the sign of the times and spirit of innovation, master the knowledge and ability about management, economy, law and finance, as well as money management. Also, the students will be senior talents who will deal with financial affairs, finance control, as well as teach and research work in the units about industry and commerce, financial enterprises, government organizations and institutions. Revolve around the talent cultivation target; the design of training system should be systematic, coherent, and integrative.

Under the cultivation of 'theory + 'Pyramid-shaped' Undergraduate Training System', the undergraduates should master advanced ability of apply knowledge comprehensively, self-study, and innovation, furthermore; they should have a preferable social adaptivity, professional quality, and they also could actively adapt the development of local social and economic development.

2.4 Design the training system bases on school-running features

The school-running features is different among different universities, for example, the school-running features of Guangxi University of Technology is 'Blend the, corporate with enterprise', and looking for a development way which is distinct from the accounting (financial management) major of other colleges. Particularly, Guangxi University of Technology is a local college, which means we should focus on the reality and local area, cater to the demand of talent in Pan-Pearl River Delta (PPRD) and China-ASEAN Free Trade Area (CAFTA), as well as the social and economic development of Liuzhou.

Therefore, we must take 'emphasize on high-quality' and 'value practice' as guidance, base on the school-running features to structure the training system, pay attention to cultivate the practice ability of students who will be advanced applied talents of accounting (financial management) major, they are different from comprehensive university and economic & business schools, and show the school-running features.

3 THE FRAME AND CONTENTS OF THE 'PYRAMID-SHAPED' UNDERGRADUATE TRAINING SYSTEM WHICH BASES ON CULTIVATION OF EXCELLENCE ACCOUNTANTS (FINANCIAL MANAGERS)

3.1 The frame of the 'pyramid-shaped' undergraduate training system which bases on cultivation of excellence accountants (financial managers)

Show in Figure 1.

3.2 The contents of different level of the 'pyramid-shaped' undergraduate training system which bases on cultivation of excellence accountants

3.2.1 Single courses training system
Single courses training system means the trainings are offered to each single course's contents. The trainings aim to make sure that students can

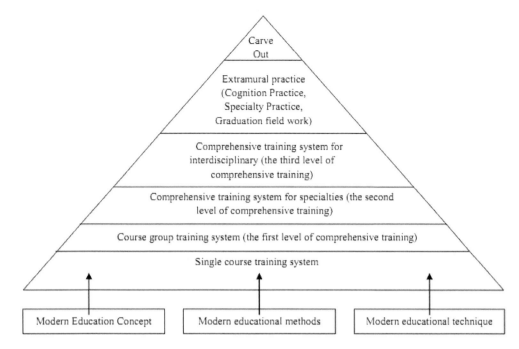

Figure 1.

apply the theory after the courses, and achieve the target that improve the students' ability which refers to intergrades theory with practice and enhance the practice capability. The contents of the single course training should include the major knowledge of the course. For example, the course training of Financial Management should contain the basic content of raising funds, using funds and allocating funds. The whole system is constructed by all the courses in the specialty.

3.2.2 *Course group training system*
The first step of designing the course group training system is to separate the courses to different course groups base on certain relationship between the courses. Then, establish the first level of comprehensive training base on the basic theory of each course group and cover all the major knowledge of each course group. The table shows the course groups structure of accounting (financial management) major (Table 1).

3.2.3 *Comprehensive training system for specialties*
The training is the second level of comprehensive training base on the situation that students finish has finished all the major courses, for example, the intramural analog simulation comprehensive training of accounting (financial management) major.

The purpose of this level's training is to cultivate the student's professional skills; therefore, the content should cover the basic knowledge and theory of the whole major.

3.2.4 *Comprehensive training system for interdisciplinary*
Comprehensive training system for interdisciplinary is the third level of comprehensive training that fits to the society demands of the professional, inter-disciplinary and innovative talents and with the help of 'Finance and Economics Interdisciplinary Analog Simulation Platform'. The purpose of the training is to enhance students' comprehensive quality, cultivate the students' ability of practice and solving actual problems, as well as communication and coordination, gather experience indirectly and ground for the work after graduate. The training takes manufacture companies and manufacturing service companies as business entities, takes the actual transactions, capital flow and information in enterprise operation as thread, constructs the real internal and external environment of enterprise operation, then, leads the students to assemble teams and play their roles, imitate enterprise strategy decision, operational decision, administrative decision and mutual communication with external service. The train of thought is showed in Figure 2.

Table 1.

Course group	Training proposal	Courses contained	Contents
Economic Analysis Course Group (includes 5 courses)	Regional Economic Analysis	Microeconomics Macroeconomics The Economics of Money, Banking and Financial Markets Statistics Financial Markets and Institutions	(External) economic circumstances, financial policy, Status and development trend of the industry
Management Practice Course Group (includes 7 courses)	Enterprise Management Analysis	Principles of Management Management Information System Marketing Oracle Strategic Enterprise Management Operation Research B Human Resources Management Electronic Commerce	(Internal) organizational structure; system design; human resource; strategy management; informatization level
Accounting Practices Course Group (includes 5 courses)	Accounting Information Processing	Principles of Accounts Intermediate Accounting Cost Accounting Application of Financial Softwares Taxation Law	(Accounting information) financial accounting; cost accounting; taxation law; accounting computerization
Financial Analysis and Decision Course Group (includes 7 courses)	Capital Operation Proposal	Investment Financial Analysis Financial Management Management Accountant Financial Risk Management Capital Operation	(Financial projections and decision) enterprise financial diagnosis and governance; evaluation and decision-making; analysis of rotation of capital

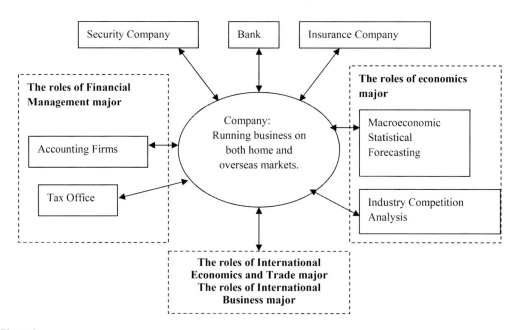

Figure 2.

3.2.5 *Extramural practice*

Extramural practice refers to extramural field work in companies or training bases in summer & winter vacation, specialty practice and graduation field work period, is the prelusion before them starting a career. This aims to cultivate the student's adaptive capacity and professional ability. Extramural practice includes cognition practice, specialty practice and graduation field work.

3.2.6 *Carve out training*

Carve out training means the college cooperated with companies to support the students' entrepreneurship. This aims to cultivate the ability of carving out and innovation.

4 SUPPORTING MEASURES OF PRACTICING 'PYRAMID-SHAPED' UNDERGRADUATE TRAINING SYSTEM WHICH BASES ON CULTIVATION OF EXCELLENCE ACCOUNTANTS (FINANCIAL MANAGERS)

4.1 *Construction of teachers*

The training model of excellence accountants (financial managers) is coalition of schools and enterprises, obviously, it is important to construct an adaptive construction of teachers. First of all, cultivation of accountants (financial managers) acquires the teachers have certain accountant or financial working experiences. Therefore, appoint the teachers who have the teach competency, at the same time, send the teachers who do not have account and financial working experience to the companies to take part in the practical work. On the other hand, the accountants and financial managers come from companies must be appointed, specially the teachers who have Certified Public Accountant qualification.

4.2 *Training bases*

In order to construct the 'Pyramid-shaped' undergraduate training system, the relative training bases should be prepared. The training bases can be considered from both on or off the campus. First, construct intramural practice platform, which is showed in Figure 3.

Furthermore, extramural training bases should be established. These training bases should be permanent and have good facilities, specific training target and contents, also, the scope of the bases should be fit for the requirements of different

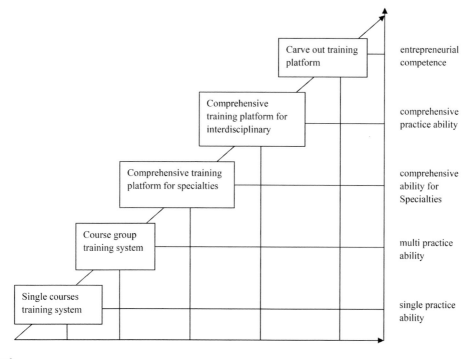

Figure 3.

507

majors, and offer a great training education environment for cultivating the innovation and practice ability.

4.3 *Certain level of inputs and cooperation*

The training work of excellence accountants (financial managers) needs the policy support from different departments which means the communication among government departments, departments of the trades and enterprises, universities and enterprises, universities and education departments. We should adopt measures, address the issues, and academic world should strength the cooperation with enterprises, university should organize the plan thoroughly.

Local government should make the relative policies, promote the enterprises to take part in the 'Excellent Program', and support the local universities that join the program.

REFERENCES

[1] Mo Lingna, Shi Lingfang. Some thoughts on the transformation of traditional industry knowledge. Financial and economic, 2005(7).
[2] Mo Lingna, Li Zhongqian. The market characteristics of the knowledge economy. Modernization of shopping malls. 2005(8).

Education Management and Management Science – Zheng (Ed.)
© 2015 Taylor & Francis Group, London, ISBN 978-1-138-02663-6

A study on improving college students' creativity by constructing "innovation centers"

M.Z. Liao
Business School, Sichuan Normal University, Chengdu, Sichuan, China

D.Y. Yao & C.Y. Jiang
Fundamental Education College, Sichuan Normal University, Chengdu, Sichuan, China

ABSTRACT: Colleges and universities in China try to promote the cultivation of college students' creativity through a variety of scientific and technological innovation activities and academic competitions. But college students' creative thinking and capability still lack independence, inheritance, development and practicality. To solve these problems, this paper proposes constructing two kinds of "innovation centers"—an innovation management center under the Department of Academy of the University Youth League Committee, and an innovation guidance center under every institute, thereby forming a multi-layered, omni-directional, and three-dimensional model to cultivate college students' independence, inheritance, development and practicality in creativity.

Keywords: college students; creativity; innovation centers

1   INTRODUCTION

Xi Jinping says, "We should open vaster sky for the youth to think freely, build up larger stages for the youth to make innovations, provide more opportunities for the youth to shape their life, create more favorable conditions for the youth to make achievements and contributions." However, as bases of cultivating creative college students, colleges and universities are still facing many problems: students depend too much on teachers, having poor independent creativity; teachers' guidance models are invariable, rigid in management and regulation, and students' developmental creativity lack motive force; students' scientific and technological innovation teams are restricted by their majors and grades, so team creativity has no inheritance; innovation practice platforms provided by colleges are far from enough, thus many innovation projects and talents lack opportunities to apply theories in practice, ending up with idle theorizing. Therefore, it is urgent that special organizations or institutions enforce right management, regulation and guidance.

2   CONSTRUCTING "TWO CENTERS"—AN INNOVATION MANAGEMENT CENTER UNDER THE UNIVERSITY YOUTH LEAGUE COMMITTEE, AND AN INNOVATION GUIDANCE CENTER UNDER EVERY INSTITUTE

In order to improve college students' creative spirits and capabilities, the University Youth League Committee and every institute can establish special "innovation centers".

2.1   *Innovation management center under the University Youth League Committee*

The innovation management center specifically established under the Department of Academy of the University Youth League Committee is responsible for organization, management and supervision of the series of competitions with "Challenge Cup" as the leader. Its work mainly includes.

2.1.1   *Classified management and organization*
Scientific and technological innovation competitions can be classified into different categories

(Wang & Li 2012). They can be classified into those of national level, provincial level, municipal level, university level, institute level; or into those of creativity and plan, subject research and essays, modeling, scientific and technological inventions (Chen 2010), venture investment. Correspondingly, the innovation management center can design, arrange, and coordinate the system, process, and organization of different competitions, thus constructing platforms with specialization, feature and difference to cultivate creative talents, academic talents, technical talents, entrepreneurial talents, and to improve college students' creativity step by step in many multi-layered and distinctive ways.

In addition, the innovation management center can regularly organize and hold experience exchanges and achievement exhibitions for the techniques, products or creativity projects that students have developed. It can invite experts in and outside the university to view and direct, and can invite some research institutions and commercial organizations which are interested in these creativity achievements to invest or sponsor. But it is not involved in the actual operation; instead, it is only responsible for organization and promotion.

2.1.2 Enhancing training and practice

In training, the innovation management center should arrange lectures by experts, academic forums, exchanges of technological innovation experience, outdoor team training, entrepreneurship contest training camp and so on to specifically cultivate creativity. In practice, the innovation management center needs to take full advantage of the superior resources to build more and better practice bases of college students' creative projects with school-run factories and logistics group as the core; and outside the university, the innovation management center can depend on university innovation funds as material support, cooperate with design institutes, research institutes, college students' venture incubation parks, scientific and technological innovation parks, industrial and commercial enterprise to build the series of practice bases and practice platforms, combining production, learning and research, providing basic conditions and vast opportunities for college students to turn their planning projects, inventions, technology patents into actual products, and thus improving college students' practical creativity.

2.1.3 Perfecting evaluation and supervision

The innovation management center should also organize experts and teachers to make a scientific evaluation and supervision system to improve college students' creativity, putting students' creativity of independence and inheritance under the protection of system (Casar 2000). It can establish

"creativity" credits, and giving comprehensive quantitative scores to students for all kinds of creative activities that they lead or take part in. The incentive policies are implemented according to the results of comprehensive quantitative scores, and these scores are one of the important indexes which are considered in scholarships, honorary titles, further study recommendation, and employment recommendation.

2.2 Innovation guidance center under every institute

The innovation guidance center under every institute is established by the Institute Sub-Committee Youth League, and the Youth League Sub-Committee Secretary or a student can be in charge of it. Its work mainly includes.

2.2.1 Information communication

By making display panels, establishing websites, or communicating with reporters of Student Union, the innovation guidance center conveys the purpose, content, requirements, time, place and other information about creative projects and activities, explaining such activities within the university. And what's more, the innovation guidance center actively communicates with business groups or individuals outside the campus for sponsorship for the center or its activities.

2.2.2 Training of teachers and students

The innovation guidance center should construct a database of tutors of creative activities and a database of excellent creative works and projects, the former including the corresponding tutors' majors, research fields, papers, books, and guidance of students' scientific and technological innovation, the latter including outstanding award-winning scientific and technological innovation achievements both in and outside of the university. Both databases are open to students of all majors and grades, and students can refer to the related information freely and independently. Meanwhile, with the institute as the platform, the guidance center should invite experienced innovation experts in and outside of the university to train the tutors of the activities or competitions, introducing new guidance concepts and models, and consequently cultivating students' developmental creativity.

2.2.3 Inspection

The guidance center should regularly check whether the institute is opening its computer rooms, reading rooms, laboratories, meeting rooms so that it is convenient for students to practice on computers in computer rooms, consult document data in reading rooms, make experiments by using the instruments and equipment in laboratories, hold conferences in meeting rooms. Thus, students' creative consciousness and practical abilities can be improved in an open cultivation model (Sak & Ayas 2013). If the institute has not opened these rooms or has only opened some of them, the innovation center will immediately contact the person in the institute in chare of this to solve the problem and ensure that all the students who are willing to innovate can have access to reasonable recourses.

3 IMPROVING COLLEGE STUDENTS' FOUR KINDS OF CREATIVITY OMNI-DIRECTIONALLY—INDEPENDENT, INHERITABLE, DEVELOPMENTAL AND PRACTICAL CREATIVITY

3.1 *Improving college students' independent creativity*

The academic knowledge (Cho & Ahn & Han 2005) that tutors convey and the innovation projects and subjects that tutors choose mostly originate from their own research interests and orientations, which only provide students with the existing knowledge and information, and don't drive students' curiosity and creativity to explore the unknown fields (Munakata & Vaidya 2013). Nevertheless, to cultivate college students' independent creativity, tutors should first foster students' courage to discard old ideas, challenge existing knowledge, overthrow stereotype paradigms, and then actively encourage students to develop their own subjects independently, explore their own interests, likes, research orientations, and propose their own research subjects and projects independently. The innovation management center and guidance center should give the college students who have independent consciousness and can think and act independently more "creativity" credits and start-up fund of scientific and technological innovation as a reward.

3.2 *Improving college students' inheritable creativity*

There are so many classes and competitions in colleges and universities, and many academic elites, creative talents, and entrepreneurial pioneers have been cultivated. The two centers can also invite these successful students to the school platform to share their experience with the other students. This can not only promote and inherit the experience of success, but also accumulate force and lay extensive mass foundation to produce new projects and teams. Meanwhile, within the institute, led by the innovation guidance center, teams of interest or teams of creativity can be set up according to specialty characteristics of each institute.

These teams are not set up for a certain competition, and they won't be dismissed after a certain competition, either. Instead, new members will add themselves in these teams every year, and these teams are organizations which have strong cohesion, common goals, dominant culture, enough back-up talents, which can provide continuous intellectual guarantee for time-consuming research projects and subject learning, and which are the effective ways to improve college students' creativity of inheritance.

3.3 *Improving college students' developmental creativity*

For one thing, teachers should alter the too rigid cultivation model in the past and adopt heuristic guidance, listening to students' thinking and opinions, encouraging them to probe into problems from different angles and perspectives and try new schemes and new ways. Teachers should develop students' creativity by enhancing their self-confidence in creative thinking (Xu & Chen 2010). For another, the innovation center and the innovation guidance center can establish special learning websites, providing some books, materials, courseware, and videos of training courses, which introduce some basic theories, principles, thinking methods (Moneta & Siu 2002), skills and techniques of innovation on the websites. Students can download, browse, and independently learn these materials anytime, continuously exercising and improving their abilities to process information, acquire new knowledge, read and write by these materials and courses on the websites.

3.4 *Improving college students' practical creativity*

To cultivate college students' creativity of practicality, the university must seek for more training opportunities and build more practice platforms for them (Mcwilliam & Dawson 2008). The existing bases of practical training can't satisfy the needs of college students' creative practice (Yuan & Wang 2013). The innovation management center under the University Youth League can train college students' creativity by cooperating with school-run factories, logistics group, and internship bases outside the school, or can develop some large innovative projects with ventures outside the school. Or, the university can set up venture fund for college students, encourage college students to practice and start their own businesses (Garcia-Cepero & Maria Caridad 2008). Furthermore, the innovation guidance center can even cooperate with universities abroad to develop and construct practical courses to improve creativity, closely combining

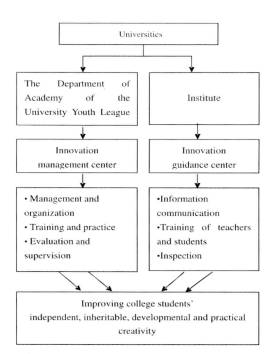

Figure 1.

lecturing with activities, competitions, projects and practice.

4 CONCLUSIONS

This paper proposes constructing two centers, an innovation management center under the Department of Academy of the University Youth League Committee, and an innovation guidance center under every institute, as is shown in Figure 1.

And this thinking has formed in the frontline work of cultivating and improving college students' creativity. The model of constructing innovation management center under the department of academy of the University Youth League Committee has produced preliminary results in some universities, greatly encouraging teachers' and students' enthusiasm and independence in innovation and venture, thus increasingly enlarging innovation teams on campus; however, the concept and proposal of constructing an innovation guidance center under every institute is yet needed further practice and application. In conclusion, only if the University Youth League Committee and every institute play their roles and embody their features

respectively, and the above-mentioned two centers coordinate with each other and work jointly, can they lay solid foundation for the further improvement of college students' independence, inheritance, development and practicality in creativity.

ACKNOWLEDGMENT

This research was financially supported by the National Nature Foundation of China (Grant No. 71202166), the Youth Fund of Educational Department in Sichuan Province (Grant No. 12SB105).

REFERENCES

[1] Chen, Y.R. 2010. A research on the cultivation system of innovative talents at colleges. *International Education Studies* 3(2): 143–147.
[2] Cho, S. & Ahn, D. & Han, S. 2005. Preference for challenging tasks: the critical factor on late academic achievement and creative problem solving ability. *KEDI Journal of Educational Policy* 2(2): 57–78.
[3] Casar, J.R. 2000. Encouraging students' attitude of innovation in research universities. *European Journal of Engineering Education* 25(2): 115–121.
[4] Garcia-Cepero. & Maria Caridad. 2008. The Enrichment Triad Model: nurturing creative-productivity among college students. *Innovations in Education and Teaching International* 45(3): 295–302.
[5] Mcwilliam, E. & Dawson, S. 2008. Teaching for creativity: towards sustainable and replicable pedagogical practice. *Higher Education* 56(6): 633–643.
[6] Moneta, G.B. & Siu, C.M.Y. 2002. Trait intrinsic and extrinsic motivations, academic performance, and creativity in Hong Kong college students. *Journal of College Student Development* 43(5): 664.
[7] Munakata, M. & Vaidya, A. 2013. Fostering creativity through personalized education. *Primus: Problems, Resources, and Issues in Mathematics Undergraduate Studiesn* 23(9): 764–775.
[8] Sak, U. & Ayas, M.B. 2013. Creative Scientific Ability Test (C-SAT): A new measure of scientific creativity. *Psychological Test and Assessment Modeling* 55(3): 316–329.
[9] Wang, L.P. & Li, M.Q. 2012. On the cultivation of automation majors' research innovation ability based on scientific research projects. *Higher Education Studies* 2(4): 137–141.
[10] Xu, Z.H. & Chen, H.L. 2010. Research and practice on basic composition and cultivation pattern of college students' innovative ability. *International Education Studies* (3)2: 51–55.
[11] Yuan, L. & Wang, C.H. 2013. Strengthening Scientific Research Ability of Undergraduates, Cultivating Practical and Innovative Talents. *International Journal of Information and Education Technology* 3(6): 648–650.

Education Management and Management Science – Zheng (Ed.)
© 2015 Taylor & Francis Group, London, ISBN 978-1-138-02663-6

How does the establishment of CAFTA affect CO_2 emissions of China?

Yuhuan Zhao
School of Management and Economics, Beijing Institute of Technology, Beijing, China

ABSTRACT: The relationship between global climate change and international trade is a hot topic in recent years. Foreign trade contributes a lot to CO_2 emissions of China. The establishment of CAFTA affects foreign trade of China, and so affects CO_2 emission of China. By using data from 1990 to 2010, this paper studies the effects of CAFTA on CO_2 emission of China empirically. The result shows that CAFTA's establishment will have negative influence on China's CO_2 emission in comprehensive level. So in order to gain most from CAFTA, corresponding measures should be taken by the government to control the possible negative effect on environment.

Keywords: CAFTA; CO_2 emission; China

1 INTRODUCTION

In recent years, global climate change has drawn more and more attention internationally. With the rapid growth of economic development and foreign trade, China faces huge pressure of CO_2 emission reduction. According to IEA's report, China has become the largest CO_2 emission country in 2007. China became the largest exporting country in 2009. Studies showed that foreign trade contributed a lot to China's CO_2 emission. CAFTA, which was signed by China and ASEAN, is the first Free Trade Agreement for China. The establishment of CAFTA affected China's foreign trade and CO_2 emission theoretically. Using data from 1990 to 2010, this paper studies the effects of CAFTA on CO_2 emission of China empirically.

2 MODELING

In this paper, we use a simple model, which is revised from the ACT model, to estimate China's environmental effects caused by its trade with ASEAN. The basic empirical model is as following:

$$TREM = c_1 \ln EXPORT + c_2 CACLP + c_3 CADPR + c_4 D1 + c_5 time + c_6 D1 * time + \omega \tag{1}$$

where c_i for $i = 0, 1, 2, 3, 4$ are coefficients to be estimated, and ω is random error term. TREM is China's CO_2 emission caused by its export to ASEAN. Import is not taken into consideration because it is commonly believed that import could reduce importer's CO_2 emission. Official data of trade caused CO_2 emission is not available. So in order to get China's CO_2 emission caused by its export to ASEAN, original CO_2 emission data need to be processed. The processing method is given below:

$$TREM = EXPORT \times CO_2/GDP \tag{2}$$

As export is a part of one country's GDP, so export related CO_2 emission could be got by using export to multiply the country's CO_2 emission intensity, which is the latter part of equation (2).

EXPORT stands for China's export to ASEAN. Here $\ln EXPORT$ is the natural logarithm form of EXPORT. China's export saw a relatively large change, so in order to make the series steadier, we use export's natural logarithm in the estimation process. We can see that the new series is steadier than the original series.

GDP is used to represent trade's scale effect on China's environment. In other empirical literatures, GDP is always used as indicator of the scale effect. However, as China's GDP is a result of domestic consumption, investment and foreign trade, it could be vogue when GDP is used to indicate the scale effect. In other words, trade related scale effect should be separated from investment and domestic consumption. In addition to that, ASEAN is only one of China's international trade partners, so even if GDP is used as indicator of the scale effect, we still need to separate China's GDP change caused by trading with ASEAN, which could be inapplicable as result of shortage of related data. So China's export to ASEAN is the most direct indicator of the scale effect.

CACLP is the ratio of China's capital to labor ratio to ASEAN's capital to labor ratio. It is used to indicate the composition effect. In recently studies, export structure is the most frequently used indicator of composition effect. However, as a shortage of data of China's detail export structure to ASEAN, this measure could hardly be implemented. In fact, capital to labor ratio can reflect the composition effect in an indirect way. Antweiler and Copeland's (1998) also use this indicator to reflect the composition effect and they make a detailed theoretical derivation to figure out the logic inside it. But there is a point should be noticed that in this paper, the labor to ratio is different from that of Antweiler and Copeland's (1998). In their paper, physical capital stock per worker represents capital labor ratio. In this paper, the capital is the gross capital formation in current USD. This change is mainly caused by a shortage of statistical data. Beside, in order to reflect trade's composition effect, we use the ratio of China's capital labor ratio to ASEAN's to reflect the change of two sides' comparative advantage in factor abundance. In this way, we try to evaluate how this change could draw influence on China's CO_2 emission.

CADPR is the ratio of China's GDP per capita to ASEAN's GDP per capita. We use this term to indicate the technical effect. In other related literature, GDP per capita or national income and corresponding revised form is used as indicators. The ACT model has given a theory description to explain the how this indicator draw indirect effect on CO_2 emission. Here we use the ratio value to replace the absolute value is because we want to reflect the influence of ASEAN. In this way, we try to evaluate the effect of technique difference on CO_2 emission. CO_2 emission intensity is another indicator could reflect the technical effect, but as it has been part of the dependent variable, so we decide to not use it to limit the possible multicollinearity.

In our model, *time* is stand for the time trend term. We try to evaluate the possible influence of time trend on CO_2 emission because the environment is generally deteriorated with time by taking the time trend terms into model. And *D1* is a dummy variable, which is used to reflect the change of intercept term in the two evaluated periods. From analysis, we can see that there is a notable change in China's CO_2 emission resulted from trade with ASEAN. And the turning point is between 2002 and 2004. In fact, in 2004, the "Early Harvest Program" is being smoothly implemented. So we try to evaluate if the trade caused CO_2 emission is significant change at this point, which is reflected by *D1*time* term.

3 DATA DESCRIPTION

In this paper, China's CO_2 emission data is from World Bank. Here CO_2 emissions are those stemming from the burning of fossil fuels and the manufacture of cement, they include carbon dioxide produced during consumption of solid, liquid, and gas fuels and gas flaring. This paper evaluates trade's environment effect from 1990 to 2010, so World Bank's data is not adequate enough for our evaluation period because the latest available data is for 2008. In order to get emission data of 2009 and 2010, we use the estimation results of CDIAC, from where the World Bank gets the definition and partial data of CO_2 emission. In CDIAC's (Carbon Dioxide Information Analysis Center) databank, all emission estimates are expressed in thousand metric tons of carbon. So in order to convert these estimates to units of carbon dioxide, we need to multiply the original data with 3.667. In this way we get China's emission data of 2009 and 2010. GDP data is get from the World Bank's databank as well and expressed in current US dollars.

China's export data to ASEAN countries is not available for years earlier than 2000. The UN comtrade databank has China's trade data with world's most countries from 1984 to 2011. Fortunately, all ASEAN's ten countries are included in this range, except for the import data from Brunei in 1993 (In SITC.1). So here we get China's export data to ASEAN by aggregating the export data of China to ten specific countries. We compare the data we got in this way with the official available data, and find no significant difference. Besides, data got by this way ensure the consistency of data source. The export data is expressed in billions of current US dollars.

China's capital to labor ratio is calculated with the gross capital formation (current US dollars) and total labor force data of World Bank. For ASEAN's corresponding data, as there is no related data for the total group in such a long series, we calculate it by using ten countries' aggregated gross capital formation (current US dollars) to divide the corresponding total labor force data. With the above two data series, we can get the ratio of China's capital to labor ratio to the ratio of ASEAN's capital to labor ratio. The explanation of gross capital formation is given by the World Bank: "Gross capital formation (formerly gross domestic investment) consists of outlays on additions to the fixed assets of the economy plus net changes in the level of inventories. Fixed assets include land improvements (fences, ditches, drains, and so on); plant, machinery, and equipment purchases; and the construction of roads, railways, and the like, including schools, offices, hospitals, private residential dwellings, and commercial and industrial buildings. Inventories are stocks of goods held by firms to meet temporary or unexpected fluctuations in production or sales, and "work in progress." According to the 1993 SNA, net acquisitions of valuables are also considered capital formation. Data are in current U.S. dollars".

And the definition of labor force is "Total labor force comprises people ages 15 and older who meet the International Labor Organization definition of the economically active population: all people who supply labor for the production of goods and services during a specified period. It includes both the employed and the unemployed. While national practices vary in the treatment of such groups as the armed forces and seasonal or part-time workers, in general the labor force includes the armed forces, the unemployed and first-time job-seekers, but excludes homemakers and other unpaid caregivers and workers in the informal sector."

CADPR is the ratio of China's GDP per capita to ASEAN's GDP per capita. China's GDP per capita is from the International Monetary Fund's (IMF) World Economic Outlook Database. But there is no corresponding data for ASEAN. In order to get ASEAN's data, we calculate it by aggregating individual country's GDP and then divide it with total population of ASEAN. With the above two series, we can get the series of *CADPR*.

And finally, for the dummy variable, in order to evaluate if the "Early Harvest Program" is the turning points of China CO_2 emission trend, 2004 is defined as the turning point. And from 1990 to 2003, the value of *D*1 is 0; from 2004 to 2010, the value is 1. 1990 is selected as the starting point is mainly the result of two factors: first of all, related data before 1990 is not complete enough and have many missing value; besides, it is from 1990s that China began having frequent interaction with ASEAN countries. The summary data table is presented in appendix A.

4 RESULTS ANALYSIS

In this paper, the estimation results are evaluated with Eviews 6.0 software. Table 1 presents the estimation results for the basic model. If there is no special note, all the default regression method is OLS.

From Table 1, we can get the estimation equation for the Basic Model which is shown below:

$$TREM = 48.00 \ln EXPORT + 115.17 CACLR$$
$$- 180.50 CADPR - 129.14 D1$$
$$(4.39)\,(3.32)\,(-2.99)\,(-2.71)$$
$$- 6.29\,time + 12.10 D1 * time$$
$$(-2.55)\,(4.23) \tag{3}$$

The estimated coefficients for all the five variables are significant at 0.05 confidence level. For $\ln EXPORT$, *CACLR*, *CADPR*, and $D1 * time$, their estimated coefficients are significant at 99% level. The estimated results show that the variables in this paper's model have significant influence on the dependent variable.

Specifically, scale effect's indicator $\ln EXPORT$ has a coefficient of 48.00. Holding other independents variable to be fix, then one unit change of the scale effect will result in 48.00 unit change on CO_2 emission. This trend is consistent with the theoretical forecast of the scale effect. If we fix other independent variable at the 1990 level, and taking 2010's $\ln EXPORT$ value into calculation process, we will find that the estimated CO_2 emission will be 171.19 million tons higher than the estimated emission of 1990, and 169.95 million tons higher than the actual emission of 1990, from which we could see the significant influence of the scale effect on CO_2 emission.

The composition effect, indicated by *CACLR*, actually draws negative effect on China's CO_2 emission. In the evaluated period, the value of *CACLR* shows an increasing trend, which implies that China is more abundant with capital factor com-

Table 1. Estimation of the basic model.

Variable	Coefficient	Std. Error	t-Statistic	Prob.
LNEXPORT	48.00359	10.93437	4.390157	0.0005
CACLR	115.1749	34.72084	3.317171	0.0047
CADPR	−180.5040	60.43471	2.986760	0.0092
D1	−129.1427	47.60642	−2.712716	0.0160
@TREND	−6.294768	2.473003	−2.545395	0.0224
D1*@TREND	12.10443	2.860047	4.232249	0.0007
R-squared	0.970449	Mean dependent var		88.97321
Adjusted R-squared	0.960599	S.D. dependent var		71.91475
S.E. of regression	14.27480	Akaike info criterion		8.389825
Sum squared resid	3056.550	Schwarz criterion		8.688260
Log likelihood	−82.09316	Hannan-Quinn criter.		8.454593
Durbin-Watson stat	2.397103			

pare to ASEAN. So China enjoys advantage in exporting commodities with higher capital to labor ratio. This may imply that China has the trend of specializing in producing emission intensity products after several years of free trade if no other policies to influence it. Also, it verifies the common phenomenon that capital abundance products tend to generate more pollution compare to labor abundance products.

And for the technical effect, the negative coefficient implies that technical advancement can impede the increase of CO_2 emission, which also goes agree with the theoretical forecast. Again, we set other independent variables at the 2010 level, and take 1990's technique level into calculation, we could find that the estimated CO_2 emission will be 178.68 million tons higher than the estimated emission of 2010 and 207.90 million tons higher than 2010's actual CO_2 emission. From these results, we could see that technique effect does make a difference in CO_2 emission reduction.

The regression results shows that China's CO_2 emission is indeed change with time as the estimated coefficient of *time* is significant. However, surprisingly, from 1990 to 2003, the changing trend is not that CO_2 emission increase with time goes on, but decrease with time goes on. Several factors may account for this abnormal phenomenon and we believe that China's increasingly strict environmental regulations, which we fail to take into consideration in this paper's simple model, should charge for much of it. In the second period (2004~2010), CO_2 emission actually increase with time goes on. This implies that in this period, the negative effect of economic development surpassed the positive effect of related environmental regulation. The significant coefficient of *D1*time* proves that China's CO_2 emission caused by trading with ASEAN really change its increase speed at the point of 2004. This implies that the liberalization process of the trade between China and ASEAN is indeed accelerating the CO_2 emission increasing rate, which could be verified by the bigger coefficient of the time trend term in the latter period of equation (4)

$$\begin{cases} 48.00 \ln EXPORT + 115.17\, CACLR - 180.50\, CADPR \\ -6.92\, time \hspace{3.2cm} (1990 \sim 2003) \\ -129.14 + 48.00 \ln EXPORT - 115.17\, CACLR \\ -180.50\, CADPR + 5.81\, time \hspace{1.2cm} (2004 \sim 2010) \end{cases}$$

$$(4)$$

5 CONCLUSIONS

CAFTA's establishment actually draws negative effects on China's CO_2 emission. At a comprehensive level, after 2004, in which year the "Early Harvest Program" is smoothly implemented, China's CO_2 emission trend is deteriorated. We believe that the more and more freer trade between China and ASEAN should play an important role for this change. Specifically, the trends of the three effects are all changed after CAFTA's establishment. For the scale effect, CAFTA's establishment indeed play negative role in affecting China's CO_2 emission trend, because its indicator's coefficient change from 57.73 to 152.25. For the technique effect, CAFTA's establishment actually intrigues positive influence, which could be known by comparing two coefficients of *CADPR* in the two evaluated periods. The composition effect actually plays negligible role in the period from 1990 to 2003, but after the "Early Harvest Program" was being implemented, it began draw negative effect on China's CO_2 emission.

REFERENCES

[1] Antweiler, W., Copeland, B.R., & Taylor, M.S.. Is free trade good for the environment?. National Bureau of Economic Research, 1998.
[2] Atici, C.. Carbon emissions, trade liberalization, and the Japan-ASEAN interaction: A group-wise examination. Journal of the Japanese and International Economies. 2011.
[3] Copeland, B.R., & Taylor, M.S. International trade and the environment: a framework for analysis. National Bureau of Economic Research, 2001.
[4] Douglas, S., & Nishioka, S. Emissions Intensity and Global Patterns of Trade and Development. West Virginia University, 2011.
[5] Gallagher, P. and Y. Serret, "Implementing Regional Trade Agreements with Environmental Provisions: A Framework for Evaluation". OECD Trade and Environment Working Papers, 2011/06, OECD Publishing.
[6] George, C. and Y. Serret, "Regional Trade Agreements and the Environment: Developments in 2010", OECD Trade and Environment Working Papers, No. 2011, OECD Publishing.
[7] Grossman, G.M., & Krueger, A.B. Environmental impacts of a North American free trade agreement. National Bureau of Economic Research. 1991.
[8] IEA (International Energy Agency). World Energy Outlook 2011. 2011. 205–209.

Education Management and Management Science – Zheng (Ed.)
© 2015 Taylor & Francis Group, London, ISBN 978-1-138-02663-6

On Outward Bound for college students conducted by China University of Geosciences (Wuhan)

Lun Li & Xiao Hong Niu
Physical Education Division, China University of Geosciences, Wuhan, China

ABSTRACT: This paper makes a brief introduction to Outward Bound in respects of its origin, development, roles, significances and program objectives by studying literature and interviewing experts. It also puts forward proposes for the development of Outward Bound among students at China University of Geoscience (CUG) in Wuhan.

Keywords: Outward Bound; college physical education; program objectives

1 INTRODUCTION

Technological advance has brought more convenience and comfort to people's daily life since the coming of the 21st century. However, as it is a double-edged sword, it has also caused some social problems. For example, students in today's colleges and universities have gradually lost the spirits of hard work and teamwork, as well as the initiative to break their own limitations. While Outward Bound has its unique advantages in cultivating people's teamwork spirit and conquering themselves, PE Class boasts favorable conditions for training students to be adventurous, innovative, and competent to challenge their limits which cannot be matched by other courses. PE Class and Outward Bound have various close relations in content and form. Therefore, incorporation of Outward Bound into PE Class can realize objectives set for the PE Class and the Outward Bound as well as enrich the PE Class without increasing students' course load. This paper studies the Outward Bound for students conducted in China University of Geosciences (Wuhan) (CUG), attempting to provide reference for other colleges and universities to carry out Outward Bound.

2 METHODOLOGIES

2.1 *Literature research method*

This paper has referred to relevant books and journals, as well as collected a lot of data on Outward Bound from websites related to sports in China.

2.2 *Expert interviews*

Authors of this paper has interviewed experts and professionals in this field and sought their opinions and suggestions about the development of Outward Bound.

3 RESULTS AND ANALYSIS

3.1 *Origin and development of Outward Bound*

As an extreme sport which integrates survival, adventure, thrill, recreation and education, Outward Bound originated in Britain during the World War II. At that time, many ships had sunk into the Atlantic Ocean due to attacks, and groups of sailors fell into the water. The cold water and far distance from land led to the death of many young seamen. But from the survivors, a shocking phenomenon was found that those people who survived the disasters were not those well-armed men full of energy but those with the firmest belief and the strongest desire to survive. Then Kurt Hahn from Germany proposed that, some activities and programs should be conducted for young mariners with some natural conditions and man-made facilities to cultivate and improve their psychological quality. Later, his friend Lawrence Holt and German educators made in-depth analysis and study on the phenomenon. A survival training school (the first Outward Bound school) was open in Wales in 1941 at Hahn's proposal, to provide special trainings for people on belief to survive. This was the prototype for Outward Bound.

After World War II, a type of management training named OUTWARD BOUND emerged in Britain. This type of training cultivated managers and entrepreneurs from aspects of psychology and management by using outdoor activities to simulate real management context. Due to the novel form of Outward Bound and positive effects, this type of training became popular in the field of educational training in Europe, and spread to other parts of the world in the next 50 years. Up to the present, 48 schools under the uniform name Outward Bound have been established in 28 countries and regions around the world. Those Outward Bound schools have formed an international organization, whose headquarters is located in Ottawa, Canada.

In 1995, Outward Bound entered China. Due to lack of understanding toward Outward Bound and media publicity, Outward Bound had been waged only in big cities like Beijing, Shanghai, Shenzhen and Guangzhou. In addition, the form and content of Outward Bound was fixed and monotonous with little innovation due to the number and quality of trainers. As people change their thinking and have better understanding about Outward Bound, people have acknowledged the functions of Outward Bound as well as its innovation and development. Therefore, Outward Bound has been conducted in schools, enterprises and public institutions at different levels, which has created good social effects and won popularity.

3.2 Roles and significances of Outward Bound

Outward Bound allows students to participate in various tasks, discuss and share experience with others, reflect on themselves and draw upon valuable lessons that can be applied into their future job. After the whole training, students will be able to realize their potentials and boost their confidence; overcome the psychological inertia and hone their perseverance in order to weather the storm; adjust their attitude toward life and work and be optimistic about challenges; realize the power of a group and have the awareness of participate in a group and shoulder responsibilities; stimulate their imagination and creation and enhance their problem-solving abilities; improve interpersonal relationships by learning to care about others and work more closely with others and be able to appreciate, care for and protect Nature.

Outward Bound is more than the sum of sports and recreation. Rather, it is a refining of formal education and a useful complement. Often when we talk about quality improvement, we think about gaining certificates, diplomas, and degrees such as MBA. As a matter of fact, knowledge and skills are just visible assets while will and spirit are invisible power. Under what circumstances can our limited knowledge and skills release the maximum energy? How to tap into your deepest potentials that you have barely noticed yourself? How to make out how far you can go in conversations and to what extent you can trust each other? Answers to these questions are exactly what Outward Bound is designed for.

The main idea of introducing Outward Bound into PE teaching is to offer well-targeted and tailor-made guidance and training to students in ideological and moral education, sports skills development, innovative ability improvement, teamwork spirit development, community and volunteer services and jobs and club activities so as to develop human resources among college students. These six aspects are a basic reflection as well as important evaluation criteria of the overall quality of a college student. Implementation of Outward Bound in higher education must center on the development of college students' ideological and moral concepts and promote their all-round development of other qualities at the same time. Students who have received the training program are expected to be equipped with relatively high sports culture attainments and sound general knowledge; able to track and identify new technologies and concepts; have the awareness and habit of creative thinking; capable of dealing with different types of people and have a good social mentality; good at organizing, managing and coordinating in future jobs and have the ability to care for others and serve the society.

3.3 Course objectives of Outward Bound

Individual task objectives: help students realize their potentials and develop good mental qualities, such as courage, tenacity and perseverance.

Group task objectives: enhance students' understanding of the power of a group, develop their sense of belonging and teach them the importance of communication, cooperation, simplicity, harmonious relationships and obeying orders.

Leadership program objectives: enhance their leadership and coordination skills by engaging them in specially designed role-play tasks.

Orientation program objectives: develop team spirit of new employees and enhance their communication and understanding among themselves through specially designed tasks; instill corporate values into these new employees at the same time so as to enable them to fit into normal operations of a company within the shortest possible time.

Youth program objectives: develop their sense of collectivity, correct interpersonal skills, and awareness of creative thinking and environmental protection by allowing them to live together as a group for a specific period of time and experience the warmth of the group.

3.4 *Outward Bound programs suitable for CUG*

1. Individual and pair tasks include the blind running through a minefield, untie knots among others.
2. Communication tasks include walking as a blind, build a square of the blind, tell numbers by gestures, tearing paper, lost on a hidden island, message passing among others.
3. Ice-breaking tasks include face-to-face introduction, body passing, cry out loud, human wave among others.
4. Group tasks include quickly pass the ball, link hands, spider web among others.
5. Base tasks include pole in the air, rock climbing, sulfuric acid pool, broken bridge in the air climb over a wall among others.

3.5 *Requirements of Outward Bound teachers, sports field and facilities*

The key to launching Outward Bound on campus is to have qualified teachers who are excellent at organizing, supervising, guiding and summarizing. PE teachers with rich teaching experience and a basic understanding of management, organizational behavior, and psychology will make perfect trainers at Outward Bound programs after receiving simple training and studying. All training tasks in Outward Bound are just like games, so the rules and arrangements of the playing field are easy to master. Furthermore, most trainers of Outward Bound in the earliest days were lecturers in universities and colleges or teachers in middle and high schools, and even today many teachers are doing part-time jobs in outdoor activity clubs as trainers. Therefore, it is possible for PE teachers to offer reliable organization and guidance in Outward Bound programs. CUG has already enrolled undergraduates majoring in Outdoor Sports for two years. And it proves that these students are able to be assistants to PE teachers after acquiring a certain amount of professional knowledge and skills. This internship experience will also help them raise their level of professional skills by helping understaffed PE teachers serve more students of different majors, and lay a solid foundation for entering the outdoor sports industry in the future.

CUG is the only one with Outward Bound field in Hubei province, which covers a land area of 13 mu (about 0.87acre) and capable to host over 40 tasks. This field is equipped with complete facilities and maintenance personnel, making it very convenient to conduct Outward Bound at CUG. PE classes offered at CUG including wilderness survival experience and some open elective courses all include Outward Bound programs.

4 CONCLUSION AND SUGGESTIONS

4.1 *Conclusion*

Outward Bound makes you feel a strong sense of achievement and great pleasure in fighting together with your partners by creating non-ordinary situations where you discover your weaknesses, blind spots and potentials that are hardly noticed in everyday life. This experience will have a positive transfer on your future job and further study, and help you develop a more active attitude toward life. Although PE classes have proven effects in fostering students' fighting spirit, motivation, and cooperative ability, they have less impact on one's mental quality compared with Outward Bound. Other activities in traditional PE classes, such as ball games, martial arts and gymnastics have a certain degree of Outward Bound training, but they lack scenarios. Therefore, using Outward Bound as a complement to and an extension of PE classes will definitely be greatly valued, for it not only has a large room for development, but also meets the requirement of being healthy today, namely an all-round development of physical conditions, mental status and social adaptability.

4.2 *Suggestions*

1. Think out of the box and create conditions to allow everyone to participate and understand the value behind the program.
2. PE teachers should naturally combine teaching with Outward Bound training by setting up scenarios creatively, correctly understand what happy health education is about, and modernize the current simplified and boring organization in PE classes. They should apply some contents and tools of Outward Bound to make PE classes more vivid and attractive. By doing this, they need to make students to truly understand that no pains, no gains.
3. Increase usage frequency of Outward Bound field. First, introduce club systems of various types that are managed and arranged wholly by PE teachers and concerned leaders with the aim to protect students' safety during training. Second, open the Outward Bound field to the public moderately on the condition that all PE classes have sufficient playing field available when it is needed, which will not only generate revenue for the university but also serve the society at the same time. This effort will give full play to the function and value of the Outward Bound field.
4. Integrate Outward Bound into PE classes. This is more demanding for PE teachers. They should not only be able to set up scenarios accordingly

but also capable of guiding students to internalize their training experiences and apply what they have learned into their future life and work.

REFERENCES

[1] Mao Zhenming, Wang Changquan. Outward Bound and psychological development at school. Beijing: Beijing Sport University Press; 2004.

[2] Zhou Fang, Fei Yingqin. Application of vocation-orientated training in higher education. China higher medical education; 2003(4).

[3] Zhang Yaqi. Thoughts on the introduction of Outward Bound into PE classes. Liaoning sport science and technology; 2008(4).

[4] Wang Jieer. Application of Outward Bound in developing college students' mental qualities. Theory and practice of education; 2004(2).

[5] Li Guoyan. Discussion on the future popularization and development of Outward Bound in terms of its characteristics and functions. Shandong sports science & technology; 2005(4).

Education Management and Management Science – Zheng (Ed.)
© 2015 Taylor & Francis Group, London, ISBN 978-1-138-02663-6

A survey of current status of self-efficacy concerning college students' emotion regulation

Huiying Liu, Bian Du, Xueke Huang & Yanjie Hou
Department of Education, Zhengzhou University's, Zhengzhou, Henan, China

ABSTRACT: The purpose of this study is to understand the current status of self-efficacy concerning college students' emotion regulation. The regulatory emotional self-efficacy scale was adopted to conduct questionnaire investigation on 1321 college students in Henanand statistical analysis, which was conducted by SPSS software. The results show that: the overall level of self-efficacy concerning college students' emotion regulation was high; the male and female college students in regulatory emotional self-efficacy subscales were significantly different, but there was no significant difference in the scale total score; the only-child students and students with siblings had no significant difference in regulatory emotional self-efficacy; students of different grades also had no significant difference in regulatory emotional self-efficacy; in the POS dimension, medical students scored significantly higher than students majored in arts and engineering; in the ANG dimension students majored in science scored significantly higher than students majored in arts, engineering and medicine with students majored in engineering having a higher score than the medical students.

Keywords: college students; self-efficacy; emotion regulation

1   INTRODUCTION

Emotion is subjective experience of human to objective things. As a complex psychological phenomenon, it has important impact on interpersonal communication, behavioral motive, physical and psychological health and other aspects; and emotion is a barometer to show one's mental health condition. Therefore, emotion regulation has great significance as it is essential for the social competence and mental health (Dongling Tang et al. 2010). The contemporary college students, especially, will be faced with many challenges when encountering negative emotion incidents. People with low regulatory emotional self-efficacy are lack of confidence on self-emotion regulation competence; they merely take action to regulate negative emotion or just fall into despair when the first regulation becoming invalid. Among educational practice, knowledge and skills about emotion regulation shall be conveyed to college students to guide them taking the generation of emotion as an opportunity for enhancement of self-awareness and improvement while accepting the emotion, and to think about their cognition on the basis, gradually trying to use diversified attitudes to face emotion and gradually promote regulatory emotional self-efficacy. Regulatory emotional self-efficacy indicates the confidence

level of an individual concerning whether he/she can regulate his/her own emotion effectively. Since the study of it by Caprara G.V., an Italian psychologist, regulatory emotional self-efficacy has a great research prospect as a new research field (Ellen Heuven et al. 2006). Strengthening of training regulatory emotional self-efficacy has great significance (Ren Li, 2012; Guiqin Liu et al. 2012; Caprara G.V et al. 2008) to improve individual's pressure handling, quality of relationship, social adaptation level, subjective well-being, etc. This research investigates the current situation of self-efficacy concerning college students' emotion regulation and gives an reference to the mental health counsel for the college students.

2   RESEARCH METHOD

2.1   *Research object*

By adopting the method of random sampling, 1321 college students were selected for this research to do questionnaire survey; 1400 pieces of questionnaire were issued, and 1317 pieces of valid questionnaire were recovered; the valid recover rate is 94.36%. Among the research objects, there are 629 male students and 688 female students; 311 are only-child and 1006 are children with siblings; there are 256 freshmen, 654 sophomores,

303 junior students and 104 senior students; 528 are majored in arts, 194 are majored in science, 417 are majored in engineering and 178 are majored in medicine.

2.2 Measuring tool

The measuring tool used is the Chinese edition of Regulatory Emotional Self-Efficacy Scale[6] revised by Caprara. This scale has 12 questions being calculated in 5 levels from completely not match to completely match. The scale is divided into three dimensions. Perceived self-efficacy in expressing positive affect, POS, perceived self-efficacy in managing despondency/distress, DES, and perceived self-efficacy in managing anger/irritation, ANG. Every dimension includes four questions. Related research and analysis indicate[7] that, coefficient α of this scale is 0.82; coefficient α of the POS subscale is 0.74; coefficient α of the DES subscale is 0.74; coefficient α of the ANG subscale is 0.76, and this scale has better validity.

2.3 Statistical method

SPSS is adopted to sort valid data; and methods of descriptive statistics, independent sample t-test and one-way analysis of variance are adopted to perform statistical analysis.

3 RESULTS

3.1 Overview of self-efficacy concerning college students' emotion regulation

Results of Table 1 show that the minimum values of POS and ANG are all 4, and the maximum values are 20; average value of POS is 15.63 which is larger than its theoretical median 12; average value of ANG is 13.37 which is larger than its theoretical median 12; minimum value of DES is 5, the maximum value is 20, and the average value is 13.90 which is larger than its theoretical median 12; minimum value of aggregate score of RES is 16, the maximum value is the full score 60, and the average value is 42.91 which is larger than its theoretical median 36.

3.2 Difference comparison in different demographic variables of self-efficacy concerning college students' emotion regulation

Table 3 shows that in regulatory emotional self-efficacy, only-child students have greater scores in every dimension score and in the total score than

Table 1. Scores of self-efficacy concerning college students' emotion regulation (M ± SD).

Dimension	Minimum	Maximum	Mean	Std. Deviation
POS	4	20	15.63	2.62
DES	5	20	13.90	2.59
ANG	4	20	13.37	2.83
RES	16	60	42.91	6.15

Note: ***$P < 0.001$, **$P < 0.01$, *$P < 0.05$.

Table 2. Difference comparison in gender of self-efficacy concerning college students' emotion regulation.

Dimension	Boys	Girls	t
POS	15.19 ± 2.70	16.04 ± 2.48	−5.991***
DES	14.09 ± 2.58	13.72 ± 2.59	2.621**
ANG	13.56 ± 2.82	13.19 ± 2.83	2.408*
RES	42.85 ± 6.23	42.96 ± 6.07	−0.314

Note: ***$P < 0.001$, **$P < 0.01$, *$P < 0.05$.

Table 3. Difference of self-efficacy concerning college students' emotion regulation between only-child and child with siblings.

Dimension	Only child	Child with siblings	t
POS	15.83 ± 2.79	15.57 ± 2.57	1.47
DES	14.05 ± 2.69	13.85 ± 2.56	1.15
ANG	13.38 ± 3.21	13.36 ± 2.70	0.11
RES	43.27 ± 6.67	42.79 ± 5.97	1.13

Note: ***$P < 0.001$, **$P < 0.01$, *$P < 0.05$.

students with siblings; however, the difference is not great between these two categories of students.

Table 4 shows that, for the POS score, junior students get the highest and senior students get the lowest; for the DES score, junior students get the highest and senior students get the lowest; for the ANG score, freshmen get the highest and senior students get the lowest; for the RES score, sophomore get the highest and senior students get the lowest. But there is no great difference between different grades.

Table 5 shows that there is great difference in different majors in the POS dimension score and the ANG dimension score ($P < 0.01$). Among the scores, for the POS score, students majored in medicine get the highest and students majored in engineering get the lowest; for the DES score, students majored in science get the highest and students majored in arts get the lowest; for the ANG score, students majored in science get the

Table 4. Grade difference of self-efficacy concerning college students' emotion regulation.

Dimension	Grade (M ± SD)				F
	Freshman	Sophomore	Junior	Senior	
POS	15.37 ± 2.79	15.70 ± 2.56	15.81 ± 2.63	15.36 ± 2.53	1.81
DES	13.91 ± 2.87	13.90 ± 2.51	14.01 ± 2.65	13.52 ± 2.18	0.92
ANG	13.53 ± 2.85	13.48 ± 2.82	13.11 ± 2.94	13.00 ± 2.43	2.06
RES	42.82 ± 6.58	43.09 ± 6.06	42.94 ± 6.22	41.89 ± 5.28	1.15

Note: ***$P < 0.001$, **$P < 0.01$, *$P < 0.05$.

Table 5. Major difference of self-efficacy concerning college students' emotion regulation.

Dimension	Major (M ± SD)				F
	Major of arts	Major of science	Major of engineering	Major of medicine	
POS	15.62 ± 2.60	15.71 ± 2.80	15.38 ± 2.51	16.18 ± 2.70	3.964**
DES	13.71 ± 2.56	14.17 ± 2.62	13.98 ± 2.50	13.97 ± 2.86	1.775
ANG	13.24 ± 2.77	13.94 ± 2.76	13.45 ± 2.79	12.94 ± 3.09	4.511**
RES	42.58 ± 6.06	43.83 ± 6.40	42.81 ± 5.89	43.10 ± 6.64	2.047

Note: ***$P < 0.001$, **$P < 0.01$, *$P < 0.05$.

highest and students majored in medicine get the lowest; for the RES score, students majored in medicine get the highest and students majored in arts get the lowest. Through the multiple comparisons afterwards, it shows that in the POS dimension, the students majored in medicine get a much higher score than students majored in arts and engineering; in the ANG dimension, students majored in science get a much higher score than students majored in arts, engineering and medicine with students majored in engineering getting a much higher score than students majored in medicine.

4 DISCUSSION

4.1 *Gender difference of self-efficacy concerning college students' emotion regulation*

The result of this research shows that for the POS score, female students have a much higher score than male students (t = −5.991, P < 0.001); for the DES score, male students have a much higher score than female students (t = 2.621, P < 0.01); for the ANG score, male students have a much higher score than female students (t = 2.408, P < 0.05); for the RES score, there is no great difference between male students and female students. The reason may be: male college students neglect and restrain more positive emotion while female college students pay more attention to and unbosom[6] more positive emotion.

Due to the difference between females and males on level of sensitivity in daily life, females can often sense tiny positive emotion, while rough males will neglect some tiny positive emotions, instead they pay attention to greater positive emotion, therefore, males' self-regulation ability on positive emotion is lower than females'; and due to sensitivity, it is easy for females to sense the negative emotion. Strategies of females and males used to regulate negative emotion show that females are more emotion-focused, while males use cognitive strategy (Jiajin Yuan et al. 2010) more. Since females are sensitive to negative emotion and weak in regulation as well, therefore, females have low regulatory emotional self-efficacy on regulating negative emotion (depression/pain, anger/rage); due to that males and females are respectively good at regulating negative and positive emotion, the difference between males and females on aggregate score of regulatory emotional self-efficacy is not obvious.

4.2 *The difference of self-efficacy concerning college students' emotion regulation in the two variables of whether the student is only child or not and grade is not great*

With development of economy and change of parents' mind, the environment difference (especially on the aspect of economy and education) during the growing process between child with siblings and only-child is getting smaller, which makes difference between the two groups getting smaller

523

and causes that the variable of whether being only-child or not becomes indistinctive. The self-efficacy concerning college students' emotion regulation is mature and tends to be in a stable status with a slow development process (Ping Zhang, 2010). This may be the reason why the difference between college students of different grades with small age span is not great.

4.3 *In the variable of major, there is great difference of self-efficacy concerning college students' emotion regulation*

The result of the study shows that in the POS dimension, the students majored in medicine get a much higher score than students majored in arts and engineering; in the ANG dimension, students majored in science get a much higher score than students majored in arts, engineering and medicine with students majored in engineering getting a much higher score than students majored in medicine. The reason for this result may be: academic stress for medical students are relatively greater than students of other majors, so medical students pay more attention on taking advantage of positive emotion in life to regulate their own life, thus they have higher assessment of self-regulation ability on positive emotion; and since medical students have great academic stress, they experience negative emotion (depression/pain, anger/rage) often; the situation that they could not control their negative motion happens a lot; therefore, they have lower assessment of self-regulation ability on positive emotion.

5 CONCLUSION

In all dimensions, there is great difference of regulatory emotional self-efficacy between male and female students while the difference between the total scores in the scale is not that great. There is great difference in students of different majors of self-efficacy concerning college students' emotion regulation.

ACKNOWLEDGEMENT

This article is one of the phased objectives of soft science project (project number: 122400450116) of Henan Science and Technology Authority.

REFERENCES

[1] Caprara G.V., Giunta L.D., Eisenberg N., et al. Regulatory Emotional Self-efficacy in Three Countries, *Psychological Assessment*, pp. 227–237, 2008.
[2] Dongling Tang, Yan Dong, Guoliang Yu, Shufeng Wen, The Regulatory Emotional Self-Efficacy: A New Research Topic, *Advances in Psychological Science*, vol. 18, pp. 598–604, 2010.
[3] Ellen Heuven, Arnold B. Bakker, Wilmar B. Schaufeli, Noortje Huisman, The Role of Self-efficacy in Performing Emotion Work, *Journal of Vocational Behavior*, pp. 222–235, 2006.
[4] Guiqin Liu, Meiyu Wang, Relationship of Junior High School Students Emotional Adjustment Self-efficacy and Subjective Well-being, *The Education of Psychological Health*, pp. 49–52, 2012.
[5] Jiajin Yuan, Yu Wang, Enxia Ju, Hong Li, Gender Differences in Emotional Processing and Its Neural Mechanisms, *Advances in Psychological Science*, vol.18, pp. 1899–1908, 2010.
[6] Miner Huang, Dejun Guo, Emotion Regulation and Depression of College Students, *Chinese Mental Health Journal*, vol.15, pp. 438–441, 2001.
[7] Min Zhang, Jiamei Lu, An Analysis of the Results of the Regulatory Emotional Self-efficacy Scale in Chinese University Students, *Chinese Journal of Clinical Psychology*, vol.18, pp. 568–570, 2010.
[8] Ren Li, Introduction to the Cultivation of Emotion Regulation Self-efficacy, *Course Education Research*, pp. 63, 2012.

Education Management and Management Science – Zheng (Ed.)
© 2015 Taylor & Francis Group, London, ISBN 978-1-138-02663-6

The exploration of the integrated teaching mode of interpretation under the perspective of Piaget's theory of cognitive

Xiao-ya Qin
The English Department of the School of Humanities and Law, North China University of Technology, Beijing, P.R. China

ABSTRACT: Interpretation has been paid more and more attention as it has become one of the compulsory courses for English major students. However, since the course has a relatively short history in China, "the research on interpretation teaching theory and methodology has been dropped behind. As a course, interpretation has many weaknesses in China, such as the opening time, teaching principle, teaching content and training method and so on, and it has a large margin to improve. (Fang Jian-zhuang, 2002)". This study attempts to analyze the necessity and feasibility of establishing an integration model of interpretation by way of digging out the inner relationship between interpretation and other courses, with the perspective of Piaget's theory of cognitive structure, aiming at offering reference to improve the teaching quality of the course of interpretation.

Keywords: theory of cognitive structure; interpretation teaching; integration model

1 THE INTRODUCTION OF PIAGET'S THEORY OF COGNITIVE STRUCTURE

J. Piaget is the founder of the Cognitive Structure Theory, Swiss psychologist and philosopher. This theory holds that the cognitive structure is the knowledge structure in the learner's mind. It is the content and organization of all the ideas or concepts of the learner or within a particular field. Learning joins new materials or new experiences and the old materials or experience, thus forming an internal knowledge structure, that is the cognitive structure which is in the forms of assimilation, accommodation and balance (J. Piaget; 1981). American psychologist J.S. Bruner improves and deepens the theory of cognitive structure. He believes that learning is not a course of passive reaction, but a process to form initiative cognitive structure. It is composed by a series of processes, so in the process of learning, learners should pay attention to the basic structure and the inherent association among all the disciplines.

Schema is a key conception in J. Piaget's Cognitive Structure Theory. It is the starting and key point of the cognitive structure. The formation and change of schema is the essence of cognitive development and while the development of cognitive is affected by three processes: assimilation, accommodation and balance.

1.1 *Assimilation*

In the theory of cognitive development, assimilation is the process of filtering or changing of the individual towards the stimuli input. That is, when individuals feel stimulated, they put them into the mind of any original schema, making it a part of itself.

1.2 *Accommodation*

Accommodation means regulating the internal structure of the body to adapt to the specific situation of a process. When an individual meets the new stimulus that cannot be assimilated by the existing schema, it will have to modify or rebuild the original schema to adapt to the new environment, which is the process of accommodation. Objectively, without accommodation, there is no development. No knowledge can exist without the assimilation and accommodation of knowledge schema. Knowledge is not only the cognitive schemata accommodating to the external objects, but also outside objects assimilating to the cognitive schema (Yu Wensen; 2009).

1.3 *Balance*

Piaget believes that individual's cognitive schema is to develop by way of assimilation and accommodation and to adapt to the new environment. A general rule is that whenever an individual encounters a new stimulus, he always tries to assimilate the new stimulus with the original schema. If the assimilation succeeds, a temporary balance can be achieved. If the original schema cannot assimilate the environmental stimuli, the individual will have to accommodate it. That is to adjust the original schema or rebuild a new schema, until it reaches a new equilibrium. But the state of equilibrium is not an absolute state. A low-level equilibrium will be lifted to a higher-level of equilibrium through the interaction between individual and environment. This continuous development of equilibrium is the whole process of cognitive development (Piaget; 1981).

2 THE TEACHING CHARACTERISTIC OF INTERPRETATION

2.1 *It is multi-dimensional*

Interpretation is a language operating activity based on the well-established abilities in listening, speaking, reading, writing, translation and other skills. It is the highest level of language applications. Interpretation Course is usually opened in the high grades at college. It is a useful tool for foreign language teaching (Bao Chuan-yun; 2004). This suggests that the effect of interpreting training courses mainly relies on the speaking, reading, writing, translation and other language skills that the learners have acquired.

2.2 *It is comprehensive interdisciplinary*

The interpreters have a high demand for social position. They are not only required to have superb bilingual skills, but also to have a good applied linguistics, socio-linguistics, psycho-linguistics, neurological linguistics, cognitive science, artificial intelligence and other knowledge. They also need to have the political, economic and cultural wide range of knowledge, history in both language countries (Guo Lanying; 2007). Interpretation teaching should be human-centered. It should help students build knowledge systems that interpretation needs and train their bilingual thinking ability (Liu Heping; 2005).

2.3 *It is intercultural communicative*

As a cross-cultural communication activity, Interpretation is not a decoding from the original language to the target language, but the information transformation from the original language to the target language. Referring to Yingelunmu's interpretation semiotic communication model, Coach Hoff proposed a dual system of communication. In this system, information is composed of verbal and nonverbal system. The information sender (S1) encodes the information in the particular social and cultural context, and then sends to the receiver (R1). The conclusion of this mode is that the interpreter's understanding of the B language largely determines the output effect.

3 INSPIRATION ON ESTABLISHING THE CORRECT INTEGRATION OF UNDERGRADUATE TEACHING MODE OF INTERPRETATION

3.1 *The target of undergraduate interpretation teaching is the skillful bilingual conversion*

Piaget believes that new knowledge is more easily absorbed only when it is put into the existed knowledge. Interpreting course is a comprehensive language output behavior for English majors. It is usually opened at the fifth or sixth academic semester. The beforehand language skills it needs include the ability of vocabulary, listening, speaking, translation and reading comprehension. Only when students really obtain these early language skills, can they achieve the output freedom in the later work, and improve the quality of the output. Therefore, the goal of interpretation teaching is not to expand the new language knowledge, but rather to help students' integrated pre-order knowledge and improve students' understanding and application capabilities of the existed knowledge, thus achieving the skillful output of the second language.

3.2 *The process of interpretation teaching is to constantly break the equilibrium state of students' prior knowledge and to establish a balance at the higher level*

Piaget emphasizes that cognitive development is process of constant construction of balancing. The cultivation of capacity is through the co-effect of assimilation and accommodation, and through the constant repetition by way of imbalance—balance—imbalance to develop and enrich from a lower level to a higher level. Therefore, in the interpretation teaching process, teachers must constantly break the balance of students' prior knowledge to help students establish a new balance, so as to promote the continuous development of students' cognition and to obtain better interpretation ability.

4 LINK MODE BETWEEN INTERPRETATION AND OTHER BASIC COURSES OF ENGLISH MAJORS

Kathryn Bock's cognitive structure of human language holds that the information collection of interpretation starts from auditory sense. "Auditory

Analysis System" is the key in the process of auditory cognition. This reveals the great importance of listening in interpretation. Therefore, in the class of Elementary Listening at Grade One at college, the teachers should not only cultivate the students' listening ability, but also train the students ability of "One mind, two uses" by way of shadow exercise. Gerver (Gile, 1994) effort models theory thinks that TR (Total Requirement) in interpretation must be smaller than TA (Total Capacity Available). Therefore, the larger the vocabulary bank an interpreter has, the easier he can find the target vocabulary in the process of interpretation. Seleskovitch says that an interpreter's language ability should meet many criteria: He must comprehend it the same time when he hears it. He must have an intuitive knowledge, articulate, has a large vocabulary (Seleskovitch, 1978). Based on this, courses like *Elementary English, Elementary English Reading* and *English Grammar* are all focus on enlarging the students' reading amount and vocabulary. While *English Speaking* majors are focusing in cultivating the students' expression ability. This suggests that Grade One is the elementary phase in interpretation training.

Over a year's study, students' listening and speaking ability have both achieved a significant progress. And their vocabulary has also been significantly expanded. As the students are getting in touch with the new knowledge, they immediately involve this into their old schema. And this new knowledge has become part of their old knowledge. This is the process of assimilation. However, the old schema cannot explain all the new stimulus. Human mind needs to accommodate the original schema. Students are improving themselves by way of convoluted process. As far as the courses at Grade Two, *Elementary English* and *Advanced English Listening* belong to the assimilation in cognition. But in many universities, English majors start to have *Writing* course. This aims at cultivating students' ability of bilingual conversion on the basis of sentences and paragraphs. This new stimulus also needs the process of assimilation and accommodation because it links with both the new and old knowledge. Robinson believes that interpretation operation is a continuous learning process. Translators constantly deepen their knowledge according to the mode of "Extension—induction—interpretation of" in order to obtain a potential second nature.

According to Gile's cognitive theory and translation mode, in the process of interpretation teaching, teachers should, as early as possible, make the students know that the process of foreign language learning is a process of conducting the information that the original language conveys. Compared with the results of language learning, the process is more important for students. Language learning is a dynamic process. It requires the learners to actively participate in the learning activity. Only by this can it mobilize and motivate their learning ability, and generally strengthen students' cognitive ability on language and on the information it bears (Guo Lanying; 2007). Therefore, in the third year of college, through the "Advanced English" course, the students continue to enhance their language skills. And at the same time, students start to be exposed to the original foreign books through "*Anglo-American Newspaper Reading*". This can enhance students' cognitive ability besides language, such as culture, history, politics, economy and so on.

In summary, the first year's study at college for English majors gives priority to the cultivation of language skills through the courses of *Elementary English, Elementary English reading, Listening* and some other basic courses. In the second year, *Elementary English, Elementary English reading, Listening* can further enhance students 'listening, speaking, vocabulary, especially the *English writing* course focuses on developing students' second language expression. All these help to establish students' English thinking ability. In the third year at college, *Advanced English, Anglo-American Newspaper Reading, Translation* give students an opportunity to form higher second language ability. They make the students deepen their understanding of the Western culture, economy, politics and so on. And all these enlarge their understanding of the English society. In the fourth grade, students have already owned a large vocabulary, have a good bilingual conversion capability, they are possible to achieve a breakthrough in their interpretation training. After the process of assimilation and accommodation, these courses have achieved the appropriate balance. However, such balance is not absolutely still.

5 CONCLUSION

According to the statistics of Mu Lei, there are more than 400 universities in China which include English as a major, and this number is still on a rise. According to the teaching syllabus for English majors, *Translation* is a course which is usually opened at the third or fourth year at college. It is a comprehensive display of students' language capability. However, the course of *Interpretation* is opened later than this time. In China, there has no a comprehensive mode in interpretation teaching. This attributes to the low development level of interpreting in China (Liu Heping; 2006). *Interpretation* is a course which closely bases on students' language ability including listening, speaking and translation. But the same time, there are sharp distinction between them (Fang Jianzhuang; 2002). Therefore, to establish an integrated mode in Interpretation teaching can not only urge the great development of Interpretation teaching, but also improve the teaching for English majors.

REFERENCES

[1] Fang Jianzhuang, On the Characteristics and Compiling Principles of Interpretation Teaching books: Chinese Science and Technology Translation Volume 1, 2002.
[2] Zhou Hongmin, On the Analysis of Translation Schema: Shanghai Science and Technology Translation Volume 3, 2003.
[3] Jean Piaget, The Principes of Genetic Epistemology, Commercial Press, 1997.
[4] Bao Chuanyun, The Positioning and Consideration on China's College Interpretation Teaching: Translation of China, Volume 5, 2004.
[5] Guo Lanying, The Research on Interpretation and the Training of Interpreters: Science Press House, 2007.
[6] Liu Heping, Challenge and Discussion on French Translation Theory: Translation of China, Volume 4, 2006.
[7] Liu Heping, Interpretation Theory and Teaching: China Translation and Publishing Corporation, 2005.
[8] Fang Jianzhuang, Research on the Teaching Books in China between 1988 and 2008: Shanghai Translation, Volume 2, 2009.
[9] Wu Bing, The Characteristics, Criteria and Requirements of Interpretation: Teaching Theory and Practice, 2004.
[10] Liu Boxiang, The Observation and Consideration upon Interpretation: Foreign Language and Foreign Language Teaching, 1998.
[11] Fang Jianzhuang, Discussion on the Reform of Interpretation Teaching, Chinese Science and Technology Translation 1998.
[12] R.A. Wilson, and F.C. Keil, The MIT Encyclopedia of the Cognitive Sciences. Cambridge: The MIT Press., 1999.
[13] Gile, D. Opening up in Interpretation Studies, In Snell Homby, M., F. Pochhacker & K. Kaindl, 1994.
[14] Danica Seleskovitch, Interpreting for international conferences: problems of language and communication, Pen and Booth, 1978.

Education Management and Management Science – Zheng (Ed.)
© *2015 Taylor & Francis Group, London, ISBN 978-1-138-02663-6*

Does export growth increase CO$_2$ emissions of China?

Yuhuan Zhao

School of Management and Economics, Beijing Institute of Technology, Beijing, China

ABSTRACT: In recent years, CO$_2$ emissions of China are increasing very fast, and China became the largest exporter of goods in the world. Does export growth increase CO$_2$ emissions of China? In order to answer this question, this paper conducts an empirical study on the relationship between export and CO$_2$ emissions of China using data of export and CO$_2$ emissions from 1978 to 2009. The results show that there are long-run and short-run equilibrium relationships between the two variables. The short-term elasticity of LnEX on LntCO$_2$ is 0.127, while the long-term elasticity of LnEX on LntCO$_2$ is 0.321. It means that, during the period of 1978–2009, export plays an important role in driving the increase of CO$_2$ emissions of China not only in the short run but also in the long run.

Keywords: export growth; CO$_2$ emissions; China

1 INTRODUCTION

T As the biggest developing country, China enjoys an incredible economic growth over the past three decades. China is the largest exporter and second-largest importer of goods in the world. In 2010, the total foreign trade value of China is US$3.64 trillion, which US$1.89 trillion in exports and US$1.74 trillion in imports. At the same time, the CO$_2$ emission of China also has seen a rapid growth. In 1994, CO$_2$ emissions of China were 3.07 GT, while in 2004, CO$_2$ emissions rose to 5.7 GT. According to an IEA report, China replaced the USA and became the largest CO$_2$-emitting country in the world in 2007. Acceleration of carbon emissions in China has given rise to appeals for carbon reduction and images of China as a threat.

Many people think that export growth is an important factor in driving CO$_2$ emissions in China. As the "world factory," China's energy consumption is not all used for domestic consumption. The rapid growth of exports has greatly increased energy consumption and CO$_2$ emissions of China.

Based on the above analysis, in order to make clear the role of export in China's CO$_2$ emission, it is necessary to conduct an empirical study on the relationship between export and CO$_2$ emissions of China.

Many scholars have done research on trade and environment, in recent years, some of them focus on international trade and climate change. Because of the increasingly importance of China in world economy, China became one of the focus of the related research. Here, we only take several researches for example. Lan (2004) uses panel data of China's 30 provinces and cities during the period

of 1995–2001 to study the relationship foreign trade and CO$_2$ emissions of China, he found that trade liberalization reduced the CO$_2$ emissions of China. Li and Zhang (2004) employed CO$_2$ emissions as the indicator to analyze impact of export growth on China's environment, they found the evidence of EKC inverted U shape, and the results show that with the expansion of exports, CO$_2$ emissions growth rate decrease. Song et al. (2008) investigated the relationship between environmental pollution and economic growth in China based on the EKC hypothesis using Chinese provincial data, the results showed that there is a long-run co-integration relationship between per capita emissions of three pollutants (waste gas, waste water, and solid wastes) and per capita GDP. Furthermore, the results show that all three pollutants are inverted U shape in China. Jalil and Mahamud (2009) employed the time series data of China from 1975 to 2005 to examine the long-run relationship between China's CO$_2$ emissions and energy consumption, the long-run relationship between income and foreign trade, the results showed that there is a negative relationship between trade and CO$_2$ emissions, but not statistically significant.

On the basis of the previous studies, using data from 1978–2009, this paper conducts an empirical study on the relationship between export and CO$_2$ emissions of China.

2 DATA SOURCES AND MODELING

C To analyze the relationship between export and CO$_2$ emissions of China, we need export and CO$_2$

emissions data of China from 1978–2009. Export data is from *China Statistic Yearbook 2011*, and CO_2 emissions data is from *World Bank Database* and *IEA Key World Energy Statistics*.

From the data, we can see that the trend of the export and CO_2 emission are roughly the same with the time, they are both on the rise in general.

Before China entered WTO in 2001, the export increased slowly, while after 2001, it leads to a sharp rise in export.

In order to analyze the relationship between export and CO_2 emissions, we establish model as the following:

$$LnTCO_2 = \beta_0 + \beta_1 LnEX + u_t \qquad (1)$$

where TCO_2 is the total CO_2 emission, EX is the export, t represent year, β_0 is constant term, ε_t is chance error.

3 UNIT ROOT TEST

In order to have effect results, it is necessary to do unit root test on the variables. The critical value of each test model is calculated based on the table of critical value which provided by Mackinnon (1991), using the formula: $C_{(\alpha)} = \varphi_\infty + \varphi_1 T^{-1} + \varphi_2 T^{-2}$ to get the values. Where α is test level, this paper adopts the test level of 5%, i.e. $\alpha = 0.05$; T is the sample size.

We do the ADF test, the results show that, at five percent level of significance, the variable LnEX is unit root at level and stationary at first difference. In case of 1st difference and the model is (c, 0, k), t-statistics is −3.54 which is lower than the critical value −2.96, so the LnEX is integrated of order of one, I(1). Similarly, when the model is (c, t, k) in case of first difference, t-statistics value of $LntCO_2$ is −3.79, less than the critical value of −3.57, therefore, this time series is non-stationary at 1st difference too. In conclusion, all the variables (LnEX and $LntCO_2$) are unit root (non-stationary) at levels and stationary at first difference, in other words, they are integrated of order of one, I (1).

The necessary condition of the variables co-integration is that the two variables have the same order of integration. Since the variables of the model are both integrated of order of one, it satisfies the premise requirement of the co-integration test, we can proceed to the next stage: co-integration test.

4 CO-INTEGRATION TEST

Time-series data of LnEX and $LntCO_2$ from 1978 to 2009 are employed to the Co-integration test. The following are the results.

Firstly, employ OLS method to estimate the equation (1).

$$LntCO_{2t} = 1.126 + 0.318 \, LnEX_t \qquad (2)$$

(18.43)   (37.20)

$R^2 = 0.979$ DW = 0.577

Since the P-value of intercept is zero, it means that it is not zero significantly, and the intercept term should be kept. On the basis of this judgment, the regression model is:

The assumptions of OLS method tell that: the existence of autocorrelation will make the regression coefficients do not have minimum variance, and the variance of the error will also be underestimated, so that the OLS regression can be not accurate. As a result, it is important to examine whether the regression has autocorrelation. This paper employs Durbin-Watson (DW) test to analyze it.

According to the discriminant rules of DW test, since the DW value equals to 0.577 is quite low, and between (0, dL), the H_0 is refused, it means that there is a strong correlation of the residual term. According to this, an appropriate lag term should be added to the equation (2). Having the distributed-lag model of LnEX and $LntCO_2$:

1. Distributed One-lag model:

$$lnTCO_{2t} = 0.209 + 0.127 \, LnEX_t + 0.788$$

(1.392)   (1.760)

$$+ \, lnTCO_{2t-1} - 0.052lnEX_{t-1} \qquad (3)$$

(6.186)   (−0.704)

$R^2 = 0.992$ DW = 0.998

The result shows that, the DW value is 0.998, between the range (0, dL), then refuse the original hypothesis and recognize there is a one-order positive autocorrelation of the residual. In other words, the autocorrelation is not eliminated. For this reason, after this process, a distributed two-lag model should be estimated.

2. Distributed two-lag model:

$$lnTCO_{2t} = 0.426 + 0.095lnEX_t + 1.292lnTCO_{2t-1}$$

(3.162)   (1.533)   (8.037)

$$-0.023lnEX_{t-1} - 0.675lnTCO_{2t-2} + 0.051lnEX_{t-2}$$

(−0.218)   (−4.154)   (0.718)

$R^2 = 0.995$ DW = 1.821   (4)

The results show that the value of DW equals to 1.821. The observation is 32: T = 32, number of explanatory variables is 5: $k = 5$, from the Table 3 we can find the associated critical value are: $dL = 1.11$, $dU = 1.82$, so the DW value between the range is $(dU, 4-dU)$, therefore the H_0 should be accepted and recognize the residual does not have autocorrelation.

The above analysis shows that the autocorrelation is eliminated. Thus it can conclude that there is a long-run equilibrium relationship between LNEX and $LNtCO_2$ preliminarily.

We do the ADF test for the residual; the results are presented in Table 1. Table 1 shows that, at five percent level of significance, the variable Residual is not unit root at level and stationary at level. In case of the $(0, 0, k)$ model, t-statistics is -2.35 which is lower than the critical value -1.95, so the Residual is stationary at level.

The co-integration test results tell us that, there is a long-run equilibrium relationship between the two variables ($LnEX$ and $LntCO_2$), in other words, a long-term stability of the positive correlation is existing between $LnEX$ and $LntCO_2$. At the same time, from the equation (4–4), the long-run elasticity of $LnEX$ can be calculated:

$$\text{The elasticity} = (\beta_1 + \beta_3 + \beta_5)/(1 - \beta_2 - \beta_4)$$
$$= (0.095 - 0.023 + 0.051)$$
$$/(1 - 1.292 + 0.675) = 0.321$$

We can see that there is a long-run equilibrium relationship between the two variables. But whether it has a short-run equilibrium relationship between variables? Under the premise of the existence of co-integration, the error correction term should be introduced to make an error correction model, in order to infer the question more effectively. At the same time, the error correction model can examine short run relationship between the two variables ($LnEX$ and $LntCO_2$).

Hence, we employ error correction techniques to build model and analyze.

5 ESTABLISH ERROR CORRECTION MODEL

The long run equilibrium does not mean the end of the analysis; the error correction test should

Table 1. ADF results of residual.

Variable	Residual		
Level	Test model (c, t, k)	t-statistic	Critical value
	(c, t, 0)	−2.178466	−3.562882
	(c, 0, 0)	−2.309892	−2.960411
	(0, 0, 0)	−2.345932	−1.952066

be carried out to deal with the short-run fluctuations. An Error Correction Model (ECM) can be built in the case of the variables are non-stationary and have co-integration relationship, to study the relationship between variables. Due to the emergence of the error correction term, ECM may also study the short- and long-term causality in the meantime.

Compared with co-integration techniques to investigate long run relationship among the variables, error correction techniques are used to examine short run relationship respectively. Engle and Granger (1987) permit long-run components of variables to conform long-run equilibrium relationship to the short-run components having a flexible dynamic specification.

ECM has many obvious advantages: (1) The use of first-order differential can eliminate the existence of possible trend factor of variables, thus avoiding the problem of spurious regression. (2) The use of the first-order differential can also eliminate the multicollinearity of the model. (3) The introduction of error correction term makes sure that there is no variable level value of information has been neglected. (4) The stationary of error correction making the model can be estimated by classical regression method, in particular, the T test and F test can be used to select the differential of the model. Except these, there are also many other advantages of ECM.

Time-series data of LNEX and $LNtCO_2$ from 1978 to 2009 are employed to the test. The following are the empirical results:

$$\Delta \ln TCO_{2t} = 0.209 + 0.127\Delta \ln EX_t - 0.212 \ln TCO_{2t-1}$$

$$(1.392) \quad (1.760) \quad (-1.663) - 0.068 \ln EX_{t-1} \quad (1.879)$$

$$R^2 = 0.965 \qquad DW = 1.736 \qquad (5)$$

We write the equation (5) to the form of error correction model as following:

$$\Delta \ln TCO_{2t} = 0.127\Delta \ln EX_t - 0.212 \, (\ln TCO_{2t-1}$$
$$- 0.986 + 0.321 \ln EX_{t-1}) \qquad (6)$$

The equation (6) shows that the two variables ($LnEX$ and $LntCO_2$) have not only long-term but also short-term relationships. The short-term elasticity of $LnEX$ on $LntCO_2$ is 0.127, while the long-term elasticity of $LnEX$ on $LntCO_2$ is 0.321.

6 CONCLUSIONS

This paper employs the data of export and CO_2 emissions of China during 1978–2009 to analyze the relationship between export and CO_2 emissions of China. The results show that there are long-run and short-run equilibrium relationships between the two variables. The short-term elasticity of LnEX on LntCO$_2$ is 0.127, while the long-term elasticity of LnEX on LntCO$_2$ is 0.321. It means that, during the period of 1978–2009, export plays an important role in driving the increase of CO$_2$ emissions of China not only in the short run but also in the long run. We can see that export growth increases CO$_2$ emissions of China. To reduce CO$_2$ emissions of China, government should pay attention to the role of export.

REFERENCES

[1] Ang, J.B. CO$_2$ Emissions, Research and Technology Transfer in China. Ecological Economics, 2009, 68(10): 2658–2665.

[2] Aaron, K., Mary R. A further inquiry into the Pollution Haven Hypothesis and the Environmental Kuznets Curve [J]. Ecological Economics, 2010, 69: 905–919.

[3] Cole, M., Elliott, R, and Wu, S. Industrial Activity and the Environment in China: an Industry-level Analysis. China Economic Review, 2008, 48(1): 393–408.

[4] Jalil, A., Mahamud, S.E. Environment Kuznets Curve for CO$_2$ Emissions: A co-integration Analysis for China. Energy Policy, 2009, 37(12): 5167–5172.

[5] Khalil, S., Inam, Z. Is Trade Good for Environment? A Unit Root Cointegration Analysis. Proceeding of the National Academy of Sciences of the USA, 2006, 104(24): 10288–10293.

[6] Kearsley, A., Riddel, M. A further Inquiry into the Pollution Haven Hypothesis and the Environmental Kuznets Curve. Ecological Economics, 2009, 69(4): 467–491.

[7] M. Auffhammer, R.T. Carson. Forecasting the Path of China's CO$_2$ Emissions using province-level information. Journal of Environmental Economics and Management, 2008, 55(3): 229–247.

[8] Managi, S., Jena, P.R. Environmental Productivity and Kuznets Curve in India. Ecological Economics, 2008, 65: 432–440.

[9] Plassman, F., Khanna, N. Accessing the Precision of Turning Point Estimates in Polynomial Regression Functions. Econometric Reviews, 2007, 26(5): 503–528.

Education Management and Management Science – Zheng (Ed.)
© 2015 Taylor & Francis Group, London, ISBN 978-1-138-02663-6

Tourism informationization teaching design and practice of tourism management professional in university

W. Zhang, S.Q. Tian & L.L. Gou
Business School, Shandong Normal University, Jinan, Shandong, P.R. China

ABSTRACT: With the arrival of the tourism informationization wave, the transformation process which is driven by informatization from tourism to modern service industry is accelerated day by day; the tourism informationization teaching design and reform of tourism management professional in university are imminent. This paper is based on development of tourism informationization, analyzing the situation of tourism informationization teaching construction of tourism management professional in university at home and abroad, and pointing out the existing problems. Finally combining with the practice, we put forward to the tourism informationization curriculum teaching design plan of tourism management professional in university, in order to make the students adapt to the development of tourism informationization and improve students' tourism informationization literacy.

keywords: tourism management professional; tourism informationization; teaching design and practice

1 SIGNIFICANCE AND SOCIAL CONTEXT ON THE DEVELOPMENT OF TOURISM INFORMATIZATION

1.1 Concept and significance of tourism informatization

Focusing on the structure of tourism industry, tourism informatization takes initiative in making use of information technology to integrate various tourism information resources and transform them into productive forces supporting the development of tourism industry as well as the booster driving the improvement of development and management level of tourism industry. The tourism informatization helps realizing effective management of tourism resources and e-business, better serves tourism planning and decision-making through finding out the information deep in the tourists' behavior. Furthermore, it plays a crucial role in the management of domestic tourism industry, cultivation of tourism market, promotion of information communication between tourism destinations and related enterprises and potential tourism consumers, rendering of an effective and convenient exchange environment in favor of tourism transaction, creation of innovative operation model in tourism industry as well as boosting the development of tourism industry.

1.2 Social background for the development of tourism informatization

The 21st century comes into an era of information, when rapid development of information technology has great impact on all industries in the society. Information infuses the development of tourism industry with infinite vigor and energy and transforms operation and management models in this sector. With booming DIY tour and self-driving tour, the tourists have much higher demand on tourism information. Provisions on optimization of tourism consumer environment, marketing of tour destinations and prevention and control of false information advertisement are specified in the *Tourism Law of the People's Republic of China*, Clauses 26, 32 and 48. In 2012, basic approach for information driven transformation from traditional tourism to modern service industry was established by China National Tourism Administration, which expressly embodied the development thinking on the tourism by focusing on and relying on information in macro level. It unveils the fact of accelerating course of transformation of tourism informatization from traditional model to modern service industry, which is reflected in the higher education in the development of tourism professionals that it's imperative to design and reform in the teaching of tourism courses informatization for the students of tourism management major.

2 SETTINGS OF TOURISM INFORMATIZATION COURSES FOR STUDENTS OF TOURISM MANAGEMENT MAJOR FROM HOME AND ABROAD AND PROBLEMS ARISING FROM TEACHING THEREOF

2.1 Situation about the settings of tourism informatization courses for students of tourism management major from home and abroad

Foreign situation about tourism informatization experiences five stages, namely, computerization, automation, networking, integration and synergy. Foreign colleges have a long history in teaching students of tourism management major courses relating to tourism informatization, together with richer course settings and more operability than domestic counterparts. Take Plymouth University in UK as an example, curriculum for students of tourism management major consists mainly of information system, information technology management, mathematics and computer based information processing, multimedia computer based information processing, webpage development and etc.

All domestic colleges with famous settings of tourism management major attach great importance to the construction of the courses concerning tourism informatization. Sun Yat-Sen University is a good example in terms of early initiation and sound development of such courses. Tourism informatization and e-business oriented are set by the Department of Tourism Management & Planning of School of Tourism Management to meet the needs of developing industries of Internet and digital information, and core professional courses such as tourism informatization, tourism e-business, operation of tourism website and etc. are developed accordingly. The students are developed professional abilities with regard to operation and management of tourism website, enterprise engaging in tourism informatization and design of tourism website and etc. so that such students, after graduation, are component to start and develop their career in enterprises in the fields of internet-based tourism, informatization, tourism administrations, network operator, travel agency, etc.

As to general situation about the settings of courses concerning tourism informatization for students of tourism management major in domestic colleges, two pictures are in view, of which one picture is that lecturers of tourism management major in a handful of colleges fully realize the importance of tourism informatization, leading to more courses oriented to tourism informatization for students of tourism management major; and another picture is that for most college students of tourism management major, only one course,

tourism management information system focusing on management system of hotels and travel agencies, is taught, in addition to such public courses as introduction of computer technology and etc.

2.2 Problems arising from teaching of tourism informatization to college students of tourism management major

2.2.1 Dislocation between old-fashioned course settings and teaching contents and intensive social need for application of tourism informatization

Except for well-established courses concerning tourism informatization being taught for students of tourism management major in some colleges, only traditional courses, such as tourism management information system and etc, are taught in most colleges, furthermore, such courses focus only upon design, function and exercise of relevant software of management information system in connection with hotels, travel agencies and tourism resorts, all of which are easily understandable by such students from routine training after starting their career. At present, courses or contents relating to intelligent tourism reflecting the features of micro consumption, dynamic tour and cloud services as well as network-based marketing and tourism e-business characterized by networking developed in the context of big data and mass information are seldom expressed.

2.2.2 Mutual independence in the design of courses concerning basic knowledge about tourism informatization and computer technology courses

Currently, the course settings of basic knowledge about tourism informatization and computer technology are designed separately in the design of courses concerning tourism informatization for students of tourism management major in most colleges, thus two independent teaching systems are developed. For instance, courses (generally professional ones) relating to tourism informatization for students of tourism management major in many colleges deal with management information system, while the courses of computer technology are designed independently as public courses in these colleges, which lead to overall superficial understanding of the students about tourism informatization as well as poor ability in real operation and settlement of real problem.

2.2.3 Practical courses and theoretical courses out of step

Although courses of basic knowledge about tourism informatization and those of computer technology are design in step for students of tourism management major in some colleges, insufficient teaching hours are spent in teaching the courses of database principles and advanced language program, and the course of management information system is taught in theory before practice of computer operation. In recent years, increasing efforts are made to introduce hardware facilities of tourism informatization for college students of tourism management major, however, practice oriented teaching of tourism informatization should not only focus on objective environment, but also be embodied in sound combination of practice and theory by developing the students' ability of real operation to reduce the gap between theory and practice.

2.2.4 *Insufficient faculties for tourism informatization and simplex practice mode*

The tourism management major in many colleges wants for qualified teachers of tourism informatization, thus it's hard to give effective tutorship to the students in practice contents. In some cases, existing practice in tourism informatization only pays attention to classroom explanation and allows the students with insufficient after school time for review the knowledge studied in the classroom, thus no effective practice mode is available.

2.2.5 *Illogical allotment of teaching hours with curriculum design*

In the curriculum design of tourism informatization for students of tourism management major in most colleges, theoretical courses account for most teaching hours, however, tourism informatization is featured by vast contents and information technology difficult to comprehend, which, due to less time being allotted for practical exercise, leads to undesirable teaching effect beyond doubt.

3 DESIGN OF TEACHING CONTENTS FOR THE COURSES OF TOURISM INFORMATIZATION

The rapid development of global information technology is an opportunity and a challenge to tourism. The key to promote development of tourism informatization is development of information technology professionals. Therefore, it's an urgent subject for immediate solution for the education of tourism major in colleges by improving information-based development of tourism professionals to adapt to the development situation of tourism. However, the most fundamental practice of tourism management major in colleges is to enrich the setting contents and strengthen practical operation in the design of courses of tourism informatization, take initiatives to gear to the needs of the development of tourism informatization for tourism professionals as well as foster professionals of innovation and application oriented.

In our curriculum settings of all previous development plan of tourism management major, courses relating to tourism informatization, such as computer based management of hotels, tourism cartology, Photoshop image processing, management information system and etc., are toughed in succession in addition to computer related public courses. To keep up with the era of information, we set course of tourism informatization in revising 2012 edition of development plan by including classroom teaching and computer operation. Main contents of classroom teaching are an overview of tourism informatization, Tourism Information Management System (TMIS), tourism information mapping & e-map, application of 3S technology to the survey of tourism resources, tourism multimedia technology, tourism administration informatization, tourism e-business, intelligent tourism (including application of novel tourism media). The part of computer operation includes tourism management information system software (Hotels, travel agencies, and scenic spot), exercise of graphical software CorelDRAW, exercise of image processing software Photoshop, exercise of geographical information system software MapInfo, applications of major tourism websites and etc.

Above teaching contents are extensive, of which the biggest problem is that, as each module is in itself a subject, we assure that it's hard to be proficient in none should all of them be put into one box of tourism informatization with limited teaching hours. Nevertheless, this design meets the requirements for the development of versatile tourism professionals, moreover, the students are allowed extensive study before independently choosing various modules for in-depth learning by their interests and career requirement in the future, which is much more useful than simply teaching of management information system. In addition, some of these modules are superposed with each other, with which scientific and logic system needs to be constructed to further define the mainline for consistency and integration. More importantly, tourism informatization can be deemed as all-inclusive, for which effective sifting and integration is necessary on the basis of profound study with regard to design of modules and teaching of contents in the limited teaching hours.

4 COUNTERMEASURE FOR IMPROVING THE STUDENTS' QUALITY OF TOURISM INFORMATIZATION

4.1 *Orientation to the demand of tourist market, clear in development goal*

Whether curriculum design of tourism informatization for students of tourism management major in colleges is adapted to rapid development of tourism industry, whether the students developed are innovative information technology professionals, upon which tourism professionals developed in the colleges are transformed to real productive forces, depend. Scientific and logic design of courses of tourism informatization is the fundamental task of the colleges for realizing such development goal the core aspect for curriculum design of tourism informatization is connecting the structure of courses and the social and market demands. The curriculum design of tourism informatization should be performed by focusing on the development goal and in light of actual situation of the college, so as to improve the students' quality in tourism informatization.

4.2 *Logically arrange teaching hours, expand realm, strengthen adaptability*

Modern tourism enjoys raid development and demands professionals of higher quality. Together with fierce competition in employment, curriculum design of tourism informatization not only imposes logical allotment of teaching hours, but also expansion of realm and better adaptability for adapting to market demand and in light of the students' interest. For instance, reduction of teaching hours for theoretical courses of tourism informatization allows the students to choose reading books relating to tourism informatization by their interest, so as to expand their realm of information technology and strengthen their momentum for sustainable career development.

4.3 *Strengthen exercise of practice ability, promote application of novel technology*

In the curriculum design of tourism informatization, strengthened exercise of practical ability and application of novel technology should be crucial to strengthen the students' ability for versatile application of tourism information technology. Modern college students are active in thinking, they're young people with strong interest in information technology, also, female students account for the majority of tourism management professionals, they're less interested in information technology than male students. This requires that curriculum design of tourism informatization should take into consideration to social context for the development of tourism informatization, actual demand of tourism industry on information technology, the students' interest, emphasis on strengthening the students' practical ability, that is, acquaint them with application of mature and practical technology of great momentum to tourism industry as well as strengthening application of novel technology in their career as tourism professionals.

ACKNOWLEDGEMENTS

The Third Batch Experimental Teaching Reform Projects of Shandong Normal University (Project number: SYJG302150).

The Innovative Undertaking Training Program of University Students (Project number: 2014211).

REFERENCES

[1] Cai Hong, Li yunpeng, Tourism Development & Tourism Education in the Context of Information Technology & Networking. Beijing: Economic Press of china, 2013.
[2] Diao Zhibo, Theoretical Study and Practical Innovation in Tourism informatization Industry. Beijing: Tourism Education Press, 2012.
[3] Geng Songtao, Preliminary Exploration of Curriculum Settings & Teaching of Tourism Management Major. Market Modernization, 2012(14):103~104.
[4] Li Yunpeng, Chao Xi, Shen Huayu et al, Intelligent Tourism-From Tourism informatization to Intelliegtn Tourism. Beijing: China Travel & Tourism Press, 2013.
[5] Wang Guoming, Tu Lihong, Accelerate Information-based Construction of Courses, Boost Construction of Open Colleges. Journal of Hubei Radio & Television University, 2012, 32(7):7~8.
[6] Wang Yanxiang, Study on Curriculum Settings for College Tourism Management Major. Sotuerwest University: 2012, 22~34.

Education Management and Management Science – Zheng (Ed.)
© 2015 Taylor & Francis Group, London, ISBN 978-1-138-02663-6

Leave career predicament, promote self-improvement—to reduce the college teaching secretaries' decompression

Hong Wang

School of Economics, Shanghai University, Shanghai, China

ABSTRACT: This article firstly analyzes the psychological problems in college teaching secretaries group: emotional sub-health, lower sense of achievement, and negative relationship with others. Then it points out that the causes of these problems are the traditional quality requirement, belittled service identification, heavy work load and great psychological pressure. Finally, it proposes that this special working group establish noble educational notion, master excellent professional skills, foster healthy body and mental quality and develop positive living attitude. This is the effective strategy for them to make self-improvement.

Keywords: predicament; college teaching secretary; pressure; self-improvement; quality

1 INTRODUCTION

College teaching secretaries are the grass root component of school undergraduate teaching management, in which they have undertaken a great deal of basic work. Their main responsibility contains helping the department head organize and revise various professional teaching plan and syllabus; assisting the daily teaching management routine and conducting investigation research for teaching quality improvement and teaching reform.

2 THE PROFESSIONAL PREDICAMENT OF COLLEGE TEACHING SECRETARIES

The work of college teaching secretaries belongs to typical helping professions and has assistant and subordinate characteristics. They mainly serve the teachers and students, also care for the teaching department and college leaders. Under the long term working pressure, they are likely to fall into the following professional predicament.

2.1 *Emotional sub-health*

Due to the low professional reputation, teaching secretaries' working value is hardly acceptable and they have unreasonable salary and treatment. Naturally, their self esteem and evaluation will be lower. They are likely to feel tired. Their anxiety and irritability will make them difficult to engage in their work, short of tolerance and understanding with teachers and students. Consequently, they are lack of the curiosity to explore the new knowledge, which leads to working efficiency.

2.2 *Lower sense of achievement*

In teaching management, lower working position and complicated personal relationships result in the slow process of secretarial work. Sometimes their management cannot be understood, their enthusiasm to serve the teachers and students are treated coldly. It will make them feel incapable to cope with their job. Some teaching secretaries begin to be perfunctory, late to come and early to leave. And they are less confident to their future professional development.

2.3 *Negative personal relationships*

The emotional sub-health and lower personal sense of achievement enables teaching secretaries to have problems in personal relationships. They are likely to have no trust and sympathy to the leaders and colleagues. They are unwilling to deal with people and take part in group activities. Some teaching secretaries hold negative attitude in the communications with teachers and students. They use sarcasm to upset the new teachers who do not completely adapt to their work. They are impatient to the students who fail to comply with the teaching management regulation. They find various reasons to avoid work, pass the buck and even produce extreme interpersonal friction.

3 REASON ANALYSIS

The college teaching secretaries' work has the professional features of both teachers and secretaries. Their work scope is rather big and they face with unique professional predicament. The reasons are as follows.

3.1 *Traditional quality requirement*

As far as the personal quality requirement of teaching secretary is concerned, the previous educational degree and capability requirement was not high. For example, in one university, before the education ministry teaching evaluation for undergraduate students in 2006, 80% of the teaching secretaries come from transferring faculty or the relatives of the talent introduction. The main reason is that the teaching secretary position is usually regarded as collecting documents and simple transactional work with no technical content. Additionally, some disqualified teaching secretaries work so carelessly that teaching management accidents take place. This causes negative influence. Consequently, they are despised within the college (Zhang, 2010).

3.2 *Belittled service identification*

Teaching secretaries' work feature is to serve others. This makes their right of management to be diminished. In one hand, the teachers who pursue academic autonomy are unlikely to comply with the teaching management. As the academic authorities who grasp advanced science and technology, ideology and culture, the college teachers are admired by the students and supported by leaders, they sometimes become the great block for the teaching secretaries to execute their work. On the other hand, students are lack of the awe attitude toward teaching secretaries, who have no right to participate in evaluating their academic reward and political performance. Therefore, students will find various reasons to delay the work informed by the teaching secretaries.

3.3 *Heavy work load*

The teaching secretarial work is tedious and complicated. Usually, they are busy from the very beginning to the end of the term. The routine work comes one after the other, the working schedule is very compact. Frequently, they have to answer the questions from the teachers and students every day. At the same time, they have to deal with some sudden management problems. For example, after the temporary information from the teaching department, they will possibly make more than ten phone calls to the related teachers or students to execute the superior arrangement. And the repeated work which explains the same issue again and again will not be recorded in their performance evaluation.

3.4 *Great psychological pressure*

The daily routine of teaching secretaries ranges from class arrange, class adjustment, examination informing, performance recording to submitting various statistics, etc. A slight negligence will cause slips. For example, if the secretary fails to arrange a cause or misspell the classroom number, it will cause a teaching accident, which brings negative influence to college teaching management. Simultaneously, the various complaints from teaching department, college leaders, teachers or students will keep teaching secretaries highly nervous. It will continue their concerning consciousness.

4 SELF-IMPROVEMENT OF COLLEGE TEACHING SECRETARIES

Due to the low quality requirement, belittled service identification, heavy work load, and great psychological pressure, college teaching secretaries often encounter their professional predicament. How to help them improve personal quality, foster professional identification and win the respect from teachers and students is really a task, which can be accomplished in the following aspects.

4.1 *To establish noble educational notion*

The 18th Chinese Communist Party Conference further emphasized the priority status of education in the national development strategy. Currently, the national strategy, which aims to build well-off society completely and deepen reform and opening-up comprehensively, puts forward newer and higher requirements for education. The construction of industrialization, informatization, urbanization and agricultural modernization, the change of economic development method, and the implementation of innovation-driven development strategy need strong support of talents from high quality education (www.12371.cn). This requires the teaching management staff march with the times, and set up "people-oriented" educational notion. With the notion of scientific development as the guideline in various work, the teaching secretaries should set the noble-ethics-for-quality-people as the working target, actively serve for the teachers and students whose need is their working foundation, and contribute themselves for the teaching quality improvement.

4.2 To master excellent professional skills

Teaching management is the teaching research, which regards teaching and learning as the centre, high quality teaching and learning as the target, scientific management as the main line. It also focuses on the subjective law of organizational management. A qualified teaching secretary should have certain theoretical foundation of high education, educational management and educational psychology. At the same time, he/she should care for the development law of high education, have a good command of "High Education Law" and teaching management system, etc. Thus his/her management can be up to specification.

Teaching secretaries should master the teaching management platform operation. They should make overall plan and foresee the periodic problems of teaching management. Also they should effectively cope with sudden accidents. Meanwhile, teaching secretaries should be well aware of the college professions, causes, teaching material construction, and other information, such as faculty condition, teaching circumstances and students' quality. Therefore, they can provide the person in charge with accurate detailed information at any time. Additionally, teaching secretaries should learn to skillfully use modern office equipment, data processing software and application of computer aided system in order to improve work efficiency.

One of the important links of teaching management is to communicate and coordinate. Teaching secretaries should build a bridge to smooth the relations among teaching department, college and department. At the same time, they should communicate with college teachers and students frequently and help them coordinate teaching management subjects in various aspects. Also, they could establish their own QQ group, in which they can share the working experiences and personal ideas with each other promptly.

4.3 To foster healthy body and mental quality

The basic condition for the qualified teaching secretaries is good health and steady psychological quality. The increasingly heavy management work in high education increases the pressure of teaching secretaries physically and mentally. Therefore, they must actively strengthen physical exercises and keep energetic state. At the same time, good mental quality will help them understand and deal with various conflicts correctly, endure all kinds of pressure and misunderstanding and far from the worries of trouble. Someone has done such a statistics (see Fig. 1): 40% of worry never happens, 30% of worry concerns decision to belong to the

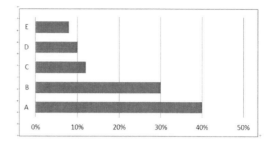

Figure 1. The percentage of worry sources.
Notes: A: worry which never happens
B: worry concerns decision to belong to the past and can not be changed
C: worry comes from others, who make in appropriate criticism out of inferiority
D: worry relates to health
E: worry belongs to the reasonable scope.

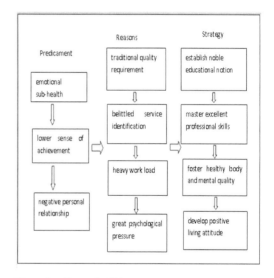

Figure 2. Toward self-improvement.

past and can not be changed, 12% of worry comes from others, who make inappropriate criticism out of inferiority, 10% of worry relates to health, only 8% of worry belongs to the reasonable scope (Zhao, 2012).

In other words, what really worries us is insignificant in one's life. A lot of pressures are just unrealistic things that come from one's own imagination. To take positive action is the best way to cope with stress.

4.4 To develop positive living attitude

During spare time, teaching secretaries should take part in some significant activities, such as singing, photography, gardening, etc., and develop positive living interests. Living interests are not only the ornament of life, but a living attitude, a kind of value-orientated philosophy of life (Zhou, 1994). If they face the lack of material with the satisfaction of spirit and the environment pressure, sense the carefree life with aesthetic emotion, the teaching secretaries will accept themselves happily and maintain a good state of mind. In the presence of teachers, teaching secretaries will have dedication as "the willing ox with bowed head"; In the presence of students, they will have the influence ability to "moistens everything silently". Through such a harmonious interpersonal relationship, the teaching secretaries will gain the respect of teachers and students.

5 CONCLUSION

To conclude, the college teaching secretaries' work involves in various basic links of teaching management and belongs to the important force of the teaching management system. As Figure 2 shows,

due to the previous professional prejudice, heavy work load and lack of identification from teachers and students, teaching secretaries have great physiological pressure. This will easily leads to their sub-health, lower sense of working achievement and negative personal relationships. To establish noble educational notion, master excellent professional skills, foster healthy body and mental quality and develop positive living attitude are the effective countermeasures for them to leave career predicament and promote self-improvement.

REFERENCES

[1] www.12371.cn (website of Chinese Communist Party members), 18th National Representatives of Chinese Communist Party Conference Report by Hu Jintao.
[2] Zhang Zheng, Research on the Job Burnout and Countermeasures for the College Teaching Secretaries, June 2010.
[3] Zhao Yan, How to Face with Pressure of Secretaries, *Secretaries*, February 2012.
[4] Zhou You, *Life of Interests*, Haitian Press, 1994.

Education Management and Management Science – Zheng (Ed.)
© *2015 Taylor & Francis Group, London, ISBN 978-1-138-02663-6*

Large shareholders' participation, issuing discount and tunneling in private placement

C. Xie & Q. Xu
College of Business Management, Hunan University, Changsha, Hunan, China

ABSTRACT: A discount factors regression model and tunneling comparison regression models are established to explore the private placement discount and tunneling of the large shareholders' participation and the specific factors of tunneling. The result shows that discount are widespread among listed companies that conduct private placement. Discount is not the only factor influence large shareholders' tunneling behavior. Company and private placement's characteristic all have significant influence on wealth transform.

Keywords: private placement; large shareholders' participation; discount; tunneling influence factors

1 INTRODUCTION

Private placement and rights issue have always been the main way for equity refinancing of listed companies. With the development of China's capital market, particularly the reform of non-tradable shares and the implementation of "*Administrative Measures for the Issuance of Securities by Listed Companies*", private placement was immediately applied by lots of companies for refinancing and restructuring. Statistics show that in 2011 and 2012 the number of private placement listed companies was 179 and 148 respectively, among which 436 companies were implementing or had implemented the private placement plan in 2013, but only 42 listed companies chose rights issue to refinance during these years. Private placement is widely used by listed companies mainly because it can make the company avoid the impact of the secondary market, and finance enough fund from the large shareholders, institutional investors or strategic investors directly. In addition, private placement reduces the information disclosure requirements and audit time, thus saving a lot of financing cost.

The core consideration of private placement is the price, which is directly related to the interests of the old and new shareholders, particularly the minority shareholders'. In general, the discount can reflect whether the private placement price is appropriate. Numerous studies have shown that, discount and tunneling behavior is widespread in private placement of large shareholders' participation. For example, Wruck analyzed the data of 37 American list companies which issued private placement to particular objects and found that sample companies' issuing price is average discounted by 13.5% relative to the trading day prior to the announcement. Hertzel mainly analyzed the private placement discount from the view of information asymmetry, and stated that the greater the degree of information asymmetry is, the higher discount the large shareholders request. Krishnamurthy believed that the main reason for the high discount rate is the compensation of large shareholders' liquidity within a time limit, because the private placement' shares of listed companies cannot be sold within a prescribed period of time. Barclay divided investors into positive and negative investors and analysis their private placement discount respectively. It was concluded that the discount to the negative shareholders (20.8%) was higher than a discount to the active shareholder (1.8%).

According to existent literature, some researchers proposed that private placement with large shareholders participation demonstrate tunneling behavior. Wruck and Wu's research showed that large shareholders consolidate their control rights by private placement to gain their own interests. He and Zhu proposed that large shareholders achieve the goal of diluting the small and medium-sized shareholders' equity through the discount. Zhang and Guo supposed the controlling shareholders of listed companies transfer the company wealth by using their rights of control of private placement. Discount level and large shareholders' subscription proportion codetermine the wealth tunneling amount. Wang and Zhang divided the listed companies samples that conduct private placement into three groups such large shareholders, institutional investors, and those does not belong to the two groups referred before. The results showed the existence of wealth transfer that the large shareholder gain by discount in the process of private placement.

In a word, current literature focuses on the investigation of private placement discount of the listed company, as well as the wealth transform with discount and encroached on the interests of small and medium-sized shareholders. However, there is hardly research on influencing factors of the large shareholders' tunneling behavior. This research suggests more factors should be taken into account other than discount to explore large shareholders wealth tunneling behavior. When will large shareholders transfer wealth? What factors impact it? Whether large shareholders to participate in the private placement will be accompanied by the tunneling or not? Whether large shareholders' tunneling can be judged by the discount level or not? These questions must be answered.

This paper selects listed companies which issued private placement between 2010 and 2012 as the sample to explore whether China's listed companies' private placement have discount phenomenon and what factors of large shareholders' tunneling from the angles of company characteristics and private placement characteristics. We will provide suggestions for the supervision agency on private placement and whether large shareholders transfer wealth.

2 THEORY AND HYPOTHESES

2.1 Large shareholders' opportunistic behavior and agency problem

There is a widespread ownership concentration problem in China's listed companies, and large shareholders tend to "a dominant". The issue objects of listed companies' private placement are usually large shareholders or their affiliated parties, so they can often control the price to some extent. To maintain their own control rights not be diluted, large shareholders will choose to expropriate minority shareholders' interests to realize the tunneling. In addition, Chinese private placement started late, the imperfection of the related rules and regulations, and the relatively simple regulators' approval process all offer a potential motive for large shareholders tunneling through the discount, so the opportunistic behavior of large shareholders' information opaque is the important factor that affect the listed companies' private placement.

Jensen and Meckling stated the agent theory that the greater the separation degree of the agent and the principal's interests, the more serious the

agent problem is. In private placement, the interests between large shareholders and minority shareholders tend to be separated. There are serious agency problem in listed companies, which demonstrate as the conflicts of interests between large shareholders and minority shareholders. At the same time, because of China's less developed capital market and the high degree information asymmetry, large shareholders are more likely to transfer wealth through the private placement discount. As such, we expect:

Hypothesis 1: The issuing discount of private placement widespread distributes, and it will be higher with large shareholders participation. Large shareholders have the use of the discount for tunneling.

2.2 Large shareholders' wealth tunneling influence factors

When there is discount with large shareholders participation in listed companies' private placement, large shareholders transfer wealth by purchasing more shareholding ratio than what they already have. The larger the difference is, the more amount of profits is. Beak studied the South Korean's family companies and found that the major way to transferring interests is high degree discount rate. However, the large shareholders' purchasing ratio in the private placement may not greater than the original stake. Also, when listed companies issue non-public shares to the large shareholders, institutional investors, or strategic investors, or a large shareholder injected high quality assets to a listed company, the purchase discount will not necessarily associated with the wealth tunneling with the participation of large shareholders. Accordingly, we expect that:

Hypothesis 2: Whether and how much large shareholders to transfer wealth cannot simply explained by private placement's discount, it is determined by company features and private placement characteristics at the same time.

3 RESEARCH DESIGN

3.1 Sample selections

This article selects companies which issued private placement in A-share market between 2010 to 2012 as the research sample. The ST enterprises, enterprises with abnormal extreme value and the enterprise whose financial disclosure is not complete were excluded because the extreme value will influence on statistical results. Eighteen companies, which conducted private placement twice in the sample period, were included. Finally, there are 344 companies' data as sample in this paper.

The data mainly comes from CSMAR, and also references the listed companies' private placement announcement Shanghai and Shenzhen Stock Exchange disclosed.

3.2 Determination of calculating variables

3.2.1 Private placement's characteristics and private placement

Discount: It measures the degree of issuing discount. Existing literatures' calculating methods of private placement discount is not consistent. In this article, we adopt Beak's method that used the private placement announcement trading day's closing price minus the purchase price, and then divided by the closing price of private placement announcement day.

Tunnel: It measures the amount of large shareholders' wealth transfer. we adopt Zhang and Guo's method which used the balance between large shareholders' purchasing ratio in the private placement and the original ratio. But there is also a special case that the large shareholders' purchasing ratio is less than their original ratio, or private placement's shares is issued at a premium when the large shareholders do not transfer wealth from the listed companies. Therefore, we define another variable *Tunnel1* to explain the wealth tunneling when shares issued at a discount and the large shareholders' purchasing ratio is larger than the original ratio.

Offertype: Private placement's objects can purchase the listed companies' shares through cash and assets. High-quality assets injection by large shareholders can optimize the assets structure of the listed companies and promote its benign development; but the private placement is a kind of private offering. Lack of strict regulation and the difficulty determination of the assets' real value may cause large shareholders transfer listed companies' wealth rely on the way of injecting non-performing assets or overvalued the assets.

Participate: Large shareholders' participation in the purchase of listed companies' private placement has direct influence on the discount. In general, the private placement discount tends to be higher with large shareholders participation.

Fraction: It is measured by the ratio of the amount of private placement shares to the total number of shares after issuance. Bigger the fraction indicates the higher proportion shareholders or institutional investors purchased listed companies' assets, and the greater risk the private placement's objects need to assume. The higher discount is, the more likely that transfer wealth will happen.

Time: Because the private placement's issue price set by the CSRC (China Securities Regulatory Commission) shall not be less than 90% of 20 trading days' stock market average shares price before the announcement day, we use the average yield of 20 trading days before private placement's announcement day of Shanghai Composite Index or Shenzhen Composite Index to state private placement's external market characteristics. Market timing has effect on the discount of private placement, we expect the discount is larger as the stock market is flourishing than as the stock market is depressed, so large shareholders will consider the wealth tunneling combine with the market timing.

3.2.2 Company features and private placement

Lnsize: It measures the degree of information asymmetry by the logarithm of listed companies' total assets. The bigger the company size is, the more the company's information can be got from the market, and the degree of information asymmetry is lower. Additionally, the discount large shareholders request will be lower, and the possibility of the tunneling will be small.

Level: It is measured by the asset-liability ratio, which is an important indicator for measuring the quality of the company. High asset-liability ratio and financial leverage may lead to companies' financial trouble, at that time listed companies will want to finance fund at a lower price which is more quick, as a result, the degree of discount will be high. In addition, the issue objects for their own interests protection require high discount.

Growth: Companies' growth ability is related to PB (price to book value). The growth ability represent the company's sustainable development in the future. Companies with high growth ability can attract more investors and large shareholders will probably not transfer interests from the companies. All variables mentioned above are listed clearly in Table 1.

3.3 Construction of regression models

According to above analysis, this paper introduces Equation (1) to test the hypothesis 1, and construct Equation (2), Equation (3) to test hypothesis 2. Among them, the Equation (3) is Equation (2) joining the *Discount* variable.

$$Discount = \beta_0 + \beta_1 \, Participate + \beta_2 \, Offertype$$
$$+ \beta_3 \, Fraction + \beta_4 \, Level + \beta_5 \, Time$$
$$+ \beta_6 \, Growth + \beta_7 \, Lnsize + \beta_8 \, Tunnel$$
$$+ \beta_9 \, Tunnel_1 + \varepsilon. \quad (1)$$

$$Tunnel\,1 = \beta_0 + \beta_1 \, Participate + \beta_2 \, Offertype$$
$$+ \beta_3 \, Fraction + \beta_4 \, Lnsize + \beta_5 \, Level$$
$$+ \beta_6 \, Growth + \beta_7 \, Time + \varepsilon. \quad (2)$$

Table 1. Variables and definition.

Variable	Variable's definition
Discount	Private placement announcement trading day's closing price minus the purchase price, and then divided by the closing price of announcement day
Tunnel	The balance between large shareholders' purchasing ratio and their original ratio
Tunnel 1	Dummy variable, under discount circumstance, and large shareholders' purchasing ratio larger than the original stake is 1, otherwise is 0
Offertype	Dummy variable, 0/1 = assets purchase/cash purchase
Participate	Dummy variable, 0/1 = not have/have large shareholders participation
Fraction	The ratio between the amount of private placement shares and the total number of shares after issuance
Lnsize	Listed companies' total assets' logarithm
Time	The average yield of 20 trading days before private placement's announcement day of Shanghai Composite Index or Shenzhen Composite Index
Level	Total assets divided by total liabilities
Growth	Price to book value

Table 2. Variables' descriptive statistics.

Variable	Minimum	Maximum	Mean	S.D.
Discount	−0.8188	0.8328	0.2137	0.2201
Tunnel	−60.73	100.00	8.5674	25.3431
Tunnel1	0	1	0.29	0.455
Offertype	0	1	0.69	0.462
Participate	0	1	0.73	0.496
Fraction	0.0145	0.8915	0.2033	0.1612
Lnsize	18.1623	25.9911	21.706	1.2783
Level	0.0046	1.6957	0.5336	0.2117
Growth	35.0614	40.5902	4.9271	4.8483
Time	0.0098	0.0089	−0.0002	0.0034

$$Tunnel\ 1 = \beta_0 + \beta_1\ Participate + \beta_2\ Offertype$$
$$+ \beta_3\ Fraction + \beta_4\ Lnsize + \beta_5\ Level$$
$$+ \beta_6\ Growth + \beta_7\ Time + \beta_8\ Discount + \varepsilon.$$
$$(3)$$

4 EMPIRICAL RESULTS AND ANALYSIS

4.1 Descriptive statistical analysis

Among the 344 private placement samples, only 35 issue at premium. From Table 2, we can see the mean value of Discount is about 21%, and the value reach 25% when excluding premium samples. This suggests that the private placement of China's listed companies is generally issued at a high discount and confirms the hypothesis 1 that discount is widespread distribution. Wealth tunneling percentage variable Tunnel's average value was 8.56%, which indicates that the purchase percentage of private placement shares large shareholders purchased is 8.56% more than the shareholding ratio before the purchase in average. But the mean value of the dummy variable Tunnel1 is only 0.29, and the standard deviation is 0.455, which shows that under the condition of discount wealth transfer's scale is not high and discount is not the only way

large shareholders used for tunneling. It also supports that we cannot judge large shareholders' wealth transfer behavior only by the discount that proposed in hypothesis 2. Because the Participate and the Offertype variables' value can only be 0 or 1, and their mean value is 0.73 and 0.69 respectively, which indicates that private placement with large shareholders participation has a high proportion, and issuing objects tend to subscribe for the purchase shares of listed company in cash.

4.2 Regression results analysis

The regression results of the discount influence factors' model are in Table 3: (1) Large shareholders' participation and Discount has a significant positive correlation at the 10% level, which is consistent with the hypothesis 1 that private placement discount with large shareholders is higher. (2) Tunnel1 also has a significant positive correlation at the 10% level with Discount, and it means that large shareholders typically require a high discount, which also supports the hypothesis 1. (3) The Fraction variable and the Level variable is all significantly related to Discount at 1% level. It indicates that the larger the large shareholder's subscription ratio is and the higher the asset-liability ratio is, the higher the discount level required will be, and this is the same as we expected. (4) The Offertype and the Discount has a significant negative correlation, which means discount with cash subscription will be lower than the others way. (5) Moreover, market timing also directly impact discount at 1% level, which suggests that when the market is flourish, the market average return is high, and it will boost investors' market sentiment, in that case, large shareholders will require a higher discount rate to protect their shareholding ratio.

In Table 4, it can be seen that from the results of Equation (2) that the equation fits better when not including the discount factor in which the large shareholders' participation, the Offertype,

Table 3. Discount influence factors model's estimate result.

Variable	Equation (1)
Constant	0.247 (1.382)
Participate	0.047 (1.974)*
Offertype	−0.074 (−3.006)***
Fraction	0.371 (5.619)***
Level	0.137 (2.761)***
Time	11.234 (4.118)***
Growth	0.003 (1.325)
Lnsize	−0.009 (−1.074)
Tunnel	0.001 (1.384)
Tunnel1	0.079 (1.785)*
$\bar{R}^2$	0.382
F-value	24.526
P-value	0.0000

*Numbers in bracket is t statistic; $^*p < 0.1$, $^{**}p < 0.05$, $^{***}p < 0.01$; the same below.

Table 4. Wealth tunneling's influence factors model's estimate result.

Variable	Equation (2)	Equation (3)
Constant	58.640 (2.667)	0.451 (1.514)
Participate	0.607 (17.319)**	0.571 (15.998)***
Offertype	−0.195 (−4.974)***	−0.158 (−3.981)***
Fraction	0.028 (0.251)	−0.099 (−0.857)
Lnsize	−0.024 (−1.707)*	−0.019 (−1.409)
Level	0.219 (2.600)***	0.161 (1.922)*
Growth	−0.003 (−1.012)	−0.004 (−1.238)
Time	−6.000 (−1.289)	−9.537 (−2.048)**
Discount		0.344 (3.836)***
$\bar{R}^2$	0.574	0.591
F-value	67.109	62.956
P-value	0.0000	0.0000

the *Level* and the *Lnsize* is significantly correlated with the *Tunnel1* variable at 5%, 1% and 5%, 10% level respectively. That means besides the discount, companies' private placement characteristics and company features have obvious influence on large shareholders' wealth transfer behavior. It is consistent with the hypothesis 2. At the same time, if comparing the regression results of Equation (2) and Equation (3), it can be found that there is no big increase of the goodness of fitting and the adjusted R square of the model when including the *Discount* variable, which also supports hypothesis 2 that we can not judge the behavior of large shareholders' wealth transfer only according to discount.

Equation (3) mainly analysis factors influence the large shareholders' wealth tunneling behavior in private placement. The result is as follow: (1) The *Offertype* has a significant negative correlation with wealth tunneling at the 1% level, which indicates that the wealth transferred amount large shareholders purchase in cash is less than in asset, because subscription in cash can avoid large shareholders injecting bad assets or miscalculating companies' assets value. This is consistent with our expectation. (2) The *Participate* and the *Discount* is all obviously correlated with the *Tunnel1* positively at 1% level, which shows that large shareholders participation and high level discount contribute to more interests delivery. (3) The *Timing* and the *Tunnel1* have a significant negative correlation at the 5% level, and this is because when the market is in a downward cycle, the stock market is also in a downturn, which will influence the public's investment enthusiasm. In order to keeping their assets continue to increase, the large shareholders are likely to use private placement to transfer interests.

4.3 Robustness testing

In order to test the reliability of results, we use the *Discount * Tunnel1* variable to replace the *Tunnel1* to represent wealth tunneling, and exclude listed companies' data which issuing private placement twice during the research period and the new regression's results still support the hypothesis 1 and hypothesis 2.

5 CONCLUSION

In this research, we selected listed companies that have conducted private placement during 2010 to 2012 of the A-shares market as samples to explore the private placement's discount phenomenon as well as influence factors of wealth tunneling under large shareholders' participation empirically. The empirical results suggest that discount is widespread in China's listed companies' private placement, and the discount level is higher when large shareholders participate in the purchase. But discount is not the only factor affects large shareholders' tunnel behavior. Other factors like the offer type, the listed company's assets scale and financial condition, as well as the companies' growth ability and market timing are all important factors large shareholder considered. For supervision authority, it is important to strengthen market price constraints and large shareholders restraint mechanism. This study has theoretical and policy implications of protecting the interests of public shareholders.

ACKNOWLEDGEMENTS

This work was financially supported by the National Natural Science Foundation of China (71373072), the Foundation for Innovative Research Groups of the National Natural Science Foundation of China (71221001) and the National Soft Science Research Project of China (2010GXS5B141).

REFERENCES

[1] Barclay M.J. et al. 2007. Private placements and managerial entrenchment. *Journal of Corporate Finance* 13(4): 461–484.
[2] Beak J.S. & Kang J.K. 2006. Business groups and tunneling: Evidence from private securities offerings by Korean chaebols. *Journal of Finance* 61(5): 2415–2449.
[3] Hertzel M.G. & Smith R.L. 1993. Market discounts and shareholder gains for placing equity private. *Journal of Finance* 48(3): 459–485.
[4] He. X.-J. & Zhu. H.-J. 2009. Tunneling, information asymmetry and private placement discount. *China Accounting Review* 7(3): 283–296.
[5] Jensen M.C. & Meckling W.H. 1976. Theory of the firm: Managerial behavior, agency costs, and ownership structure. *Journal of Financial Economics* 3(4): 305–360.
[6] Krishnamurthy S. et al. 2005. Does investor identity matter in equity issues: Evidence from private placement. *Journal of Financial Intermediation* 14(2): 210–237.
[7] Liu. B.-L. et al. 2012. Private placement, agency conflict and small and medium-sized shareholders' protection. *Financial theory and practice* (4): 76–81.
[8] Wang. Z.-Q. et al. 2010. A study on the tunneling of private placement in Chinese market. *Nankai Business Review* 13(3): 109–111.
[9] Wruck K.H. 1989. Equity ownership concentration and firm value: Evidence from private equity financings. *Journal of Finance Economics* 23(1): 3–28.
[10] Wruck K.H. & Wu Y. 2009. Relationships corporate governance and performance: Evidence from private placement of common stock. *Journal of Corporate Finance* 15(1): 30–47.
[11] Zhang, M & Guo, S.-Y. 2009. Private placement under the control of major shareholder and wealth tunneling. *Accounting Research* (5): 78–83.

Education Management and Management Science – Zheng (Ed.)
© *2015 Taylor & Francis Group, London, ISBN 978-1-138-02663-6*

Constructing the practical teaching platform of the pan-engineering and whole process to develop the excellent engineers in frontline of production

Shunpeng Zeng, Zhilin Qi, Kanhua Su & Jingcheng Liu
Chongqing University of Science and Technology, Chongqing, China

Xiuwen Yang
Logistical Engineering University, Chongqing, China

ABSTRACT: According to the Chinese petroleum industry proposed new talent standard of petroleum engineering, the experimental teaching platform of the pan-petroleum engineering and whole process was structured. Our university also deepened the school-enterprise "six common points" education mechanism, strengthened the pan-engineering consciousness, ability of engineering practice and comprehensive quality in order to cultivate excellent engineers in the production frontline.

Keywords: experimental teaching; pan-engineering; whole process; excellent engineer

1   INTRODUCTION

In order to keep up with the national strategic transformation, contribute to the development of the oil industry, provide services to the development of shale gas in Chongqing, Chongqing University of Science and Technology promotes the training plan of excellent engineers education, implementation of National Experimental Teaching Demonstration Center construction strongly. Our university also rebuild the necessary practical teaching system and platform for training the frontline Petroleum Engineers, established the education mode innovatively which applied talent integration of production and education, deepened the school-enterprise cooperative mechanism, strengthened the training of the pan-engineering awareness, the ability of professional practice and the overall quality.

2   THE REQUIREMENTS OF FRONT LINE ENGINEERS IN PETROLEUM ENTERPRISE

Our school takes China petroleum industry's talent demand as the guidance, develop the front line engineers' training standards combined with the three oil group experts, and establish the educational mode and the experimental teaching idea by combining production and teaching, coordinating school and enterprise.

2.1   *Update the application personnel training standards*

The training standard of front line engineers in china petroleum enterprise: to adapt to the development of petroleum engineering technology and the demands of oil industry international market, master the theory of petroleum engineering and professional skills, master petroleum engineering theory and professional skills, can be competent oil and gas drilling engineering, exploitation engineering and reservoir engineering design, engineering supervision, application technology R & D and modern reservoir management work etc. with the remarkable ability of engineering practice, pan-petroleum engineering consciousness and petroleum enterprise culture.

2.2   *Adhere to the advanced experiment-teaching ideas*

Experimental teaching is a key link of cultivating engineering application capabilities, must adhere to the student oriented, students' knowledge, capabilities and qualities standards as the main line, focusing on the collaborative of theory and practice teaching, outstanding experiment-teaching ideas of engineering practical ability and the cultivation of the sense of innovation.

2.3   *Establishing school-enterprise cooperative mode*

Persisting educational through depending on oil industry; Forming education mode of "2561" practice-oriented talents: the double subject of enterprise training and school education, adhering to the "five combined" to achieve innovation for experiment-teaching, school and enterprise "six common points" to education. Adhere to professional ethics education, the participation in the implementation of talents training through the entire process.

3 THE PRACTICE TEACHING PLATFORM'S BUILDING OF THE PAN-ENGNIEERING AND WHOLE PROCESS

Petroleum Engineering Specialty in our college take the cultivation of the enterprise' front line excellence Engineer as the goal, to implement excellent engineer national development plan as

a guide, accelerate the construction of national experimental teaching demonstration center, build the practice teaching platform of the pan-petroleum engineering and the whole process.

3.1 *Experimental teaching system*

In view of petroleum engineering technology intensive and specialized subject crossed characteristics, adaptation "geological—drilling-oil production—gathering and transportation" the whole process and oilfield service system-wide needs of modern oil and gas production, building covers a large number of professional personnel training objectives Petroleum Engineering need "Four Levels, Five Module" experimental teaching system.

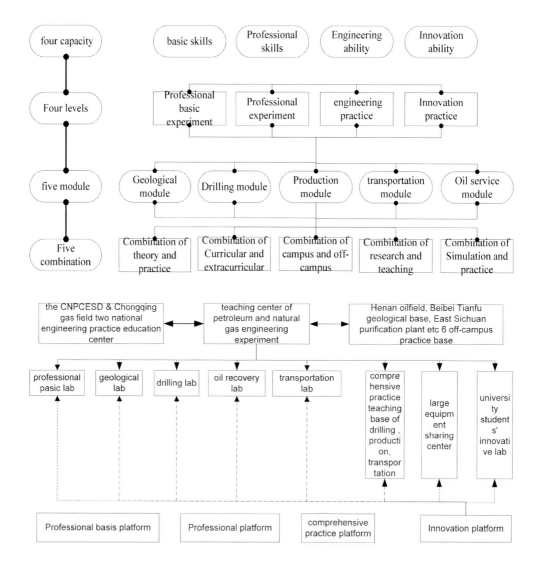

3.2 Set up the experimental teaching platform of school enterprise cooperation

According to "promote the transformation of students' knowledge and practical ability, cultivate students' engineering awareness and innovation" principle, set up a experimental teaching platform of pan-petroleum engineering "geological—drilling—oil production—gathering and transportation" of whole process and oil service system.

3.3 Cultivate international and engineering type's teaching team

A variety of ways to strengthen the ability of engineering practice and international awareness of teachers: oil field testing exercise, the international academic exchange, foreign cooperative research, to go abroad or domestic universities to study science and technology, oilfield services and personnel training, the formation of a team with the experimental teaching system, platform to match, a teaching staff with the ability to live the high level of practical guidance.

3.4 Omnibearing improvement of experimental conditions

Designed and built the "drilling and transportation comprehensive practice teaching base", to take real and simulation combine to reproduce the whole process of production of petroleum engineering, but higher than the actual production (simulate all kinds of production engineering accident), can carry out 100 kinds of technology experiment, teaching, scientific research test and oil culture exhibition; at the same time, school and enterprise cooperation 3 National Engineering Practice Education Center, 6 of out of school practice teaching base, being combined with CNOOC construction of offshore oil platform training base.

4 EXPERIMENTAL TEACHING FOR EXCELLENT ENGINEER TRAINING

Since 2012, Petroleum Engineering has been continuously optimizing talent training standards and program and strengthening experimental teaching reform according to the requirements of excellent engineer education training plan of Ministry of Education.

4.1 Enrich experimental teaching resources

According to the concept of pan-petroleum engineering and modern oil industry production technology, Petroleum Engineering has established five experimental teaching modules (i.e. geological, well drilling, oil recovery, gathering and transportation, oil service) and organized 63 course experiments, 37 experimental courses and 120 engineering practical training items. Additionally, in accordance with the four levels of plans (i.e. basic professional experiment, professional experiment, engineering practice, innovative practice), Petroleum Engineering has scheduled progressive, step by step professional engineering practice, including two weeks of understanding practice, five weeks of production practice, five weeks of graduation training and sixteen weeks of graduation design.

4.2 Systemic reform of the experimental teaching content

Fully considering the intersection between different subjects and modern petroleum engineering technology, the Petroleum Engineering has compressively and systemically reformed the experimental teaching contents according to the talent training standards. While arranging teaching contents of each module, much attention is paid on organic integration and promotion among basic experiment, comprehensive experiment, engineering experiment and innovative practice. Meanwhile, the Petroleum Engineering also lays emphasis on reasonable distribution of traditional teaching and modern teaching contents; strengths the close ties between scientific researches, engineering and the practical applications in the enterprise, timely absorbs results of scientific and technological innovation as well as achievements of experiment teaching reform, and upgrade and compensate relevant experimental projects; strengths the teaching contents of imitation and emulation and engineering training; and emphasizes the improvement of students' engineering practical abilities and research innovation abilities.

4.3 Reform teaching methods

In consideration of the experiment teaching characteristics of Petroleum Engineering, following measures are taken:

1. Heuristics and discussion teaching methods are applied order to cultivate students' learning interest and participation consciousness.
2. The practical teaching methods of imitation training and field practice are used to cultivate students' ability in engineering practice.
3. The combination of internal and external practice is applied order to transform students' knowledge to ability and quality effectively.

Goal-driven experimental teaching methods are applied order to cultivate students' innovation ability and comprehensive quality.

4.4 *Improving the experimental evaluation system*

According to different experiment course or experimental project, we applied various forms of assessment (including the experimental operation, the report the respondent, the experiment report or the paper), paid more attention to the combination of the process of evaluation and results evaluation, and evaluated students' experimental results comprehensively. We have made the scoring rules and measures of general experimental project, design experiments, innovative experiment project and the practice course (on campus and off campus).

5 GREAT PROGRESS HAS BEEN MADE IN CONSTRUCTION

The experts of China's three major oil companies has spoke highly of the cooperative education in our school education mode, admission first choice 100% nearly five years, graduate employment rate has remained at more than 90%; Double-professionally-titled teachers at over 80%, Oilfield technical services more than 4000 million RMB per year; More than 100 domestic colleges and universities visited our engineering practice platform and teaching system and give the full affirmation; Annual training of domestic and foreign oil companies in engineering and technical personnel more than 2,000 people per year.

Our school will further improve the pan-petroleum engineering "the whole process, the whole system" in experimental teaching platform and experimental teaching system, deepen the school-enterprise "six common" education mechanism, comprehensively promote the teaching experiment to the Experimental teaching with the "four transformations", as well as training the popular excellent engineer to the oil and gas production line.

AUTHOR BRIEF INTRODUCTION

Shunpeng Zeng (June 1965), male, Chongqing, doctor's degree, The professor (level two), Director of oil and natural gas National Experimental Teaching Demonstration Center, mainly engaged in teaching and research of petroleum development engineering, responsible for the professional training, e-mails:z-boshi@163.com.

REFERENCES

[1] Shixian Zhang & Yongping Li, Summary of the research on the reform of the training mode of Applied Undergraduate Talents, Higher Education Forum, 2010(10).
[2] Maoyuan Pan, The positioning, features and quality of China higher education, Chinese University Teaching, 2005(06).
[3] Xinyu Zhao, The analysis of the cultivation mode that used in Engineering Training session of application-oriented universities, Policy Research & Exploration (The second half), 2011(11).
[4] Jingming Cai & Zhubao Wei, Strategic thinking of Applied Undergraduate Talents, China Higher Education, 2008(12).
[5] Guoying Qian & Gang Wang & Liqing Xu, The characteristics of Applied Undergraduate Talents and the construction of the training system, Chinese University Teaching, 2005(9).
[6] Xuejun Fang & Tiecheng Ma & Xiumei Gu, Research and Practice on the reform of the training model of Applied Undergraduate Talents, Liaoning Education Research, 2008, 12:65–67.

Education Management and Management Science – Zheng (Ed.)
© 2015 Taylor & Francis Group, London, ISBN 978-1-138-02663-6

Operating mechanism construction of university technology transfer organizations

Xiaoli Li

Wuhan Textile University, Wuhan, China

ABSTRACT: University technology transfer organizations in China have different forms. However, based on the number and amount of the technology transfer contracts and patent technology transfer contracts, the efficiency of university technology transfer organizations in China could be much improved. This research proposed the operating mechanism of China's university technology transfer organizations from the perspectives of income distribution, human resource management and technology transfer diffusion.

Keywords: technology transfer; technology transfer organization; operation

1 INTRODUCTION

In order to promote the development of China's university technology transfer and train specialized personnel of technology transfer, the relevant state ministries, universities or combined with companies have established all kinds of technology transfer organizations. The forms of China's university technology transfer organizations vary, but most of them are immature, and the corresponding functions haven't also reached the necessary level. Generally, the domestic types of technology transfer mechanism can be divided into the types officially recognized by national state department and none-state department. Since September, 2009, the former State Economic and Trade Commission and the Ministry of Education have recognized 10 national technology transfer centers, 7 of which are in university. The State Ministry of Science and Technology and the Ministry of Education have identified 86 national university science parks in eight batches as a support platform for university technology transfer since 2001. In addition, the country has approved and identified 266 national technology transfer demonstration institutions since 2008. Though the technology transfer organizations have been much set up, China's current transfer rate of scientific and technological achievements is only about 25%, the real industrialization is less than 5%, which shows the efficiency of China's university technology transfer organizations is expected to be improved.

2 STATUS OF UNIVERSITY TECHNOLOGY TRANSFER CONTRACTS

2.1 *Number and amount of technology transfer contracts*

Figure 1 shows in 2005–2010, the number and amount of China's university technology transfer contracts are relatively stable. The number of technology transfer contracts reached the highest of 9,159 in 2010; the contracts in 2005 amounted to the highest value of RMB 2,215 million as shown in Figure 2. Regarding the growth of China's substantial research funding, university technology transfer didn't increased fast enough. Thus, the development of university technology transfer organizations at this stage is not enough specialized. University technology transfer organizations should focus on the timely delivery of information to guide the development of market demand for

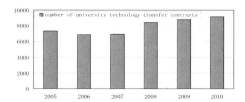

Figure 1. The number of university technology transfer contracts from 2005 to 2010.
Source: China's University Science and Technology Statistical Yearbook from 2006–2011.

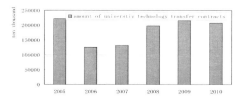

Figure 2. The amount of university technology transfer contracts from 2005 to 2010.
Source: as Figure 1.

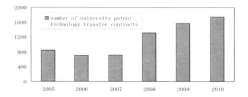

Figure 3. The number of university patent technology transfer contracts from 2005 to 2010.
Source: as Figure 1.

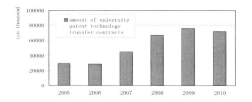

Figure 4. The amount of university patent technology transfer contracts from 2005 to 2010.
Source: as Figure 1.

the direction of the university applied technology development, while promoting the development of technology transfer from various channels.

2.2 Number and amount of patent technology transfer contract

The number and amount of patent technology transfer increase obviously. In 2005, the number of university patent contract is 842. Although in 2006 and 2007, the number of patent contracts declined, it started its rapid growth in 2008 and reached 1,745 in 2010 while patent contracts accounted for the entire technology transfer contracts increased from 11.5% in 2005 to 19.1% in 2010 as shown in Figure 3. Figure 4 shows the amount of patent technology transfer contract demonstrated an obvious growth in the observed period except in 2006. And the value of patent technology transfer contracts accounted for the entire technology transfer contracts also increased from 13.3% in 2005 to 34.6% in 2010, which reveals that university technology transfer organizations have enhanced the awareness of intellectual property and commercialization strategies.

However, the growth rate of patent grants reached 54.4% in 2010, which shows that the growth rate of the university patent technology transfer falls far behind the patent grants, and compared with the overall technology transfer, the proportion of patent technology transfer is still relatively low. University patent technology is an important resource for the university and the community, but the current patent technology transfer contracts accounted for only 34.6% of the total technology transfer contracts, which is partly resulted from the immaturity of university technology, and also shows there is still much room for operations' improvement of technology transfer organizations, so government, university and industry should advance the technology intermediary functions from all aspects.

3 OPERATION STRATEGIES OF UNIVERSITY TECHNOLOGY TRANSFER ORGANIZATIONS

Based on the number and amount of the technology transfer contracts and patent technology transfer contracts, China's university technology transfer organizations are making progress, but the absolute amount and number are still limited. Based on overseas experiences, the operating mechanism of China's university technology transfer organizations is constructed as follows:

3.1 Strategy of funding source and income distribution

In the first four years that the university transfer organizations set up, the government and universities should be important supporters of funding, and the proportion of direct government investment should strive for 40%–50%. Technology transfer organizations should allocate 30%–40% of the technology transfer revenue to maintain the operation and patent applications for university technology transfer organizations. And then, net income should be allocated among inventor, inventor's faculty, inventor's university, and also be deposited directly to their accounts to ensure all the technology transfer entities' interest. As the fixed allocation ratio of income distribution is concerned, the distribution principle in the United States that each party receives one-third of the income could be learnt. In addition, in order to ensure inventor's interest to a greater extent, the fixed ration for inventor could be increased, such as 40%, 30% and 30% respectively for inventor, inventor's faculty and inventor's university. However, net income of technology transfer can be also distributed as a progressive diminishing way that sets distribution ratio based on a different range of income. When the income is low, the income distribution ratio for inventor is higher, such as inventor, inventor's faculty and inventor's university could be allocated to 35%, 30% and 35% when technology transfer net income is less than RMB 200 000, but 25%, 35% and 45% is distributed respectively to them when that income is more than RMB 200 000. Under this circumstance, inventor's interest could be much ensured when the gross income is not enough high.

3.2 Strategy of human resource management

As technology transfer is directly related to the market operation, the selection criteria of technology transfer personnel must be clear. The qualified technology transfer staff should have relevant technology, finance, management, law backgrounds with business experience. Technology transfer manager selection should be based on the technical advantages of supporting university. The training of technology transfer staff can be carried out in two ways: one is to take the intensive training for the existing staff. Some staff should be sent to technology transfer organizations of well-known universities abroad. Through learning by doing, they could work independently after returning home. The other is to cultivate talents of intellectual property management, scientific and technological achievements transfer, which is a long-term staff development plan. Training center for talents of intellectual property and technology transfer should be established under grants in university with certain foundation to train a number of compound talents that are not only with engineering expertise, but also familiar with technology transfer procedures.

Besides, the responsibilities of technology transfer managers, and the relevant evaluation criteria must be clear. The primary duties of technology transfer staff are to determine the technical needs of industry, look for potential licensees, and sign the transfer contracts with enterprises to promote the commercialization of the technology. The number of spin-off, patent disclosure, patent applications, technology transfer contracts, enterprises and researchers that are contacted with should be taken as evaluation indicators for technology transfer managers. (Xiaoli Li, 2011). Meanwhile, in order to strengthen internal personnel management, all positions in technology transfer organizations should have appropriate job descriptions and job specifications as criteria on recruitment, selection, training, performance appraisal of technology transfer managers.

3.3 Strategy of technology transfer diffusion

3.3.1 Alliance strategy
Horizontal and vertical network alliances of university technology transfer organizations help expand the scope and possibilities of technology transfer. Through vertical integration of industry alliances, regional alliances and horizontal integration of technology transfer association alliances, the degree of stability and trust in both supply and demand of technology transfer could be strengthened. Thus, university technology transfer organizations could get more opportunity to interact with the company, and expand the technology transfer.

3.3.2 Patent portfolio and sub-contract strategies
In 1998, German scholar, Holger Ernest first proposed the concept of "patent portfolio". As the value of the patent portfolio is often greater than the value of a single patent, patent portfolio is widely used in the corporate R & D strategy planning, patent layout and technology merger. But university technology transfer organizations will not be so lucky to have a complete technology patent portfolio. Granieri (2003) proposed a set of technical and cost-effective technology transfer strategy. The basic idea is to evaluate the existing patent portfolio and find out the potential one. Despite the relatively high value, the patent portfolio lacks flexibility in the transfer process as some companies may only be interested in some patents of the patent portfolio. In order to expand the scope of technology transfer, patent portfolio and sub-contract strategies ensure the demand elasticity of technology acquirer, and integrated value of the patent portfolio. For just a single piece or several pieces of patent technologies are required, university technology transfer organizations take sub-contracting strategy of patent portfolio, and directly licenses. Meanwhile, when the patent portfolio reaches the level of commercialization, university technology transfer organizations directly license the portfolio to the transferee. If the technology patent portfolio is not complete, university technology transfer organizations seek complementary patents from other institutions; if the technology within the existing patent portfolio is immature, university technology transfer organizations needs to re-organize the research team to conduct research so that the patent portfolio could be put into use. When the technology is immature or incomplete, university technology transfer organizations should coordinate the research of technical complementarities and cooperation across organizations. By flexibly using the patent portfolio and sub-contract strategies, university technology transfer organizations expand the technology transfer and increase the likelihood of success, thus improve the performance of the technology transfer organizations.

3.3.3 Spin-off strategy
In order to increase the diffusion of technology transfer, besides the sales of technology transfer from university to enterprises, spin-off strategy could be used to expand the scope of technology transfer, as shown

in Figure 5 (Kano, 2001). When enterprises' evaluation capability for receiving technology is higher than β values in Figure 5, university technology transfer organizations directly transfer the technology to the target through its basic operation, namely the "connection 1" as shown in Figure 5, which is the core of university technology transfer organizations' design system. When enterprises' evaluation capability for receiving technology is less than or equal to β value in the Figure below, in which case the enterprises are unable to take in the technology or new technology is lowly related to the main business especially when a large degree of uncertainty or a major technology is concerned, it is unrealistic for university to transfer new technology to the existing enterprises (Fred, 2011; Müller, 2010), so there is a gap of technology transfer between the two sides to break through. It is time to consider establishing spin-off enterprise to commercialize university research achievements, and spin-off also becomes the important model of technology transfer. University technology transfer organizations could help the target enterprises to spin off or university could do this itself as the "connection 2 or 3" shown in the Figure below. University technology transfer organizations could possibly negotiate with governmental risk investment agencies, university science parks, researchers to help target enterprises incubate new business. 20/80 principle could be used to efficiently look for the target enterprises (Hoye, 2009) as repeat buyers of a new technology are often the most powerful to incubate the technology with uncertainty. For some groundbreaking, forward-looking and practical scientific research, university technology transfer organizations may consider the establishment of university spin-off enterprises in the absence of a target, and try to get the support from the government.

But there is a challenge for university technology transfer organizations to set up a new spin-off because organizations should have the specialized department, and be able to integrate social capital and solve the cultural conflict (Rasmussen, 2010). When the university spin-off is mature, the university can sell the shares. In this way, university technology transfer is achieved and the opportunities of internships and employment are provided for students.

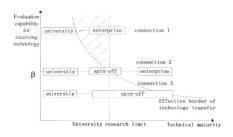

Figure 5. Spin-off strategy.

4 CONCLUSIONS

Based on the status of China's university technology transfer contracts and overseas experiences, the operating mechanism of university patent technology transfer organizations could be re-constructed with strategies of funding source and income distribution, human resource management, technology transfer diffusion. In the funding source and income distribution, the government and universities should be important supporters of funding for technology transfer organizations, and to ensure inventor's interest to a greater extent, the fixed ration for inventor could be increased. For human resource management strategy, the selection criteria of technology transfer personnel, the responsibilities of technology transfer managers, and the relevant evaluation criteria must be clear. Alliance strategy, patent portfolio and sub-contract strategies and spin-off strategy could be taken for technology transfer diffusion.

ACKNOWLEDGEMENT

This project is granted by National Social Science Foundation of China (Grant No.12CGL012).

REFERENCES

[1] Fred P., Paul G. Commercializing inventions resulting from university research: analyzing the impact of technology characteristics on subsequent business models Original Research Article. Technovation, 2011, 31(4): 151–160.
[2] Granieri M. Beyond traditional technology transfer of faculty-generates inventions: building a bridge toward R&D. Les Nouvelles, 2003, 38(4): 167–175.
[3] Hoye K., Pries F. Repeat commercializers? The habitual entrepreneurs of university industry technology transfer. Technovation, 2009, 29(10): 682–689.
[4] Kano S. Introduction and Comparison of Technology Transfer Models in University-Industry Relations: The Concept of Technology-Transfer Effectiveness Frontier and Its Application. Business Model (Electronic Journal of Japanese Society for Business Model), 2001, 1(1): 1–10.
[5] Müller K. Academic spin-off's transfer speed: analyzing the time from leaving university to venture. Research Policy, 2010, 39(2): 189–199.
[6] Rasmussen E., Borch O.J. University capabilities in facilitating entrepreneurship: A longitudinal study of spin-off ventures at mid-range universities. Research Policy, 2010, 39(5): 602–612.
[7] Xiaoli Li. Research on Dynamic Evolution of University Technology Transfer Organization in U.S.A based on the Model of Triple Helix. Library and Information Service, 2011, 55(14): 36–41.

Education Management and Management Science – Zheng (Ed.)
© *2015 Taylor & Francis Group, London, ISBN 978-1-138-02663-6*

The analysis on cognition and behavioral intention of adventure tourists to Chengdu

Tingting Xiao & Mei Li
Sichuan Agricultural University, Ya'an, Sichuan, China

Yong Yu
Beijing Davost Intelligence Group, Chengdu, Sichuan, China

Chaohong Ma
The Ecotourism Association of Sichuan, Chengdu, Sichuan, China

Chao Yu
Sichuan Agricultural University, Ya'an, Sichuan, China

Yun Liao & Xiaohua Yang
The Ecotourism Association of Sichuan, Chengdu, Sichuan, China

ABSTRACT: The research on relationships among tourists' perceived value, tourists' satisfaction and tourists' behavioral intention has been a topic of concern in the field of tourism consumption behavior. Taking domestic tourists to Chengdu as an example to make a sampling survey, this article made use of the questionnaire survey and model analysis methods to study dimensions of adventure tourists' perceived value and their relationships, and their influence on tourists' satisfaction and tourists' behavioral intention. The results indicate that function value, currency value and social value have significantly positive effect on emotion value; function value, social value and emotion value have significantly positive effect on tourists' satisfaction; social value, emotion value and tourists' satisfaction have significantly positive effect on tourists' behavioral intention. The research results show that domestic adventure tourism development and management should pay more attention to the quality of supporting services as well as the design and development of tourism projects.

Keywords: adventure tourism; tourists' perceived value; tourists' satisfaction; tourists' behavioral intention

1 INTRODUCTION

Adventure tourism is that the tourism subjects with individuals or teams go to a relatively remote and original place, for a variety of outdoor tourism activities. Generally, it needs professional guides and special sports equipment. The tourists can have exciting and unusual experience in the activities[1]. Nearly 30 years, foreign related studies focus on safety and risk of adventure tourism[2,3], adventure tourism market[4], the influence of adventure tourism[5,6], etc. Domestic researches mainly concentrate on the introduction of overseas studies, the description of domestic related activities and the exchange of experience. The purpose is that the scientific development and standardized management of adventure tourism has sufficient basis, the research on the tourists' perception of adventure tourism perceived value and tourist satisfaction is necessary.

2 DEFINITION AND MODEL HYPOTHESES

2.1 *Definition*

Combined with the results of previous studies[7,11], this text make the following definitions of tourist perception, tourist satisfaction and tourist behavior intention: The perceived value can be defined as an overall assessment of the tourism product utility, including four dimensions, namely, the functional value, emotional value, cognitive value and social value. Tourist satisfaction is a subjective feeling of visitors through the formation of tourism product perception and expectation effect.

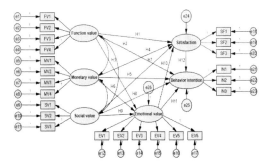

Figure 1. The original path graph of the ideal mode.

Tourist behavior intention is the specific activity and behavior tendency what tourists may be taken after they experience tourism product and service, there will be "revisiting willingness" and "recommendation intention" as measures of tourist behavior intention.

2.2 The hypotheses of theoretical model

Take Emotional value, satisfaction and behavior intention as endogenous latent variables; Function value, monetary value and social value as exogenous latent variables; ei as error, this thesis uses structural equation model analysis software AMOS18.0 to build path graph of theoretical model, containing 12 model assumptions, what is shown in Figure 1.

3 RESEARCH METHOD

3.1 Sample selection

Chengdu is located in the western Sichuan Plain, by virtue of its strategic location, convenient transportation and pleasant climate, it attracts numerous adventurers. Selecting Chengdu as the destination for the investigation has a great typical and representative meaning. The study focuses on Chengdu tourists relatively concentrated in the major travel agencies and domestic tourists concentrated in the major attractions.

3.2 Questionnaire design and survey

Generalizing foreign adventure tourism options in their cognitive research questionnaires, 28 indicators are classified for the assessment; after rewriting the indicators one by one, and using a Likert 7 scale, the measurement problem sets come into being; in November of 2012, a total of 330 copies of questionnaires were handed out in Chengdu Temple Jinsha, Chengdu Comfort Travel CYTS

to do the pre-survey, and SPSS18.0 software were used to analyze the validity of these questionnaires and to modify them. Excluding 5 indexes of lower reliability and validity, ultimately, 23 evaluation questions set is attained (see Table 1).

3.3 Data analysis methods

Factor loading for every latent variable observation index reaches the lowest load standard 0.5, and it ensures that three indicators in each dimension have reached the standard, without deleting. The KMO value is 0.892, which shows that the adequacy of sample is good. Bartlett test approximate chi square was 1389.748, df = 253, sig = 0.000, which shows that there are some differences between correlative coefficient matrix and unit matrix and it is suitable for scale data to make factor analysis. The author has made reliability test for the 23 measurement items, finding that the table overall Alpha coefficient is 0.917, and each of the 6 latent variables Composite Reliability (CR) Alpha coefficients is higher than 0.7, which has met the suggested standards, and the latent variables have internal consistency.

Indexes for fitting can be derived from theoretical models computing: χ^2/df is between 2~3, and SRMR as well as RMSEA is less than 0.08, which achieves the desired standards; But NFI, NNFI, CFI and GFI are less than 0.90 which do not meet the desired requirements (see Table 2). From the perspective of

Table 1. Variables and specific testing items.

Latent variables	Observation variables
Function value	Consistency
	Well-designed
	Quality standard
	Services and organization
Monetary value	Good returns
	Good value for money
	High cost performance
	Equitable price
Social value	Recognized by his teammates
	Accept his teammates
	Meet the target
Emotional value	Happy feeling
	Worthy of appreciation
	Freshness
	Curiosity
	Adventure and challenge
	Sense of reality
Satisfaction	Overcome the difficulties and challenges
	Desired journey
	Satisfied with the decision
Behavior intention	Recommendation intention
	Revisiting intention
	Other journey

Table 2. The second-order confirmatory factor model fit indices and standard parameters.

Fit indices					
Original	Improved	Path	Gamma value	Path label	Expected signs (test results)
		Function value→Satisfaction	0.231	H1	+(+)
$\chi^2 = 628.524$	$\chi^2 = 457.174$	Function value→Behavior intention	/	H2	+(−)
df = 215	df = 210	Function value→Emotional value	0.173	H3	+(+)
$\chi^2/df = 2.924$	$\chi^2/df = 2.177$	Monetary value→Satisfaction	/	H4	+(−)
SRMR = 0.0495	SRMR = 0.0452	Monetary value→Behavior intention	/	H5	+(−)
GFI = 0.875	GFI = 0.907	Monetary value→Emotional value	0.337	H6	+(+)
CFI = 0.897	CFI = 0.938	Social value→Satisfaction	0.196	H7	+(+)
NFI = 0.853	NFI = 0.903	Social value→Behavior intention	0.239	H8	+(+)
NNFI = 0.879	NNFI = 0.926	Social value→Emotional value	0.325	H9	+(+)
RMSEA = 0.071	RMSEA = 0.056	Emotional value→Satisfaction	0.522	H10	+(−)
		Emotional value→Behavior intention	0.302	H11	+(+)
		Satisfaction→Behavior intention	0.275	H12	+(+)

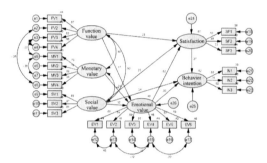

Figure 2. The path diagram of the improved model.

effectiveness and credibility, the MI value (correction factor) is selected, the largest and most realistic one which is also the most reasonable and effective one to modify the model. The improved model is shown in Figure 2. After the improvement, the indicators have reached the ideal standard[12], indicating that the model has a good fit and good constructive validity (see Table 2). All T test values are greater than 1.96, P values are less than 0.05, which means that all measuring indicators are significantly associated with each potential variable.

The number of travelers who participate in adventures differs in every season. Due to the weather in winter tourists are relatively fewer. To make this questionnaire survey more complete and reliable, this research was carried out in January, April, July, November, 2013 to Chengdu Xiling Snow Mountain, Kam, Dujiangyan Qingcheng Mountain and CYTS, China. Official investigation was conducted formally 4 times, in which 2097 valid questionnaires were returned out of 2200 questionnaires, with the effective rate as high as 95.3%.

4 RESULTS AND ANALYSIS

4.1 Demography characteristics

The sample population remains mostly male, higher education and young age. These people have controlled the budget of tourism consumption. These are related to male physical condition and psychological needs, professional knowledge and skills, income level and so on (see Table 3).

4.2 The theoretical model has stronger explanation ability

Table 2 shows that, the validity of adventure tourists' cognition and behavior intention model has been confirmed, the reliability and validity of the survey data has been tested, and it has good dependability. From the test of model hypotheses, in addition that the "functional value→behavior intention", "monetary value→satisfaction" and "monetary value→behavior intention" are not supported by the empirical study, the remaining 9 model assumptions have been well verified, indicating that the model has stronger explanation ability.

4.3 Function value, monetary value and social value has a positive effect on emotional value

Table 2 and Figure 2 show that, function value, monetary value and social value have positive effects on emotional value (Gamma values were 0.17, 0.34, 0.33, respectively). The function value of the trip is higher, tourists get quality at lower cost. The social value of the tourists who participate in the adventure in social groups is higher, tourists can feel more pleasure, excitement

Table 3. Demographic profiles of tourist samples (N = 2097).

Item	Classification	Copies	Rate
Sex	Male	1248	59.5
	Female	849	40.5
Age	18–25	1097	52.3
	26–45	1000	47.7
	46–60	0	0
	More than 60	0	0
Education background	Junior high school or lower	0	0
	Senior school or technical school	113	5.4
	Bachelor degree	1644	78.4
	Master degree or higher	340	16.2
Daily budget	Below 100 yuan	472	22.5
	101–200 yuan	774	36.9
	201–300 yuan	491	23.4
	301–400 yuan	151	7.2
	More than 400 yuan	209	9.9

and happiness. In the analysis of path, the four attribute of function value have a positive impact on tourists' emotional evaluation; the relationship between the reasonable price of tourism products and emotional value is significant; that adventure tourists get social recognition and improve self image in exploration experience will affect tourists' emotional experience. In contrast, function value has weaker influence. Monetary value and social value has stronger influence. Therefore, H3, H6, H9, these three hypotheses have been verified.

4.4 There are positive effects between function value and monetary value, monetary value and social value, function value and social value

Table 2 and Figure 2 show that, there are positive effects between function value and monetary value, monetary value and social value, function value and social value (Gamma values were 0.59, 0.44, 0.60, respectively), Which shows that the three of functional value, monetary value and social value are inseparable. That is, a factor changes certainly will affect the changes of the two other factors. It promotes each other among the three.

4.5 Emotional value, functional value and social value have significantly positive effects on satisfaction

Table 2 and Figure 2 show that, emotional value, functional value and social value has positive

effects on satisfaction (Gamma values were 0.52, 0.23, 0.20, respectively). The after-tour emotions of adventure tourists will directly affect satisfaction; effective organization, elaborate design, quality standards and supporting services of this trip have significant impacts on satisfaction; tourists win the team recognitions and establish the social relations by adventure tourism, that social value can be increased significantly to improve the satisfaction; monetary value indirectly affect the tourist satisfaction, tourists tend to care more about the stimulation, adventure and challenge of adventure travel, so the cost isn't the key that tourists pay attention to. The test results support H1, H7, H10, the three hypotheses, but they do not support H4.

4.6 Social value, emotional value and satisfaction have significantly positive effects on behavior intention

Table 2 and Figure 2 show that, social value, emotional value and satisfaction have positive direct effects on behavior intention (Gamma values were 0.23, 0.30, 0.28, respectively). The tourists who participate in the process of adventure tourism in social groups social value of is higher, and the higher emotional evaluation of tourists, tourists are more satisfied with the travel, the tourist revisit rate can be increased much better, the assumptions of H8, H11, H12 have been verified. Function value and monetary value of the trip have indirect influences on behavior intention of adventure tourists, adventure tourists pursue breathtaking, exciting, irreplaceable and fleeting adventure experience, the consistent design of the journey, quality standards, supporting services and the cost of money are not the key factors that adventure tourists consider, therefore, the two assumptions of H2, H5, are false.

5 CONCLUSIONS AND DISCUSSIONS

5.1 The after-tour emotional evaluation of adventure tourists are affected most by the economic cost

Monetary value has been an important evaluation variable of attraction competitiveness[13]. Perceiving the economic costs not only has an important role in the purchase decision-making stage, but at the end of the tourism activities it affects tourists' emotional evaluation on this travel experience[14], which is consistent with the findings by Havel et al. Therefore, the adventure tourism operators should take full account of the economic cost of the product while developing in order to increase the cost-effective.

5.2 Emotional value has most significant impact on the satisfaction of adventure tourists

Rojas and Camarero took the heritage tourism destination for example, analyzed the impact of emotional factors on emotional experience for tourists and noted that the emotional factors had an important adjusting influence in the formation of tourist satisfaction[15]. American marketing scholars Westbrook did the empirical research on the relationship between customer satisfaction and consumer emotion for the first time, and the results showed that consumer emotions directly affected customers' satisfaction[16]. The results of this study also show the same phenomenon. Development of domestic adventure tourism should pay attention to the design and development of adventure tourism projects. Good tourism projects can give the adventure tourists a thrilling, exciting and challenging experience, thus meeting the need of tourists' seeking adventure experiences, so as to enhance visitor's satisfaction.

5.3 The behavior of adventure tourists is easy to be affected by the combined effects of social value, emotional value and satisfaction

Emotional evaluation of the visitors on this whole travel schedule is considered to be the endogenous motivation of revisiting willingness[17], the model in this paper analyzes that emotional value displays a positive impact on the formation of visitor loyalty. Visitor satisfaction will affect many aspects such as choosing travel destinations, travel products and services consumption, or whether to revisit, whether to recommend, etc. Many studies have confirmed that visitor satisfaction will produce a good reputation and lead to more loyalty to the attractions, and high satisfaction usually triggered positive behavioral intention, which is crucial for Tourism marketing[18,20]. Adventure tourism operators should focus on the post-tour emotions of visitors, so as to enhance the development of products based on the needs of tourists' satisfaction, and to prompt visitors to make recommendations to the surrounding population, and establish revisiting induction strategy to promote sustainable development of adventure tourism.

5.4 Emotional evaluation, satisfaction and behavioral intention are tend to be affected by social value

Unlike sightseeing, travel process of adventure tourism is not an activity which can be done alone, but a more dependent teamwork. Through adventure, tourists build friendships and social relationships and gain the social recognition and a sense of belonging, so that visitors are satisfied to promote revisiting willingness. Therefore, the development of adventure tourism products should exert a special focus on whether the product can meet the expectations of visitors on social value.

ACKNOWLEDGEMENTS

A project supported by scientific research fund of Sichuan Provincial Education Department in 2011, No.11ZA078.

REFERENCES

[1] Buckley, R. 2006. *Adventure Tourism.* Wallingford, UK: CAB International, 1.

[2] Bentley, T., Meyer, D., Page, S., & Chalmers, D. 2001. Recreational tourism injuries among visitors to New Zealand: An exploratory analysis using hospital discharge data. *Tourism Management,* 22(4): 373~381.

[3] Bentley, T.A., Page, S.J., & Macky, K.A. 2007. Adventure tourism and adventure sports injury: The New Zealand experience. *Applied Ergonomics,* 38(6): 791~796.

[4] Loverseed, H. 1997. The adventure travel industry in North America. *Travel and Tourism Analyst,* (6): 87~104.

[5] Shackley, M. 1996. Community impact of the camel safari industry in Jaisalmar, Rajasthan. *Tourism Management,* 17(3): 213~218.

[6] Silori, C.S. 2004. Socio-economic and ecological consequences of the ban on adventure tourism in Nanda Devi Biosphere Reserve western Himalaya. *Biodiversity and Conservation,* 13(12): 2237~2252.

[7] Zeithaml, V.A. 1988. Consumer Perceptions of Price, Quality and Value: A Means-End Model and Synthesis of Evidence. *Journal of Marketing,* 3(52): 2~22.

[8] Chen, C.H., & Tsai, D.C. 2007. How destination image and evaluative factors affect behavioral intentions. *Tourism Management,* 28(6): 1115~1122.

[9] Kim, Y.K., & Lee, H.R. 2010. Customer satisfaction using low cost carriers. *Tourism Management,* 31(1): 1~9.

[10] Woodside, A.G., Frey, L.L., & Daly, R.T. 1989. Linking service quality, customer satisfaction and behavioral intentions, *Journal of Health Care Marketing,* 9(4): 5~17.

[11] Seines, F. 1993. An examination of the effect of product performance on brand reputation, satisfaction and loyalty. *European Journal of Marketing,* 127(9): 19~35.

[12] Wen, Z.L., Hau, K., & Herbert, W.M. 2004. Structural equation model testing: Cutoff criteria for goodness of fit indices and chi-square test. *Acta Psychological Sinica,* 36(2): 186~194.

[13] Shi, C.Y. 2006. An empirical study on the relationship between the scenic image and tourists' perceived value, satisfaction and loyalty. *Human Geography,* (3): 72~77. (in Chinese).

[14] Javier, S., Luis, C., Rosa, M.R., & Miguel, A.M. 2006. Perceived value of the purchase of a tourism product. *Tourism Management,* (27): 394~409.

[15] Rojas, C.D., & Camarero, C. 2008. Visitors' experience, mood and satisfaction in a heritage context: Evidence from an interpretation center. *Tourism Management,* 29(3): 525~537.

[16] Westbrook, R.A. 1980. Intrapersonal affective influences on consumer satisfaction with products. *Journal of Consumer Research,* 7(1): 49~54.

[17] Deng, M.Y. 2004. Discussion of Mount Emei tourism image orientation. *Journal of Southwest University for Nationalities,* (4): 177~179. (in Chinese)

[18] Li, Y. 2008. The analysis on tourists' satisfaction of tourism destination and its influencing factors. *Tourism Tribune,* 23(4): 43~48. (in Chinese)

[19] Lee, S.Y., Petrick, J.F., & Crompton, J. 2007. The Roles of quality and intermediary constructs in determining festival attendees' behavioral intention. *Journal of Travel Research,* 45(4): 402~412.

[20] Tracey, M., Vonderembse, M.A., & Lim, J.S. 1999. Manufacturing technology and strategy formulation: Keys to enhancing competitiveness and improving performance. *Journal of Operations Management,* 17(4): 411~428.

Education Management and Management Science – Zheng (Ed.)
© *2015 Taylor & Francis Group, London, ISBN 978-1-138-02663-6*

Impact of risk propensity and entrepreneurial self-efficacy on entrepreneurial intention

Botao Yan
North China Institute of Science and Technology, Hebei, China
China University of Mining and Technology (Beijing), Beijing, China

Kunshu Ma
Southwest University for Nationalities, Chengdu, China

ABSTRACT: The research on impacts of entrepreneurial intention is an important area in the field of entrepreneurship research which is widely concerned by scholars because it is an important index variable to explain entrepreneurial behavior. On the basis of related theories of risk propensity, entrepreneurial self-efficacy and entrepreneurial intention, the paper probes into the relationship among them by means of regression analysis on questionnaire data. This empirical study demonstrates that: i) risk propensity has positive effect on entrepreneurial self-efficacy and entrepreneurial intention; ii) entrepreneurial self-efficacy has also positive effect on entrepreneurial intention; and iii) entrepreneurial self-efficacy partially intermediates the impact of risk propensity on entrepreneurial intention.

Keywords: risk propensity; entrepreneurial self-efficacy; entrepreneurial intention; mediating effect

1 INTRODUCTION

Why do some people start new business ventures? Scholars are always trying to answer this question. Entrepreneurship is not instinctive reaction of human being, but goal-directed and premeditated actions (Bird 1988, Katz & Gartner 1988). Intention is the best variable to facilitate the understanding and predicting such kind of purposed-unconventional behavior (MacMillan & Katz 1992). Most of current theoretical models (Krueger et al. 2000, Ajzen 1991, Shapero1982) that well explain entrepreneurial intentions and behaviors significantly emphasize the factor of self-efficacy. Obviously, as the key variable impacting entrepreneurial intentions, self-efficacy is worth to be discussed because of its important roles. Moreover, risk propensity has been attracting scholars' attention because of its impact on entrepreneurship. Its impact on entrepreneurial intention and entrepreneurial behavior is often discussed by scholars (Sitkin & Pablo 1992, Barbosa et al. 2007), and its impact on self-efficacy is now a new focus (Barbosa et al. 2007, Zhao et al. 2005). The paper probes into the impact of risk propensity and entrepreneurial self-efficacy on entrepreneurial intention by means of empirical study in order to further explain the generation mechanization of entrepreneurial intention.

2 HYPOTHESES

Individual risk propensity is an inevitable factor when starting a new business adventure. Individual risk propensity (Barbosa et al. 2007) and willingness to undertake it (MacCrimmon & Wehrung 1990) influence, to some extent, the process of entrepreneurial intentions. The individual with higher risk propensity is generally willing to take more entrepreneurial risks for higher returns, more willing to start a new business, and even inclined to start new business adventures with more risks (Forlani & Mullins 2000). Hence the first hypothesis (Hypothesis 1) is formulated as following.

Hypothesis 1(H1): Risk propensity is positively correlated to entrepreneurial intention. The higher one's risk propensity is, the higher his entrepreneurial intention is; the lower one's individual risk propensity is, the lower his entrepreneurial intention is.

As an important factor influencing intention (Bird 1988, Krueger 1993) and behavior (Bird 1988), self-efficacy plays an essential role. Self-efficacy is a person's own beliefs about their ability to do a task or activity (Bandura 1977), which is a fundamental concept of social cognitive and learning theory, and has directive influence on mindset, intention and behavior (Bandura 1986). The perceived self-efficacy reflects different behaviors and tasks because of different requirement for capabilities, so corresponding domain feature exists (Bandura 1977). Deriving from this point, entrepreneurial self-efficacy emphasizes the self-confidence for an individual to undertake entrepreneurial actions, which is usually used to explain entrepreneurial behavior (Zhao, Hao. et al. 2005, Forbes 2005, Hmieleski & Corbett 2008). According to self-efficacy theory, the more a person is confident to a certain action, the more likely he has the willingness to take this action (Bandura 1986). Taking the domain feature of self-efficacy into account, it is the best to adopt entrepreneurial self-efficacy rather than the common sense of it when talking about influencing factors of entrepreneurial intention. Hence the following hypothesis (Hypothesis 2) is formulated.

Hypothesis 2(H2): Entrepreneurial self-efficacy is positively correlated to entrepreneurial intention. The higher one's perceived entrepreneurial self-efficacy is, the higher his entrepreneurial intention is. The lower one's perceived entrepreneurial self-efficacy is, the lower his entrepreneurial intention is.

The impact of risk propensity to entrepreneurial self-efficacy has been discussed previously; however it was based on risk propensity, a stable feature variable (Barbosa et al. 2007). The current research mainly discuss risk propensity when probing into personal risk propensity because that feature cannot be used to well explain entrepreneurial behavior (Gartner 1989). The person with higher risk propensity is more convinced with his entrepreneurial role (Zhao, Hao. et al. 2005), thus he is more confident to conduct his entrepreneurial behavior. The hypothesis (H3) is derived from this.

Hypothesis 3(H3): Risk propensity is positively correlated to entrepreneurial self-efficacy. The higher one's risk propensity is, the higher his entrepreneurial self-efficacy is. The lower one's risk propensity is, the lower his entrepreneurial self-efficacy is.

As can be concluded that risk propensity has impact on both entrepreneurial self-efficacy and intention, entrepreneurial self-efficacy has impact on entrepreneurial intention as well. The relationship among them is formulated as Figure 1.

Figure 1 clearly shows the relationship of three variables. A person with higher risk propensity is

more likely to choose risky entrepreneurial action as his own career development routine possibly because he is more confident with successful startup (Zhao, Hao. et al. 2005), that is, he is more confident with his capability to undertake entrepreneurial action. Given this point, entrepreneurial self-efficacy serves as a bridge of the correlation between risk propensity and entrepreneurial intention, which is the mediating variable. Considering the influence of risk propensity to entrepreneurial intention, the paper proposes that entrepreneurial self-efficacy doesn't completely mediate the relationship between them, but partially mediates it. Thus the following hypothesis (H4) is formulated.

Hypothesis 4(H4): The influence of risk propensity to entrepreneurial intention is partially mediated by entrepreneurial self-efficacy.

3 METHODOLOGY

3.1 Measurement of related variables

The paper adopts individual behavior "propensity" as the definition of risk propensity and uses Sitkin and Weingart's risk propensity measuring scale (Sitkin & Weingart 1995), 5 items. Zhao Hao et al.'s scales with 4 items (Zhao, Hao. et al. 2005) are respectively adapted to measure entrepreneurial self-efficacy and entrepreneurial intention. All scales are 10-Point Likert Scales.

After a two-week test-retest, the test shows that all items meet the requirement of internal consistency reliability and retest reliability. After pretest, some business managers and entrepreneurs were invited to check the scale and then adjust some phrasing for a better understanding. The formal scale with a certain face validity and content validity was formulated after revising many times.

3.2 The sample

The research samples consist of entrepreneurs, business managers, seniors and third-year grade students from Sichuan, Guangdong, Fujian, Anhui, Jiangsu, Shanghai and Hubei, etc. 400 questionnaires were issued while 352 of them were retrieved. 337 questionnaires are valid with the recovery of 84.25% except those were incompletely or incorrectly filled in. Men and women account for 50.45% and 49.55%, respectively. The age of 68.53% samples ranges from 25 to 44 years old, which is the same as the age distribution of Global Entrepreneurship Monitor.

3.3 Scale test

The statistical analysis on the data in this research was accomplished by means of SPSS13.0. Firstly,

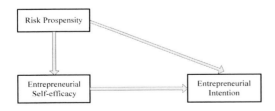

Figure 1. The relationship among risk propensity, entrepreneurial self-efficacy and entrepreneurial intention.

the retrieved questionnaires were used to test all scales. Reliability test was implemented to test the reliability of scales via Cronbach α. The values of Cronbach α of risk propensity scale, entrepreneurial self-efficacy scale and entrepreneurial intention scale are 0.82, 0.84 and 0.82, respectively. Then the validity of all variables was tested. The values of KMO of measure items in risk propensity scale, entrepreneurial self-efficacy scale and entrepreneurial intention scale are 0.84, 0.81 and 0.78, respectively.

4 ANALYSIS AND RESULTS

4.1 *The relationship among risk propensity, entrepreneurial self-efficacy and entrepreneurial intention*

Regression analysis was adopted to separately test the hypotheses on impact of risk propensity and entrepreneurial self-efficacy on entrepreneurial intention, the dependent variable. Table 1 shows the test results of regression analysis to all hypotheses.

Hereinto, the result of regression analysis shows that there is a positive correlation between risk propensity and entrepreneurial intention ($r = 0.494$, $P<0.001$), which indicates that risk propensity positively influences entrepreneurial intention. Therefore H1 is supported: The higher one's risk propensity is, the higher his entrepreneurial intention is; the lower one's individual risk propensity is, the lower his entrepreneurial intention is.

The result of regression analysis shows that there is a significantly positive correlation between entrepreneurial self-efficacy and entrepreneurial intention ($r = 0.425$, $P<0.001$), which supports the H2: Entrepreneurial self-efficacy is positively correlated to entrepreneurial intention. The higher one's perceived entrepreneurial self-efficacy is, the higher his entrepreneurial intention is. The lower

one's perceived entrepreneurial self-efficacy is, the lower his entrepreneurial intention is.

The result of regression analysis shows that there is a positive correlation between risk propensity and entrepreneurial self-efficacy ($r = 0.353$, $P<0.001$), which supports the H3: Risk propensity is positively correlated to entrepreneurial self-efficacy. The higher one's risk propensity is, the higher his entrepreneurial self-efficacy is. The lower one's risk propensity is, the lower his entrepreneurial self-efficacy is.

4.2 *Mediating effect of entrepreneurial self-efficacy*

According to the test method of mediating effect, the relationship between risk propensity (an independent variable) and entrepreneurial intention (a dependent variable) must be tested firstly. As shown in Table 1, the unstandardized regression coefficient (B_1, $P<0.001$) of H1 is 0.494, and its coefficient of determination (R_1^2) is 0.240. Then the relationship between risk propensity (an independent variable) and entrepreneurial self-efficacy (a dependent variable) needs to be tested. As shown in Table 1, the unstandardized regression coefficient (B_2, $P<0.001$) of risk propensity to entrepreneurial intention is 0.353, and its coefficient of determination (R_2^2) is 0.108. The relationship of risk propensity and entrepreneurial self-efficacy (two independent variables) and entrepreneurial intention (a dependent variable) is to be tested finally. The unstandardized regression coefficient (B^3, $P<0.001$) of risk propensity to entrepreneurial intention is 0.385; the unstandardized regression coefficient (B_4, $P<0.001$) of entrepreneurial self-efficacy to entrepreneurial intention is 0.308, and its coefficient of determination (R_3^2) is 0.337. As shown in the data analysis, all corresponding regression coefficients are significant; when entrepreneurial self-efficacy was introduced into the

Table 1. The results of regression analysis.

	H1 Independent variable: Entrepreneurial intention		H2 Independent variable: Entrepreneurial intention		H3 Independent variable: Entrepreneurial self-efficacy		Independent variable: Entrepreneurial intention	
	r	t value	r	t value	r	t value	r	t value
Risk propensity	0.494***	10.295			0.353***	6.362	0.385***	8.104
Entrepreneurial self-efficacy			0.425***	9.334			0.308***	6.97
F value	105.994		87.132		40.478		84.812	
R^2	0.240		0.206		0.108		0.337	
Adjusted R^2	0.238		0.204		0.105		0.333	

Remark: *** $P<0.001$; r = Correlation Coefficient.

last regression equation on the regression of coefficient of risk propensity to entrepreneurial intention decreases, which indicates that entrepreneurial self-efficacy mediates the influence of risk propensity to entrepreneurial intention. Moreover, the regression coefficient of risk propensity to entrepreneurial intention is still significant implicating that entrepreneurial self-efficacy only partially mediates the influence of risk propensity.

The change of regression coefficient intuitively illustrates the mediating effect. Besides that, Sobel Test may be adopted to accurately confirm the mediating effect. The equation of Sobel Test is:

$$Z = ab \div \sqrt{b^2 S_a^2 + a^2 S_b^2}$$

where a = the unstandardized partial regression coefficient of independent variable to mediator variable; b = the unstandardized partial regression coefficient of mediator variable to dependent variable in the regression equation including independent, mediator, and dependent variables; and $S_{a(b)}$ = the standard error corresponding to $a(b)$, respectively.

After computing, $Z = 4.683$ ($P<0.001$), which implicates that entrepreneurial self-efficacy has significant mediating effect between risk propensity and entrepreneurial intention. The product of a (-0.353) and b (0.308) is 0.109 that is mediating effect and the regression coefficient B_1 (0.494) of risk propensity (independent variable) to entrepreneurial intention (dependent variable) is the total effect, which implicates that entrepreneurial self-efficacy partially mediates the influence of risk propensity (independent variable) on entrepreneurial intention (dependent variable).

5 CONCLUSIONS AND DISCUSSIONS

This study presents a much clearer profile of the relationship of risk propensity, entrepreneurial self-efficacy and entrepreneurial intention. From the viewpoint of the relationship of two variables, risk propensity has positively impact on entrepreneurial self-efficacy ($r = 0.353$, $P<0.001$) and entrepreneurial intention ($r = 0.494$, $P<0.001$), respectively. In addition, entrepreneurial self-efficacy may be used to significantly predict the formulation of entrepreneurial intention ($r = 0.425$, $P<0.001$). From the viewpoint of relationship of three variables, the mediator effect of entrepreneurial self-efficacy was confirmed and the influence of risk propensity on entrepreneurial intention is partially transmitted by the media of entrepreneurial self-efficacy. Specifically, an individual with higher risk propensity is more likely to start a new adventure, which

helps him recognize a start-up as an opportunity rather than a threat and then promotes the formulation of his entrepreneurial intention. Entrepreneurial self-efficacy, as a bridge, partially transmits the influence of risk propensity on entrepreneurial intention.

This paper only investigates partial influences of entrepreneurial intention. The related variables will impact entrepreneurial intention differently if their quantity or quality is changed. In the future research on entrepreneurial intention, the relationship among these three factors maybe changed if other variables are introduced into it, which is the future research focus of the author.

REFERENCES

[1] Ajzen, I. 1987. Attitudes, Traits, and Actions: Dispositional Prediction of Behavior in Social Psychology. *Advances in Experimental Social Psychology* 20(1):1–63.
[2] Ajzen, I. 1991. Theory of Planned Behavior. *Organizational Behavior and Human Decision Processes* 50(2):179–211.
[3] Bagozzi, R. et al. 1989. An Investigation into the Role of Intentions as Mediators of the Attitude-behavior Relationship. *Journal of Economic Psychology* 10(1): 35–62.
[4] Bandura, A. 1977. Self-efficacy. Toward a Unifying Theory of Behavioral Change. *Psychological Review* 84(2): 191–215.
[5] Bandura, A. 1986. *Social Foundations of Thought and Action: A social-cognitive View.* Englewood Cliffs: Prentice-Hall.
[6] Barbosa, S.D. et al. 2007. The Role of Cognitive Style and Risk propensity on Entrepreneurial Self-efficacy and Entrepreneurial Intentions. *Journal of Leadership and Organizational Studies* 13(4): 86–104.
[7] Baron, R.M. & Kenny, D.A. 1986. The Moderator-mediator Variable Distinction in Social Psychological Research: Conceptual, Strategic, and Statistical Considerations. *Journal of Personality and Social Psychology* 51(6): 1173–1182.
[8] Bird, B. 1988. Implementing Entrepreneurial Ideas: The Case for Intentions. *Academy of Management Review* 13(3): 442–454.
[9] Boyd, N.G. & Vozikis, G.S. 1994. The Influence of Self-efficacy on the Development of Entrepreneurial Intentions and Actions. *Entrepreneurship: Theory and Practice* 18(4): 63–77.
[10] Forbes, D.P. 2005. The Effects of Strategic Decision Making on Entrepreneurial Self-efficacy. *Entrepreneurship Theory and Practice* 29(5): 599–626.
[11] Forlani, D. & Mullins, J.W. 2000. Perceived Risks and Choices in Entrepreneurs' New Venture Decisions. *Journal of Business Venturing* 15(4): 305–322.
[12] Gartner, W.B. 1989. Some Suggestions for Research on Entrepreneurial Traits and Characteristics. *Entrepreneurship: Theory and Practice* 14(1): 27–37.

[13] Gist, M.E. & Mitchell, T.R. 1992. Self-efficacy: A Theoretical Analysis of its Determinants and Malleability. *Academy of Management Review* 17(1): 183–211.

[14] Hmieleski, K.M. & Corbett, A.C. 2008. The Contrasting Interaction Effects of Improvisational Behavior with Entrepreneurial Self-efficacy on New Venture Performance and Entrepreneur Work Satisfaction. *Journal of Business Venturing* 23(4): 482–496.

[15] Katz, J. & Gartner, W.B. 1988. Properties of Emerging Organizations. *Academy of Management Review* 13(3): 429–441.

[16] Krueger, N. 1993. The Impact of Prior Entrepreneurial Exposure on Perceptions of New Venture Feasibility and Desirability. *Entrepreneurship Theory and Practice* 18(1): 5–21.

[17] Krueger, N.F. et al. 2000. Competing Models of Entrepreneurial Intentions. *Journal of Business Venturing* 15(5–6): 411–432.

[18] MacCrimmon, K.R. & Wehrung, D.A. 1990. Characteristics of Risk-taking Executives. *Management Science* 36(4): 422–435.

[19] MacMillan, I. & Katz, J. 1992. Idiosyncratic Milieus of Entrepreneurship Research: The Need for Comprehensive Theories. *Journal of Business Venturing* 7(1): 1–8.

[20] Shapero A. 1982. The Encyclopedia of Entrepreneurship. In C. Kent, D. Sexton and K. Vesper, eds. *Social Dimensions of Entrepreneurship*. Englewood Cliffs: Prentice-Hall.

[21] Sitkin, S.B. & Pablo, A.L. 1992. Reconceptualizing the Determinants of Risk Behavior. *Academy of Management Review* 17(1): 9–38.

[22] Sitkin, S.B. & Weingart, L.R. 1995. Determinants of Risk Decision-making Behavior: A Test of the Mediating Role of Risk Perception and Propensity. *Academy of Management Journal* 38(6): 1573–1592.

[23] Zhao, Hao. et al. 2005. The Mediating Role of Self-Efficacy in the Development of Entrepreneurial Intentions. *Journal of Applied Psychology* 90(6): 1265–1272.

Education Management and Management Science – Zheng (Ed.)
© 2015 Taylor & Francis Group, London, ISBN 978-1-138-02663-6

A research on approaches and methods of cultivation of the professional core competence for higher vocational colleges

Lihong Yang
Mechanical and Electrical Engineering Department, General Party Branch,
Hebei Chemical and Pharmaceutical College, Hebei, China

Qingfeng Chai, Qiang Zeng & Zhaohua Yu
Hebei Chemical and Pharmaceutical College, Hebei, China

ABSTRACT: In the social and economic transformation period, colleges should have professional core competencies, such as, solid professional knowledge, interpersonal communication, information processing, business innovation, application of digital and so on. This paper analyzes the connotation of core competence and the significance of occupation training core competence for vocational college students. Through the creation of occupational core ability training courses, reforming teaching methods, the counselor precept, strengthening the construction of the community, we can cultivate the core ability of occupation for higher vocational students who can adapt the development of society and economy.

Keywords: higher vocational colleges; approaches and methods of cultivation; core competence

1 INTRODUCTION

Employment is vital to people's livelihood, and employment of college students is currently an urgent social problem. If the university wants to cultivate graduates who adapt to society and industry, the cultivation of core occupational competence is valuable and vital breach.

2 THE CONNOTATION OF THE CORE OCCUPATIONAL COMPETENCE

"Occupational core competence" is put forward and used in foreign countries. It has different names in different countries; its significance is the ability beyond of professional skills. In our country, the Ministry of labor and social security first proposed about the core competence system development occupation ideas in 1998, and in the Ministry of labor and social security "as the important project of national skills development strategy" in accordance with occupation classification will be divided into specific ability of occupational ability, industry universal ability and core competence. We divide the occupation of core competence into communication with the people, digital applications, information processing, cooperating, solving problems, self-learning, innovation, and foreign language application, which are known as "the eight core competencies"[1]. Occupation specific ability is each kind of occupation itself, and applies only to a specific occupation jobs[2]. The ability and quality we need when we engaged in a certain industry called general ability, the former two are professional ability. The core competence is the ability to engage in any occupation and industry, it is an universal ability that has applicability, interdisciplinary and mobility, so it is a kind of general ability, and suitable for any occupation and industry, sustainable development can promote the individual, is the goal of the current quality education, is able to succeed in a job search, business development and the necessary ability. With the workplace changing, the professional man can adapt to different jobs, and quickly work into the role and achieve, the most important factor is the occupational core ability.

---

[1] The Ministry of labor and social security occupation skill identification center group. Occupation of social competence training handbook. Beijing: People's publishing house, 2008, 8.
[2] Zhang Xiling. On the cultivation of core competence management in higher vocational colleges students occupation. Journal of Henan Mechanical and Electrical Engineering College, 2011, 9.

3 IMPORTANT OCCUPATIONAL CORE COMPETENCE OF STUDENTS IN HIGHER VOCATIONAL COLLEGES

3.1 *Occupational training of core competence is the needs of the society for talents*

At present, it is the age of knowledge economy and social transformation period, the demand of social and enterprise on the talent is no longer limited to strong professional knowledge and skills, more emphasis on talent group team spirit, innovative consciousness and ability, information collection and processing power, and communication and coordination ability, analysis and solve the question ability, foreign language level and so on, which are our core competencies that occupation. With the emergence of new occupation and the increasingly fierce competition, the work flow of people is also accelerating, weather we can adapt to the requirements of different positions, the key lies in the core competence of personal occupation. The high-grade high schools should follow the development of society and educational situation, and pay more attention on the occupational core ability of students, then cultivate the graduates who adapt to society.

3.2 *Occupational training of core competence is the needs of sustainable development*

The core competence is the effective occupation ability, basic ability for everyone's success, and it is not only the key ability of competition in the workplace, but also the golden key to open the door to success. Students can participate in the teaching reform in the school by accepting school training, and actively participate in school curriculum, students will participate in community to strengthen the cultivation of core competence of their own occupation, such as the strong professional knowledge and skills, group team spirit, innovative consciousness and ability, information collection and processing power, and communication and coordination ability, then the students will realize the life value and sustainable development.

3.3 *Occupational training of core competence is the internal requirement and inevitable trend of the development of higher vocational education*

"Employment oriented, service for the purpose, ability standard" is as the school mission of higher vocational education. The Ministry of education "about improving the quality of teaching in the higher occupation education of a number of opinions" (high [2006] No. 16) (hereinafter referred to as the "16 document") on "strengthening quality education, strengthen the occupation morals, and points out that the cultivation goal" this problem clearly: "be aimed at the character of students in higher occupation colleges, the cultivation of students' social adaptability, educate students to establish the concept of lifelong learning, improve learning ability, learn to communication and teamwork, and improve the students' practical ability, creative ability, employment ability, train socialist builders and successors of all-round development." The No. 16 document of the "capacity" is the core of occupation ability mentioned above, this is the social position of individual comprehensive quality requirements[3]. Thus, occupation training core competence of vocational college students is the internal requirement and inevitable trend of the development of Higher Vocational education. Higher vocational education should not only cultivate high qualified practical talents for the society, but also make every student into a man, do things, cooperation, study of human occupation.

4 APPROACHES AND METHODS OF CULTIVATING OCCUPATIONAL CORE COMPETENCE IN HIGHER VOCATIONAL COLLEGES

4.1 *Set up occupation core competence curriculum*

At present, the national core competence have occupation specialized training, qualified success will obtain the corresponding occupation core ability qualification certificate, after that they can carry out the relevant training to the students in higher vocational colleges, and some have already begun to implement. The core competence of occupation school can send some excellent teachers and counselors to participate in the national formal training, to learn the essence, obtain the corresponding occupation qualification certificate to return to school, build the core competence of the occupation training team, set up the core competence of the students occupation course, which carries on personal cultivation combining the reality of students, through the different grade points and levels.

4.2 *Reform the classroom teaching*

We should pay more efforts to practice teaching in teaching. For example, first we can use situational

---

[3]Liu Hongxing, Li Weiguang. Analysis on present situation and Countermeasures of occupation core ability of vocational college students. Journal of Huanggang Polytechnic College, 2010, 8.

teaching approach, which could let students listen and operate together for integration of teaching. Second, we can use task driving method which divides students into different groups for task decomposition in order to cultivate the team spirit, communication, cooperation, analysis and so on. Third, the school should increase the intensity of university-enterprise cooperation, expand the construction of enterprise workstation in school which is able to help students access into production environment for cultivating them to solve some actual problems, so it could lay a foundation for employment.

4.3 The precept and example of instructors

College counselors are the backbone of Ideological and political work for students, full-time engaged in ideological education and behavioral management, is an important part of teachers. They can also be used as an important team of college students occupation of core competence. In one hand, counselors can through the theme class meeting, theme activities, invite the graduate exchange, organizing various class activities, to carry out targeted occupation core ability education and guidance. On the other hand, instructors should often go to the dormitory, talk with students, friends and students, so it is easy to pull closer between instructors and students' psychological distance, students are more willing to accept the education and guidance counselors. Counselors can in the usual study, work and life give students seepage type education, transfer the relevant information needed in the influence character by environment, competition in the workplace of the key occupation core ability training to students, it also helps students occupation of core competence.

4.4 Strengthen the construction of the community

School is equipped with professional associations and interest community, the students' quality development is implemented based on the credits quantity, then carry out the personal cultivation, students are encouraged to undertake choosing according to their ability and interest, let the student to develop their advantage in the community, and enhance self-confidence.

4.4.1 Strengthen the construction of teachers and funds guarantee

Higher vocational colleges should attach great importance to improve the innovation ability in the student community, remit special funds to ensure the stability of the necessary funds investment. At the same time, colleges should equip the tutor team members who have the certain professional background, who guide carefully college students' associations into the healthy development.

4.4.2 Strengthen the construction of community connotation

The college should strengthen the construction of community connotation, elaborately design different types, different practical activities of the project. According to the principle that is rich, closed to the reality, lively, healthy and tasteful. These approaches are able to improve the quality of the college activities, and create the high-quality college activities, which can let the college activities quality to be improved, and attract the attention of the students, and can provide students with the stage to show themselves, let the students communication skills and problem solving ability get promoted.

4.4.3 Establish and improve the characteristics of community management system

In the construction of community management, we can plan the community management system according to the feature of vocational education, combining with "the factory in the school, the school in the factory, the cooperation between them". We can launch the student task allocation by company mode, which can improve their foreign language application ability, information management and mathematical operation.

4.4.4 Improve the system of education and training assessment

The higher vocational colleges should pay more attention to the education training system, establish and improve the related training. At the same time, the school is obliged to pay attention to the appraisal for members, motivate students to strive for a higher goal, and carry out the "excellent member and excellent community" comparison, stimulate students' initiative and enthusiasm, and further promote students improve the self-realization.

5 STATEMENT

This paper is the research achievement of the Planning Project (134076177D) of 2013 Hebei Science and Technology, and the project's name is the Research on Approaches and Practices of Cultivation of the Professional Core Competence for Vocational College Students.

AUTHOR BRIEF INTRODUCTION

Yang Lihong, Mechanical and Electrical Engineering Department Deputy Secretary of general

Party branch, Vice Professor, Hebei Chemical & Pharmaceutical College.

Research direction: the ideological and political education and the employment of College Students.

Chai Qingfeng, ZengQiang, Vice-president, Professor, Hebei Chemical & Pharmaceutical College.

Yu Zhaohua, instructor, Hebei Chemical & Pharmaceutical College.

REFERENCES

[1] Xie Zhiping. China's occupation of core competence development, Guangzhou occupation education forum, 2012, 10.

[2] Liu Hongxing, Li Weiguang. Analysis and Countermeasures of the occupation, the core competence of higher vocational students, Journal of Huang gang Polytechnic College, 2010, 8.

[3] Zhang Yu. The occupation of core competence of vocational training of empirical research, model, vocational education forum, 2012, 8.

[4] Li Ying, Ceng Xianwen, Zhang Xichun. Higher vocational education is embedded in the core competence of the occupation of work value exploration, modern education science, 2011, 6.

[5] Feng Peijie, Zhang Peng. Research teaching mode to cultivate the core competence of Higher Vocational Students Occupation, New West, 2010, 12.

[6] Deng Feng, Wu Yingyan. Higher vocational students' occupation core competency training system construction and practice, education and occupation, 2012, 9.

[7] Huang Wen, Li Shuzhen, Wu Hong. College Students' occupation of core competence research, Journal of higher education finance, 2012, 9.

[8] Qiu dongxiao. Occupation the connotation of core competence analysis and in the higher vocational education training, Guangzhou province, Journal of Career Technical College, 2011, 4.

Education Management and Management Science – Zheng (Ed.)
© *2015 Taylor & Francis Group, London, ISBN 978-1-138-02663-6*

Can investors' attention affect earnings management? Evidences from China

Hao-jing Guo & Qian-wei Ying
Business School of Sichuan University, Chengdu, Sichuan, China

ABSTRACT: This paper empirically tests the relationship between earnings management and investors' attention. And it also investigates how the firms' type of ownership affects the relationship between earnings management and investors' attention. Using the data of Baidu searching frequency of the nonfinancial Chinese listed firms from 2007 to 2011, we construct a measure of investor attention in this paper and find that firms with higher investors' attention are exposed to higher level of earnings management. Compared with state-owned enterprises, investors' attention has a larger effect on the non state-owned enterprises. This paper enriches our understanding of the corporate earnings management behavior and the role of investor attention in corporate governance.

Keywords: earnings management; investors' attention; ownership

1 INTRODUCTION

To improve the quality of accounting information is one of the important goals of corporate governance. However, earnings management leads to poor quality of accounting information (Cohen and Zarowin 2010). Earnings management means that the operators alter financial reports using accounting means or arranging transactions to mislead stakeholders' understanding of the company's performance. Literature shows that corporate governance has an effect on earnings management. Corporate governance mechanism can be divided into external and internal mechanisms. Internal corporate governance mechanism includes the establishment of the board, the appointment of independent directors, the arrangement of appropriate ownership structure and so on (Beasley 1996). External governance mechanism includes the legal system, media, analysts, investors, etc. Most existing researches on the topic of earnings management focus on the effect of internal corporate governance mechanisms, such as the audit quality (Becker et al. 1998), the board of director characteristics (Klein 2002), the managers' equity incentives (Cheng and Warfield 2005), and the CEO turnover (Hazarika et al. 2012). A few studies have also discussed the effect of media and analyst coverage (Yu 2008; Qi et al. 2014). However, the external effect of investor attention on earnings management has been ignored by the literature. As a matter of fact, investors' attention plays an important role in the performance of a firm's stock

price which not only determines its financing cost, but also serves as an important indicator of the performance of its executives. Therefore, managers' incentives to manage earnings may be closely related to investors' attention.

This paper uses the Baidu searching volume index as a proxy for investors' attention and find that investors' attention will cause higher capital market pressure on the managers, which motivate managers to manage earnings. Further results show that investors' attention has less impact on state-owned enterprises than on private-owned ones.

The contribution of this paper mainly lies in two aspects. First, it enriches the existing research on the determinants of earnings management by investigating the role of investor attention, and help us better understand how external pressure affect the executives' motivations of earnings management. Second, this paper find a different effect of investor attention on state owned enterprises and non state owned enterprises, which implicates inherent differences on the motivations of earnings management between state owned enterprises and non state owned ones.

2 THEORY AND HYPOTHESES

The separation of ownership and control leads to a conflict of interests between managers and shareholders. Managers have the motivation to conduct earnings management for his/her interests which may be inconsistent with the interests of the shareholders and investors. One of the most popular ways of Chinese investors to collect a firm's information is to use Baidu search engine which takes up more than 80% of the search engine market share. Therefore, the Baidu searching volume index can reflect the degree of investors' attention on each stock in Chinese stock market. Investors will make investment decisions based on the financial statement information of a firm. Their attention on the positive or negative information of a firm may directly affect the extent of upward or downward fluctuations of the firm's stock price, which causes market pressure on the managers. One of the main motivations of managers to manage earnings is to receive a favorable feedback from the capital market. Faced with high pressure from the capital markets, managers will take strategies to manage the earnings (Healy and Wahlen 1999). The increasing degree of investors' attention will bring managers higher market pressure, and thus stronger motivation of earnings management. We can conclude the above analysis with the following hypothesis:

H1: High degree of investors' attention will lead to greater extent of earnings management by creating more market pressure on managers.

Earnings management motivations are different between state-owned enterprises and private-owned enterprises (Bo and Wu, 2009). The managers of the private-owned companies care more on the firms' stock market performance since their salaries and bonuses are usually closely related to it. On the other hand, salaries and bonuses of the managers in the state-owned enterprises are less correlated with the stock market performance of the firm, i.e., their income is relatively stable regardless of the change of the firm's stock price. Therefore, they care less on the firm's stock market performance. Instead, the managers of the state owned enterprise may pay more attention to building up a good relationship with the superiors and the government so as to get a promotion more easily (Bo and Wu, 2009). Consequently, the managers of state owned enterprises have a weaker incentive to do earnings management, and the investor attention also has little effect on their behavior of earnings management. In addition, when faced with high investor attention on the negative information of the state-owned enterprises, the government may use political power to support state-owned enterprises to improve its performance t. Therefore, pressure from capital market on state-owned enterprises is relatively small. In summary, we can conclude that investors' attention in market has less impact on state-owned enterprises' earnings management.

H2: Investors' attention imposes less capital market pressure on state-owned enterprises than on private-owned enterprises.

3 DATA AND MODEL

3.1 Variable definitions

3.1.1 Earnings management

Following Dechow et al. (1995) and Aboody et al. (2005), this paper adopts the modified Jone's model. Total Accruals (TA) are calculated by equation (1) below, in which ΔCA stands for the changes in current assets, ΔCL stands for the change in current liabilities, $\Delta CASH$ stands for the change in cash and cash equivalents, $\Delta STDEBT$ stands for the change of short term liabilities maturing within a year, and DEPN stands for depreciation and amortization.

$$TA_{jt} = \left(\Delta CA_{jt} - \Delta CL_{jt} - \Delta CASH_{jt} + \Delta STDEBT_{jt} - DEPN_{jt}\right) \quad (1)$$

Then we run the following cross-sectional OLS regression shown in model (2) within each industry classified by the Chinese Security Regulatory Commission. $Asset_{it-1}$ represents the book value of total assets for firm i at year $t-1$. ΔREV_{it} stands for the change in main operational revenues for firm i at year t, and PPE stands for the book value of fixed assets for firm i at year t.

$$\frac{TA_{it}}{Asset_{it-1}} = K_{1t}\frac{1}{Asset_{it-1}} + K_{2t}\frac{\Delta REV_{it}}{Asset_{it-1}} + K_{3t}\frac{PPE_{it}}{Asset_{it-1}} + \varepsilon_{it} \quad (2)$$

Using the estimates of K_1, K_2, and K_3, from the regression above, we can calculate the Non-discretionary Accruals (NA) with the formula (3), where ΔAR stands for the change in accounts receivable.

$$NA_{jt} = K_{1t}\frac{1}{Asset_{jt-1}} + K_{2t}\frac{\Delta REV_{jt} - \Delta AR_{jt}}{Asset_{jt-1}} + K_{3t}\frac{PPE_{jt}}{Asset_{jt-1}} \quad (3)$$

Finally, the Discretionary Accruals (DA) can be calculated by the difference between Total Accruals (TA) and Non-discretionary Accruals (NA), shown as

$$DA_{jt} = \frac{TA_{j,t}}{Asset_{j,t}} - NA_{j,t} \quad (4)$$

3.1.2 Investors attention

Baidu searching volume index is used to indicate the degree of network exposure and users' attention of any key words during a defined period. Baidu index can reflect the "user attention" of different key words in a period of time. Investors will use the companies' name or stock code to search for it if they are interested in the company. This paper uses the average of the Baidu searching volume index on both of the two keywords, and then take its logarithm as the measure of investor attention, denoted by Matt.

3.1.3 Control variables

According to existing literatures on the determinants of corporate earnings management, we choose a set of control variables including Company Size (*Size*), the return on assets (*Roa*), financial leverage (*Lev*), sales growth (*Growth*), book to market ratio (*Bm*), auditor's reputation dummy variable showing whether the auditor is one of big four or not (*Big4*), the share of institutional ownership, and a state ownership dummy variable showing whether the ultimate controller of the firm is state owned or not (*dum_gy*).

3.2 Data sources and sample selection

The sample of this paper includes all the non-financial A share Chinese listed firms in Shanghai and Shenzhen exchanges from 2007 to 2011. Except for the data of Baidu searching volume index manually collected from the Baidu's website, the data of all the other variables are collected from CSMAR data base. To exclude the effect of outliers, all the variables are winsorised at 1% level. The summary statistics is presented in Table 1.

3.3 Empircal model

To test the hypotheses proposed in part 2, this article employs the following benchmark regression model

$$DA_{it} = \alpha_0 + \alpha_1 matt_{it-1} + \alpha_2 Size_{it} + \alpha_3 Lev_{it}$$
$$+ \alpha_4 ROA_{it} + \alpha_5 Growth_{it} + \alpha_6 MB_{it}$$
$$+ \alpha_7 Big4_{it} + \alpha_8 Inst_{it} + \alpha_9 dum\_gy_{it}$$
$$+ \sum_{j=1}^{n} \beta_j IND_j + \sum_{k=1}^{m} \gamma_k year_k + \varepsilon_t \qquad (5)$$

where DA_{it} represents the discretionary accruals for firm i at year t, $matt_{it-1}$ represents the measure of investor attention for firm i at year t. Besides of controlling the effect of firm size (*Size*), financial leverage, Return On Assets (*ROA*), sales growth (*Growth*), market to book ratio (*MB*), auditor's reputation (*Big4*), share of institutional ownership (*Inst*) and state ownership (*dum_gy*), we also control the industry (*IND*) and year fixed effects. Based on the benchmark regression, we further divide the sample into two groups, state-owned enterprises and non state-owned ones to run subsample regressions.

Table 1. Summary statistics.

Variable	Mean	Median	Sd	Max	Min
Da	0.168	0.123	0.160	0.802	0.002
Matt	5.262	5.199	0.697	7.228	3.568
Bm	0.629	0.620	0.261	1.193	0.090
Growth	0.230	0.159	0.503	3.588	−0.687
Size	21.76	21.62	1.251	25.41	18.69
Lev	0.505	0.503	0.257	1.968	0.054
Roa	0.040	0.037	0.065	0.239	−0.289
Inst	0.181	0.122	0.180	0.728	0.000
Audit	0.063	0.000	0.243	1.000	0.000
Dum_gy	0.608	1.000	0.488	1.000	0.000

Table 2. Earnings management and investor attention.

	(1)	(2)	(3)
	All sample	Non state owned enterprises	State owned enterprises
	DA	DA	DA
Matt	0.020***	0.040***	0.004
	(0.005)	(0.008)	(0.005)
Bm	0.033***	0.111***	−0.025*
	(0.013)	(0.023)	(0.014)
Growth	0.061***	0.054***	0.070***
	(0.006)	(0.010)	(0.009)
Size	−0.020***	−0.048***	0.000
	(0.004)	(0.006)	(0.004)
Lev	0.058***	0.048**	0.063***
	(0.017)	(0.024)	(0.019)
Roa	−0.009	0.155*	−0.149**
	(0.051)	(0.080)	(0.060)
Inst	−0.013	−0.004	−0.010
	(0.012)	(0.025)	(0.012)
Audit	−0.007	−0.015	−0.015*
	(0.007)	(0.021)	(0.008)
Dum_gy	−0.030***		
	(0.005)		
constant	0.410***	0.990***	0.151**
	(0.062)	(0.114)	(0.070)
N	6222	2436	3786
r2	0.086	0.108	0.091

Note: The standard errors are shown in parentheses. ***p < 0.001. **p < 0.05. *p < 0.01.

4 EMPIRICAL RESULTS

The regression results are shown in Table 2.

It can be seen from the result that state-ownership is negatively related to the degree of earnings management, indicating that the earnings quality of state-owned enterprises is higher than private-owned enterprises'. The first column shows that investor attention has a positive and significant effect on the company's earnings management behavior, reflecting that investors can produce market pressure on the company. Hypothesis 1 is thus verified. The second column and the third column show the regression results on the group of non state-owned enterprises and state-owned enterprises respectively. It can be found that investor attention has a significant and positive effect on earnings management in the non state-owned enterprises group, while its effect on earnings management in the state-owned enterprises group is not significant. This result indicates that Investors' attention imposes less capital market pressure on the earnings management of state-owned enterprises than on the earnings management of private-owned enterprises. This evidence supports Hypothesis 2.

5 CONCLUSION

Using the data of Baidu searching volume index of the non-financial Chinese listed firms from 2007 to 2011, this paper constructs a measure of investor attention on each firm and investigates the relationship between investor attention and earnings management. According to the regression results, investor attention can produce a market pressure on the company. Higher the degree of investor attention is, the higher is the extent of earnings management. Further empirical evidence found in sub-sample regressions shows that the ownership type of the firm plays a key role in the relationship between Higher investor attention causes market pressure and leads to significantly higher earnings management in non state-owned enterprises, while it has no significant impact on earnings management in state-owned enterprises. The conclusions of this study indicate that outside investors cannot play an efficient supervisory role in corporate earnings management behavior but may instead reinforce it. This means that the market needs more powerful and efficient compelling force to restrict the corporate earnings management and convey true information of a firm. Therefore, it is necessary to improve regulatory agencies' monitoring efficiency and market supervision mechanism, and strengthen law enforcement to promote the formation of an efficient capital market.

ACKNOWLEDGEMENTS

The corresponding author, Qianwei Ying would like to acknowledge the financial support from the National Science Foundation of China (NSFC71373167; NSFC71003108), the social science research fund of Sichuan University (skqy201312), and the Fundamental Research Funds for the Central Universities (Excellent Young Scholars Research program in Sichuan University, 2013SCU04 A32).

REFERENCES

[1] Aboody, D., J. Hughes, and J. Liu. 2005. Earnings quality, insider trading, and cost of capital. Journal of Accounting Research 43(5): 651–673.
[2] Beasley, M.S. 1996. An empirical analysis of the relation between the board of director composition and financial statement fraud. Accounting Review 71(4): 443–465.
[3] Becker, C.L., M.L. DeFond, J. Jiambalvo, and K. Subramanyam. 1998. The effect of audit quality on earnings management. Contemporary Accounting Research 15 (1):1–24.
[4] Bo, X. and L. Wu. 2009. The Governance Effects of State Ownership and Institution Investors: the Aspect of Earnings Management, Economics Research Journal (6): 77–93.
[5] Cheng, Q., and T.D. Warfield. 2005. Equity incentives and earnings management. The Accounting Review 80 (2):441–476.
[6] Cohen, D.A., P. Zarowin. 2010. Accrual-based and real earnings management activities around seasoned equity offerings. Journal of Accounting and Economics 50(1): 2–19.
[7] Dechow, P.M., R.G. Sloan. 1995. Detecting Earnings Management, Accounting Review 70 (2):193–225
[8] Dechow, P.M., I.D. Dichev. 2002. The quality of accruals and earnings: The role of accrual estimation errors. The accounting review 77(s-1): 35–59.
[9] Hazarika, S., J.M. Karpoff, and R. Nahata. 2012. Internal corporate governance, CEO turnover, and earnings management. Journal of Financial Economics 104(1):44–69.
[10] Healy, P.M., J.M. Wahlen. 1999. A Review of the Earnings Management Literature and its Implications for Standard Setting. Accounting Horizons 13(4):365–383.
[11] Hermalin, B.E, M.S. Weisbach. 2001. Boards of directors as an endogenously determined institution: A survey of the economic literature. National Bureau of Economic Research Working paper.
[12] Klein, A. 2002. Audit committee, board of director characteristics, and earnings management. Journal of Accounting and Economics 33 (3):375–400.
[13] Li, T, L. Sun and L. Zou. 2009. State Ownership and Corporate Performance: A Quartile Regression Analysis of Chinese Listed Companies, China Economic Review 20:703–16, 2009.
[14] Qi, B., R. Yang, and G. Tian. 2014. Can media deter management from manipulating earnings? Evidence from China. Review of Quantitative Finance and Accounting 42 (3):571–597.
[15] Yu, F. 2008. Analyst coverage and earnings management. Journal of Financial Economics 88 (2):245–271.

Education Management and Management Science – Zheng (Ed.)
© *2015 Taylor & Francis Group, London, ISBN 978-1-138-02663-6*

Exploration of the effectiveness strategy in university physical education

Ke Shen
Sports Institute of Hunan University of Technology, Zhuzhou, Hunan, China

ABSTRACT: The effectiveness of university physical education is the basic quality protection of it. This article bases on the function and current situation of university teaching to discuss the basic characteristics of university physical education. It includes three aspects: the learner-centered, discipline system as the goal, harmonious teacher-student relationship as the foundation to explore the strategy of university physical education, which provides reference for the construction and teaching practice of university physical education.

Keywords: university; physical education; effectiveness; strategy

1 EFFECTIVENESS IN PHYSICAL EDUCATION

The definition of effective education is that the teacher invests as little time, energy and material forces as possible in organizing activities in order to achieve efficiency of teaching as high as possible and to meet the requirements of social and personal development in accordance to rules and regulations of teaching. It is a teaching concept combining efficiency, effectiveness and benefit. Physical education is no exception. It advocates effective teaching as well. With the implementation of new curriculum standards for physical education, the reform in this area has been carrying out in full swing. However, new problems arise in the actual teaching process. Many seemly good courses achieve unsatisfactory results. The effectiveness of physical education is embodied that students succeed in developing themselves physically and arouse a strong sense of experience and a deep interest in learning. Meanwhile, students receive physical education and feel the fun of sports in an vigorous, vivid, joyful and pleasant atmosphere. An effective physical education means that students' physical quality, sports skills, participation in sports, mental health and social adjustment are all being improved.

2 FACTORS RESTRICTING THE EFFECTIVENESS OF UNIVERSITY PHYSICAL EDUCATION

2.1 Teachers lack of reasonable teaching objectives

New course requires university physical education in sports participation, sports skills, physical health, mental health, social adaption and other areas to adapt to a multi-way goals. However, some teachers in order to reflect the diversity of the target, they make the teaching objectives seem big and empty, far-fetched and unfocused. Therefore, it is necessary to strengthen the teaching objectives awareness, establish the guiding ideology of health ahead, and focus on the development of university students' personality, to make the target layered, progressive, clarified, operable and scalable. This is the most important prerequisite for effective teaching.

2.2 Teachers lack of respect for university students

The new course advocates that students individuals should receive adequate development, which is a contradiction with the traditional forms of classroom organization. Some teachers are self-centered, and lack of consideration of individual students in the choice of teaching contents and methods. Students are often difficult to understand in the classroom or they are not interested in the contents, or the interesting contents does not appear in the classroom. As a result, the effectiveness of teaching will reduce without the active participation of students, which will affect the formation of lifelong interest in university sports development in the long term. Therefore, it needs to adopt the layered teaching method and give students options. According to the current status of university teaching, it is not operational for students to choose teaching contents freely. But it is a positive measure to enhance the effectiveness of teaching through choosing their own receptive learning methods by students.

2.3 Teachers lack of effective teaching knowledge and skills

University physical teachers have profound professional knowledge, but knowledge about learning theory and technology teaching, strategies and methods are very limited. Most professional soports teachers have adopted the teaching strategies and methods, followed their feelings as a student or traditional teaching, and do not reflect whether it can help students learning effectively and improve teaching quality. Therefore, the university physical teachers should have some related theory and technology, and pay attention to the research of teaching and improve the effective teaching skills, otherwise, they cannot receive the effective teaching as they wish.

3 THE STRATEGY OF IMPROVING THE EFFECTIVENESS OF UNIVERSITY PHYSICAL TEACHING

3.1 Learner-centered strategy

To learn is to master. Effectiveness of teaching is mainly reflected in the students' progress and development. Therefore, effective university physical teaching should treat students as the center of teaching. It requires teachers to consider students in teaching, arouse student's study enthusiasm to promote students to learn actively. Specific requirements are as follows:

1. Getting to know the actual situation of the students. Knowing the master degree, technical level and cognitive ability of them on a certain sport-skill can be benefit for designing plans meeting the needs of their activities, methods and progress.
2. The target must manifest the skills and abilities, and it also should pay attention to students' emotion, attitude and value views. Designing teaching plan should consider if it is suitable for both excellent students and poor-quality students. And this requires the teachers to grasp the teaching material well and set up an appropriate degree.
3. Students-centered when design skills teaching. Designing different ways of learning according to different learning content in the process of teaching design, letting every student has the opportunity to try and understand. Considering the individual differences of students comprehensively, leaving time for mutual discussion or practice of student in order to solve the skills actions that are not yet clear. Therefore, in the process of teaching design, the variety ways of learning can also let students to organize instead of teachers. Only the students'

intrinsic cognitive level and grasping degree are considered and their study situation are taken into account, can the actual teaching effect be reached.

3.2 Subject system-oriented

University physical education is a transmission and innovation process of physical knowledge, which aims at making students master the subject content and improve their technical competence. Thus the effective teaching of physical education curriculum should centre on the subject. The discipline-oriented teaching requires teachers should focus on teaching aims and pay attention to the subject contents with detailed explanation to enhance students' understanding, mastering and creative application of the subject. The specific requirements are as following:

(1) It requires physical exercise-oriented with proper basic knowledge learning of university physical education, especially the study of health consciousness. (2) There must be a reasonable exercise intensity and density of physical education. Appropriate exercise load is the demand of working out, mastering sports skills and promoting the psychological development of university students. In the design of learning, the time and the number of practice are expected to be made more specific. (3) Teachers try to enhance the teaching of technical skills. In the design of learning, the teacher should focus on motor skills and physical exercise and take sports participation, mental health and social adaptation into consideration. (4) The teacher should arrange the teaching according to the law of physical exercise strictly, using the least time to help the students master the physical knowledge and skills with scientific and effective method. Teaching should be carried out according to the law of education to enhance teaching effect. (5) Safety education should be enhanced and the sense of responsibility should be cultivated in physical education. The safety education is an important guarantee of physical education since it is a physical practice with sport equipments, which may be dangerous.

3.3 Harmonious teacher-student relationship as foundation

A harmonious teacher-student relationship is very essential to fulfill the teaching task successfully. Therefore, effective university physical education should base on harmonious teacher-student relationship, which requires gym teacher to love, care, trust and respect students through education. The intimate friendship between teachers and students would form a harmonious relationship, making students to get close to their teachers, trust and follow their guidance, enjoy their teaching and education. The specific requirements are as following:

1. Teachers should love and care about students, which is teachers' professional ethics. In university physical education, loving students means caring about their welfare and interests, teaching them with all our knowledge without reservation, providing with more opportunities and creating the best conditions for students' development and growth.

2. Teachers and students must build a harmonious mental environment. This is mainly controlled by the relationship between teachers and students. However, teachers' teaching attitude and style is the determinant, so teachers should actively construct a passionate learning atmosphere with a good teaching attitude. What's more, teachers should fully promote democracy and give students an eager "expectation", at the same time, they should help students to establish a friendly group to help with each other, offering them a strong emotional appeal.

3. Teachers should communicate with students timely and effectively. Education, teachers' praise and classmates' encouragement would have a direct positive stimulation on the students. Whereas, teachers' rudeness, autocracy and distrust, classmates' mock would result in the students' depression, or even appearing to a terrified and negative mental state. At the same time, teachers should often guide the students to discover other classmates' advantages, leading all the students to applause for someone who gets progress and arousing students' initiative and enthusiasm. Only a harmonious teacher-student relationship, loving and caring between classmates could activate the classroom learning atmosphere.

It is because of the specificity and complexity of university physical education, to give a good PE lesson and be a qualified gym teacher, one not only has to improve his or her comprehensive quality, but also be full of noble virtue, solid professional knowledge, strong abilities, rich and wide cultural qualities, and healthy physical and mental quality. This should be a direction for our university physical teachers to pursue progress. Therefore, in order to promote PE lesson forward the target of curriculum reform, enable teachers to grow together with students and realize the effective education of PE class, teachers have to summarize and explore new teaching methods, build confidence, change idea, and at the same time, implement teaching creatively in physical education.

REFERENCES

[1] Yuan Ling & Xie Chi. 2006. On the Problems in the Evaluation of Teachers' Teaching Effect in universitys and the Strategies. *University Education Science* (1):44–46.

[2] Yao Limin. 2000. Thinking on the relationship between Teaching and Research. *China Electric Power Education* (1):9–11.9.

[3] Gu Zheng. 2005. Research on Effectiveness of Class Teaching in University. *Culture and Teaching Materials* (32).

[4] Wu Jian. 2007. Effective Classroom Teaching and Healthy Physical Education under the new curriculum. *China School Physical Education* 7.

[5] Fu Jianming. 2005. Educational Principle and Technology. Guangzhou: Guangzhou Educational Publishing House.

[6] Zhang Jiajun & Wu Haiou. 2005. Discuss Valid P.E. Teaching and Its Tactics. *Sichuan Sports Science* (12): 108–110.

[7] Sunpeng & Wang Shudong. 2004. Research on the Validity and Strategy of Physical Teaching. *Sports Culture Guide* (1):54–55.

Education Management and Management Science – Zheng (Ed.)
© 2015 Taylor & Francis Group, London, ISBN 978-1-138-02663-6

Research on safety and appliance of mobile-phone QR code payment tool based on QFD

Cai Xia Chen, Chun Shi & Xian Fen Zhang
Hainan Normal University, Haikou, China

Qi Xiang Lan
Central University of Finance and Economics, Beijing, China

ABSTRACT: With the development of global communication technology, the number of mobile application keeps increasing; this includes applications with QR code payment. In this paper, a model of mobile-phone payment is built by using QFD product research and development technology to present a brief structure of mobile payment and provides guidance of product development and innovation.

Keywords: mobile-phone payment; QR code; QFD

1 BACKGROUND OF MOBILE-PHONE QR CODE PAYMENT

The 34th report (Statistics by Jun, 2014) provided by China Internet Network Information Center (CNNIC) shows that mobile-phone network user had already reached 527 million and outnumbered PC network user. The number of mobile-phone payment rose by 63.4% in half year (Data sources: Kuaipai Cloud). In Mar, 2014, the number of new mobile QR code registered in China is 25.74 million, and QR code scanning add up to 9.08 million times (Data sources: Imageco).

Mobile-phone QR code payment is coming into people's lives and also making a great influence. However, existing providers of QR code product lack a systematic approach when designing and operating QR code application, which result in inefficiency in design and security problem. In this paper, a model of mobile-phone payment is built by using QFD product research and development technology. The model would present a brief structure of mobile payment to show different concepts of several levels, and so as to provide guidance of product development and innovation.

2 INTRODUCTION OF QR CODE PAYMENT

2.1 *Definition of QR code*

A QR code consists of black modules (square dots) arranged in a square grid on a white background, which can be read by an imaging device.

The technology of mobile-phone QR code is the technology of encoding and decoding of QR code on mobile phone.

2.2 *Application of QR code in mobile payment*

There are 3 main ways of QR code application in China: Information Acquisition, Network Access, and Verification.

Information Acquisition: to scan the QR code by mobile-phone camera, the information would be given by decoding software. This saves time for users and would be quicker and easier. Example: Quick pay app.

Network Access: to scan the QR code, a URL link is given and user could click into the link to visit the website and download data. This is mostly used in advertisement on bus. Example: Weixin scanning.

Verification: to scan the QR code, the data is uploaded to the server, and the server would verify if the product or service is available. This is mostly used in e-ticketing and anti-counterfeit. Example: QR code coupon, etc.

3 PAYMENT TOOLS STRUCTURAL MODEL

3.1 *Definition*

Payment tool: a generic name of the technology, software and procedure that support the transfer of value.

A traditional payment procedure includes: Start a payment, Interchange of payment order and liquidation, Close the account. The last two steps are always finished by inner banking system.

Internet payment tool: Except banking system, a lot more ways, for example internet or WAP, are available for fund transfer.

Mobile-phone QR code payment tool: Adopt more new ways, for example wireless payment, than internet payment tools. It brings great convenience to people but meanwhile causes more safety problems. And its product also faces a homogeneity problem.

The model tries to make the resource utilization more efficient and provide several advices to prevent payment risk. This would be also available for other application on the model.

3.2 Classification of mobile-phone QR code payment tool

There are several kinds of payment tool classified by their different producer.

Produce by Third-Party Payment Platform: Alipay QR code receipts service;

Produce by Bank Payment system: Mobile banking service;

Produce by APPs: Quick Pay APP;

Produce by Mobile Service Provider: QR code ticket by China Mobile.

3.3 Structure of mobile-phone QR code payment tool

A standard mobile-phone QR code payment tool consists of 7 basic parts.

First, the person, the one who starts the transaction, the value transfer.

Second, the QR code creator, the machine or software that creates a QR code.

Third, the payment terminal, where the orders be sent out and receive.

Fourth, the way that orders transfer.

Fifth, the one who deal with these orders and finish the deal.

Sixth, the payment account, where the value sends out.

Seventh, the receipt account, where the value goes to.

These basic parts could be linked together, and each part of it would be influenced and limited

by the basic features of the tool. There are 3 basic features of the mobile-phone payment tool: Safety, Regularity and Amount. And there are also addictive features such as convenience, reliability, popularity, etc. These addictive features and the 3 basic features make a payment tool working and influencing it all the time.

4 RISK ANALYSIS OF MOBILE-PHONE QR CODE PAYMENT MODEL

4.1 Hardware risk analysis

Figure 2 shows that there is more than one QR code creator in the market, which results in:

Lack a standard code system. Original code system in China is translated from U.S. PDF417 and Japan QR. However, existing QR code do not share a same code system, which show a different code quality and may cause compatibility problems.

Make it easier for virus invasion. QR code reader is not included in mobile phone in China, so phone users have to download another one. This provides a chance for virus to spread through QR code and there are several real examples to prove it.

Terminal equipment is weak. The decoding of QR code relies on the ability of camera and the support of operating system. Today many mobile phones in China do not optimize for QR code decoding within 100 mm or further than 10 m, which causes a lot of mobile phones could not decode QR code successfully.

Private information leaks. A QR code without encryption may leads to information leakage.

To Resolve These Problem, the writer suggests the government department to set a QR code service platform and cooperate with QR code technology department, code provider, equipment

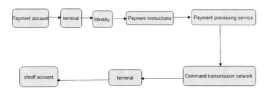

Figure 1. Describing the basic structure.

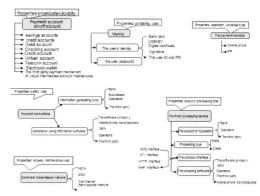

Figure 2. Describing the sub-structure and listing the application in the end of each branch.

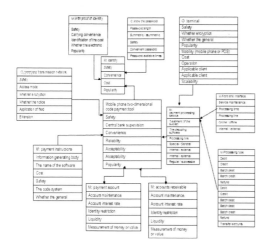

Figure 3. 'M' indicates a necessary structure, 'O' indicates an optional structure.

producer to standardization the code system and service of QR code.

4.2 Payment structure risk analysis

Figure 2 shows that there are more than one institution which could do orders dealing work, this causes a safety problem.

Traditional POS payment is based on credit card. Through the connection between China Unionpay Service System and NPC system in Central Bank, the liquidation among different banks could be done. All of the liquidation object should have a liquidation account in the accounting department PBC.

However, mobile-phone QR code payment, as a new wireless payment system, replaces POS payment by QR code decoding. The final fund transfer could be done without the bank account but within the virtual account created by Third-Party Payment Platform, and this account cannot be supervised by any bank. This leads to a big safety problem.

The writer suggests that: connect other mainstream third-party payment platform to payment platform in central bank as a supplement; and set a series of payment standard for central bank to easily supervise virtual account.

4.3 Legal risk analysis

From Figure 1, it could be revealed that there are several differences between the procedures of mobile-phone QR code payment and *the Rule of Credit Card Operation Management*.

Article 14 in *the Rule of Credit Card Operation Management* claim that department have

responsibility to manage the operation of capital gains of stores. But, the mobile-phone QR code payment adopts online payment rather than offline payment, which is hard to define.

The writer suggests that the PBC should set a new rule with more detail and a standard charge requirement for the whole country to fit the new condition of the market.

5 APPLICATION OF THE MODEL AND INNOVATION ANALYSIS

5.1 To guide the production of new QR code product

When there is a model of mobile-phone payment, there was no need to do new research when a new product is expected. An existed product line and a complete module inventory were available for product producing. This would increase efficiency and save resources.

5.2 To guide the design of payment tool

Chart 1 shows a QFD matrix. The user requirement feature is showed on the left; the product

Chart 1.

structure based on the model is showed on the top. The correlation degree is set according to 3-level method. A high correlation degree leads to 9, normal leads to 3, low leads to 1.

By analyzing this matrix, producer could optimize the design of the product. For example, more than one password is needed when users try to modify their pay order. But all this identifying methods are finished on the mobile phone, and this would cause lost if the mobile phone of user is controlled by other people.

To use QFD technology to improve the structure and parameter of the product, different user's need would be satisfied.

5.3 *To guide the combination of different way of payment*

Person, Terminal Software and Pay Order match different payment way. Start from the structure of the model, a combination of different payment methods would be achieved and reach a new development. For example, ATM QR code is a more efficiency way to withdraw cash.

5.4 *To guide the innovation research of payment product*

The innovation is not limited to the model but could also according to the feature of different product. For example, it is possible to replace ID card or credit card by QR code. Or, department may create an APP with the support of bank, service provider and third-party payment platform where users could make QR payment successfully no matter which platform he/her choose. A new financial product is also possible.

6 CONCLUSION AND EXPECTATION

The QR code is a mature technology, but when it combine with mobile phone it create a new space of payment. Explosive demand for QR code apps pushes the development of payment method and value-added service. The model would help to product innovation and the plan making of management, and achieve healthy improvement of mobile payment.

ACKNOWLEDGEMENT

Supported by the natural science foundation of Hainan Province Project number: 714276.

REFERENCES

[1] China Electronic Commerce Association. Beijing: China Standard Press, 2008.
[2] Linxi Zhu. *E-Payment and Internet Bank*. Tsinghua University Press. Beijing Jiaotong University Press. 2010.
[3] Qing Yang. *Electronic Finance*. Fudan University Press. Feb. 2009.
[4] Lei Wang. *Online Payment and Settlement*. Zhejiang University Press. Aug, 2007.
[5] European Committee for Banking Standards. Electronic Payment Initiator (ePI).
[6] Qiusheng Liu. *New Product Development*. Tsinghua University Press. 2001:127–161.
[7] Xinping Pan. *Structural Model of Payment Tool*. 2010, 1:62–64.

Education Management and Management Science – Zheng (Ed.)
© 2015 Taylor & Francis Group, London, ISBN 978-1-138-02663-6

Analysis on using performance and optimization of excellent resource-sharing course

Guihua Wang, Debo Ding & Tingxian Qian
College of Trade and Finance, Jiangsu Institute of Economic and Trade Technology, Nanjing, China

ABSTRACT: Excellent resource-sharing course represents the direction of course reform in high vocational education. It is the important orientation of the prospective course construction in vocational colleges. However, there are some problems severely limited the openness and sharing of course resources, such as low utilization rate and sharing difficulty, which reduce its important function in teaching and personnel training. This paper selects 200 teachers in 80 vocational colleges, investigates the openness and sharing situation of excellent resource-sharing course. Based on the analysis of the present status of cognitive channels, visiting frequency, visiting purpose, access mode, access content, using performance and access barriers of excellent resource-sharing course, some strategies have been put forward finally to optimize the construction and sharing of excellent resource-sharing course.

Keywords: excellent resource-sharing courses; construction; sharing; survey

1 INTRODUCTION

Since 2003, excellent course has been basically formed a national-provincial-school level platform. In 2012, the ministry of education indicated that the national excellent course should be transferred to excellent resource-sharing course, which brought more challenge to course sharing application. Openness and sharing of excellent course not only can improve college teaching method between the exchanges, shorten the gap between the students, but also can promote the new knowledge innovation and the quality of teaching. How to expand the sharing degree of excellent resource-sharing course with modern information technology and educational technology is an imperative problem remained to be solved.

2 CONSTRUCTION SITUATION OF EXCELLENT COURSE

Excellent resource-sharing course is an important part of teaching quality and teaching reform project in Colleges and Universities. It not only has extremely strong specific aim, but also has strategic significance. With the course construction, many scholars pay more attention to the course construction study. Shan F.R. (2003) pointed out that, according to the characteristics of higher vocational course reforming requirements and management, advanced teaching ideas and clear construction ideas should be established to grasp the construction of excellent course. Wang X. & Chen X.B. (2007) pointed out that excellent course web increased year by year, which would further improve the assessment standard. Chen D. (2010) suggested that higher vocational colleges should implement the school enterprise cooperation mode to construct and specialize courses in line with the requirements of the higher vocational education course based on working process model. With the course construction, many scholars studied the sharing problem of course resources. Wang Z.R et al. (2012) pointed out that there were many obstacles in sharing quality resources, such as resources are difficult to obtain and the channel obtain is not smooth and so on. How to fully understand the status of course sharing, put forward the effective optimization way and promote the sharing of educational resources, is the biggest challenge for course reform of higher vocational education.

3 CURRENT SITUATION SURVEY OF EXCELLENT RESOURCE-SHARING COURSE

To gasp the popularization and sharing status of excellent resource-sharing course, we selected 80 vocational colleges throughout the country, issued 200 pieces of questionnaire in this survey, and 160 pieces of questionnaire were recovered. From the analysis of survey data, the current course sharing is not optimistic.

3.1 Cognitive channel analysis

The cognitive channel survey shows that the course website accounted for 22.22% of the total sample, other people 28.57%, search engine 26.98%, meetings 14.29%, files 4.76%, media and other means each 1.59% (Fig. 1). Most teachers get information through internet search and they obtain excellent resource sharing course.

3.2 Visiting frequency analysis

The survey shows that, more than 80% of the teachers have visited the national excellent resource sharing program. The visiting frequency is different from person to person. As Figure 2 shows, 51.6% of the teachers visit it once a month, 14.89% once a week, 12.77% once a year, only 2.13% every day, and 19.15% never visit or use the resource sharing program. Therefore, most of the teachers visit it once a month. Only 17.02% of the teachers insist on visiting the excellent resource-sharing course every week. 19.15% of the teachers know nothing about excellent resource-sharing course. This explains the lack of popularity of quality resources sharing course.

3.3 Visiting purpose analysis

There are several different purposes that teachers visit excellent resource-sharing course. As Figure 3 shows

that 30.63% of the teachers visit it for lesson preparation, 25.23% for downloading the information from its website, 21.62% for improving their teaching level, 13.51% for the studying of the construction of their own courses, 9.01% for the purpose of developing new courses. It shows that most teachers' visiting is for the purpose of teaching and course construction. This is also the original construction purpose of our country's excellent resource-sharing course.

3.4 Access mode analysis

The survey shows that there are several different access ways that teachers visit the excellent resource-sharing course. As Figure 4 shows, 21.7% study the sharing courses as reference, 29.25% for the understanding of the relevant information and materials of the course, 7.55% for the experiment and practice to improve their ability, 1.89% for taking the online test and access the course forum in the internet. Most of the teachers only browse the courses sharing resources, only a few of the teachers really use them in the practice. It is urgent to let teachers and students use it and take parts in it in their everyday work.

3.5 Access content analysis

The survey of access content shows that, 34.51% of the teachers mainly access the data of the courses, such as the teaching plan and the courseware,

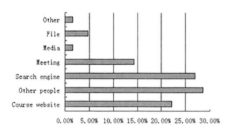

Figure 1. Cognitive channels of excellent resource-sharing course.

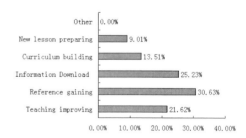

Figure 3. Visiting purpose of excellent resource-sharing course.

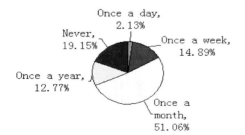

Figure 2. Visiting frequency of excellent resource-sharing course.

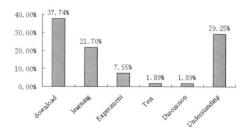

Figure 4. Access mode of excellent resource-sharing course.

23.01% access the teaching video, such as the video of classroom teaching, 22.12% acquire the exercises of the course, such as the examination questions and the simulation tests, 19.47% visit the sharing courses for the course introduction, such as the course syllabus, time distribution and course announcements as well. There are other 0.88% of the teachers who expect to get the other contents of the courses, such as the media material (Fig. 5).

3.6 Using performance analysis

The survey shows that the using performance of excellent resource-sharing course is not good enough. As Figure 6 shows, 65.91% of the teachers think it is not good enough, 15.91% are satisfied, 6.82% are very satisfied, 6.82% are dissatisfied, 4.55% are very dissatisfied. The satisfaction of the excellent resource-sharing course is only 22.37%. This illustrates that there is a big gap between the course construction and expected result.

3.7 Access barriers analysis

Teachers meet some problems during visiting of excellent resource-sharing course. As Figure 7 shows, the problems of the content not available

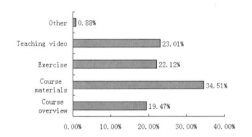

Figure 5. Access content of excellent resource-sharing course.

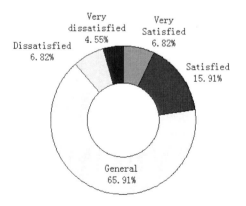

Figure 6. Access mode of excellent resource-sharing course.

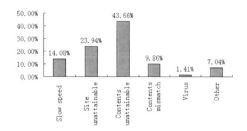

Figure 7. Access barriers of excellent resource-sharing course.

account for 43.66%, the site not available accounts for 23.94%, slow speed, long response and waiting time account for 14.08%, the mismatch of link refer and display accounts for 9.86%, the virus problems account for 1.41%, there are other problems such as fees, not free to use, account for 7.04%. Most of the teachers give up visiting excellent resource-sharing course because of their traditional teaching mode and particular stress on theoretical teaching. The whole teaching contents are lack of attractiveness.

4 THE OPTIMIZING APPROACHES OF EXCELLENT RESOURCE-SHARING COURSE

4.1 Construct the benefit mechanism of sustainable development

For the course builders, the immediate task was to get recognized, and the recognized project is a one-time task. Once the course is recognized, the task has been completed for the builder. The follow-up constructing and updating become a passive task, and some course builders even stop constructing, which leads content's poor usability. To change this situation, innovative benefit mechanism must be constructed, such as strengthening the development of paid resource and the implementation of market-oriented operation. Only with enough economic power, the builders will continue to build and update the course resource.

4.2 Build the teaching resources sharing platform of school level

The low visiting frequency is closely related to the single channel. Now the sharing platform of excellent resource sharing course is only limited to national platform. The lack of the school platform is the main obstacle unable to penetrate the teaching. The current builders of national excellent resource sharing courses are mostly professional and well-known teachers from the few forefront of the country school. However, the course users are teachers,

585

students and other learners from the Higher Vocational Colleges all over the country. Building the teaching resources sharing platform of school level is the way to play its due role of excellent resource sharing for excellent resource sharing course. So the popularization and application of excellent resource sharing course depend on the construction of school sharing platform, which can make the excellent resource sharing course really go deep into the classroom, and benefit teachers and students greatly.

4.3 Increase the training and enterprise practice content

As the national quality of teaching resources, the excellent resource-sharing course should lead the teaching reform trend both in the teaching mode and teaching content. Especially for the teaching content, the builder must increase the training and the practice of enterprises with fully considering the characteristics of specialization and occupation, which can realize the full integration of the teaching content and the enterprise practice. Enterprises' practice contents are helpful to realizing the course and the practice of enterprise, and also conducive to improve the practicability and attraction of course and change the status of single access and boring content. At the same time, rich practice teaching helps students familiar with occupation characteristics and operation of enterprises, and helpful to prepare the subsequent occupation career.

4.4 Optimize the course content and related resources with teaching service consciousness

At present, the main objective for teachers to get the resources sharing is to collect more resources and materials for teaching. For students, the purpose is to constantly improve them with the autonomous learning. Although the purpose of construction of resource sharing is not limited to the above, but the first purpose is to meet the needs of the majority of the teachers and students in teaching and learning with service consciousness. In one hand, builders must make the content refinement, such as using short and fine micro course, which can improve the interest of learning and the effectiveness of users; on the other hand, builders must choose matching resources and media material, such as the latest case base, industry development policy, test questions library, self-testing system and so on. The builder must provide more practical and pertinent related resources for teachers and students.

4.5 Strengthen the investment and software upgrade of network equipment

Although more than 80% of the teachers have visited excellent resource-sharing courses, most visiting frequency is low. The important reason is that the course access and download are not easy to get. It leads to Webpage failed to open, and teachers waste so much time waiting. Over time, teachers lose confidence to the excellent resource-sharing courses, at the same time the excellent resources are not fully utilized. So it urgently need to strengthen the investment and upgrading of network hardware equipment and software, such as the use of cloud management, which can make sharing the access program and download more convenient and smoother.

5 CONCLUSION

The construction of excellent resource-sharing course has made certain achievement, but the whole constructing and sharing situation is not optimistic. The popularization of excellent resource sharing course is not enough because of many reasons. The objective and the way of the teachers visiting the course are limited to download data, mostly for school use. The technical and connotation construction still exist some problems, which result in a less-than-ideal effect. If the excellent resource-sharing course wants to achieve the expected effect, some measures must be taken to solve the problem of sharing of excellent resource-sharing course fundamentally.

ACKNOWLEDGEMENTS

The research work was sponsored by Qing Lan Project of Jiangsu Province, supported by Higher education reform projects of Jiangsu Province under Grant No. 2013JSJG082 by the name of "Study on the construction and sharing of excellent resource-sharing course".

REFERENCES

[1] Shan F.R. 2003. On management courses construction in Higher Vocational Education. Journal of Liaoning Vocational College. (6): 62–64.
[2] Wang X., Chen X.B. 2007. Present evaluation situation and problems of Course construction in our country. Research of Educational Development. (10): 83–85.
[3] Chen D. 2010. Study on the construction and achievement of excellent courses in Higher Vocational. Vocational and Technical Education. (32): 9–11.
[4] Wang R.D., Li E., Zhao D.N. 2012 Application status, problems and countermeasures of course resources sharing. Higher Education Forum (2): 20–23.

Education Management and Management Science – Zheng (Ed.)
© 2015 Taylor & Francis Group, London, ISBN 978-1-138-02663-6

The research and practice of the undergraduate graduation design based on the national excellent engineers training plan

Jiang-chuan Liu, Zhi-gang Yin & Wei Ji
*School of Water Resources and Environmental Engineering, Changchun Institute of Technology,
Changchun, China*

ABSTRACT: Graduation design is an important link of the undergraduate teaching, it is analyzed by all kind of the existing factors which is affecting on the quality of undergraduate graduation design, and discuss the mode of graduation design which is based on the excellent engineers training by improving several aspects, just as training mode, instructors, supervision mechanism, appraisal mode and so on, and improve the quality of the students' designs. From the point of practical effect, it can offer a new thought for the guidance work which is developed by the Water resources and hydropower engineering, and it can also be referred by other similar engineering majors.

Keywords: undergraduate graduation design; teaching mode reform; quality of graduation design

1 INTRODUCTION

Graduation design is a comprehensive, summary, and practical teaching link in the last teaching plan, a continuation, deepening and expansion of the each teaching link in the four years at the university, a comprehensive training for engineering design to cultivate students' comprehensive application of the basic theory, professional knowledge and basic skills, an important stage of the students' comprehensive quality and engineering practice ability [1][2].

By the practice of graduation design, the ability of Students' independent work can be cultivated, the professional fields can be broadened, and the studied speculative knowledge can be deepened. Therefore, Graduation design can't be replaced by other teaching links as a practice in the role and status in the whole teaching of undergraduate course [3][4][5].

The ministry of education "Excellence engineers training plan" (Referred to as the "excellence initiative") is the implementation of a major reform program of the "national medium and long-term education reform and development plan outline (2010–2020)" and "national medium and long-term talent development plan outline (2010–2020)", Our school of water resources and hydropower engineering was listed in the "excellence initiative" in 2011. This program aims to nurture a large number of innovative ability, which can adapt to the economic and social development needs of high quality engineering and technical personnel.

According to the general standard and industry standard for cultivating engineering talent, the students' engineering ability and innovation ability is strengthened. This requires that we must develop a graduation design practice to adapt the demand of excellent engineer talent training. From the fundamental purpose of undergraduate education as the starting point, the reform and practice of the undergraduate course graduation design is discussed based on the training objectives of national excellence engineers training plan.

2 THE INFLUENCE ON THE GRADUATION DESIGN OF UNDERGRADUATE COURSE

2.1 *The selected topic difficulty in graduation design*

Upon to the topic selection of graduation design, most universities require that the selection of students' graduation design topic cannot be the same as each other, and shall not be repeated within a few years. This brings a great deal of difficulty to teachers. According to their own research direction and currently engaged in the research content, the teachers select the students' topic. The different sessions of students, even the same session, their graduation design is often the same; just the topic or the specification of the parameters is different. So the selection of the graduation design content is only decided by the teachers without considering the students' interests and the employment direction in the future. Due to this, many students don't do their graduation design, to work perfunctorily on the graduation, they will copy others' design content on the eve of final project defense.

The topic selection of the graduation design and the training orientation of the excellence engineers are inconsistent. There are many problems in the topic selection. First, it is not based on solving practical problems in engineering area, second it is seriously out of line with the actual research and production, and third it can't be combined with the enterprise and local economic development. All these reasons made quite a number of students' graduation design be limited to theoretical research, without combined with production practice, this and the cultivation of outstanding engineers serious deviation. This is a serious deviation from the training position of the excellence engineers.

2.2 *The loss of the graduation design process monitoring*

In the graduation design process, although the schools put forward specific requirements for instructors and students, the lack of effective monitoring for the whole course of graduation design make it very hard to execute.

1. The instructors' reasons
   With the rapid increase in the number of College enrollment, the comparatively lagging behind contradiction of the construction speed of the teaching staff is increasingly prominent. In one hand, an instructor takes too many students in one time; he can't completely master every student's design process. On the other hand, for the senior teachers, they play multiple roles, their time and energy are limited, and they won't have time to instruct student's graduation design; for the introduction of young teachers, due to the deficiency of the Engineering practice and individuals 'capability, at the same time, many of them also are head teachers, they cannot complete and monitor the graduation design task independently. All these may directly affect the quality of graduation design.
2. The students' reasons
   During the graduation design, students invest in a shortage of time and energy on the design. In one hand, most of the college graduation design is arranged in the eighth semester (The last semester of senior year), during this period, most of the students are busy looking for a job or go to other places to participate in the recruitment and interview, because of the serious employment pressure, Students don't

have a mind to do the graduation design. On the other side, some excellent students have to review the postgraduate entrance examination course for postgraduate re-examinations during the time; they also don't have time on the graduation design. In terms of school and teacher, it is impossible to ask the students to force more energy on the graduation design instead of looking for a job and postgraduate re-examinations, this also make the monitor of graduation design process out of control.

2.3 *The problems of intermediate inspection*

The purpose of the opening report is to make students consult the literature extensively, understand the whole process of graduation design, master each link of graduation design, and complete the design based on the given task. The instructors and institutes should make a detail review on the design ideas, research method, and technological means, and make suggestions for revision. However, most colleges and universities are lack of this link. Every student has the opening report for the reply, but most of the students hand in the reports after the opening report, which is just to deal with the graduation reply and lost the meaning of the opening report itself.

The intermediate inspection of graduation design is to look at the student's whole design process and verify preliminary work; it is an important part of the content. However, most colleges and universities have ignored the link, even if some colleges and universities have the mid-term examination in the teaching task, but this is only an illusion, not really relating the process to the graduation performance evaluation.

2.4 *The general evaluation system of graduation design*

The current graduation design evaluation system is a general assessment about the students' learning attitude during graduation design, the design results, the defending thesis, according to the result of assessment, evaluating the excellent, good, moderate, pass and fail. This is a one-time appraisal way, which can be difficult to control the phenomenon of individual students' plagiarism. Firstly, the marker and exam-markers tend to be influenced by instructors; they will not deviate from the instructors' given level. In this way, some students have a good relationship with the instructors; they will get high grades although their design results and performance defense are poor. All these make the graduation design achievement injustice and affect the quality of graduation design.

3 THE DISCUSSION ABOUT THE EXCELLENT ENGINEER TRAINING ORIENTED OF THE UNDERGRADUATE GRADUATION DESIGN

3.1 On the basis of manufacture, learning and research, based on employment, optimizing subject

The Graduation design topic selection is on the basis of manufacture-learning-research, carrying out three way selection in the instructors, students and enterprises. Based on the students' employment direction, aimed at solving the teachers' scientific research project and the enterprise's actual engineering problems. In one hand, teachers make their own research projects or academic issues decomposed into the appropriate design topic for the students who has admitted graduate students and the students who participate in the college students' science and technology innovation; on the other hand, the cooperative units should put forward the practical problems they encounter in the engineering and have a consultation with the teachers to form an appropriate design topic for the students. For the students who are going to work, they will enter the working mode in advance and strengthen the consciousness of engineering and engineering quality based on the Engineering practical problems. By this graduation design, students' practical ability is stronger and they can be competent for a job directly. In addition, schools can get the content of the graduation design combined well with experimental equipment based on the existing conditions of the laboratory, which cannot only deepen and master the understanding of professional knowledge, but also can cultivate a series of students' comprehensive practice ability, such as independent design, installation and debugging.

3.2 On the basis of professional direction, building up a guidance model of graduation design team

According to the goal of national excellent engineers training plan and the requirements and regulations of the school for the excellent training experimental specialty, taking the scientific research and engineering practice, students' employment and college students' scientific and technological innovation as the starting point, forming a number of teachers' team cooperation guidance mode.

1. The instructors' Graduation design team should consist with a head of the subject direction and several teachers of the direction; the structure model of professional title is Professor, associate professor and lecturer. In the process of cultivating students, each of teachers has their own specialty and advantages, which can complement each other's advantages. The head of the subject direction is in the center of the team, his role is to coordinate the whole team in all kinds of things, such as: the topic selection, the solution of the key technology, defense, etc; Other members are responsible for specific guidance to students. They should form the guide mode with assessment, answering questions, the meeting.

2. The students' team
   The students' team should be built up according to their interests, the employment direction, and the service ability; the forming process should be based on the principle of two-way choice; in one hand, the students can select the instructors, on the other hand, the teacher can also select students according to the nature of the subject and the direction of the team. Once the team built up, teachers should know well each student's learning, living, working and various aspects, they should also gasp the students' knowledge level, and choose the coordinated ability and organizational ability of the students to be the team leader which is responsible for the coordination work.

3.3 Strengthening the supervision of course

From the temporal view of point, the graduation design process is carried out in accordance with the "3 + 1" mode, that is the early stage of the design, middle, late, and finally rejoin process, each stage has specific guidance and requirements.

The students who do the graduation design at school should be in the uniform location (such as library, design classroom, etc.), the group leader should be responsible for attendance assessment, and assure the daily design time. At the same time, teachers should go to the location. in one hand, teachers can look at the students' designs; on the other hand, teachers can guide students solving puzzles. The students who do the graduation design outside school should be required to report to the team leader once a day and to report to the instructors the progress of the design once a week. At the same time, Business mentors and teachers should communicate once a week; in one hand, they can confirm whether the design is carried on in time; on the other hand, they can examine students' design progress.

3.4 Reconstruct evaluation system of graduation design based on students' achievement

The guidance, review, and defense three evaluation links should be optimized; the Graduation design achievement should be quantitative evaluation in every stage, they are made to 40%, 30% and 30% according to the proportion of scoring criteria. In order to truly reflect the students' actual level, the inspection of graduation design process should be focused on.

The student's design requires periodic examination, on the one hand, students' attendance should be examined; on the other hand, according to examine the Literature review, the phased objectives, the whole students' designs can be known by the teachers. To quantify the every detail, the marking teachers are mainly responsible for the examination of students' design achievement, which involves the topic selection, the workload, design drawings, calculations, and other aspects. The defense result should reflect the student's design level, and the grasp of the basic knowledge.

4 THE PRACTICE AND RESULT

We take the excellent engineers training plan oriented graduation design teaching mode embodies the industry and the requirements of times. By optimizing the graduation design topic selection, building a guidance of graduation design team mode, strengthening the monitor of the graduation design process, reconstituting the evaluation system of graduation design. According to combine the teaching of the graduation design with the students' employment direction, teachers' research subject and engineering practices, the students' design ability and practice ability can be strengthened.

4.1 The quality of graduation design is improved greatly

In one hand, by the monitor of the graduation design process, the time of graduation design can be ensured, rather than just on the last week or two weeks to deal with the design, because of this, student can grasp of the whole design more comprehensively, and their design results can be improved. On the other hand, according to the construction of team model, the connection of the professional class teachers, professional basic course teachers, especially young teachers can be strengthened. At the same time, the understanding of the different teachers 'relative courses can be deepened and the teaching quality can be improved further.

4.2 The students' practical ability and employment quality can be improved

Students select the teachers' research subjects, employment direction and engineering practice as their graduation design topics, they must devote fully into the overall design, they can solve the problems according to mutual cooperation and the integrated use of learned knowledge.

The students who do the design outside school generally go to the company they have signed or the company they want to sign; the enterprise appoints the instructors fully participating in the graduation design process, which can make the instructors have plenty of time to examining the students' professional proficiency, practical ability and personal quality, by doing this, the instructors can gasp the students' specific circumstances. In the process of enterprise design, students can understand the task and the nature of the work ahead of time. Through practice, students' employment rate and employment quality can be greatly improved.

4.3 Professional competence of the young teacher is promoted

The young teachers who don't have scientific research subject can join in the direction of the professional team through the graduation design, they can monitor the whole graduation design under the academic leaders' instruct. By doing this, their scientific research ability and teaching ability can get improved. In one hand, they can get some funding from schools and colleges for scientific research based on the scientific research project team. On the other hand, to be a good teacher, they can absorb the good teaching methods and teaching means from other teachers of the team.

ACKNOWLEDGEMENTS

This work was financially supported by Jilin Province education scientific planning.

REFERENCES

[1] Chujie Jiao, Junping Zhang, Shanhu Wu, et al. Teaching reform on graduation design of civil engineering major in local universities and colleges. *Journal of Architectural Education in Institutions of Higher Learning*, 2010, 19(5):112–116. (in Chinese).

[2] Wenjing Zheng, Huiqin Wu. Implementation of university-enterprise cooperation in graduation design of civil engineering specialty. *Journal of Architectural Education in Institutions of Higher Learning*, 2012, 21(5):139–141. (in Chinese).

[3] Zhen Luo, Min Deng, Yanbing Ye. Practice and measures to graduation design quality of civil engineering specialty in local universityies and colleges. *Journal of Architectural Education in Institutions of Higher Learning*, 2010, 19(3):123–126. (in Chinese).

[4] Jianzhong Xia, Jianhua Wu. Study on practice teaching in education of excellent civil engineering engineer. *Journal of Zhejiang University of Science and Technology*, 2010, 22(5):387–391. (in Chinese).

[5] Yanbin Li, Feng You. Probe into the effects of engineering graduation design based on cooperative teaching. *Journal of North China Institute of Water Conservancy and Hydroelectric Power (Social Science)*, 2012, 28(5):162–164. (in Chinese).

Education Management and Management Science – Zheng (Ed.)
© *2015 Taylor & Francis Group, London, ISBN 978-1-138-02663-6*

Evolution and introspection on the allocation of resources in higher education since the founding of new China

Bo Peng
School of Education of Huaibei Normal University, Huaibei, Anhui, China

ABSTRACT: This study focused on the evolution and introspection about the allocation of resources in higher education since the founding of new China. The characteristics of the allocations of resources in higher education under both the background of planned economy and market economy were firstly analyzed to compare the management and economic system differences of allocation under the two different historical backgrounds. Finally, this study summarized and introspected on the allocation of resources in Chinese higher education. The innovation of this work is the analysis on inefficient resources in higher education based on the binary difference theory.

Keywords: Chinese higher education; marketization; allotment standard; financial allocation

1 INTRODUCTION

The higher education resources have high comprehensiveness. They can be divided into two categories—material resources and non-material resources, which can be also called two domains—hardware or software. The higher education in China started from the new-type education of Guangxu years in Qing Dynasty. Based on the imperial examination system and examination hall in old China, the higher education then adopted the western higher education pattern, gradually forming its higher education with unique characteristics. Since the founding of new China, the higher education in China gradually formed new patterns [1]. However, influenced by the planned economy, the new China had adopted planned economy patterns to conduct resources management in higher education for a long time since the founding. In new stage of this new era, the development of Chinese higher education should rely on the summary and introspection of management and teaching experiences in traditional higher education, thus providing references to the improvement of management quality of Chinese higher education and promoting the long-term advancement of Chinese higher education. Not only is it the developing requirement of Chinese higher education, but also the rejuvenation of China [1].

2 EVOLUTION OF ALLOCATION OF RESOURCES IN CHINESE HIGHER EDUCATION SINCE THE FOUNDING OF NEW CHINA

2.1 Concept of the allocation of resources in Chinese higher education

The education resources in universities have both broad concept and narrow concept. In the broad sense, the education resources in universities include human resources, financial resources, material resources and incorporeal property, such as brand assets. In the narrow sense, they only refer to the financial resources and corresponding resources allocation of universities, namely, capitals input from the country. This study mainly focused on the evolution of the allocation of financial education resources [2].

2.2 Allocation of resources in higher education under the background of planned economy

After the founding of new China, the allocation of resources in higher education is based on the potency and its future development of a university. The allocation of resources in higher education mainly relies on the relevant departments of universities and financial department of government, who defines the amount of education fund of a university in view of its scale, daily expenses and costs. Then, the financial allocation of this year is based on the sum of the education fund of last year and the one of this year, finally achieving the allocation of education fund.

This kind of allocation is of planned economy, which regards universities as affiliates of government; and all the financial funds of universities should be allocated through central finance or provincial finance. The funds are earmarked for its specified purpose only, the residual funds should be returned to treasury departments at the end of the year [3]. Thus it can be seen that the allocation of higher education before 1985 was a planned economy pattern, which could handle the allocation to a large extent. However, the planned economy pattern also largely constrained the creativity on the allocation of resources in higher education, exerting negative influence on the development of universities and their creativity.

2.3 Allocation of resources in higher education under the background of market economy

Since 1985, China has been gradually developing its market economy. Correspondingly, the Chinese treasury departments have adopted the comprehensive allotment standard with special subsidies for the allocation of resources in higher education. The comprehensive allotment standard is derived from the absolute allotment standard of the planned economy. The biggest difference between the two allotment standard is that the comprehensive allotment standard is decided by the Treasury Department and relevant administrations, with funding expenses based on the number of students and the difference between different majors and educational background. All the capitals from different items add up to the fund allotment. Another part of the allocation is special subsidy, which is based on the development situation of different universities and is the supplement of the allocation of comprehensive universities. The special subsidies are offered by Treasury Department to solely subsidize universities. Since the fund is earmarked for its specified purpose only, it is called a special subsidy.

The biggest characteristic of this pattern is the transformation from financial expense return model to allotment standard, which means there is no more fund if universities overspend their funds and the residual funds belongs to universities. However, this kind of allotment still has some problems. First is the narrow source of education funds for universities, which can cause capital shortage of higher education. Second is the low efficiency of education funds. The two problems largely restrained the allocation efficiency of resources in higher education. Therefore, it is

necessary to deeply study the combination of comprehensive allotment standard and special subsidies, evaluating its advantages and disadvantages.

3 DIFFERENCES OF TWO KIND OF ALLOCATION MANAGEMENT AND ECONOMY SYSTEM IN HIGHER EDUCATION

3.1 Significance and evolution reasons of the allocation of base plus development

The planned allocation, namely, the base plus development, is mainly carried out according to the national conditions of planned economy, which is related to the development of national economy in the early stage of new China. In this stage, the financial allocation is the only source of fund for universities, and the government undertook the role of both administrator and manager of universities. Meanwhile, universities did not have clear property rights and so did the powers and duties of the administrator, leading to the single pattern and low efficiency of higher education. In 1980s, Chinese universities still relied on the financial allocation. However, the new model of comprehensive allotment standard plus special subsidies better solved these problems.

During the Ninth Five-year Plan, Chinese universities gradually tended to raise their education funds from multiple channels and supplement their funds further. The education funds of higher education gradually showed a pattern of comprehensive funding from the society, government and individuals. Currently, the allocation of resources in higher education is no longer the full-amount financial allocation, but the central plan as a whole and decentralized management of local education administrations. Afterwards, some universities are supervised from directly under the central government to the co-management of central government and local government.

3.2 Significance of allotment standard plus special subsidies

The influences on the development of universities brought by the two kinds of allocation of education resources are totally different. Under the background of traditional planned economy, the expenses of universities mainly relied on the full amount funding of Treasury Department, with the country covering all the education expenses of universities. Therefore, universities became the affiliates of the government, so there were no competitions on resources or expenses among different universities and crisis awareness wouldn't be aroused. Meanwhile, the development of universities became more natural and students could also get some subsidies. At that time, higher education was totally free of charge, so university students wouldn't be poor for receiving higher education.

However, based on the comprehensive allotment standard of higher education resources, the finances undertake more resources burdens for universities and different universities are in competitions for financial allotment. Universities need to be checked and inspected by education administration for financial support since not all universities can get full-amount of funds they need. If universities want to seek more standard allotment and scientific funds, they should have breakthroughs on scientific researches and teaching activities, which include three tendency of influence.

First, the education service offered by universities tends to become industrialized and commercialized. Universities have transformed from passive status to active status, no longer receiving full-amount financial support from central finance. Therefore, some universities gradually step on a track of industrialization. For example, some universities may offer scientific researches for value and some universities may introduce administration and logistics management into socialization domains. Especially, many private capitals are pouring into higher education.

Second, the gaps among different universities are becoming larger. In order to gain advantages over their counterparts, some universities are finding their teaching characteristics, forming their unique teaching products, teaching advantages and attractive majors. All these efforts create the competition situation among universities.

Finally, on the allocation model of resources in higher education, the tuitions of some universities are relatively higher, leading to the fact that some university students may become poor for receiving higher education, which protrudes the issue of education fairness.

4 SUMMARY AND INTROSPECTION ON THE ALLOCATION OF RESOURCES IN CHINESE HIGHER EDUCATION

4.1 Evolution and comparison of two kinds of allocation of resources in higher education

The two different kinds of allocation patterns, base plus allotment standard and comprehensive allotment standard plus special subsidies, have played positive role in different historical times in China. The first kind of allocation has high efficiency and simple procedure; influenced by planned economy, universities and treasury departments only need to negotiate on the amount of funds. However, this kind of allocation is appropriate to the fact that the number of universities is small and the structure of universities is single. In addition, this kind of allocation mainly refers to the past development situation of universities. Therefore, when conducting adjust accounts on the costs of universities, there is often no scientific accounting method. If a university has large financial expense in the past year, the university can have more funds in this financial year. Therefore, under the allocation pattern of planned economy, misbehavior can stimulate extravagance and waste. The 1985 allocation pattern, namely, base plus special subsidies, can better control the disadvantages of planned economy allocation of resources in higher education.

4.2 Summary and introspection on the evolution of allocation of resources in Chinese higher education

First, the evolution of the two kinds of allocation can be attributed to the adaptation of the development of higher education. In order to adapt to the development of market economy, especially the transformation from planned economy to market economy, the allocation of resources in Chinese higher education evolved from planned allotment standard to allotment standard plus fund-raising. This evolution deserves appreciation. The demand on higher education is growing while the education resources are relatively short, thus creating a kind of contraction. Therefore, if public finance continues to conduct full-amount allocation, the financial order will be disturbed. Therefore, the popularization of higher education needs the combination of fund-raising plus allotment standard pattern.

Secondly, market innovation is optimizing the allocation of resources in higher education. The evolution of allocation from planned base plus development to comprehensive allotment standard plus special subsidies involved market as a fundamental role in allocating resources. The limitation of special subsidies require universities to strive for the subsidies, which stimulates universities to take more efforts on improving the teaching levels to face competition, enhancing the education efficiency.

Thirdly, the intervention of market requires the equal protection of resources allotment in higher education. Since universities have certain extent of public welfare, the total marketization will make the tuitions become unaffordable for a large number of students. Therefore, the allocation of resources in higher education still needs the coordination between plan and market. In addition, the right to education of university students should be protected financially to promote the equality of education chances.

The development of higher education cannot be separated from continuous allocation pattern of resources. With the development of times, the allocation of resources in higher education also needs some adjustments. The education administration should offer policies to change the allocation of resources according to the social development, promoting the allocation structure of resources to be appropriate to the requirements of social development.

ACKNOWLEDGEMENT

The work was one of staged achievements of Study on the Allocation and Cultivation of Higher Education Resources from an Ecology-Oriented Perspective (CIA110149), the youth issue of National Social Science Fund (Education).

REFERENCES

[1] Li Ping. *Consideration on the Optimization of Investing Reform of Higher Education*. Journal of Higher Education. 2006(10)54–57.

[2] Liang Dongqing, Hu Zhongfeng. *Problems of Reform of China's Investment in Higher Education issues and Solutions*. Higher Education Exploration. 2006(02)52–53+83.

[3] Zhang Min. *On the Rational Allocation of Resources of Higher Education*. China Adult Education. 2007(02)18–19.

Education Management and Management Science – Zheng (Ed.)
© 2015 Taylor & Francis Group, London, ISBN 978-1-138-02663-6

Influencing factors of MICE tourism based on experience economy—case of Guilin in China

Ying Tang & Yun Zhao
Guilin Institute of Tourism, Guilin, Guangxi, China

ABSTRACT: MICE tourism has become an important way of changing Guilin's traditional development mode of tourism, enhancing the competitiveness and development potential. Thus, the acceleration of MICE tourism becomes an inevitable trend to guide scientific development of Guilin tourism. Experience economy advocating consumption experience, as an extension of service economy, has been affecting all walks of life in current society. Based on theory and background of experience economy, it was combined with MICE tourism. Related research was conducted on influencing factors of MICE tourism in Guilin with questionnaire, thus searching a new development direction of MICE tourism in Guilin.

Keywords: experience economy; MICE tourism; influencing factors; Guilin

1 BACKGROUND OF EXPERIENCE ECONOMY

1.1 *Meaning of experience economy*

Experience economy is the economic form that enterprise creates series of consumers-centered activities to satisfy consumers' requirements of personal experience and validation in experience period. Positive effect can be formed after meeting consumers' psychological demand, thus promoting the long-term development of enterprise [1].

1.2 *Characteristics of experience economy*

1.2.1 *Emotionality*
Emotionality of experience economy refers to that individual will generate happy or sick feelings while experiencing events with emotion as a starting point.

1.2.2 *Subjectivity*
The implementation of experience economy relies on cognitive and practical activities of experiencer, which cannot be replaced by the third party. Experiencers obtain distinctive understanding and feeling through their own need, value orientation, cognitive structure, emotional structure and personal experience [2].

1.2.3 *Intangibility*
Experience economy was developed based on manufacturing and service economy. Enterprise attracts consumers to purchase tangible goods, thus generating actual consumption while bringing experience products with services as point of increase. The quality of experience products cannot be evaluated without consumers' experience. Thus, experience products cannot exist alone without subject and object; instead, it is attached to products and services [3]. The intangibility of experience economy is similar to characteristics of tourism products.

1.3 *Correlation between experience economy and MICE tourism*

The experience economy has impacted the development of all walks of life. There is a natural inner correlation between MICE tourism and experience economy due to characteristics of tourism products. MICE tourism is an economic phenomenon growing with the development of exhibition industry, relying on the good environment of exhibition development. It is a new type of tourism activities conducted during or after the exhibition, with tourism resources as the complement. Four basic types of MICE tourism include conference tourism, exhibitions, incentive travel and festival tourism. Taking exhibition projects and tourism resources as attractions, MICE tourism satisfies the demand of participants or exhibitors in exhibition activities and tourism activities, with the starting point of creating unique experience of consumption. The comprehensive degree of experience in MICE tourism is influenced by many factors, including atmosphere of exhibition scene, conference scene, uniqueness of incentive travel, local culture, supports of local people experience, etc. These are in accordance with characteristics and ideas of experience economy.

2 INFLUENCING FACTORS OF GUILIN MICE TOURISM

2.1 Factor analysis

Based on previous relevant researches of MICE tourism, influencing factors of Guilin MICE tourism were analyzed combined with Guilin's actual situation.

2.1.1 City reputation
City reputation refers to the degree known by public, the breadth and depth of social impact. The degree of visibility makes influence on the effectiveness of large-scale exhibition activities.

2.1.2 Urban environment and tourism resource
The overall environment and tourism resource of a city is the core power of tourism development. Though MICE tourism is different from traditional tourism, entities tourism resource such as historical sites and natural resources is still a very important component.

2.1.3 Economic attractiveness and industrial base
Different from other forms of tourism, MICE tourism of various themes has higher requirements for development level of related industries. Only with strong industrial base can the sustainable development of theme MICE tourism be supported.

2.1.4 Popularity and publicity of exhibition
Brand popularity of exhibition should be enhanced—only with high popularity the exhibition can become preferred object of exhibitors and visitors. Meanwhile, exhibition with high popularity attracts high-quality exhibitors and buyers. The enhancement of visibility should be achieved through reasonable propaganda.

2.1.5 Transportation convenience
As material carrier of tourism and business activities, traffic is an important link connecting destination and starting point. Tourist traffic has become an important part of tourism product, which has a significant impact on development and operation of MICE tourism products.

2.1.6 Attraction and services of exhibition theme
MICE tourism possesses dual nature of exhibition activities and tourism activities, thus the quality of exhibition theme determines whether exhibitors and visitors visit the exhibition. Tourism activities cannot be implemented without exhibition activities. Meanwhile, service quality of exhibition

affects next participation of exhibitors. The lasting and benign MICE tourism cannot be maintained without good themes and exhibition services.

2.1.7 Government incentives
Various policies and measures of cities have been introduced for the local development of exhibition industry to seize exhibition market. Thus, the development of exhibition industry can be regulated, which lays foundation for the operation of MICE tourism.

2.1.8 Participation fee
Under the trend of increasing participation fee, enterprise chooses proper exhibition within the budget according to their financial resources. Participation fee should not result in additional burden on enterprises, especially for small and medium enterprises with cautious expenditure. The level of participation fee affects customer's choice of exhibition, which is an important factor in project selection.

2.1.9 Urban reception force
Exhibition activities will bring lots of people and logistics, thus city should have sufficient accommodated ability to undertake large-scale exhibition projects, which is a precondition for the development of MICE tourism. Urban reception capacity is mainly reflected on dining, shopping, entertainment, lodging, attractions, service capabilities, local transportation, and support degree of residents.

2.2 Sample, data and method

Random sampling survey was adopted in the research. Questionnaires were designed to research Guilin International Tourism Expo for study analysis. The survey was conducted in September 2012, with 200 questionnaires distributed and 198 recovered, and the recovery rate was 99%. There were 192 valid questionnaires, so the valid rate was 96%. Participants were selected from exhibitors and professional visitors in the Third Guilin Tourism Expo.

The importance degree of variables that influence participants to choose MICE tourism was divided into "extremely important", "more important", " general", "less important", and "extremely unimportant"; other factors adopt the form of individual choice.

AHP (Analytic Hierarchy Process) method of quantitative analysis method was used in relevant statistical data and research factors. Compared to other quantitative analysis methods, AHP can clearly reflect the proportion relationship and mutual relevance among various factors. Besides, the objectivity and scientificity of MICE tourism in Guilin can be greatly enhanced with recommendations from exhibitors and professional visitors in the survey.

2.3 Analysis on survey results

There are 140 of 192 participants in the survey are from Guangxi, accounting for 72.9%; number of participants from other provinces in China is 43, accounting for 22.3%; 9 participants are from foreign countries, accounting for 4.6%. Thus, participants in Guilin MICE tourism are mostly from Guangxi.

Based on the influence of experience economy on enterprise, exhibitors are still willing to pay fee to participate in exhibitions and conduct appropriate tourism activities despite the rapid development of Internet technology. The purposes include establishing new business contacts, strengthening existing business, promoting company's new business, looking for new agents, developing new markets, increasing fame, and knowing company's market share. However, in case of importance division, exhibitors—who consider the establishment of new business contacts to be very important—occupy the majority, followed by the improvement of tourism reputation and publicity of company's new travel business. Therefore, for most exhibitors, finding new customers and increasing their fame is most important for participating exhibition, indicating the significance of visitor number in exhibition.

The researchers conducted the investigation of influencing factors affecting Guilin MICE tourism. The results found that the main influencing factors affecting exhibitors to select exposition of Guilin's tourism and MICE tourism activities included Guilin's fame (15.6%), urban environmental tourism resources (14%), popularity and propaganda of exhibition (13%), preferential policy of government (12.5%), attraction of exhibition theme (11.9%), economic attraction and industrial base of Guilin (10.4%), exhibition service (8.8%), reception capacity (8.7%), benefits of exhibition (8.3%).

3 SUGGESTIONS ON MICE TOURISM UNDER BACKGROUND OF EXPERIENCE ECONOMY

Through empirical analysis on Guilin International Tourism Expo, combined with the connotation of experience economy and the developing trend of times, suggestions were proposed on planning and design of exhibition and MICE tourism as follows:

3.1 Experience and innovation of MICE tourism product theme

In experience economy times, the experience theme of exhibition tourism activities should be designed from consumers' perspective, thus changing passive acceptance to active participation of exhibitors and visitors. Taking Guilin International Tourism Expo as an example, organizers paid more attention to the use of experience in exhibition planning, designing Guilin "Experience Travel". Best scenic spots and paths were selected from Guilin tourism resources of high-quality as the tour line of exposition of "Experience Travel". Merchants, buyers and travel agency principals were arranged to obtain live experience, thus more exhibitors and buyers could know tourism and fine scenery of Guilin through experience. Arab Theme Pavilion in Shanghai World Expo was another successful case of MICE tourism. Visitors could fully experience the culture of various aspects in Arab countries without being in United Arab Emirates through the unique design modelling and high-tech means of pavilion. Therefore, visitors could understand the theme and essence of pavilions in the experience.

3.2 Matching service of experience in MICE tourism

MICE tourism possesses the intangibility of tourism product, thus the quality of MICE tourism products can be experienced through the consumption of exhibitors and visitors. Service is the most direct experience for consumers, which can effectively help to achieve optimization of MICE tourism products. In experience economy times, it's not enough to have the theme with experience and innovation, high-quality services are also necessary to establish an atmosphere of good experience. Firstly, the service image and scene should match experiencing theme, which can start from costumes of service personnel, props and design of service scenario. For example, the service site of Guilin International Tourism Expo can be designed into landscape modelling with characteristic of Guilin; service personnel can wear clothes similar to actors in the film and television of Liu Sanjie, thus providing experience propaganda while offering services.

3.3 Strengthening implement of government's preferential policy

MICE tourism has become a new sparkle of the establishment of city image and economic development of Guilin with the coming of experience economy. Government policy and support ranks fourth with 12.5% value among surveyed factors, which indicates its importance to the development of MICE tourism. Government has also fully realized this development trend—formulating relevant policies to promote the development of exhibition industry. For example, Guilin government has regarded exhibition industry as one of focused industries during "twelfth five-year" period, thus ensuring the dominance of exhibition development from government level; meanwhile, *"Interim Measures for Use and Management of Development Funds in Exhibition Industry"* was issued, stipulating in the form of documents that Guilin financial departments should arrange 10 million RMB of special funds every year to support exhibition events at all levels held in Guilin, with cultivating exhibition brand.

3.4 *Updating environment resources of Guilin MICE tourism to improve reception capacity*

As a famous international tourist city, infrastructure construction of Guilin is relatively perfect with convenient basic living facilities and tourism facilities, which is a key factor that exhibitors and buyers choose Guilin as the destination of exhibition activities. However, the infrastructure and tourism facilities have been aging with time, while the development of MICE tourism should be invested with large amount of money on city publicity, venue construction maintenance, equipment update, development of service industry, etc., thus proposing new requirements for the hardware construction of Guilin. For MICE tourism, Guilin possesses exhibition reception facilities with certain strength, such as Guilin International exhibition center, museum, art museum, etc. However, the exhibition reception facilities affect the service quality and image in practical application, such as traffic jam, insufficient catering sites, etc. Therefore, from the perspective of MICE tourism development, Guilin should invest certain financial, human and material resources for the upgrade of old reception facilities, combined with the new demand of experience economy era.

4 CONCLUSIONS

In summary, experience economy era has proposed new requirements for the development of exhibition industry and MICE tourism. Organizations and personnel of exhibition industry should accelerate self-construction to occupy MICE tourism market for the upgrade of tourism products combined with actual situation. However, the influencing factors and importance degree of MICE tourism are constantly changing with the development of exhibition industry. Therefore, exhibition workers should always focus on the development and changing trends of exhibition industry, thus improving the construction of MICE tourism products under the guidance of experience economy theory.

ACKNOWLEDGEMENTS

The work was supported by *"Integration development of Guangxi Exhibition Industry and Related Industry under the Background of Transformation in Modern Service Industry"*, Human Studies and Social Science Research Plan of Guangxi Colleges and Universities. (Grant number: SK13LX514).

REFERENCES

[1] Wu Xiaoye. *Creative Design of Incentive Travel Product from Perspective of Experience Economy*. China Collective Economy. 2008 (1): 92–94.
[2] Zhao Guoying. *Festival Tourism Development under Background of Experience Economy*. China Business & Trade. 2010 (1): 138–139.
[3] Yang Haihuan, Li Xiaohui. *Strategy of Tourism Development Based on Concept of "Experience Economy"*. Journal of Yunnan Normal University. 2005 (5): 71–74.

Education Management and Management Science – Zheng (Ed.)
© 2015 Taylor & Francis Group, London, ISBN 978-1-138-02663-6

Diversified cultivation for cultural awareness in college English teaching

Mingcai Wu
Minnan Science and Technology Institute, Fujian Normal University, Quanzhou, Fujian, China

ABSTRACT: Currently, cultural awareness cultivation is appealing more and more attention in college English teaching. However, there are cases where only language knowledge teaching is emphasized regardless of cultural awareness cultivation. Seriously affecting teaching effects, it is not conducive to the cultivation of students' ability and the comprehensive development of their quality. In this sense, the importance of cultural awareness should be fully recognized in college English teaching to cultivate students' cultural awareness on multiple angles and levels.

Keywords: college English; cultural awareness; cultural differences; diversified cultivation

1 INTRODUCTION

As a social consciousness, language has its own background and is associated with a certain culture. In language learning, only when you can master relevant cultural background and timely cultivate cultural awareness can you better use this language [1]. In fact, it has become an irresistable trend to fully understand the importance of cultural awareness cultivation and adopt diversified means to nurture the students' cultural awareness.

2 NECESSITY OF CULTURAL AWARENESS CULTIVATION IN ENGLISH LEARNING

Diverse cultures result in great differences between languages. Therefore, if you merely learning English grammars, silly mistakes or even misunderstanding that delays things can happen in actual communication.

2.1 *Differences between languages caused by diverse cultures*

There is a huge difference between Chinese and English in pronunciation and grammar, which is mainly caused by cultural differences. Thus when Chinese learn a foreign language or foreigners learn Chinese, they will come across problems in pronunciation and grammar. However, it's an isolated viewpoint to solve such problems merely based on grammars, namely "suit the remedy to the case" [2]. To radically solve this problem, people should profoundly experience and understand foreign culture and its difference from domestic culture, thus truly comprehending the causes of differences in language.

In Chinese, it's necessary that the attribute is located in the front of the subject or object to modify the subject and the object. For example, in "I have a Shanghai working elder sister", "Shanghai working" is the attribute of the object "elder sister" and can only be located in the front of the object in Chinese. Chinese barely say "I have an elder sister who works in Shanghai"; however, when translated into English, "I have a Shanghai working elder sister" can only be "I have an elder sister who works in Shanghai". Here "who works in Shanghai", as an attributive clause, is put in the back of the object to modify the object [3]. All in all, it's easy to master grammars if you can fully understand the cultural connotations in different languages.

2.2 *Different common expressions caused by cultural differences*

Formed with certain cultural background, several adages and proverbs are frequently used in verbal communication. Therefore, only when you master such cultural knowledge can you better utilize the language in English learning. Otherwise there will be the use of some nondescript adages, causing stupid mistakes in communication.

For example, Chinese usually say "ài wū jí wū", which can only be translated into "Love me, love my dog" rather than "Love the house, love the crow (word-for-word translation)". What's more, "rù xiāng suí sú" can only be "When in Rome, do as the Romans do" in English.

In China, "dog" is a bad thing—people would say: "You dog thing!" when cursing someone. In

English-speaking countries, dog nonetheless is not a bad thing but a good word used to praise others. People there think dogs are loyal and often use dogs as compliments. For instance, "You are a lucky dog." doesn't mean "You are a dog." but "You are so lucky."

When Englishmen tell you: "I wonder if I can go somewhere?" Their real intention is to ask you: "Can I go to the bathroom?" rather than actually going somewhere. At this moment, they will be confused if you say: "Yes, you can go anywhere." When a foreigner says you are a blue blood, they mean you have noble descent other than your blood is blue. However, once they say you were once in a blue moon, they think you are a risqué person filled with nasty thoughts.

2.3 Influence of cultural differences on daily communication

In addition, cultural differences can generate a significant difference in daily communication. For example, both Chinese and English use "Hello" as the greetings in the first meeting. But they differ greatly in greeting acquaintances. Generally, Chinese use "What are you going to do?" or "Have you had your meal?" when greeting others, which is inappropriate for foreigners. Foreigners will feel uncomfortable if you say: "What are you going to do?" to him; a foreigner will mistakenly believe that you intend to invite him for dinner if you ask him: "Have you had your meal?" In English-speaking countries, the topic of greeting would usually be weather, health, etc. For instance, if a foreigner says: "Lovely day, isn't it?" to you, there's no need to discuss weather conditions with him—he was just greeting you.

In Chinese, asking the ages of different objects has various versions: "fāng líng" for girls, "guì gēng" for young people and "gāo shòu" for old men. However, all those versions become a single sentence "How old are you?" when translated into English. Thus never compliment a foreign old man by saying "You are in good health at such an old age." Because he would be very angry once you said that—he assumed you want him to die and said "You old fool!" to him.

3 CURRENT STATUS OF CULTURAL AWARENESS CULTIVATION IN COLLEGE ENGLISH TEACHING

3.1 Emphasis on cultural awareness cultivation

At present, it has become a consensus that cultural awareness cultivation is important in English teaching and conducive to mastering and using English knowledge. In fact, various universities are beginning to develop students' cultural awareness through a variety of means. And English teaching is combined with the education and cultivation of English-speaking countries' cultures.

3.2 Problems in cultural awareness cultivation

Although various universities have made lots of efforts in cultivating students' cultural awareness, there're unclear viewpoints and targets in implementation. As a result, it's severe that the use of equipments becomes a mere formality. In fact, many teachers still merely emphasize grammars and syntax in English teaching regardless of cultural awareness cultivation. Consequently, the goals and means of teaching become seriously disconnected. Besides, there're cases where only English-speaking countries' cultures are emphasized instead of cultivating students' cultural awareness. And the cultural awareness cultivation based on the difference between foreign and domestic culture has not been achieved.

4 DIVERSIFIED CULTIVATION FOR CULTURAL AWARENESS

In college English teaching, cultural awareness cultivation for students should be diversified through various means, nurturing students' pluralistic cultural awareness.

4.1 Explanation of words and phrases with cultural awareness cultivation

It's necessary to combine cultural awareness with the explanation of words—the basic elements constituting English. Thus the explanation of words can't be stuck with the meaning, property and usage of words. It should always have relevant cultural awareness such as the generating and evolving background knowledge of words. Accordingly, students can more deeply appreciate cultural differences via methods such as polysemy and multiple uses of one word. For example, the word "dog" is often used as abusive expressions in China: "gǒu nú cái (lackey)" and "gǒu tuǐ zǐ (henchman)"; However, dog is not a swearword in English. On the contrary, it's often used for praise: "You are a lucky dog." and "You are a clever dog." The both are praises other than bad languages.

Phrases, formed by multiple words, are often used in social intercourse. They are not the simple piling up of words but common expressions based on grammatical knowledge and cultural background. For this reason, it's difficult to understand many phrases without the cultural background of English-speaking countries. "Borrowing power to do evil" means "hú jiǎ hǔ wēi (the fox assuming the majesty of the tiger)" or "gǒu zhàng rén shì (a dog threatening people on the strength of its master's power)" in Chinese. And "rob Peter to pay Paul" is equal to "chāi dōng qiáng bǔ xī qiáng" in Chinese. In a word, words and phrases teaching should timely expand knowledge to help students develop cultural awareness.

4.2 Cultural awareness cultivation in reading lessons

English reading materials often involve cultural traditions of English-speaking countries such as religion, marriage, etc. Sequentially, only when students master such cultural background can they better understand the reading materials in reading lessons. Meanwhile, teachers should encourage thought diffusion and knowledge expansion when it comes to relevant reading materials. With the aid of the cultural background in reading materials, they can help students master relevant cultural knowledge and develop cultural awareness.

4.3 Grammar teaching with cultural awareness cultivation

Grammars often reflect cultural knowledge, while cultural knowledge generates different grammars. For example, the previously mentioned "Lovely day, isn't it?" is not a question but a simple greeting. What's more, English imperative sentences are often used in the form of interrogative sentences. For instance, the imperative sentences expressing requests usually don't directly use "please" but adopt interrogative sentences guided by "would you" or "may I". In sum, these are grammatical knowledge reflecting different cultural knowledge.

4.4 Create situations for cultural awareness cultivation

Why Chinese students don't understand western culture? The important reason is that they hasn't grown up in that environment and can't feel that atmosphere. Thus it's feasible to create situations that enable students to experience Western culture. The ways of creating situations can and should be diverse, so teachers should actively brainstorm to create situations for students.

The first kind of creating situations: students can experience Western culture by participating into classroom activities combined with the Western culture mentioned in classes. During this process, teachers are neither strict supervisors nor onlookers. They are timely guidance and make summaries later.

Situations can be created via hosting some activities. For example, students can design several

real-life situations as creatively as their imagination allows. Afterwards, they can perform these scenarios, while other students act as observers. Then teachers make comments and concluding remarks. During such activities, students can enhance the understanding of Western culture and comprehend different cultural environments suitable for various languages.

4.5 Extra-curricular activities for cultural awareness cultivation

Students should be encouraged to participate into extracurricular activities to enrich knowledge and enhance their understanding of Western culture. For example, they can take part into summer camps to contact with foreigners. Then they can form firmer and deeper impression since they personally experienced more Western knowledge.

Besides, Western Culture Knowledge Contests can help students collect information about Western culture as much as possible. Students' enthusiasm can thereby be mobilized through contests, thus making active preparations. The preparations are to understand Western culture, while the contests are to test cultural knowledge. In this game, competition mechanism stimulates students' initiative as much as possible. Finally, students fully dedicate themselves to the learning via contests, greatly improving learning efficiency.

4.6 Offer specialized public courses in Western culture

Without personal experience, most students' knowledge about Western culture is limited to books. Thus the specialized public courses in Western culture can help students experience and understand Western culture, cultivating their cultural awareness. The major role of such courses is to allow students to feel things, while teachers are primarily responsible for interpreting some confusing parts. For example, Western films enable students to experience Western culture and the origin of the cultures from other English-speaking countries. At the same time, Western films can help students master the specific application scenarios of language and exercise English listening and speaking ability. Over time, cultural awareness will go deep into students' brain—they can freely use it for communication.

4.7 Cultural awareness cultivation in the comparison between Chinese and Western culture

The cultural awareness cultivation for students is not merely experiencing Western culture but comprehending the cultural differences. Comparing Chinese and Western culture, students can thereby more deeply experience the different usages of languages in different cultures and know the importance of cultural awareness in language learning; and such comparison can enable students to have a deeper understanding of different cultures. Meanwhile, they won't end up with forgetting Eastern culture after mastering Western culture.

Chinese and Western culture can be compared in various ways: everyday language, greetings and proverbs and so on, enabling students to know the differences between different countries; in addition, students can better master various cultures by comparing cultural allusions; besides, comparing the historical development of China and the West allows students to better grasp the similarities and differences between Chinese and Western cultures as well as their own strengths.

The key of cultural awareness cultivation is to enable students to understand the relationship between culture and language, rather than studying Western culture while losing domestic culture. This process requires the proper guidance of teachers. Otherwise, the allusion "Handan toddler (imitate others and thus lose one's own individuality)" will repeat itself in English teaching.

5 CONCLUSIONS

In a word, mastering different cultural knowledge and understanding the cultural differences between Chinese and Western culture is very important in English learning. Therefore, college English teachers should cultivate students' cultural awareness through various means. Then students can combine language learning with culture to avoid nondescript communication. In addition, students' cultural awareness cultivation should avoid blind xenophilia and nurture students' cross-cultural awareness, rather than studying Western culture while losing domestic culture.

REFERENCES

[1] Peng Yi, *Strategies for Cross-cultural Awareness Cultivation in the English Teaching of Higher Vocational Colleges*, Time Education, 2014 (11): 18–19.
[2] Pu Shengche, *Causes of Students' Lack of Cultural Awareness in the Reading Instruction of High School*, Future Talent, 2014 (07): 171.
[3] Fu Wen, *First Exploration on Weak Cultural Awareness in High School English Teaching*, Course Education Research, 2014 (13): 92.

Education Management and Management Science – Zheng (Ed.)
© 2015 Taylor & Francis Group, London, ISBN 978-1-138-02663-6

Influence on English translation caused by culture difference between China and the West

Yijun Huang
Minnan Science and Technology Institute, Fujian Normal University, Quanzhou, Fujian, China

ABSTRACT: With the development of multi-culture, people today have more requirements in learning and reading desire. So many famous foreign articles need to be translated in writing. However, great differences in using words, thinking method and culture background make written translation uncertain to some degree. The first thing of accomplishing a good translation is to know how these differences affect translation, so the essence of translation in Chinese and English articles can be well understood. Analysis on influence in English translation caused by culture difference between China and the west is made in this work by adopting some examples.

Keywords: Chinese and western culture; culture difference; English translation in written

1 INTRODUCTION

The purpose of studying language is to widen vision and learn the advance mind and culture from many kinds of civilization. Nowadays, it is not simply transferring words or misinterpretation out of context. Large-scale translation and brand new language structure not only satisfy reading requirement of people, but also present wisdom of translators through different cultures. There are many hard-working famous translators in China such as Yang Jiang, Zheng Yonghui and Fu Lei, who show people advanced translation methods of getting through ancient and modern, and unifying local and abroad. Although having no historical obligation at present, translators have the same society responsibilities after abundant reading and selecting correct articles [1]. Translators with such responsibilities can push forward the whole society, enhancing the national spiritual power.

2 CURRENT SITUATION OF ENGLISH TRANSLATION IN WRITING

With the rapidly globalization, more attentions are paid to English study and application ever than before. People are required to master certain English knowledge to achieve activities, such as foreign trade, culture communication and competition, daily using of imported goods, reading foreign articles, and watching foreign films etc. Translations are needed in daily life and social economical development, thus achieving better culture communication and proper delivery of messages. There are great amount of translation personnel covering all areas. However, the situation of lacking excellent translators happens, and it is hard to get the accurate judgment because of different translation level [2]. Continuous self-reflection and promotion are needed for universities, training groups and translators, so English activities can be continuously improved.

As an important part, English written translation alternates and delivers Chinese-English message through language expression. Activities such as contract, product description can be easily solved, and also Chinese and foreign articles can be enjoyed and understood. It requires no extremely high response and instant recalling ability like oral translation. However, faithfulness, expressiveness and elegance are definitely required when translating foreign articles. Thoroughness in using words, setting language condition and separating the emotion is required to translator, and these factors can be used to judge the capability of translators, thus showing professional ability of translators [3].

There are many translators coming from English or translation major of university. Anyway, they still need to accumulate translation skills and experiences by large-scale written translations, although they accepted professional education in university which is certain systematic and authoritative. Silly mistakes also happen in above period, which influence their working confidence and passion, even impacting future job plan. Translation, as a bridge of two cultures, requires translators to be quite familiar with two cultures and deep culture understandings, thus making translation fluently and bring reader the enjoying and better understanding. Translation of articles is not only simply language alternation, but also the valuable second-time creation. However, some normal translators, although they translate articles with complete content, and it seems okay, will overplay and be laughed by professional personnel. Their translations are far from original meanings because of inappropriate expression and Chingish grammar.

3 INFLUENCE IN ENGLISH TRANSLATION CAUSED BY CULTURE DIFFERENCE BETWEEN CHINA AND THE WEST

Colorful culture and civilization between China and the west diversifies the activities of communication, thus forming mixture and boundary of different areas. Since now, translation is not only keeping the original meaning of articles with foreign symbol, or showing Chinese characteristics, but also is required by spiritual need. In particular, written translation can provide abundant reading pleasure to people by integrating with China and the West, so creative reform can be achieved through translating progress.

3.1 *Influence on written English using different words*

Translation is not simply checking dictionary, but literature creation integrating multi-resources. It is beyond question that translation is like a production line. Translators familiarly alternate the words of original article using abundant vocabularies in their minds, thus keeping the same language environment. It is the basic test for translator in this stage. Plenty of vocabularies support them in translation. In the same time, they might have difficulties in selecting correct words. How to meet the original requirement of article; how to make it easy understand by local readers and even selected words make the structure reasonable or not—all these issues need to be considered by translators. Successful translation should not only focus on familiarly using and applying of words, but also make readers trace the original thoughts of article. Translation words can be refreshed by reviewing translator's native language from special aspect.

Throughout the history of China, the most representative culture is Confucianism. The five ethics, charity, justice, propriety, wisdom and loyalty, are presented by Confucianism, which are continuously delivered by Chinese people in the progress of changing world through thousands years.

And life faith is also differing greatly from other nations and countries, which can be reflected from literature and articles. There are many humble addresses like "Ning Zun", "Ning Tang", "Quan Zi" and "Han She" etc., showing humble thoughts and manners. And these words greatly support the description of figures in articles.

Comparatively, influenced by western renaissance, more focuses are paid on protection and liberty of human. Humanity is priority in the historical progress. "Heroes" and "Strong people" are worshiped by popular, and these features are shown in daily life. The difference of culture between China and the west is that two habits and emotions are presented separately. In China culture, words are more conservative and sensitive, while words are straightforward in western culture. For example, the words "ambition" is more aggressive in Chinese opinion, while western people prefer to think it "aspiration" more positively. Also, Chinese prefer "old" as reputation people with long age, while western people think it as useless. These culture differences make translators take time to select proper words, and they try to use suitable words to make readers accept and like, thus getting phrase of articles.

3.2 *Influence on written English with different thinking method*

The difference of thinking method between Chinese and western people is caused by different location and history. The "brain" is also formed during the interactive among peoples, which provides fixed way for personal thinking mind. Although people around world can communicate with each other easily by website and transportation tools, this original thinking method still has its fixed difference, which can help people learn from each other.

With proper understanding of the difference in thinking method, translator's works can be accepted by more reader, so readers can get the same feeling with translator after reading their translation. The phenomenon of personal privacy protection in western culture is also rooted in respect for humanity in history. Nowadays, foreigners will refuse to answer instead of saying "That is a secret", when Chinese people ask about their age, work and family. This is because western people prefer to act energetic and passion rather than publishing the true age or answering family situation to protect their family. For example, Chinese people like to call themselves as "descendant of the dragon", while the dragon is treated as a horrible creature by western people. They can't understand why Chinese people treat themselves as a bad creature. Therefore, there are two translations about dragon: *Loong* and *Dragon*. *Loong* is for the real dragon in Chinese meaning, while *Dragon* is the terrible dragon. So the problem of misunderstanding can be solved by this method.

Translators should try to use this "Refuse" method properly when translating English to Chinese. Then local readers can be far from misunderstanding and unpleasant when reading. It is known that there is difference of thinking method between China and the west. Anyway, translators with experiences are still needed to solve the problems of reader's misunderstanding and unpleasant, thus ensuring reader's comfort in reading and thinking.

3.3 Influence on written English with culture background between China and the West

Ancient civilization of China and the west brought great fortune to both sides, including allusion, incident, or principle. These are important parts of human civilization. And they are important materials that should be conserved by human beings. Anyway, theses precious materials with culture background are like two sides of a coin for translators. They can translate the article for readers with familiar local language background, while its limit can obstruct the translation because of "conventions" formed thousands year ago.

There are many familiar conventional languages based on local history and culture, which make written translation hard to improve. For example, the word "land" is understood as normal and cheap thing because of vast territory in China, while land is very precious resource in Britain Island. There is one Chinese saying, "one who spends money like land", while in England, they quote seawater as vast thing. Also in China, they say Buddha bless you while western people say God bless you. Then all the problems in translation can be easily solved by fully understanding the different culture background.

History of each nation might influence their people's knowledge and judgment to a certain extent. People may have no ability to predict future thing, but can fill in their brains with knowledge by analyzing incidents happened in the past. The allusion quoted in literature is the most representative knowledge. And these allusions are known by people with same culture background. They seem confused but people can actually understand its principle by deep thinking. Foreigners can't understand the meaning of "cao chuan jie jian", "bei gong she ying" and "wen ji qi wu", while Chinese cannot understand the meaning of "jump the gun", "stick in the mud" and "spill the beans". Professional translators should pay more attentions to these slangs. Proper usage of these slangs can make perfection of translation for translators.

4 CONCLUSIONS

Difference between China and the west provides the chance of second creation for written translator. Culture collision inspires deeper language charming, and this is the reason why readers prefer translated articles. People need to use more wisdom to feel the article's spirit and language feature. Written translation shouldn't be limited by culture difference, and a translator can be a good messenger through proper quotation, understanding and utilization, thus better spreading the civilization of human being.

REFERENCES

[1] Ping Ping, Ma Tiechuan. *Translation Limitation Caused by Culture Difference and Its Countermeasure.* Net Friends. 2014 (3): 57–57.

[2] Peng Tingting. *Discussion and Research about the Difference between China and the West.* China Technology Overview. 2014(3): 296–296.

[3] Wu Zhiwei, Han Ying. *"Brief of Oral and Written Translation in Global Background on International Conference.* East Journal of Translation. 2014(2): 90–92.

Education Management and Management Science – Zheng (Ed.)
© 2015 Taylor & Francis Group, London, ISBN 978-1-138-02663-6

Themes of education philosophy in modern society

Guiling Han
Faculty of Education Science, Shangqiu Normal University, Shangqiu, Henan, China

ABSTRACT: For education in modern society, corresponding ideology is needed to guide proper and efficient educational work. It is positive for educational work to enhance study of education philosophy in modern society. In terms of current state of domestic education development, problems emerge mainly from objective reality and psychology of teachers and students. Education philosophies and thoughts tend to have several similarities, where humanization, expansibility and practicability of education are the most representative.

Keywords: modern education philosophy; humanization; expansibility; practicability

1 INTRODUCTION

Education philosophy is a field where philosophy is used to interpret and influence human educational activities. Dating back to ancient times, some philosophies gave significant enlightenment to educational thoughts and ideas, such as Confucian thought of equal education for various students, and early thought system of materialism advocated by legalists. In modern times, people with lofty ideals paid more attention to education, with emancipation of minds and eastward spread of western culture [1]. This contributed to the formation of advanced and enlightened education mode which laid a necessary foundation for today's education prosperity. Due to inseparable relationship between education philosophy and educational activities, coordination of them helps achieve the goals—to impart knowledge and educate people and to rejuvenate China. On the contrary, without appropriate philosophy, teaching performance is likely to walk on a wrong path, ultimately trapped in the vicious circle of talent training and giving adverse guide and negative impact to students. Therefore, schools and teachers need some ideas of philosophies as guidelines to ensure their work in a right direction. Guided by education philosophies, teachers are able to grasp main contradictions in educational work, consider and balance all factors and prolong vitality of teaching practice.

2 CURRENT STATE OF EDUCATION IN MODERN SOCIETY

As a Chinese saying goes, strong youths lead to strong country. Education is the hope of a country and a nation, with next generation education regarded as a main approach to change future social life. Throughout history, education forms developed from old-style private school and imperial examination in ancient times to modern popular prenatal education, preschool education, compulsory education and higher education [2]. Whatever their purposes are, people commonly hope children to be well-educated for a better future.

Before studying in school, people, in most cases, are innocent, ignorant and pure. Enjoying those characters, they may fail to cope with kinds of challenges from cultures, interpersonal communication and complex society when they grow up. Education in fact is a process for a person to be physically and mentally matured, and form and perfect outlook on life and values. Both educational activities and philosophies displaying in activities will profoundly and significantly influence the innocent children.

Covering a long history of education, Chinese extracted precious experience and lessons from teaching practices in a long time, such as equal education without classification of students and striving with one's own efforts under teachers' guidance. Chinese education now faces a huge amount of educated people, various types of education and spotty education levels, showing great differences in terms of time and space [3]. It is easy to find schools enjoying good equipment, kinds of curriculums and strong teaching staff, while there are schools with few students, blocked traffic, one or two rural teachers, no fine blackboard and desks and chairs. Due to the huge objective differences, educational activities and qualities also diverge greatly from other schools.

In addition, students are forced into fierce competition by gradual saturation of education capacity. Students are compelled to study excessively due to issues of difficulties of going to school or entering higher schools; besides, heavy school work always squeezes out-of-class time. Such cruel learning mode will accompany students from kindergarten to senior high school and university for over ten years. For domestic students, grueling study has turned into a nightmare. In foreigners' eyes, Chinese students always study hard, though they are not born like this. Compared with American children aged around ten who receive liberal education and light homework within one hour, Chinese students have already been imbued concepts about score, test and enrollment rate by schools and parents. As results, student's nature and creative thinking are constricted, contributing to failure to dominate their growth and study even with strong learning inertia. This kind of education will expose its defects rather than improvement in the long run, leading to retrogression of scholastic ability and other students' abilities without promoting domestic education. Advanced educational concepts handed down for years are still significant models for modern education to learn from, such as incorporating things of different diverse nature proposed by Cai Yuanpei and emancipation of minds. However, facing improper distribution of educational resources and partial and rational educational concepts, school, together with teachers and parents loses composure. Proper and effective modern education cannot be achieved without recognition of social reality and stable promotion of education forwardly.

3 THEMES OF EDUCATION PHILOSOPHY IN MODERN SOCIETY

3.1 *Humanization of modern education*

Humanization practically represents an enlightened attitude and inclination, upholding that respecting every single person is essential prerequisite for certain social change. The renaissance in early Europe showed that recognition of people's views or thoughts and liberation of human nature can not only ensure correct and scientific guiding thoughts for activities but also accumulate strong spiritual strength for a team or group to promote people's behavior.

Education is a bright field where creative teaching and guiding work are engaged. Long-term psychological construct is required in teachers' imparting knowledge and training students. As with the concept of human innate goodness proposed by Mencius in The Doctrine of the Mean, everyone has a sense of sympathy, sense of shame and sense of right and wrong, which are indicated in modern education. Teachers are required to replace reprimand and punishment in class with proper and timely communication and psychological instruction in line with the efforts of function transformation of teachers and schools. For example, teachers are supposed to walk down rostrum into students, make friends with them for psychological intimacy, communicate with them and educate them. The change of attitude and identity not only helps with teachers' and schools' smooth transformation, but also psychologically relax students to carry out learning plan freely. Meanwhile, a bilateral communication mode between teachers and students is formed to gain knowledge and abilities.

In addition, humanization of modern education is also shown in specific teaching activities. Single and old teaching concepts cannot cope with a large number of students with various characters, and students aged around ten or twenty who think even more actively and boldly. Therefore, previous thoughts need transforming to improve education activities spirally and dynamically as soon as possible. At present, firm adherence to familiar pipelining teaching and duck-stuffing teaching hardly stops the teaching dilemma but escalate contradictions between students, teachers and teaching, ultimately destroying teacher-student cooperation. It is more significant to classify students' learning needs and treat students differently. With foci both on group and individual students, teachers are able to get a whole picture of students' learning to make use of scale advantages, and instruct individual student with best efforts for comprehensive development. Though this approach is bound to demand more manpower, and material and financial resources, it complies with its purpose—all for children. With different characters, students who have huge potential will infinitely enlarge positive effect of the investment. Therefore, talents in various fields will be discovered through education, making educational activities charming.

3.2 *Expansibility of modern education*

Facing chances and challenges brought by new social elements, domestic educational work now still needs refinement and improvement. Deng Xiaoping has ever pointed out that education shall be oriented to the needs of future, of the world, of the modernization, which now are keys to keep constant development of education.

Current educational work is different from that of two or three decades ago, mainly attributing to development of hardware devices. Teachers now take advantages of multimedia and remote resources in class while students replace their handwritten notes with multimedia courseware and electronic devices. In addition, Internet revolution provides both convenience and temptation for teachers and students through changing traditional class mode with textbook and notes in educational field—the peak of knowledge activities. Despite the easy access to knowledge brought by internet, educators are reminded to think about preservation of education value. Meanwhile, educators are supposed to turn students' volatility in information era into learning desire. Educational expansibility will be retained, with the dynamic tracking of education tendency and social talent demands and the timely understanding of current affairs' influence on talent training. Therefore, demands adjustment is likely to be conducted by catering to the policies.

Khan Academy mentioned in Salman Khan's book persistently carries out free online education which relies on strong internet function of resource transport and broad display platform. It wins good comments and reputation from the world mainly by its use of internet resources along with internet revolution and realization of its value proceeding from public benefit. Besides the development space brought by internet, educators are supposed to pay attention to hot issues, international interactions and educational innovations, which are fresh to people and have strong vitality from birth to death with possible changes. Therefore, people need to dig them up in depth within a very limited time to help with teaching activities, for example, topics may be about air accident of missing Malaysia Airlines flight 370, food safety issues about KFC and McDonald's, FIFA World Cup in Brazil. With topic's involvement in teaching activities, teachers are enabled to guide students to think about reasons of the accidents and instruct their seeing and hearing. Education constant development and innovation need endless materials to provide new ways and references, and various lessons from other parts of society to succeed.

3.3 *Practicability of modern education*

Different from the fixed pattern formed by imperial examination for selection of governors in ancient times, modern education particularly has practical significance. Specialized operation with high standards demands of people sufficient basic education and professional education. With the concept of education serving economy and society which is not customized for vocational education, kinds of education are supposed to combine purposes and aims with the demands of talent training. By the combination, education emphasizes more on some particular aspects according to social needs of talent for social and economic construction, in order to overthrow the "useless" bias of education and create fortune for humankind. As domestic education benefited from society should in turn benefit the society, modern education philosophy seems to follow this trend to bring powerful impetus for domestic education which stepped into a plateau.

As teaching plan of practical education is based on adequate practical surveys, teachers are guaranteed to change teaching focus by reliable data and practical experience. For example, in Chinese teaching, students' practical states may drive enhancement of practical writing, reading comprehension and proper curriculum adjustment. In terms of examination, corresponding adjustment may be conducted according to different practical value of different chapters, in order to examine and train students' strain capacity. All these adjustments require schools and teachers to break through the traditional exam-oriented education to optimize structure of teaching contents and ensure maximum benefits for students. Additionally, to discriminate and select good teaching plans also indicates practicability of education, with some impractical and idealistic innovations excluded from educational practice due to grandstanding and flashy contents. Though education is encouraged to stretch thinking and apply innovative concepts, the essences of education cannot be neglected. Education may achieve false prosperity by some strange titles and teaching performance as well as impractical methods which hardly represent the direction of education development, scarcely improve teaching quality but puzzle and disgust students. In measurement and evaluation of teaching innovation, people ought to be cautious to strictly balance practicability and feasibility. Teaching practice is essential to evaluate even fantastic teaching concepts, contributing to the prevention of distortions and the deterioration of education and the formation of precise understanding of education philosophy.

4 CONCLUSIONS

Under the background of a complex and volatile society, constant development of domestic education essentially requires precise understanding of education philosophy. Studies and discussions conducted by specialists and scholars, along with educators' practices concluded the more clear themes of education: human concern for students, dynamic understanding based on internet and social affairs and continuous enhancement of practicability, all of which provide right direction for education development and space for educators' self-improvement. Meanwhile, other advanced education philosophies and corresponding ideologies will be examined and improved in teaching practices.

REFERENCES

[1] Wen Hengfu, Yang Li, Significance of Constructive Postmodern Educational Philosophy on Education Reform: Summary of International Academic Symposium on Constructive Postmodernism and China, *Philosophical Trends*, 2012 (11): 111–112.

[2] Pan Xiwu, The Fall of Modern Educational Philosophy and its Consequences, *Journal of Schooling Studies*, 2011 (4): 5–10.

[3] He Junhua, Construction of Modernized Educational Philosophy with Axiology: Comments on Wang Qingkun's Book of Educational Philosophy—A Study of Philosophy with Axiology, *Jounal of Higher Correspondence Education (Philosophy and Social Science)*, 2008 (6): 79–80.

Education Management and Management Science – Zheng (Ed.)
© *2015 Taylor & Francis Group, London, ISBN 978-1-138-02663-6*

Graphic design and visual processing of computers

Luhua Zhao
Henan Quality Polytechnic, Pingdingshan, Henan, China

ABSTRACT: It's vital for graphic designs to have aesthetics. In the later period, computer graphic designs can rely on technical platform and graphic software. The designs are closely connected with visual impressions and impacts. Therefore, it's studied in the work on the connections between them with the graphic designing theories, concepts and visual impressions.

Keywords: computer image; graphic design; visual processing; analysis

1 INTRODUCTION

Computer image is the most widely used in the field of graphics. Based on the technique and computer software, the graphic processing in the later period will remedy the photography to realize the visualization and modification of images [1]. Therefore, it's possible for the computer graphic designs. Nowadays, there are quite a few fields and styles based on computer graphic designs. It will directly influence people's visual impressions on how to realize the best design effects. There is a potential connection between graphic design and visual processing. Therefore, the real graphic visualization and visual diffusion will come true through the relationship coordination.

2 CONCEPTS OF COMPUTER GRAPHIC DESIGNS

The computer graphic design is a relatively modern form. Transferring from papers to computers, images are becoming more intelligent and visual. As the purpose is to satisfy people's needs and realize aesthetic feelings. For graphic images, the designs are very important both in the early and late stage. New feelings can be shown in old images through processing in the later period. The premise of designing computer graphic images is based on many basic factors [2]. Therefore, it takes advantage of computer techniques and software to make brand-new graphic images. However, the concept based on the graphic images is the most important in design. The designing concept is the soul, and the expression of graphic meaning will not come true until concepts are conveyed.

2.1 Design of concepts of computer graphic images based on aesthetics

The short-term memory in nature shows the significance of pictures. As pictures come out due to people's wish to remember the past. Graphic images are given more contents due to the development of techniques and occurrences of photography. Beauty is a kind of pursuit, as well as a kind of enjoyment. It's very important to show beauty in the field of graphic images. It's successful to have pictures showing people aesthetics. Meanwhile, aesthetic transmission should be based on designer's aesthetics while designing computer graphic images [3]. It's relatively easy for computer graphic images to have aesthetic designs. Can we realize the transmission of aesthetic designing concepts with computer technology? To achieve these goals, it's needed to analyze as follows:

First, transmission of designing concepts of subjective aesthetics; based on designer's awareness, subjective aesthetics add aesthetic feelings to designing concepts, and create pictures of subjective aesthetics. Subjective aesthetics is a relatively abstract designing concept, showing an individual's understanding. It means the designed computer graphic images will not be accepted by the majority. Therefore, subjective aesthetics is a designing concept belongs to the minority. However, it enjoys both necessities and developing potentials for computer graphic designs to convey the concepts of subjective aesthetics. Although the objective aesthetics are popular, someday people will have aesthetic fatigue. Therefore, it's worthwhile to advocate concepts of subjective aesthetics.

Second, designing concepts of objective aesthetics; computer graphic images will mostly tend to objective aesthetic designs, because it can be recognized widely with the help of objective aesthetics. It's easier for people to accept direct aesthetics. While abstract aesthetics need further understanding to be accepted. Meanwhile, the concepts of graphic designs, which are based on the computer technology, will tend to objective aesthetics. As the traditional graphic designs are deeply rooted, designers are willing to express their feelings with brushes. While the computer emphasizes more on the glaring colors and standard lines, it has more advantages in showing objective aesthetics.

2.2 Design of concepts of computer graphic images based on ideology

Any picture shows an artistic conception, as well as a theme. The expression of a theme conveys the ideology. It's easier to connect with subjective aesthetics when studying graphic ideology transmission. However, some differences actually exist. It takes advantage of technology to finish the design of computer graphic images. Computers are stricter and more accurate in controlling lines and colors. However, designing of graphic images is not so accurate. Sometimes the anomaly and defect will highlight the graphic image. Therefore, advantages of computer graphic images also have defects. There're two possibilities in expressing ideology with computer graphic images.

First, the design of computer graphic images fully utilizes the computer technology. Therefore, it's accurate in drawing lines and using colors, especially in showing hard lines with brush tools. All these are important advantages. With the help of relevant software and operation platform, it shows a hale or beautiful picture in the most accurate and beautiful way. Sometimes, the transmission of ideology needs the accurate expression, and the computer graphic images will do a good job.

Second, it's obvious that computer graphic images have advantages in painting intelligently with the principle of sensor. However, due to the need of something irregular, powerful and unconstrained style, computer graphic designs will be deficient. The high accuracy and intelligent imagination affects the expression of images to a large extent. It's difficult for computers to draw some messy and tonal differences, or something not following the rules. Therefore, computer graphic images will not always demonstrate the ideology perfectly.

3 PRINCIPLES OF COMPUTER GRAPHIC DESIGNS AND COLOR MATCHING

Designs of computer graphic images should follow some principles, including the designer's ideology

and theme. However, something needs to be solved will affect the design in the process. As the theme is the soul of a picture, the design needs to satisfy the most basic elements. A high quality design fully expresses a thought to make audiences understand the designer, so that the graphic design pays.

Besides, it emphasizes color matching more in graphic design. Therefore, it influences the graphic comfort to a large extent. In order to ensure the integrity of the image, not only the artistic conception should satisfy the designer, but also the color matching should cater to people's appreciation. Therefore, an image of high quality should meet the following requirements:

First, it's required that the color matching is reasonable, as well as relatively full difference of warm and cool color. Meanwhile, soft degrees of whole picture should be guaranteed. The realization of soft degree lies in the rational use of color to make painting style aesthetic.

Second, standard of painting style should cater to the requirements of modernization, the too dark or depressed one should not appear. Modern society always pursues the positive energy. The simple and sunshine style appeals more to people's aesthetics.

In all, the computer graphic design emphasizes the improvement of design efficiency, real reduction and accuracy of the color.

4 VISUAL PROCESSING EFFECTS OF COMPUTER GRAPHIC IMAGES

In computer graphics, mainly we design initial pictures, which not influenced by outside environment. However, in real jobs, pictures are the premises, and people do the later processing. We mainly deal with the photographed pictures in later processing, and it's quite common for computer graphic later processing. It emphasizes visual effects to make colors and tonal appeal more to requirements, so that it can be used in different conditions. The later processing and visual adjustments of computer graphics are based on the editable digital image and the principle of minimum unit pixel. In daily lives, it's quite common to do visual processing of computer graphics. Therefore, it's necessary to analyze from the following aspects.

4.1 Adjustment of color in improving the visual effects and the contrasts

Color expresses the way to emphasize the sensory effects of image. It's aimed to show visual impacts and visual aesthetics with contrasts. Color is always adjusted in the process of computer graphic design. For example, cool color is added to bright images to create silence and mystery. It's a relevant color adjustment related to real needs to form special visual effects and strong visual impacts. Besides, visual contrast is widely applied in the computer graphic images, aiming at adjusting the dual tone in the opposite way to form computer graphics based on anti tones. The creation and later process of these images will form a strong visual contrast. Meanwhile, they will improve the picture's potential infection to the largest extent and add many characteristics to the whole picture.

4.2 Later processing of adjusting the sharpening and fuzzy visual effects

It needs later processing because some details in the picture will not be demonstrated or covered while using. For example, the picture of a car, especially the private ones, need to cover the plate. The way to deal with the plate will be associated. So that sharpening and fuzzy are the two most common ways in visual processing and later processing of pictures.

4.2.1 Visual effects of fuzzy implementation

The blurring represents fuzzy process of certain district. However, there are many different ways of fuzziness in the actual use. Gaussian blur is the most commonly used software, and it mainly deals with subtle changes with blur effects. It has great advantages in the process, covering without seeing less. Therefore, the gaussian blur is commonly used. Especially with the help of the fuzzy radius of relatively small diameter, it's more capable of achieving the blur effects.

4.2.2 Realization of effects of clear visual texture by sharpening

The later process of some pictures, especially the small size ones, will adopt the method to make the textures clearer. However, the principle of sharpening is not to deepen the texture, but to make the bright one brighter, darker one darker. The whole picture will be in order, and visual texture cleaning will appear in strong contrast. Therefore, to make later process more appeal to visual aesthetics, it follows regulations to achieve the prediction, based on real needs.

5 CONCLUSIONS

The development of computer technology pushed forward the techniques of graphic images. It allows the application of higher level software and techniques in the graphic field, changing the traditional viewpoint of image. Meanwhile the graphic image processing based on computer technology can design images satisfying real needs with the help of excellent later processing. And the key of designing is to ensure your picture has subjective awareness and demonstrates the theme and ideology with colors and lines. However, the later process deals with visual effects according to real needs. The adjustment of color is to make images have impact forces after improving visual shocking sense. However, common operations such as the sharpening and guassian blur are to make excellent imaging. In all, it's relatively easy and highly efficient in dealing with computer graphic images. Designer with subjective awareness can design pictures he/she enjoys. The later process further promotes the overall composition of pictures and realizes the real intelligent designing.

REFERENCES

[1] Feng Xun, Tang Xiaohua, Liu Meilian. *Analysis of Thread Tension Factors of Computer Embroidery Machine Bolt on Graphic Image Technology*. Development and Innovation of Mechanical and Electrical Products. 2013(01):85–86.
[2] Wei Hui, Liu Chun. *Method of Displaying Aircraft Track on Image Processing*. Computer Engineering and Design. 2013(02):87–91.
[3] You Yuhu, Liu Tong, Liu Jiawen. *Summary of Autofocus Technology on Image Processing*. Laser and Infrared. 2013(02):132–136.

Education Management and Management Science – Zheng (Ed.)
© 2015 Taylor & Francis Group, London, ISBN 978-1-138-02663-6

Ideas and advantages of Sino-foreign cooperative education

Guimin Feng

E&A College, Hebei Normal University of Science and Technology, Qinhuangdao, Hebei, China

ABSTRACT: Under the background of education reform in China, the effects of Sino-foreign cooperative education should be facilitated to improve Chinese education level. By analyzing current debates on the ideas and advantages of Sino-foreign cooperative education and some necessary idea changes, this study reached several conclusions on the developing trend of Sino-foreign cooperative education. With the comprehensive discussion on the ideas and advantages of Sino-foreign cooperative education, this study proposed some plans for the sound development of Sino-foreign cooperative education so as to promote the positive meanings of Sino-foreign cooperative education.

Keywords: Sino-foreign cooperative education; teaching ideas; teaching advantages

1 INTRODUCTION

With the continuous education reform in China over the recent years, Sino-foreign cooperative education has been common in higher education. For example, some Chinese education institutions and foreign education institutions are exchanging students. However, due to the continuous increase of returned students and low rate of economic return on current higher education, many families become careful to studying abroad. Through Sino-foreign cooperative education, the education cost can be reduced while the education level can be improved [1]. However, there are some differences in teaching ideas, which need to be updated in current Sino-foreign cooperative education, so as to facilitate the deeper development of Sino-foreign cooperative education and further demonstrate its advantages.

2 ANALYSIS ON THE IDEAS OF SINO-FOREIGN COOPERATIVE EDUCATION

2.1 Debates on the ideas of Sino-foreign cooperative education

Sino-foreign cooperative education can be regarded as an important way for China to internationalize its higher education. The current education internationalization is demonstrated in two forms. One is higher education internationalization and the other is open national education market. The first form aims at promoting mutual cooperation among international education institutions. The second form adheres to market support, and its education model and activities mainly use economic development to facilitate education cooperation, regarding higher education as an international commodity [2]. The two forms of education internationalization reflect the differences about education philosophy. The core debate on this issue includes the following aspects: the first is whether higher education needs to adhere to market competition; the second is whether higher education can be consumption product for students; the third is whether higher education can be commodities exchanged in different countries. Some scholars believe that higher education is a kind of social cultural heritages and an important part to study social structures. Therefore, higher education, in essence, should be public property. In a sense, it can be even a kind of public responsibility. For developing countries, higher education is an important channel for them to promote their social productivity.

Other scholars argue that apart from the consideration on the education essence, there are other factors on the differences of cooperative education ideas. On the operational level, many traditional famous universities have been integrated with local culture. Therefore, the teaching ideas of these famous universities, like Oxford and Cambridge, are hard for other countries to copy successfully [3]. Those successful Sino-foreign cooperative universities are newly found, and these universities haven't been constrained by academic traditions or systematic regulations. Therefore, they can leisurely face the pressure from academic circle and satisfy market demands, which is also an important way for the innovation of cooperative education.

Currently, there are a few of OECD countries that oppose education marketization. However, education marketization has been the developing trend for cooperative education. Many higher education exporting countries are conducting education cooperation with undeveloped areas for economic benefits rather than education assistance. Therefore, the ideas of Sino-foreign cooperative education should accept commercialized education and observe market-oriented operating ideas and models, which ought to be applied in enrollment of students, major setting, education goals and management.

2.2 Innovation on the changes of Sino-foreign cooperative education ideas

Over the recent years, the financial crisis caused by American subprime crisis has spread to the whole world. Both the eastern and western world are considering the future trend of economic development under the backdrop of economic globalization. Politicians and scholars are seriously study the challenges to human commercial morals brought by liberal economy and virtual economy. In a sense, this is not merely a simple financial issue, but the nature of economic conducts served by this financial system. Therefore, when people are pursuing for wealth and benefits, even the most advanced liberal economy system can be manipulated and changed during its development.

As demonstrate by economic development, in the process of Sino-foreign cooperative education, we need to review the relationship between education support and education commerce under the background of over marketization. We also need to ask that whether the profit-oriented education ideas have deviated from the essence of education and the goals of Sino-foreign cooperative education. It is questionable whether this profit-oriented cooperative education will cause financial loss to China. If the answer is yes, how we can use national policies to guide cooperative education, thus establishing correct ideas and readjust the direction of cooperative education. All these questions are not aimed at reducing the profits of cooperative education, but aimed at promoting the development of cooperative education with certain economic returns based on the model of non-government funded education. If we suddenly reduce or constrain the profits of cooperative education, the foreign investment would be severely influenced, thus hindering the introduction of foreign education resources. In addition, with the guidance of national policies rather than reliance on market only, the profit-oriented trend of cooperative education can be rectified in a certain extent and the scientific development direction can be clear and definite.

3 ADVANTAGES OF SINO-FOREIGN COOPERATIVE EDUCATION

3.1 Beneficial to attracting foreign investment

The Sino-foreign cooperative education is the necessary choice for Chinese education in the international environment. Seen from the different type of Sino-foreign cooperative education conducted by higher education institutions, an important advantage of Sino-foreign cooperative education is that it can attract large amount of foreign investment, which plays an important role in making up the shortage of higher education funds. Since China adopted its reform and opening up in last century, the whole country have been emphasizing on attracting foreign investment and have drawn up many preferential policies to attract foreign investment. Over the past several decades, China has achieved a lot in attracting foreign investment. Large amount of practices show that introducing foreign investment is important in quickening the development of education. The Sino-foreign cooperative education is the best form in education field to introduce foreign investment.

3.2 Beneficial to attracting foreign high-quality resources

The Sino-foreign cooperative education plays an important role in introducing foreign advanced school-running ideas and thoughts, management experiences and teaching methods. The guidance of these advanced teaching ideas is also beneficial to the cultivation of high-level internationalized talents. On the one hand, the Sino-foreign cooperative education is beneficial for the enhancement of the comprehensive education level in China. Under the influence of foreign advanced teaching ideas, we can conduct corresponding comparisons and introspections on domestic teaching systems and policies, thus promoting the deep reform and further development of Chinese education system. On the other hand, with the help of Sino-foreign cooperative education, the shortage and empty of some majors in Chinese universities can be made up. For example, the introductions of some majors like logistics, sports management and leadership have not only cultivated correlative professionals needed urgently, but also played a very important role in facilitating the enhancement of the comprehensive quality of the whole industry. The Sino-foreign cooperative education can lay a vital foundation for China to have continuous development in these industries and even reach the global advanced level.

In addition, Sino-foreign cooperative education can effectively use international education resources to enhance international competition level of domestic teachers through mutual integration and influence. The teacher partnership system adopted by Sino-America international management school can also facilitate the formation of close relationship between Chinese and foreign teachers, thus having mutual development in scientific research projects and teaching. In cooperative education, when foreign professors are lecturing, they can actively set up public gallery for Chinese young and middle-aged teachers, which can deepen the integration of both parties and provide superior conditions for the improvement of strong teaching staff in China.

3.3 Beneficial to the enhancement of Chinese education development

One of the goals of Sino-foreign cooperative education is to achieve sustainable development, which needs increasing funds input and innovation on running model and type of operation, thus forming its own characteristics. Sino-foreign cooperative education is supposed to promote its competitive power in fierce market competition so as to effectively activate Chinese education market. Then, different forms of Chinese education can continuously learn from Sino-foreign cooperative education. Meanwhile, these forms of education can bring threat and competition to existing education institutions. However, it is competitions that can better facilitate the reform and development of Chinese education and promote the innovation on education systems and ideas, thus improving the education institutions in China.

3.4 Beneficial for the development of international trade in China

In a sense, current Sino-foreign cooperative education is market-oriented. It becomes part of the international education trade in China, so developing Sino-foreign cooperative education to facilitate the education development in China, in essence, is contributions to the development of international trade. Since education industry has a basic and leading status with conditions progress and development, Sino-foreign cooperative education can play an important role in facilitating the sustainable development of international education trade and society in China.

4 DEVELOPMENT PLANS OF SINO-FOREIGN COOPERATIVE EDUCATION

4.1 Choosing cooperative countries in view of the general situation of national education development

Choosing proper countries to cooperate is a vital premise for Sino-foreign cooperative education. The choice needs to follow the strategic goal of national high-quality education development and some countries that can meet the future vocational requirements and students requirements are needed. The Chinese the Twelfth Five-year Plan regards vocational education as one of an important direction for the future development of Chinese higher education. Currently, German vocational education institutions have advanced management experiences, so China can choose to conduct cooperation with Germany according to the higher education demands in China. In addition, the US has excellent graduate education, so China can have cooperative education with the US in order to have first-class higher education and facilitate the continuous development of research-based universities. Through the platform of cooperative education, China should co-establish academic innovation platform with other countries and further strengthen the academic exchanges, thus seeking funds support from national policies to optimizing the Sino-foreign cooperative education.

4.2 Focusing on eliminating unbalance of education

Due to regional unbalance of higher education in China, when cooperating with foreign education institutions, China should lean to undeveloped areas to eliminate the unbalance situation of education, which can make up the shortage of insufficient education resources in these areas. To achieve this goal, China needs to have strong national guidance and education development plans since during the development of marketization, the strong will always strong. Only through policy intervention and guidance can China effectively optimize its education layout. For example, government can subsidize Sino-foreign cooperative education institutions and allow them to gain reasonable economic return, thus eliminating the unbalance problems of education resources.

4.3 Optimizing disciplines and majors

When conducting Sino-foreign cooperative education, China should actively optimize majors and even introduce some burgeoning disciplines and imperative majors. Therefore, education institutions need to analyze which major is needed. For example, many majors about financial management have been set up by Sino-foreign cooperative education over the recent years. With the occurrence of financial crisis and the gloom of Chinese security market, graduates of financial management and economy can hardly find good jobs. These majors are in saturation and needs to be pruned. Comparatively, with the continuous increase of infrastructure investment, talents of infrastructure management are rare, especially graduates of light rail construction and engineering management. So, China can deeply cooperate with Germany and Japan in these areas to enhance the quality of professional talents. It will have positive effect on Chinese economic development.

5 CONCLUSIONS

Although there are some divergences on the ideas of Sino-foreign cooperative education, it can play an important role in facilitating the development of education industry and the reform of education system in China. However, Sino-foreign cooperative education cannot be profit-oriented, and government should optimize the forms of education and strengthen the cooperation of week majors and disciplines through the guidance of policies. The government is also supposed to support the education resources to lean to undeveloped areas, thus effectively helping the development of education industry and facilitating its sustainable development. For the society, only through the sustainable development of education can we achieve social sustainable development. In this point, the advantages of Sino-foreign cooperative education will be more obvious.

REFERENCES

[1] Wang Daoxun. *Dilemma and Outlet of SFCRS in Chinese Central Regions, Education and Vocation.* 2014 (14):52–53.
[2] Yan Xiao. *Basic Principles and Influence Factors on the Introduction of Foreign Good-quality Education Resources of Sino-foreign Cooperation Education,* Jiangsu Higher Education. 2014 (01): 120–122+155.
[3] Qi Xiaodan, Zhang Xiaobo. *Running Mode and Development Path of Sino-foreign Cooperative Education Institutions,* Beijing Education (Higher Education). 2013 (12):14–16.

Education Management and Management Science – Zheng (Ed.)
© 2015 Taylor & Francis Group, London, ISBN 978-1-138-02663-6

Exploration for new rural social pension insurance system in China

Zou Zhang
Beijing Normal University, Zhuhai, Guangdong, China

ABSTRACT: With the rapid economic development, the gap between rich and poor is increasing, which is inconsistent with the goal of common prosperity. So, protecting the rights of poor and vulnerable groups has become the primary means of regulating social conflicts. China's new rural social pension insurance system, the latest model focusing on farmers' pension, aims at guarantee farmers' basic livelihood in their old-age periods. Therefore, research on the premise and objective of implementation of new rural social pension insurance system seems urgent. It is of positive significance in improving the new rural social pension insurance and narrowing the gap between urban and rural areas.

Keywords: new rural society; pension insurance; population aging

1 INTRODUCTION

New rural social pension insurance, including farmer personal fees and the subsidy of government and society, is deposited into farmers' individual accounts to guarantee farmers' daily life. Such insurance system, combined with land security, social support and assistance policies, has become an important part of China's insurance system. The implementation and improvement of China's new rural social pension insurance system can provide certain financial support to farmers, reducing their economic pressure [1]. That is also the main part of narrowing the gap between rich and poor and easing social conflicts. To some extent, new rural social insurance system is the key measures to solve farmers' supporting problem. New rural social pension insurance can improve the income level of farmers, and promote social harmony and progress. However, in the specific implementation process of this system, reasonably design is needed, for only reasonable system can comprehensively improve farmers' economic ability. New rural social pension insurance system can not only promote the realization of China's socialist target, but also solve the problem of farmers' basic supporting problem practically [2]. Therefore, to perfect China's new rural pension insurance system, emphasis should be given on the premise and objective of implementation, as well as the tasks that need completed in the process.

2 PREMISE OF IMPLEMENTATION OF CHINA'S NEW RURAL SOCIAL PENSION INSURANCE SYSTEM

2.1 *Traditional agricultural country, the foundation of implementing new rural social pension insurance*

Large land with rich resource and complexity of climates make China suitable for the growth of various plants. Therefore, it is the superior natural condition that makes China a traditional agricultural country, and agriculture has long been the pillar industry of China. According to China's State Statistics Bureau, till February 22, 2013, the total population of China in mainland is 135,404. Among them, the rural population is about 64,222, 48.8% of the total population, and has improved 1.1 percentage points compared with former year. Rural population in China occupies a large proportion, so famers' living guarantee is the major social factor affecting social stability. Famers' living guarantee has a direct impact on the development of grain production and other food processing industries, so farmers are of importance in agriculture and economy. Maintaining the stability and steady development of society, as well as gradually realizing a well-off society, is the requirements of China's socialist system [3]. They are also the main objective of this vulnerable group—farmers. China's new rural social pension insurance system, starting from agriculture, rural areas and farmers, is an important measure to really service "three agriculture". Therefore, the proportion of rural population more than a half is the premise of implementation and necessary requirement of China's new social pension insurance system.

2.2 Deepening of population aging degree

As the most populous country in the world, China has a relatively large population base. Over the recent years, affected by family planning policy, and the improvement of people's cultural level and the overall quality of society, people have gained a deeper understanding of education and responsibility. On the whole, people's attitude towards fertility has undergone a new change, leading to a steady decline in the number of newborns. While the population of the elderly is relatively large, leading young people have to shoulder the increasing pressure of supporting the elderly. At present, many newly married couples are the only children, so they need to work to support the elderly, or even grandparents both sides. Based on statistics of State Statistics Bureau, till February 22, 2013, the population of 60 years and above has reached 19,390, accounting 14.3% of the total population, with a 0.59 percent increase compared with former year. The population-aging problem in China is emerging, and is seriously affecting the normal life of society. And the excessive pressure generated from support for the elderly has brought a direct negative effect to young people's work and life. Therefore, the deepening of population aging degree is the main driving factor of implementing new rural pension insurance system.

2.3 Adjustment of economic structure

China has been in primary stage of socialism, with low level of economic development, when economic development structure needs continuous adjustment. Labor demand is constantly expanding, especially in the eastern coastal areas, where economic level is relatively high and with a lot of small and medium enterprises. Moreover, China's labor market appears an imbalance. The surplus labor in the west is relatively excessing, while the economic development level not high enough, leading to a weak demand for labor. So lot of western people are more accustomed and familiar with the flowing life. And it is very common for many young people to go to metropolitan to develop. Most young people, with low education, will not follow their parents to engage in agriculture production. They will choose to work in cities to broaden their horizons, thus changing their destiny. While the college students, who can create wealth through knowledge, seldom return to the countryside again after graduation. This vicious circle makes lonely elder people and

left-behind children become more and more. Data from the State Statistics Bureau show that the national total population of Family Separation is 279 million. Among them, 236 million are the flowing population. Young people depart their homes, worrying their children and elderly parents. In the countryside, the economy is relatively backward and medical conditions relatively poor. The elderly people mainly depend on the grain, which cannot fully guarantee their life in the case of illness and attending children. China's new rural pension insurance system gives the elderly a real guarantee and sustenance, thus making young people work outside attentively. In this way, young people can develop themselves while caring their families, thus enabling the society to develop more stable and harmonious. Therefore, the flowing population caused by the adjustment of economic structure is also the premise and requirement of quick execution of the new rural pension insurance system.

3 PURPOSE OF CHINA'S NEW RURAL PENSION INSURANCE SYSTEM

Construction of socialist new countryside is an important part of China's socialist construction. Rural pension insurance system, a software system in rural areas, can provide a powerful guarantee for socialist new countryside characterized by Chinese features. Meanwhile, it is also an important method to reflect the superiority of socialist system. Implementing new rural pension insurance system can alleviate the problem of the aging population and provide more opportunities and spaces for young people. So that young people can better devote himself or herself to work, thus making their own contribution for socialist modernization. It is also significant to solve the problem of support for the elderly people.

Implementation of new rural pension insurance system is an integral part of China's socialist pension system, and also a major breakthrough of the traditional pension model. Subsidies for farmers' pension insurance will change the self-supporting model, alleviating their financial burden. It can also provide farmers with some financial support, thus excluding the worries of young people and reducing their pressure. To some extent, it can increase farmers' distributable income levels and improve their living standards. Besides, it is of historical significance in the promotion of harmonious development, common progress and common prosperity. China's socialist system shows strong advantage. From the social protection degree perspective, the social pension insurance is an important reflection. Pension insurance has a strong practical significance in the aspects like narrowing the gap between rich and poor and realizing social justice. From the tax perspective, it can maintain the country's overall balance of payments, collect more social capital and increase revenue. In this way, more farmers will be provided with better services, with their basic living improved and their legitimate interests protected. Besides, the society will be more stable for the rapid and smooth development of China's economy, thus achieving better progress.

4 SPECIFIC TASKS OF CHINA'S NEW RURAL PENSION INSURANCE SYSTEM

4.1 Strengthening propaganda of new rural pension insurance system

The overall quality of farmers is not high, and their understanding of the pension insurance system is inadequate. So, explaining this new system in detail is necessary. For it enable farmers to have a real understanding of it, thus eliminating their reluctance to payment and reducing their blind resistance to pension insurance. Government should guide farmers to learn the advantages of pension insurance, including the benefits and normalization of it. Then farmers will be active in participating and cooperating, and the pension system be perfected. Make farmers know the vulnerability of land, then farmers will accept the new system and form, thus better understanding the new social insurance system. Encourage farmers to participate actively and making their paying fees more transparent, so that farmers will build a sense of trust to relevant departments and government, making the government well supported. Moreover, certain government propaganda is required to strengthen the new rural pension insurance system, thus ensuring better promotion and development.

4.2 Strengthening the support dynamics

Capital is the foundation to ensure the smooth implementation of pension insurance system. Government should visit farmers' life to understand their actual situations. And government also needs to manage the capital, income and insurance system well, increasing fiscal spending to improve the living standards of farmers. Capital is the best support for farmers. While strengthening the support dynamics, some targeted measures are also needed. For example, government can use different standards for different income levels, and different charging methods for villagers' different economic situations. Moreover, management of the capital is also an important step in pension insurance. With capital increase, certain supervision and control of capital flowing is necessary, so that more capital

can be better used. During the operation and support of capital, government should supervise the capital and pension system to protect farmers' legitimate rights and interests. Finally, a strict supervision system is crucial to avoid the emergence of corruption.

4.3 Promoting rural economic development

While increasing the capital support, rural resources should also be developed actively. That includes developing the economy with local conditions, stimulating employment and increasing technical support for rural economic development. In this way, the living standards of farmers will be fundamentally improved. In the process of becoming rich, government needs to strengthen rural infrastructure construction, like the construction of highway. Transport is the key link of grain transportation, and only smooth road can seize the better sales period of vegetables and grain. Rural economic development is the foundation of improving famers' disposable income. Government, while providing support and assistance, should focus on the local economy and exploit the development potential of rural areas, thus protecting the basic living of farmers. Providing technical support is also to promote economic development, since it is the key to increase grain production level. Guiding farmers to acquire more advanced production technology enable rural grain and cash crops to better play its economic value, so as to protect famers' life and guarantee their pension.

4.4 Perfecting related systems

New rural social pension insurance combines government subsidies and farmers' contributions. Farmers' contributions are a burden for every family, and they maybe hesitant. So, reliable systems and strict execution measures are necessary to prove the feasibility of pension system, giving farmers a sense of trust. In addition, issues such as the restitution of pension that most valued by famers also need reasonable research and improvement according to insurance system. At this point, government and relevant departments should put restitution in the first place, ensuring the prompt restitution and the amount. Perfecting supervision and restitution enable farmers to maintain a harmonious relationship with the government and relevant departments, achieving the long-term cooperation. Then, farmers' real benefits and pension will be guaranteed. Sound system is a guarantee for the implementation of rural pension insurance, because it can provide guidance and direction for the government and relevant departments. However, to better implement and enforce it, more institutional and regulations are needed to safeguard the legitimate rights and interests of people. From another perspective, a sound social pension insurance system is also the requirements of protecting farmers' benefits.

5 CONCLUSIONS

New rural social pension insurance has achieved great success in rural areas throughout China, protecting the benefit of farmers well. However, shortage of many subsidies still exists. That means the pension system is still not perfect, and need the efforts of all aspects, including the mutual cooperation and supervision of the government and family. After comprehensive analysis of the premise and objective of implementation, as well as the tasks need to be completed in the process, we should firmly adhere to the socialist road and the road of common prosperity. Only concerning the vital interests of farmers from actual conditions,

farmers' pension will be achieved and social justice realized. Meantime, the pressure on young people to support the elderly will be alleviated, thus they will have more time and energy to work and make a better contribute to the socialist construction. The superiority of socialist system needs to be reflected in every policy and system. As believing the party and government, people should be confident and hopeful about the future life.

REFERENCES

[1] Li Xingxing, *Study on rural social pension insurance legal from urban and rural harmonious perspective,* Modern Business Trade Industry. 2013 (9): 145–146.

[2] Cong Yun, *Improvement of the necessity and thought of China's rural social pension insurance system,* Science and Technology and Enterprise. 2013 (8): 222–223.

[3] Qi Min, *Policy effect of lemon market theory and social pension insurance system in rural areas of China,* China Economist. 2012 (10): 25–27.

Education Management and Management Science – Zheng (Ed.)
© 2015 Taylor & Francis Group, London, ISBN 978-1-138-02663-6

Cultivation of vocational students' English practical ability based on multimodal theory

Jinrong Shu
Hunan Railway Professional Technology College, Zhuzhou, Hunan, China

ABSTRACT: Disadvantages have been gradually exposed in the fixed mode formed by vocational English teaching during a long period. With the deep exploration of teaching reform, educators continuously enhance students' practical ability for their value realization in future professional post. As an advanced element with multi-angle and profundity in vocational English teaching activities, multimodal theory will provide new ideas in the aspects of teaching condition and network technique as well as college-enterprise combination. Meanwhile, it can also significantly improve English learning of vocational college students.

Keywords: multimodal theory; vocational students; English practical ability

1 INTRODUCTION

With the popularization of higher education, teaching activities have widespread promotion in various vocational colleges. They act as the bond of schools and society, aiming to provide tremendous young professionals for social production. However, considering weak school condition and teaching resource of most vocational colleges, there is a doubt about the effect of vocational education as well as the professional quality of their graduates, which also concern students, parents and enterprises. Although currently the government has increased funding and policy support for vocational education, more efforts of schools and educators are required in complex teaching links in order to really turn the input into reasonable educational output [1]. It is necessary to consolidate vocational education advantages and introduce new teaching concept and mode for the improvement of teaching work and the quality optimization of talent cultivation during the long-term practice and run-in period. The teaching exploration, with social responsibility and significance, needs frontline teachers' breakthroughs from fix thinking mode as well as strong coordination and support of students and their colleges [2]. These measures will ensure smooth process of the improvement activities and bring good achievements.

2 STATUE OF VOCATIONAL STUDENTS' ENGLISH PRACTICAL ABILITY

As a fundamental subject, English has a growingly consolidated place in education field. Increasing proportion of English course is considered in various teaching reform and examination adjustment, indicating the cultivation of students' English ability is still the key emphasis in future educational work. It needs attention and more efforts from schools, teachers and students. With the rapid internationalization of society and economy, there will be enormous job openings, which will give new challenges for internal students' English learning. In the meanwhile, various industries require plenty of special vocational employees with related professionals knowledge, and more expect that these employees can contribute for the perfection of departmental work system and the internationalized transformation with their stronger English practical ability and high-level professional quality [3].

Like most English major teaching, vocational English education presents certain limitation with its fast development in China. Seen from basic teaching condition, there is a lack of normalization in English textbooks adopted by vocational college students. In the early days, these colleges with short running time had no special English textbooks, adopting the same textbooks of undergraduate education. This behavior ignored the characteristic and special social responsibility of vocational education, and caused trouble to teacher-student English course activities during the practice. Students failed to adapt course difficulty and schedule, which influenced their learning quality and the enthusiasm in English subjects. In the next years, the government unified teaching textbooks for the regulation of teaching activities and teacher behavior in colleges. However, there are cases where some colleges still use self-compiled

English textbooks to meet students' further demand of specialization. Only the complexity in teaching textbooks, consequently, has already made vocational education in trouble of management and development. Meanwhile, with some colleges' limitation in infrastructure construction as well as teaching and teacher resource, it is worrisome for English ability of vocational students. In addition, teachers are helpless for cultivation of students' practical ability due to the fixed mode of exam-oriented education in schools. With monotonous inflexible curriculum provision and traditional lecture-based learning, students have become line products of vocational college English major. Lacking certain application and management ability of English knowledge, they still stay at the initial stage of English learning.

From the above situation, graduates, educated by this vocational English education, are hardly appreciated by employers and are difficult to realize their career ambition. For economic specialty students, they have to master English knowledge related with commodity information, transportation, trade negotiation, contract correspondence, etc., and can apply the English ability of listening, speaking, reading and writing to work. Only in this practical way can vocational English embody its practical significance and help vocational college students to realize their career planning and life value.

3 APPLICATION OF MULTIMODAL THEORY IN CULTIVATION OF ENGLISH PRACTICAL ABILITY

Multimodal theory is modal symbols including language, image, characters, behavior, etc. It is applied in daily communication among social members to optimize common knowledge information transfer and feeling expression with its meticulous quantized standard. The purpose of this mode is to change traditional single-mode theory and to fully explore individual inner potential in senses of slight, smell, hearing and touch. With its application in different practical activity, the more scientific and reasonable theory can be generally adopted in various fields.

3.1 *Expanding teaching forms with equipments*

Today's English education is more than words, sentence pattern and grammar. In vocational fields with high professionalism, especially, corresponding teaching activity reform cannot be limited to the increase of teaching section or the adjustment of teaching requirement. Pointed by *On A Synthetic Theoretical Framework for Multimodal Analysis*,

multimodal exploration in aspects of culture, language environment, content and expression is influenced by specific behavior of educators. And to some extent, it is also affected by non-language teaching attachments and implements.

It is suggested that English education teaching be a pioneer in utilizing various advanced technologies. With certain material and economic basis, these technologies from education are applied in education for the huge demand of society. Diversified teaching tools are necessary to keep teaching activities up to date. Especially in the application of multimodal theory to vocational English, schools and teachers are more supposed to actively find various possibilities with diversified equipments and teaching form to realize multimode of students' English practical ability cultivation.

Currently, widely used teaching tools have possessed multimedia devices such as projector, PPT and voice prompt system, helping students visually see course content. Besides, vision-oriented teaching form can easily present fuzzy language and ideology with images or characters. This provides free choices for students' absorption and comprehension, and further emancipates students' mind. Students are encouraged for the thought and attempt of practical English in language, movement and expression in order to arouse their stronger learning desire.

The breakthrough of traditional teaching pattern provides teachers a greater unlimited space. For each student's professional English competence, teachers' deep inquiry and survey can increase the pertinence and effectiveness of their teaching activities. With learning platform possessed high-tech elements, students can clearly know basic information such as major requirements of English competence, emphasis of major-related jobs and their present English practical ability, and then rationally schedule their school life. This reform is due to teachers' accurate choice of multimode as well as their scientific control of sensorial utilization and environmental factor. Consequently, the main idea of multimodal theory is conveyed, which is maximally beneficial to the teaching and learning.

3.2 *Efficient teaching system formed by network*

Currently, work of various industries is widely integrated into information network where data and information can be easily collected and collated and even be quickly called and conveyed if needed. Education field, cradle of these knowledge and skill, is more supposed to utilize integrated and networked hi-tech tools for the upgrade of specific work.

Vocational English teaching can find the most suitable curriculum provision and teaching content with reference to network textbooks. According to English learning effect of vocational students at different stages, timely adjustment and upgrade of such textbooks can avoid the discrepancy of teaching activities and textbooks. Multimodal theory emphasizes teachers' cultivation of students' practical ability in various English teaching forms. Teachers are required to focus on whether students have mastered enough practical English vocabulary and grammar, whether they can skillfully apply what they learned in reality and whether they can quickly adjust to new environments and interpersonal relationships. These proposed new contents and standards, which are far beyond initial teaching requirements, not only indicate more comprehensive emphasis of modern society on talent quality, but also can gradually adapt to the wide existence of integration and network technique in daily life. Teachers can properly balance and classify each student's ability of listening, reading, speaking and writing and even their practical ability, creative ability and divergent ability, and then integrate related data into special information database to form archives. The database is a direct record and reflection of students' English learning and can effectively urge their active learning. Additionally, the database also provides reference data for training programs of vocational colleges, connecting teaching activities with modern technology.

As the core role in multimodal teaching pattern, information concretely includes course content, teachers' language expression and students' feedback and interaction. After the integration of these activities into network system, it can be qualitatively and quantitatively evaluated about teaching results, input-output ratio and inherent potential. And the pattern operation can even provide reliable recommendations for teaching improvement. Consequently, only the network of teaching activities has already largely reduced teachers' working strength and decreased their futile action in teaching process. Meanwhile, it forms a virtuous circle between function and effect of vocational English with high teaching standard. On this basis, vocational colleges can expand network technique to other majors in teaching reform, student information management and teacher assessment and even schools' internal affairs management, to realize the modernization transformation of vocational college education as soon as possible.

3.3 Increasing students' practice opportunities based on multimodal theory

Effective connection between teaching activities and social positions determines the success of talent cultivation in vocational colleges as well as the realization of purpose to serve the economy and society. "Cooperation of colleges and enterprises, combination of work and study" is the personnel training mode proposed by Ministry of Education recently. It requires vocational colleges to adjust traditional lecture-based teaching, and emphasizes the consistency of society requirement and teaching requirement in college-enterprise cooperation to guarantee pertinence and effectiveness of students' learning. The best method to practice multimodal theory is to increase students' practice opportunities by all means. It will provide necessary reference to students' future career if they can comprehensively know about position statement, job specification and corresponding stimulation treatment of their major-related jobs.

In Hubei Province, vocational unified examination will retreat from its historical stage and be replaced by knowledge & skill assessment. The attempt to differentiate vocational education and general education not only further emphasizes the professionalism of teaching and learning in vocational colleges, but also is able to realize the important value of "From the society and to the society". With extensive use, vocational English has rich content including business English, travel English, scientific English, etc. It requires students have sufficient ability in aspects of spoken English, cultural background, reading and translation, daily writing, etc. These skills can make students high standard technical employees who are qualified for their future positions. And what's more, good English practical ability will increase additional value for these young workplace freshmen, providing them with more independent choice right. With negotiation and contracts, vocational colleges can establish stable and reasonable win-win cooperation mode with work units such as enterprises and departments. It is effectively helpful to open students' eyes by the organization visits of specific workflow related to practical English, the participation in lectures made by foreign specialists and scholars and even the transportation of quantitative outstanding internships to foreign enterprises or famous domestic enterprises. Through these measures, students will have a further and comprehensive understanding of actual requirements in own majors and realize the distance between their learning effect and job requirements, which can urge them to lay more emphasis on English competence and self-improvement.

4 CONCLUSIONS

For the students in our vocational schools, both schools and teachers should pay more attention to its professionalism as well as students' corresponding practical ability and strain capacity considering specific actual requirements in subjects of vocational colleges. Multimodal theory, compared with traditional teaching theory, has unparalleled superiority due to its visible and comprehensive perspective. Meanwhile, it has unique application value to students' different senses in teaching activities as well as various fields related to teaching. With the support of schools and related departments, teachers can constantly expand teaching forms in this mode and integrate diversified teaching activities into network by information technology. With these measures and further college-enterprise cooperation, finally, students' comprehensive quality including practical ability will be improved.

ACKNOWLEDGEMENT

This work was one of periodical achievements of Hunan provincial level project *Cultivation of Vocational Students' English Practical Ability Based on the Input and Output of Multimodal Theory* (NO. XJKO13CZY003) and one of initial outcomes of Hunan Railway Professional Technology College school-level course *Connection of Public English course and Professional Education in Higher Vocational Colleges* (No. K201414).

REFERENCES

[1] Yang Wenge. Several Problems of Vocational English Course Teaching. *Time Education*. 2014 (11): p. 238.
[2] Liu Shuqi. Preliminary Discussion on Employment-oriented Vocational English Teaching. *The Merchandise and Quality*. 2014 (3): p. 170.
[3] Luo Ying. Application of Multimodal Pattern in Translation Teaching of Vocational Business English. *Management & Technology of SME*. 2014 (1): p. 277.

Education Management and Management Science – Zheng (Ed.)
© 2015 Taylor & Francis Group, London, ISBN 978-1-138-02663-6

Construction design and maintenance measures for underground concrete construction

Fenglan Tian
Department of Material and Engineering, Inner Mongolia Vocational College of Chemical Engineering, Hohhot, Inner Mongolia, China

Donghai Bai
Vocational Counsel, Inner Mongolia Vocational College of Chemical Engineering, Hohhot, Inner Mongolia, China

ABSTRACT: The coal production operations are mainly concentrated in the mine shafts under the ground. For the architectural design of the mine shafts, many factors should be considered to ensure safety and normal production of coal. Currently, as regards China's coal production environment, the design of mine shaft is mainly on basis of its firmness and corrosion resistance. However, in the daily course of construction, maintenance measures for underground building still need to be considered. Therefore, the focus of this article is to analyze the construction design for underground concrete building and explore the main measures for the maintenance and operation of the mine.

Keywords: concrete; building; construction design; maintenance measures

1 INTRODUCTION

The level of industry is an important symbol of a country's comprehensive strength. Therefore, the importance of industrial development is obvious. The energy resources such as coal and oil etc., as the basis for industry guarantee the development of industries. The production status of energy resources can directly affect the industrial level. In addition, many rare metal mineral resources also play an important role in industrial development. Yet mining of coal and rare metal minerals usually require mine shafts. In the past, there was not enough understanding of mine construction to a certain extent, affecting mining efficiency in actual production. In the early times, due to lack of funding, foreign companies and investment were attracted, in hope of promoting China's economic development with foreign funds and the rich mineral resources were the main driving force for foreign companies to come into China [1]. Until now, the exploitation of resources through the mine shafts remains the main driving force of economic growth, but by the survey it's found that companies attach more importance to productivity and economic benefits than to the construction of mines, which didn't only influence mining efficiency, but also resulted in some safety incidents.

2 CURRENT STATE OF CONSTRUCTION DESIGN FOR CHINA'S UNDERGROUND CONCRETE BUILDINGS

2.1 Features of underground concrete buildings

Underground buildings compared with ordinary concrete buildings are particular. Especially for some deep shaft buildings, the concrete needs to withstand greater pressure. The mine shafts as auxiliary facilities provide the channel for exploitation and transportation, besides, with some special buildings, the efficiency of mining can be improved. In actual process, the underground concrete buildings should vary according to different resources. For instance, in the mining of coal resources, the safety measures like fire prevention must be taken into account. As for mining of rare minerals, the mining efficiency must be paid more attention [2]. The deeper the resources are stored the greater load should the building bear, thus the correspondingly thicker concrete. China is vast in territory and rich in resources. There are large numbers of mine facilities. Adjustment should be made according to different geological and geographical conditions. In the process of the excavation of the mine shafts, if the favorable geological and geographical conditions could be made use of, it can largely improve the efficiency of construction, such as the use of layer with larger bearing capacity, which could reduce the bearing load of the respective building. So the most conspicuous feature of underground concrete buildings is the bearing capacity.

2.2 Factors affecting construction design of underground concrete buildings

Many factors need to be considered, with regard to the underground concrete buildings, such as the resources to be exploited, geological and geographical conditions, etc. Factors should be taken into full consideration in construction design. The first is the geological condition. As the discovery of mineral resources require certain exploration methods. There will be a detailed geological and geographical information report in the exploration process and this report contains the information of layer formation and resource reserve, etc. In addition, the construction design would differ according to different resources [3]. Especially for some flammable or toxic resources exploitation, it's quite different from the ordinary underground buildings, and no flammable materials should be used. Also corrosion resistance must be considered so that the construction quality can be ensured to maximum extent. In order to improve the mining efficiency, the mechanical equipment should also be considered in the construction design, such as placement of equipment facilities. If the design and construction could be based on the existing equipment, the efficiency in use of equipment could somehow be improved. In addition to these factors, the quality level of a designer is also an important factor which influences the construction design of underground concrete buildings. Only the personnel with high quality can reach a scientific and rational construction design on the basis of these above factors. Survey shows that the average quality of construction designer in China is still low and foreign designers are employed for many large underground concrete buildings.

3 MEASURES FOR CONSTRUCTION DESIGN OF UNDERGROUND CONCRETE BUILDINGS

3.1 Quality improvement of the designers

In order to improve the situation of underground concrete building fundamentally, it's critical to ensure that the designers are of adequate professional quality and well aware of all the factors influencing the design. However, the quality of a designer is decided by the education. With the current situation of China's education system, the quality of designers is low. Obviously the education system reform can't be completed in a short period. Under such circumstance, companies can train their designers to learn from the western countries. Surveys show that the construction designs of underground concrete buildings in developed countries are comprehensively considered. Besides the factor of production efficiency, the safety for personnel and production is also paid much attention. Designers gained comprehensive knowledge in colleges or universities, but they didn't have any design experiences. When they first entered their company, they learned design from experienced staff, and then they started to design some small part of the construction. In contrast, many enterprises in China have adopted this model, it is difficult for new entrants who just graduated from college to calm down to learn from the experienced staff, thinking that they could design underground buildings independently. It causes the age of China's high-quality designers to be relatively high. In order to change this situation, a reward mechanism might be introduced to give a certain reward to those who complete a stage of learning.

3.2 The system for design check

To ensure the construction design of underground concrete building is scientific and perfect, not only should the designer be of certain level of quality, taking into account various factors, but also the design should be checked after it is completed. If the conditions permit, advanced computer-based simulation technology is advised, and the construction shall be carried out when no problems are found in the simulation. In the phase of design check, in addition to find out the existing problems in the design, it's also helpful to optimize the design. If the inspectors for design check are not the designers, there would be differences in understanding of the design, and the inspectors can optimize the design according to their own understanding. So ensuring the design is reasonable to the maximum. Especially, the design of large underground concrete building is a complicated work, requiring long period, during which, some advanced construction technologies may appear and these technologies can be used in the design for optimization in the check stage. It's obvious that the design check weighs heavily in ensuring it is scientific and reasonable. For a comprehensive check of the design, a perfect checking system should be formed. Firstly, the inspectors should be themselves designers with rich experience, and then they should be divided into several groups and each group reviews a part of the whole design. During the check, they could propose their remedy, and discuss with the designers during modification.

4 MAINTENANCE MEASURES FOR UNDERGROUND CONCRETE BUILDINGS

4.1 *The problems in the maintenance of underground concrete buildings*

After completion of the concrete building, certain precaution must be carried out to ensure the longest service life and the maximum bearing capacity of the building. And when quality problems occur, maintenance is required. As underground buildings are particular, especially with the use of large mechanical equipment, the concrete buildings are often seriously worn yet the companies in China have not paid enough attention to the maintenance of the buildings. When problems happen, the bearing capacity of the whole building shall be reduced if they are not timely maintained. In some serious cases, accidents were caused. Facts show that China's underground concrete buildings were not timely repaired when they were worn. Especially the less important buildings are repaired only during regular large-scale maintenance. However, in western countries, the underground concrete buildings are immediately repaired as long as the damage is discovered. For some sensitive parts, even experiments in mechanics are carried out and better materials are used to repair damages to ensure the safety. It's clear that maintenance not timely carried out is the main problem in China's underground concrete buildings at present. Although for some important buildings, the damages were repaired promptly, the materials and the technique used are relatively simple. Buildings could not even meet its original strength, thus resulting in bad influence on the safety of the whole building and re-damage of these repaired parts in future use. Therefore, in the process of maintaining underground concrete buildings, problem identification must be carefully carried out to ensure that measures are taken accordingly to solve the problems. And the safety of underground buildings is thus guaranteed. As for the problem solving process, comprehensive analysis on basis of the design and the actual construction conditions are required to ensure successful problem solving.

4.2 *Maintenance measures for underground concrete buildings*

If timely maintenance needs to be performed when problems happen to the underground concrete buildings, there must be a perfect maintenance system. And it must be guaranteed to be practiced with a strict reward and punishment mechanism. The first thing of the system is the discovery of problems, and currently the problems of underground concrete buildings are mainly found out by two means, one is the report from staff and the other by maintain personnel from regular inspection. After a problem is found, close analysis should be conducted on cause of the damage of buildings and the effect on the mine production. According to the influence to the production, different maintenance mechanism is used, for example, if the problem is of considerable importance, it should be responded to in the shortest time. In addition to keeping the maintained part with sufficient strength, in order to ensure the safety of underground concrete buildings, regular maintenance should be conducted and potential hazards should be identified to keep the underground concrete buildings at their optimum state thus the prolonged service life. It can be seen from above that a good maintenance system, do not only ensure the integrity of the building, but also, at the same time, greatly improves production efficiency and safety of mine production and personnel. And the implementation of maintenance system relies on the corresponding reward and punishment mechanism. Punish those who damage the building and reward those who protect it. A perfect reward and punishment mechanism is the key to successful implementation of maintenance measures. As in the process of maintenance by a team, it is very important to mobilize their initiative. Only when the construction personnel are positive and initiative could they deal with problems of their own accord in the maintenance process, and the total independent operational mode is achieved to some extent.

5 CONCLUSIONS

Mine shafts as auxiliary facilities play an important role in exploitation of underground resources. And the constructions of underground concrete buildings do not only affect the production efficiency directly, but also weighs heavily in the safety of production. We could know from the article that, as regards production in mine in the past, lacking of knowledge of the underground buildings, some accidents happened to the personnel and equipment greatly affected the company benefits. In order to avoid this problem fundamentally, all factors must be considered and a perfect maintenance system should be made for the construction design of underground concrete buildings.

REFERENCES

[1] Li Jing, *Crevice prevention in mass concrete construction, Transport Standard,* 2012 (21): 21–23.

[2] Ren Jinlong, Zhang Jian, Zhang Long. *Organization and management of mass concrete casting in overlength foundation slab, Architecture Technology*, 2012 (11): 977–980.

[3] Cheng Hong, Lin Shuai, Wang Shibing, Lu Qiwei, Yang Zijing, Xintian. *The design and realization of a highly reliable emergency power supply for auxiliary fan of mine shaft, Coal Engineering*, 2012 (05): 15–17.

[4] Zhang Xiaojie, Xu Jianxin, Du Shuo, Xu Chuanzhi. *Stability evaluation of foundation for coal mine gob site constructions, Journal of Hebei Polytechnic University: Social Science Edition*, 2012 (3): 137–140.

Education Management and Management Science – Zheng (Ed.)
© 2015 Taylor & Francis Group, London, ISBN 978-1-138-02663-6

Computer information safety analysis based on the network database data storage principle

Yanjie Zhou & Min Wen

School of Mathematics and Computer Sciences, Jiangxi Science and Technology Normal University, Nanchang, Jiangxi, China
School of Civil and Structural Engineering, Nanchang Institute of Technology, Nanchang, Jiangxi, China

ABSTRACT: In the Information Age, Internet is the main communication tool for people. Safety of computer information gradually becomes the concern of Internet users, who deeply analyze the protection and prevention of the computer information safety based on the principle of current computer network technology and the storage of network database. Through interpreting the storage principle of network database, this work analyzes the protection measures against the computer information safety based on the network database as well as makes a corresponding summary of the future development of the computer information safety.

Keywords: computer network; information safety; network database; safety analysis

1 INTRODUCTION

The development of computer network technology makes the Internet become the main tool and means of communication for people's life and work. It could be considered that the Internet has changed the traditional world to a certain extent. People's basic necessities have been closely linked with the Internet while their relevant information has been brought into the Internet as well. However, as the Internet brings conveniences to people's life and work, it also has somewhat impact on people's information safety [1]. Currently, with the arrival of the Age of Big Data, people are paying an ever-increasing attention to the safety of information as people become more and more dependent on the Internet. Network database storage is a data information storage method of computer network, aimed at saving corresponding data cache. Computer information security will have problems because of that. Therefore, this work is focused on analyzing the computer information safety based on the network database storage principle.

2 ANALYSIS OF THE NETWORK DATABASE DATA STORAGE PRINCIPLE

The interaction and transmission of network information is mainly realized by the network hardware equipment such as the mainframe, server, and switch. In traditional sense, network data mainly refers to the transmission between data packets and then using interpretation to decode, trying to reduce the method which users could recognize and read. Yet, switching equipment is usually needed during the interaction of information and data. Since the information transmission will cause some information to cache, the transmission equipment is needed to be equipped with the function of storage [2]. With the information interaction increasing continuously, the hardware buffer equipment couldn't meet the current requirements any more. Therefore, some research has been done in the computer software. Some network database based on the Internet has begun to possess bigger storage space, which expands the storage space directly to a certain degree. Then what is the storage principle for different storage methods? What comparison relationship does these two have in the advantage of storage?

2.1 *Analysis of traditional hardware cache storage principle*

Before analyzing the memory cell of the hardware, the basic principle and component of the network hardware equipment should be specified first of all. Though computer network is virtual, and still needs the hardware to be the basic platform, so as to realize the establishment of virtual network [3]. The server is the major equipment in the hardware constitution. The server is the base which computer network could login in normally. The login of the network is be completed by the server, which enable people to browse the links with other domains. While in data processing, the server needs to set the cache device for the purpose of storing a large amount of information temporarily so that it could ensure the high efficiency of data information processing. Beside that is the switch equipment. In the network system, the switch is used to allocate the resources better, thus realizing a reasonable allocation of resources. The switch also possesses the cache equipment, the principle of which is the same with that of the server. Both are aimed to process the information and data more effectively.

Through the above analysis, people could have some understanding of the storage means and equipment of the hardware equipment. Then what is the storage principle? In the processing procedure of network exchange information, how to realize the cache and transmission?

First of all, node control of hardware storage. No matter the mainframe, server or switch, they are all used to provide an effective network operation platform, which allows users to exchange information in a long range. Then when a large amount of information emerges within regions, there will be such problem—how to process the information. The hardware equipment of the network is used to transmit the information more quickly and accurately, so as to realize an efficient network environment. Therefore, when the hardware is caching the information, the cache nodes need to be set and determined so that the most accurate node for information cache could be calculated out. Normally, the setting of cache nodes of network equipment should take the size of data packets as the standard. Take the space, which could store under the normal network environment as the cache nodes.

Secondly, the analysis of hardware storage principle. Transferring the network data in the way of signal via optical fibre or base station and persisting suspendedly in the hardware equipment is the basic principle of hardware storage. In a series of data transmission processes, characteristic signal should be decrypted and decoded targetedly, thus ensuring that the signal transmitted every time could be transmitted to the terminal in the way of data and realizing the interaction of information.

2.2 *Analysis of the data storage principle of network database*

The hardware equipment could store some data, but the storage amount is very small. What's more, in theory, its storage amount is not the real storage. Instead, it is a kind of cache mechanism, aimed at releasing the efficiency and stress of data

processing through cache to ensure the patency of the whole transmission network. Therefore, it can be analyzed, in a certain sense, that the storage of network database is the real storage space.

2.2.1 *Network database gradually replacing traditional hardware storage*

Mobile Hard disk is often used accessory equipment which is related with computer. In the Computer Age and Information Age, data is the most important resource of all. Furthermore, some operations about storing and processing the data, to a certain degree, could only be finished with the assistance of some auxiliary equipment. In this way, the mobile hard disk is mainly to be able to transfer the local data. However, it is very easy to be damaged in the using process and its storage space is also limited. And there are some inconveniences when storing and transmitting the data. Hence, in order to enable users to better make use of the storage and usage of resources and data, network database is developed based on the Internet to store the data. The storage mode of network database is very simple. Only with the network, data resources could be uploaded and then backed up into personal SkyDrive. This is a kind of data storage with no transfer property. For this reason, the emergency of network database makes it more convenient and simple to store the data.

2.2.2 *Analysis of the storage principle of network database*

Network database, also known as network dish, is a kind of virtual memory space based on the Internet. With very obvious advantages, the network dish could establish its own SkyDrive space. Moreover, it is free for people to apply and possess their own network spaces. The realization and establishment of network space could largely avoid the inconveniences of peripheral equipment and enable data to be uploaded and backed up in the first time as well as handled timely. The storage space of network database is formed on the basis of the layering mechanism of the Internet and the network space principle. Since the saving and reading is all processed in the form of data, it will be more effective and fast for data processing and analysis.

3 ANALYSIS OF THE INTERACTION PRINCIPLE OF COMPUTER NETWORK

The interaction of computer network, in fact, is to make a reasonable allocation of network resources and transmit the corresponding information according to the instructions. Interaction is a two-way instruction mode, which could make use of the network equipment within the range and carry out the command to guarantee the effective transmission of information. In fact, the interaction principle of computer network, built upon the mutual cooperation between hardware and software, is to guarantee the transmission of data layer in hardware by building while in network aspect, layered structure could be used to allocate all content and resources in a reasonable way and then make corresponding handling finally.

4 INFLUENCING FACTORS AND SAFETY PROTECTION OF NETWORK DATABASE ON COMPUTER INFORMATION SECURITY

From the above analysis of network database and storage principle, it can be known that it is very easy and convenient to use network database and it could replace the traditional mobile hard disk drive. However, even though network storage has many advantages, there are still risks for the computer information safety. As a result, attention must be paid to the safety protection of computer information over a wide range of using the network database.

4.1 Analysis of the influencing factors of computer information security

Computer information safety is a very important problem. With the popularization of network, more and more personal information will be published to the website, which becomes a potential risk. Besides, to provide a safe network environment, a summary and analysis of the influencing factors of computer information safety should be made by aiming at some protection measures.

4.1.1 Information safety problems easily caused by the network vulnerabilities

Network vulnerabilities exist absolutely. Since the computer network adopts advanced language programmers to build the network layer, some network vulnerabilities is thus given birth to. That said, there will be some loopholes in the establishment of logical language inevitably, which would bring about the safety loopholes in the later using process of computer internet. Therefore, network loopholes are also a very important hidden danger for the information safety of computer.

4.1.2 Information safety threats brought by the Age of Big Data

The Age of Big Data could be thought as a new time of the Internet. The emergence of Big Data has also changed the original Internet mode, affecting the information safety of computer to a certain extent. In fact, Big Data is a technical term aimed at professional fields. Yet, in the Internet Age, the quantity of data is much huger and fast, which gives rise to Big Data. The influence of Big Data on the information computer safety could be analyzed in the following directions.

On the one hand, Big Data makes some related information become the key point to decode personal private information, which may cause the disclosure of personal information and directly affects the personal information security. Under the background of the Big Data Age, it is impossible for personal related information not to be totally exposed in the Internet environment. Therefore, it is these related information that causes our information to be disclosed and gives rise to some information safety crisis. This kind of crisis would cause our information to loss or make personal network life get interfered.

On the other hand, blind research is more convenient for the retrieval of the overall information in the Age of Big Data. Some other information could be retrieved by some related information. Currently, some search modes of magnet links on the network would retrieve all related information according to the current related information. Therefore, compiling some query and retrieval programs could help retrieve and position more quickly, thus being able to find out the information that the searchers want very quickly.

4.2 Influence of the network database on computer information safety

As network database provide conveniences for the storage and processing of data, it also has some impact on the computer information safety. There are very many problems in the actual network use and life, which would easily cause some safety problems. Generally, those data and resources that need to be stored will usually be backed up. Maybe there are a lot of resources that need our key protection and keep private. Though there is a unique and independent account and password in the network database, some criminals would also be able to steal the account and password through some evil means, who will enter your SkyDrive and stole the content inside. It affects the computer information safety to a certain degree. Moreover, some unqualified companies will set up a SkyDrive, too, because the threshold for application is very low. However, since there is no way to identify its normality, it is still risky for the content you store to be stolen. Undoubtedly, the most serious problem is the information leakage problem in the Age of Big Data. In the Age of Big Data, careful research is not carried out according to specific keywords any more. Instead, it is realized by some relativity. In reality, it is a very dangerous thing. Since your resource and data is uploaded to your own SkyDrive, once the SkyDrive gets attacked or is with any problem, the information will be leaked. And blind research by Big Data would influence its safety, too.

5 CONCLUSIONS

Through analyzing the principle of network database data storage and the influencing factors of the computer information safety, this work, to a certain extent, could enable us to how to reduce the hidden dangers of information by reducing the exposure of personal information when using the Internet. The information safety under the Age of Big Data, therefore, takes on added importance. Though the network database has a lot of advantages, there are still some potential problems in the management and protection of information security. And what's more, this kind of potential problem is very serious, which demands the computer network users to adopt a safe operation mode when using the network. And the application of network database could also be divided by the security levels to make data storage targetedly. Try not to transmit the most private content into the network database, in order to avoid very serious loss if the database is stolen. All in all, a reasonable using of network database, storing data targetedly, reducing the overdue exposure of personal information would all help reduce the potential safety problems of the information.

REFERENCES

[1] Li Guoqing, Li Xianguo. Research on Mobile Platform-oriented Web Information Interaction Model, Computer Applications and Software, 2011 (01): 176–180.
[2] Zhang Xian, Song Yuchuan, Liufei, Dan Bin, Deng Jianxin. Information Interaction and its Implementation for Virtual Enterprise Based on P2P, Computer Integrated Manufacturing Systems, 2004 (05): 579–584.
[3] Zhang Zimin, Zhou Ying, Li Qi, Mao Xi. Emergency Response Information Model Based on Information Sharing (Part II): Model Calculation, China Safety Science Journal (CSSJ), 2010 (09): 158–165.

Education Management and Management Science – Zheng (Ed.)
© 2015 Taylor & Francis Group, London, ISBN 978-1-138-02663-6

Advantages of integration of multimedia network and interactive English teaching

Juan Lei

School of Foreign Languages, Leshan Normal University, Leshan, Sichuan, China

ABSTRACT: Multimedia technology has been widely used in current teaching with a certain progress in the application. But there are still some misunderstandings. Teaching effect has been undesirable for a long time due to improper application and heavy dependence on multimedia. With the advancement of science and technology, multimedia is bound to be more and more used in teaching. Multimedia can play a better role only with current problems overcome. In this work, advantages and disadvantages of multimedia network teaching were analyzed at the start. The advantages of integration of multimedia network and interactive English teaching were then thoroughly researched for the innovation of teaching mode.

Keywords: multimedia network; interactive teaching; English teaching; integration

1 INTRODUCTION

As a tool of modernization, multimedia has brought a lot of convenience for teaching. But misunderstanding existing for a long time has led to some deficiency in teaching with multimedia. Efficient classroom can be promoted with these problems overcome and current interactive teaching integrated [1].

2 ADVANTAGES AND DISADVANTAGES OF TEACHING WITH MULTIMEDIA NETWORK

2.1 *Analysis based on classroom capacity*

Large capacity is the most obvious characteristic of multimedia teaching. Large capacity teaching with multimedia is helpful for control of classroom rhythm and increase of knowledge quantity. However, there are also disadvantages of blind pursue for large capacity in current multimedia teaching. For example, "The Olympic Games" and "Computer" can be finished in just one class. But with ignorance of main body role and acceptance ability, students cannot actually absorb all, although teacher passes on the knowledge in great quantities [2].

2.2 *Analysis from interesting in teaching*

With a large number of videos, images and cases, multimedia teaching stimulates the students' senses, greatly mobilizing students' interest in learning. Students' learning enthusiasm and efficiency can be improved to some extent. However, more attention is paid on the ornamental value of courseware currently. In order to attract students, some teachers make the courseware gaudy. What really attracts students is the courseware instead of imparted knowledge. The order is reversed. In another words, students often concentrate on the interesting pictures, rather than knowledge.

For example, some practical cases, videos or pictures of winning, fun in The Olympic Games are used in teaching "The Olympic Games". Students' enthusiasm of learning can be mobilized with advanced multimedia [3]. But too much multimedia materials can make the secondary supersede the primary. Students can ignore real learning of language in English teaching, only pondering The Olympics Games.

2.3 *Analysis based on interaction and students' participation*

The chief requirement of current curriculum reform is to exert the principal role of students, making students fully involved in the classroom. The classroom efficiency greatly can increase with the help of multimedia courseware. But interaction between teachers and students is partly limited for the difficulty of courseware production and modification. Teachers are hard to timely solve the emergencies with stiff multimedia courseware, bringing a rigid classroom.

Modern multimedia classroom or voice classroom can improve some interaction to a certain extent. But with disconnection to the Internet, students are difficult to obtain new knowledge by using multimedia technology. If connected, students trend to do something irrelevant to the class.

3 ADVANTAGES OF INTEGRATION OF MULTIMEDIA NETWORK AND INTERACTIVE ENGLISH TEACHING

Multimedia network is an advanced teaching method, while interactive teaching is a new teaching concept. Integration of the both takes their respective advantages, promoting efficiency of English classroom.

3.1 Better situation establishment

A situation is often needed in interactive English teaching for the requirements of the new curriculum reform. Students can better understand English in a situation in which teacher and students participating together. For the big difference between English and Chinese, there is a certain difficulty on establishment situation sometimes. But with the aid of multimedia network, English situation can be relatively easy to create. Student can better learn English knowledge through data collection and interaction.

For example, in the class "I have a dream", students are hard to understand Martin Luther King's speech for little comprehension of American racial issues. With multimedia network, racial discrimination is easy to reappear, making students immersed in that time. Thus, students can better realize meaning of the speech through identification and communication.

3.2 Eliminating communication barriers between teacher and students

Being shy or afraid to communicate with teachers directly makes a big communicational obstacle between students and teacher. Especially in English learning, students are usually shy for unskilled grammar and oral expression defects, fearing to communicate with teacher. It is unfavorable to improve classroom efficiency for teacher knowing just a little about students' learning. More encourage from teacher contributes to cultivate students' sense of participation and courage. Multimedia network helps students eliminate obstacles of interaction.

When talking about "Friendship", teacher asks some questions, such as "You want to see a very interesting film with your friend, but your friend cannot go until he/she finishes cleaning his/her bicycle, what he do?" Uncomfortableness of students can be avoided with multimedia network platform when students talk directly facing teacher and other students. Thus they can communicate better.

3.3 Teacher and students playing their roles

It has come to a common view of the roles that teacher and students play in class under curriculum reform. Students are viewed as main body, while teacher is leading factor. Teacher and students should fully play the roles, respectively. With multimedia network applied in classroom teaching, teacher can easier to control classroom rhythm, playing roles of guide, inspirator and dominantor.

With intense intellectual curiosity, students have very strong interest in new things. Students' curiosity and desire for knowledge can be stimulated by using advanced multimedia network in interactive English teaching. So that students are more willing to be involved in classroom, better learning knowledge and skills.

3.4 Learning from more teachers

Interactive classroom is not just confined to interactions of teacher-to-students and students-to-students. Interaction with more people through network is also a connotation of interactive classroom. New Oriental School and Crazy English are pacemakers in English education field in China. But not everyone has the chance to contact with them. Multimedia network sets up a platform for improvement of teachers and students.

Teacher can learn peers' experience on multimedia network platform, enriching his own teaching contents. Besides, with communication to peers, teacher can get resource sharing and effective solution for teaching. Students can expand their horizons and enrich knowledge with learning from English classes in other schools and regions on this platform.

3.5 Facilitating students to "go abroad"

The earth has become a global village with development of network. Students can obtain "face-to-face" contact with people in foreign countries on multimedia network platform. Thus, students can really understand English. "Face-to-face" communication is helpful to overcome phenomenon of Chinglish (Chinese English) in daily learning. Students can learn English more authentically.

English cannot be studied without understanding of western culture. For long time affected by Chinese culture, students know little about the western culture. It makes students' level of English hard to improve. Establishing friendship with western students, students in China can know more about western culture and local conditions and customs. Thus, English level of students can be ascended quickly. Meanwhile, interaction between young people is beneficial for learning knowledge, as well as deepening friendship between Chinese and western countries. Spreading Chinese culture and learning from western cultural do great favor to students themselves.

3.6 *Promoting students' quality and ability*

It is an era of quality-oriented education currently, except for English teaching. English knowledge should be passed on, such as grammar and syntax. Besides, all-round development of students should be promoted through teaching. Interactive English teaching with multimedia network can arouse students' interests—actively participating in class. As a result, students can better master knowledge of English. Furthermore, students are more willing to participate and to cooperate, increasing their thirst for knowledge.

Students' communication ability can be cultivated on interactive platform, laying a foundation for their future development. Students' ability of expression can also be improved, so that they can better adapt to society in the future. IQ and EQ of students can be ascended at the same time.

Voice communications with western students can exercise Chinese students' speaking and listening, while written communications can improve their writing. Skills of equipment using can also be better trained. For example, the word "can't" is usually typed as "can' t" for misusing space key. With more practice of computer, the mistake can be avoided. Besides, through contact with network and multimedia, students can correctly understand and treat network.

4 INTEGRATION WITH MULTIMEDIA NETWORK AND INTERACTIVE ENGLISH TEACHING

Multimedia without network is like water without source, wood without root. Without necessary equipment, function of multimedia can be weakened, either. Multimedia network should be well combined with interactive teaching in accordance with the requirement of quality education.

4.1 *Guaranteeing necessary equipment*

No housewife can cook a meal without a food and vegetables. Even with strong ability of theory and practice, one can be restricted to development without necessary equipment. The abilities have to be shelved. For better function in teaching and combination with interactive teaching, multimedia network should be guaranteed with necessary equipments. Schools and related departments should actively construct modern classrooms, such as network multimedia classrooms and speech classroom, with allowed conditions. Multimedia network classroom is necessarily equipped with computer and broadband network. These computers should be connected in a Local Area Network (LAN), with installed software of demonstration and control. Thus, teaching becomes lively and controllable, convenient for interaction and rein.

4.2 *Mastering necessary skills*

Tool is only a reflection, not an equation to productivity. Necessary skills are the support for equipment for modernization and interaction. Otherwise, equipment can only be ornament.

Staff of multimedia classroom should undoubtedly master configuration and maintenance of hardware and installation of software. It is the precondition for interactions of teacher-to-students and students-to-students. Interactive communication cannot be cut off with advanced software and hardware technology and unobstructed network. Classroom rhythm can also be timely controlled.

Teacher should correctly use modern equipment in teaching, developing the role of advanced equipment to the maximum. Necessary communicational skills should be handled well to interact with students. Courseware production and oral English skills should be improved as well.

With multimedia network for interactive teaching, English teachers first of all should practice skills of courseware making, modification and using. Input of English words is another necessary skill. Teacher should also expertly master knowledge of network and use of software and hardware for smooth teaching. These skills will be precious abilities and wealth of teacher himself.

Students, as the main body of class, are required to grasp equipment operation skills, ensuring to timely communicate with teacher and students. Besides, use of some software, online learning and English expression should also be mastered so as to communicate with teacher and classmates and online friends (especially British and American friends).

4.3 *Setting up correct understanding and idea*

Only with correct understanding can practice be guided correctly. As implementer in class, teacher with correct understanding can implement correct effective class. Teacher should change the old teaching idea—teacher is the dominator in classroom. Classroom should be enriched with equal and interactive communication of teacher and students, rather than teacher feeding students. Furthermore, traditional understanding of multimedia in English teaching should be changed for better combination of multimedia and interactive English teaching.

Students should set up correct understanding of actively participation. With more interactive communication with teacher and classmates, students should improve participation awareness and team cooperation consciousness. Despite of shyness or fear, students should actively participate in English classroom teaching. Moreover, students should treat multimedia network as a tool for learning knowledge, instead of a means of amusement.

5 CONCLUSIONS

Integration of multimedia network and interactive English teaching is different from traditional teaching with multimedia. The front can overcome disadvantage of multimedia teaching. Meanwhile, the integration can also bring a lot of benefits for teaching and all-round development of students. Integration of multimedia network and interactive English teaching needs joint efforts of schools, teachers and students, with improvement in two aspects—software and hardware.

REFERENCES

[1] Zhao Tiantian. *Interactive principle of mode of multimedia network English teaching in university* (China E-Commerce, 2014 (6): 142).
[2] Lei Hongyu. *Empirical research on effect on university English listening and speaking ability with network teaching—case in Qinzhou university* (Journal of Qinzhou University, 2013 (13): 54–57).
[3] Song Weichao. *Application of multimedia network teaching in economic law teaching in secondary vocational colleges* (Wen Li Dao Hang, 2014 (3): 6).

Education Management and Management Science – Zheng (Ed.)
© 2015 Taylor & Francis Group, London, ISBN 978-1-138-02663-6

Strategy of college students' psychological health education based on position requirements

Huabei Hu

Anhui Audit Vocational College, Hefei, Anhui, China

ABSTRACT: College students' psychological health education, faced with new situation and challenge, has a great influence on selection, achievement and other requirements of position. For employment of college students, strategy of college students' psychological health education based on position requirements has been the key point of research. The problems of college students' psychological health education are analyzed and solved. The new idea of psychological health education is established, thus improving college students' psychological health and achieving employment.

Keywords: college students; psychological health education; position requirements

1 INTRODUCTION

With development of reform and open, the level of national economic development has been improved a lot, and the whole society has been deeply reformed. The development of national economy has a great influence on the development level of the whole society, with employment situation of college students greatly changed. Social development affects thought and behavior of college students a lot. Various thoughts are constantly introduced to China because of international cooperation, directly affecting college students. However, College students have less contact with society, and weak ability to differentiate. Besides, some temptation and negative thoughts affect development of college students' psychological health. So, attention should be paid to college students' psychological health, thus affecting development of college students and requirement of society. Recently, the social competition has been more and more intense [1]. College students, without working and social experiences, rest all their hope on finding suitable jobs in order to obtain approval of family and society. In the process, great pressure can make college students controlled by negative thoughts, thus causing kinds of danger. Consequently, college students' psychological health guidance is comprehensively carried out by colleges, thus keeping them healthy, and helping them find suitable jobs and realize values themselves.

2 COMMON COLLEGE STUDENTS' PSYCHOLOGICAL PROBLEMS OF EMPLOYMENT

2.1 *Anxious psychology because of position competition*

Anxious psychology of college students is a kind of emotion because of not achieving original employment target when college students are faced with intensive competitions. As a kind of common emotion, anxious psychology is a kind of anxiety that value of something will be possible to decrease in future [2]. This emotion consists of tension, anxiety and worry, greatly affecting physiology and psychology.

Recently, financial crisis has resulted in great decrease of employment, thus causing difficulty for employment of college students. Any college student desires to get a suitable job and realize his own value. However, college students will worry about whether they can find ideal jobs, compare actual conditions themselves with objective employment situation, and sway between employment and postgraduate entrance examination. E.g. some enterprises employ only males; some require that the applicants have experiences in student cadres; some students are not reconciled. The above requirements make pressure to the applicants, and the applicants will be anxious when they are possible to lose the job [3]. This anxious emotion has obvious performance in the students with introversion nature, common achievement and no famous university. It is easier for these students to form inferiority and anxiety.

2.2 Conceit in process of employment

Conceit, as a negative emotion, means that someone overestimates himself, expressed as blind obedience and crankiness. With development of society and influence of national policy, the ratio of only children in college students is higher. Only children can easily form positive and optimistic nature because of superior family environment and unique favor given by parents. They have advantages of large curiosity and activity. However, incorrect education will cause distortion of personality and conceit. E.g. some parents overindulge their children. In addition, lots of college students, especially excellent ones consider that after decades of knowledge accumulation they have had enough knowledge and ability to practice and realize their own values. However, some college students are unsatisfied with requirements of enterprises in the process of employment, and only interested in jobs with good environment and generous income to realize their own values. This leaves bad impression for recruiters, and makes it difficult to find suitable jobs and achieve scheduled target, thus causing anxiety and vicious circle.

2.3 Dependent psychology

People with dependent psychology are short of independent consciousness and thought, and unwilling to undertake responsibility and make decision independently. Used to rely on parents and friends, they are unwilling to dominate their own destiny and short of enterprising spirit. College students, with dependent psychology, are afraid of social competition and job application. Instead of actively pursue the jobs they like and improve abilities and knowledge, they rest their hopes on parents and wait for the job. This kind of dependent psychology will make college students miss optimal time for application and good opportunity of training and improving ability. If this dependent psychology is unsolved, it will affect the whole life of college students and cause serious results.

2.4 Mass-following psychology

Mass-following psychology of college students' employment is that college students follow other people or blindly pursue hot occupation, instead of selecting jobs according to their own interests. This phenomenon is common in the process of college students' employment. The graduates cannot deeply analyze employment market because of lack of social experiences and independence. The information, usually got from people around them, will make great influence on college students. The students, with mass-following psychology, don't have sufficient self-cognition and definite occupation target to undertake huge social pressure. So, without enough determination, they give up their interest and choose another occupation.

3 STRATEGY OF COLLEGE STUDENTS' PSYCHOLOGICAL HEALTH EDUCATION BASED ON POSITION REQUIREMENTS

3.1 Changing idea of psychological health education

College students' psychological health education and humanistic care should be promoted to make all aspects of society and students to recognize its importance to employment of college students. Meanwhile, educational workers should be strict with themselves based on psychological health standard and develop active psychology in work and life, thus building good example. Besides, educational workers of college students' psychological health should correctly guide value outlook of college students and analyze all reasons of psychological health, thus establishing an effective system, integrating all effective theories, and finally forming a perfect psychological health education system and new educational idea. The new idea should help students form noble quality, so psychological health education can provide position requirements.

3.2 Constant psychological health education

At present, little attention is paid to students' psychological health education in most of colleges. Psychological health education is not set as required course, or set in a certain term instead of the whole college learning process. Sometimes, Psychological health education is set as selective course, learned by only students with interest, thus restricting its influence range and ideal achievement. The setting modes of the above two courses, with characteristic of less class time, is bad for changing ideological concept. Course setting of psychological health education should run through the whole learning in college and cover all the college students. With increase of time for psychological health education, knowledge of the course will gradually affect college students, thus forming good quality. So, college students can accept comprehensive psychological health education and accurately reply the coming pressure of occupation.

3.3 Psychological health education surrounding employment psychological problems of college students

Based on research on psychological pressure of college students in the process of employment, it is discovered that the negative emotions easily to form are inferiority, conceit, dependence, mass-following, indifference, etc., thus making negative influence on psychological and physiological health of college students. Educational workers should actively deal with these problems according to requirements of society and students. The concrete problems should be concretely analyzed to find the way to solve all kinds of negative emotions, thus forming concrete solution and systematic theories. The theories are applied to psychological course to give active guidance of psychological health education and help college students understand, thus forming good spirit quality and ability.

3.4 Helping students recognize frustration and improve courage to undertake frustration

In the process of employment, there will be much pressure and frustration, which is nothing in the eyes of some students. However, different characteristics make different performance. It is difficult for excellent college students to undertake frustration. The superiority complex formed by praise of teachers and students will be changed into frustration by the rejection of enterprises. Then correct guidance of psychological health is needed. Firstly, the students should not deny themselves but face up to frustration, and recognize that frustration is inevitable in life. Frustration is the problem each graduate will come across. Secondly, the students should be helped to analyze the reasons causing frustration from two aspects: objectivity and subjectivity. Only to find the crucial reason in active attitude can the students solve the problems based on actual situation themselves. At last, the students should be directed to face up to frustration calmly. Failure is the mother of success, which means that early failure can pave the way for greater success in future.

3.5 Correct attitude of employment

Employment is a way to test comprehensive ability, including knowledge reserve, ability level, personnel quality, etc. Firstly, college students should objectively evaluate themselves based on their own advantages and disadvantages to find suitable position. In addition, they should adjust employment expectation according to change of actual condition and employment situation, thus decreasing psychological pressure because of large difference. Secondly, college students should keep optimistic psychology to face up to frustration correctly, thus avoiding losing more opportunities because of too anxiety in the key period of employment. At last, college students should reject unfair competition, temptation and unreasonable requirements of some enterprises, and keep sense of justice. The employment of college students is a cruel and fierce process. Fair competition will promote location and progress of college students. However, unfair competition will land college students in great difficulties, which is not beneficial to development of college students. So, college students should take a correct attitude and find suitable jobs through fair competition.

4 CONCLUSIONS

With employment situation more and more rigorous, college students are faced with new situation and larger challenge. Colleges should pay attention to college students' psychological health education and further analyze employment requirements. Based on the analysis results, psychological health education is carried out to help college students establish correct attitude and view of employment, improve their personnel quality, etc. Besides, psychological health education should run through the whole process of learning and life of college students, thus providing more comprehensive and intensive guidance. However, the concrete problems should be concretely analyzed, and the theory should be connected with practice, thus improving the timeliness of psychological health education.

College students' psychological health education based on position requirements is carried out for successful employment of college students. With competitive pressure more and more larger, education in China is constantly changing, although certain achievement have been obtained, thus promoting development of psychological health education along with change of time.

REFERENCES

[1] Tian Cong, Way of Psychological Health Education Based on Employment of College Students, Latter Half of Month, Folk Art and Literature, 2012 (10): 268–268.
[2] Xie Jingwen, Psychological Problems and Influence Factors Based on Employment of College students, Time Education, 2012 (23): 75–75.
[3] Zhang Dan, Psychological Health Education Based on Employment of College Students, Heilongjiang Science and Technology Information, 2010 (16): 152–152.

Education Management and Management Science – Zheng (Ed.)
© 2015 Taylor & Francis Group, London, ISBN 978-1-138-02663-6

Legislation principles and operation efficiency of European and American ecotourism

Buyao Guo
College of Applied Science and Technology, Hainan University, Haikou, Hainan, China

ABSTRACT: Along with the global problem of deteriorating ecological environment, human gradually establishes clear and definite norms and constraints on their activities. Ecotourism legislation has always been a focus in European and American countries. During decades of exploration, human beings constantly strengthened the legality, scientificity and humanity of ecotourism, obtaining certain achievements and leading to a brighter and clearer prospect of ecotourism legislation. Based on the study of legislation principles in European and American countries, the work creatively proposed some strategies to improve the operation efficiency of ecotourism.

Keywords: European and American ecotourism; legislation principles; operation efficiency

1 INTRODUCTION

Tourism travel is realized on the basis of fully satisfied material life. In the 1980s, global tourism activities gradually affected and changed ecological environment, and then human's "changing the world" entered a new stage. Since then, the concept of "ecotourism" was put forward, and people were actively or passively influenced by this concept in their travel and entertainment activities. Furthermore, national legislation of many countries begins to emphasize this field. At present, worldwide ecotourism is showing a booming situation, and its economic income accounts for twenty percent of that of the entire tourism industry, presenting a rapid growth. In European and American countries, modern ecotourism is manifested in more abundant forms, such as original villages, undecorated buildings and unsophisticated folkways, exactly satisfying ecotourism lovers' pursuit. Those "alternative" new forms require special protection from tourism departments and necessary support from local residents [1]. Actually, ecotourism concept aims to call on human to reduce their interference with the original form of tourism. The key to ensure authentic attraction of scenic spots is to keep or restore its original attraction to the largest extent. Thus, people's attitude and purpose of tourism have some distinct modern characteristics.

2 CONCEPT OF ECOTOURISM

The concept of ecotourism appeared in 1980s and was put forward by the International Union for Conservation of Nature (IUKN). It emphasizes that people should not unilaterally pursue the entertainment effect of tourism during any travel activity, ignoring the sustainable development of scenic spots. Instead, it requires people to pay attention to local ecological environment and residents' living conditions as well. This claim still shows its correctness and foresight today. If human continues unlimited exploration of scenic spots, largely changing national environment and violating the objective laws of the development of things, it will lead to exhaustion of tourism resources and damage of ecological environment. As a result, natural landscape cannot realize its value of tourism and appreciation. Furthermore, the contradiction between human and nature may be further intensified, thus resulting in irreparable loss and damage to human society. According to current rapid developing situation of global tourism industry, the exploration and utilization of natural resources is close to saturation [2]. In fact, nearly every place on the earth is marked with human's exploration footprints. Besides, with the rapid development of tourism, transportation, catering and other service industries, there will be an increasing passion of tourism travel. In terms of China, most scenic sports are crowded with tourists every holiday and golden week. In fact, every natural landscape has its lifetime and tourist-tolerance capacity. In addition to the regular professional maintenance from scenic-spot personnel and the regular closure of scenic spots, more efforts are needed to realize a long-standing and better tourism service for people. In the 2014 International Ecological Civilization BBS held in Guiyang, Hans Gelman, the

Swiss Federal Parliament Speaker, proposed the idea of mutually beneficial cooperation for ecological environment protection. Hans expressed the wish to realize the complementation and communication with China in the aspects of transportation, forestry and energy. Therefore, protecting ecological environment and realizing ecotourism has become a global wish. Efforts from different fields and countries around the world are needed to make the wish come true, thus guiding tourism industry towards sustainable development as soon as possible [3].

3 LEGISLATION PRINCIPLES OF EUROPEAN AND AMERICAN ECOTOURISM

3.1 *Legality*

The legislation concept and related work about ecotourism in European and American countries appeared earlier than in other areas, and the main protection content advocated in their work mainly referred to the protection of the primitiveness and integrity of natural landscape. That's why there are lots of scenic spots of primitive forests and ancient buildings in European and American countries. In fact, relevant tourism-supervision departments transformed their emphasis from creation into retention and inheritance; such behaviors of respecting history and time explain why worldwide tourists are so keen on the scenic sports in European and American countries. The emphasis of ecotourism in European and American countries is firstly manifested in the early and comprehensive legislation of ecotourism. Such serious attitude and rigorous rules regulate people's behaviors and firstly arouse enough attention from relevant departments and personnel, making convenience for the smooth work of legal compliance and law enforcement in the next step.

As early as in 1916, America began to apply legal procedures in tourism management, and National Park Service was the initial management and law-enforcement department for tourism. In 1993, UK established *National Park Protection Law*, mainly aiming to regulate the exploration of natural landscape and to protect ecological environment. It can be seen that European and American countries refined and deepened the regulations and laws of tourism very early in order to realize the restriction of tourism industry and tourist behaviors, thus easing human's unlimited utilization of natural resources. Moreover, tourism administrations in some countries also tried to separate scenic-spot management from business management so as to achieve supervision based on relevant international laws and regulations. Therefore, the authority of

relevant laws and regulations can be strengthened through various means.

3.2 *Scientificity*

With the rapid development of advanced and scientific management concept and information technology, human spares no effort to link the work in different fields with science, hoping to promote sustainable development of this field through advanced science. Based on a survey from the International Ecotourism Society, people's tourism-travel choice gradually develops towards "protectional tourism". Meanwhile, some foundations or research institutes organize "exploration activities", taking a group of tourism lovers to some special regions that they are not familiar with or have not been to before. Thus, they can investigate local ecological resources and environment as well as people's living conditions, conducting scientific and reasonable evaluation through the efforts of all members. Therefore, they can get much knowledge and information for future protection and development of natural landscape. Such exploration activity is based on professional knowledge and exploration spirit. It not only provides reliable reference for the development of tourism resources, but also expands people's ideas on ecotourism legislation, ensuring the correctness of specific legislation work. For example, a British "Blue Adventure Foundation" organized some people to do a one-month field investigation for the original region of marine organism in Madagascar. In the end, they obtained the award from the United Nations and built a biological protection base there. Therefore, they realized the most effective protection to local organisms, laying necessary foundation for human scientific research and ecotourism-resource development in the future. European and American countries have fully used their abundant science resources as necessary foundation of ecotourism resources protection and legislation. Such combination of science and ecotourism not only proves enough accuracy, but also provides good examples for the common development of tourism industry and scientific careers for other countries.

3.3 *Humanity*

Ecological legislation itself is a phased measure produced with population growth and resource exhaustion. Thus, there should be some reasonable explanations for the necessity of ecological legislation, but people's recognition and understanding is another important reason for the generation of ecological legislation. It is obvious that some imbalanced-development problems still exist in tourism industry. However, the key for people to accept ecotourism concept is whether those imbalanced-problems affects their self-interests. In fact, in order to make ecotourism legislation no longer embarrassed, "human" immediate interests should be considered from the beginning. The value of ecotourism is not to help tourism practitioners gain maximum profits but to ensure the integrity and lasting vitality of ecotourism resources. Therefore, the investigation of local people's living conditions should also be considered as the reason for tourism legislation. Thus, local people's rights and interests can be fully protected, making future rational utilization and development of scenic resources possible. For example, contaminated land is called "brown land" in the United States. In 1960s and 1970s, America treated the brown land of Gas Work Park in Seattle with 2,000 tons of chemical materials filling and leveling up that local "river of love", realizing so-called "treatment". However, the water quality survey in 1976 discovered that the "river of love" contained lots of chemicals including some carcinogens. At that time, 364 residents lived near by and more than 400 students studied in a primary school there. In the following 20 years, the U.S. government continued pollution treatment for that river until it was finally "treated". After this incident, the United States enacted related laws in 1980 to improve the inhumane management of ecological environment. According to the law, people whose living area were polluted or occupied can get corresponding compensation and policy supports from government.

4 OPERATION EFFICIENCY OF EUROPEAN AND AMERICAN ECOTOURISM

4.1 Improving efficiency through "differential treatment"

Ecotourism legislation actually is faced with various challenges. Natural landscapes in each country have different characteristics and specific environment. Thus, it is difficult to apply the same rules or methods to treat different natural landscapes. Besides, it is also not easy to realize a scale management for those natural landscapes. Therefore, "differential treatment" is necessary in the specific work of ecotourism management. People's distinction of those tourist attractions and natural resources are firstly established on the basis of field investigation, because the differences of various objects can only be accurately defined through field experiences and investigations. Therefore, ecotourism legislation and corresponding management on such basis can realize specific treatment according to different levels, requirements and goals. Such treatments can appropriately reduce management

cost and generate more efficient response between complementation plans and objects, largely improving the pertinence and effectiveness of ecotourism legislation. The treatment of "brown land" in Germany accurately interpreted the concept of "differential treatment". On one hand, large-scale shopping centers and corresponding urban facilities can be constructed in areas with lower level of pollution to attract tourists, thus creating new value for those areas. On the other hand, heavily polluting factories requiring large cost of treatment should be shutdown for purification treatment, and then transformed into a factory-related history museum. This kind of treatment plan, fully considering economic factors, is a win-win solution for environment and economy benefits. Such treatment plans can also be extended to the specific implementation of ecological-protection laws and regulations. Thus, human can strictly abide by the laws and regulations. Meanwhile, they can also creatively achieve a long-term development prospect for natural resources and certain economic values, thus realizing the harmony between human and nature.

4.2 Strengthening the participation of scientific and technological ideas

Many countries pay more and more attention to the role of science and technology in ecotourism legislation, but more in-depth exploration efforts should be paid to better solve the problems and difficulties in actual work. Such scientific efforts in ecotourism legislation are not limited to certain scientific inventions or some advanced technologies. In fact, some ideological concepts such as advanced management mode and scientific working methods also contribute to a better ecotourism legislation. For example, in the aspect of ecological environment protection, fund sources and its function ways present a new direction. America Travel Bureau gets special funding for ecological environment protection from the government; at the same time, it is also supported by taxes from the profitable units inside the scenic spots and donations from some social welfare organizations. Therefore, this kind of funding system increases the special funds at travel bureau's disposal and utilization. Meanwhile, it also optimizes the structure of main behavior bodies of such protection work, thus further strengthening the status of ecotourism and environment protection in public mind. In addition, international cooperation of ecotourism is developing gradually. Therefore, the tourism administrations and ecotourism legislation departments in some countries can share certain resources in tourism management. Furthermore, they can also exchange their experience and lessons of ecological protection and tourism management through information-network platform, providing some constructive suggestions for the management mode of future international tourism and ecological management. Meanwhile, the effect of ecotourism legislation can be significantly improved.

4.3 *Enhancing the operability of legislation*

The evaluation of legislation work largely depends on its implementation effects and feedbacks. European and American countries have achieved commendable practices in this respect. Besides, the standardized application of relevant laws and the flexibility in corresponding treatment cases help western countries to achieve smooth and successful ecotourism management. The United States enacted *Animal Protection Act, Safety Drinking Water Act, Clean Air Act* and some other series of laws and regulations as the norms of people's behavior, thus enabling American to construct good environment-protection consciousness. For example, it has been a common rule that certain green area is necessary for a house no matter where it is built. In summer, tourists should make a reservation in advance when planning to visit Yellowstone, indicating the good feasibility and operability of a perfect legislation in the United States. In addition, more and more serious ecological-environment situations and problems also contribute to establishing relevant laws and regulations and realizing the concept of environment protection. The declining air quality, global warming issue and extreme climate disasters draw public's attention to the practice of ecological protection.

Everybody will go on vacations, and their ways to get along with environment as well as the results increase their senses of responsibility and mission. People with different social roles in different fields all make better and more obvious effects of ecotourism legislation, thus achieving greater practical significance. Laws and regulations related to ecotourism activities play the charm through everyone's effort of compliance with those regulations and rules. In return, people's long-term recognition and practice of the laws will lead to a huge difference to their living environment.

5 CONCLUSIONS

After years' of efforts from professional tourism-management personnel and legislation staff, the ecotourism legislation in some European and American countries has achieved great results. Nowadays, natural landscapes are no long considered as only tourism resources. Besides, the historical reference value, cultural value and scientific value of natural landscapes are gradually discovered and appreciated by the public. This kind of more scientific and comprehensive recognition of ecotourism is not only beneficial to tourism management but also plays a positive role in the sustainable development of global ecological environment.

ACKNOWLEDGEMENT

The work was funded by 2012 Hainan philosophy and social science planning project "*Research of Hainan Ecotourism Legal Issues*". Project No.: HNSK(GJ)12–32.

REFERENCES

[1] Xu Xiaoling, Comparison of the Legislation of Wetland Conservation in China and in America. Science & Technology Information. 2010(6): 80–81.
[2] Yao Xiaoyan, Economic Thinking of Ecotourism Development. Commercial Times. 2014(13): 122–123.
[3] Zou Qi, Comparison Study of Western and Eastern Ecotourism. Management & Technology of SME. 2014(05): 154–155.

Education Management and Management Science – Zheng (Ed.)
© 2015 Taylor & Francis Group, London, ISBN 978-1-138-02663-6

Strategy of e-commerce development in chinese clothing retail

Zhuo Chen
Anshan Radio and TV University, Anshan, Liaoning, China

ABSTRACT: The development of e-commerce has a huge impact on Chinese clothing retail industry. Enterprises should continue exploration in the highly competitive market environment to adapt to new situations. The work began with the status quo of e-commerce applications in Chinese clothing retail, followed by solutions of the problems in e-commerce development. New composite channel should be constructed with reasonable positioning, thus achieving sustainable development of enterprises through continual improvement of various aspects in e-commerce application process.

Keywords: clothing; retail industry; e-commerce; strategy

1   INTRODUCTION

Retail industry plays an important role in Chinese economic development, reflecting on the following aspects: large number of retail enterprises, absorption of much employment, and large proportion of sales in GDP. Clothing retail occupies a high proportion in retail industry. Wang Zhimin, chairman of Baodao Glasses Ltd, thinks that traditional retailers should start the transfer to online retail because of the increasing cost pressures and the changing consumer habits of younger generation. They will be substituted by the market if not transformed to e-commerce. By December 2013, the number of Internet users in China had reached 618 million, and the Internet penetration was 45.8%, 3.7% higher than that of 2012 [1]. Under the slowdown situation of Chinese retail industry, network retail keeps thriving. According to the report of China Internet Association, the total trade amount of network retail in China reached 1.8 trillion RMB in 2013, indicating that network retail gradually has entered a mature period of steady growth [2]. Thus, it is worth exploring how clothing retail with rapid growth develops e-commerce to achieve online and offline integration.

E-commerce was developed on the basis of computer technology, network technology and database. It is a new business model representing the future direction of trade development, using digital electronic forms for data interchange and implementation of business development. The success and promotion of e-commerce models such as Alibaba, Taobao, Lynx, Dangdang, has brought great and fast impact beyond people's anticipation on traditional retail. It is an opportunity as well as a challenge for Chinese clothing retail with great sales; thus it should continue further exploration [3].

2   CURRENT SITUATION OF E-COMMERCE APPLICATION IN CLOTHING RETAIL

Clothing retail is a traditional cyclical industry with a strong seasonality—there are significant slack seasons and peak seasons. The sale on weekends and holidays is much more than usual time, and sales in the second half is also far more than in the first half. Therefore, businesses should make long-term strategic planning. Own characteristics of Chinese clothing retail should be paid attention in its application of e-commerce, designing rational new business models to adapt to the changing markets [4].

E-commerce in China has been applied in various industries since 1990s, presenting prosperity of the overall development of e-commerce retail. Clothing retail is one of the top ten industries. Chinese e-commerce sales surpassed the United States in 2003 as the world's largest network retail markets. According to *China Electronic Commerce Research Center Report*, the transaction of Chinese clothing online shopping market in 2013 was about 407.6 billion RMB, with an increase of 33.6%. The growth rate was slightly higher than that of the overall online shopping. Clothing retail was still the largest category of online shopping. A growing number of apparel retailers have joined e-commerce, constantly seeking better marketing models.

3 PROBLEMS IN APPLICATIONS OF E-COMMERCE IN CHINESE CLOTHING RETAIL

Many companies, such as "Li Ning", "were" started by e-commerce. However, after several years of trial and business operations, more and more enterprises have recognized that it is necessary to break current restrictions of some models and concepts for further extension of e-commerce concept. For example, integration of online and offline channels should helps to better consolidate the market, achieving higher sales growth. E-commerce broadens the marketing channels of clothing retail business. Thus, costs can be effectively reduced through the full play of comprehensive advantages of online and offline fusion, improving market responsiveness and creating a good corporate image. Meanwhile, the efficiency of information issue and feedback can be improved for better competition in the market. Many companies are still in the exploration in e-commerce applications; thus deficiencies in many areas need to be solved.

3.1 Challenges of clothing retail transformation

Clothing retail has made considerable progress after the reform and opening up as well as rapid development of computers and networks. However, there are still some problems including less overall competitiveness and shortage of electronic extent. Some large enterprises directly station in e-commerce platform, while the competitiveness of self-employed online shops is in decline. Thus, enterprises of different sizes cannot take the same path. Differences competition such as special services and personalized products is the key for online shops to survive. Companies should exert strengths to seek their own development direction, adapting to the new era of e-business requirements and overcoming the transition challenges. Traditional clothing retail needs to complete an independent e-commerce platform. However, lots of platform design is not reasonable due to lack of professional e-commerce talent and operating assistance, complex interface operation, low information processing abilities, poor logistics and inaccurate positioning, thus hindering the development of e-commerce.

3.2 Problems in positioning of online and offline sales products of clothing retail

Many clothing retailers lack overall strategic plan and long-term strategic planning, mainly selling their outdated and over-quarter on-line products. The e-commerce platform is regarded as a channel for clearing inventory and dumping goods at low-cost. Consumers will lose interest after a long time, resulting in the loss of customer base. Most companies lack specialized product design and network ordering for segment of network consumer groups. Moreover, they have no reasonable product portfolio positioning and overall sales strategy, losing customers' loyalty and failing in full use of respective advantages and characteristics of online and offline fusion.

3.3 Problems in transactions trust and payment risk of e-commerce

Security has become a core issue of e-commerce development and the basic condition of transaction. It is difficult to assess the information of online transactions, thus buyers and sellers cannot confirm the identity of each other under e-commerce environment. The separation of capital flow and logistics leads to certain risks in both sides. Online shopping faces transaction-trust problems and security issues while bringing convenience, influencing the consumer behavior. The depth of e-commerce development is inseparable from the improvement of online trust and payment system, which should be taken seriously for resolution.

3.4 Hysteresis between logistics development and electronic commerce

Fashionability of clothing makes consumers have higher requirements for logistics services of online shopping. Customers always hope to get the goods in the shortest time. Users' satisfaction of logistics express is still low, according to "*Report on User Experience and Complaints Detection Chinese E-commerce in the First Half of 2013*". The rapid development of e-commerce requires consistent modern logistics network, while current logistics presents a low automation of equipment, poor timeliness of delivery due to imperfect network systems, and low customer satisfaction. Clothing retail enterprises should attach importance to the contradictions between e-commerce development and logistics hysteresis.

3.5 Imperfect e-commerce laws and regulations

Chinese e-commerce laws and regulations are still under construction, which are mainly regulations of industries, sectors and localities, lagging behind other e-commerce-developed countries. Though there are regulations such as "*Electronic Signature Law*", "*Regulations of China Internet Network Domain Name*", "*Management Regulations of Electronic Authentication Service*", and "*Management Approach of Third Party Payment*" in China, it still lacks specific details of transaction for detailed specifications and guidance. E-commerce laws and regulations also require further standardization and improvement, as well as rational application in practice.

4 STRATEGY FOR E-COMMERCE DEVELOPMENT OF CHINESE CLOTHING RETAIL

E-commerce has brought new opportunities while bringing tremendous impact on traditional clothing retail. It is imperative for the integration of online and offline, exerting its advantages to create more economic benefits. E-commerce development strategy for Chinese clothing retail should be made based on its own characteristics and existing problems.

4.1 Strengthening construction of network platform to promote customers' purchasing behavior

Online shopping can improve the purchasing efficiency by saving time costs. Customers can also compare the clothes to be purchased by electronic equipment, seeking the most satisfactory products with highest performance price ratio. Therefore, e-commerce businesses should strengthen infrastructure construction, building personalized online shopping space. Branded apparel products can be displayed with full range on e-commerce platform, comprehensively introducing product information, improving information processing capability to attract more potential customers. The trading range can be infinitely expanded into international market. Excellent network platform is essential for the improvement of visibility and sales increase.

4.2 Establishing composite channel models to achieve online and offline integration

With rapid development of e-commerce, traditional clothing retail companies can establish their own websites, optimizing the allocation of resources. Meanwhile, they can establish development strategies to achieve online and offline integration. The composite channel model of physical store and e-commerce store can be built to increase competitive advantage of enterprise market. Traditional clothing retail has irreplaceable advantages including field experience, apathy, and intuitiveness of commodities. The advantages of e-commerce are that customers can buy products anywhere and anytime. Meanwhile, they can compare a large amount of product information to make a purchase decision. More and more consumers prefer

to try on clothes in physical stores and buy them online. Some consumers browse clothing information online and expect to experience in physical stores; thus the store becomes customer's dressing room. In other countries, the trend of online and offline integration is being further strengthened. Physical store can be made as extension of online shop, finally developing as the base for customers to take or replace goods as well as after-sales service. Consumers can achieve online and offline free shopping space and flexible purchasing behavior according to personal preferences.

4.3 Providing efficient, simple, and flexible means of payment

Clothing retailers should regulate platform environments of online payment. Credit problems in e-commerce transaction process can be solved through the use of "Alipay" and other payments, enhancing its safety to make payment efficient, simple and flexible. Consumer's worry on transaction confidence and payment risk can be reduced through improving security and usability of Internet payment. Relevant laws and regulations should be introduced to improve network security system, strengthening supervision of electronic payments. Training of technology and management should be conducted for all kinds of personnel. Safe and effective payment should be ensured for the smooth conduct of e-commerce, avoiding various risks of "electronic pickpockets", online scams, online hackers and information leaks.

4.4 Establishing three-dimensional network distribution system for protection of distribution services

Three-dimensional and perfect network distribution system is necessary for e-commerce development of clothing retail. Faced with the fact that logistics development seriously lags behind e-commerce development, existing logistics equipment of logistics and allocation should be upgraded, achieving three-dimensional network construction using scientific management methods. Clothing retailers can change their physical stores as logistics nodes and platform of electronic commerce for taking or replacing goods, thus satisfying customer's needs for field and intuitive experience. Physical stores and online businesses can be implemented as two systems: customer purchases clothes online, and then couriers can deliver the goods to the designated address or stores. The control of distribution costs plays an important role in the reduction of product's total cost, increasing competitive advantage and improving customer's satisfaction. Therefore, clothing retail enterprises should make proper logistics distribution system based on their own development strategies. Independent logistics department can be established to centrally arrange the delivery goods. Meanwhile, clothing retail enterprises can work with third-party logistics companies, thus providing a full range of convenient and efficient logistics and distribution services.

4.5 Improving e-commerce laws and regulations

The development of e-commerce is inseparable from transparent and harmonious legal environment. In order to reduce the risk resulted from law hysteresis, e-commerce crime should be constrained through safety legislation. Meanwhile, the scope of legislation should be adjusted in accordance with the requirements of economic development. Network clothing retail industry should be standardized to adapt to the new e-commerce forms with constant development and progress, ensuring normal trading of e-commerce activities. E-commerce market can be standardized through the establishment of government regulation including security management systems of computer information and certification institutions management, thus perfecting laws and regulations of clothing retail industry. The authenticity of sales information can be ensured to make customers avoid being misled through the standardization of users and members' registration information, and behaviors of electronic signatures, electronic payment, promotion and sales services. Unfair competition and imparity clause cannot be allowed between enterprises.

4.6 Developing computer online shopping into mobile online shopping

According to Based Statistics of China Internet Network Information Center, the number of Internet users in China has reached more than 600 million by the end of 2013; mobile phone users reached 500 million, representing an increase of 80.09 million than that of 2012, which is a tremendous force to support the development of electronic commerce. In addition, mobile phone user is the base for the transformation from computer network shopping to mobile online shopping. Internet users preferring online shopping are usually young with high education degree, higher income and higher consuming ability. In addition, they will form a strong support for mobile online shopping.

With the rapid development of mobile devices and rapid increase of users, mobile online shopping has become a general trend. Mobile users obtain information and related services of apparel product through mobile devices without restriction of time and place. Moreover, the overall related shopping activities such as payment and feedback can be completed through mobile devices. Mobile online shopping will be the future direction of e-commerce; thus clothing retail enterprises should seize this new model with breakthrough innovation, achieving new growth of sales.

5 CONCLUSIONS

Current mature consumers cannot just live online or stay offline. Their online and offline consuming demands cannot be completely separated, which requires clothing retailers to achieve real integration to satisfy the market requirements. The successful completion of integrating online and offline communication and marketing requires cooperative effect of external and internal factors for an excellent enterprise. Meanwhile, companies should take use of advantage to achieve their goals.

As the development direction of trade means in the 21st century, e-commerce has the undisputed superiority and advancement. Chinese clothing retail e-commerce, as the role of pioneer, will keep rapid growth in the coming years. Various retail enterprises should raise awareness to create new business models. Consumer psychology should be understood due to the changing consuming behaviors under network environment. Long-term development plan should be made for enterprises to meet the requirements of market development, thus better completing sales growth and sustainable development of enterprises.

REFERENCES

[1] Statistics Report of the 33rd China Internet Development, 2014-01-16. http://www.199it.com.
[2] Zhou Jianxia, Zhao Bingxin. Strategy for New E-commerce Model of Small Appliance Industry. Shandong University, 2008, (09): 35–36.
[3] Zhang Zhijun. Using e-commerce Strategy to Accelerate Transformation of Traditional Retail Business Beijing. China Market, 2011, (19): 79–81.
[4] Huang Shiqiang, Fan Bingsi. Strategies of Online Market in Clothing Retail Industry. East China Normal University, 2011, (03): 52–55.

Education Management and Management Science – Zheng (Ed.)
© 2015 Taylor & Francis Group, London, ISBN 978-1-138-02663-6

Evaluation index system and model for the construction of school ethos in university classes

Guanghai Shang, Qiankun Yang & Shiqi Liu
The School of Management, Hefei University of Technology, Hefei, Anhui, China

ABSTRACT: The evaluation of constructing school ethos in classes has been emphasized in many universities to guarantee teaching quality and learning atmosphere. Currently, the common score assessment, special factors assessment and other new methods are being tested by practice in those universities. Besides, the advantages, application scope and development direction of these models have been investigated and analyzed. On this basis, the evaluation of constructing school ethos in universities will become more rational and effective. Thus the construction of school ethos in university classes can be better promoted.

Keywords: construction of school ethos; evaluation index; model

1 STATUS OF THE CONSTRUCTION OF SCHOOL ETHOS IN DOMESTIC UNIVERSITIES

School ethos, as the core ideology and goals of a school, has always been an important campus work emphasized by various universities. Thus the works related to the construction of school ethos have become more and more significant. Nowadays, people's evaluation standard for a university is not merely limited to faculty, students' test scores, graduation rates and employment rates but favor a more profound level of constructing school ethos [1]. As the major bases of cultivating young talents, universities are very similar in curriculum and teaching. Therefore, it's necessary to emphasize the construction of school ethos to enable college students to acquire the edification of higher education. Such construction is, of course, based on students' learning. Students and teachers can improve teaching methods through teaching programs. Students' learning interests can thereby be continuously stimulated to improve teaching quality and learning atmosphere. On this basis, current students carefully study professional knowledge and continuously enhance the pragmatic learning attitude. Such campus activities through "teaching" and "learning" can achieve the mutual promotion of teaching and learning. And they're the widely used method of constructing school ethos in most universities.

At present, most domestic universities are actively promoting their own construction of school ethos and have accomplished several achievements. For example, Jiangnan University has divided the construction of school ethos into three powers: the leading role of backbone teachers in daily teaching, the exemplary role of student leaders in learning process and the principal role of most students in the whole activity. The coaction of the three can not only ensure the construction of school ethos covers the whole school but also achieve a clear responsibility division [2]. As a result, various works can be launched based on evidences, guaranteeing both quality and quantity.

In addition, some universities have innovatively created the "characteristic" construction of school ethos consistent with their actual characters, achieving the blossoming of all styles. For instance, National Defence University PLA China mainly emphasizes military construction. Therefore, it focuses on the strict management of tests in construction of school ethos. In addition to the conventional four levels: "inferior, qualified, good and excellent", the evaluation of cadets has even rigorously set up different proportions of recommending outstanding students. And the links of college examination have introduced not only the advanced evaluation system for national defense students' comprehensive quality and the annual appraisal plan but also the more humanized mutual assessment and recommendation between cadets. In this way, more young talents can be appropriately discovered and comprehensively cultivated. With such model of constructing school ethos, students can be more comprehensively and scientifically urged. And the purposes of cultivating college talents can become more distinctive and well presented. Plus, it's an important factor for universities to carry on the past and open a way for future.

2 COMMON EVALUATION INDEX SYSTEMS FOR THE CONSTRUCTION OF SCHOOL ETHOS IN UNIVERSITY CLASSES

2.1 Traditional scoring evaluation system

Scoring evaluation has been most widely used in constructing school ethos and school's annual assessment. In this method, various evaluation contents, code of conduct, rewards and punishments are represented by different weights based on their different status. An object of evaluation will obtain the corresponding score after actually achieving the set standard. And the final results are judged and determined through each individual's score [3]. Such method, with less possible disputes, is the most simple and straightforward one among a number of assessment methods. Thus it's very persuasive for objects. Simple and crude as it may be, this "all or nothing" way of judgment is very effective. If people want to flexibly grasp the different factors in such method, they can improve the setting of evaluation contents or the allocation of scores to increase adaptability and practicality. Let's take the scoring evaluation rules of constructing school ethos in a university from Zhuhai as the example. Its evaluation can be divided into five areas: planning and implementation for constructing school ethos (25%), students' learning (25%), class honors and punishments (20%), characteristic works (20%) and others (10%). These areas have rather completely covered the main goal of constructing school ethos. As the important parts of the system, the corresponding weights not only reflect the different values presented by different contents but also provide the basic specifications for the following scoring. In addition, the system also includes specific assessment contents, scores, scoring standard, assessment basis and remarks. A complete evaluation system of constructing school ethos requires the evaluation team to rigorously compare and judge the behaviors of participants. If a participant reached a certain standard, he/she will obtain the corresponding scores. Thus the obtained final data is well documented and can provide a reasonable explanation regarding the results of constructing school ethos for all participants. Currently, most universities mainly adopt such method to evaluate work efficiency and appraise excellent talents.

2.2 Evaluation system with special awards

The evaluation system of constructing school ethos with scoring standard has been widely used in practical college works. In addition, some universities evaluate the construction of school ethos based on the outstanding performance of classes.

Adopted by some universities, such evaluation can be accurately implemented on class or individual level. Thus they can obtain good supervision and encouragement. In fact, this kind of evaluation model is linked with the reward system: as long as a class has an outstanding performance in one link of constructing school ethos activities, it will obtain the corresponding honorable tiles. Such assessment results can often be converted to material rewards. And the effects have certain advantages compared with scoring evaluation and are more applicable in modern universities. Although such special titles have many categories, they have to focus on the construction of school ethos with comparable and distinctive characteristics. For example, the titles for individual college students: "Learning Model", "Learning Individual Excellence" and "Best Progress Award". And the titles for classes: "Mutual-help Class" and "Good Construction of School Ethos Award". Multiple as they may be, these awards are not flashy. Diversified evaluation methods can not only better retain the advantages and characteristics of each individual but also increase flexibility and enjoyment in the serious evaluation of constructing school ethos. Consequently, teachers and students can get rid of the previous prejudice that the school ethos only emphasizes academic scores. In this way, they can envisage the construction of school ethos and actively widen thinking. Besides, teachers and students' learning-based understanding in knowledge, feelings, mentality and behavior is optimized to promote sustainable development of constructing school ethos in universities.

2.3 Election and self-recommendation evaluation system

As the forefront of constructing school ethos, universities are required to reasonably accomplish the related supervision and organization and achieve meaningful formal innovation. Such gathering of high-level intellectuals should generate progressive and divergent thinking to better promote the implementation of school's activities. Focusing on the initiative and enthusiasm of participants, class elections and self-recommendation has broken the previous patterns where classes passively accept the evaluation of constructing school ethos. Such "rose to the challenge" method can not only reflect classes' confidence in the ability of their own teachers and students, but also show the good state of preparation for tests and great cooperation. Thus the evaluation regarding students' academic scores, learning attitude, class atmosphere and interpersonal relationships can be smoothly carried out. Besides, its authenticity and reliability can be adequately protected, which will make relevant college works even better. Once such creative evaluation methods were widely promoted in universities, most classes, teachers and students will have more stringent requirements for their own behaviors. Accordingly, a good academic atmosphere can be formed in teaching and learning, presenting an enthusiastic situation in a long period—comparing, learning, catching-up, helping and surpassing. To some extent, adding such humane elements to school evaluation system can alleviate the situation where students, teachers and classes excessively pursue the so-called scholarships, material rewards and honorary titles. Thus the deterioration of school ethos can be avoided on campus. In reality, the real purpose of constructing school ethos is not the victory of one individual but the improvement of atmosphere and quality around the entire university.

3 STUDY ON THE EVALUATION MODEL OF CONSTRUCTING SCHOOL ETHOS

3.1 *Relatively stable fixed-factors evaluation model*

The construction of school ethos in universities has always been an issue emphasized by all the teachers, students and education departments. Its timely implementation, effectiveness of evaluation and reasonability of feedbacks are important factors influencing its effects. In the evaluation model of constructing school ethos, students, teachers, schools and families are indispensable basic factors. Included in construction activities, they can determine the success or failure of the evaluation of constructing school ethos. Among them, the teaching environment formed by teachers and students are playing the most obvious role—the state, ideology and behavior of the both can affect the entire school ethos. As a result, the Butterfly Effect can be reflected in appearances such as student achievement, the strength and reputation of universities. Meanwhile, universities, as the main body of activities, need to control the collection, selection and processing of information as well as the final results during the evaluation. Thus they are required to have a safeguard—firm evaluation system and reliable evaluation team members. In addition, universities need to control the forming and developing trend of school ethos and take the reality and ethos of the school into account so that the both can harmoniously develop. Besides, parents can provide strong supports, real-time feedbacks, timely supervision, constructive comments and suggestions for the evaluation. In a word, these four aspects are complementary and indispensable in the construction of college ethos.

3.2 *Relatively flexible special-factors assessment model*

The evaluation of constructing school ethos in domestic universities commonly uses some winning titles or awards to reflect its successful completion. Why does this way exist until now? The reason is that it reflects the real value in people's practice and has a certain mass base—people's recognition and affirmation. In fact, the setting of various so-called special prizes can, to some extent, reflect the priorities in constructing school ethos and people's expectations for this work. Rigorous learning attitude, excellent teaching environment and strong academic atmosphere are the purposes that schools, teachers and students want to achieve. Thus "Excellent Class" and "Advanced Individual" need to meet the requirements of constructing school ethos and make outstanding contributions during the process. Besides, the school can create the following evaluations: the uniform evaluation on the whole class that investigates whether the learning atmosphere inside the class is strong and whether the overall level of students is high; the personal capacity evaluation based on the individual ability or contribution in each class; the progress award that compares the performance of each class in different periods to achieve vertical reference. The above approaches can not only make the conventional evaluation of constructing school ethos flexible and operable, but also conduct personalized evaluation based on the psychological characteristics of classes and individuals. Such diversification will inject greater vitality into the construction of college ethos. Thus a series of proactive self-improvement and perfection can be stimulated, promoting the optimization of school ethos.

3.3 *Other evaluation models of constructing school ethos*

Compared with the schools on other levels, domestic universities have sufficient hardware and software support for constructing school ethos. With high quality, educational level and degree of ideological emancipation, college teachers and students can brilliantly absorb new ideas and methods. On this basis, constructing school ethos and its evaluation can obtain effective implementation and feedbacks. In addition, the study of college students requires more self-restraint and self-learning. Therefore, once formed and becoming mature, the school ethos commanding the whole campus will become a valuable intangible asset of the school. In such atmosphere, both teachers and students can develop their resources and personal abilities to the fullest degree while benefiting a lot.

In addition to the steady development on the original track, the construction of school ethos requires further innovation and expansion in order to "retain youthfulness". Combined with the public expectations for college talents, the school evaluation team can add several new standards and directions to the project design: the evaluation of constructing excellent courses and the quality evaluation for academic conferences (from the perspective of teachers); a student's research project results, social acceptance and self-management classes (from the perspective of classes). Based on the previous academic-atmosphere-centered situation, such new evaluation methods introduce factors coming from academic authority and practical needs of society. As a result, the evaluation of constructing college ethos has become more comprehensive in structure. And the internal development of universities has acquired a striving direction. In other words, these methods are acting as both assessment and guidance. All in all, the school leaders and the management team of constructing school ethos should pay adequate attention to "quality" and "quantity", mastering effective methods and advanced concepts of constructing school ethos. Only when the both complement each other can the objective of talents training and teaching bring out the best in each other inside universities.

4 CONCLUSIONS

The superior evaluation of constructing college ethos is playing a significant positive role: optimizing academic atmosphere, improving teaching quality and establishing a good image of college and campus culture. Thus universities will obtain good results in constructing school ethos through clearly considering the evaluation indicators, optimizing and upgrading evaluation models as well as the coaction of teachers and students. School ethos is the soul of a school. Therefore, the gradual improvement of the evaluation will act as the pilot for various academic activities on campus. And the positive energy it brought will be continuously effective in a long period.

ACKNOWLEDGEMENTS

This work was funded by the National Undergraduate Training Programs for Innovation and Entrepreneurship "Study on the Evaluation Index System and Model of Constructing School Ethos in University Classes Based on SPSS Factors" (No. 201310359050).

REFERENCES

[1] Chen Zhen, *Analysis on the Evaluation System of Constructing College Ethos based on Two-factor Theory*, Journal of Hubei Radio & Television University, 2013.33 (8): 116–117.

[2] Wu Yan, Zhang Wenping, Cui Ruiling. *Introspection about College Students' Construction of School Ethos based on Pedagogy*, Charming China, 2014 (13): 179–179.

[3] Li Haibo, *Exploration on Problems in the Contemporary Construction of College Ethos*, Business, 2013 (52): 157–157.

Education Management and Management Science – Zheng (Ed.)
© 2015 Taylor & Francis Group, London, ISBN 978-1-138-02663-6

Relationship between innovation training of social sports instructors and ecological sports

Bin Wang
School of Physical Education, Baicheng Normal University, Baicheng, Jilin, China

Jianrui Yu
Baicheng Medical College, Baicheng, Jilin, China

ABSTRACT: Social sports refer to an effective way to achieve national fitness. Social sports instructors play a special and key role in social sports development and are the dynamic and guider. Currently, the shortage of social sports instructors, poor professional skills and irrational talent structure has seriously hampered China's social development and the implementation of national fitness. To change the status of the social sports development, two aspects should be emphasized. On one hand, the existing difficulties and problems should be solved to achieve the innovation training of social sports instructors. On the other hand, the harmonious relationship between human and nature should be emphasized, and the ecological idea should be introduced to the guidance of social sports. In this way, the highest level of sports—the integrity of man and nature—will be achieved, thus improving the physical and mental quality of people. Besides, the public consciousness of lifelong exercise will be established, and the development of China's sports cause will enter in a virtuous cycle. So, it can constantly improve people's quality of life and physical and mental quality.

Keywords: social sports; instructor; significance of training; problem; ecological sports

1 INTRODUCTION

Social sports are both an important part of sports cause and an important form of mass participation in sports. And it is of importance in improving people's physical and mental quality. Social sports instructors are the leader and core driving force of social sports development. They can be combined with public comprehensive physical quality to effectively guide individuals to participate in physical exercise, scientifically improving their health and spiritual style. That is of significance for promoting social harmonious development and social moral construction and improving the quality of people's life. At present, social sports instructors' training in China has many shortcomings, which have restricted the development of sports cause [1].

2 SOCIAL SPORTS INSTRUCTORS' TRAINING

2.1 *Significance of social sports instructors' training*

The objectives of social sports instructors are social mass, millions of ordinary people. First, social

sports instructors are beneficial to the development and implementation of national fitness. For one thing, social sports instructors can effectively guide people's exercise, making sports activities secure and healthy. Then people can exercise healthy and happily under the scientific guidance, thus enjoying the benefits of sports to the greatest degree [2]. For another thing, social sports instructors can provide technical support for the scientific and correct exercise of people, thus making social sports safe, health and scientific. Second, social sports instructors can promote the enhancement of national quality. Social sports are comprehensive sports activities beneficial for the physical and mental health of people. With the guidance of social sports instructors, the training purposes of sports fans will be clearer. Then, social sports participants can have a scientific understanding of their own health and physical characteristics, and take corresponding exercise to improve their physical and mental qualities. Third, social sports instructors are of positive significance to improve people's quality of life. With the development of economy in China, people's spiritual demands are higher than material life. Social sports are beneficial to strengthening of people's physical and mental.

It can also enrich people's spiritual life, which is an indispensable part of the evaluation of modern living standards. Thus, social sports are significant to improve people's quality of life. Fourth, social sports instructors can effectively promote the comprehensive development of China's sports cause. On the one hand, social sports are an important part of China's sports cause and can improve the qualities of people [3]. On the other hand, social sports instructors not only play the role of sports skills guidance, but also shoulder the responsibilities of sports propaganda, organization and management. Therefore, they are important to bring China's sports cause a stable and healthy development.

2.2 Status and problems of social sports instructors' training

Currently, the professional orientation of social sports instructors is still relatively vague. The salary of social sports instructors is neither unified nor stable, and most of them provide voluntary service or only charge a small reward, impacting the social sports instructors' training. The main features of social sports instructors' training are as follows. First, the scale of social sports instructors' training is so small that trained people cannot meet the needs of society. Second, most of the social sports instructors have only got high school qualifications with low cultural foundation and educational background. Thus, their professional knowledge of sports and fitness is not complete enough to accomplish the organization, management and onsite guidance of social sports. Moreover, they cannot make good use of professional practice, sum up the law society of sports development and solve problems by scientific theory in the development of social sports. In a word, the current occupation level of social sports instructors is difficult to meet the needs of social sports development. Third, most social sports instructors are old, in which retirees accounting for more than 90%. They can only drive the elderly to exercise instead of the young people. In addition, the age structure of social sports instructors is unscientific, which will lead to the acceptance of new things becoming conservative and slow. It is not conducive to the promotion and development of advanced and personal sports, which may lead to the uneven development of social sports. Fourth, big difference exists in the development of urban and rural social sports. On the one hand, there is not enough rural sports infrastructure to meet the needs of people. On the other hand, most of the social sports instructors are concentrated in urban areas, so rural participants are difficult to get a professional sports guide to stimulate their sports interest. Fifth, the related legal system of social sports is not perfect, affecting the standardization of social sports instructors' training. In addition, government support is insufficient, and the training methods are relatively single, so it is difficult to exert the potential of training institutions.

2.3 Training objectives of social sports instructors

With the development of social sports, the training requirements of social sports instructors are constantly increasing. Firstly, social sports instructors need to systematically and comprehensively grasp of the guiding theory for social sports and extend the scope of knowledge to fitness, health, psychology and other fields. Only in this way can social sports instructors better guide the development of social sports. Secondly, social sports have strong comprehensiveness. Therefore, it requires social sports instructors to have excellent professional knowledge, strong comprehensive ability. Thirdly, social sports instructors should have the awareness of national fitness and develop scientific fitness plans according to social policies and appeal, improving the health and fitness of people.

3 INNOVATION TRAINING OF SOCIAL SPORTS INSTRUCTORS FROM ECOLOGICAL SPORTS PERSPECTIVE

National fitness is the trend of China's social development. Influenced by this trend, people have greatly changed their attitude toward life and had a deep understanding of the importance of health. They prefer to spend money on health rather than on medicine. Thus, most people would like to spare some time to join the social sports. With the updated of people's values for sports, social sports instructors' training also needs innovation and reform to update their concept.

3.1 Positioning of social sports instructors' training from ecological sports perspective

Ecological sports were proposed while human beings facing the serious problems of ecology and resources. Faced with ecological imbalance, the extinction of plants and animals and frequent natural disasters, humans begin to rethink the relationship between nature and human beings. Ecological sports emphasize the harmonious unity and coexistence and co-prosperity between human and nature. It also takes sports as the relationship between culture and ecology, seeking to a harmonious development. On this basis, the primary task of social sports instructors' training is to correctly and scientifically position the functions of social sports instructors. First of all, social sports instructors should respect, love and contact nature. Instructors should position their functions combined with the thought of "harmony between human being and nature", so as to make contribution to the development of social sports. Secondly, social sports instructors should be combined with the idea of freedom, amusement and volunteer to train their awareness of serving society. Afterwards, they will use an attitude of serving society to communicate with others, thus promoting the harmony of social sports and strengthening the body and heart of people. Also, they can make contributions to the development of social sports with a spirit of devotion. Thirdly, social sports instructors should emphasize humane care, concern natural and the individual and love life and health to ensure the harmonious relationship between social sports development and ecological protection. Moreover, social sports instructors should comprehensively understand their positive significance for social development and purposively train their abilities. In addition, they should carry out differential guidance for people in physical exercise, and give full play of the human-oriented idea to highlight the positive significance of social sports in human health and entertainment.

3.2 *Approach analysis of social sports instructors' innovation training*

There are countless ties between social sports and ecological sports. Social sports attach importance to the fitness, heart health of heart and entertainment in sports regardless of competitiveness. So, it is a physical fitness exercise in game and entertainment. The idea of ecological sports is approaching to this. Ecological sports emphasize on human initiative in sports, stressing health first and pleasure sports. It also emphasizes on the relationship between human and nature, the harmonious unity and common development between human and sports. So, the purpose of two sports is highly consistent. Firstly, great attention should be paid to the structural optimization of social sports instructors and the training of young instructors. And social sports with modern concepts should be closely combined to achieve the goals of national fitness and personal lifelong sports. Secondly, the diversification of instructors' guiding skills should be emphasized, and the content of social sports instructors' training should be constantly enriched. Integrating the nature, sports and culture to raise social sports instructors' awareness of social ecology, then it enable instructors to spread healthy and scientific sports skills. Instructors should constantly enrich people's amateur life and effectively

adjust their psychology. Then perfect personality of people will be shaped, promoting a harmonious and stable development of society. Thirdly, the open model of social sports instructors' training should be perfected to make instructors contact with a wider range of training content, thus exerting their learning initiative and creativity. Meanwhile, instructors should diversify their own career planning, so as to providing more reasonable and scientific services for people's health. In this way, a wide range of sports knowledge and ecological information will be provided for people to enhance their intellectual capacity and physical and mental health. Fourth, social educational resources should be fully used to make the routes of social sports instructors' training diversified. On the one hand, the single way of social sports instructors' training should be gradually improved to expand the scope and scale of training. Then related training institutions and units will train more professional and high-quality guiding talents to meet the needs of the social sports development, thus providing better service for China's sports cause. On the other hand, university, with rich teaching resources and good environment, is an important foundation to train guiding talents and should be scientifically used. University can also scientifically cultivate sports talents according to the demand of participation in sports. Thus, making best use of the university, it also can promote the development of China's nationwide fitness programs.

3.3 *Optimization of social sports instructors' training environment*

Environment is an important factor for talents training. Training social sports instructors should pay high attention to the optimization of environment and the awareness of environment and ecology. Firstly, experiencing the relationships between sports and ecology should be emphasized in instructors training to make them realize the importance of environment on human health in real and intuitive experience. Sports instructors should establish a correct development concept of social sports and integrate the harmonious relationship between human and nature into social sports. Then people can experience and enjoy the nature and have the awareness of protecting natural ecology and yearn for ecological sports experience. In this way, the development of social sports and ecological sports can be associated, thus effectively improving the physical and mental health of people. Secondly, legal construction of participation and guiding of social sports should be emphasized to fully exert its social and economic benefits under the protection of law. It can also strengthen the determination and confidence of social sports instructors to serve social sports development. Thirdly, the training process of social sports instructors should be improved to make them extensively absorb knowledge. Then instructors can combine theory and practice with theory and thought in serving social sports and equip themselves, taking rational and keen response for various social issues. Therefore, instructors can pave the way for the rapid and stable development of social sports.

4 CONCLUSIONS

Innovation training of social sports instructors is essential for the development of social sports. The combination of the concept of social and ecological sports is helpful to break through traditional sports guiding and decrease the requirements of sports skills. It can fully show the effect of strengthening body and heart health, so that the participants can actively participate in social sports under the guidance of ecological awareness and health notion. Then the role of sports for fitness will be enhanced. If public awareness of healthy and lifelong exercise was established, the overall quality of people will be improved.

Thus, people can make greater and longer contributions for socialist construction.

ACKNOWLEDGEMENT

The work is within the Research of Social Science Found Project of Jilin Province "Countermeasure Research of Innovative Training Social Sports Instructors by University Resources" (No. 2013B256).

REFERENCES

[1] Zhu Ling, Classroom Construction of Ecological Sports in the View of Ecological Civilization, Contemporary Sports Technology, 2013 (09): p. 126.
[2] Xi Li, Yang Zhimin, Factors Influencing the Development of Social Sports Instructors and Training Strategies—a Case Study of the City Of Hefei, Journal of Hefei University of Technology (Social Science), 2011 (04): pp. 156–106.
[3] Chen Wei, Xu Fuzhi, Zhu Yalin. Construction of Social Sport Instructors Training Model in High School P.E Major, Science & Technology Vision, 2013 (25): pp. 19–20.

Education Management and Management Science – Zheng (Ed.)
© *2015 Taylor & Francis Group, London, ISBN 978-1-138-02663-6*

Dividend policy and corresponding stock market reaction in China

Meiling Liu
Wuxi Institute of Commerce, Wuxi, Jiangsu, China

ABSTRACT: With the development of socialist market economy and economic globalization, the growth of stock market in China has been tested in many aspects. Enterprises, shareholders and relevant regulators need to have joint efforts in facilitating the harmony and prosperity of Chinese stock market, especially in dividend policy and information disclosure. The warming trend of Chinese stock market will have a retroaction on dividend policy. Given the policy adjustments of some enterprises and more accurate information disclosure, stock market in China still has a large development potential.

Keywords: dividend policy; stock market reaction; information disclosure

1 INTRODUCTION

Stock market in China started in 1989. Although the early development of Chinese stock market was not smooth, it has come to act on international convention and develop comprehensively after China became a member of WTO and had its market extended to the whole world. Regarding to the fast changing stock market, some experts raised several kinds of bold predictions and thorough analysis on the economic growth rate, the amplification of inflation rate, reform of IPO and law enforcement efforts. In order to gain a positive position in stock market competition, enterprises need strong management and reasonable dividend policy to earn the trust and support from shareholders. The sustainable development of Chinese stock market also needs the exploration of China Banking Regulatory Commission (CBRC) and China Securities Regulatory Commission (CSRC) with the assistance of perfect and scientific policies to guarantee the benefits of concerned parties and the coordinated development of economy [1].

2 CURRENT SITUATION OF DIVIDEND POLICY IN CHINA

Dividend policy is a standard of dividend appropriation proposed by general meeting of stockholders or board of directors. It often contains the ways and time of dividend appropriation and influences enterprises' distribution and plan of profits. Therefore, Chinese enterprises are cautious of dividend policy and different enterprise has its own characteristic of dividend policy.

2.1 No or rare dividend appropriation in many enterprises

As investors of companies, shareholders have the rights to proportional distribution of dividend, which is explicitly stipulated in Company Law of the PRC. According to legal procedure, when a company has allocable profits, general meeting of stockholders or board of directors should first draft profit distribution plans for general meeting of stockholders to discuss [2]. If these plans are passed, the company should make a dividend according to the plans. Dividend appropriation ought to be the operating principle for companies to observe, and it can also attract the trust and further investment of shareholders.

Facts show that dividend appropriation can help the stock price rise of enterprises in some industries. For example, China Mobile Communication Corp (CMCC) and China Unicom began dividend appropriation in March, 2003. Then, the stock price of the two companies stopped falling tendency and began to rise. CMCC had great rise of its stock price for its good market performance and stood out over the same period of stock. Thus it can be seen that dividend appropriation is good for the development of enterprises in the stock market. However, current dividend policy in China can hardly reach the situation of healthy and steady dividend appropriation, which needs people to conduct further study. Before the year of 2002, dividend appropriation was not common in all kinds of companies. About 220 listed companies never had dividend appropriation, and some of them even never had any form of profit appropriation [3]. The shortage of knowledge about listing standard and imperfect mechanisms can explain the situation in early stock market of China. As China became a member of WTO, it began to positively participate in global market competition. The stock market, where most competition exists, requires more normative and reasonable dividend policy. Although the number of listed companies that never had dividend appropriation is declining, the phenomenon of rare appropriation still exists, partly because of the low rate of return and profitability of Chinese companies.

2.2 *Partial adoption of residual dividend policy*

The history of stock market in China is short and China has been learning from the stock markets in developed countries. As far as China listed companies are concerned, their quality assessment was stipulated by China Banking Regulatory Commission in 2005. The main idea of this assessment is that the quality of companies should be established on the foundation of maximizing shareholder interests, including management ability, profitability, financial liquidity and information disclosure. Therefore, the dividend policy is affected by the operating conditions of companies to some extent.

The residual dividend policy has been applied by some Chinese listed companies. However, this application is periodic, long-term adherence to this policy will restrict the interests of companies and their develop capabilities. The residual dividend policy is generally applied to those companies with many investment opportunities and higher rate of investment return, so these companies can distribute residual profits after the internal capital transfer. This model can not only guarantee the steady operation of companies, but also help some initial listed companies or even newly established companies to form reasonable capital structure, ensuring their steady development at early stage. Some Chinese listed companies have small scale of capitals, so the residual dividend policy can help them relieve the capital pressure of reinvestment. However, short-term of application can support the growth of young companies while the long-term application can only bring the disjunction between stock prices of companies and their business circumstances to be more obvious. The detrimental factor of dividend is not stock price, and the weak dividend awareness can make shareholders suspicious of the business circumstances of companies. Seriously, some shareholders may even withdraw shares from their companies. A series of malignant consequences can have bad influence on the debt paying ability and investment ability of enterprises. If the managers who adopt this model can only seek to earn lot of profits but not to allocate them, the company will finally be isolated.

2.3 *Disturbance from untrue information*

The dividend policy of rare or no dividend demonstrated by some listed companies in China comes down to nonstandard operation and even illegal behaviors of managers. The imperfect of regulations and systems of stock market allow some companies to infringe the legal rights of many shareholders and they even are kept in the dark.

Firstly, in the process of making profits, the accounting and supervision of companies often have some exaggerations. Although both the government and enterprises are calling on information disclosure and fair competition, in fact, nonstandard calculation of assets debts and cash flow, the information got by the government, CBRC, enterprises and shareholders has been contaminated at source. The profits distribution based on false information has already deviated from honest goals. Therefore, the dividend policy has no longer been profitable for shareholders. The misappropriating of some listed companies ignored Company Law and Accounting Law as well as some industry norms, increasing the suspicion of reaping huge profits with other institutions and endangering the Chinese dividend policy and stock market. In order to solve these problems, the Accounting Department of Jinhua, Zhejiang once tried Centralized Office to have the financial work conducted in one place where auditing departments can timely supervise and check the financial work to ensure the normalization of accounting information and the truth of accounting report offered by companies. Some similar controlling methods are constantly being proposed to deal with the untrue information disclosure and non-valid transactions among enterprises, guaranteeing the steady of stock market and the legal interests of shareholders.

3 ANALYSIS ON THE REACTION OF STOCK MARKET IN CHINA

After the long-term depression, stock market in China gradually shows signs of a turnaround. The good news of central government policy, the overall prosperity of economy and relaxed market environment give hope to many enterprises and their managers. Many reforms will continuously emerge in stock market. For the comprehensively economic development, China had steady economic development with controlled indexes in reasonable scope, laying a good foundation to further economic development and policy reforms. Huatai-Pinebridge Fund points out that the optimistic situation will continue and there will be obvious increase in manufacturing industry, architecture industry and consumption industry, which is a good news and chance for stock market. However, for listed companies, whether their interests can be guaranteed or not depends on the appearance of relevant dividend policy. Judging from the trend of recent years, the number of companies with no or rare dividend appropriation will be smaller and smaller. The national policy and regulations of CBRS also don't allow the existence of unreasonable form of dividend appropriation. The prosperity demonstrated by stock market will act on the dividend appropriation policy and standard of some listed companies in a certain period. Some companies will seek to strengthen the structural allocation and have their dividend appropriation on the basis of fairness and justice. The number of dividend should be defined by clear evidence and the time of dividend appropriation should be decided by the general meeting of shareholders. In addition, some companies draw up corresponding plans for the confirmation of business operation, capital transfer as well as the benefits protection of both the companies and their shareholders, scientifically avoiding risks to the large extent.

3.1 *Further protection of shareholders' interests*

The protection of shareholders' interests has long been emphasized by national policy. The current prosperity of Chinese stock market should be attributed to the support of shareholders whose capitals enable enterprises to continuously enhance their flow of fund and expand their business scales. The cooperation between enterprises and their shareholders is becoming closer and closer, making stock market in China gradually become mature over the decades of development. Protecting the legal interests of shareholders is not only the guarantee of cooperation, but also the key link of maintaining the order and steady of stock market. With the increasing supervision and management on the behaviors of enterprises, the functions of shareholders are becoming more regulatory and reasonable. One share one vote rule and the principle of fairness, justice and open have been deepened, guaranteeing the rights and dividend appropriation of shareholders from the whole company. Meanwhile, the enactment of policies on the protection of shareholders' interests legalized the behaviors of different parties, motivating the dynamic of the whole stock market.

The status of shareholders changed with the adjustment of enterprises' policies. Shareholders have come on the stage rather than stay behind the stage, and their rights of information and inquiry have been strengthened further. At the stage of dividend appropriation, companies should take priority of shareholders' interests and they cannot have internal absorption of dividend with unreasonable reasons. The supervision department has set up reporting institutions to supervise the illegal behaviors of some listed companies. Then, the operation of stock market should more rely on the shareholders' interests and listed companies ought to strictly regulate their business and profits allocation, facilitating the harmonious development of stock market.

3.2 *Larger scale of information disclosure*

The information disclosure of enterprises has long been the focal issue because the data and materials of the enterprises' business and their reality not only concern the image and fortune of an enterprise, but also arouse the supervision of the whole stock market and economy. With the increasing expansion of business scales of some enterprises, the successful conduction of enterprises' activities can be the result of the cooperation among several enterprises and government departments. Therefore, once the information collection is inaccurate, the whole business chain would be endangered and the investigation of enterprises or officials would arouse uproar among the public. In order to guarantee the legality of internal operation of enterprises and make information or procedures demonstrated before shareholders more persuasive, expanding information disclosure is the focal work of current stock market and economic field in China.

The problems of information disclosure in China can be summarized as untruth of information, nonstandard form of information disclosure and out of date information disclosure. The resolution of these problems should find reasons from enterprises and relevant departments. Firstly, the supervision development should enact uniform regulations for the information management of enterprises and establish complete information process to replace the nonstandard operation of some enterprises. In addition, enterprises should strengthen internal control and set up prestigious supervision and audit panels to have necessary supervision over the daily work of enterprises. Special emphasis is supposed to be given to accounting behaviors to prevent the alteration and omission of information at source. The information management of enterprises is their vital fortune, and good information disclosure can earn the recognition of more shareholders and other enterprises, helping to create good atmosphere in the whole stock market.

4 CONCLUSIONS

Stock market in China has undergone tests from different aspects over the past decades. In the unsmooth practices, some immature behaviors have been demonstrated by some listed companies in stock market. Compared with other countries, the dividend policy of China is not perfect and has not formed a set of code of conduct. No or rare dividend appropriation, unreasonable dividend policy and poor information disclosure greatly hindered the development of stock market in China, damaging the legal rights and benefits of shareholders and enterprises. Nowadays, enterprises and supervision departments are trying to take measures to bring the stock market to high-efficiency development track. On that basis, dividend policy and stock market quality can be optimized and upgraded.

REFERENCES

[1] Qian Lili, Sun Yujun. *Analysis on Dividend Distribution of Ningbo Listed Companies*. Commercial Accounting. 2014(9):76–77.

[2] Wang Chen. *Correlation of Listed Companies' Capital Structure and Cash Dividend Policy*, Heilongjiang Science and Technology Information. 2014(7):295–295.

[3] Zhang Junmin, Wu Fengmin, Fu Shaozheng. *Financial crisis, Ownership Concentration and Cash Dividend Policy—Based on Empirical Evidence from A-share Listed Companies*. Journal of Nanjing Audit University. 2014(3):39–48.

Education Management and Management Science – Zheng (Ed.)
© 2015 Taylor & Francis Group, London, ISBN 978-1-138-02663-6

Analysis on creativity application in painting teaching

Lihua He
School of Art Design, Wuchang University of Technology, Wuchang, Hubei, China

ABSTRACT: With development of education and talent training, creativity application of students has been the main direction of teaching innovation. Painting field, with characteristic of creativity and artistry, has been the direction of teaching innovation. Based on single, rigid and nervous classroom atmosphere in painting teaching, teachers should fully exert subjective activity and search for the corresponding improvement scheme, thus achieving creativity excavation and application of students.

Keywords: painting teaching; creativity; application

1 INTRODUCTION

In teaching, domination of Chinese, mathematics and English has been past, and painting, music and performance are gradually on the right track, thus making an important meaning of improving comprehensive quality of students. In painting course, the creation of students cannot be comprehensively evaluated by "like or not" and "beautiful or not". The teaching target of this course is gradually close to development of students. This change can be suitable to the trend of talent training, and the characteristics of children including opening thought and flexible mind at the present stage [1]. Teaching and learning mutually affect, so the teachers can find suitable teaching method, and adjust teaching target and plan, thus making the modified effect more and more popular.

2 PRESENT SITUATION OF PAINTING TEACHING IN CHINA

2.1 *Single form of painting teaching*

As a creative behavior, painting course plays a role of arousing divergent thinking and creativity of students in the process of training. At present, insufficient attention has been paid to painting course, reflected in weak teacher resource, lack of specialty of teachers and occupation of painting class in most of the schools. This situation is caused by laying stress on course of Chinese, mathematics and English and achievement for examination for a long time in China. It is difficult for school and teachers to realize the importance of painting teaching to creative adaption and intelligence development [2]. So, the teaching activity of

painting is restricted. On the one hand, although some schools set up painting course and provide the students with books, teaching target and better teaching method are not researched by teachers. On the other hand, painting learning is restricted in composition and color filling with brush and white paper in some schools, thus causing lack of understanding and necessary practice of other painting forms. The essential reason is narrow teaching idea and vision. Stay in basic painting form for a long time will restrict creative thinking and manipulative ability and make painting teaching as the eight-part essay.

Painting, as a kind of diversified art, can effectively liberate human nature and divergent thinking. At the stage of mental maturity and flexible mind, the students should be provided with wide stage. Oppositely, painting teaching are tied up with fixed form and thinking, thus making painting teaching no sense and losing a chance of individualized training for students [3]. Consequently, the disadvantages are more than advantages.

2.2 *Fixed teaching target*

Painting course in some responsible schools is better than that in other schools, such as rigorous teaching procedure, teachers with professional painting qualification, scheduled teaching target, etc. However, in order to catch essence of painting teaching, flexible control should be achieved in the teaching process. In general, the teaching plan teachers draw up is compulsory. E.g. the students are required to learn brushwork of some kind of little animal in a few class hours, and exert themselves in the latter part of the course, etc. However, because of limit of teaching time and opportunistic teachers, most of painting classes are the learning process of copying works of teachers, and the students can hand over a beautiful work to school and parents. The teaching target and procedure actually violate the meaning of painting teaching and job requirement of teachers, although seeming perfect. Such teaching mode makes creative painting education fall into old teaching idea and mode, thus seriously affecting active and optimistic teaching ethos.

The reaction ability, flexibility and inventiveness of students, the strongest in their infancy and boyhood, should not be ignored. Once missed the period, it will be difficult for creative teaching activity of teachers to achieve good effects, and the potential development of students will be meaningless. Teaching target is stable. However, if teachers cannot control the degree, the stability will change into rigid restriction, thus limiting thought of students and the teaching activity of teachers.

3 NERVOUS ENVIRONMENT CAUSED BY TEACHERS AND PARENTS

For a long time, learning effect of students has been a target for teachers and parents, and the students have become larger group under pressure in society. The children are expected to get excellent achievement by the teachers, and perform better than others by parents. However, with weak ability, the children require comfortable environment provided by teachers and parents, thus promoting maximization of all abilities of students. In fact, the expectation of teachers and parents has unconsciously evolved into exaction and compression. In addition, immature mind, unsolved pressure and heavy schoolwork burden result in deformity of their learning behavior. So, it is difficult to present excellent achievement teachers and parents expect, thus restricting development of children themselves. In painting teaching, which should have been easy and joyful, some teachers evaluate the paintings of students with ugly words because of the reason of themselves, or unintentionally compare the paintings, thus making the students nervous. Besides, the students are more afraid of strict or arbitrary evaluation by parents. Psychology of nervousness, fear and anxiety can produce negative emotion and big shadow in the next painting class. E.g. the students only create images that parents and teachers like or ever praised using normal color configuration, and they are afraid of expressing what they are thinking actually. It is not good for students. As the so-called normal aesthetic formed in this process, their thought or creativity is blotted out. Parents and teachers should calmly think about which is more important.

4 SEVERAL WAYS OF CREATIVITY APPLICATION IN PAINTING TEACHING

4.1 Expanding diversified painting teaching

Actually, the form of painting can be divided into abstract painting and realistic painting according to properties, and into painting on the glass, canvas, board and container according to painting tools. Kinds of expressions can be transferred to painting class. School should increase financial support, because painting teaching, which is the same as other art teaching, requires certain hardware support. This good effect to be formed in the future should be expected, thus deeply affecting creativity development and aesthetic training of students.

Besides paintbrush, painting tool including crayon, scrubbing brush, pigment, nicking tool, etc. should be used to expand art vision of students. So, they can understand artistic convey and performance with different forms, thus promoting diversification of expression in painting and other fields. In painting teaching, the way teachers teach and students learn will directly affect development level of students' creativity. Diversified painting teaching, according with thought of modern education, has an advantage of artistry compared with Chinese, mathematics and English. The teachers are provided with space of teaching innovation and constantly try the new teaching form the students learn interesting content for their research in order to increase ability and efficiency of autonomous learning. So, the creativity of students can be deeply developed by a higher platform provided by diversified forms of art, and the teaching activity of teachers can be easy.

4.2 Actively excavating creative thought of students

Regular teaching mode cannot be suitable to the requirement of modern talent training to teaching activity. In this teaching mode, students will be slow to respond and short of flexible ability, instead of regular pattern of behavior and thought. Research shows that a baby of eight and nine months has two special abilities including distinguishing different faces of apes and slightly different voices. However, nine months later, the two abilities will disappear. So, the training of children should be controlled in specific periods. Once missed, it will directly affect training of ability and quality. Likewise, the training of students' creative thought should be controlled in a certain period. In this period, a more reasonable education scheme should be established by teachers or parents to effectively improve creativity of students. In early learning, the students are the most active in thought, and the teachers can help them to detect similarities and differences between things and expand thought according to striking pattern, brilliant color and different expressions. So, the students can be encouraged to present their own thought and original view by using untried expressions. The teachers should be more active to make the students overcome fear and pressure, thus forming intelligent way of dealing with things and flexible thinking. Besides, the teachers, as the guider in teaching activity, should help students get actual achievement through persuasion of thinking and language. Meanwhile, without intervening behavior of children, the teachers should give them with suitable help, thus improving their autonomous ability and activity of learning behavior.

4.3 Establishing comfortable teaching environment

Teaching environment has an influence on learning attitude, emotion and creativity of students. So, comfortable psychological environment, the same as hardware facilities and class environments, should be provided for the students. Training activities of creative thought should be added as follows: exhibition of creative works of calligraphy and painting can be developed; the students can be organized to watch documentary of art and masterpiece; the students who got achievements in artistic field like painting can be commended. The teachers should try to approve and praise thought and works of students in daily painting teaching as much as possible. They should tactfully evaluate some unsatisfactory works, instead of criticism and blind denial. So, the students can be happy to achieve more diffusion and association of thought and make full use of their creativities in creative process.

In addition, comfortable learning environment and active psychological state can form polymerization advantage, and good atmosphere of the whole team or class can benefit teachers and students from cooperation and healthy competition. Quality an artwork is not intelligence of somebody. The group or collective formed in teaching can make an effect that one plus one is more than two, thus increasing ratio of good works in class. Once the collective advantage is formed, intense thinking inertia will take shape in students. By using creative thought in daily life, the students can develop the habit of observation and thinking, which is beneficial to painting, each aspect related and students the whole life.

5 CONCLUSIONS

In fact, painting teaching is the easiest to develop creativity training of students, and its advantage in morphology and property can easily attract students. Likewise, the creative thought of students can be reasonably used in painting field. This relationship provides creative application of painting teaching with right timing and place. Besides, more feasible attempt should be performed by schools and teachers. In painting class, certain modern education characteristics can be presented by introducing diversified forms of painting teaching, constructing learning environment without pressure and improving original teaching method. So, training and application of creativity can be promoted, and this kind of teaching mode can be permeated into other teaching fields, thus pushing comprehensive development of modern education.

REFERENCES

[1] Zhang Xiaohong, Analysis on Painting Teaching Merged into Various Fields, Examination Week, 2014 (31): 192–192.
[2] Hu Lichang, Learning to Enjoy Paintings of Children, Little Writer Selection: Teaching Communication, 2013 (12): 483–483.
[3] Zhu Yuefeng, Return of Children's Happiness of Doodle, Forum on Education Research, 2011 (11): 27–28.

Education Management and Management Science – Zheng (Ed.)
© 2015 Taylor & Francis Group, London, ISBN 978-1-138-02663-6

Cooperative marketing game based on industrial cluster

Yijun He
Department of Economics and Management, Shaoyang University, Shaoyang, Hunan, China

ABSTRACT: Economic benefit needs mutual cooperation and coordination of characters in non-isolated business activities. Cooperative marketing model of industrial cluster emerges with the enhancement of economic level, offering a new model to economy development. The cooperation of production, packaging and marketing can be achieved. It contributes to conservation of resources and improvement of economic benefits of cooperative partners. In this work, backgrounds, impact factors and specific strategies of cooperative marketing of industrial cluster were analyzed with Game Theory.

Keywords: industrial cluster; development background; impact factor; cooperative marketing strategy; game

1 INTRODUCTION

Industrial cluster cooperation is a common and distinctive model in modern economics. It contributes to connections of enterprises and other economic organizations, strengthening relations and communications among various enterprises. Cooperative marketing of industrial cluster can promote development and progress of whole region and establishment of regional brands. Thus, the marketing model has been accepted by many regions and enterprises. More tight relations and cooperations emerge with the implementation of industrial cluster cooperation in one region. But competitive relationship within these enterprises, especially similar enterprises, is fierce and brutal. Coexistence of competition and cooperation is the true relation state among enterprises [1]. Therefore, competition and cooperation should be paid same attention to well balance the relationship based on industrial cluster. The "win-win" can be achieved with Game Theory. More economic value can be created based on the guarantee of economic interests of cooperative partners.

2 BACKGROUNDS OF COOPERATIVE MARKETING OF INDUSTRIAL CLUSTER

Socialist economic level in China is relatively low since it is in the early stage. Economic development speed increases continuously. Environment for China's market development is not stable compared with other developed countries. Meanwhile, small and medium-sized enterprises play an important role in the socialist market economy, but with characteristics of less capital, small scale

and weak vitality. They are too feeble to survive in the unstable market in China. Thanks to industrial cluster cooperation, sale cost and transportation expense can be decreased for internal marketing of cluster [2]. With labor division and cooperation, industrial cluster has adapted to macro economics in China and micro economics and trade environment of small and medium-sized enterprises. Industrial cluster provides more opportunities and a broader space for cooperation of enterprises.

From the perspective of present status in China, there has been a lot of industry cluster cooperation in relatively developed southeast coastal areas. For example in Wenzhou, collaboration between enterprises has become more and more common in economic development. But cooperation management of industrial cluster is still affected and restricted by many factors. Industrial cluster in China is not a reasonable system. Seamless regional cooperation and pipeline connection are rarely seen. Research on cooperative marketing of industry cluster is still superficial without special scientific theoretical guidance [3]. Moreover, others factors also affect cooperative marketing, such as trends of economic development, different resources of partners, guide of market, personal preferences of managers.

3 IMPACT FACTORS OF COOPERATE MARKETING UNDER INDUSTRY CLUSTER BACKGROUND

3.1 *Trends of economic development*

Enterprises and economic organizations are strengthened continuously with the rapid increase of economy and the sustainable development in China. Business relationship between enterprises emerges for the purse of economic benefits. Joint efforts are made to achieve economic benefits. Such development state provides more opportunities for cooperation of enterprises. Similar enterprises in the same region tend to cooperate for marketing of industrial cluster. Compared with other marketing methods, cluster cooperative marketing endows strong competitiveness of quality, features and cost performance of products. Better yet, a certain degree of monopoly can be formed, such as featured brand in one area. Therefore, backgrounds and trends of economic development are important factors for promoting cooperative marketing.

3.2 *Different resources of partnerships*

Economic interest is the certain goal of all enterprises and economic organizations. Mutual using of resources is the premise of cooperation between enterprises, for resources of enterprises are the foundations for creating economic benefits. Similar with other cooperation, cooperative marketing should take full advantages of partners' resources for better value and economic benefits. Partners should make their own contributions for marketing, achieving "1 + 1 > 2" – meaning a successful cooperation. Therefore, different resources of partners are also impact factors for cooperative marketing.

3.3 *Guidance of market*

In market with widespread cooperation, there emerges an economic organization for creating and strengthening cooperation, called cooperative marketing intermediaries. The intermediaries offer careful planning for marketing strategy of enterprises. Cooperation between intermediaries and customer enterprises can be achieved, as well as cooperation with more partners for customer enterprises. A kind of freemasonry delicate relationship is formed between enterprises of competition and cooperation. Feedback of economic phenomenon along with market can be acted back on market. More enterprises and economic organizations should develop a broader market for expanding business, so as to attract more economic partners. The level of cooperative marketing can be improved, thus forming a benign circle of economy.

3.4 *Personal preference of manager*

Manager plays a decisive role in one enterprise's plan and decision, directly affecting healthy development and persistent operation. Manager's personality has an obvious intervention in cooperative marketing decision. If manager has a positioning of business object and idea, together with an ambition, the enterprise can develop for the distinct object on the basis of cooperation.

Superficially, there will be more business expansion and cooperation and continually excavated market. Instead, if a rational manager has a goal of stability and an absolute control of enterprise, the enterprise is in obscurity to a certain extent. Old customers are the main sources of income. Therefore, personal preference of manage has a direct impact on enterprise development, as well as cooperative marketing decision.

4 SPECIFIC STRATEGY FOR COOPERATIVE MARKETING OF INDUSTRIAL CLUSTER

4.1 *Cooperative pricing in industrial cluster*

In terms of same kind of product, internal enterprises of industrial cluster own great advantage in pricing. Regionalism is an obvious characteristic of cooperation in industrial cluster. Cost and spending can be saved in raw materials flow. Large-scale cultivation and manufacturing of raw materials are popularized in cluster enterprises for smooth production. Cost of product can be saved with scientific standardized management and obtaining materials locally. In the collaboration of related enterprises, proper internal price is to be adjusted and negotiated for more economic benefits. Products with such cost and price are very competitive in industrial cluster, compared with other similar products. Of course, low price is more attractive for customers. Enterprises are able to earn more profits.

Low price in industry cluster is achieved on the basis of guaranteeing enterprises' profits. Meanwhile, such a low price has no damage on legitimate rights and interests of consumers. With saving some expenditure in a certain extent, enterprises can obtain recognition, support and trust of customers. Economic goals of cooperation partners can be achieved at the same time. As a result, the pricing is entirely different from price in "price alliance" of market—intentionally driving down price. Industrial cluster can realize cooperative target of "killing three birds in one stone". Product of higher cost performance can be achieved with scientific and reasonable management and tight cooperation. On the basis of product value and price advantage, pricing is to be agreed for occupying a stable market. Moreover, prolonged trust relationship between enterprises can be maintained for better long-term cooperation.

4.2 *Planning of marketing activity in industrial cluster*

With income guaranteed and profits achieved, product sale is a key component of enterprise's business. Planning of professional skills is needed. For example, discount promotion has many different modes. Cooperation promotion has obvious advantages over personnel promotion, advertising and other forms. Media agencies, public relation companies and websites, along with the market, provide a good external environment for industrial cluster marketing cooperation. By using these conditions, internal enterprises of industrial cluster can make mutual cooperation with joint promotion activities. The risk can be appropriately dispersed. Promotion messages can be widely spread through media and network, so as to attract more consumers' attention. E-commerce can also be run on website. More market shares can be achieved with "small profits and good sales" by connecting with international market directly. In the promotion activities, enterprise and brand should establish a good image. Furthermore, regional economic development should be boosted with increase of sales efficiency and decrease of time.

Regional economy and brand are signatures for promotion activities. One kind of product in the region is a conspicuous representative. With collaboration between enterprises in cluster, many unknown small enterprises can obtain exchange opportunities on product exhibitions. Thus, small and medium-sized enterprises can cooperate with strong partners for its attractive excellent product quality and performance. Reform of product quality of whole industry is to be driven. Under mutual supervision of fair competition, more and more opportunities are provided for small and medium-sized enterprises. With market expanding, small and medium-sized enterprises can achieve more economic benefits by taking advantages of regional economy.

4.3 Diversification of sales channels for industrial cluster enterprises

Sales channel is an essential sales method, which is different from promotion and exhibition. With great significance, it brings the most stable income for enterprises. Industry cluster enterprises have a certain advantage in product's price, as well as in after-sales service. It is convenient for the service and maintenance of product. By taking advantages of industrial cluster, a whole sales network can be built with cooperation of similar enterprises. The scope and channels of sales can be expanded for the sharing of partners. "Enhancing advantage and avoiding disadvantage" can be achieved. More than low cost of product, cost of market development can be shared with cooperation of enterprises in industrial cluster. Thus, overall cost can be saved, and risk of investment can be avoided. More enterprises are able to co-found well-known regional brands in the same broad market with mutual supervision and improvement. For strong sense of trust, regional economy can attract more domestic and foreign large-trading partners. Success rate of foreign cooperation increases continuously. Enterprises can progress together with cluster partners in one region.

Meanwhile, industrial clusters and regional economic integration development need government's support and assistance. Government should actively advertise local cluster in the cooperation and communication with external economic entity. Regional economy can be more competitive and reputable. Diversity of sales channels can be developed, as well. Government should provide industrial cluster for beneficial policy support, so as to better serve local economic development and enterprises' value achievement.

5 CONCLUSIONS

Industry cluster is a trend of future economic development. Obvious competitive relationship is among similar enterprises. With inadequate cooperation and competition methods, problems and obstacles will occur in the development of enterprises. Analyzed from backgrounds, impact factors and strategies, cooperative marketing of industrial cluster should be promoted with well balanced cooperation and competition. Based on Game Theory, management of internal enterprises in industrial cluster can better deal with profits and continuous development. Economic interests can be achieved with rational benign cooperation. Enterprise can obtain long-term and sustainable development in fierce competition, contributing to enhancement of regional economic level. Economy in China can be developed soundly and rapidly. With the guidance of enterprise leaders, industry organizations and national government, enterprise in industry cluster can progress with a better and broader prospect.

ACKNOWLEDGEMENT

This research was funded by Education Department of Hunan Province "Innovative network of cluster based on knowledge management" with Grant number 11C1144.

REFERENCES

[1] Zhang Xiaobao, Yin Fangzi, Chen Cong and Gui Ling. *Marketing models of garment industry cluster in China—based on data of Humen clothing industry* (Modern Business Trade Industry, China, 2012).
[2] Yang Baojun. *Marketing models of industry cluster in western regions in China—Moslem food in Ningxia* (Heilongjiang National Series, China, 2007).
[3] Huang Benxiao, Luo Qin. *Characteristics and trends of development of manufacturing industry cluster—case of textile industry cluster in Shaoxing of Zhejiang Province* (Science-Technology and Management, China, 2005).

Education Management and Management Science – Zheng (Ed.)
© 2015 Taylor & Francis Group, London, ISBN 978-1-138-02663-6

The analysis of wisdom for Chongqing tourism public information service system

Yan Liu & Wentian Jiang
School of Tourism and Service Management, Chongqing University of Education, Chongqing, China

ABSTRACT: The wisdom for Chongqing tourism public information service system can improve the comprehensive competitiveness of tourism destination. Because the government attaches great importance, Chongqing has constructed travel industry information management system independently and set the wisdom tourism business foundation platform fully. Chongqing has enhanced the level of the wisdom of tourism marketing gradually. But compared with developed areas, the technology wisdom of Chongqing tourism information service level is not high and wisdom of county is not enough. To coordinate regional development for security and meet the demand of tourists, in some of wisdom information for tourists demand, need to increase intelligence technology as the foundation and accelerate the realization of tourism public information service real wisdom.

Keywords: Chongqing; tourism public information service; wisdom; wisdom tourism

1 INFORMATION AND WISDOM

Information makes full use of information technology, development and utilization of information resources. It promotes the exchange of information and knowledge sharing, improves the quality of economic growth, and promotes the transformation of the historical process of social and economic development. Since the 1990's, information technical revolution of information technology driven to Chinese economy, society and so on various aspects of the changes have been agreed by the world. And what's more, the 18th National Congress of the CPC more puts forward new requirements on our country's information construction. That is must be fully aware of the new situation of informatization development, new opportunities and new tasks. Here of the new situation of informatization is a new stage of informatization development, or we call it wisdom [1].

Wisdom is the third phase in the development of informatization, from digital to networked to wisdom. It is the integration of information technology application. The main application of the Internet of things technology, Cloud computing technology, Intelligent terminal technology, Big data technology. It products form the wisdom of the data through a large amount of data analysis of the information world, integration, mining, processing [2]. And then return to the real world, to its development to optimize the huge role of ascension. Its essence is to establish a virtual brain system for entities. It makes smarter, simulation has the wisdom. Nowadays, wisdom development is very broad. Its application is in all aspects of society. The key application is embodied in three areas: Wisdom city, Wisdom industry and Wisdom enterprise. As needed with the help of information technology means to improve tourists travel environment and experience in terms of the tourism industry, leads to the ecological, social and economic comprehensive value maximization. How can achieve wisdom is a concern topic by scholars, enterprises and government in recent years.

Chongqing has the first policy support of national wisdom city and with the construction of the dream of a world famous tourist destination. Chongqing should grasp the development trend of future tourism new formats, comprehensive grasp tourism public information service system construction; enhance the connotation of the wisdom of the tourism industry. It has great significance to improve the level of tourism information service and comprehensive competitiveness greatly, accelerate the development of become a tourism highland in the west of China and create high-end tourism city brand [3].

2 OVERVIEW

Tourism public information service is the core elements of public services, widely in content of tourism in developed countries and regions such as

Europe and the United States. For example, in tourism electronic business and information services, British tourist public information service system is very perfect electronic government affairs public and information service function. In many tourist website will also provide the perfect tourism literature delivery service. The tourist information center is government travel service agency engaged in tourism consulting services. It provided free of charge at least hundreds of single page or folding information for tourism and the content involves travel every aspect of life. The tourism public information service of Australia is also very convenient. No matter in domestic anywhere, even if the language barrier. Visitors can through a free interpretation services and translation hotline 131450 get a satisfactory answer. More dotted with Australian tourist information consulting services, you can see all kinds of public information service entity forms in attractions, streets, railway stations, airports, ports, shopping malls and parks. Tourists in Australia can see advertising everywhere and can get free travel promotional information from a variety of sources. It improve the overseas tourists know degrees of accommodation line tour for entertainment information, transparency greatly. And also guarantee the overseas tourists in Australia consumer satisfaction and retention fully [4].

However, domestic research on tourism public information service is a relatively new topic. As early as 2006 to 2008, the central related conference has been put forward to speed up the construction of tourism public service system in our country. But due to various places have different level of economic development and government sources have different degree and emphasis. Lead to the government, market awareness and put into tourism public service system construction are also different. In 2011, China national tourism administration determines the importance of tourism public service construction [5]. The public information service as an ascending tourism public service to start and put forward to develop the tourism public information service clearly. Strive for in 2015 basic perfect tourism information consultation service system and so on. In order to speed up the implementation of domestic tourism wisdom laid the foundation policy.

3 THE ACTUALITY

3.1 Constructed tourism industry information management system independently

At present, the city's tourism industry information management system has entered the stage of trial operation by the municipal tourism administration develop independently. The system can detect report and statistics by each district and county tourism bureau in the city, resources of tourism enterprise and tourism information data automation. Keep track of the type, level and quantity distribution dynamic change. And the dynamic data acquisition, not only embodies in the number of macro maintenance, such as the guide, the dynamic monitoring and management of tourism vehicles, will solve for a long time for the black guide of tourism, tourist vehicles overload, speeding, the phenomenon such as to change the route. And also will through the wisdom of information management, guide and support chongqing tourism enterprises through its own e-commerce applications, business process reengineering, and how to carry out comprehensive toward a more technical field of wisdom forms, greatly enhance the efficiency and effectiveness of the industry management.

3.2 Set the wisdom tourism business foundation platform fully

The biggest beneficiaries should be tourist in wisdom tourism. Regular visitors through smart phones or computers can obtain all travel service agency to provide information services. This is to improve the tourist satisfaction and increasing the important power of tourist attraction. These services need to build to achieve information operation and support the basis of application service platform system. At present, the wisdom of chongqing tourism platform construction is in the country is still in its preliminary stage of development. But it has been gradually online, to promote the wisdom of chongqing tourism development laid a solid foundation such as "the Three Gorges International Tourism Festival", in May 2014, this is wisdom of chongqing tourism platform comprehensive online node. It is reported that in the service of the Three Gorges International Tourism Festival wisdom of chongqing tourism platform were built four branch platforms. One is the first regional provincial tourism wisdom network platform (www. fortripbook.com). The second one is the travel booking service hotline, the phone is 12301. The third one is scenic and tourist code statistics. The fourth is the chongqing tourism information release of the official letter. Through these four branches improve gradually, tourism enterprises and tourism at the next higher level units will be able to more timely understanding of the tourist site. Visitors will also be able to view and understanding of the destination through mobile phones, and formulate corresponding schedule. In the end, all data will be consolidation and summary. Through the analysis for tourism enterprises to provide strategy and marketing direction of the next reference.

3.3 Enhance the level of the wisdom of tourism marketing gradually

Tourists travel mode and qualitative changes have taken place in the tourism products in the Internet age. Wisdom tourism can provide visitors with a brand-new tourism experience. As a result, the basic goal of its development is to enhance the tourism perception as the center, with a variety of marketing methods to achieve higher efficiency of information processing and personalized configuration.

Before the National Day in 2013, chongqing Yuzhong district tourism bureau with sina the cooperation, launched the official microblog. By sina, sina microblog and sina 25 local station centralized propaganda tourism theme. In has witnessed a huge marketing effect under the background of wisdom, "Chongqing tourism" micro official also followed by a formal letter online operations. Service elements of comprehensive integration of chongqing tourism resources, to arrange released the latest tourism information and tourist in the marketing platform. It provides the most practical travel guide and tourist information query. People can use mobile Internet, with the aid of all kinds of portable terminal device; more easily get all kinds of tourist information by online. At the same time, people also travel service providers to adopt diversified marketing provides more reference.

4 THE PROBLEM

4.1 Technical wisdom degree is not high

Wisdom tourism is based on highly developed Internet of things from the definition of the wisdom of tourism we can clear to appreciate. Through the Internet to realize the wisdom of the tourism industry and tourists travel wisdom, and is not a simple computer and the Internet plus technical software is wisdom. Wisdom is a continuation of the informatization and intelligent.

Although chongqing is now focus on exploring the public information service system for intelligent building has a clear direction and goals, but limited by the objective conditions. At present, we basically still stay in the informatization and intelligent level. Travel wisdom of the relationship between the entities has not been truly established. Travel service provider still pays more attention to tangible products, ignore the information service. Technical equipment backward, information release slowly. Abroad and even domestic coastal areas such as shenzhen, Shanghai has been widely used model of travel APP, intelligent technology such as RFID, wireless smart cover and even the city within the territory of chongqing many blocks are still not popular. And GDS distribution, GPS positioning, and the wisdom of the commonly used foreign tourism marketing means such as e-commerce has not received enough attention. Even the tourism information posted by running mainly relies on manual collection and experts design, unable to realize automatic updates. In tourism informatization revolution is full of the world at the same time, we should see chongqing tourism wisdom the status quo of public information service system and there are large gaps when compared with developed areas. Thus promoting wisdom tourism construction, strengthen the tourism information exploration, the chongqing tourism public information service actively as soon as possible into the mainstream of the market is very necessary.

4.2 Area county wisdom popularization is not enough

Wisdom tourism development is system engineering, covering multiple formats, more technical its content is complex, involved in all aspects, to the top floor design and construction of low-rise synchronization carry on, to provide, how, in combination, combining both up and down. Building in chongqing city national wisdom of objective support, the wisdom of the tourism industry design should pay attention to both the macroscopic level of unified platform, unified standard, unified framework, etc. We should also speed up the county level, standardization, refinement, procedural, digitization and informatization infrastructure construction, and make full use of the wisdom of tourism construction achievements, convert it to promote the overall tourism destinations, public service and management ability.

However local tourism administration in chongqing is still under the system of regional segmentation of reality, some counties such as Jiulongpo District, Nanan District, Jiangbei District, Liangjiang New Area really carried out different levels of wisdom to explore, and walk in the place of them. However there are still some counties affected by local financial resources and policy, tourism information of intelligent building is still on the design ideas and goals slogan. For local information design in the whole tourism industry lack of unified deployment and coordinated action. Many resources have not been effective sharing of software and hardware. A variety of software data docking port is not open to each other. And redundant construction is more severe.

4.3 Wisdom of micro design is not much

Wei Xiaoan, director of the Chinese tourism research academic committee, said that if just put all kinds of resources on the Internet, it only instrumental sense, tourism is not wisdom. Wisdom tourism need to pay attention to micro: micro consumption, micro tourism, micro core, micro life, micro response. Especially must pay attention to young people's feelings, needs and so on. Here refers to micro are mainly for tourists to the tourist information design details. From the role of the early stage of the tourists, in the middle of the purchase, to use the process and to the back of the comments on feedback system, the wisdom of the details should be based on the tourist experience. It is designed on the basis of micro information intelligence. Wisdom is the core of the construction of the tourism; it is the key to tourism ultimately win the market. However, some local governments and businesses in the building of the wisdom of tourism excluded the tourists, ignoring the needs of tourists, completely from technology to technology cycle, it is not true wisdom.

From the perspective of the building of chongqing tourism public information service system, whether it has built online travel industry information management system, and set up perfect the wisdom of tourism infrastructure, affected by traditional service concept guide, are more inclined to point to the "wisdom" data management, rather than for tourist information requirements of information design. For some local governments and enterprises, intelligent control is more important than wisdom of tourists' perception.

However the core of the wisdom of tourism is the tourists as the center, the ultimate purpose is to facilitate visitors to tourist destination information sharing and exchange. If you ignore the destination of tourists' perception and experience, it is difficult to achieve the delivery of information, will lose the real value of tourism development of wisdom. This is also the construction of tourist destination problem urgently to be solved in the public information service system. Therefore, the public information service system of Chongqing for the future of the wisdom development should take the tourists as the core, in such aspects as information analysis, collection and integration should be

close to its needs change. To travel for clues, construction network technology infrastructure supporting complete. At last realize information and resource sharing, so as to ultimately achieve tourists experience and needs.

5 CONCLUSION

With the rapid development of information technology and tourism, realize the tourism enhanced the wisdom of the public information service system has become a tourist experience perception, an important means to promote the competitiveness of the tourism destination. Chongqing to realize the goal of build the world famous tourist destination, must be based on enhanced intelligence technology and to coordinate regional development for security. We also on the premise of meet the demand of tourists, to speed up the realization of information service for tourism destination, tourism enterprises, and tourists of true wisdom.

ACKNOWLEDGEMENT

Project Supported by Scientific and Technological Research Program of Chongqing Municipal Education Commission (Grant NO. KJ1401401).

REFERENCES

[1] *Chongqing tourism new era for wisdom* http://cqrbepaper.cqnews.net/cqrb/html/2014-05/19/content_1743954.htm, 2014-05-19.
[2] *Development of wisdom tourism needs to focus on some questions* http://www.hnta.cn/Gov/dongtai/fzcl/2014-01/24069054735.shtml, 2014-01-15.
[3] *Zhang Lingyun. The basic concept of wisdom tourism and theoretical system.* Travel journal. 2012,(5):66–73.
[4] *Liu Junlin. The composing of the wisdom of tourism value and the development tendency* Chongqing academy of social sciences. 2011,(10):121–124.
[5] *Wisdom tourism, Meet the demand of individual character* http://www.cq.xinhuanet.com/2012-12/17/c_114048463.htm, 2012-12-17.

Education Management and Management Science – Zheng (Ed.)
© 2015 Taylor & Francis Group, London, ISBN 978-1-138-02663-6

Application of humanistic management in sports clubs

Yanni Zhang
Yangtze University College of Technology and Engineering, Jingzhou, Hubei, China

ABSTRACT: With the comprehensive development of sports causes, the improvement of management mode is an urgent task for sports clubs of all sizes in China. As an important characteristic of modern management, humanistic management is suitable for organizations, especially like sports clubs, that combine management with service. It has become the general trend in current sports clubs. Considering current core aging and under-performing of domestic clubs, the application of humanistic management will bring them unprecedented vigorous development.

Keywords: humanistic management; clubs; application

1 INTRODUCTION

As Mao Zedong said, "material elements need man's subjective initiative for their exploitation", which reveals the important significance of people-oriented management. Years of practice has improved the content and extension of humanistic management. And its related specific management channels are also expanded by incessant exploration of people. Operation state and development prospect are affected by the transformation of people, environment, culture and value in the entire organization. And this is the reason why increasing people are looking for breakthroughs in state of people's mind and interpersonal relationship as well as organization forms in order to realize the improvement of management quality.

2 CURRENT MANAGEMENT SITUATION OF SPORTS CLUBS

2.1 *Unreasonable management mechanisms of clubs*

In China, professional and amateur sports clubs are in combined form of the government and enterprises. Similar to most clubs, sports clubs profit from supplying professional sports equipment, facilities and stadium, or from competitions and performances by employed professional athletes. With various sports programs, clubs of football, basketball and taekwondo are the most common, making up a large proportion of clubs [1]. And these clubs have fixed forms and scales of management. Meanwhile, the spring-up of other professional clubs (i.e., clubs of shooting, cycling and motorcycle) diversifies the development of

professional and amateur sports cause in China. Most domestic sports clubs, unlike sales industry or manufacturing industry, seldom have great demands and fixed distribution channels. Faced with certain management stress, their operating profit only relies on membership fee, venue rental fee and ticket receipts. Uncertain operation and the lack of finance from public companies have broken the industry operation chain of most sports clubs. Clubs in small and medium size and clubs of second-tier or third-tier cities are on the verge of closing down. Obviously, the single outdated management has shown its limitation. As the source of club activities, club management urgently requires necessary improvement and reformation for continued and prosperous development of sports clubs.

2.2 *Lack of professional spirit in athletes*

Athletes in training or competitions are an important component of professional and amateur sports clubs. With an excellent match or performance, athletes can build favorable team image and develop positive aggressive sportsmanship for belonged clubs [2]. For clubs with precise systems, normative discipline of training and life for athletes is required and so does professional competence in athletes, which both can improve overall image of teams and clubs. However, poor management in some clubs and the lack of fairness and justice in competitions suppress the expected enthusiasm of athletes, leading to the degradation of their public image. Besides, instead of delivering positive impression to coaches and the public, some of athletes are lax in discipline and unpunctual in training, and they even contradict coaches and referees, insult opponents, etc [3]. These behaviors are against sportsmanship and damage the reputation and image of clubs, which will cause a vicious circle in the future development.

2.3 Weak relationships between sports clubs and the public

With strong professionalism, most professional sports clubs realize their strategic importance and value by closed training and profitable matches. In China, there are several famous clubs such as such as Guangzhou Evergrande, Shandong Luneng and Henan Jianye who have compact training and matches. Many sports enthusiasts can well know club schedules, competition arrangements and athlete conditions. And closely crowed with these big fans, famous clubs are gradually far away from common people. More people know about these clubs only by Chinese Super League and Soccer Game of National Games, without other reliable accesses. Therefore, for most people, sports clubs are still high-class or independent organizations, and there is nothing known about their training venues and professional daily performance. Lacking organization planning, some amateur sports clubs or leisure clubs also have weak relationship with the public, failing to realize the purpose to "develop physical exercise and build up people's health". Theoretically, on the basis of the masses, such organizations could never exist without the support of common people. Most current sports clubs, however, have ignored their social responsibility and twisted their position in the entire sports cause, gradually abandoning their characteristic of socialization.

3 APPLICATION OF HUMANISTIC MANAGEMENT IN SPORTS CLUBS

3.1 Optimization of human resource allocation in management layer

Similar to other enterprise units, the management of sports clubs needs professional financial staff, committee, department managers, etc. Humanistic management concept emphasizes the role of people in management of enterprises or organizations. For example, for club management, managers have to optimize the staffing in organization. "People orientation" mentioned in humanistic management requires that chairmen or supervisors of clubs should have reasonable evaluation and positioning of individual condition, working capability and potential capability of each staff. And on this basis, post allocation and function setting can be gradually carried out. Wise managers know the way to realize the maximal function of limited human resource. Despite of the system of government-enterprise combination, sports clubs still require internal staff have strong professional qualification and handling ability of interpersonal relationship as well as

organizational skills, which are beneficial for further simplification and efficiency of club activities. Especially commercialized sports clubs that are hot recently are more supposed to positively introduce young and popular consumption patterns as well as leisure sports forms. And then with scale effect and chain management, these clubs can change the public fixed opinion of sports clubs and promote mutual development of the entire field.

In addition, expanded financing channels and enterprise cooperation can also help some clubs out of trouble. Club managers have to abandon traditional management and positively participate in market competition and resource sharing. Such modern management will renew the enthusiasm of entire work groups in clubs to make up the deficits and get surpluses. Similar to most organizations, sports clubs are work teams composed of individuals, where is suitable for the application of humanistic management. Qiaodan Sports, for example, has staff rooms in working areas, well-equipped facilities and perfect service. And the company organizes staff for various outdoor training and provides further study opportunities. Meanwhile, in order to maximally eliminate staff's concern and help employees focus on their jobs, the company even offers special rooms for working couples and builds bilingual kindergartens for staff's children. Such humanistic management gains people's praise and reliance as well as promotes employees' loyalty and efficiency. Cautious and conscientious work in large scale will produce huge power for the transition of difficult period. For sports clubs with bumpy development, people-oriented staff management will generate enormous internal drive for clubs to ensure their permanent development.

3.2 Development of athletes' potential and spiritual strength

The image and spirit of athletes are significant symbols of clubs. Excellent athletes in clubs can be involved in the national team of corresponding events. Training of elite athletes is always carried out in clubs. It is indicated that there is a close and meaningful relationship between athletes and their clubs. In consequence, clubs should place athletes in an important position in management.

By necessary means, humanistic management mode can realize the functions of cultivation, exploitation and encouragement in management of athletes. Athletes with pertinent evaluations will form intensified team spirit. Such athlete team is more likely to treat routine training seriously and achieve good performance in competitions. In fact however, poor management and lax athlete management system in domestic clubs have led to frequent athlete flow and made it difficult for clubs to retain talents. Destabilizing effect of team is against the development of athletes and the entire clubs. Additionally, focusing on the quota or ranking, clubs widely utilize the system of "survival of the fittest", which increases more mental stress on athletes. Sports supposed to inspire fighting spirit have become a tool to limit and suppress athletes. Such condition without timely reversal will cause serious vicious circle in athletes and prevent further development of athlete career and clubs. Therefore, chairmen and coaches have to put individual development of athletes in the first place, and respect and rely on athletes, looking for the perfect balance in cultivation of athletes. Meanwhile, the "balance" requires coaches' strict checking of daily training and corresponding disciplines of training and life. Athletes are required to be punctual and obedient in the training. And far way from things harmful to image of athletes and clubs, athletes have to discipline themselves to gain reorganizations of the pubic.

In addition, athletes need self-judgment and self-selection in their training without the interference of coaches. Having enough space for freedom and development, athletes can improve their training methods in controllable range, which gradually establishes personalized management mode. The development of athletes' potential is provided with strong support to promote their performance in training or competition for renewed records. The relationship between the club and athlete is like that of a ship and water. Water can keep the ship afloat or sink it. With work emphasis on the development and cultivation of athlete abilities, clubs can not only meet the demand of self-development of numerous athletes, but also increase own strength and competitiveness. Finally, the positive effect will bring dominant position to clubs in corresponding field.

3.3 Establishment of sports clubs full of humanistic concern

Following "people orientation", the management of sports clubs also has to present its humanistic concern. On the basis of good internal management system and athlete development, clubs have to provide intimate service to the masses for their support of sports cause and sports clubs for years. And in the external management of clubs, service optimization should be treated as the long-term working principle of staff in order to improve the quality of club service. Clubs, especially some commercialized or leisure clubs, have to provide corresponding perfect service for people with payment to gain favorable impression and recognition of customers. For hardware facilities, clubs are required to change the damaged and obsolete sport equipment in venue, increase a certain amount of fitness equipment, and ensure the neatness and well function of equipment. All these measures can provide comfortable environment and enjoyment

for customers and meet their hard requirements, ultimately realizing the basic value of clubs.

Besides, clubs can set up some innovative training programs or conduct some interesting activities for customers. With closed relationship, clubs can more effectively communicate with customers. Some of big clubs even involve typical service programs such as catering and entertainment. Managers have to increase the cultivation of employee services and standardize their specific behavior in customer reception and transaction processing. High quality of service will become a club's business card to attract more customers. In addition to this, sports clubs are supposed to regard fitness of the masses as own duty and actively carry out series activities that benefit people. For example, clubs can regularly open the venue for free, provide instruction of physical exercise for common people, and support the cultivation and selection of youth athletes in school. The activities not only respond to the call of nationwide fitness programs, but also obtain funds subsidization from the government and enterprises. What's more, good relationship with the masses will establish positive social image for clubs. With support of the public, clubs will gradually move towards a better future.

4 CONCLUSIONS

With the management form of modern enterprises, sports clubs shoulder the important social responsibility of sports cause development in China. Therefore, there is great significance of the upgrading in their management patterns and concepts. With exact emphasis on the role of humans in organizations and society, humanistic management requires improvements in internal personnel management, athlete development, and social image. Domestic sports clubs will come to a high level under humanistic management, and humanistic care will spread all over the management of the entire clubs. Such improvement can not only improve clubs' current condition of weak development and destabilizing workforce, but also help clubs establish modern operation mode, which will bring long-term benefit for various sports clubs in China.

REFERENCES

[1] Qiu Lewei. Analysis on the Relationship of Sports Competition and Economic Competition in Professional Sports Clubs. *Hubei Sports Science*. 2013 (8): 663–665.
[2] Zhao Qiutao. Preliminary Study on Humanistic Management in New Economy Era. *Young Society*. 2014 (4): 292–293.
[3] Ni Mengmeng, Deng Xinran. Analysis on Humanistic Management of Enterprises in China Based on Foxconn Workers Suicide. *Business*. 2013 (47): 134–134.

Education Management and Management Science – Zheng (Ed.)
© 2015 Taylor & Francis Group, London, ISBN 978-1-138-02663-6

Education mode of applied innovative talents in higher vocational medical colleges

Jianrui Yu

Baicheng Medical College, Baicheng, Jilin, China

ABSTRACT: Effect of talents cultivation in high vocational medical college has a direct influence on medical achievement and level of medical worker in future China. The public hopes for advanced and professional health care. Applied innovative talents are urgently needed. Many higher vocational medical colleges attach importance to cultivation of talents through various means. Facing with many troubles, colleges and teachers boost cultivation effect with development of students' source, reinforcement of faculty, improvement of courses and initiative cooperation. Thus, a scientific integrated mode can be formed for cultivation of applied innovative talents.

Keywords: medical college; innovative talent; education mode

1 INTRODUCTION

With better employment prospect and social status of medical worker, medical colleges are popular with examinees and patriarch. However, higher vocational medical colleges has received cold shoulder from the public. Long-term bias and some flaws of colleges make it ineffective for medical higher vocational education. Corresponding talents cultivation is difficult to orderly carry out. With intention of image-changing, higher vocational colleges need deep going exploration of running situation and education mode [1]. They should make efforts to achieve the goal of cultivating professional high-quality talents for society.

2 STATUS OF EDUCATION MODE OF TALENTS CULTIVATION IN HIGHER VOCATIONAL MEDICAL COLLEGES

2.1 Substandard enrollment of higher vocational medical colleges

With decrease of students' number, some colleges recruit very limited students every year, although colleges in China has greatly lowered threshold of admission. Moreover, only one child in family is common in China. Number of students recruited in any college is hard to achieve expected level. Some are even difficult to meet with eligibility criteria. In the field of medical colleges, enrollment in some well-known undergraduate universities is fever for their reputation and regular employment provided. Especially in universities in first-tier cities, enrollment of students is over-fulfilled for relatively huge demand [2]. On the contrary, higher vocational medical colleges are lack of visibility and reputation for located in the second-tier or third-tier cities. Additionally, with unsatisfactory employment, higher vocational colleges are hard to recruit students. For example, only a quarter of higher vocational medical colleges in Hebei province completed their enrollment plans in 2013. Most of higher vocational colleges are in trouble with admission.

The causes are complicated. Faculty and hardware conditions cannot satisfy most students and their parents. Long-term prejudice drive students selectively ignore higher vocational colleges. Little attention will be paid, even though colleges make sufficient preparations for new coming students and their future developments. Most of higher vocational colleges attract students with advertising in senior high school or contacting with students proactively. But these methods are not long-term solutions. Colleges should find internal reasons and effective solutions, so as to increase enrollment [3].

2.2 Single teaching mode in medical college

Teachers in medical professional are required to combine practice with theory in teaching. Students should learn theoretical knowledge, such as basic medicine, clinical medicine, preventive medicine and health medicine. Additionally, students should learn certain clinical practice skills with overcoming psychological obstacle in practical. Single teaching mode is common in most of higher vocational medical colleges presently. It does not differ significantly from teaching methods of other theoretic disciplines. Actually, higher vocational medical colleges need more combination of theory and practice in class. In the practice, students can master professional skills later used in the jobs, and can form earnest rigorous attitude of medicine.

For example, graduates of clinical medical can work on clinical treatment in community health service and basic medical places in the future. Graduates are required to detail understanding of pathogenic condition, physical condition and demand of patients. Besides, patients should be treated with a generous heart and good professional ethics. Especially for current tense doctor-patient relationship, students should be trained of reaction and psychological adjustment ability by teachers. Thus, students can better adapt to future medical work. Cultivation of students in medical colleges involves several aspects such as profession, society and psychology. Such knowledge or experience cannot be simply passed on by teachers' oral presentations. But most of higher vocational medical colleges still conservatively carry on this teaching mode. Students can contradict to or be tired of fixed mode, thus causing non-ideal teaching effect. Above all, adjustment and reform of teaching method are necessary in medical colleges.

2.3 *Separated from practical problems in teaching*

For a long time, conservative and relatively stable medical education has cultivated excellent medical personnel with solid basic skill and earnest rigorous attitude for society. However, such relatively static learning environment and mode imprison creativity and imagination of students to a certain degree. As a practical skill, saving lives is the responsibility of each medical staff. However, conventional methods are often limited in practical diagnosis and treatment, affecting efficiency of healing the wounded and rescuing the dying. For example, some basic medical service center is equipped with rare devices and insufficient treatment. Critical patients can only accept conservative treatment or temporarily pain relief. Nidus of patient is not able to be removed as quickly as possible. With some emergencies, patients cannot be placed reasonably due to limited scale and quantity of medical infrastructure. Misdiagnosis or medical accidents can be caused, let alone decrease of doctor's efficiency.

The above are practical problems faced by teachers and students in higher vocational medical colleges. But most of them pay no attention to the intractable problems. These problems are not discussed among students in the teaching, causing no creative solutions and countermeasures. Teaching is separated from practical problems existing in the professional field, with an effect of null. So-called teaching reform and innovation become castles in the air.

3 CULTIVATION MODE OF INNOVATIVE TALENTS IN HIGHER VOCATIONAL COLLEGES

3.1 *Expanding strength of teachers and optimizing resource of students*

Optimization of students is to be achieved by every college. Sufficient absorption of students represents strong strength of college. With sufficient students, more advanced teaching improvement and practice can be implemented, thus improving teaching quality and forming good school ethos. Cultivation of students can be smoothly boosted with more obvious scale effects.

In recent years, higher vocational medical colleges have made efforts to train applied innovative talents. Enough high-quality students and backbone teachers are required as foundation of all education reform at first. More attention should be paid to improve strength of teachers. As a core competitiveness of colleges, strong faculty can be formed by detention of outstanding graduates, external recruitment, internal preferment and further education of teachers. Young, vibrant, potential teachers can contribute to better innovation in higher vocational medical colleges.

Teachers with outstanding performance in medical education should be excavated seriously. By using good salary or benefits, some teachers with excellent ability and innovation can be attracted as leaders of innovative talents cultivation. Model teacher should be broadcasted. Team of teachers can be encouraged for bold innovation of students' basic medical study, clinical practice and career planning. Teachers should create valuable theoretical tips and operation methods by combination with actual situation and cast off traditional framework. Additionally, teachers can be collectively trained for changing old thinking and accepting modern teaching methods and ideas. Teachers' disconnect with young students should be avoided, in order to better guide students to develop new idea into scientific and standardized innovation. Cultivation of young teachers should not be ignored for little generation gap with students. Teachers should provide students for free loose thinking space. Students' innovative ability can be displayed and exchanged, promoting cultivation of applied innovative talents.

3.2 *Opening mind and flexible teaching*

Inherent idea should be changed by higher vocational colleges to cultivate innovative talents at first. Colleges should try to accept new ideas and concept popular in modern society. Non-objective views should be changed. With realistic positioning of a cradle of talents, college can concentrate on cultivating students without distractions. Talent training is a proactive and certain advanced job. Higher vocational medical colleges are shouldered with multiple pressures from society, students and parents. Colleges should open their minds to explore directions for future teaching.

Change of teaching mode is an important link in cultivating applied innovative talents. Communication of students and teachers is mainly involved in the process of teaching. Students' ways of thinking are largely affected by professional teachers. Especially in medical class, certain experience or sure conclusions can be passed on by teachers. But it is not beneficial for flexible teaching for innovative talents training, although such knowledge is certainly valuable. Students should present their own opinions or ideas after completion of teaching content. Besides, students can gradually form habits of independent thinking with more time for autonomous learning. Thus, they can explore more professional knowledge so as to deal with various events in future jobs.

Daily class can be transferred into lab, where students are encouraged to practice. Videos and lectures can be used to deepen students' understanding of professional knowledge and future work. For necessary practice of students, colleges can achieve contract or agreement with hospital or other medical institutions. Theoretical or practical innovations during internship should be recommended, helping students master a complete theoretical system or standard operating procedures. Teachers can also increase assessment points for encouraging students' authority awards or innovations. With the direct stimulation, cultivation of students' innovation can be boosted, thus offering society high-quality medical talents.

3.3 Providing a wide range of support and improving applied innovation

With weak ability and embarrassing situation in recent years, higher vocational should look for support and cooperation from all sides. External assistance can improve talents cultivation. With emphasis on colleges' resources, higher vocational medical colleges should focus on supportive policies of China. These policies contribute to more convenient cooperation with some counterpart enterprises. Cultivation agreement can be achieved between colleges and hospitals and medical and health services center. Relatively stable and "win-win" cooperation models can be formed, providing more opportunities for students' internship and employment.

In addition, colleges and teachers can be inspired by many academic seminars for talents cultivation.

On the platform, teaching methods can be communicated among teachers to make common progress. Sharing of diversified teaching resources and exchanges with across colleges can be achieved with intentional colleges during seminars. Scope of cooperation and influence of colleges can be enlarged.

Some skills contest among higher vocational colleges and national professional skills contest can contribute to cultivating students' innovation ability. Competitions can urge students to more actively explore skills in theory or practice, stimulating their enthusiasm and motivation of applied innovation. Joint effort of teachers and students is the key of cultivation of applied innovative talents. Innovation of education mode requires collaboration of teachers and students, as well. Traditional knowledge in book is important. But surefooted practice guided by teachers is also needed for students' open thinking and bold innovation concept. Teachers should offer timely answer or assistance when students have questions or ideas. Faced with academic requirements beyond abilities, teachers should ask for more powerful help from colleges. Academic institutions are to enrich students' innovative idea or method, shaping their senses of accomplishment and self-confidence.

4 CONCLUSIONS

With obvious career orientation, cultivation in medical colleges is related to future career and employment of students. They should master professional medical skills and practice for patients. Colleges and teachers should gradually emancipate their minds, paying more attention to faculty improvement, teaching mode reform and introduction of more partners. Students, especially in medical major, can accept most valuable education. Thus, the modern education goal of applied talents cultivation can be achieved. And also education mode of cultivating talents in higher vocational colleges can acquire a completely new outlook.

REFERENCES

[1] Li Qin, Ren Liping, Dong Huixia. *Understanding of teaching method reform in higher vocational medical colleges.* (Modern Medicine & Health, 2013).
[2] Zhu Huifang. *Strengthening students clinical practice abilities in higher vocational medical.* (Du Shi Jia Jiao, 2014).
[3] Wang Feng. *Four problems of employment of higher vocational medical graduates.* (Academy, 2011).

Education Management and Management Science – Zheng (Ed.)

Cultural governance mode of response mechanism and national identity of ethnic-minority sport culture policy

Yunong Wu

The Sport Department of Baise University, Baise, Guangxi, China

ABSTRACT: As a multi-ethnic country, China still needs to make efforts to strengthen the national identity of each ethnic and improve specific response mechanism. Thus, a strong sense of belonging and cohesion among all ethnics can be ensured to achieve common progress. At present, the economy of ethnic minority areas relatively lags behind other areas, forming a severe contradiction with the urgent desire of development. On such basis, government departments and local organs of self-government should actively rely on economy to develop cultural governance plans according to local conditions. Therefore, the scientificity and effectiveness of ethnic work can be ensured, thus promoting the development and prosperity of the whole country.

Keywords: ethnic minority; response mechanism; cultural governance; economic foundation

1 PROBLEMS IN THE CULTURAL GOVERNANCE OF ETHNIC MINORITIES

1.1 *Relatively backward economy of ethnic minority areas*

Since the reform and opening up, the economy and culture undertakings of ethnic minority areas have achieved constant development and prosperity, showing a big change compared with previous situation. Meanwhile, various preferential policies have been implemented to ethnic minority areas. Besides, all ethnic groups in China are working hard together for the prosperity of the country, ensuring the work and policies in ethnic minority areas implemented step by step. Therefore, the policy objectives of ethnic equality, unity and common prosperity become more and more clear. However, in order to promote economic, political and cultural prosperity of ethnic minority areas, more efforts and attention should be paid to the deficiencies of national governance. Besides, government should start from the key and difficult issues to overcome the main contradictions, optimizing governance level and quality [1]. Since the economy of ethnic minority areas is still a key issue, a solid economic foundation will directly affect the living quality of ethnic minority areas as well as the practice of various ethnic policies.

At present, the economic level of some ethnic minority areas in China is still lower than national or regional average level. Although economy in ethnic minority areas develops rapidly, it still cannot get rid of relatively backward economic development mode. As a result, the development of those ethnic minority areas becomes difficult. Geographic locations and historical factors of those areas mainly lead to such difficulties. For example, ethnic minority areas are mostly featured with fragile ecological environment and dangerous terrain, directly affecting the development of agriculture, food and clothing as well as the economy development modes. Besides, the transportation that contacts with outside world is also constrained, making it difficult for minority people to travel out of the mountains and accept advanced education and talent trainings. As a result, it is difficult for ethnic minority areas to smoothly integrate into modern labor forms and business activities. Therefore, ethnic minority areas can hardly develop their regional economies based on above constraints [2]. Those factors are reflected in some economic indexes, such as per capita GDP, per capita total industrial output, per capita consumption and saving level. In fact, all those indexes have certain space for improvement. Therefore, government should make more efforts to develop Chinese ethnic economy, thus realizing ethnic policies of economic and cultural development in those areas.

1.2 *Ethnic people's pursuit of urban lifestyle*

Human is born with desires and demands that cannot be ignored. The poor hope to gain wealth through efforts. The young want to stand out among others through persistent striving. All sorts of demands come from the gap and differences between the ideal and the reality. For a long time, population migration and people's persistent self-improvement are all for the purpose of finding a better living environment, making a rich life and realizing certain ideal.

Efforts from government, organs of self-government and all Chinese people are needed to diminish the gap and differences of development level between national average and ethnic minority areas. Such power comes from ethnic people's yearning of urbanization and urban lifestyle. Regardless of the limitation influence of ethnic minority development, ethnic people all want to get equal education opportunities, wider choices of professions, and more chances to know people and fresh things [3]. Therefore, more and more young minorities are willing to integrate into the operation mode of modern society and accept more advanced society elements. After long-time wars between nationalities and national integration, minority people now prefer to walk out of past activity areas and life circle with the help of national support policies, convenient information transmission and efficient transportation. They gradually adapt to high-efficient and fast-pace urban life. Meanwhile, the encouragement of local governments and warm welcome from society also contribute to the smooth transformation and management work in ethnic minority areas, which helps to find a new direction for cultural governance at the same time.

1.3 Strong development demands of ethnic-minority sport culture

Chinese ethnic minorities possess a strong cultural connotation, and these national cultures become more precious after the long-time historical precipitation. There are 55 ethnic minorities in China, among which most ethnics have their own unique costumes, customs and lifestyles. Some ethnics still keep their special native spoken and written languages as well as histories. Chinese people's idea of tracing the source is reflected in the protection and inheritance of those cultural heritages. Ethnic minorities are keen to make their unique ethnic culture accepted, respected and protected by people.

Ethnic-minority culture includes two aspects: material level and spirit level, both possessing a long history and high-degree national identity. For example, grand song of Dong minority is a unique way of singing in Dong minority area. Originated from the minority singing form of the spring and autumn period, such singing form generates harmonious melody through participants' own harmony. Grand song of Dong minority once won widespread praise in the 1986 Paris Autumn Festival in France. Obviously, these precious ethnic cultures not only provide profound supports to Chinese multi-civilization, but also arouse a worldwide resonance, further proving the saying "of a nation, of the world". In addition, people usually have a strong desire to develop their national sports cause, especially local traditional sports. Sports cause can ensure the social stability, national unity and border security of China. Besides, it can also provide systematic professional trainings and national-fitness environment for minority people to maintain good health and rich amateur cultural life at the same time, thus forming a strong sense of belonging and cohesion.

2 CULTURAL GOVERNANCE MODE OF RESPONSE MECHANISM AND NATIONAL IDENTITY OF ETHNIC-MINORITY SPORT CULTURE POLICY

2.1 Cultural governance necessarily relying on economy

Well-developed regional economy is the foundation of cultural governance in ethnic minority areas. After all, the economic basis determines superstructures such as cultural-governance policies and measures. Besides, economic foundation provides material insurance and financial supports for a series of cultural governance projects. In terms of work assignments, governments need to control the main direction of economic construction, and publish policies conducive to economic construction of ethnic minorities. Meanwhile, government personnel should be meticulous and serious about this work to make sure those preferential policies accurately implemented to minorities peoples. For example, Luoyang Ethnic Affairs Commission in Henan conducted a comprehensive survey of ethnic-minority economy, sorting out and summing up the concrete operation situation of three large-scale Moslem food enterprises and two ethnic minority villages. Afterwards, corresponding opinions, suggestions, development solutions and supporting policies were developed, laying a solid foundation for the implementation of following policies.

In addition, China adheres to regional autonomy of ethnic minorities, so local organs of self-government also have corresponding rights of independent development. Thus, national cultural advantages and characteristics of different regions and ethnics can be considered to achieve adaptive and personalized economic development. Furthermore, local organs of self-government can take advantage of the special natural resources to develop ecological economy in the region. Thus, economic development projects such as characteristic agriculture and ecological tourism gain striving power and force through protecting and restoring the living environment of that area. Once the industrial chain is formed with certain standard and scale in ethnic minority areas, it will directly bring huge economic benefits for the region. With sufficient funds support, corresponding cultural governance activities can be implemented smoothly. Increasing income also promotes the consumptions of mobile phone and Internet. The development of information network helps people to quickly understand current national affairs and policies. Meanwhile, it also brings convenience for national governance policies to be spread and understood. Therefore, various cultural governance activities get more effective practice channels, strengthening governance purpose and effects.

2.2 *Establishing response mechanism based on respecting and protecting ethnic culture*

Response mechanism, associated with social, economic and cultural constructions, is a further sublimation of government functions. It not only emphasizes the actual demands of social public, but also develops clear divisions and regulations for the treatment methods, response effects and ideas tend of these demands. However, proper adjustment should be made to better conform to the special culture and compatriots emotions of ethnic minorities when applying response mechanism. Governments and related staff should firstly carry out detailed research work in regional governance to understand its policy and humanistic environment as well as regional development history of this area, thus avoiding impractical plans or obstacles in policy implementation.

Ethnic minorities, as living fossils of human history evolution in China, play the roles of enriching Chinese national cultures and forming magnificent national traditions and features of ethnic minorities. Torch Festival and Lesser Bairam are two typical national traditions. Adult Ceremony, Marriage Customs and Ancestor-Worshipping Celebration are common ceremonies with ethnic characteristics. Those ethnic elements, containing emotional sustenance of ethnic minorities, are the wisdom crystallization of predecessors. Therefore, great importance should be attached to those ethnic elements in specific work. As for the policy response of culture and sports development in ethnic minority areas, government should listen to the opinions of local people and focus on the protection and respect of national traditional cultures. Meanwhile, measures and policies that hurt national emotions or do not conform to local customs and cultures should be firmly rejected.

Therefore, in order to achieve a "wise" response mechanism, government should care more about ethnic minorities and rely on ethnic people. Meanwhile, it is also important to get a better understanding of ethnic culture. Furthermore, general national policies and regulations should be refined to feasible measures in line with ethnic emotions and regional environment, thus timely and effectively meeting economic and cultural needs of ethnic minority compatriots. Therefore, government functions can be deepened and renovated through response mechanism, turning good ideas and starting points into visible achievements and effects.

2.3 *Conducting cultural governance of national identity based on local conditions*

National identity has always been a serious issue concerning ethnic unity and national prosperity. It involves people's recognition level of the country and sense of belonging. Governments' efforts on economic construction and cultural governance in ethnic minority areas have achieved significant effects. Since the founding of the People's Republic of China, governments have provided great financial supports to ethnic minority areas. Especially after the implementation of the Western Development Strategy, the fixed assets investment in ethnic areas has reached 3.0204 trillion RMB during the "Tenth Five Years" period. Nowadays, characteristic governance and supports have been implemented with better plans to help minority people get rid of poverty and poor living conditions. In addition to necessary financial supports, local conditions should also be considered to further deepen ethnic people's national identity. More specifically, government should analyze specific issues according to local conditions, applying different measures of economic construction and cultural governance to different regions and ethnics. At present, eastern coast regions are developing faster than western interior areas in China. Therefore, government brought up a counterpart support plan—one eastern coast city supports one ethnic minority area or village through sharing and transmitting economic resources and talents. Thus, new development motive force can be generated in relatively backward minority areas. Meanwhile, the counterpart support plan also helps people from two areas to forge deep friendship, thus enhancing national cohesion and realizing national unity and common prosperity.

Ethnic groups in China live together over vast areas while some live in individual concentrated communities in small areas. Therefore, governance work should firstly develop an overall planning. Then, government should come up with specific plans in accordance with local characteristics after field investigation of minority settlement areas. For example, ethnic minorities in southwest area mainly include Yi, Dong, Dai and Zhuang nationalities. Most of those ethnic people live in the areas of well-preserved ecological environment with some dangerous but important terrains or wonderful natural landscape. In addition to agricultural farming, they also live on ecological tourism. Thus, governments and organs of self-government at all levels can rely on this form to strengthen traditional culture of ethnic minorities such as languages, words, painting, clothing, dancing and other aspects. It is possible to make profits while protecting ethnic culture. Such kind of cultural governance mode combined with local economy enables traditional cultural governance to effectively integrate with modern economic operation mode. Meanwhile, it also provides a certain amount of financial supports for specific governance projects. Thus, civilization of minority areas can be protected and retained more systematically. People's desire for developing their own history and culture can be satisfied at the same time. In return, ethnic people's emotions and national identity of the country will also be improved.

3 CONCLUSIONS

As a multi-ethnic country, China's governance of economy and culture in ethnic minority areas is closely related to domestic development, social stability, national unity or even the country's survival. The cultural governance mode of response mechanism and national identity needs the guidance of correct national policy and scientific work methods. More importantly, the dialectical relationship between economic foundation and cultural governance should be well controlled to make them support and promote each other, ensuring scientific cultural governance. Scientific management mode plus joint efforts of all ethnic groups will eventually contribute to smooth ethnic development, ethnic unity and common prosperity of all ethnic groups.

REFERENCES

[1] Chen Lufang, Response Mechanism and Cultural Governance of Ethnic-Minority Cultural Demands. Social Sciences in Yunnan. 2010(4): 121–125.
[2] Wang Zhaoqian, Modernization Development of Traditional Ethnic-Minority Sports Based on National-Game Performance Characteristics in China. Journal of Chifeng University: Natural Science Edition. 2014.30(12): 143–145.
[3] Zhang Jing, Zhang Boyang, National Unity Creates Brilliance; Mutual Help and Support Creates Future—To Commemorate the 30th Anniversary of the Promulgation and Implementation of "Law of the People's Republic of China on Regional National Autonomy". Northern Economy. 2014(4): 12–14.

Education Management and Management Science – Zheng (Ed.)
© 2015 Taylor & Francis Group, London, ISBN 978-1-138-02663-6

Function and value of cultural communication in national folk sports events

Bikai Dong

Guangxi Normal University for Nationalities, Chongzuo, Guangxi, China

ABSTRACT: China is a unified multinational country with vast territory, abundant resources and long history of national culture. Likewise, the culture of national folk sports events has also been developing continuously and kept pace with the time in the history river. As an important part of Chinese traditional culture, the culture of national folk sports events is of profound significance for Chinese cultural communication and inheritance. Based on the analysis of the background of national folk sports events and existing problems in the development, this work gave a detailed description about the specific contents of its cultural communication. Following that, the important value of Chinese folk sports culture were further discussed.

Keywords: national folk; sports events; cultural communication; background; communication value

1 INTRODUCTION

Traditional life-style and living habits have direct impacts on the formation and development of culture. China is a multinational country with complex historical culture and traditional customs. Characteristic cultures of various nationalities and regions have also promoted the diversified development and integration of national folk sports events. Without doubt, the institutions and rules of national folk sports events have strong features and regional characteristics. This not only helps to spread local culture and spirit through sports, but also allows people to exercise in the entertainment, thereby enhancing their physical quality. Diverse national folk sports, including dragon-boat race, playing diablo, kicking shuttlecock, tug-of-War, horse racing and sports dance, are all Chinese national folk sport projects. These projects, representing national characteristic culture and reflecting the local spirits, have played an important role in the long history of Chinese sports [1]. Additionally, the communication function of national folk sports events is irreplaceable by other forms, and it is the best part of cultural heritage deposited in special historical periods. Therefore, national folk sports events are of important significance in promoting the communication and mixture of Chinese national culture, enhancing national pride and patriotism spirit, and strengthening national unity.

2 BACKGROUND OF CHINESE FOLK SPORTS EVENTS

Since the reform and opening-up, the economic strength of China has been increasing and its international status also has changed a lot. Obviously, the development of sports and communication of culture plays supporting function to such achievement. The sports cause of China has had a qualitative leap, and the past "sick man of East Asia" has become a sports power in Asia. In the development of sports, the communication of sports culture and presentation of sports spirit show a more far-reaching significance than gold medal. So, sports power should set promoting the communication of sports culture, spreading sport spirit and enhancing its international influence as the lifelong development goals. In fact, China is making constant efforts toward this direction. The entire sports culture in China is relatively thick, and great importance has been attached to the development and communication of sports culture [2]. During Chinese traditional cultural inheritance, national folk sports events have also been strongly supported and constantly encouraged by government and society. Such background has provided a fertile soil for the cultural development of Chinese folk sports events. This enables the advocators, successors and laborers of Chinese traditional culture to foresee the culture prospect of Chinese folk sports. However, compared with competitive sports, national folk sports are still in a passive position. Chinese nationalities reflect various features, which is unique in the world. So, it is necessary to take advantage of these features as well as Chinese national culture to influence world culture [3]. This is the cultural background and development goal of Chinese folk sports events.

3 PROBLEMS IN CULTURAL COMMUNICATION OF CHINESE FOLK SPORTS EVENTS

3.1 *Chinese national folk culture being ignored compared to competitive sports*

Competitive sports culture takes up a great proportion in sports culture. In competitive sports, the pursuit of results, rankings and medals presents a distinct bias. Chinese competitive sports have made outstanding contributions to improve the international status and comprehensive strength of China. However, too much emphasis on medals and rankings has led to a big deviation in the development of sports culture. While promoting the development of competitive sports, Chinese folk sports culture has been constantly ignored, thus hampering its development.

3.2 *Small-scale of national folk sports events*

Chinese media are still willing to or even habitually focus on various sports contests while attaching less attention to national folk sports. Indeed, many media directly refuse or neglect the broadcast of national folk sports events, making a lot of national folk events rest at local and become a national folk entertainment. In recent years, health care sports, such as shadowboxing, are popular among people. However, there is no formal organization or institution to create platforms for propagating shadowboxing culture. In short, compared with competitive sports culture, Chinese folk sports are at a disadvantage in communication and promotion.

3.3 *Social responsibility of media not strong enough*

There are few sports broadcasting channels in China, and CCTV sports channel is the major broadcasting platform for all sporting events. In numerous sports activities and competitive events, traditional projects such as table tennis, diving and football occupy the superior position. Since these projects attract more viewers and fans, TV stations can improve their audience ratings, thus getting more economic benefits. Additionally, many competitive sports projects originated from China have

a long history and great significance of cultural inheritance. However, some of those projects such as curling have been excluded by other projects in the development of sports. Further, similar sports projects and events are gradually being marginalized, thus making the communication of national sports culture appear a deviation.

In summary, there are many problems in the communication of Chinese folk sports culture. In the future, it is needed to solve the above problems and defects so as to ensure Chinese traditional culture and sports cause makes co-coordinated progress.

4 IMPORTANT VALUE OF THE CULTURAL COMMUNICATION OF NATIONAL FOLK SPORTS EVENTS

4.1 *Cultural communication function of national folk sports events as the requirement of the times*

With continuous development of economy, people's disposable income is gradually increasing. Since more and more people begin to focus on the need of spirit level, cultural development has become the inevitable requirement of the times. Chinese sports cause develops rapidly and has a strong influence at the international level, which can be clearly reflected from the number of medals and the successfully hosting of Olympic Games and Paralympic Games. However, compared to the Olympics or World Cup, Chinese folk sports events seem very bleak. Furthermore, many people tend to have prejudice against the cultural development of Chinese folk sports events, thinking it impossible to make the goal of becoming sports power. In fact, this concept—of the national, of the world—accurately express the importance of national folk sports events. Some national folk sports events, such as shadowboxing, playing diablo and square dance, have formed a relatively complete contest rules in the long-term studies. These activities provide local people a way for self-amusement, and also help people to understand local cultural characteristics through such meaningful national activities. Even more, it can attract international friends to better understand China and recognize different nations. To make the world culture more splendid, the social concepts of cultural integration and pace of globalization should be better implemented. Thereby, more national culture will enter into world stage, and more meaningful and valuable culture will be excavated and developed. While learning the history, the communication and inheritance of national culture should keep pace with the times. Only national culture exists in large spaces for a long time, its international influence can be truly enhanced. In this way, national folk sports culture will have effect on world culture, thus better meeting the requirements of new era.

4.2 *Communication function of national folk sports events culture as requirements of national unity and security*

China is a unified multinational country with fifty-six nationalities. Even with large number of nations, Han population occupies the leading position. So the relationship between Han and other minorities should be better handled to maintain social stability. For different nations, their own beliefs and unique culture are different. In the long history, cultural forms that left behind always have highly competitive power and special significance. Almost every national folk sport has a history or a beautiful legend. Such cultural connotation in activities will undoubtedly leave people in every dynasty the national cultural beliefs, and then people inherit classic in entertainment. Therefore, every national folk sport is of important significance, and the emphasis of national culture represents the respect of the entire nation.

National folk events are rare opportunities that bring the minorities from different areas of China together. And every nation has its own unique cultural form and expression method. Subtle movements, small symbols or the totem in the corner of clothing are all of national significance. Such sports events are golden opportunities to show the national mien—people can make use of those opportunities to exchange cultures, thus making contribution to the communication of national culture. And in the exchange of different nations, people will find every national culture the same miraculous, full of strength and worthy of respect and protection. Furthermore, people of different nationalities will virtually feel kind, and possess national self-esteem, self-confidence and pride of the motherland, thus promoting national unity and maintaining social stability. For the development of culture, national folk sports events can provide more opportunities to attract funding. That is, more resources and opportunities can be used to promote the development of cultural cause. For the government, national problem is a thorny problem. But the development of national folk sports events can solve the national problem to some extent, thus promoting the coordinated development and common progress of nations.

4.3 *Exchange value of cultural communication function of national folk sports events*

For a nation, there are countless ties from the ancient time to today, especially for ethnic minorities with small population. Normally, a traditional national folk sport has strong national spirit, and during the inheritance, people of the nation have very strong sense of identity. Such cultural form is an important way to unite the nation and gather people. As a traditional modern country, China still suffers from the compression of feudalism. However, Chinese deep-rooted culture form still affects many contemporary people.

In ancient China, women were required to follow "the three obediences and four virtues", without much discourse power and liberty in life, including their marriage. Apart from "parents' order, match-maker's canvasses", the "competition for marriage in martial skills" and "throwing a ball made of rolled colored silk" have also appeared on TV play of historical theme. Indeed, "throwing a ball made of rolled colored silk" is a traditional national folk sport. In ancient, if there was a contradiction or conflict within nation or between tribes, the majority would take the same cultural forms (e.g., dancing, wrestling) to reach the reconciliation. Using the identity of same cultural form to offset the hostility of each other, the distance between each other can be narrowed through national cultural activities. A better exchange and communication environment for both sides guarantees the nation's peace and stability. These activities are typical cases of good communication using national folk sports activities by ancient people. For modern people, the society is relatively stable, but terrorist activities and East Turkistan assault happen frequently. Due to these phenomena, there are some misunderstanding and prejudice among some social people against the compatriots in Xinjiang who are kindhearted and diligence. Even, some people reject and contradict this nation to some extent. In fact, victims are the compatriots in Xinjiang. When national conflicts emerge, perhaps national activities are the alleviation means, such as Xinjiang dance. In this way, people will appreciate the good faith of other nations though communication and exchange, thus promoting the cultural communication of national folk sports. In the inheritance of Chinese culture, only unity and cooperation among nations can safeguard the stability and harmony of motherland.

5 CONCLUSIONS

National folk sports events have obvious national identity and unique cultural connotation. Their traditional cultural feature has strong appeal of native people's spirit, and local customs can be seen from all the details. However, with the rapid rise of competitive sports, many people have ignored the soul of national culture, constantly marginalizing the form and culture of national folk sports. Therefore, for Chinese national folk sports events, it is necessary to analyze their background and the problems in cultural communication. Additionally, the requirements of the times for Chinese national folk sports events should also be analyzed. Then it is possible to better promote the cultural communication of national folk sports events, thus showing their cultural value. National folk sports events are an important part of Chinese sports development, possessing important significance in cultural communication, entertainment, fitness and promoting exchanges. Therefore, the joint efforts of all aspects are required to protect this valuable intangible cultural heritage and inherit it, realizing the nation is the world.

REFERENCES

[1] Zhang Junping, National Folk Sports Development Strategy and Path Selection from the Perspective of Regional Culture, Wushu science, 2012 (12): 92–95.
[2] Yang Bin, Connotation and Modern Value of Tujia Traditional Sports—Based on the Field Work of Wantan Town in West Hubei Province, Journal of Yangtze Normal University, 2012 (08): 5–9.
[3] Yan Jinfang, Culture Communication of Sports Events—Cultural Feast of Olympic, The Guide of Science & Education, 2014 (12): 101–101.

Author index